Advancements in Zeolites and Micro–Meso Porous Hierarchical Materials

Fabian N. Murrieta-Rico
Universidad Politécnica de Baja California, Mexico

Joel Antúnez-García
Universidad Autónoma de Baja California, Mexico

Vitalii Petranovskii
Universidad Nacional Autónoma de México, Mexico

IGI Global
Scientific Publishing
Publishing Tomorrow's Research Today

Published in the United States of America by
IGI Global Scientific Publishing
701 East Chocolate Avenue
Hershey, PA, 17033, USA
Tel: 717-533-8845
Fax: 717-533-7115
Website: https://www.igi-global.com E-mail: cust@igi-global.com

Library of Congress Cataloging-in-Publication Data

Names: Murrieta-Rico, Fabian, 1986- editor | AntuÌ nez-GarciÌ a, Joel, editor | PetranovskiiÌ†, V. P. (VitaliiÌ† Pavlovich) editor
Title: Advancements in zeolites and micro-meso porous hierarchical materials / edited by: Fabian N. Murrieta-Rico, Joel AntuÌ nez-GarciÌ a, and Vitalii Petranovskii.
Description: Hershey PA : IGI Global Scientific Publishing, [2025] | Includes bibliographical references. | Summary: "This book aims to offer extensive research in the latest advancements in Zeolites and Micro-Meso Porous Hierarchical Materials, this, by providing a general introduction to this topic and exploring its diverse applications in various fields"-- Provided by publisher.
Identifiers: LCCN 2025005790 (print) | LCCN 2025005791 (ebook) | ISBN 9798337303208 hardcover | ISBN 9798337303215 paperback | ISBN 9798337303222 ebook
Subjects: LCSH: Zeolites--Case studies | Mesoporous materials--Case studies
Classification: LCC TP245.S5 A295 2025 (print) | LCC TP245.S5 (ebook) | DDC 661/.068324--dc23/eng/20250519
LC record available at https://lccn.loc.gov/2025005790
LC ebook record available at https://lccn.loc.gov/2025005791

British Cataloguing in Publication Data
A Cataloguing in Publication record for this book is available from the British Library.

Table of Contents

Preface xvii

Chapter 1
Optical Properties of Mordenite-Type Zeolite in VUV-UV-Vis and IR Regions . 1
Vitalii Petranovskii, Center of Nanoscience and Nanotecnology, Mexico
S. V. Kholodkevich, St. Petersburg Federal Research Center of the Russian Academy of Sciences, Russia
Joel Antúnez-García, Institute of Advanced Studies of Baja California, Mexico
A. V. Cvetkov, Pskov State University, Russia
K. V. Gusev, Pskov State University, Russia
Marina S. Ivanova, Pskov State University, Russia
Tamara G. Matveeva, Pskov State University Branch in Velikiye Luki, Russia
V. G. Solovyev, Pskov State University, Russia
S. V. Trifonov, Pskov State University, Russia
Roberto Núñez-González, Universidad de Sonora, Mexico

Chapter 2
Nuclear Magnetic Resonance to Probe Micro-/Mesoporous Zeolites 23
Marina G. Shelyapina, Saint Petersburg State University, Russia

Chapter 3
FTIR Studies of Adsorption Properties of Zeolites 65
Alexey Tsyganenko, St. Petersburg State University, Russia

Chapter 4
Influence of Natural Zeolites on the Crystalline Properties of ZnO: X-Ray Diffraction Technique and Its Application in Characterization 101
Karin M. Álvarez Gómez, Benemérita Universidad Autónoma de Puebla, Mexico
Miguel Ángel Hernández-Espinosa, Benemérita Universidad Autónoma de Puebla, Mexico
Luis Pérez-Cabrera, Universidad Nacional Autónoma de México, Mexico
Viridiana Evangelista-Hernández, Universidad Nacional Autónoma de México, Mexico
Vitalii Petranovskii, Universidad Nacional Autónoma de México, Mexico

Chapter 5
Ion Exchange and Zeolites: Sustainable Approaches to Water Pollution Mitigation.. 151
Amita Somya, Amity University, India
Amit Prakash Varshney, Independent Researcher, India

Chapter 6
Comparison of ZSM-5 and $MgO-Al_2O_3$ Nanosheet Supports and Their Performance in Hydrodesulfurization 181
Luis Perez-Cabrera, Centro de Nanociencias y Nanotecnología Universidad Nacional Autónoma de México, Mexico
Victor Alfredo Reyes-Villegas, Centro de Investigación Científica y de Educación Superior de Ensenada, Mexico
Jesus Isaías De León-Ramírez, Centro de Investigación Científica y de Educación Superior de Ensenada, Mexico
Joel Antúnez-García, Universidad Autónoma de Baja California, Mexico
Rosario Isidro Yocupicio-Gaxiola, Instituto Tecnológico Superior de Guasave, Mexico
Miguel Angel Hernández-Espinosa, Benemerita Universidade Autónoma de Puebla, Mexico
Viridiana Evangelista-Hernández, Universidad Autónoma de Baja California, Mexico
Armando Reyes-Serrato, Universidad Nacional Autónoma de México, Mexico
Sergio Fuentes-Moyado, Universidad Nacional Autónoma de México, Mexico

Chapter 7
Clinoptilolite Perspectives as a Zeolite for Removing Pollutants From Water: A Review 215
Manuel Alberto Flores-Hidalgo, Universidad Juarez del Estado de Durango, Mexico
Diana Barraza-Jiménez, Universidad Juárez del Estado de Durango, Mexico
Heimi V. Ríos-Arámbula, Universidad Juárez del Estado de Durango, Mexico
Hugo I. Flores-Hidalgo, TecNM, Jiménez Institute of Technology, Mexico
Sandra I. Torres-Herrera, Universidad Juárez del Estado de Durango, Mexico

Chapter 8
Are NEREA® Zeolitic Nanostructured Materials Equal to NPK Fertilizers Mixed With Natural Zeolites? 265
Gerardo Rodríguez Fuentes, Instituto de Ciencia y Tecnología de Materiales, Cuba
Inocente Rodríguez-Iznaga, Instituto de Ciencia y Tecnología de Materiales, Cuba

Chapter 9
Post-Synthesis Surfactant-Induced Mesoporosity in Zeolite Crystals: State-of-the-Art, Open Questions and Perspectives 315
Joaquín Pérez-Pariente, Instituto de Catálisis y Petroleoquímica, Spain
Manuel Sánchez-Sánchez, Instituto de Catálisis y Petroleoquímica, Spain
Julia Aguilar Pliego, UAM-Azcapotzalco, Mexico

Chapter 10
Recent Advances in Sonochemistry for the Synthesis and Modification of FeY Zeolite Catalyst 359
Victor Alfredo Reyes, Centro de Investigación Científica y de Educación Superior de Ensenada, Mexico
Jesus Isaias De Leon Ramirez, Centro de Nanociencias y Nanotecnologia UNAM, Mexico
Luis Perez-Cabrera, Centro de Nanociencias y Nanotecnologia UNAM, Mexico
Arturo Estolano-Cobian, Facultad de Ciencias Quimicas e Ingenieria, Universidad Autonoma de Baja California, Mexico
Jose Roman Chavez-Mendez, Facultad de Ciencias de la Salud, Universidad Autonoma de Baja California, Mexico
Vitalii Petranovskii, Centro de Nanociencias y Nanotecnologia UNAM, Mexico

Chapter 11
Ultrasonic Coating of LTA Zeolites by Antibacterial and Antibiofilm Nanoparticles 395
Jesus Isaias De Leon Ramirez, Centro de Nanociencias y Nanotecnologia UNAM, Mexico
Victor Alfredo Reyes-Villegas, Centro de Nanociencias y Nanotecnologia UNAM, Mexico
Sergio Perez Sicairos, Instituto Tecnologico de Tijuana Centro de Graduados, Mexico
José Román Chávez-Méndez, Universidad Autónoma de Baja California, Mexico
Luis Pérez-Cabrera, Centro de Nanociencias y Nanotecnologia UNAM, Mexico
Verónica González Torres, Universidad Autónoma de Baja California, Mexico
José Constantino González Crisóstomo, Universidad Autónoma de Baja California, Mexico
Ethiel Zavala-Flores, Universidad Autónoma de Baja California, Mexico
Fabian N. Murrieta-Rico, Universidad Politécnica de Baja California, Mexico
Vitalii Petranovskii, Centro de Nanociencias y Nanotecnologia UNAM, Mexico

Chapter 12
Theoretical Advances in Zeolites 427
Dipan Kumar Das, Centurion University of Technology and Management, India
Padmaja Patnaik, Centurion University of Technology and Management, India
Nibedita Nayak, Centurion University of Technology and Management, India
Sudip Kumar Das, Dr. C.V. Raman University, India

Compilation of References 457

About the Contributors 559

Index 563

Detailed Table of Contents

Preface xvii

Chapter 1
Optical Properties of Mordenite-Type Zeolite in VUV-UV-Vis and IR Regions . 1
Vitalii Petranovskii, Center of Nanoscience and Nanotecnology, Mexico
S. V. Kholodkevich, St. Petersburg Federal Research Center of the Russian Academy of Sciences, Russia
Joel Antúnez-García, Institute of Advanced Studies of Baja California, Mexico
A. V. Cvetkov, Pskov State University, Russia
K. V. Gusev, Pskov State University, Russia
Marina S. Ivanova, Pskov State University, Russia
Tamara G. Matveeva, Pskov State University Branch in Velikiye Luki, Russia
V. G. Solovyev, Pskov State University, Russia
S. V. Trifonov, Pskov State University, Russia
Roberto Núñez-González, Universidad de Sonora, Mexico

Zeolites are porous crystalline solids, which can be used as host matrices to create nanocomposite materials by infiltrating their voids with guest substances (clusters of metals, semiconductors, dielectrics). The guest nanospecies are characterized by uniform sizes (~1 nm), regular arrangement of particles repeating the system of voids in crystals, and high concentration (up to ~5×10^20 cm-3). The main benefit of zeolites as matrices for the stabilization of semiconductor clusters is their optical transparency, which makes it possible for light to interact with the entire volume of the composite crystal. To analyze the spectral properties of zeolite-based guest@ host systems, it is very important to know the spectral features of the empty matrix. In this work, the optical spectra of mordenite in a wide region (2.0 –12.7 eV for the VUV-UV-Vis spectra and 450 – 4000 cm–1 for the IR spectra) have been studied in detail. The data obtained are compared with available literature data.

Chapter 2
Nuclear Magnetic Resonance to Probe Micro-/Mesoporous Zeolites 23
Marina G. Shelyapina, Saint Petersburg State University, Russia

Zeolites are widely used in industry as molecular sieves, ion exchangers and catalysts. According to recent studies mesoporosity is beneficial to the rational design of catalysts with controlled product selectivity and the improved catalyst lifetime due to the efficient mass-transport properties. To tailor structures and morphology of zeolite catalysts for specific reactions, it is necessary to understand the effects of various modifications (either during synthesis or post-synthetic treatments) on physico-chemical properties of the target material. Nuclear magnetic resonance (NMR) has proven to be a reliable method for studying zeolite structure, surface acidity and dynamics of adsorbed molecules. This Chapter reviews recent NMR studies of zeolites with hierarchical porosity mainly focusing on, structural studies by ^{27}Al and ^{29}Si NMR spectroscopy, 2D T_1-T_2 correlation and T_2-T_2 exchange experiments and NMR cryoporometry.

Chapter 3
FTIR Studies of Adsorption Properties of Zeolites 65
Alexey Tsyganenko, St. Petersburg State University, Russia

The paper summarizes the results of FTIR studies of specific properties of zeolites, compared with metal oxides, in the adsorption of simple molecules. An attempt is made to explain the unusual properties of zeolites, such as linkage isomerism of adsorbed CO, extremely high Lewis acidity of exchangeble cations, Brønsted acidity of H-zeolites and basicity of zeolite framework oxygen atoms. Acidity of zeolitic hydroxyl groups, choice of test molecules for acid and basic sites, advantages of carbon monoxide and fluoroform for testing the strength of surface centers, correlation between their frequency shifts and absorbance are discussed, as well as the influence of environment upon the electron-accepting properties of cations and their ability for isomerism.

Chapter 4
Influence of Natural Zeolites on the Crystalline Properties of ZnO: X-Ray Diffraction Technique and Its Application in Characterization 101
Karin M. Álvarez Gómez, Benemérita Universidad Autónoma de Puebla, Mexico
Miguel Ángel Hernández-Espinosa, Benemérita Universidad Autónoma de Puebla, Mexico
Luis Pérez-Cabrera, Universidad Nacional Autónoma de México, Mexico
Viridiana Evangelista-Hernández, Universidad Nacional Autónoma de México, Mexico
Vitalii Petranovskii, Universidad Nacional Autónoma de México, Mexico

ZnO-zeolite composites were synthesized and characterized by sol-gel method and X-ray diffraction (XRD). Different proportions of clinoptilolite, mordenite, erionite and chabazite with ZnO were analyzed. The results show that the addition of zeolites affects the crystal structure, particle size (32.3-66.5 nm), dislocation density and microdeformations. Rietveld refinement confirmed variations in the lattice parameters. It is concluded that these materials can improve the dispersion and stability of ZnO, being promising for catalysis and gas purification.

Chapter 5
Ion Exchange and Zeolites: Sustainable Approaches to Water Pollution Mitigation.. 151
Amita Somya, Amity University, India
Amit Prakash Varshney, Independent Researcher, India

Water pollution is a critical global issue requiring sustainable treatment methods. Ion exchange and zeolite-based processes are effective solutions due to their efficiency in pollutant removal and alignment with sustainability principles. This chapter explores their role in mitigating water pollution. Ion exchange is widely used for removing heavy metals, nitrates, and harmful ions due to its high selectivity and regenerability. Zeolites, naturally occurring aluminosilicate minerals, have a porous structure, large surface area, and high cation exchange capacity, making them ideal for water treatment. The present chapter examines ion exchange mechanisms and zeolite properties that enhance pollution control. It also highlights advancements in zeolite modification for improved contaminant removal and discusses environmental benefits like reduced chemical usage and lower energy consumption. Finally, future prospects of these technologies in addressing emerging pollutants are explored, emphasizing their role in sustainable water management.

Chapter 6
Comparison of ZSM-5 and MgO-Al_2O_3 Nanosheet Supports and Their Performance in Hydrodesulfurization 181
Luis Perez-Cabrera, Centro de Nanociencias y Nanotecnología Universidad Nacional Autónoma de México, Mexico
Victor Alfredo Reyes-Villegas, Centro de Investigación Científica y de Educación Superior de Ensenada, Mexico
Jesus Isaías De León-Ramírez, Centro de Investigación Científica y de Educación Superior de Ensenada, Mexico
Joel Antúnez-García, Universidad Autónoma de Baja California, Mexico
Rosario Isidro Yocupicio-Gaxiola, Instituto Tecnológico Superior de Guasave, Mexico
Miguel Angel Hernández-Espinosa, Benemerita Universidade Autónoma de Puebla, Mexico
Viridiana Evangelista-Hernández, Universidad Autónoma de Baja California, Mexico
Armando Reyes-Serrato, Universidad Nacional Autónoma de México, Mexico
Sergio Fuentes-Moyado, Universidad Nacional Autónoma de México, Mexico

This chapter developed pure and iron-modified ZSM-5 by soft template synthesis—CTABr functions as a layered director. ZSM-5 was post-pillaring by TEOS to create interlamellar mesopores. Pillared ZSM-5 and iron-modified ZSM-5 were efficient catalytic supports for NiMoS used in the HDS of DBT reaction, which is significant for clean fuel production. From both NiMoS-supported, ZSM-5 and MgO-Al_2O_3, ZSM-5 exhibited better HDS performance. Al_2O_3-MgO oxides were obtained from hydrotalcite with different Al/(Al+Mg) ratios. Characterization of the catalysts by XRD, UV-Vis, and TEM indicated that a good dispersion of the sulfide phase was obtained due to the specific acid-base properties of the supports. The most active catalyst in the reaction of desulfurization of dibenzothiophene was the one presenting the highest acidity along the MgAl-x series, which was alumina due to improved dispersion of the nickel-promoted MoS_2 crystallites (low stacking and shorter lengths). All catalysts on Al_2O_3-MgO binary oxides exhibited better selectivity to the hydrogenation pathway than the alumina catalyst.

Chapter 7
Clinoptilolite Perspectives as a Zeolite for Removing Pollutants From Water: A Review.. 215
Manuel Alberto Flores-Hidalgo, Universidad Juarez del Estado de Durango, Mexico
Diana Barraza-Jiménez, Universidad Juárez del Estado de Durango, Mexico
Heimi V. Ríos-Arámbula, Universidad Juárez del Estado de Durango, Mexico
Hugo I. Flores-Hidalgo, TecNM, Jiménez Institute of Technology, Mexico
Sandra I. Torres-Herrera, Universidad Juárez del Estado de Durango, Mexico

A review on clinoptilolite, a natural zeolite, which is widely used in water remediation due to its porous structure and ion exchange properties. It effectively removes pollutants like heavy metals, ammonium, nitrates, and organic compounds from water sources through selective adsorption, offering a cost-effective, eco-friendly solution for treating industrial wastewater, groundwater, and drinking water. Its efficacy varies with factors such as origin and modification, requiring tailored optimization. Research employs advanced characterization (e.g., electron microscopy, X-ray diffraction, etc.) and computational simulations to enhance understanding of its pollutant removal mechanisms. With its versatility, affordability, and environmental benefits, clinoptilolite is a promising tool for sustainable water management. This review synthesizes its strengths, limitations, and future potential, emphasizing the need for continued innovation to address global water quality challenges.

Chapter 8
Are NEREA® Zeolitic Nanostructured Materials Equal to NPK Fertilizers Mixed With Natural Zeolites? 265
Gerardo Rodríguez Fuentes, Instituto de Ciencia y Tecnología de Materiales, Cuba
Inocente Rodríguez-Iznaga, Instituto de Ciencia y Tecnología de Materiales, Cuba

The most recommended use of natural zeolite in agriculture is the addition to the soil as amending agent or in mixture with fertilizers, but the results are not observed in short time, and its reproducibility is not entirely positive. A NPK chemical fertilizer requires water to be solubilized and make nutrients available for plants. If zeolite particles are present can retain part of these nutrients, other part remains in the soil increasing the pollution of the environment. New nanostructured zeolitic materials named NEREA® have changed the paradigm of natural zeolites in agriculture. The distinction is to introduce steadily and homogeneously all the nutrients that plants require into the zeolite particles, before incorporating them into the soil. Only the plants take up the nutrients modifying the architecture and morphology of their roots to involve NEREA particles. If the mechanical mixing of zeolite mineral with NPK fertilizers has represented an important contribution, NEREA products solve its shortcomings, representing an innovation for agriculture and zeolite science.

Chapter 9
Post-Synthesis Surfactant-Induced Mesoporosity in Zeolite Crystals: State-of-the-Art, Open Questions and Perspectives 315
Joaquín Pérez-Pariente, Instituto de Catálisis y Petroleoquímica, Spain
Manuel Sánchez-Sánchez, Instituto de Catálisis y Petroleoquímica, Spain
Julia Aguilar Pliego, UAM-Azcapotzalco, Mexico

In this chapter, the state-of-the-art on the generation of intracrystalline mesopores in zeolite crystals by post-synthesis treatments with cationic surfactants in alkaline medium is critically reviewed. So far, five 12R large pore FAU, BEA, MAZ, LTL and MOR, and one medium pore 10R MFI zeolites have been subjected to this treatment. The main aspects of the process and the chemical and structural factors of the parent zeolites that trigger the mesostructuring process have been thoroughly examined. Particular attention has also been paid to discuss the influence of the treatment on the total acidity and acid strength of the resulting materials. Finally, some general considerations have been elaborated to serve as guidelines for future work, with the aim of helping to develop the most suitable materials for specific catalytic processes.

Chapter 10
Recent Advances in Sonochemistry for the Synthesis and Modification of FeY Zeolite Catalyst 359
Victor Alfredo Reyes, Centro de Investigación Científica y de Educación Superior de Ensenada, Mexico
Jesus Isaias De Leon Ramirez, Centro de Nanociencias y Nanotecnologia UNAM, Mexico
Luis Perez-Cabrera, Centro de Nanociencias y Nanotecnologia UNAM, Mexico
Arturo Estolano-Cobian, Facultad de Ciencias Quimicas e Ingenieria, Universidad Autonoma de Baja California, Mexico
Jose Roman Chavez-Mendez, Facultad de Ciencias de la Salud, Universidad Autonoma de Baja California, Mexico
Vitalii Petranovskii, Centro de Nanociencias y Nanotecnologia UNAM, Mexico

Sonochemistry is recently used to improve the zeolitization process and post-synthetic treatments of potential zeolite-based catalysts. This work explores the use of ultrasound to synthesize a faujasite zeolite with a Y-type structure and the reported interconversions associated with FAU. Reviewing the effects of sono-assisted treatments of Y zeolite for the formation of mesopores and modification with alkaline metals, alkaline earth metals, and metals. Based on this specified literature review, preliminary work was conducted, employing ultrasound irradiation to synthesize a selective catalyst for oxidizing benzene to phenol (OBP). The sono-assisted synthesized zeolite Y was sonochemically ion-exchanged and modified in an ammonia buffer solution with dissolved Fe. The catalyst activity for the OBP was measured by UV-Vis spectroscopy and GC/MS, with no significant difference between the techniques, and an enhanced selectivity. This preliminary study highlights the importance of the sonochemical method for potentially designing a catalyst for the benzene-to-phenol reaction.

Chapter 11
Ultrasonic Coating of LTA Zeolites by Antibacterial and Antibiofilm Nanoparticles 395
Jesus Isaias De Leon Ramirez, Centro de Nanociencias y Nanotecnologia UNAM, Mexico
Victor Alfredo Reyes-Villegas, Centro de Nanociencias y Nanotecnologia UNAM, Mexico
Sergio Perez Sicairos, Instituto Tecnologico de Tijuana Centro de Graduados, Mexico
José Román Chávez-Méndez, Universidad Autónoma de Baja California, Mexico
Luis Pérez-Cabrera, Centro de Nanociencias y Nanotecnologia UNAM, Mexico
Verónica González Torres, Universidad Autónoma de Baja California, Mexico
José Constantino González Crisóstomo, Universidad Autónoma de Baja California, Mexico
Ethiel Zavala-Flores, Universidad Autónoma de Baja California, Mexico
Fabian N. Murrieta-Rico, Universidad Politécnica de Baja California, Mexico
Vitalii Petranovskii, Centro de Nanociencias y Nanotecnologia UNAM, Mexico

Sonochemistry uses ultrasound to initiate chemical reactions under extreme conditions, difficult to achieve with conventional methods like hydrolysis or condensation. Ultrasound generates mechanical, transport, and chemical effects in liquids: microjets can erode surfaces, enhancing them for coatings, while localized high temperatures and radicals break molecular bonds. Ultrasound has been employed as an innovative approach to enhance the bactericidal properties of Zeolite A (LTA) by supporting nanoparticles. These improvements are attributed to cavitation effects, which promote adhesion and homogeneous distribution of nanoparticles onto the zeolite surface. Raising attention for environmental and medical applications by reducing the need for additional chemicals and shortening preparation time. The efficiency of ultrasound allows to explore how these modifications influence the bactericidal action of zeolitic materials. Ultrasound presents a more efficient, cleaner, and quicker method for enhancing zeolite-based materials.

Chapter 12
Theoretical Advances in Zeolites 427
Dipan Kumar Das, Centurion University of Technology and Management, India
Padmaja Patnaik, Centurion University of Technology and Management, India
Nibedita Nayak, Centurion University of Technology and Management, India
Sudip Kumar Das, Dr. C.V. Raman University, India

Zeolites, crystalline microporous materials with high thermal and chemical stability, have garnered significant research interest due to their diverse applications in catalysis, adsorption, and ion exchange. Recent theoretical advancements have expanded our understanding of their structural, electronic, and adsorption properties. Studies have demonstrated novel insights into CO_2 adsorption in non-Löwenstein zeolites, revealing enhanced uptake efficiency. Computational approaches, such as QM/MM methods, have provided accurate descriptions of diffusion behaviour within porous frameworks. Investigations into cation-loaded Mg^{2+}-zeolite layers have furthered developments in energy storage applications. Theoretical and experimental characterizations have elucidated the adsorption mechanisms of heavy metals and sulfur in zeolites. Advancements in zeolite-templated carbon structures have led to improved materials for batteries and gas separation. Theoretical studies on Pb-exchanged zeolites and phosphorus-modified frameworks have refined our understanding of their electronic interactions.

Compilation of References 457

About the Contributors 559

Index 563

Preface

The remarkable versatility of zeolites and hierarchical porous materials continues to inspire cutting-edge research across disciplines—from catalysis and environmental remediation to nanotechnology and energy systems. The growing complexity of these materials and their applications demands a multidisciplinary approach that combines advanced characterization techniques, theoretical modeling, and innovative synthesis routes. This edited volume, *Advancements in Zeolites and Micro-Meso Porous Hierarchical Materials*, provides a comprehensive snapshot of current scientific advances and challenges in the field. It brings together contributions from experts across the globe, each presenting their recent research, critical reviews, or technological developments.

The chapters in this book cover a wide array of topics—from the fundamental understanding of zeolite structures and their optical, adsorption, and catalytic properties to emerging applications in water purification, sustainable agriculture, and materials science. The book also addresses recent advances in post-synthetic modifications, theoretical modeling, and the use of sonochemistry to enhance zeolite performance. This diverse yet focused compilation offers both foundational insights and forward-looking perspectives, making it valuable for researchers, graduate students, and professionals working in materials science, environmental engineering, and catalysis.

CHAPTER OVERVIEW

Chapter 1: Optical Properties of Mordenite-Type Zeolite in VUV-UV-Vis and IR Regions

Vitalii Petranovskii and co-authors present a comprehensive analysis of the optical spectra of mordenite, providing valuable baseline data for future guest@host optical materials. Their work spans a broad spectral range and emphasizes the importance of the zeolite matrix's transparency in nanocomposite applications.

Chapter 2: Nuclear Magnetic Resonance to Probe Micro-/Mesoporous Zeolites

Marina Shelyapina explores advanced NMR techniques for characterizing the structure and dynamics of hierarchical zeolites. The chapter highlights how various NMR experiments can elucidate the effects of synthesis and post-synthesis treatments on zeolite properties.

Chapter 3: FTIR Studies of Adsorption Properties of Zeolites

Alexey Tsyganenko compares the adsorption behavior of zeolites and metal oxides using FTIR spectroscopy. The study discusses unique surface features such as Lewis and Brønsted acidity, and the isomerism of adsorbed molecules, enhancing our understanding of zeolitic surface chemistry.

Chapter 4: Influence of Natural Zeolites on the Crystalline Properties of ZnO

Karin Álvarez Gómez and colleagues investigate how various natural zeolites affect the crystal structure and properties of ZnO composites. Using X-ray diffraction, they demonstrate the potential of zeolite-ZnO materials for catalytic and gas purification applications.

Chapter 5: Ion Exchange and Zeolites – Sustainable Approaches to Water Pollution Mitigation

Amita Somya and Amit Varshney review the role of ion exchange and zeolite materials in sustainable water treatment. The chapter outlines mechanisms of pollutant removal and discusses how modified zeolites align with eco-friendly and low-energy consumption goals.

Chapter 6: Comparison of ZSM-5 and MgO-Al2O3 Nanosheet Supports in Hydrodesulfurization

Luis Perez-Cabrera and co-authors present a comparative study on zeolitic and mixed oxide supports for HDS reactions. They detail synthesis methods, catalyst performance, and structural properties, identifying ZSM-5 as a superior support for clean fuel production.

Chapter 7: Clinoptilolite Perspectives as a Zeolite for Removing Pollutants from Water – A Review

Manuel Flores-Hidalgo and his team provide a comprehensive review of clinoptilolite's effectiveness in water purification. They cover its adsorption behavior, structural characterization, and the role of modification techniques in enhancing its environmental performance.

Chapter 8: Are NEREA® Zeolitic Nanostructured Materials Equal NPK Fertilizers Mixed with Natural Zeolites?

Gerardo Rodríguez Fuentes and Inocente Rodríguez-Iznaga introduce NEREA®, a novel zeolitic material that integrates nutrients into the zeolite matrix for improved agricultural performance. The chapter contrasts traditional zeolite-fertilizer mixtures with these innovative composites.

Chapter 9: Post-Synthesis Surfactant-Induced Mesoporosity in Zeolite Crystals

Joaquín Pérez-Pariente and colleagues critically examine post-synthesis strategies using surfactants to induce mesoporosity in large-pore zeolites. They discuss factors influencing mesostructuring and propose guidelines for optimizing materials for catalytic applications.

Chapter 10: Recent Advances in Sonochemistry for the Synthesis and Modification of FeY Zeolite Catalyst

Victor Reyes and co-authors explore the use of ultrasound in the synthesis and modification of FeY zeolite catalysts. Their work illustrates how sonochemical methods enhance catalytic activity and selectivity, particularly for benzene oxidation reactions.

Chapter 11: Ultrasonic Coating of LTA Zeolites by Antibacterial and Antibiofilm Nanoparticles

Jesus Isaias De Leon Ramirez and his team apply sonochemistry to develop antibacterial coatings on LTA zeolites. The chapter demonstrates how ultrasound improves nanoparticle distribution and bonding, with potential applications in environmental and biomedical fields.

Chapter 12: Theoretical Advances in Zeolites

Dipan Kumar Das and co-authors present recent theoretical developments in the study of zeolites. Covering topics from CO_2 adsorption to zeolite-templated carbon materials, this chapter underscores the growing importance of computational methods in material design.

This collection represents the synergistic progress in experimental and theoretical domains of zeolite science. By combining fundamental studies with applied research, this book aims to serve as both a reference and a springboard for future innovation in the field.

Chapter 1
Optical Properties of Mordenite-Type Zeolite in VUV-UV-Vis and IR Regions

Vitalii Petranovskii
https://orcid.org/0000-0002-8794-0593
Center of Nanoscience and Nanotecnology, Mexico

S. V. Kholodkevich
St. Petersburg Federal Research Center of the Russian Academy of Sciences, Russia

Joel Antúnez-García
https://orcid.org/0000-0003-3668-1701
Institute of Advanced Studies of Baja California, Mexico

A. V. Cvetkov
Pskov State University, Russia

K. V. Gusev
Pskov State University, Russia

Marina S. Ivanova
Pskov State University, Russia

Tamara G. Matveeva
Pskov State University Branch in Velikiye Luki, Russia

V. G. Solovyev
Pskov State University, Russia

S. V. Trifonov
Pskov State University, Russia

Roberto Núñez-González
https://orcid.org/0000-0001-7202-2549
Universidad de Sonora, Mexico

ABSTRACT

Zeolites are porous crystalline solids, which can be used as host matrices to create nanocomposite materials by infiltrating their voids with guest substances (clusters of metals, semiconductors, dielectrics). The guest nanospecies are characterized

DOI: 10.4018/979-8-3373-0320-8.ch001

by uniform sizes (~1 nm), regular arrangement of particles repeating the system of voids in crystals, and high concentration (up to ~5×10^20 cm-3). The main benefit of zeolites as matrices for the stabilization of semiconductor clusters is their optical transparency, which makes it possible for light to interact with the entire volume of the composite crystal. To analyze the spectral properties of zeolite-based guest@ host systems, it is very important to know the spectral features of the empty matrix. In this work, the optical spectra of mordenite in a wide region (2.0 –12.7 eV for the VUV-UV-Vis spectra and 450 – 4000 cm–1 for the IR spectra) have been studied in detail. The data obtained are compared with available literature data.

INTRODUCTION

Zeolites are well-known regular porous crystalline solids, which are also recognized as «molecular sieves» (Breck, 1974; Rabo, 1976). They owe this name due to the fact that the cavities and channels of zeolite voids are elements of their crystal structure, and, accordingly, are calibrated in size within the errors of the positioning of atoms in the crystal, while their sizes are close to the sizes of molecules. Accordingly, molecules of larger sizes, exceeding the apertures of pore entrances, are not adsorbed by zeolites. Zeolite-based molecular sieves are essential for many industrial processes such as catalysis, adsorption, ion exchange, and many others. All new forms and applications of these materials are rapidly developing (Lin et al., 2012). The advancement of “green processes” to preserve the environment is forcing the replacement of traditional acidic and basic homogeneous catalysts with solid substances. Among these heterogeneous compounds, zeolites are environmentally friendly to nature, and very important in heterogeneous catalysis due to their homogeneous microporous structure, and the associated size and shape selectivity (Martínez & Corma, 2011). Zeolites and zeolite-based materials, through their use in numerous traditional and sustainable applications, are very important to facilitate the necessary transition to a more cyclic society. A large number of applications and properties of these materials have been discovered over the last decades (van Vreeswijk & Weckhuysen, 2022).

The aluminosilicate frameworks of zeolites are constructed of tetrahedrally coordinated silicon and aluminum ions. TO_4 tetrahedra (T = Si, Al) are connected through the vertices by bridging oxygen ions. Each of the oxygen ions in the Si–O and Al–O bonds is connected to two T cations (Breck, 1974). It is well known that regular tetrahedra do not tile the three-dimensional Euclidean space completely (Conway & Torquato, 2006). Dense crystal structures involving tetrahedrons can be obtained only by a combination of tetrahedrally and octahedrally coordinated ions. Therefore, the formation of pores in such structures is quite natural, and the lace structures of zeolites confirm this by their very existence (Breck, 1974). The cavities and channels of their open structures calibrated by the crystal lattice are regularly arranged in space. These voids, to compensate for the negative electrical charges of the aluminum tetrahedron $[AlO_4]^-$ in their framework, contain off-lattice exchangeable cations (e.g., Na^+, K^+) as well as water molecules. Both cations and water molecules move freely within the zeolite voids. The general chemical formula of these crystalline aluminosilicates (*Formula 1*) (Breck, 1974) is:

$$Me_{x/n}\left[Al_x Si_y O_{2(x+y)}\right] \bullet zH_2O \quad , \qquad (1)$$

where square brackets show the crystal framework and Me^{n+} denotes the ion-exchange cation. The silica to alumina atomic ratio AR, that is, Si/Al = y/x depends on the type of zeolite. For some structures, in particular for mordenite, this ratio may vary depending on the method of its synthesis.

Although the bonds in the single tetrahedra of $[AlO_4]^{5-}$ and $[SiO_4]^{4-}$ exhibit tetrahedral coordination, they vary in length, strength, and reactivity due to differences in atom size (T–O bond length) and electronegativity. Also, the O–T–O bond angles (the ideal O–T–O angle is 109.5°) and T–O–T bonds vary. The average Si–O–Si angle in zeolite frameworks is 152.9°, varying from 127.0° to 180°; the average value for Al–O–Si is 141.9°, varying from 124.8° to 179.3° (Baur & Fischer, 2023). For the most common bond lengths, these authors obtained values of 1.603(11) Å for tetrahedral Si-O and 1.736(8) Å for Al–O. The bond angles in Si–O–Al groups can vary depending on the ratio of SiO_2 to Al_2O_3. The larger size of the $[AlO_4]^-$ tetrahedron creates strain in the crystal structure. As the SiO_2/Al_2O_3 ratio increases, the peak width of the Si–O–Si bond angle distribution decreases. This indicates that if more of the same $[SiO_4]^{4-}$ tetrahedra are bonded together, the structure becomes more ordered.

Zeolites are classified as dielectrics, which means that they have a large bandgap. Silica, as well as alumina, are known for their optical transparency in a wide range of wavelengths; this property is also inherent in aluminosilicates. Accordingly, there is an opportunity to utilize zeolites (as well as microporous silica (Bogomolov, 1978)) as light-permeable host matrices to design a wide variety of unique nanocomposite materials by infiltrating their voids with guest substances (Kumzerov & Vakhrushev, 2004; Solovyev et al., 2009; Seff, 2007). The systems of guest nanoparticles in zeolite host matrices are characterized by small sizes that cannot exceed the nanometer dimensions of the cavities or channels containing them (on a scale of about 0.3 – 1.5 nm), precise arrangement of guest species in accordance with the symmetry of the host zeolite, and their high concentration (up to ~5×10^{20} cm^{-3} for the A-type zeolite structure).

The combination of all these properties makes zeolite voids unique environments for chemical reactions and stabilization of unusual nanostructures (Bein, 2007). Their pore system can consist of one-dimensional, two-dimensional, or three-dimensional channel systems. The pore system can be charged, which implies a population of charge-compensating cations that can be exchanged for other ions if necessary. The crystal structures of pores and channels can act as attractive hosts for the assembly of numerous host-guest systems that can have entirely new properties compared to the parent substances. To control host-guest nanoengineering, the study of non-covalent interactions such as electrostatic interactions, hydrogen bonding, and metal-ligand coordination (zeolite can be considered as a solid polydentate ligand) is required. The confined environment of a nanoscale pore system has a strong influence on electron transfer processes and chemical reactions; both the spatial arrangement of reactants and the limited nanometer-scale reaction space can drastically alter reaction pathways compared to a homogeneous solution. Spatial constraints are also key to the stabilization of nanoscale metal, semiconductor, or dielectric clusters (Braschinsky & Steed, 2022). Confinement chemistry is a growing field with many potential applications in synthetic, catalytic, and especially pharmaceutical industries. Restriction of materials can reveal new properties compared to their bulk (Zhao et al., 2024). Encapsulation of a guest in a host cavity can open new pathways for guest reactivity or halt reaction modes normally observed in solution (Breiner et al., 2011).

Among the currently known 258 types of topologically different zeolite structures (Database, 2025), the so-called "big five" zeolites, FAU (USY), BEA, MOR (mordenite), MFI (ZSM-5), and FER, occupy a special place – five structures that have found the widest application in industry. These zeolites are stable, highly industrious, and their synthesis is easily scalable, making them inexpensive to produce. Therefore, there has been an increased interest in studying these particular types of zeolites.

"*Empty spaces are abhorred by nature, which immediately rushes in to fill the void*" (Tabacchi, 2018). The approach to "ship-in-a-bottle" type syntheses based on zeolite matrices allows the preparation of self-organizing low dimensional nanostructures from 0D clusters and 1D filaments; such materials have remarkable properties that they acquire. due to the synergism of the *guest@host* system (Murrieta-Rico et al., 2021). The combination of zeolites and guest nanoparticles in a single catalyst allows fine-tuning of the properties of such tandem catalysts (Konnov et al., 2024). Examples of stabilization of metallic 0D clusters in such zeolites as mordenite, β-zeolite, Y-zeolite, and ZSM-5 (Petranovskii et al., 2005; Lopez-Bastidas et al., 2012; Petranovskii et al., 2013; Lopez-Bastidas et al., 2013), 1D metallic filaments in mordenite matrix (Romanov, 1993), semiconductor 0D clusters in mordenite and ZSM-5 (Raymond et al., 2007; Susarrey-Arce et al., 2011; Jaime-Acuña et al., 2014), and dielectric nanoparticles dispersed in matrices of zeolites of types A and X (Matveeva et al., 2024; Balabinskaya et al., 2005) have been published.

Numerous data indicate the absence of chemical interactions between zeolite matrices and molecules, clusters, and nanoparticles stabilized in their voids. Zeolite matrices act as inert substrates, spatially isolating the guests and preventing their interaction with each other and/or with the environment. The overall properties of the system in the first approximation are reduced to the sum of the properties of its individual components (Solovyev et al., 2009).

The aim of this work was to study in detail the optical properties of the unfilled aluminosilicate matrix of mordenite-type zeolite and to compare the obtained data with the previously published properties of some other zeolites and zeolite-like aluminophosphates (including NaA and AFI materials studied by us earlier (Solovyev et al., 2009)) in a wide spectral region.

STRUCTURE OF ZEOLITES

Mordenite exists both as a natural mineral and as a synthetic material. *Formula 2* represents the idealized composition of routinely synthesized mordenite in its sodium form (zeolite NaMOR):

$$Na_8\left[Al_8Si_{40}O_{96}\right] \bullet 24H_2O$$

$$(AR = 5). \qquad (2)$$

Figure 1 schematically shows its structure. Mordenite belongs to the orthorhombic crystal lattice family, with space group *Cmcm* (Database, 2025). It possesses a two-dimensional channel system. Two straight channels are distinguished in its structure, whose major axis is parallel to the *c*-direction. In cross section, these elliptical channels are formed by 8- and 12-membered rings of tetrahedra (8MR and 12MR), with cross sections of 0.67 × 0.70 *nm* and 0.29 × 0.57 *nm*, respectively. The channels are interconnected by so-called “side pockets”. A detailed description of the positions of the Si, Al, and O atoms in the crystal framework, as well as the positions of the ion-exchangeable sodium cations, can be found in (Antúnez-García et al., 2022; Antúnez-García et al., 2024a). Three of these positions (see (Devautour et al., 1999; Maurin et al., 2002)) are shown in Roman numerals in Figure 1.

Figure 1. MOR-type zeolite (mordenite) framework. Roman numerals I, IV, VI denote the most populated positions of Na^+ cations

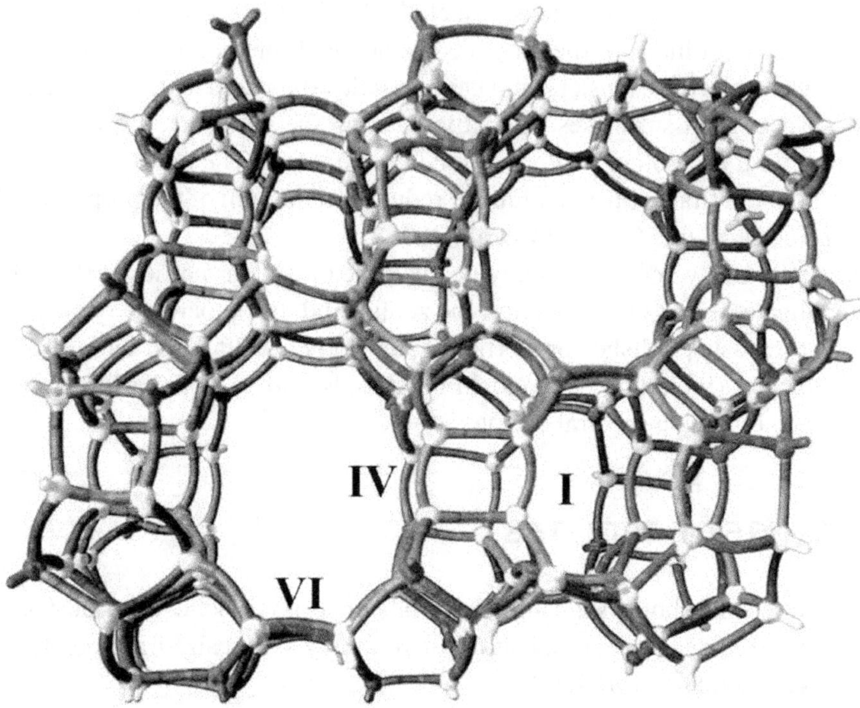

(Solovyev et al., 2009; Devautour et al., 1999; Maurin et al., 2002).

The framework of synthetic NaA zeolite belongs to the cubic crystal lattice family, with space group $Pm\bar{3}m$ *(Database, 2025), with composition* (Formula 3):

$(Na_{12}[Al_{12}Si_{12}O_{48}] \cdot 27H_2O)$ *(3)*

Its crystal structure contains two types of cavities – α– cages (supercages) and β– cages (sodalite cages), with inner diameters of 1.14 nm and 0.66 nm, respectively (Figure 2). The diameter of the windows between adjacent α– and β– cages is 0.22 nm; the α– cages themselves are interconnected by windows with a diameter of 0.42 nm. The lattice of voids of zeolite LTA is cubic with a lattice constant of 1.23 nm (for the pseudo-elemental cell). Thus, the concentration of α– or β– cages in zeolite LTA is $(1.23\ nm)^{-3} \approx 5\times10^{20}\ cm^{-3}$.

Figure 2. (a) LTA-type zeolite framework and (b) structural assembly between an α– and a β–cage.

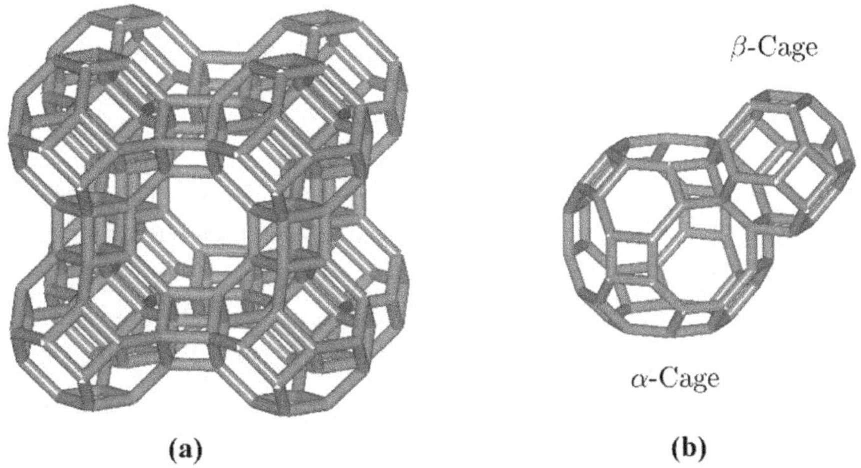

Aluminophosphates represent a special case compared to aluminosilicates. The TO_4 tetrahedra connected by bridging oxygen atoms have a total charge equal to zero for SiO_4, and -1 for AlO_4. Phosphorus, whose valence is +5, provides the PO_4 tetrahedron with a total charge of +1. Since the concentration of aluminum atoms in the AFI ($AlPO_4$–5) aluminophosphate is equal to the concentration of phosphorus atoms, the total charge of the crystal structure is zero, and therefore its pores are free of exchangeable cations present in aluminosilicate structures.

Figure 3. Zeolite-like aluminophosphate AFI framework.

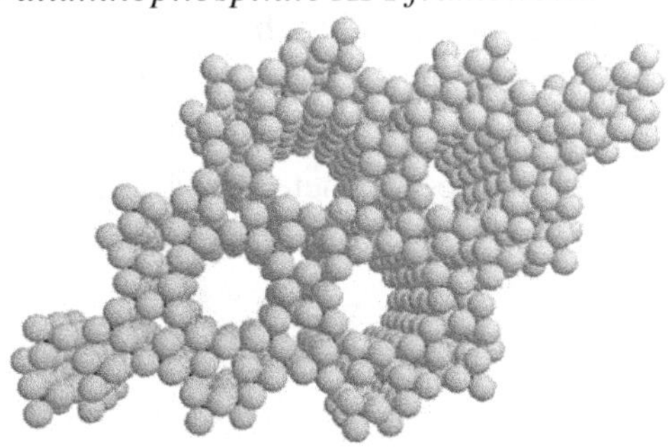

The image was taken from (Solovyev et al., 2009).

The parallel channels of AFI crystal with an inner diameter of 0.73 nm are formed by 12-membered rings of AlO_4 and PO_4 tetrahedra and are oriented along the **c**-axis of the crystal (Figure 3). These pores are very convenient for infiltration by guest substances. The ideal formula of the AFI unit cell framework (Formula 4) is:

$$Al_{12}P_{12}O_{48} \tag{4}$$

and the unit cell parameters are a = 1.37 nm, c = 0.84 nm. The framework of synthetic AFI zeolite belongs to the hexagonal crystal lattice family, with space group P6/mcc (Database, 2025).

EXPERIMENTAL PROCEDURE

Commercial synthetic mordenite produced by the Experimental Laboratory at Kazan Petrochemical Plant, Russia, was used in the present work. Diffuse reflectance spectra $R_d = f(\hbar\omega)$ of powdered samples were determined according to the standard method (Figure 4). Magnesium oxide served as a reference sample. An SF-16 spectrometer equipped with an integrating sphere PDO-1 for diffuse reflectance measurements was used in the UV-Vis spectral range of photon energies $E = \hbar\omega$ from 2.0 to 5.6 eV. Vacuum ultraviolet (VUV) spectra ($4.4\ eV \leq E \leq 12.7$ eV) were obtained on a Beckman spectrometer. To measure the VUV, the powder was pressed into tablets, dehydrated at 350 °C for 4 h, and then transferred to the vacuum chamber of the spectrometer. The pressure in the chamber in the range of about 10^{-2} Pa during the measurement was provided by a diffusion vacuum pump. Thus, this portion of the spectrum was obtained for dehydrated mordenite. Hydrated mordenite powder at atmospheric pressure was used for the UV-Vis measurement. Despite this difference, a good agreement of values was observed in the overlapping region of both spectra. Infrared (IR) spectra in the wavelength range $450\ cm^{-1} \leq \lambda^{-1} \leq 4000$ cm^{-1} were recorded on a Fourier spectrometer FSM 1201, for hydrated mordenite, also at atmospheric pressure. All spectra were measured at room temperature.

Figure 4. Experimental setup for measuring diffuse reflectance spectra.

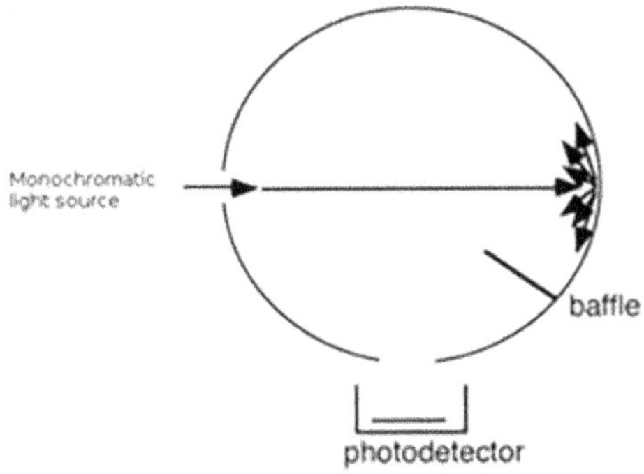

There exist many theoretical approaches allowing to convert the experimentally measured diffuse reflectance spectra $R_d = f(\hbar\omega)$ into the optical absorption of the sample in relative units. The most commonly used is the classical Gurevich – Kubelka – Munk function (*Equation 5*) (Gurevich, 1930); Kubelka & Munk, 1931; Vogt et al., 2023), assuming that the sample thickness is semi-infinite:

$$F_{GKM}(R_d) = \frac{(1 - R_d)^2}{2R_d}$$

. (5)

In some cases, the Simmons function (Simmons, 1975) (*Equation 6*) is also used for the same purpose:

$$F_S(R_d) = \frac{(ln R_d)^2}{2}$$

. (6)

Figure 5. Gurevich – Kubelka – Munk (1) and Simmons (2) functions converting measured reflectance spectra$R_d(\hbar\omega)$ into the optical absorption in relative units.

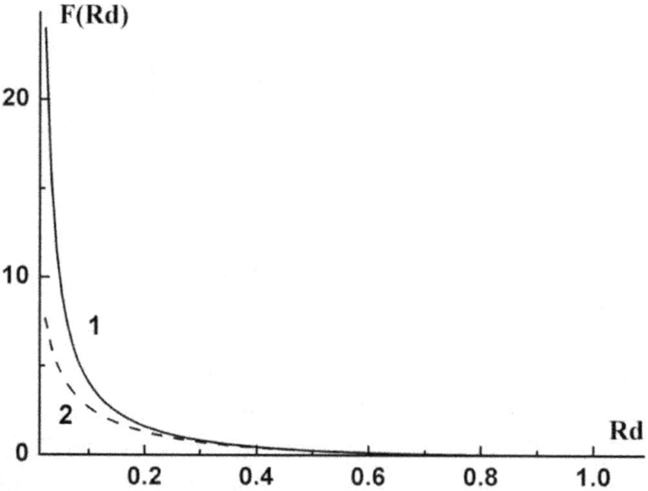

At large values of R_d ($x = 1 - R_d \ll 1$) these functions coincide ($F_{GKM}(R_d) = F_S(R_d) \approx \frac{x^2}{2}$), but at large absorption $F_{GKM}(R_d) > F_S(R_d)$ (Figure 5).

RESULTS AND DISCUSSION

Figures 6 and 7 represent the plots of the Gurevich – Kubelka – Munk and Simmons functions, respectively, extracted from the diffuse reflectance spectra for mordenite (Formula 2) measured in this work. Both plots show a pronounced absorption edge (its spectral position is $\hbar\omega$ = (7.5 – 8.0) eV), as well as four absorption bands near 4.9, 6.5, 8.5, and 10.5 eV. The width of the forbidden zone for the studied mordenite corresponds to the intersection point of the straight line; according to Figures 6 and 7 they are 6.9 and 6.7 eV, respectively.

Assuming a parabolic approximation to the valence and conduction bands, one can obtain the value of the energy required for the transition of a valence electron into a conduction band. The minimum energy is named E_g, the band gap, or the absorption edge. In the case where the vertices of the parabola have the same wave-vector or momentum, this is the direct bandgap, and it can be shown that the linear absorption coefficient depends on E_g (*Equation 7*):

$$\alpha\hbar\omega = A\left(\hbar\omega - E_g\right)^n$$

,   (7)

where α is the linear absorption coefficient, ω is the frequency of light, ħ is the Planck constant, and A is the proportionality constant. The degree of the parenthesis, n, is taken to be equal to ½ for materials with a direct band gap. When the incident radiation scatters in a perfectly diffuse manner, the absorption coefficient K becomes equal to 2α. In this case, considering the scattering coefficient S as constant with respect to the wavelength, the Kubelka – Munk function is proportional to the absorption coefficient α (Gurevich, 1930), and applying eq. (7), we obtain the Relation 8:

$$\left[F_{GKM}\left(R_{\infty}\right)\hbar\omega\right]^{2} = A\left(\hbar\omega - E_{g}\right). \tag{8}$$

The high-energy absorption peaks near the absorption edge (Figures 6 and 7) are probably due to the formation of excitons. Note that various types of excitonic absorption of SiO_2 in the spectral range of photon energies from 8.5 to 10.7 eV were also found in (Trukhin, 1992). It should be noted that the excited states of AlO_4 and SiO_4 tetrahedra in the zeolite structure can also appear at 9 eV and 10 eV, respectively (Devautour et al., 1999).

Optical absorption of the dangling oxygen bonds was observed in the vacuum ultraviolet spectral region of 6.5 – 6.8 eV (Hosono et al., 2002; Skuja et al., 2011). The authors suggested that this absorption band is due to transitions to a higher-energy excited state in NBOHC in amorphous SiO_2. The absorption band near 7 eV in MOR-type zeolite (Figures 6 and 7) may have the same origin. Recent theoretical studies (Antúnez-García et al., 2024b) show that for sodium mordenites with Si/Al ratio equal to 7 and equal to 5, the calculated band gaps are 6.7 eV and approximately 7.0 eV, respectively. It is noteworthy that for Si/Al=5, the theoretical band gap closely matches the value obtained using the Kubelka-Munk function for the mordenite used (Fig. 6). Moreover, the theoretical results show that the band gap arises predominantly from the tetrahedral Si-O and Al-O interactions. McComb and Howie (McComb, & Howie, 1990) report a band gap of 8.0 eV for protonated mordenite (also with a Si/Al ratio of 5), emphasizing the influence of the exchangeable cation on the zeolite properties. Thus, the E_g value in a particular mordenite sample is a function of the chemical composition of its crystalline matrix (Si/Al value), and the type of ion-exchangeable extraframework cations. For comparison, a gap energy of about 6.5 eV in NaA zeolite, which differs from mordenite in lattice topology and composition, with a Si/Al ratio of 1, was theoretically estimated in (Núñez-González et al., 2023). On the other hand, the electronic absorption of OH^- groups and H_2O molecules can also occur in the spectral range of 6.5–8.0 eV. Note that the broad IR absorption band at about 3500 cm^{-1} (Figure 8) corresponds to vibrations of OH^- groups (Breck, 1974; Rabo, 1976; Luque-Alvarez et al., 2024; Pechar & Rykl, 1983; Rahman et al., 2009). This suggests that the range of transition energies for oxygen in these systems is relatively narrow.

The absorption at 4.8 eV observed in amorphous and crystalline SiO_2 is usually attributed to nonbridging oxygen hole centers (NBOHCs) (Trukhin, 1992; Skuja et al., 2011). We hypothesize that similar defects with absorption near 4.9 eV can be found in the aluminosilicate structure of mordenite with a sufficiently high Si/Al ratio of 5.

This assumption coincides with our observation of an infrared (IR) absorption band near $\nu_1^{NaM} \approx 780\,\text{cm}^{-1}$ (Figure 8), which is probably due to local vibrations of the nonbridging oxygen atom. It should be noted that IR absorption at 850 cm^{-1} has also been attributed to nonbridging oxygen vibrations in SiO_2. Similar IR absorption in the spectral region of $(790-800)\,\text{cm}^{-1}$ was also observed in NaY-type zeolite (Titova et al., 1993). An alternative explanation for this IR absorption was given in (Núñez-González et al., 2023; Król et al., 2019; Olla & Carbonaro, 2024), where bands at about $(700-815)\ \text{cm}^{-1}$ are explained by external symmetric stretching vibrations of Si-O-Si bond linking two tetrahedra.

A broad IR band at about $\nu_2 \approx 1000\,\text{cm}^{-1}$, which has been attributed to asymmetric stretching vibrations of $T\text{–}O$ bonds in TO_4 tetrahedra, is characteristic of all zeolites and zeolite-like materials (Breck, 1974; Rabo, 1976; Antúnez-García et al., 2025; Luque-Alvarez et al., 2024; Rahman et al., 2009; Król et al., 2019). However, this band in aluminophosphate crystal AFI ($AlPO_4$–5) is shifted by ~100 cm^{-1} towards a higher frequency ($\nu_2^{AFI} \approx 1080\,\text{cm}^{-1}$) in comparison with aluminosilicate zeolite NaA (where $\nu_2^{NaA} \approx 980\ \text{cm}^{-1}$) (Solovyev et al., 2009). This considerable high-frequency shift is due to the difference in T–O bond lengths: the P–O bond distance (0.154 nm) in zeolite-like aluminophosphate AFI is shorter than either Si–O (0.161 nm) or Al–O (0.175 nm) bond lengths in zeolite NaA.

On the other hand, the aluminosilicate zeolite mordenite exhibits this broad absorption band in the IR range at $\nu_2^{NaM} \approx 1060\,\text{cm}^{-1}$(Figure 8). Thus, it can be concluded that $\nu_2^{NaA} < \nu_2^{NaM} < \nu_2^{AFI}$. This effect is due to the difference in the lengths of the above-mentioned bonds as well as to the *Si/Al* ratios: $R^{NaA} \approx 1 < R^{NaM} \approx 5$. A similar trend was also observed for the stretching modes in sodalites: substituting a silicon atom with aluminum causes both asymmetric and symmetric stretching modes to shift to lower wave numbers (Olla & Carbonaro, 2024). Note that ratio $\nu_2^{NaM}:\nu_1^{NaM} \approx 1060\,\text{cm}^{-1} : 780\,\text{cm}^{-1} \approx 1.36 \approx \sqrt{2}$; this value correlates with a simple theoretical estimation for the vibrational frequencies of bridging and nonbridging oxygen.

Figure 6. Gurevich – Kubelka – Munk function plot extracted from the diffuse reflectance spectra for mordenite.

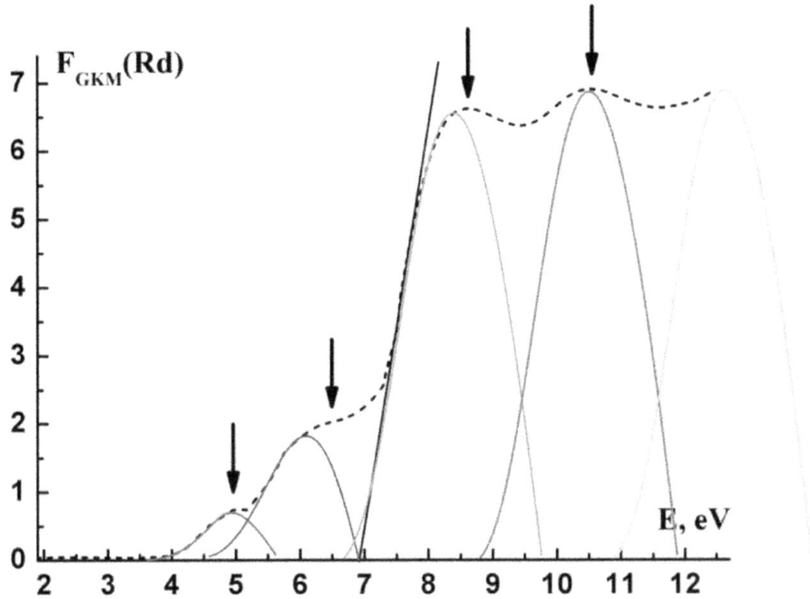

The black arrows indicate different absorption maxima. The red arrow marks the line derived from the Tauc plot method, corresponding to the first prominent absorption peak. In this case, the method yields a band gap of 6.9 eV.

Figure 7. Simmons function plot extracted from the diffuse reflectance spectra for mordenite.

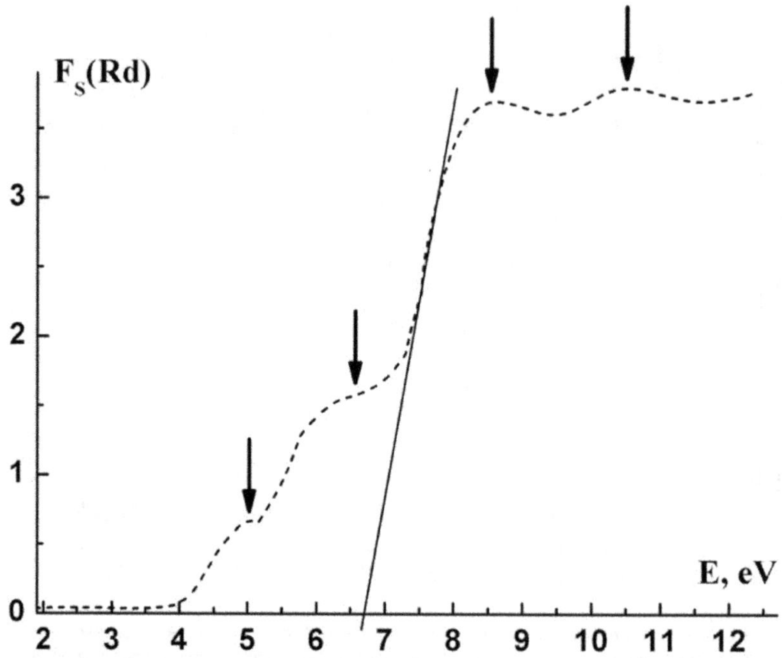

The black arrows indicate different absorption maxima. The red arrow marks the line derived from the Tauc plot method, corresponding to the first prominent absorption peak. In this case, the method yields a band gap of 6.7 eV.

Figure 8. Infrared spectrum of mordenite.

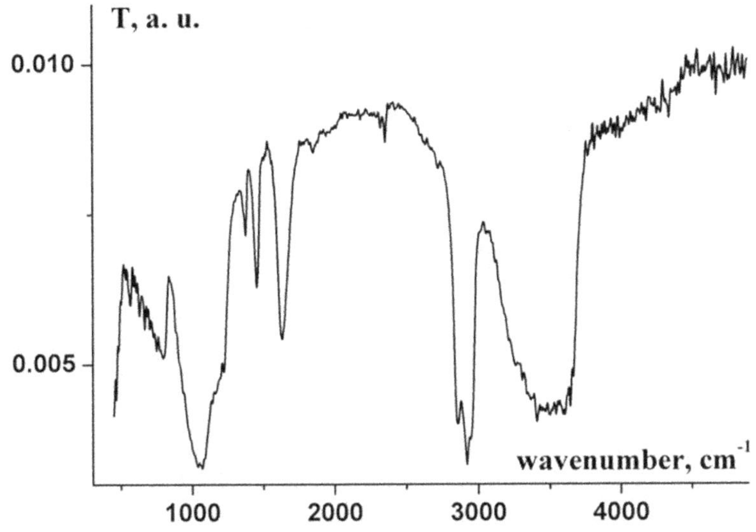

CONCLUSION

In the present work, optical absorption spectra of mordenite of standard composition (sodium form, AR equal to 5) were obtained in the visible, ultraviolet, and vacuum ultraviolet regions, in the photon energy range of $2.0\,\mathrm{eV} \leq \hbar\omega \leq 12.7\,\mathrm{eV}$*, as well as infrared transmittance in the spectral range of* $450\,cm^{-1} \leq \nu \leq 4000\,cm^{-1}$. The data obtained were compared with previously known published data for other aluminosilicate structures and for zeolite-like aluminophosphate AFI. The optical absorption of the zeolite framework as well as that of the defects in the structure was identified. The band assignments are compared with those published in the literature.

ACKNOWLEDGEMENTS

The authors are grateful to Dr. Yu.A. Kumzerov and Dr. R. Machorro for inspiring and fruitful discussions during the preparation of this article. Also, the authors thank the Department of Zeolite Research of the Institute of Science of the BUAP (ICUAP), and its staff, for fruitful discussions of the results obtained. Computing time was provided by the National Supercomputing Laboratory of Southeast Mexico (Project 202301020C). The financial support from DGAPA-PAPIIT Project IG101623 is gratefully acknowledged.

REFERENCES

Antúnez-García, J., Galván, D., Petranovskii, V., Murrieta-Rico, F. N., Yocupicio-Gaxiola, R. I., Shelyapina, M. G., & Fuentes-Moyado, S. (2022). Aluminum distribution in mordenite-zeolite framework: A new outlook based on density functional theory calculations. *Journal of Solid State Chemistry*, *306*, 122725. DOI: 10.1016/j.jssc.2021.122725

Antúnez-García, J., Núñez-González, R., Petranovskii, V., Hmok, H., Reyes-Serrato, A., Murrieta-Rico, F. N., Xiao, M., & Zamora, J. (2025). Theoretical study of the electronic and optical properties of a composite formed by the zeolite NaA and a magnetite cluster. *Beilstein Journal of Nanotechnology*, *16*, 44–53. DOI: 10.3762/bjnano.16.5 PMID: 39834743

Antúnez-García, J., Petranovskii, V., Murrieta-Rico, F. N., Yocupicio-Gaxiola, R. I., Shelyapina, M. G., & Fuentes-Moyado, S. (2024a). Investigating patterns in aluminum distribution and stability in mordenite zeolite structure: Analysis and theoretical conclusions from existing experimental data. *Journal of Solid State Chemistry*, *331*, 124548. DOI: 10.1016/j.jssc.2024.124548

Antúnez-García, J., Ponce-Ruiz, J. L., Núñez-González, R., Petranovskii, V., Murrieta-Rico, F. N., Serrato, A. R., Shelyapina, M. G., Xiao, M., & Zamora, J. (2024b). Theoretical investigation of Fe and Cu cations hosted within the MOR zeolite framework. *Materials Today. Communications*, *110418*, 110418. Advance online publication. DOI: 10.1016/j.mtcomm.2024.110418

Balabinskaya, A. S., Ivanova, E. N., Ivanova, M. S., Kumzerov, Y. A., Pankova, S. V., Poborchii, V. V., Romanov, S. G., Solovyev, V. G., & Khanin, S. D. (2005). Investigation into the electrical and optical properties of sodium nitrite and sodium nitrate nanoparticles in regular porous matrices. *Glass Physics and Chemistry*, *31*(3), 330–336. DOI: 10.1007/s10720-005-0064-2

Baur, W. H., & Fischer, R. X. (2023). The flexibility of the T–X–T hinges between the coordination tetrahedra in various zeolitic frameworks: An empirical structural study. *Mineralogy and Petrology*, *117*(2), 165–179. DOI: 10.1007/s00710-023-00813-6

Bein, T. (2007). Host–Guest interactions in zeolites and periodic mesoporous materials. In Studies in surface science and catalysis (pp. 611-657, XVII-XIX). https://doi.org/DOI: 10.1016/S0167-2991(07)80806-8

Bogomolov, V. N. (1978). Liquids in ultrathin channels (Filament and cluster crystals). *Soviet Physics - Uspekhi*, *21*(1), 77–83. DOI: 10.1070/PU1978v021n01ABEH005510

Braschinsky, A., & Steed, J. W. (2022). Molecular clusters in confined spaces. *Coordination Chemistry Reviews*, *473*, 214840. DOI: 10.1016/j.ccr.2022.214840

Breck, D. W. (1974). Zeolite molecular sieves (9th ed.; Donald W. Breck, Ed.). Nashville, TN: John Wiley & Sons.

Breiner, B., Clegg, J. K., & Nitschke, J. R. (2010). Reactivity modulation in container molecules. *Chemical Science (Cambridge)*, *2*(1), 51–56. DOI: 10.1039/C0SC00329H

Conway, J. H., & Torquato, S. (2006). Packing, tiling, and covering with tetrahedra. *Proceedings of the National Academy of Sciences of the United States of America*, *103*(28), 10612–10617. DOI: 10.1073/pnas.0601389103 PMID: 16818891

Database of Zeolite Structures. (Accessed Feb. 4, 2025) (s. f.). http://www.iza-structure.org/databases

Devautour, S., Henn, F., Giuntini, J., Zanchetta, J., & Vanderschueren, J. (1999). TSDC relaxation map analysis in a Na–mordenite zeolite. *Solid State Ionics*, *122*(1–4), 105–111. DOI: 10.1016/S0167-2738(99)00048-X

Gurevich, M. (1930). Über eine rationelle klassifikation der lichtenstreuenden medien. *Physikalische Zeitschrift*, *31*, 753–763.

Hosono, H., Kajihara, K., Suzuki, T., Ikuta, Y., Skuja, L., & Hirano, M. (2002). Vacuum ultraviolet optical absorption band of non-bridging oxygen hole centers in SiO_2 glass. *Solid State Communications*, *122*(3–4), 117–120. DOI: 10.1016/S0038-1098(02)00118-7

Jaime-Acuña, O. E., Villavicencio, H., Díaz-Hernández, J. A., Petranovskii, V., Herrera, M., & Raymond-Herrera, O. (2014). Atomic and Electronic structure of quaternary $Cd_xZn_yS_\delta O_\gamma$ nanoparticles grown on mordenite. *Chemistry of Materials*, *26*(21), 6152–6159. DOI: 10.1021/cm5024585

Konnov, S. V., Bruter, D. V., Pavlov, V. S., & Ivanova, I. I. (2024). State-of-the-art strategies for the synthesis of zeolite-encapsulated subnanometric metal clusters. *Inorganic Chemistry Frontiers*, *11*(13), 3669–3706. DOI: 10.1039/D4QI00703D

Król, M., Koleżyński, A., Mikuła, A., & Mozgawa, W. (2018). Vibrational spectroscopy of zeolites. In Challenges and advances in computational chemistry and physics (pp. 301–332). DOI: 10.1007/978-3-030-01355-4_10

Kubelka, P., & Munk, F. (1931). Ein Beitrag zur Optik der Farbanstriche. Z. für techn. *Phys.*, *12*(Nr. 11a), 593–601.

Kumzerov, Y., & Vakhrushev, S. (2004). Nanostructures within Porous Materials. In Nalwa, H. S. (Ed.), *Encyclopedia of Nanoscience and Nanotechnology* (Vol. VII, pp. 811–849). American Scientific Publishers.

Lin, C. C. H., Dambrowitz, K. A., & Kuznicki, S. M. (2011). Evolving applications of zeolite molecular sieves. *Canadian Journal of Chemical Engineering*, *90*(2), 207–216. DOI: 10.1002/cjce.20667

López-Bastidas, C., Petranovskii, V., & Machorro, R. (2012). Optical response of Cu clusters in zeolite template. *Journal of Colloid and Interface Science*, *375*(1), 60–64. DOI: 10.1016/j.jcis.2012.02.036 PMID: 22440728

López-Bastidas, C., Smolentseva, E., Petranovskii, V., & Machorro, R. (2013). Plasmon features of coinage metal nanoparticles supported on zeolites. *Plasmonics*, *8*(4), 1551–1558. DOI: 10.1007/s11468-013-9571-5

Luque-Álvarez, L. A., Torres-Sempere, G., Romero-Sarria, F., Bobadilla, L. F., Ramírez-Reina, T., & Odriozola, J. A. (2024). Controlling copper location on exchanged MOR-type aluminosilicate zeolites for methanol carbonylation: In situ/operando IR spectroscopic studies. *Microporous and Mesoporous Materials*, *378*, 113258. DOI: 10.1016/j.micromeso.2024.113258

Martínez, C., & Corma, A. (2011). Inorganic molecular sieves: Preparation, modification and industrial application in catalytic processes. *Coordination Chemistry Reviews*, *255*(13–14), 1558–1580. DOI: 10.1016/j.ccr.2011.03.014

Matveeva, T. G., Ivanova, M. S., Solovyev, V. G., & Vanin, A. I. (2024). Electrophysical properties of the Rochelle salt dispersed in a porous dielectric matrix of type A zeolite. *Physics of Complex Systems.*, *5*(4), 195–201. DOI: 10.33910/2687-153X-2024-5-4-195-201

Maurin, G., Senet, P., Devautour, S., Henn, F., & Giuntini, J. (2002). Modelling of the cation motions in complex system: Case of Na-mordenites. *Journal of Non-Crystalline Solids*, *307–310*, 1050–1054. DOI: 10.1016/S0022-3093(02)01571-5

McComb, D., & Howie, A. (1990). Characterisation of zeolite catalysts using electron energy loss spectroscopy. *Ultramicroscopy*, *34*(1–2), 84–92. DOI: 10.1016/0304-3991(90)90062-Q

Murrieta-Rico, F. N., Lindner, L., Petranovskii, V., Galván, D. H., Antúnez-García, J., Yocupicio-Gaxiola, R. I., Sergiyenko, O., & Rivas-López, M. (2021) Zeolites with quantum dots as functional materials: current trends and perspectives for optical devices. *Proceedings of 2021 IEEE 30th International Symposium on Industrial Electronics (ISIE)*. Kyoto, Japan, P. 1–5. DOI: 10.1109/ISIE45552.2021.9576435

Núñez-González, R., Xiao, M., Antúnez-García, J., Ponce-Ruiz, J. L. A., Reyes-Serrato, A., Petranovskii, V., Murrieta-Rico, F. N., & Rakov, N. (2023). First-Principles Study of optical properties of Linde-Type A zeolite. *Physica Status Solidi. B, Basic Research*, *260*(12), 2300378. Advance online publication. DOI: 10.1002/pssb.202300378

Olla, C., & Carbonaro, C. M. (2024). The void side of silica: Surveying optical properties and applications of mesoporous silica. *Journal of Physics Condensed Matter*, *36*(25), 253002. DOI: 10.1088/1361-648X/ad3474 PMID: 38545962

Pechar, F., & Rykl, D. (1983). Study of the complex vibrational spectra of natural zeolite mordenites. *Zeolites*, *3*(4), 329–332. DOI: 10.1016/0144-2449(83)90177-X

Petranovskii, V., Gurin, V., & Machorro, R. (2005). Spectroscopic observation and ab initio simulation of copper clusters in zeolites. *Catalysis Today*, *107–108*, 892–900. DOI: 10.1016/j.cattod.2005.07.039

Petranovskii, V., Stoyanov, E., Gurin, V., Katada, N., Hernández, M.-A., Avalos, M., Pestryakov, A., Chávez Rivas, F., Zamorano Ulloa, R., & Portillo, R. (2013). Formation of copper nanoparticles in mordenites with variable SiO_2/Al_2O_3 molar ratios under redox treatments. *Revista Mexicana de Física*, *59*(2), 170–185.

Rabo, J. A. (Ed.). (1976). *Zeolite Chemistry and Catalysis*. American Chemical Society.

Rahman, I., Vejayakumaran, P., Sipaut, C., Ismail, J., & Chee, C. (2008). Size-dependent physicochemical and optical properties of silica nanoparticles. *Materials Chemistry and Physics*, *114*(1), 328–332. DOI: 10.1016/j.matchemphys.2008.09.068

Raymond, O., Villavicencio, H., Flores, E., Petranovskii, V., & Siqueiros, J. M. (2007). Spectral-Optic response of ZnS and mixed Cd_xZn_yS nanoclusters on synthetic mordenites. *The Journal of Physical Chemistry. C, Nanomaterials and Interfaces*, *111*(28), 10260–10266. DOI: 10.1021/jp072417y

Romanov, S. (1993). Electronic structure of the minimum-diameter Tl, Pb and Bi quantum wire superlattices. *Journal of Physics Condensed Matter*, *5*(8), 1081–1090. DOI: 10.1088/0953-8984/5/8/009

Seff, K. (2007). Doing chemistry in a one-nanometer test tube (in a zeolite). *Comments on Modern Chemistry. Part A, Comments on Inorganic Chemistry : a Journal of Critical Discussion of the Current Literature*, *28*(5–6), 173–179. DOI: 10.1080/02603590701787415

Simmons, E. L. (1975). Diffuse reflectance spectroscopy: A comparison of the theories. *Applied Optics*, *14*(6), 1380. DOI: 10.1364/AO.14.001380 PMID: 20154834

Skuja, L., Kajihara, K., Hirano, M., & Hosono, H. (2011). Visible to vacuum-UV range optical absorption of oxygen dangling bonds in amorphous SiO_2. *Physical Review B: Condensed Matter and Materials Physics*, *84*(20), 205206. Advance online publication. DOI: 10.1103/PhysRevB.84.205206

Solovyev, V. G., Ivanova, M. S., Pankova, S. V., Trifonov, S. V., & Veisman, V. L. (2009). Preparation and physical properties of zeolite, zeolite-like single crystals and zeolite-based nanocomposite materials. In Wong, T. W. (Ed.), *Handbook of Zeolites: Structure, Properties and Applications* (pp. 77–99). Nova Science Publishers.

Susarrey-Arce, A., Petranovskii, V., Hernández-Espinosa, M. A., Portillo, R., & de la Cruz, W. (2011). Optical properties of ZnO nanoparticles on the porous structure of mordenites and ZSM-5. *Journal of Nanoscience and Nanotechnology*, *11*(6), 5574–5579. DOI: 10.1166/jnn.2011.3426 PMID: 21770221

Tabacchi, G. (2018). Supramolecular organization in confined nanospaces. *ChemPhysChem*, *19*(11), 1249–1297. DOI: 10.1002/cphc.201701090 PMID: 29573368

Titova, T. I., Kosheleva, L. S., Zhdanov, S. P., & Shubaeva, M. A. (1993). IR spectroscopic study of structure-chemical aspects of the Na-Y zeolite dealumination with ethylene diamine tetraacetic acid. *Pure and Applied Chemistry*, *65*(10), 2231–2236. DOI: 10.1351/pac199365102231

Trukhin, A. N. (1992). Excitons in SiO_2: A review. *Journal of Non-Crystalline Solids*, *149*(1–2), 32–45. DOI: 10.1016/0022-3093(92)90052-L

van Vreeswijk, S. H., & Weckhuysen, B. M. (2022). Emerging analytical methods to characterize zeolite-based materials. *National Science Review*, *9*(9), nwac047. Advance online publication. DOI: 10.1093/nsr/nwac047 PMID: 36128456

Vogt, C., Wondergem, C. S., & Weckhuysen, B. M. (2023). Ultraviolet-Visible (UV-Vis) spectroscopy. In Springer Handbooks (pp. 237–264). https://doi.org/DOI: 10.1007/978-3-031-07125-6_11

Zhao, Y., Zhu, L., Kang, Y., Shen, C., Liu, X., Jiang, D., Fu, L., Guselnikova, O., Huang, L., Song, X., Asahi, T., & Yamauchi, Y. (2024). Nanoengineering multilength-scale porous hierarchy in mesoporous metal–organic framework single crystals. *ACS Nano*, *18*(33), 22404–22414. DOI: 10.1021/acsnano.4c07119 PMID: 39108023

Chapter 2
Nuclear Magnetic Resonance to Probe Micro–/Mesoporous Zeolites

Marina G. Shelyapina
https://orcid.org/0000-0003-3769-941X
Saint Petersburg State University, Russia

ABSTRACT

Zeolites are widely used in industry as molecular sieves, ion exchangers and catalysts. According to recent studies mesoporosity is beneficial to the rational design of catalysts with controlled product selectivity and the improved catalyst lifetime due to the efficient mass-transport properties. To tailor structures and morphology of zeolite catalysts for specific reactions, it is necessary to understand the effects of various modifications (either during synthesis or post-synthetic treatments) on physico-chemical properties of the target material. Nuclear magnetic resonance (NMR) has proven to be a reliable method for studying zeolite structure, surface acidity and dynamics of adsorbed molecules. This Chapter reviews recent NMR studies of zeolites with hierarchical porosity mainly focusing on, structural studies by ^{27}Al *and* ^{29}Si *NMR spectroscopy, 2D* T_1*-*T_2 *correlation and* T_2*-*T_2 *exchange experiments and NMR cryoporometry.*

DOI: 10.4018/979-8-3373-0320-8.ch002

INTRODUCTION

Zeolites are crystalline aluminosilicates of alkali and alkaline earth metals with a variable composition. Their properties, such as high ion exchange capacity, action as molecular sieves, and catalytic activity make them very attractive for use in a number of industrial processes. To tailor structures and properties of zeolites or zeolite-based materials for specific applications, it is necessary to understand the effects of various modifications (either during synthesis or post-synthetic treatments) on the physico-chemical properties of the material. Nuclear magnetic resonance (NMR) has proven to be a reliable tool for a comprehensive study of the structure and properties of zeolites (Grey, 2003; Klinowski, 1984; Shelyapina et al., 2020; Zheng et al., 2024). ^{29}Si Magic Angle Spinning (MAS) NMR spectroscopy is an important tool for determining the crystalline structure of zeolites (Ashbrook et al., 2014; Brouwer et al., 2020; Dawson et al., 2017) relying on the ^{29}Si isotropic chemical shift sensitive to local environment.

However, NMR crystallography is not limited only to isotropic ^{29}Si chemical shifts. Other nuclear spin interactions, including dipole coupling, chemical shift anisotropy, quadrupole coupling (for nuclei spin $I > 1/2$) and J coupling, provide additional structural information (Srivastava et al., 2025). ^{27}Al NMR combined with ^{29}Si NMR allows quantifying the Si/Al ratio, which is also considered an analog of the acidity of zeolite (Barthomeuf, 1987; Shelyapina et al., 2022). Moreover, ^{27}Al MAS NMR allows distinguishing framework and extra-framework aluminum species; the latter are the issue of Lewis acidity (Ravi et al., 2020; Shelyapina et al., 2022). ^{1}H MAS NMR spectroscopy provides a direct quantification of the density of Brønsted acid sites (Kanellopoulos et al., 2008). The use of various probing molecules (water (Buntkowsky & Vogel, 2020; Huo et al., 2009; Krylova et al., 2018; Shelyapina et al., 2023), phosphine oxides (Bornes et al., 2019, 2021; Lewis et al., 2018; Zasukhin et al., 2022), pyridine (Gunther et al., 2016; Wang et al., 2022), acetone (Biaglow et al., 1994; Yu et al., 2012), carbon dioxide (Yakimov et al., 2022) etc.) opens up wide possibilities of indirect characterization of zeolite acid sites. Hence, studying the strength and spatial distribution of active sites on zeolite surface multi-dimensional and multi-quantum NMR spectroscopy provides unique knowledge (Amoureux & Pruski, 2008; Ashbrook et al., 2014; Zheng et al., 2024).

Recently, interest in two-dimensional time-domain NMR has increased, opposite to the 2D Fourier transform used in 2D NMR spectroscopy, which relies upon the 2D Laplace transform. This is explained by the rather modest requirements for NMR equipment and by the development of numerical processing methods (Mitchell et al., 2014). In addition, these methods have proven to be extremely useful in the study of systems with complex pore organization, where both micro-, meso- and(or) macropores are present, the so-called systems with hierarchical porosity. 2D T_1-T_2

correlation and T_2-T_2 exchange experiments manifest the ability to determine surface affinity and explore pore connectivity in porous media, including zeolites (D'Agostino et al., 2023; Fleury et al., 2023; Robinson et al., 2021). NMR cryoporometry has been increasingly used to determine pore structure distribution on a scale from small micropores to large mesopores with an accuracy comparable or even superior to traditional gas adsorption technics or mercury porosimetry (Mitchell et al., 2008; Petrov & Furó, 2009; Webber et al., 2020a).

Since there are many worthy reviews devoted to the use of NMR for the study of zeolites (Gedeon & Fernandez, 2007; Grey, 2003; Klinowski, 1984, 1988; Stepanov, 2016; Zheng et al., 2024) in this Chapter we will focus only on some points that are important primarily for the study of mesoporous materials: namely, structural studies by ^{27}Al and ^{29}Si NMR spectroscopy, 2D T_1-T_2 correlation and T_2-T_2 exchange experiments and NMR cryoporometry in view of their application to zeolites with hierarchical porosity.

MICROPOROUS AND MESOPOROUS ZEOLITES: A BRIEF OVERVIEW

Zeolites are crystalline aluminosilicates with general formula $M_{x/n}[(AlO_2)_x(SiO_2)_y]\ mH_2O$ (where M^n is a metal cation or H^+, n is its valency), whose crystal structures can be represented as an open three-dimensional framework formed by $[SiO_4]^{4-}$ and $[AlO_4]^{5-}$ tetrahedra connected by vertices that create regular voids: cavities and channel systems extended in one, two or three dimensions. By now the 256 zeolite structures are known (Baerlocher & McCusker, n.d.) and their number is constantly expanding. For reliable orientation among the many zeolite structures, the International Zeolite Association (IZA) has adopted a three-letter coding system (Baerlocher & McCusker, n.d.), for example, MFI for ZSM-5, BEA for beta-zeolite, MOR for mordenite, FAU for faujasite etc. Selected zeolite structures are shown in Figure 1.

Figure 1. Selected zeolite frameworks and information about pore structure: channel dimensionality; maximum diameter of a sphere that can be included (D_{incl}) and that can diffuse along a specific direction x (D_x); channel size and N-membered ring it is formed by (in parenthesis).

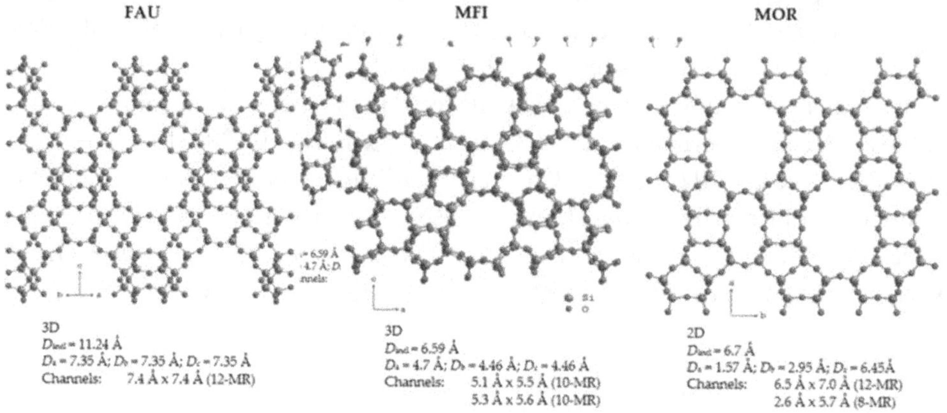

Adapted from Shelyapina, M.G. Molecules; published by MDPI, Basel, Switzerland. Creative Common CC BY license, 2024 (Shelyapina, 2024).

Depending on the framework type, zeolite can be stable in different ranges of the Si/Al ratio and the overall properties of zeolites for a specific application (activity, selectivity, durability) are closely related to the Si/Al ratios and Al distributions. Some zeolite properties as a function of the Si/Al ratio are summarized in Figure 2.

The charge compensation of the negatively charged zeolite framework that is issued by the substitution of Si^{4+} with Al^{3+}, is achieved through extra-framework cations M^{n+} (Na^+, K^+, NH_4^+, Ca^{2+} etc.) located within zeolite voids, or H^+ that form various hydroxyls on the zeolite inner surface (Breck, 1974). The presence of extra-framework cations determines the ability of zeolites to ionic exchange, which, combined with relatively high thermal and chemical stability, makes it possible to use zeolites as cation exchangers in environmental protection processes for removing undesirable impurities or for recovering valuable metals from the aqueous solutions etc. (Rodríguez-iznaga et al., 2022).

Figure 2. Changes in some properties of zeolites depending on the Si/Al ratio.

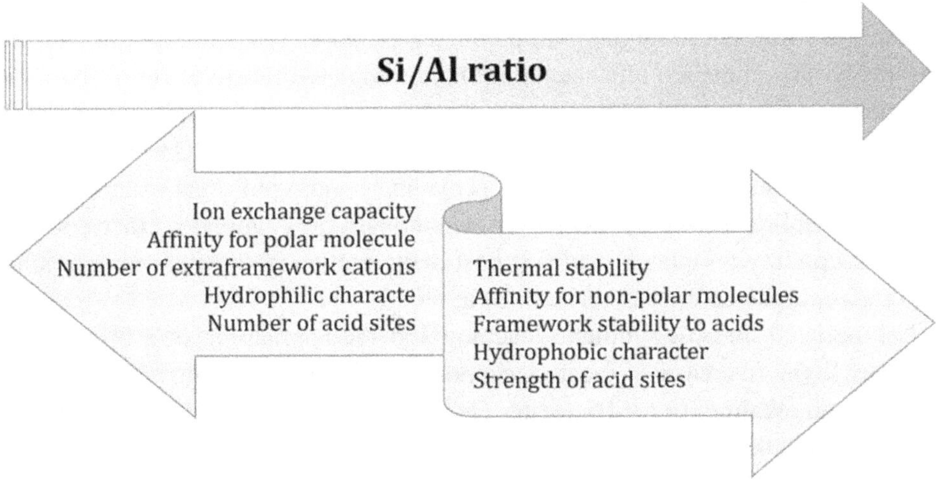

Catalysis is another important application of zeolites highly demanded in the market (Sánchez-López et al., 2021; Sun et al., 2021). Here the acid sites on the zeolite surface (their nature, strength, concentration, accessibility) play an important role. The acidity varies depending on the Si/Al ratio and the charge-compensating cations. The overall efficiency of the zeolite catalyst for a specific reaction can be tuned by regulation of both Lewis and Brønsted acidity (Ravi et al., 2021). Brønsted acidity originates from bridging hydroxyl groups (Si–O(H)–Al) (Haag et al., 1984). The issues of Lewis acidity are framework metals (such as Sn, Zr, Ti, Ga, Fe) and aluminum both extra-framework and framework-associated aluminum, although, the Lewis acidity originating from framework aluminum is not fully understood (Ravi et al., 2020). Moreover, the synergy between Brønsted and Lewis acid sites, apparently due to the close location of both sites in the zeolite voids, was reported (Yu et al., 2010, 2011).

As soon as in 3D zeolites the size of the formed voids d does not extend 1 nm, they are categorized as microporous materials (according to IUPAC, porous materials are classified as microporous ($d < 2$ nm), mesoporous ($2 < d < 50$ nm) and macroporous ($d > 50$ nm). The uniformity of the pore size allows the use of zeolites as molecular sieves. However, this limits the use of microporous zeolites in reactions involving bulky molecules as reactants or products. Moreover, several studies showed the benefit of mesoporosity in specific catalytic reactions for improving product selectivity (Abdulridha et al., 2021; Christensen et al., 2003) and catalyst lifetime (Chu et al., 2010; Milina et al., 2014) due to the efficient mass-transport properties. In addition, both the structure and the ordering of the mesopores are important (Petkovich & Stein, 2013; Qin et al., 2017).

There are different strategies to implement mesoporosity in microporous zeolites that involves (i) inclusion of a secondary hard template in form of rigid fibers, (ii) incorporation of a secondary soft template (macromolecules), (iii) application of a bifunctional template (two separate structure directing agents (SDAs) responsible for micro-and mesopore formation), (iv) covalent bonding of organosilane spacers with bulk zeolite seeds, (v) formation of nanoparticle aggregates (that does not require secondary templates) and (vi) post-synthetic leaching of alumina or silica from microporous zeolites (Möller & Bein, 2013).

Let us dwell on post-synthetic leaching and the application of a bifunctional template. In the first case, to create mesoporosity, zeolites undergo additional processing in an alkaline (or acid) medium, leading to dealumination (or desilication) and, as a result, the creation of disordered secondary porosity (mesopores), i.e., the creation of hierarchical structures containing micro- and mesopores (Čižmek et al., 1997; Oliveira et al., 2023). The main advantage of this approach is its simplicity and low cost. The size of the created mesopores can be regulated by the treatment time and/or temperature, the concentration of the solution, as well as the Si/Al ratio of the initial microporous zeolite (Čižmek et al., 1997). Thus, for ZSM-5, zeolites with a Si/Al ratio in the range from 25 to 50 must be used for optimal mesopore size and distribution (see Figure 3) (Groen et al., 2004).

Figure 3. Influence of the Si/Al ratio on the desilication of ZSM-5 zeolite.

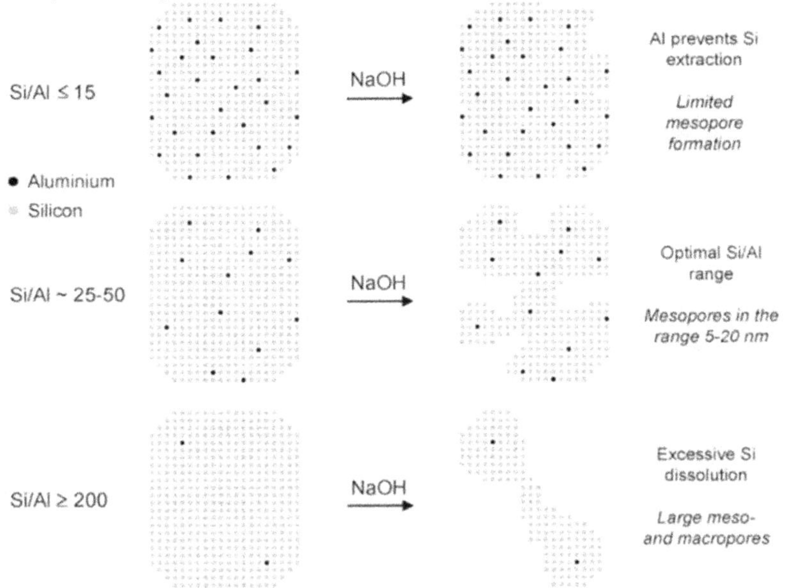

Reproduced with permission from Groen, J.C. et al. J. Phys. Chem. B published by American Chemical Society (2004) (Groen et al., 2004).

Despite the simplicity and low cost of creating zeolites with hierarchical porosity by leaching, it leads to disordered secondary porosity, which may be a limiting factor for the reaction rate. Pillaring makes it possible to create mesopores of a certain structure and size, interconnected in a certain way, which is important for optimizing transport properties (Zhao et al., 2021). However, the synthesis of pillared zeolites is a rather complex multi-stage procedure that involves (i) preparation of lamellar zeolites by self-assembling method using organic SDAs; (ii) introduction into the interlamellar space occupied by SDA forming mesopores molecules that serve as precursors for formation of pillars (e.g. tetraethoxysilane - TEOS); (iii) formation of inorganic pillars (e.g. through TEOS hydrolysis to amorphous SiO_2); (iv) calcination, to evacuate organic molecules, see Figure 4 (Na et al., 2010). More details about the synthesis and post-treatment of two-dimensional zeolites can be found in Ref. (Roth et al., 2014).

Figure 4. A schematic representation of phase transformation upon the synthesis of pillared zeolites.

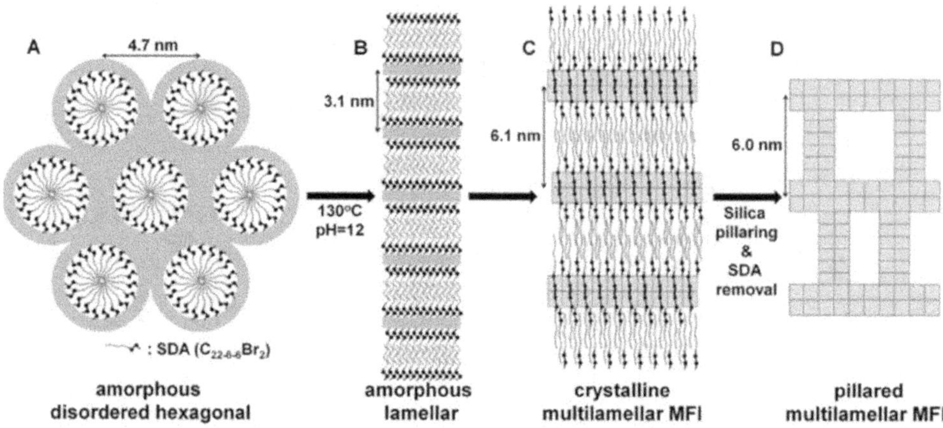

Reproduced with permission from Na, K. et al. J. Am. Chem. Soc. (2010) published by American Chemical Society (2010) (Na et al., 2010).

Modern strategies for developing new catalytic materials involve tuning their properties for specific reactions. This implies a comprehensive study of the physicochemical properties of the catalyst. NMR has proven to be a reliable method for studying microporous zeolites. Applied to such complex structures as mesoporous zeolites, NMR spectroscopy is an invaluable method for reconstructing a local structure with complex pore structure and surface organization. Moreover, as discussed further, time-domain NMR is an issue of valuable information on the mobility of guest molecules in zeolite voids, their interaction with a complex surface, and can provide access to pore size distribution and connectivity.

BASIC PRINCIPLES OF NMR PHENOMENON

The key point of NMR is that a system of nuclear spins in an external magnetic field $\boldsymbol{B}_0$ can absorb electromagnetic energy at a well-defined frequency, which is proportional to the external magnetic field and the nuclear gyromagnetic ratio γ. As soon as γ is the intrinsic property of a particular nuclear isotope, NMR exhibits the element and isotope selectivity. Beside interaction with external magnetic fields nuclear spins are involved in a number of interactions with their environment. This makes NMR a tool to probe local environment and dynamics of selected nuclei. Actually, different experimental NMR technique are widely applied in condensed matter physics, chemistry, biology, and medicine (Cardoza et al., 2004; Chizhik et al., 2014; Shelyapina, 2024; Skripov & Shelyapina, 2016).

The phenomenon of NMR is based on the property of nucleus to possess intrinsic angular momentum, called "spin". The magnetic moment $\boldsymbol{\mu}$ is directly proportional to the spin: $\boldsymbol{\mu} = \gamma \boldsymbol{I}$. Since spin I is quantized, it results in 2I + 1 discrete values of the angular momentum $m = -I, -I + 1, \ldots, +I$, and in a magnetic field $2I + 1$ levels with energy $E_m = -(\boldsymbol{\mu}, \boldsymbol{B}_0) = -m\gamma B_0$ appear due to the so-called Zeeman interaction. The distance between the neighboring levels is $\Delta E = \gamma B_0$. If one now applies an oscillating magnetic field with frequency that matches this energy distance one will observe a resonance signal.

From the point of view of classical mechanics, the NMR phenomenon can be described in terms of nuclear macroscopic magnetization $\boldsymbol{M}$ (the sum of all individual nuclear magnetic moments of the sample). In a constant magnetic field, it is an equilibrium state M_0 and is directed parallel to $\boldsymbol{B}_0$. Under the action of a radio frequency pulse (an oscillating field perpendicular to a static magnetic field $\boldsymbol{B}_0$), it deviates from the equilibrium state. The deviation angle is determined by the pulse duration. After switching off the pulse, the nuclear magnetization gradually returns to its equilibrium state (relaxes), with characteristic relaxation times T_1 and T_2, which characterize the recovering of the longitudinal component and disappearance of the transverse component of the magnetization, respectively. These times are characteristic for the system under study. During the recovery process, the nuclear magnetization vector oscillates around the field $\boldsymbol{B}_0$ and a free induction decay (FID) signal is recorded in a receiver coil of an NMR device. This FID signal can be converted into an NMR spectrum using the Fourier transform.

Internal interactions of nuclear spins with the electron shell (chemical shift), with each other (dipole-dipole coupling), interactions of quadrupole moment of the nucleus (if present) with an inhomogeneous electric field (quadrupole coupling), and others intercations, lead to shift, splitting, and broadening of spectral lines. All these interactions are anisotropic, but in liquids, due to high molecular mobility, they are averaged to the isotropic value (chemical shift) or zero (dipole-dipole and quadrupole couplings). This greatly simplifies the NMR spectrum, but some of the information is lost. NMR spectra in solids potentially contain a huge amount of information about both the structure and mobility of the substance, but it is often difficult to come by: the spectral lines are greatly broadened and some interactions mask others. Therefore, special high-resolution NMR techniques must be applied.

One of the simplest techniques is the magic angle spinning (MAS): the sample is placed in a rotor, the axis of which is located at an angle of 54°44′ to the direction of the external magnetic field $\boldsymbol{B}_0$, and is rapidly rotated (with typical rate of 10-20 kHz). This makes it possible to significantly narrow the NMR lines and partly suppress anisotropic interactions. An alternative is to use multipulse sequences. In recent decades methods of multi-dimensional and multi-quantum spectroscopy NMR have also been widely developed. More information about basic principles of NMR and its applications can be found in special monographs (Chizhik et al., 2014; Duer, 2002; Levitt, 2008).

ZEOLITE FRAMEWORK PROBED BY ^{27}AL AND ^{29}SI NMR

Zeolite samples, with rare exceptions, are microcrystalline powder. The arrangement of aluminum, although it obeys certain rules, is disordered. Therefore, the NMR method, being sensitive to the local environment of nuclei, is extremely in demand for determining the local structure of both microporous and hierarchical zeolites.

The ^{27}Al is a reliable tool for determining aluminum environments in zeolites. ^{27}Al chemical shifts are very sensitive to aluminum coordination, and AlO_n with n = 4, 5 and 6 exhibit values in the range 80-50, 40-30 and 15-0 ppm, respectively. The coordination state of both framework Al and extra-framework Al species in zeolites can be determined by applying one-dimensional ^{27}Al MAS NMR. MAS NMR technics is used for solids to suppress dipolar coupling (the main issue of NMR line broadening in solids), chemical shift anisotropy and quadrupole coupling, the latter concerns nuclei with a spin $I > 1/2$, which have a nonzero quadrupole moment. For quadrupole nuclei, which include ^{27}Al with $I = 5/2$, to resolve complex spectra with anisotropic contributions issued by chemical shift and quadrupole coupling 2D multiple-quantum magic-angle spinning (MQMAS) NMR (Pike et al., 2000) is very helpful.

^{27}Al MAS NMR chemical shift scales of zeolites are given in Figure 5. However, it should be noted that in hydrated zeolites. so-called perturbed framework Al atoms that correspond to $(SiO)_3AlOH$ groups may present. They are characterized by a broad ^{27}Al NMR resonance at about 59–62 ppm (with the quadrupole coupling constant $C_Q = 5$ MHz, and the asymmetry parameter $\eta = 0.3–0.4$) with a shoulder at 40 ppm (Brus et al., 2015). After dehydration they transform to $(SiO)_3Al$ Lewis sites, tri-coordinated to the zeolite framework. Al atoms of these sites exhibit an extremely broad ^{27}Al NMR resonance due to the large quadrupole coupling constant (the isotropic chemical shift $\delta_{iso} \approx 67$ ppm, $C_Q \approx 20$ MHz, and $\eta \approx 0.1$) that makes these Al species "invisible" to NMR (Brus et al., 2015; Xin et al., 2019).

Figure 5. ^{27}Al MAS NMR chemical shift scale of aluminum environments in zeolites.

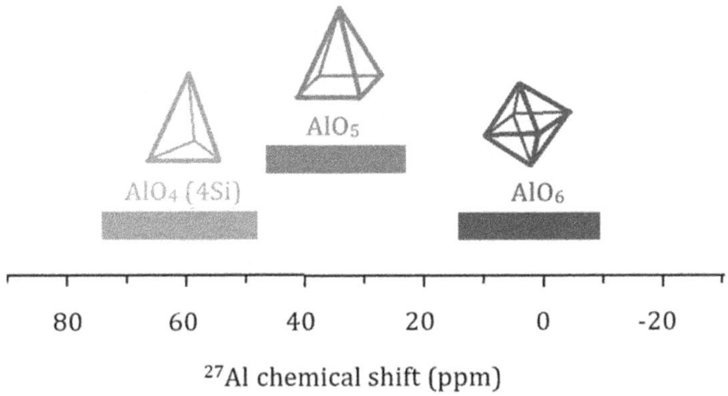

^{29}Si NMR spectroscopy is also an important tool for determining the structure of zeolites (Ashbrook et al., 2014; Brouwer et al., 2020; Dawson et al., 2017; Stepanov, 2016) that is issued by a number of factors: the natural abundance of the ^{29}Si isotope, sufficient to obtain an informative spectrum, the absence of a quadrupole moment (nuclear spin I = 1/2, which simplifies the interpretation of the spectra), as well as the sensitivity of ^{29}Si chemical shift to small changes in its local environment.

In the ideal zeolite crystal structure (siliceous microporous zeolite), all silicon atoms are located inside the oxygen tetrahedra TO_4 and are characterized by the position Q^4, see Figure 6. For pure silica zeolites, high resolution ^{29}Si solid-state NMR spectra show crystallographic resolution of the various Q^4 sites. According to a careful study of 39 different zeolite framework types done by Brouwer et al. (Brouwer et al., 2020), the ^{29}Si isotopic chemical shift ranges from –103.8 to –122.6 ppm. The authors also reported the correlations between ^{29}Si isotropic chemical shifts and various geometric parameters (Si–O–Si bond angles, Si–O bond lengths, the average s character of the oxygen orbitals in the Si–O–Si bonds). This information is extremely useful for deciphering and refinement the structure crystal of siliceous microporous (Masai et al., 2023; Srivastava et al., 2025) and hierarchical zeolites (Shelyapina et al., 2020; Usman Azam et al., 2025). The latter are normally characterized by presence of Q^3 (one neighboring hydroxyl group), Q^2 and even Q^1 (two and three neighboring hydroxyl groups, respectively) surface sites (Huang et al., 2014; Ruggiu et al., 2024), see Figure 6. The presence of Q^n with $n < 4$ also is a marker of a partial amorphization of the zeolite (Masai et al., 2023).

In zeolites with a finite Si/Al ratio some of TO_4-positions (Q^4 sites) may be occupied by aluminum and every Si atom has from zero to four neighboring Al atoms: Q^4(nAl) sites, see Figure 6. ^{29}Si NMR spectra in this case show up to five well-resolved peaks, those intensities are determined by the probability of such sites in zeolites.

Figure 6. ^{29}Si MAS NMR chemical shift scale of silicon environments in zeolites.

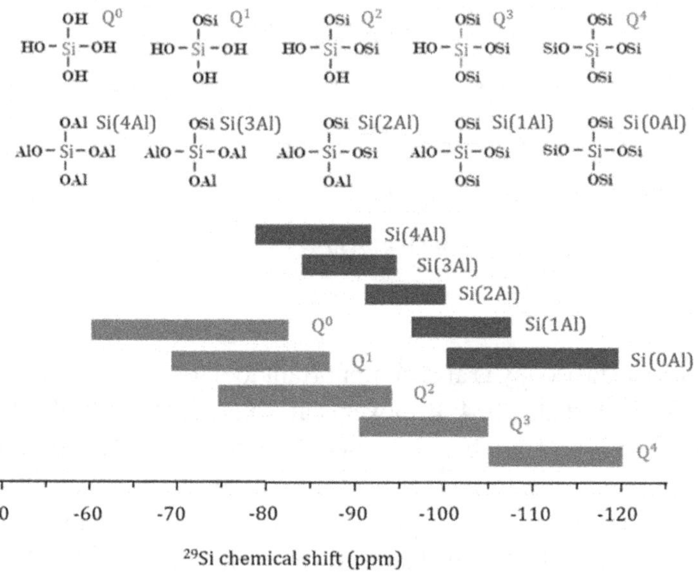

For microporous zeolites ^{29}Si NMR can be used to determine the Si/Al ratio related directly to the zeolite framework (Fyfe et al., 1991). Assuming the Löwenstein's rule (Loewenstein, 1954), which implies that the Al–O–Al bond formation is forbidden in zeolites, by analyzing integral intensities of individual ^{29}Si spectral lines one can estimate the Si/Al ratio fy the following simple expression [48]:

$$\frac{\mathrm{Si}}{\mathrm{Al}} = \frac{\sum_{n=0}^{4} I_{\mathrm{Q}^4(n\mathrm{Al})}}{\sum_{n=0}^{4} \frac{n}{4}\left(I_{\mathrm{Q}^4(n\mathrm{Al})}\right)} \tag{1}$$

where $I_{\mathrm{Q}^4(n\mathrm{Al})}$ is the peak areas of individual ^{29}Si NMR lines attributed to $\mathrm{Q}^4(n\mathrm{Al})$.

It should be noted that this method allows the most accurate determination of the Si/Al ratio in zeolites. Since standard bulk elemental analysis, such as Energy-Dispersive X-ray spectroscopy (EDX) or Optical Emission Spectroscopy with Inductively Coupled Plasma (ICP-OES) do not distinguish the framework aluminum from the extra-framework one, while tri-coordinated framework Al in dehydrated zeolites is "invisible" for ^{27}Al MAS NMR, as discussed above.

Figure 7 show the ^{29}Si and ^{27}Al MAS NMR spectra for hydrated microporous H-mordenites with the close nominal Si/Al ratio (about 7.5) but prepared by different way (Krylova et al., 2022). The samples are characterized by essentially different share of $Q^4(n\mathrm{Al})$ and one of the samples has an important number of the extra-framework Al species, as evidenced from ^{27}Al MAS NMR. This indicates dependence of Al distribution on the preparation method (Krylova et al., 2022; Shelyapina et al., 2022). Table 1 summarized the Si/Al ratio for a series of H-mordenites with the nominal (declared by a manufacture) ratio varied from 4.95 to 9.9 as determined by different methods: EDX, ICP-OES, ICP-OES combined with ^{27}Al MAS NMR (that accounts for extra-framework Al), ^{29}Si MAS NMR using Eq. (1) (Shelyapina et al., 2022). As one can see the Si/Al ratios determined from ^{27}Al and ^{29}Si MAS NMR are in reasonable agreement between each other.

Figure 7. ^{29}Si and ^{27}Al MAS NMR spectra with their decomposition for the H-mordenite samples: H-MOR-15.0 (with nominal Si/Al = 7.5) and H-MOR-15.2 (with nominal Si/Al = 7.6).

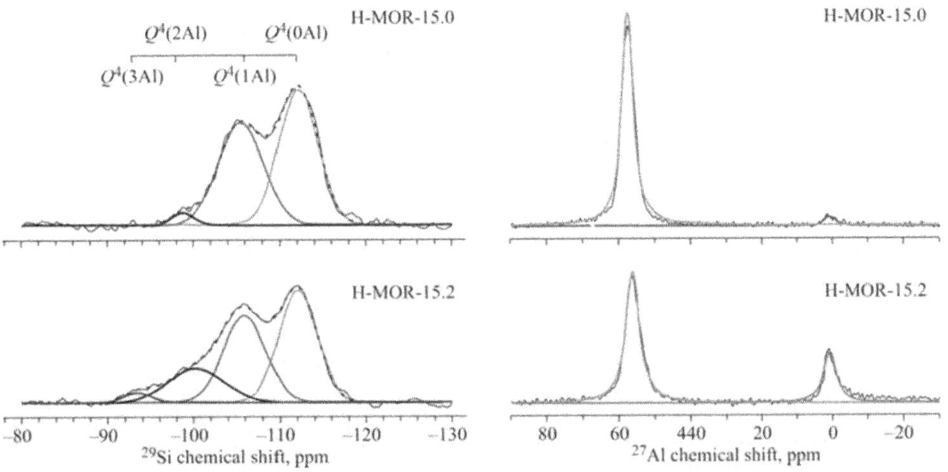

Reproduced with permission from Krylova, E.A. et al. J. Struct. Chem.; published by Pleiades Publishing, Ltd. (2022) (Krylova et al., 2022).

Table 1. Si/Al ratio for a series of H-mordenites determined by different methods (Shelyapina et al., 2022).

Sample	Nominal	EDX	ICP-OES	^{27}Al NMR	^{29}Si NMR
H-MOR-9.9	4.95	4.65	4.36	4.95	5.2
H-MOR-15.0	7.5	7.15	6.67	7.15	8.0
H-MOR-15.2	7.6	7.2	6.69	9.05	8.9
H-MOR-15.7	7.85	7.55	6.82	9.1	10.5
H-MOR-15.8	7.9	7.25	7.03	9.1	11.5
H-MOR-18.7	9.35	8.8	8.34	9.25	10.25
H-MOR-19.8	9.9	9.65	8.81	9.1	7.9

For hierarchical zeolites, it is very important to understand the influence of synthesis or post-treatment conditions on the number and location of active sites, which is determined, among others, by the local silicon environment. It is interesting to note that the Löwenstein's rule, originally developed for microporous zeolites, can be used for mesoporous zeolites as well. For example, Gomez et al. (Gómez et al., 2022) in their work aimed at the synthesis of mesoporous LSX zeolites (FAU-type structure) with the maximum number of Al sites, i.e. the minimum Si/Al ratio that is 1 according to the Löwenstein's rule. The mesoporosity induced by using sodium dodecylbenzene sulfonate during the synthesis allowed achieving Si/Al = 1.1 that is on the limit of the Löwenstein's rule. To control this parameter and the aluminum distribution over zeolite framework ^{29}Si MAS NMR was used.

Huang et al (Huang et al., 2014) applied ^{29}Si and ^{27}Al NMR to follow the evolution of the local structure of a mordenite zeolite (with Si/Al = 4.8) upon subsequent steaming-acid leaching and alkaline treatment to introduce mesoporosity. ^{27}Al MAS NMR spectra confirm the presence of extra-framework Al species in the parent microporous zeolite that are still observed for acid-leached and hierarchical samples. ^{29}Si MAS NMR spectra evidence the vanishing of Q^4(2Al) sites after acid leaching followed by their recovering after the alkaline treatment. Such a realumination of the dislodged Al species back to the zeolite framework was also reported for the alkaline-treated MFI zeolites (Verboekend & Pérez-Ramírez, 2011) and copper-exchanged mordenites (Zhukov et al., 2016).

For pillared zeolites, with pillars from amorphous SiO_2 formed due to TEOS hydrolysis, it is difficult to distinguish Q^4 and Q^3 signal arising from SiO_2 pillars and zeolite layers due to partial overlapping, see Figure 6. However, ^{29}Si NMR can be applied for monitoring formation of such complex structure. Shelyapina et al. (Shelyapina et al., 2020) followed the evolution of the ^{29}Si MAS NMR spectra upon preparation the layered zeolites with MOR and MFI topology and pillared by amorphous SiO_2. The samples were prepared using the one-pot synthesis method, as proposed by Na et al. (Na et al., 2010) (see Figure 4) with cetyltrimethylammonium bromide (CTAB) as organic SDA. The spectra for the as-synthesized layered zeolites (MOR-AS and MFI-AS) show features typical for microporous three-dimensional zeolites: Q^4(0Al), Q^4(1Al) and Q^4(2Al), see Figure 8. The Si/Al ratio calculated using Eq. (1) was found to be in fair agreement with elemental analysis done by electron dispersive spectroscopy. However, compared with 3D mordenites with a close Si/Al ratio (Kato et al., 2003; Paixão et al., 2010; Shelyapina et al., 2022) an important contributions of Q^4(1Al) and Q^4(2Al) signal were observed for the layered modification. The authors suggest that this is caused by the interaction of CTAB heads with AlO_4 tetrahedra at the surface of the zeolite layer upon zeolite formation resulting in a higher Al content in certain preferred areas of the zeolite structure. In addition, the change in T-O-T angles for Q^4(nAl) due to local structural distortions on the lamella may also contribute. Introduction of TEOS (MOR-T, MFI-T) and its further hydrolysis (MOR-TH, MFI-TH) result in the broadening of spectral lines line, and finally, the annealing (pillared zeolites MOR-P and MFI-P) results in a redistribution of the line intensities.

The partial overlapping of the Q^4(1Al) and Q^4(2Al) ^{29}Si zeolite signal with Q^4 and Q^3 of TEOS complicates the assigning of the spectral lines the typical ranges of ^{29}Si chemical shift in corresponding 3D zeolites (Brouwer et al., 2020; Fyfe et al., 1991) and TEOS (Glaser & Wilkes, 1989) are shown in Figure 8. It is worth noting that no ^{29}Si signal below 90 ppm was observed even under cross-polarization conditions. This means that the introduction of TEOS from the beginning leads to autohydrolysis and the formation of SiO_2 oligomers due to the water contained in the sample triggering the formation of a 3D network of SiO_4 tetrahedra in the interlayer space of the zeolite sandwiched between the CTAB layers. The further targeted hydrolysis finalizes the process resulting in the disappearance of the Q^3 signal and the formation of SiO_2 oligomers and then amorphous SiO_2 after annealing. For the pillared MOR-P sample, the signal at –111.2 ppm corresponds to the overlapping mordenite Q^4(0Al) site and Q^4 of amorphous SiO_2 (Brinker et al., 1988; Rainho et al., 2001). It is also interesting to note that according to ^{27}Al MAS NMR for as synthetized materials, all Al belong to the zeolite framework (the only signal near 55 ppm), neither the introduction of TEOS or its further hydrolysis affect it, but the calcination resulting in the evacuation of organics provokes a partial but not critical collapse of the local zeolite structure (the appearance of extra-framework six-coordinated Al species). This allows the authors to suggest that the organic CTA^+ cations separating zeolite lamella in the as-synthetized materials are localized near negatively charges Al containing T-units and the removal of organic cations ripping out Al from the zeolite framework.

Figure 8. (a) and (c) – ^{29}Si MAS NMR spectra for the samples at the all preparation steps with their decomposition; (b) and (d) – the ^{29}Si line maximum (left), the line width at half maximum, $\Delta\nu_{1/2}$ (center) and relative integral intensities (right) of the individual lines for the zeolite samples at the all preparation steps. Filled areas show the typical ranges of ^{29}Si chemical shift in zeolites (solid fill) and TEOS (hatched fill). (a), (c) corresponds to the MOR samples, (b), (d) to the MFI samples.

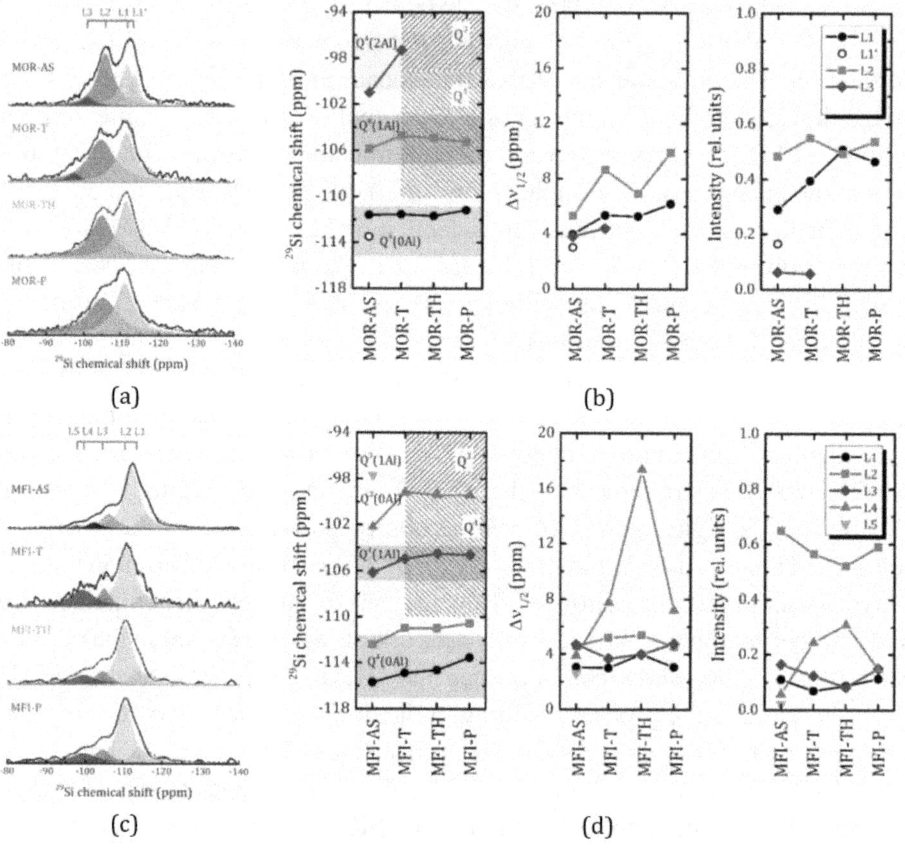

Adapted from Shelyapina, M.G. et al. Molecules; published by MDPI, Basel, Switzerland. Creative Common CC BY license, 2022 (Shelyapina et al., 2020).

Increasing of extra-framework Al in hierarchical β zeolites synthesized through one-pot hydrothermal treatment under dynamic condition with organosilane surfactant (trimethoxysilyl)propyl]ammonium (TPOAC) chloride and tetraethylammonium hydroxide was reported by Zhou et al. (Zhou et al., 2022). The decreasing of strong acid sites over hierarchical β zeolites upon TCOAC concentration increasing evidenced from NH_3-TPD measurements authors attributed to the formation of extra-framework Al observed from ^{27}Al MAS NMR.

NMR RELAXATION TO PROBE ZEOLITE ACIDITY: T_1-T_2 CORRELATION MAPS

In addition to NMR spectroscopy, which has been successfully used to study zeolites for many decades, there has been an increased interest in using NMR relaxation to probe porous media, namely to determine surface affinity and adsorbate behavior in mesoporous catalysts.

The basic idea of using NMR relaxation to study the surface acidity of zeolites can be summarized as follows. For confined liquids, in the case where it is impossible to neglect the exchange that molecules experience between surface layers with a thickness of ε, the observed relaxation rates of nuclear spins $T_{1,2}^{-1}$ differ from their bulk values $T_{1,2}^{\text{bulk}}$ (Godefroy et al., 2001; Korb, 2011; Mitchell et al., 2008):

$$\frac{1}{T_{1,2}} \approx \frac{1}{T_{1,2}^{\text{bulk}}} + \frac{1}{\frac{d_p}{2\rho_{1,2}} + \frac{d_p^2}{8D}} = \frac{1}{T_{1,2}^{\text{bulk}}} + \frac{2\rho_{1,2}}{d_p} \cdot \frac{1}{1 + \frac{\rho_{1,2} d_p}{8D}}, \tag{2}$$

where $\rho_{1,2} = \varepsilon \cdot \left(T_{1,2}^{\text{surf}}\right)^{-1}$ is the surface relaxivity, $T_{1,2}^{\text{surf}}$ are the relaxation times for molecules adsorbed on the surface of the porous media, D is the self-diffusion coefficient of bulk liquid, d_p is the pore diameter and for planar, cylindrical or spherical pores, respectively). This expression was obtained for long times t and under fast exchange condition $t \gg T_{1,2}^{\text{surf}} \gg 1/w$ (w is the exchange rate). Two extreme cases are usually of interest: the diffusion-limited ($4D/d_p \ll \rho_{1,2}$), and the surface-limited ($4D/d_p \gg \rho_{1,2}$) relaxations. In the first case, Eq. (2) can be rewritten as

$$\frac{1}{T_{1,2}} \approx \frac{1}{T_{1,2}^{\text{bulk}}} + \frac{8D}{d_{\text{p}}^2}. \tag{3}$$

For the surface-limited relaxation Eq. (2), it can be simplified by the following way

$$\frac{1}{T_{1,2}} \approx \frac{1}{T_{1,2}^{\text{bulk}}} + \frac{2\rho_{1,2}}{d_p}. \tag{4}$$

Assuming spherical pores of volume V and surface S, one obtains the simple expression

$$\frac{1}{T_{1,2}} \approx \frac{1}{T_{1,2}^{\text{bulk}}} + \rho_{1,2}\frac{S}{V}. \tag{5}$$

This means that under conditions of limited geometry, for the two considered limiting cases, the relaxation rates $T_{1,2}$ linearly depends on $1/d_p$ or $1/d_p^2$. This result is extremely important because it provides access to (i) the morphology of porous media and (ii) host-guest interaction, the measure of which is the surface relaxivities of adsorbed.

It is worth noting that the adsorption of molecules on the surface results in a decrease in their rotational and translational mobility and, hence, shortening the relaxation times: $T_{1,2}^{\text{bulk}} \gg T_{1,2}^{\text{surf}}$. Then, the T_1/T_2 ratio is equal to the ratio of the corresponding surface relaxivities

$$\frac{T_1}{T_2} \approx \frac{\rho_1}{\rho_2} = \frac{T_2^{\text{surf}}}{T_1^{\text{surf}}} \tag{6}$$

and being sensitive to surface affinity (Mitchell et al., 2013) can be used for determining the surface acidity affinity of zeolites (Robinson et al., 2021). Moreover, recent studies proved that T_1/T_2 correlates with the desorption energy of liquids confined in zeolite matrices (D'Agostino et al., 2023; Robinson et al., 2021).

It should be noted that in the case of conformed liquids, due to interaction with the surface, which is usually inhomogeneous, there is a distribution of relaxation times. Therefore, two-dimensional $T_1 - T_2$ correlation graphs are more informative and be used to identify molecular species and to probe their dynamics (Velasco et al., 2023). The resulting maps evidence the relationship between T_1/T_2 of the adsorbed liquid and the properties of the surface of a porous medium. The ideology largely echoes two-dimensional NMR spectroscopy, but instead of the 2D Fourier transform (frequency domain NMR), the 2D Laplace transform (time domain NMR) is used. In recent years, interest in such studies of porous systems has been increasing due to (i) the rather modest requirements for NMR equipment, and (ii) by the development of numerical processing methods. The experimental details can be found in Ref. (Shelyapina, 2024).

In T_1–T_2 correlation experiment for data acquisition a pulse, which is composed of an inversion recovery pulse sequence followed by a Carr-Purcell-Meiboom-Gill (CPMG) echo pulse train (Y. Q. Song et al., 2002). The acquired 2D NMR relaxation data are the normalized spin echo magnitudes depending on two independent variables: the nuclear magnetization recovery time τ and the echo time τ_e, that may be described by the following expression trough the Fredholm integral equation of the first kind (Mitchell et al., 2014; Y. Q. Song et al., 2002):

$$\frac{S(\tau, n\tau_e)}{S(\tau \to \infty, 0)}$$

$$= \iint \exp\left(-\frac{n\tau_e}{T_2}\right)\left[1 - 2\exp\left(-\frac{\tau}{T_1}\right)\right] F\left(T_1, T_2\right) dT_1 dT_2 + E\left(\tau, n\tau_e\right), \tag{7}$$

where $F(T_1, T_2)$ represents the desired 2D distributions of T_1 and T_2 relaxation times, $E(\tau, n\tau_e)$ is the experimental noise. The data analysis relies on inversion of the integral to extract $F(T_1, T_2)$ from the measured signal amplitude.

Figure 9 (a) shows the 2D $T_1 - T_2$ correlation map for pyridine adsorbed in ZSM-5 zeolites with different SiO_2/Al_2O_3 ratios (Robinson et al., 2021). Each peak on the 2D map responds to a particular pyridine/zeolite system. For the bulk liquid T_1/T_2 is close to 1 (the diagonal line). Increasing T_1/T_2 (the domain below the diagonal) can be treated as an indicator of the strength of the host-guest interaction on the pore surface, hence, the position of these peaks is determined by the relative surface affinities of pyridine in zeolite pores. This assumption is supported by the linear correlation between T_1/T_2 and enthalpy of pyridine desorption H_{des}, see Figure 9(b), the latter, in its turn, increases with increasing the number of Brønsted acid sites.

D'Agostino et al. (D'Agostino et al., 2022) used T_1/T_2 ratio to probe the host-guest surface interactions of water adsorbed in ZSM-5 and chabazite, two zeolites with different pore sizes: water confined in chabasite, that is characterized by a smaller pore size compared to ZSM-5, shows reduced mobility and a more dramatic increase in T_1/T_2 with increasing acid sites.

Figure 9. (a) – ^1H 2D T_1–T_2 correlation plots for pyridine confined in HZSM-5 with various SiO_2/Al_2O_3. The solid diagonal line corresponds to $T_1/T_2 = 1$; data for bulk pyridine are also shown; (**b**) *– $-T_1/T_2$ versus enthalpy of pyridine adsorption H_{des}; the line shows a linear fit. SiO_2/Al_2O_3 values are indicated next to each correlation peak or point.*

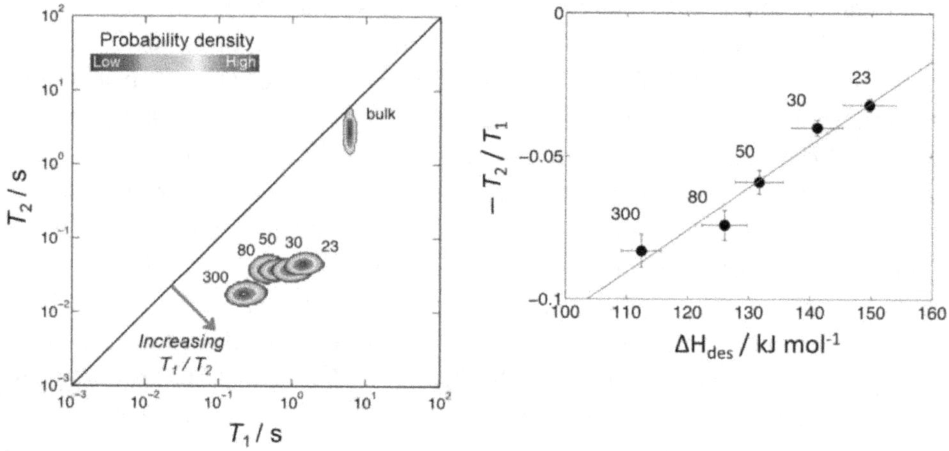

Reproduced from Robinson, N. et al. Phys. Chem. Chem. Phys.; published by the Royal Society of Chemistry. Creative Common CC BY license, 2021 (Robinson et al., 2021).

Zou et al. (2023) studied toluene adsorbed into H-Beta and Pt modified H-Beta zeolites. Enhancement of interaction between toluene molecules and the surface of zeolites after dealumination was confirmed by increasing of the T_1/T_2 ratio of toluene. Further increasing in T_1/T_2 after Pt loading the authors explained by the ability of dealuminated Pt modified H-Beta zeolites to interact with toluene, providing partial oxidation of toluene to benzyl alcohol facilitated by zeolite hydroxyls.

Forster et al. (Forster et al., 2023) applied 2D NMR relaxation maps to evaluate the effect of the reaction solvent (water, methanol and ethanol) upon the activity of Sn and Ga doped zeolite Y catalyst for the isomerization of glucose to fructose. The NMR data suggested that methanol adsorbs the weakest to the zeolite surfaces and water adsorbs the strongest with ethanol intermediate between the two substantially altering the dominant reaction mechanism. The observed lack of catalytic activity in water is issued by the strong adsorption of water molecules in the zeolite pores blocking access of reactants to Lewis acid sites active for the sugar isomerization. Ethanol being retained in the pores prevent solvated fructose from further reaction on Brønsted acid sites situated outside of the pore space.

Mensah et al. (Mensah et al., 2024) explored the performance of hierarchical H-Y zeolites for the catalytic cracking of 1,3,5-triisopropylbenzene and low-density polyethylene. Time domain NMR was employed to confirm hierarchical pore structure connectivity within dealuminated zeolite materials. A single relaxation population with short relaxation times ($\langle T_1 \rangle \approx 0.1$ s; $\langle T_2 \rangle \approx 5$ ms) observed for water interacting with parent zeolite H-Y, indicates rapid exchange between water within the zeolite micropores and within the interparticle space (bulk media), see Figure 10(a). The hierarchical zeolite also exhibits a single peak that suggests rapid exchange and hence pore connectivity; however, the increase in relaxation times ($\langle T_1 \rangle \approx 0.25$ s; $\langle T_2 \rangle \approx 26$ ms) compared to the parent H-Y material indicates that water is confined in larger pores, see Figure 10(b). The elongation of the relaxation distribution away from the parent H-Y zeolite indicates pore size heterogeneity within the dealuminated material.

Recent studies have shown that such maps allow a simple quantitative relationship between surface relaxivity and interfacial chemistry in mesoporous media (Dong et al., 2024; Forster et al., 2023; Moon et al., 2023) .

Figure 10. (a) – [1]H 2D T_1–T_2 correlation plots correlation for water and confined in parent H-Y and dealuminated DA($HNO_3$0.5h)-Y zeolites. Data for bulk water are given for a comparison. The solid diagonal line corresponds to $T_1/T_2 = 1$. (b) – NLDFT pore size distributions of Y zeolites (from Ar physisorption measurements).

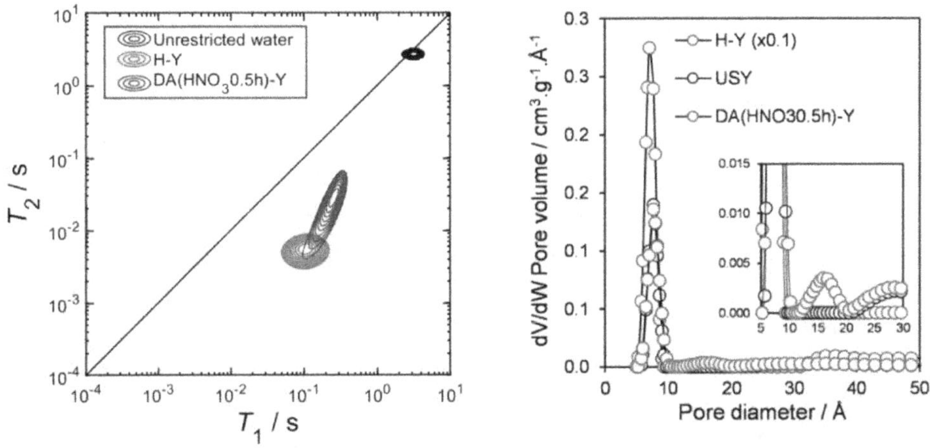

(a) (b)

Reproduced from Mensah, J. et al. ChemCatChem; published by Wiley-VCH. Creative Commons Attribution-NonCommercial-NoDerivs License (Mensah et al., 2024).

NMR CRYOPOROMETRY TO EXPLORE THE PORE SIZE DISTRIBUTION

As it was mentioned above the relaxation rates $T_{1,2}$ of confined liquids through relaxivity give access to morphology of porous media. However, along with this another NMR method, the so-called NMR cryoporometry has been increasingly used to probe pore size distribution on a large scale from sub-nanometers to several micrometers (Mitchell et al., 2008; Petrov & Furó, 2009). Contrary to more abundant gas adsorption technics that uses the Kelvin equation (constant temperature), NMR cryoporometry relies on Gibbs-Thomson Equation (constant pressure).

In this method one measures the temperature shifts of phase transitions that occurs in a matter due to confinement effects. These shifts treated in terms of pore geometry provide information on pore sizes and their distribution, and in favorable cases, about the shape of the pores.

For a matter confined in cylindrical pores with radius *r* the shift of the melting point T_m relative to its value in bulk T^0 can be expressed as following (Jackson & McKenna, 1990):

$$\Delta T_m \equiv T_m - T^0 = -\frac{4\sigma_{sl} T_m}{r \Delta H_f \rho_s}, \tag{8}$$

where σ_{sl} is the surface energy of the solid-liquid interface, ΔH_f is the bulk enthalpy of fusion and ρ_s is the density of the solid. In the similar way the shift of the freezing point ΔT_f can be defined as $\Delta T_f \equiv T_f - T^0$.

For specific cases Eq. (8) can be simplified. The corresponding shifts of melting and freezing points obtained in the large pore limits (>10 nm) are listed in Table 2. Since the $\Delta T_m / \Delta T_f$ ratio is sensitive to the pore shape it can be used for evaluation the pore geometry.

Similar to both mercury porosimetry and gas sorption a hysteresis between equilibrium cooling and heating curves is very characteristic when measuring a cryoporometry NMR signal. Several theories have been proposed to explain it (Petrov & Furó, 2006)and tested experimentally (Hitchcock et al., 2011; Perkins et al., 2008) .

Table 2. The freezing and melting temperature shift ΔT_f and ΔT_m, respectively, for various pore geometries in the limits of large pores (Petrov & Furó, 2009).

Pore shape	$\lvert\Delta T_{f}\rvert$ (K)	$\lvert\Delta T_{m}\rvert$ (K)
Sphere	$3K_c/r$	$2K_c/r$
Cylinder	$2K_c/r$	K_c/r
Slit	K_c/r	0

The signal intensity of liquid filling the pore is proportional to the pore volume $v(x)$, and the melting temperature T_m is related to the pore size distribution. Hence, the pore size distributions is accessed by differentiating and re-mapping the melting curve data using the Rahman-Smith-Strange trasform (Strange et al., 1993):

$$\frac{\mathrm{d}v(x)}{\mathrm{d}x} = \frac{K_c}{x^2}\frac{\mathrm{d}v}{\mathrm{d}T_m(x)} \cdot \tag{30}$$

In the experiment a porous material with an adsorbed probe liquid is cooled until all the liquid in the pores is frozen, after that the sample is slowly heated. The intensity of the liquid NMR signal is stored as a function of temperature. Usually, one monitors the spin-spin relaxation times of protons of adsorbed water or organic molecules, the solid and liquid phases of which are characterized by shorter (on the order of microseconds) and longer (on the order of milliseconds) T_2 values, respectively, and can be easily distinguished.

The main advantages of NMR cryoporometry over standard gas adsorption and thermo-porosimetry are a wide variety of substances that can be used as a probe liquid and that pre-drying of the samples is not essential. Webber et al. (Webber et al., 2020b) used NMR cryoporometry to probe the micro- and mesopore size distributions in USY zeolites (parent microporous zeolites with different Si/Al ratio and with induced mesoporosity) convincingly demonstrating its advantages over standard gas porosimetry. The temperature dependence (in both direction – heating and cooling) of the ^{1}H NMR signal intensity from liquid phase of water adsorbed in microporous USY zeolite is rather complex and shows several critical points that correspond to a subsequent pass through the region of supercooled water (a negative slope as a function of temperature is due to the Boltzmann Effect changing the spin populations), water around the pores that nucleates to ice, again supercooled water in pores, a rapid nucleation step with the ice formation in pores and finally complete solidification of water within zeolite pores. The melting curve at the beginning repeats the freezing curves, then exhibits a slowly rising plateau distinct from the cooling path followed by a number of rising steps indicating different porosity of different scale (from small micropores to large mesopores) finalized by the melting point.

We remind that USY zeolites belong to the FAU zeolite family, see Figure 1, and have the supercage (opening diameter of 7.4 Å), internal cage (diameter of 11.6 Å), and sodalite cages (diameter of 6.6 Å) connected via hexagonal prisms of 2.5 Å. Due to limitation of NMR cryoporometry the latter cages cannot be distinguished: even water (kinetic diameter of 2.7 Å) cannot penetrate to sodalite cages. A comparison of the pore size distribution derived from NMR cryoporometry and N_2 adsorption/desorption experiment done by the authors demonstrate a qualitative agreement between these two methods. However, NMR cryoporometry provides a more precise description in both micro- and mesopore ranges. The limitations of Barrett-Joyner-Halenda (BJH) model, used to obtain the pore size distribution from gas sorption curves, barely allow it to cover the entire peak of micropores (< 1nm). Moreover, NMR cryoporometry is free from the artifact that presents on the N_2 desorption curve (a spike near 3-4 nm). For zeolites with induced mesoporosity a pronounced decrease of microporosity is observed, however, the authors note that the micropore volume is underestimated in NMR cryoporometry experiment relative to the *t*-plot method.

As noted above, one of the advantages of the NMR cryoporometry is a wide range of probe molecules. The advantage of other molecules than water is that they can provide access to larger pores and selectively explore a certain range of pore sizes. Webber et al. (Webber et al., 2020b) along with water used n-dodecane that due to its size restrictions cannot fully enter the supercages of Y zeolite, whereas a partial insertion leads to the appearance of ~20 Å peak. Nevertheless, n-dodecane exhibits similar cumulative volumes as water, that allows the authors to conclude that this molecule occupies the same volume in micro- and mesopores as water.

It is worth noting that despite many advantages of NMR cryoporometry in studying por media it requires some care in conducting the experiment and calibration.

2D T_2–T_2 RELAXATION-EXCHANGE TO STUDY THE PORE CONNECTIVITY

Recently, along with T_1–T_2 correlations, T_2–T_2 experiments became popular for characterization of pore structure and functionality of porous materials (Velasco et al., 2023). 2D T_2–T_2 relaxation-exchange is very useful for determining the molecular transfer between different relaxation media evidencing diffusive coupling between two relaxation environments that can be used for probing pore connectivity in systems with complex hierarchical porosity (Bernin & Topgaard, 2013; Mitchell et al., 2007).

The T_2-T_2 exchange pulse sequence consists of two CPMG echo trains, *A* and *B*, during which spin-spin relaxation first evolves and afterwards is detected (Fleury & Soualem, 2009; Washburn & Callaghan, 2006). These two echo trains are separated by an exchange period τ_e (the storage interval). The acquired magnetization from two isolated reservoirs (*A* and *B*) can be written as follows (Mitchell et al., 2007):

$$\frac{S\left(t_A, t_B, \tau_e\right)}{S(\tau \rightarrow \infty, 0)} = \iint \exp\left(-\frac{t_A}{T_2^A}\right)\exp\left(-\frac{t_B}{T_2^B}\right) F\left(T_2^A, T_2^B\right) dT_2^A dT_2^B + E\left(t_A, t_B\right), \tag{9}$$

where $F\left(T_2^A, T_2^B\right)$ is the probability density of the signal components with the relaxation times T_2^A and T_2^B, $E\left(t_A, t_B\right)$ is the experimental noise, t_A and t_B are the relaxation encoding time intervals corresponding to the durations of two CPMG echo trains. Between two echo trains the nuclear magnetization relaxes with a characteristic time T_1 to its thermodynamic equilibrium along the external magnetic field. Therefore, exchange of magnetization from one site to another must be considered during all the three periods t_A, t_B and τ_e (Gao & Blümich, 2020) and the total signal amplitude is reduced by $\exp\left(-\tau_e/T_1^{AB}\right)$. Similar to T_1-T_2 correlation experiment, the data analysis is based on the 2D inverse Laplace transform to extract $F\left(T_2^A, T_2^B\right)$ from the measured signal amplitude. More technical details of the experiment can be found anywhere (Shelyapina, 2024).

A schematic T_2–T_2 map showing diagonal and off-diagonal peaks at a certain exchange time τ_e is given in Figure 11. On the T_2-T_2 map most of the signal lies on the diagonal line $T_2^A = T_2^B$. If there is an exchange between regions with different T_2 it leads to the appearance of symmetric pairs of off-diagonal peaks T_2^+ and T_2^-, the intensity of which relative to the main peaks quantifies the diffusive exchange that occurs during the mixing period. Recording a series of T_2–T_2 maps at various exchange time τ_e one can plot the amplitudes of the diagonal P^{++}, P^{--} and off-diagonal P^{+-}, P^{-+} peaks as a function of τ_e and determine the exchange rates k_A and k_B (from A to B and from B to A, respectively). The numerical expression of these amplitudes can be found in Ref. (Monteilhet et al., 2006).

Figure 11. A schematic T_2–T_2 map showing diagonal and off-diagonal peaks at a certain exchange time τ_n

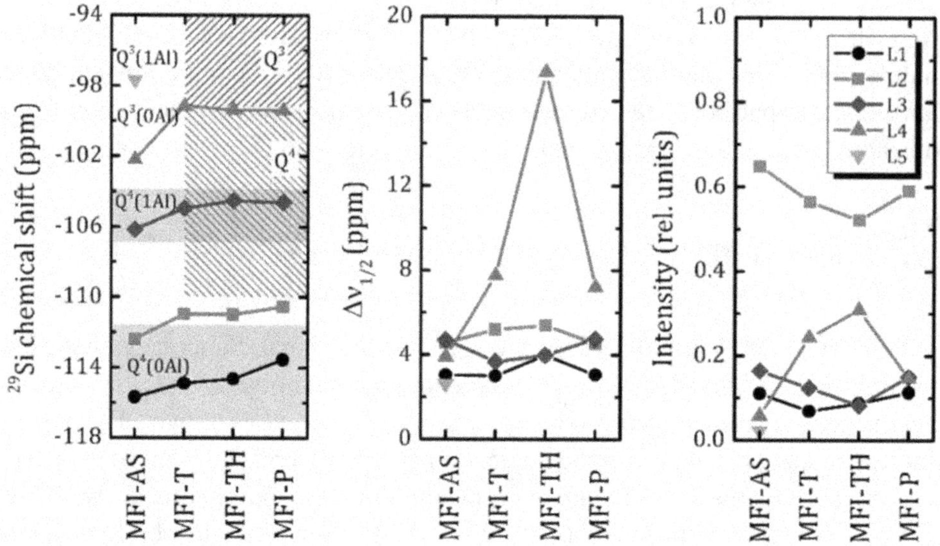

Reproduced from Shelyapina, M.G. Molecules; published by MDPI, Basel, Switzerland. Creative Common CC BY license, 2024 (Shelyapina, 2024).

Assuming a two sites model and with $k_A = k_A = k$ (two reservoirs of equal size), the exchange rate k can be found by fitting the experimental data by the following equation (Robinson et al., 2020; Washburn & Callaghan, 2006):

$$\frac{P^{+-} + P^{-+}}{P^{tot}} = A\left(1 - e^{\frac{\tau}{k}}\right), \tag{10}$$

where P^{tot} is the summed amplitude of all peaks, A characterizes amount of molecules exchanged.

The described above NMR 2D T_2-T_2 experiment provides additional information about the dynamics of liquids confined in pores with complex organization. Fleury at al. (Fleury et al., 2023) used 2D relaxation experiments to evaluate exchange rates between micro- and mesopores in hierarchical USY zeolites (FAU topology, see Figure 1). They studied two samples with essentially different mesoporosity: CBV400 (a slightly dealuminated USY zeolite with a low mesoporosity) and CBV720 (a strongly dealuminated form of USY with high mesoporosity). The authors used two different molecules as a probe: squalane and propanol. Squalane is the largest alkane molecule with a diameter of about 5 Å that allows it to penetrate into the USY micropores and exchange with the meso- or micropores. A hydroxyl group of 2-propanol makes it possible to study the porous network in a different way: due to an active interaction with the zeolite surface the mechanism of surface diffusion becomes more important. 2D T_1-T_2 maps for squalane show multiple T_1/T_2 peaks attributed to the macropores: $T_1/T_2 = 1$ (molecules in macropores), $T_1/T_2 = 4$ (molecules in mesopores), and $T_1/T_2 = 20$ and 40 for CBV400 and CBV720, respectively (molecules in micropores).

To evaluate the pore connectivity 2D T_2-T_2 exchange experiment was applied (Fleury et al., 2023). Symmetric off-diagonal peaks confirm the presence of exchange between the micro- and mesopores (with a timescale of about 6 ms for both zeolites), while a single asymmetric peak points out the molecule exchange between meso- and macropores. For 2-propanol the only exchange peaks between micro- and mesopores were observed. The difference in the exchange rate ($k = 2$ and 4 ms for CBV400 and CBV720, respectively) and the amount of exchanged 2-propanol molecules (in CBV400 is one order of magnitude greater then in CBV720) the authors associated with a high surface residence time of 2-propanol in the micropores of CBV720 due to strong interaction of polar molecules with zeolite surface.

Despite the great potential for exploring systems with complex pore organization, T_2-T_2 relaxation exchange NMR experiments have so far been little used for studying zeolites. (Fleury et al., 2023; Gizatullin et al., 2018). However, it has been successfully applied to other porous matrices with hierarchical structure: silica (Elgersma et al., 2022, 2023), polymers (Silletta et al., 2018), glasses (Mitchell et al., 2007; R. Song et al., 2016).

CONCLUSION

Zeolites with hierarchical porous structures are effective catalysts in various processes since the presence of transport mesopores ensures the accessibility of active sites and simplifies the diffusion of both reagents and reaction products. However, the synthesis of micro-/mesoporous zeolites or post-synthetic processing often leads to uncontrolled changes in their crystalline structure and porosity. As shown in this Chapter, various NMR spectroscopy and relaxation methods make it possible to monitor the structural evolution of hierarchical zeolites during synthesis, as well as to obtain information about their surface acidity, the pore size distribution and connectivity. However, despite its versatility and advantages, NMR has a number of limitations, which were discussed in the text and summarized in Table 3.

Table 3. Advantages and limitations of NMR in studies of zeolites properties.

Zeolite properties accessed by NMR	NMR techniques	Advantages of NMR	Limitations of NMR	Complementary methods
Structural characterization of the zeolite framework	^{27}Al, ^{29}Si MAS NMR	Sensitive to local structure distortion	The superposition of signals prevents the destruction of individual lines, high rotation speed in MAS experiment is required; Relatively low intensity of ^{29}Si NMR	XRD
Si/Al ratio	^{27}Al, ^{29}Si MAS NMR	Distinguishes framework Al from extra-framework Al species; provides the real Si/Al ratio	Some Al species can be NMR "invisible"	ICP-OES, EDX, XRD
Pore structure and distribution	^{1}H, ^{13}C or other nuclei of probe molecules	Provides access to a large scale of pore size distribution; different liquids can be used as a probe; provides information on the pore connectivity	Requires special mathematical treatment of the experimental data	N_2 porosimetry

Nevertheless, being a method sensitive to local changes near the probe nuclei, NMR significantly enriches our knowledge of the structure and dynamics of the substances under study. With a wide range of different experimental techniques and probe molecules and in combination with other, more familiar volumetric methods, NMR provides a deeper and more complete understanding of the structure and physico-chemical properties of such materials as zeolites with a complex pore organization.

REFERENCES

Abdulridha, S., Jiao, Y., Xu, S., Zhang, R., Ren, Z., Garforth, A. A., & Fan, X. (2021). A comparative study on mesoporous Y zeolites prepared by hard-templating and post-synthetic treatment methods. *Applied Catalysis A, General*, *612*, 117986. DOI: 10.1016/j.apcata.2020.117986

Amoureux, J. P., & Pruski, M. (2008). MQMAS NMR: Experimental strategies and applications. In Harris, R. L., & Wasylishen, R. K. (Eds.), *eMagRes* (pp. 1–31). Wiley., DOI: 10.1002/9780470034590.emrstm0319.pub2

Ashbrook, S. E., Dawson, D. M., & Seymour, V. R. (2014). Recent developments in solid-state NMR spectroscopy of crystalline microporous materials. *Physical Chemistry Chemical Physics*, *16*(18), 8223–8242. DOI: 10.1039/C4CP00578C PMID: 24675798

Baerlocher, C., & McCusker, L. B. (n.d.). *Database of zeolite structures,*http://www .iza-structure.org/databases/*; date accessed: May 30, 2024.* Retrieved May 30, 2024, from http://www.iza-structure.org/databases/

Barthomeuf, D. (1987). Zeolite acidity dependence on structure and chemical environment. Correlations with catalysis. *Materials Chemistry and Physics*, *17*(1–2), 49–71. DOI: 10.1016/0254-0584(87)90048-4

Bernin, D., & Topgaard, D. (2013). NMR diffusion and relaxation correlation methods: New insights in heterogeneous materials. *Current Opinion in Colloid & Interface Science*, *18*(3), 166–172. DOI: 10.1016/j.cocis.2013.03.007

Biaglow, A. I., Gorte, R. J., Kokotailo, G. T., & White, D. (1994). A probe of Brønsted site acidity in zeolites: ^{13}C chemical shift of acetone. *Journal of Catalysis*, *148*(2), 779–786. DOI: 10.1006/jcat.1994.1264

Bornes, C., Fischer, M., Amelse, J. A., Geraldes, C. F. G. C., Rocha, J., & Mafra, L. (2021). What is being measured with P-bearing NMR probe molecules adsorbed on zeolites? *Journal of the American Chemical Society*, *143*(34), 13616–13623. DOI: 10.1021/jacs.1c05014 PMID: 34410690

Bornes, C., Sardo, M., Lin, Z., Amelse, J., Fernandes, A., Ribeiro, M. F., Geraldes, C., Rocha, J., & Mafra, L. (2019). ^{1}H-^{31}P HETCOR NMR elucidates the nature of acid sites in zeolite HZSM-5 probed with trimethylphosphine oxide. *Chemical Communications*, *55*(84), 12635–12638. DOI: 10.1039/C9CC06763A PMID: 31580363

Breck, D. W. (1974). *Zeolite molecular sieves: structure, chemistry, and use*. John Wiley and Sons., DOI: 10.1093/chromsci/13.4.18A-c

Brinker, C. J., Kirkpatrick, R. J., Tallant, D. R., Bunker, B. C., & Montez, B. (1988). NMR confirmation of strained "defects" in amorphous silica. *Journal of Non-Crystalline Solids*, *99*(2-3), 418–428. DOI: 10.1016/0022-3093(88)90448-6

Brouwer, D. H., Brouwer, C. C., Mesa, S., Semelhago, C. A., Steckley, E. E., Sun, M. P. Y., Mikolajewski, J. G., & Baerlocher, C. (2020). Solid-state ^{29}Si NMR spectra of pure silica zeolites for the International Zeolite Association Database of Zeolite Structures. *Microporous and Mesoporous Materials*, *297*(January), 110000-1–12. DOI: 10.1016/j.micromeso.2020.110000

Brus, J., Kobera, L., Schoefberger, W., Urbanová, M., Klein, P., Sazama, P., Tabor, E., Sklenak, S., Fishchuk, A. V., & Dědeček, J. (2015). Structure of framework aluminum lewis sites and perturbed aluminum atoms in zeolites as determined by ^{27}Al{^{1}H} REDOR (3Q) MAS NMR spectroscopy and DFT/molecular mechanics. *Angewandte Chemie International Edition*, *54*(2), 541–545. DOI: 10.1002/anie.201409635 PMID: 25393612

Buntkowsky, G., & Vogel, M. (2020). Small molecules, non-covalent interactions, and confinement. *Molecules (Basel, Switzerland)*, *25*(14), 3311. DOI: 10.3390/molecules25143311 PMID: 32708283

Cardoza, L. A., Korir, A. K., Otto, W. H., Wurrey, C. J., & Larive, C. K. (2004). Applications of NMR spectroscopy in environmental science. *Progress in Nuclear Magnetic Resonance Spectroscopy*, *45*(3–4), 209–238. DOI: 10.1016/j.pnmrs.2004.06.002

Chizhik, V. I., Chernyshev, Y. S., Donets, A. V., Frolov, V. V., Komolkin, A. V., & Shelyapina, M. G. (2014). *Magnetic resonance and its applications*. Springer International Publishing., DOI: 10.1007/978-3-319-05299-1

Christensen, C. H., Johannsen, K., Schmidt, I., & Christensen, C. H. (2003). Catalytic benzene alkylation over mesoporous zeolite single crystals: Improving activity and aelectivity with a new family of porous materials. *Journal of the American Chemical Society*, *125*(44), 13370–13371. DOI: 10.1021/ja037063c PMID: 14583028

Chu, N., Wang, J., Zhang, Y., Yang, J., Lu, J., & Yin, D. (2010). Nestlike hollow hierarchical MCM-22 microspheres: Synthesis and exceptional catalytic properties. *Chemistry of Materials*, *22*(9), 2757–2763. DOI: 10.1021/cm903645p

Čižmek, A., Subotić, B., Šmit, I., Tonejc, A., Aiello, R., Crea, F., & Nastro, A. (1997). Dissolution of high-silica zeolites in alkaline solutions II. Dissolution of 'activated' silicalite-1 and ZSM-5 with different aluminum content. *Microporous Materials*, *8*(3–4), 159–169. DOI: 10.1016/S0927-6513(96)00082-X

D'Agostino, C., Bräuer, P., Zheng, J., Robinson, N., York, A. P. E., Song, L., & Fan, X. (2023). Adsorbate/adsorbent interactions in microporous zeolites: Mechanistic insights from NMR relaxation and DFT calculations. *Materials Today. Chemistry*, *29*, 101443. DOI: 10.1016/j.mtchem.2023.101443

D'Agostino, C., York, A. P. E., & Bräuer, P. (2022). Host-guest interactions and confinement effects in HZSM-5 and chabazite zeolites studied by low-field NMR spin relaxation. *Materials Today. Chemistry*, *24*, 100901. DOI: 10.1016/j.mtchem.2022.100901

Dawson, D. M., Moran, R. F., & Ashbrook, S. E. (2017). An NMR crystallographic investigation of the relationships between the crystal structure and ^{29}Si isotropic chemical shift in silica zeolites. *The Journal of Physical Chemistry. C, Nanomaterials and Interfaces*, *121*(28), 15198–15210. DOI: 10.1021/acs.jpcc.7b03730 PMID: 28751927

Dong, S., Liu, L., Ling, N. N. A., May, E. F., Johns, M. L., & Robinson, N. (2024). Toward quantifying the chemical sensitivity of nuclear spin surface relaxivity in mesoporous media. *Langmuir*, *40*(31), 16160–16171. DOI: 10.1021/acs.langmuir.4c01245 PMID: 39045723

Duer, M. J. (Ed.). (2002). *Solid-state NMR spectroscopy. Principles and applications*. Blackwell Science.

Elgersma, S. V., Sederman, A. J., Mantle, M. D., & Gladden, L. F. (2022). Measuring the liquid-solid mass transfer coefficient in packed beds using T_2-T_2 relaxation exchange NMR. *Chemical Engineering Science*, *248*, 117229. DOI: 10.1016/j.ces.2021.117229

Elgersma, S. V., Zheng, Q., Avrantinis, N., Sederman, A. J., Mantle, M. D., & Gladden, L. F. (2023). Quantifying liquid-solid mass transfer in a trickle bed using T_2-T_2 relaxation exchange NMR. *Applied Magnetic Resonance*, *54*(11–12), 1423–1443. DOI: 10.1007/s00723-023-01607-x

Fleury, M., Pirngruber, G., & Jolimaitre, E. (2023). Probing diffusional exchange in mesoporous zeolite by NMR diffusion and relaxation methods. *Microporous and Mesoporous Materials*, *355*, 112575. DOI: 10.1016/j.micromeso.2023.112575

Fleury, M., & Soualem, J. (2009). Quantitative analysis of diffusional pore coupling from T_2-store-T_2 NMR experiments. *Journal of Colloid and Interface Science*, *336*(1), 250–259. DOI: 10.1016/j.jcis.2009.03.051 PMID: 19439315

Forster, L., Kashbor, M. M. M., Railton, J., Chansai, S., Hardacre, C., Conte, M., & D'Agostino, C. (2023). Low-field 2D NMR relaxation and DRIFTS studies of glucose isomerization in zeolite Y: New insights into adsorption effects on catalytic performance. *Journal of Catalysis*, *425*, 269–285. DOI: 10.1016/j.jcat.2023.06.021

Fyfe, C. A., Feng, Y., Grondey, H., Kokotailo, G. T., & Gies, H. (1991). One- and two-dimensional high-resolution solid-state NMR studies of zeolite lattice structures. *Chemical Reviews*, *91*(7), 1525–1543. DOI: 10.1021/cr00007a013

Gao, Y., & Blümich, B. (2020). Analysis of three-site T_2-T_2 exchange NMR. *Journal of Magnetic Resonance (San Diego, Calif.)*, *315*, 106740. DOI: 10.1016/j.jmr.2020.106740 PMID: 32438312

Gedeon, A., & Fernandez, C. (2007). Solid-state NMR spectroscopy in zeolite science. In Čejka, J., van Bekkum, H., Corma, A., & Schüth, F. (Eds.), *Studies in Surface Science and Catalysis* (Vol. 168, pp. 403–XII). Elsevier., DOI: 10.1016/S0167-2991(07)80800-7

Gizatullin, B. I., Savinkov, A. V., Shipunov, T. V., Melnikova, D. L., Doroginitzky, M. M., & Skirda, V. D. (2018). Investigation of molecular mobility and exchange of n-hexane and water in silicalite-1 by 2D ^{1}H NMR relaxometry. *MRSej*, *20*(1), 18102.

Glaser, R. H., Wilkes, G. L., & Bronnimann, C. E. (1989). Solid-state ^{29}Si NMR of TEOS-based multifunctional sol-gel materials. *Journal of Non-Crystalline Solids*, *113*(1), 73–87. DOI: 10.1016/0022-3093(89)90320-7

Godefroy, S., Korb, J. P., Fleury, M., & Bryant, R. G. (2001). Surface nuclear magnetic relaxation and dynamics of water and oil in macroporous media. *Physical Review E: Statistical Physics, Plasmas, Fluids, and Related Interdisciplinary Topics*, *64*(2), 021605. DOI: 10.1103/PhysRevE.64.021605 PMID: 11497601

Gómez, J. M., Montes, I., Díez, E., & Rodríguez, A. (2022). Mesoporous low silica X (MLSX) zeolite: Mesoporosity in loewenstein limit? *Microporous and Mesoporous Materials*, *330*, 111618. DOI: 10.1016/j.micromeso.2021.111618

Grey, C. P. (2003). Nuclear magnetic resonance studies of zeolites. In Auerbach, S., Carrado, K., & Dutta, P. (Eds.), *Handbook of zeolite science and technology* (pp. 267–331). CRC Press., DOI: 10.1201/9780203911167.pt3

Groen, J. C., Jansen, J. C., Moulijn, J. A., & Pérez-Ramírez, J. (2004). Optimal aluminum-assisted mesoporosity development in MFI zeolites by desilication. *The Journal of Physical Chemistry B*, *108*(35), 13062–13065. DOI: 10.1021/jp047194f

Gunther, W. R., Michaelis, V. K., Griffin, R. G., & Roman-Leshkov, Y. (2016). Interrogating the lewis acidity of metal sites in beta zeolites with ^{15}N pyridine adsorption coupled with MAS NMR spectroscopy. *The Journal of Physical Chemistry. C, Nanomaterials and Interfaces*, *120*(50), 28533–28544. DOI: 10.1021/acs.jpcc.6b07811 PMID: 28479940

Haag, W. O., Lago, R. M., & Weisz, P. B. (1984). The active site of acidic aluminosilicate catalysts. *Nature*, *309*(5969), 589–591. DOI: 10.1038/309589a0

Hitchcock, I., Holt, E. M., Lowe, J. P., & Rigby, S. P. (2011). Studies of freezing-melting hysteresis in cryoporometry scanning loop experiments using NMR diffusometry and relaxometry. *Chemical Engineering Science*, *66*(4), 582–592. DOI: 10.1016/j.ces.2010.10.027

Huang, S., Liu, X., Yu, L., Miao, S., Liu, Z., Zhang, S., Xie, S., & Xu, L. (2014). Preparation of hierarchical mordenite zeolites by sequential steaming-acid leaching-alkaline treatment. *Microporous and Mesoporous Materials*, *191*, 18–26. DOI: 10.1016/j.micromeso.2014.02.039

Huo, H., Peng, L., & Grey, C. P. (2009). Low temperature ^1H MAS NMR spectroscopy studies of proton motion in zeolite HZSM-5. *The Journal of Physical Chemistry. C, Nanomaterials and Interfaces*, *113*(19), 8211–8219. DOI: 10.1021/jp900313y

Jackson, C. L., & McKenna, G. B. (1990). The melting behavior of organic materials confined in porous solids. *The Journal of Chemical Physics*, *93*(12), 9002–9011. DOI: 10.1063/1.459240

Kanellopoulos, J., Gottert, C., Schneider, D., Knorr, B., Prager, D., Ernst, H., & Freude, D. (2008). NMR investigation of proton mobility in zeolites. *Journal of Catalysis*, *255*(1), 68–78. DOI: 10.1016/j.jcat.2008.01.028

Kato, M., Itabashi, K., Matsumoto, A., & Tsutsumi, K. (2003). Characteristics of MOR-framework zeolites synthesized in fluoride-containing media and related ordered distribution of Al atoms in the framework. *The Journal of Physical Chemistry B*, *107*(8), 1788–1797. DOI: 10.1021/jp022227h

Klinowski, J. (1984). Nuclear magnetic resonance studies of zeolites. *Progress in Nuclear Magnetic Resonance Spectroscopy*, *16*, 237–309. DOI: 10.1016/0079-6565(84)80007-2

Klinowski, J. (1988). Recent advances in solid-state nmr of zeolites. *Annual Review of Materials Science*, *18*(1), 189–218. DOI: 10.1146/annurev.ms.18.080188.001201

Korb, J. P. (2011). Nuclear magnetic relaxation of liquids in porous media. *New Journal of Physics*, *13*(3), 035016. DOI: 10.1088/1367-2630/13/3/035016

Krylova, E. A., Shelyapina, M. G., Mazur, A., Baranov, D. A., Tsyganenko, A. A., & Petranovskii, V. P. (2022). Local structure of protonated mordenites with $SiO_2/Al_2O_3 \approx 15$ probed by multinuclear NMR. *Journal of Structural Chemistry*, *63*(6), 930–943. DOI: 10.1134/S0022476622060105

Krylova, E. A., Shelyapina, M. G., Nowak, P., Harańczyk, H., Chislov, M., Zvereva, I. A., Privalov, A. F., Becker, M., Vogel, M., & Petranovskii, V. (2018). Mobility of water molecules in sodium- and copper-exchanged mordenites: Thermal analysis and ^{1}H NMR. *Microporous and Mesoporous Materials*, *265*, 132–142. DOI: 10.1016/j.micromeso.2018.02.010

Levitt, M. H. (2008). *Spin Dynamics: Basics of Nuclear Magnetic Resonance* (2nd Editio). Wiley.

Lewis, J. D., Ha, M., Luo, H., Faucher, A., Michaelis, V. K., & Román-Leshkov, Y. (2018). Distinguishing active site identity in Sn-Beta zeolites using ^{31}P MAS NMR of adsorbed trimethylphosphine oxide. *ACS Catalysis*, *8*(4), 3076–3086. DOI: 10.1021/acscatal.7b03533

Loewenstein, W. (1954). The distribution of aluminum in the tetrahedra of silicates and aluminates. *The American Mineralogist*, *39*, 92–96.

Masai, H., Kohara, S., Wakihara, T., Shibazaki, Y., Onodera, Y., Masuno, A., Sukenaga, S., Ohara, K., Sakai, Y., Haines, J., Levelut, C., Hébert, P., Isambert, A., Keen, D. A., & Azuma, M. (2023). Siliceous zeolite-derived topology of amorphous silica. *Communications Chemistry*, *6*(1), 1–2. DOI: 10.1038/s42004-023-01075-1 PMID: 38071376

Mensah, J., Yan, P., Rawal, A., Lee, A. F., Wilson, K., Robinson, N., Johns, M. L., Kennedy, E., & Stockenhuber, M. (2024). Catalytic cracking of 1,3,5-triisopropylbenzene and low-density polyethylene over hierarchical Y zeolites and Al-SBA-15. *ChemCatChem*, *16*(3), e202300884. DOI: 10.1002/cctc.202300884

Milina, M., Mitchell, S., Crivelli, P., Cooke, D., & Pérez-Ramírez, J. (2014). Mesopore quality determines the lifetime of hierarchically structured zeolite catalysts. *Nature Communications*, *5*(1), 3922. DOI: 10.1038/ncomms4922

Mitchell, J., Broche, L. M., Chandrasekera, T. C., Lurie, D. J., & Gladden, L. F. (2013). Exploring surface interactions in catalysts using low-field nuclear magnetic resonance. *The Journal of Physical Chemistry. C, Nanomaterials and Interfaces*, *117*(34), 17699–17706. DOI: 10.1021/jp405987m

Mitchell, J., Gladden, L. F., Chandrasekera, T. C., & Fordham, E. J. (2014). Low-field permanent magnets for industrial process and quality control. *Progress in Nuclear Magnetic Resonance Spectroscopy*, *76*, 1–60. DOI: 10.1016/j.pnmrs.2013.09.001 PMID: 24360243

Mitchell, J., Griffith, J. D., Collins, J. H. P., Sederman, A. J., Gladden, L. F., & Johns, M. L. (2007). Validation of NMR relaxation exchange time measurements in porous media. *The Journal of Chemical Physics*, *127*(23), 234701. DOI: 10.1063/1.2806178 PMID: 18154403

Mitchell, J., Webber, J. B. W., & Strange, J. H. (2008). Nuclear magnetic resonance cryoporometry. *Physics Reports*, *461*(1), 1–36. DOI: 10.1016/j.physrep.2008.02.001

Möller, K., & Bein, T. (2013). Mesoporosity – a new dimension for zeolites. *Chemical Society Reviews*, *42*(9), 3689–3707. DOI: 10.1039/c3cs35488a PMID: 23460052

Monteilhet, L., Korb, J. P., Mitchell, J., & McDonald, P. J. (2006). Observation of exchange of micropore water in cement pastes by two-dimensional T2 - T2 nuclear magnetic resonance relaxometry. *Physical Review E: Statistical, Nonlinear, and Soft Matter Physics*, *74*(6), 061404. DOI: 10.1103/PhysRevE.74.061404 PMID: 17280070

Moon, H. J., Sekiya, R. S., & Jones, C. W. (2023). Probing the morphology and mobility of amines in porous silica CO_2 sorbents by ^1H T_1-T_2 relaxation correlation NMR. *The Journal of Physical Chemistry. C, Nanomaterials and Interfaces*, *127*(24), 11652–11665. DOI: 10.1021/acs.jpcc.3c02441

Na, K., Chol, M., Park, W., Sakamoto, Y., Terasakl, O., & Ryoo, R. (2010). Pillared MFI zeolite nanosheets of a single-unit-cell thickness. *Journal of the American Chemical Society*, *132*(12), 4169–4177. DOI: 10.1021/ja908382n PMID: 20136135

Oliveira, D. S., Lima, R. B., Pergher, S. B. C., & Caldeira, V. P. S. (2023). Hierarchical zeolite synthesis by alkaline treatment: Advantages and applications. *Catalysts*, *13*(2), 316. DOI: 10.3390/catal13020316

Paixão, V., Carvalho, A. P., Rocha, J., Fernandes, A., & Martins, A. (2010). Modification of MOR by desilication treatments: Structural, textural and acidic characterization. *Microporous and Mesoporous Materials*, *131*(1–3), 350–357. DOI: 10.1016/j.micromeso.2010.01.013

Perkins, E. L., Lowe, J. P., Edler, K. J., Tanko, N., & Rigby, S. P. (2008). Determination of the percolation properties and pore connectivity for mesoporous solids using NMR cryodiffusometry. *Chemical Engineering Science*, *63*(7), 1929–1940. DOI: 10.1016/j.ces.2007.12.022

Petkovich, N. D., & Stein, A. (2013). Controlling macro- and mesostructures with hierarchical porosity through combined hard and soft templating. *Chemical Society Reviews*, *42*(9), 3721–3739. DOI: 10.1039/C2CS35308C PMID: 23072972

Petrov, O., & Furó, I. (2006). Curvature-dependent metastability of the solid phase and the freezing-melting hysteresis in pores. *Physical Review E: Statistical, Nonlinear, and Soft Matter Physics*, *73*(1), 011608. DOI: 10.1103/PhysRevE.73.011608 PMID: 16486162

Petrov, O. V., & Furó, I. (2009). NMR cryoporometry: Principles, applications and potential. *Progress in Nuclear Magnetic Resonance Spectroscopy*, *54*(2), 97–122. DOI: 10.1016/j.pnmrs.2008.06.001

Pike, K. J., Malde, R. P., Ashbrook, S. E., McManus, J., & Wimperis, S. (2000). Multiple-quantum MAS NMR of quadrupolar nuclei. Do five-, seven- and nine-quantum experiments yield higher resolution than the three-quantum experiment? *Solid State Nuclear Magnetic Resonance*, *16*(3), 203–215. DOI: 10.1016/S0926-2040(00)00081-3 PMID: 10868573

Qin, Z., Cychosz, K. A., Melinte, G., El Siblani, H., Gilson, J. P., Thommes, M., Fernandez, C., Mintova, S., Ersen, O., & Valtchev, V. (2017). Opening the cages of faujasite-type zeolite. *Journal of the American Chemical Society*, *139*(48), 17273–17276. DOI: 10.1021/jacs.7b10316 PMID: 29160696

Rainho, J. P., Rocha, J., Carlos, L. D., & Almeida, R. M. (2001). ^{29}Si nuclear-magnetic-resonance and vibrational spectroscopy studies of SiO_2–TiO_2 powders prepared by the sol-gel process. *Journal of Materials Research*, *16*(8), 2369–2376. DOI: 10.1557/JMR.2001.0325

Ravi, M., Sushkevich, V. L., & van Bokhoven, J. A. (2020). Towards a better understanding of Lewis acidic aluminium in zeolites. *Nature Materials*, *19*(10), 1047–1056. DOI: 10.1038/s41563-020-0751-3 PMID: 32958864

Ravi, M., Sushkevich, V. L., & van Bokhoven, J. A. (2021). On the location of Lewis acidic aluminum in zeolite mordenite and the role of framework-associated aluminum in mediating the switch between Brønsted and Lewis acidity. *Chemical Science (Cambridge)*, *12*(11), 4094–4103. DOI: 10.1039/D0SC06130A PMID: 34163680

Robinson, N., Bräuer, P., York, A. P. E., & D'Agostino, C. (2021). Nuclear spin relaxation as a probe of zeolite acidity: A combined NMR and TPD investigation of pyridine in HZSM-5. *Physical Chemistry Chemical Physics*, *23*(33), 17752–17760. DOI: 10.1039/D1CP01515J PMID: 34190253

Robinson, N., Xiao, G., Connolly, P. R. J., Ling, N. N. A., Fridjonsson, E. O., May, E. F., & Johns, M. L. (2020). Low-field NMR relaxation-exchange measurements for the study of gas admission in microporous solids. *Physical Chemistry Chemical Physics*, *22*(24), 13689–13697. DOI: 10.1039/D0CP02002H PMID: 32525174

Rodríguez-iznaga, I., Shelyapina, M. G., & Petranovskii, V. (2022). Ion exchange in natural clinoptilolite : Aspects related to its structure and applications. *Minerals, 12*, 1628-1–39. AcademicDOI: 10.3390/min12121628

Roth, W. J., Nachtigall, P., Morris, R. E., & Čejka, J. (2014). Two-dimensional zeolites : Current status and perspectives. *Chemical Reviews*, *114*(9), 4807–4837. DOI: 10.1021/cr400600f PMID: 24555638

Ruggiu, A., Carvalho, A. P., Rombi, E., Martins, A., Rocha, J., Parpot, P., Neves, I. C., & Cutrufello, M. G. (2024). Y and ZSM-5 Hierarchical Zeolites Prepared Using a Surfactant-Mediated Strategy: Effect of the Treatment Conditions. *Materials (Basel)*, *17*(17), 4401. Advance online publication. DOI: 10.3390/ma17174401 PMID: 39274788

Sánchez-López, P., Kotolevich, Y., Yocupicio-Gaxiola, R. I., Antúnez-García, J., Chowdari, R. K., Petranovskii, V., & Fuentes-Moyado, S. (2021). Recent advances in catalysis based on transition metals supported on zeolites. *Frontiers in Chemistry*, *9*, 1–25. DOI: 10.3389/fchem.2021.716745 PMID: 34434919

Shelyapina, M. G. (2024). NMR relaxation to probe zeolites: Mobility of adsorbed molecules, surface acidity, pore size distribution and connectivity. *Molecules (Basel, Switzerland)*, *29*(22), 5432. DOI: 10.3390/molecules29225432 PMID: 39598821

Shelyapina, M. G., Krylova, E. A., Mazur, A. S., Tsyganenko, A. A., Shergin, Y. V., Satikova, E., & Petranovskii, V. (2022). Active sites in H-mordenite catalysts probed by NMR and FTIR. *Catalysts*, *13*(2), 344. DOI: 10.3390/catal13020344

Shelyapina, M. G., Nefedov, D. Y., Antonenko, A. O., Valkovskiy, G. A., Yocupicio-gaxiola, R. I., & Petranovskii, V. (2023). Nanoconfined water in pillared zeolites probed by ^{1}H nuclear magnetic resonance. *International Journal of Molecular Sciences*, *24*(21), 15898. DOI: 10.3390/ijms242115898 PMID: 37958879

Shelyapina, M. G., Yocupicio-Gaxiola, R. I., Zhelezniak, I. V., Chislov, M. V., Antúnez-García, J., Murrieta-Rico, F. N., Galván, D. H., Petranovskii, V., & Fuentes-Moyado, S. (2020). Local structures of two-dimensional zeolites – mordenite and ZSM-5 – probed by multinuclear NMR. *Molecules (Basel, Switzerland)*, *25*(20), 4678. DOI: 10.3390/molecules25204678 PMID: 33066351

Silletta, E. V., Franzoni, M. B., Monti, G. A., & Acosta, R. H. (2018). Influence of exchange in NMR pore size/relaxation correlation experiments. *Microporous and Mesoporous Materials, 269*(April), 17–20. DOI: 10.1016/j.micromeso.2017.10.025

Skripov, A. V., & Shelyapina, M. G. (2016). Nuclear magnetic resonance. In Fritzsche, H., Huot, J., & Fruchart, D. (Eds.), *Neutron scattering and other nuclear techniques for hydrogen in materials* (pp. 337–376). Springer International Publishing., DOI: 10.1007/978-3-319-22792-4_12

Song, R., Song, Y. Q., Vembusubramanian, M., & Paulsen, J. L. (2016). The robust identification of exchange from T_2-T_2 time-domain features. *Journal of Magnetic Resonance (San Diego, Calif.), 265*, 164–171. DOI: 10.1016/j.jmr.2016.02.001 PMID: 26905815

Song, Y. Q., Venkataramanan, L., Hürlimann, M. D., Flaum, M., Frulla, P., & Straley, C. (2002). T_1-T_2 correlation spectra obtained using a fast two-dimensional Laplace inversion. *Journal of Magnetic Resonance (San Diego, Calif.), 154*(2), 261–268. DOI: 10.1006/jmre.2001.2474 PMID: 11846583

Srivastava, D. J., Venetos, M. C., McCarthy-Carney, L., Baltisberger, J. H., Grandinetti, P. J., & Brouwer, D. (2025). Refining siliceous zeolite framework structures with ^{29}Si 2D J-resolved NMR spectroscopy. *Physical Chemistry Chemical Physics, 27*(1), 419–435. DOI: 10.1039/D4CP03530E PMID: 39648968

Stepanov, A. G. (2016). Basics of solid-state NMR for application in zeolite science: Material and reaction characterization. In Sels, B. F., & Kustov, L. M. (Eds.), *Zeolites and Zeolite-like Materials* (pp. 137–188). Elsevier B.V., DOI: 10.1016/B978-0-444-63506-8.00004-5

Strange, J. H., Rahman, M., & Smith, E. G. (1993). Characterization of porous solids by NMR. *Physical Review Letters, 71*(21), 3589–3591. DOI: 10.1103/PhysRevLett.71.3589 PMID: 10055015

Sun, Q., Wang, N., & Yu, J. (2021). Advances in catalytic applications of zeolite-supported metal catalysts. *Advanced Materials, 33*(51), 2104442. DOI: 10.1002/adma.202104442 PMID: 34611941

Usman Azam, M., Fernandes, A., João Ferreira, M., McCue, A. J., Graça, I., & Afzal, W. (2025). Unlocking the structure-activity relationship of hierarchical MFI zeolites towards the hydrocracking of HDPE. *Fuel, 379*, 132990. DOI: 10.1016/j.fuel.2024.132990

Velasco, M. I., Franzoni, M. B., Chávez, F. V., & Acosta, R. H. (2023). Characterization of structure and functionality of porous materials. *Journal of Magnetic Resonance Open*, *14–15*, 100099. DOI: 10.1016/j.jmro.2023.100099

Verboekend, D., & Pérez-Ramírez, J. (2011). Desilication mechanism revisited: Highly mesoporous all-silica zeolites enabled through pore-directing agents. *Chemistry (Weinheim an der Bergstrasse, Germany)*, *17*(4), 1137–1147. DOI: 10.1002/chem.201002589 PMID: 21243680

Wang, C., Dai, W., Wu, G., Guan, N., & Li, L. (2022). Application of ammonia probe-assisted solid-state NMR technique in zeolites and catalysis. *Magnetic Resonance Letters*, *2*(1), 28–37. DOI: 10.1016/j.mrl.2021.10.001

Washburn, K. E., & Callaghan, P. T. (2006). Tracking pore to pore exchange using relaxation exchange spectroscopy. *Physical Review Letters*, *97*(17), 25–28. DOI: 10.1103/PhysRevLett.97.175502 PMID: 17155481

Webber, J. B. W., Livadaris, V., & Andreev, A. S. (2020a). USY zeolite mesoporosity probed by NMR cryoporometry. *Microporous and Mesoporous Materials*, *306*, 110404. DOI: 10.1016/j.micromeso.2020.110404

Webber, J. B. W., Livadaris, V., & Andreev, A. S. (2020b). USY zeolite mesoporosity probed by NMR cryoporometry. *Microporous and Mesoporous Materials*, *306*, 110404. DOI: 10.1016/j.micromeso.2020.110404

Xin, S., Wang, Q., Xu, J., Chu, Y., Wang, P., Feng, N., Qi, G., Trébosc, J., Lafon, O., Fan, W., & Deng, F. (2019). The acidic nature of "NMR-invisible" tri-coordinated framework aluminum species in zeolites. *Chemical Science (Cambridge)*, *10*(43), 10159–10169. DOI: 10.1039/C9SC02634G PMID: 32055370

Yakimov, A. V., Sushkevich, V. L., van Bokhoven, J. A., & Copéret, C. (2022). Probing acid sites in MOR zeolite using low-temperature ^{13}C solid-state NMR spectroscopy of adsorbed carbon monoxide. *The Journal of Physical Chemistry. C, Nanomaterials and Interfaces*, *126*(7), 3681–3687. DOI: 10.1021/acs.jpcc.1c10026

Yu, Z., Li, S., Wang, Q., Zheng, A., Jun, X., Chen, L., & Deng, F. (2011). Brønsted/lewis acid synergy in H-ZSM-5 and H-MOR zeolites studied by ^{1}H and ^{27}Al DQ-MAS solid-state NMR spectroscopy. *The Journal of Physical Chemistry. C, Nanomaterials and Interfaces*, *115*(45), 22320–22327. DOI: 10.1021/jp203923z

Yu, Z., Wang, Q., Chen, L., & Deng, F. (2012). Brønsted/lewis acid sites synergy in H-MCM-22 zeolite studied by ^{1}H and ^{27}Al DQ-MAS NMR spectroscopy. *Cuihua Xuebao. Chinese Journal of Catalysis*, *33*(1), 129–139. DOI: 10.1016/S1872-2067(10)60287-2

Yu, Z., Zheng, A., Wang, Q., Chen, L., Xu, J., Amoureux, J. P., & Deng, F. (2010). Insights into the dealumination of zeolitehy revealed by sensitivity-enhanced ^{27}Al DQ-MAS NMR spectroscopy at high field. *Angewandte Chemie International Edition, 49*(46), 8657–8661. DOI: 10.1002/anie.201004007 PMID: 20931586

Zasukhin, D. S., Kasyanov, I. A., Kolyagin, Y. G., Bulygina, A. I., Kharas, K. C., & Ivanova, I. I. (2022). Evaluation of zeolite acidity by ^{31}P MAS NMR spectroscopy of Aadsorbed phosphine oxides: Quantitative or not? *ACS Omega, 7*(14), 12318–12328. DOI: 10.1021/acsomega.2c00804 PMID: 35449977

Zhao, S., He, S., Du Kim, K., Wang, L., Ryoo, R., Wang, Z., & Huang, J. (2021). Influence of hierarchical ZSM-5 catalysts with various acidity on the dehydration of glycerol to acrolein. *Magnetic Resonance Letters, 1*(1), 71–80. DOI: 10.1016/j.mrl.2021.100002

Zheng, M., Chu, Y., Wang, Q., Wang, Y., Xu, J., & Deng, F. (2024). Advanced solid-state NMR spectroscopy and its applications in zeolite chemistry. *Progress in Nuclear Magnetic Resonance Spectroscopy, 140–141*(March 2023), 1–41. DOI: 10.1016/j.pnmrs.2023.11.001

Zhou, P., Liu, M. N., Luo, Q. X., Zhang, J., Chen, H., Ma, X., & Hao, Q. Q. (2022). Synthesis of hierarchical nanocrystalline β zeolite as efficient catalyst for alkylation of benzene with benzyl alcohol. *RSC Advances, 12*(8), 4865–4873. DOI: 10.1039/D2RA00209D PMID: 35425505

Zhukov, Y. M., Kovalyov, A. N., Kultaeva, A. Y., Shelyapina, M. G., & Petranovskii, V. (2016). A comparative analysis of the protonated and copper exchanged mordenites with SiO_2/Al_2O_3 molar ratio equal to 10. *International Journal of Nanotechnology, 13*(1/2/3), 136–146. DOI: 10.1504/IJNT.2016.074529

Zou, R., Chansai, S., Xu, S., An, B., Zainal, S., Zhou, Y., Xin, R., Gao, P., Hou, G., D'Agostino, C., Holmes, S. M., Hardacre, C., Jiao, Y., & Fan, X. (2023). Pt nanoparticles on Beta zeolites for catalytic toluene oxidation: Effect of the hydroxyl groups of Beta zeolite. *ChemCatChem, 15*(21), e202300811. DOI: 10.1002/cctc.202300811

ADDITIONAL READING

Callaghan, P. T. (2011). *Translational dynamics and magnetic resonance. Principles of pulsed gradient spin echo NMR.* Oxford University Press. DOI: 10.1093/acprof:oso/9780199556984.001.0001

Duer, M. J. (Ed.). (2002). *Solid-state NMR spectroscopy. Principles and applications*. Blackwell Science.

Fujara, F., Kruk, D., & Privalov, A. F. (2014). Solid state Field-Cycling NMR relaxometry: Instrumental improvements and new applications. *Progress in Nuclear Magnetic Resonance Spectroscopy*, *82*, 39–69. DOI: 10.1016/j.pnmrs.2014.08.002 PMID: 25444698

Price, W. S. (2009). *NMR studies of translational motion: Principles and applications*. Cambridge University Press., DOI: 10.1017/CBO9780511770487

Wang, W., Xu, J., & Deng, F. (2022). Recent advances in solid-state NMR of zeolite catalysts. *National Science Review*, *9*(9), nwac155. Advance online publication. DOI: 10.1093/nsr/nwac155 PMID: 36131885

KEY TERMS AND DEFINITIONS

Chemical Shift: A chemical shift in NMR is a shift in the NMR signal depending on the chemical composition of a substance due to the shielding of an external magnetic field by atomic electrons.

Dipole–dipole coupling: Dipole–dipole coupling in NMR refers to the direct interaction between two nuclear magnetic moments.

Magic Angle Spinning: Magic Angle Spinning (MAS) in solid-state NMR refers to a technique where the sample is rapidly rotated at an angle 54°44' with respect to the external magnetic field. This technique helps to average anisotropic spin interactions, which leads to narrower spectral lines and increased resolution of NMR spectra.

Nuclear Magnetic Resonance: Nuclear magnetic resonance (NMR) is a resonant absorption or emission of electromagnetic energy by a substance containing nuclei with nonzero spin in an external magnetic field at a well-defined frequency which is proportional to the external magnetic field and the nuclear gyromagnetic ratio.

Quadrupole coupling: Quadrupole coupling refers to the interaction of quadrupolar nuclei with the electric field gradient at the nucleus.

Spin-lattice relaxation: Spin-lattice relaxation is the process by which nuclear magnetization returns to its thermal equilibrium state along a static magnetic field after excitation by a radiofrequency pulse (recovering of longitudinal component of nuclear magnetization). It is characterized by the spin–lattice relaxation time T_1.

Spin-spin relaxation: Spin-spin relaxation is the process by which the transversal component of nuclear magnetization, that appears after excitation by a radiofrequency pulse, disappears. It is characterized by the spin–spin relaxation time T_2.

Chapter 3
FTIR Studies of Adsorption Properties of Zeolites

Alexey Tsyganenko
https://orcid.org/0000-0003-0496-984X
St. Petersburg State University, Russia

ABSTRACT

The paper summarizes the results of FTIR studies of specific properties of zeolites, compared with metal oxides, in the adsorption of simple molecules. An attempt is made to explain the unusual properties of zeolites, such as linkage isomerism of adsorbed CO, extremely high Lewis acidity of exchangeble cations, Brønsted acidity of H-zeolites and basicity of zeolite framework oxygen atoms. Acidity of zeolitic hydroxyl groups, choice of test molecules for acid and basic sites, advantages of carbon monoxide and fluoroform for testing the strength of surface centers, correlation between their frequency shifts and absorbance are discussed, as well as the influence of environment upon the electron-accepting properties of cations and their ability for isomerism.

INTRODUCTION

Over the past 50 years, the number of papers on the study of zeolites by means of IR spectroscopy of adsorbed molecules has increased by an order of magnitude. Several reviews with hundreds of references were published concerning the characterization of surface active sites of zeolites using different test molecules (Zecchina & Otero Areán, 1996; Hadjiivanov & Vayssilov, 2002; Lamberti et al., 2010; Bordiga et al., 2015, Busca, 2017). In this paper, we did not plan to present

DOI: 10.4018/979-8-3373-0320-8.ch003

one more work of this kind but try to find an explanation of the unusual properties of zeolites that distinguish them from the surface properties of metal oxides based on the changes in the spectra of adsorbed molecules and surface hydroxyl groups caused by adsorption. During 15 years since the publication of our paper on low-temperature infrared studies of surface phenomena (Tsyganenko, 2011), a lot of new results appeared, both experimental and theoretical. Some new data on the problem of scattering distorting the contour of spectral bands and the ways to diminish it (Novikov et al., 218; Tsyganenko, 2013; Bogdanov et al., 2021). The role of lateral interactions in spectra formation was considered and summarized (Dobrotvorskaia et al., 2015, 2017).

HYDROXYL GROUPS

Since the early work by Angell & Schaffer (1965), a problem of hydroxyls in zeolites arose. One band at *ca* 3745 cm^{-1} almost coincides in its position with the silanol band of silica absorbing at 3749 cm^{-1}, however differs somehow in its behavior. Another one at about 3645 cm^{-1} was related to the aluminum content, but unlike the Al-OH bands, whose shift on perturbation by benzene does not exceed 120 cm^{-1} were much more sensitive to its adsorption shifting by about 330 cm^{-1}, thus demonstrating much greater acidity. It was logically concluded that the proton, which compensates for the negative charge of the AlO_4 tetrahedron, should be bound to one of four surrounding oxygen atoms already linked to silicon. Thus, the bridged Si-OH-Al groups should be formed.

Unlike bridged hydroxyls of metal oxide surfaces, according to our earlier proposed model (Tsyganenko, & Filimonov, 1972, 1973), or the bulk OH groups of metal hydroxides or minerals (Tsyganenko, 1975), where the oxygen atom is bonded to the atoms of the same metal, zeolitic OH groups are not symmetric, being bound to the atoms of different elements. We can compare such a group with the bridged group of alumina surface (type II OH using the notion of Tsyganenko, & Filimonov, 1972, 1973), or of the hydroxyls of $Al(OH)_3$. If we substitute in such Al-OH-Al group one aluminum by silicon, it means a great increase of the positive charge of cations bound to the oxygen of hydroxyl. This should lead to the OH frequency lowering from about 3740 cm^{-1} for alumina and to the increase in its acidity, what we can see in reality.

On the other hand, for silicon valence is equal to the coordination number, thus, each of the four bonds can be considered as covalent. The aluminum atom forms three bonds to three oxygen atoms, linking it to three adjacent silicons. The third bond to the oxygen of the OH group should rather be considered as a coordinate one. Thus, zeolitic bridged hydroxyls could be considered as silanol groups which form coordinate bond to adjacent three-coordinated aluminum ions. Such a bond should also lead to frequency lowering and acidity increase.

Influence of coordinate bond formation upon the properties of silanol group can be modeled by adsorption of gaseous aprotonic acids. Adsorption of SO_3 results in the reaction leading to the formation of immobilized sulfuric acid with a dangling S-OH group (Tsyganenko & Trusov, 1978). Less acidic SO_2 itself does not adsorb strongly on silanol groups, shifting the bands of silanol groups by about 90 cm^{-1} (Babaeva et al., 1984), but if it is added to a sample with pre-adsorbed pyridine or 2,5dimethylpiridine (DMP, lutidine) SO_2 becomes strongly held, while instead of the bands of H-bonded DMP at 1606 and 1585 cm^{-1} new bands of protonated base at 1655 and 1629 cm^{-1} arise (Tsyganenko et al., 2001). This phenomenon of "induced Brønsted acidity", illustrated in Fig.1, shows that as a result of SO_2 coordination to oxygen, the acidity of the silanol group increases to such extent that proton transfer to the base occurs, and the Si-OH group, whose interaction with strong bases was always limited to H-bonding, acts in this case as a real Brønsted acid site.

Figure 1. "Induced Brønsted acidity". IR spectrum of SiO_2 before (a) and after DMP adsorption (b), addition of SO_2 (c) and after short evacuation at −90°C (d). Curves (b–d) below 1700 cm^{-1} are presented after subtraction of the background silica absorption (spectrum (a)).

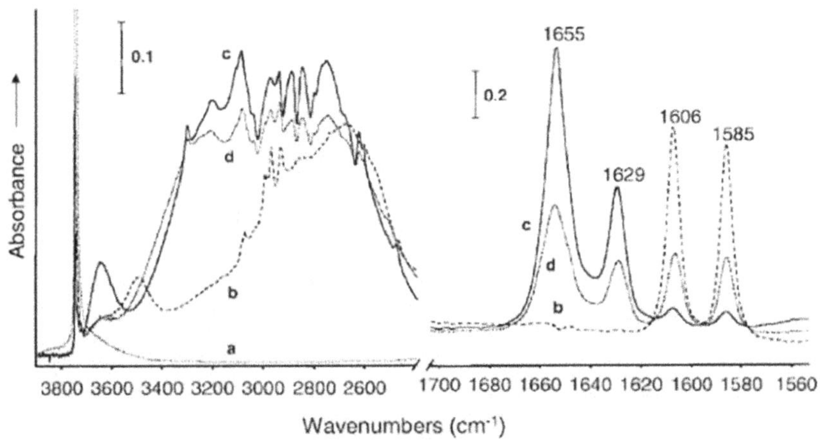

(Tsyganenko et al., 2001)

This enhancement of silanol acidity caused by SO_2 adsorption occurs in parallel with stronger adsorption of SO_2 itself. Apparently, adsorption of base increases the basicity of oxygen atom of silanol group. In fact, the spectrum of adsorbed fluoroform CHF_3 (Fig. 2) reveals the appearance of strong basic sites on silica after preliminary adsorption of ND_3, demonstrating the phenomenon of "induced basicity" (Storozheva et al., 2006). Therefore, we can speak about mutual adsorption enhancement when acid and base are coadsorbed on silica.

These data enable us to propose the answer to the question put forth by Mapes and Eischens 70 years ago (Mapes & Eischens, 1954): what is the nature of Brønsted acid sites of silica-alumina catalyst if the sample had been heated at 500°C under vacuum for 20 hours and no other hydroxyls remain, but the silanols. We now suppose that if the silanol group has in the nearest environment a three-coordinated Al^{3+} ion, then the adsorption of a base molecule should raise the basicity of oxygen atom to such extent that it forms a coordinate bond with the adjacent aluminum. The formed bridged hydroxyl, the same as in zeolites, donates its proton to the base, acting as a Brønsted acid site.

The strength of proton-donating centers, usually OH groups or sometimes chemisorbed water molecules, or even surface NH_2 groups, can be characterized by their X-H stretching frequency shifts on adsorption of a standard perturbing agent. This can be a strong base like ammonia or pyridine, but then detailed comparison is complicated because of proton transfer from Brønsted acidic sites and their simultaneous adsorption on the cations. The latter obstacle can be overcome by application of DMP (Oliviero et al., 2005) where the methyl groups prevent interaction with cationic sites.

Figure 2. "Induced basicity". IR spectrum of fluoroform adsorbed at −196°C on silica pretreated at 600°C (dash line) and on silica with preadsorbed ND_3 (solid line).

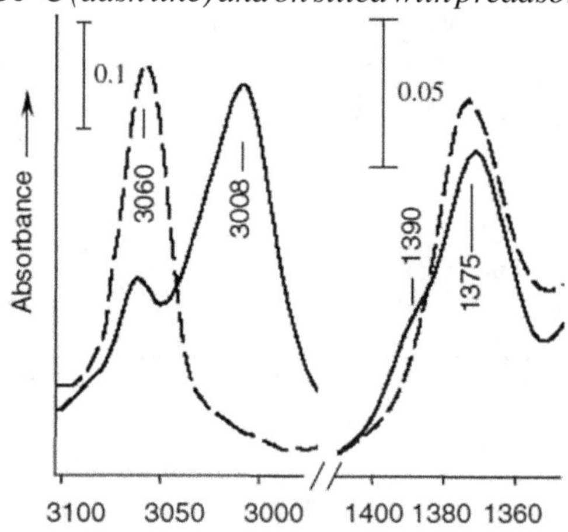

(Storozheva et al., 2006).

It was attractive, however, to find such a test that does not interact with the cations and whose interaction with hydroxyls is limited to H-bonding. Attempt to use benzene adsorption enabled the authors (Tret'yakov & Filimonov, 1972) to arrange the surface OH groups of different oxides in a sequence according to their proton-donating ability. The sequence was found to be the same if we measured the OH frequency shifts on the low-temperature adsorption of CO, nitrogen or oxygen or even on submerging the sample in liquid O_2 or N_2 (McDonald, 1958; Tsyganenko et al., 1976). It became possible in this way to distinguish the shifts of bands for different types of OH groups of the same oxides, which overlap on perturbation by stronger bases. The influence of nitrogen matrices upon the spectra of acid molecules could be compared with the effect of a sample submerged in liquid N_2 (Smirnov & Tsyganenko, 1977). For USY zeolite, using CO as the adsorbate, Makarova et al. (1994) succeeded in distinguishing three different types of OH groups with strong, medium and weak acidity characterized by OH frequency shifts of 382, 327 and 255 cm^{-1}, respectively. Zholobenko et al. (1993) have shown that in H-mordenite, the asymmetrical stretch band of bridged OH groups at 3607 cm^{-1} consists of two components, one at a higher frequency of 3612 cm^{-1} (lower effective acidity) and another at a lower frequency of 3585 cm^{-1} (higher effective acidity) with the intensity ratio of 2:l. The low-frequency OH groups are inaccessible to bulky molecules such as benzene and cyclohexane, which suggests they are located in small channels.

Chakarova, & Hadjiivanov, (2011) have studied adsorption of CO and N_2 at 100 K on H-ZSM-5. At high CO coverages, they have distinguished two kinds of CO interaction with isolated silanol groups: those that absorb at 3745 cm^{-1} are shifted to 3650 cm^{-1}, and others, which account for absorption in the 3730-3710 cm^{-1} region, move to 3580 cm^{-1}. At intermediate CO coverages, the band at 3667 cm^{-1}, assigned to AlOH groups, undergoes a shift by ~-190 cm^{-1}, ascribed to the formation of AlOH - CO complexes. The bridging OH groups with a band at 3616 cm^{-1} give two bands of perturbed hydroxyls at 3306 and 3415 cm^{-1}. As soon as perturbation with adsorbed N_2 leads to a single band of perturbed OH, it was concluded that the appearance of two shifted bands was due to Fermi resonance with the second excitation mode of δ(OH). For the reason of Fermi resonance and solvation effects, which results in additional shifts of bands with coverage, the authors concluded that CO is not convenient for measuring the acidity of the bridging hydroxyls, and in their FTIR study of H-FER (Chakarova & Hadjiivanov, 2013) they used perturbation by adsorbed N_2. This enabled the authors to distinguish three kinds of zeolite bridging hydroxyls with the bands at 3609, 3596 and 3555 cm^{-1} (3614, 3600 and 3556 cm^{-1} at 100 K). Nitrogen affects them in different ways, first, the highest-frequency hydroxyls and last, the lowest-frequency ones, shifting the bands by 90 - 111 cm^{-1}.

The application of OH frequency shifts for acidity characterization is limited by the usual H-bond. If we turn to less acidic or basic hydroxyls, we face the "abnormal" blue-shifting H-bond when the bonding increases OH frequency (Hobza & Havlas, 2000). For surface OH groups of CaO or MgO, attempts to perturb them by benzene led to zero shift (Tret'yakov & Filimonov, 1972), while submerging into liquid nitrogen results in a slight shift to higher wavenumbers (Tsyganenko et al., 1976). Although it is scarcely the case of zeolites, this means that to characterize the proton-donating ability of surface hydroxyls it is preferable to use the frequency shifts of test molecules themselves.

ACIDITY TESTING

The most popular and widely used test molecule for surface acid sites nowadays is CO, adsorbed at low temperatures (Angell, 1966; Rodionova et al.,1982; Paukshtis & Yurchenko, 1983; Zaki & Knözinger, 1987; Kustov et al., 1987; Hadjiivanov & Vayssilov, 2002; Busca, 2017, Busca & Gervasini, 2020). While the OH frequency shift on CO adsorption increases from *ca* 100 cm^{-1} for silanols up to 275-320 cm^{-1} for zeolitic bridged hydroxyls (Makarova et al., 1994; Zholobenko et al., 1993), the frequency of C-O stretching vibration increases from 2143 cm^{-1} for the gas up to 2156 - 2160 cm^{-1} for Si-OH and 2180 cm^{-1} for the most acidic hydroxyls of H-zeolites (Hadjiivanov & Vayssilov, 2002).

Among molecules whose frequency changes noticeably on interaction with surface OH groups, we must mention hydrogen. H_2 adsorption at liquid helium temperature on silica results in the O-H frequency shift by 15 cm^{-1}, while the H-H frequency lowers from 4160 down to 4140 cm^{-1} (Tsyganenko, 1980). This is in agreement with the data by Makarova et al (1994b) who have measured frequency shifts of OH groups and hydrogen adsorbed on several zeolites and has shown that acidity characterized by these parameters increases in the following order: Si-OH < SAPO-37 < H-Y < H-EMT < H-ZSM-5 < H-MOR.

Figure 3 shows IR spectra of hydrogen adsorbed on different zeolites. H_2 interaction with zeolites with high Si/Al ratio leads to a band of induced absorption shifted to lower wavenumbers from the gas phase value. In accordance with the electrostatic model, complexes with cations have T-shaped (side-on) geometry, and the magnitude of frequency shift depends on the cation (see Fig. 3a), growing with the increase of electric field for ions of smaller radius. For the zeolites with high aluminum content, H_2 forms linear complexes with framework oxygen atoms. The band position depends on the negative charge of the latter determined by the Si/Al ratio. That is why the band positions for LiNaY and CsY zeolites are almost the same (Fig. 3b), while for CsX it is much lower and close to that for H_2 on LiZSM-5.

Thus, the frequency is lowered in hydrogen complexes with both cations and anions, which can cause band assignment difficulties (Tsyganenko, 2013).

Figure 3. FTIR spectra of hydrogen adsorbed at -196°C on zeolites. a: on high-silica zeolites with Si/Al = 25: KZSM-5 (1), NaZSM-5 (2), and LiZSM-5 (3); b on high-aluminum faujasites: LiNaY (1), CsY (2) and CsX (3).

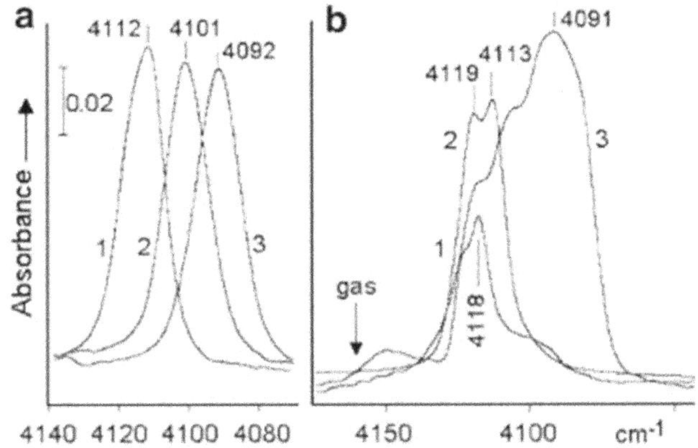

(from Tsyganenko, 2013)

Angell & Schaffer (1966) established a good linear relation for CO adsorbed on zeolites between frequency and field strength of the alkaline earth metal cations, while for transition metal cations, there were considerable deviations. This electrostatic approach describing frequency shift on adsorption as a manifestation of the vibrational Stark effect was then successfully extended to alkali cations (Storozhev et al., 2004). For each cation, the position of CO band is not the same for zeolites with different structures and Al content. The frequency is a little higher for Y than X zeolites, demonstrating the effect of environment upon the electrostatic field of cations. To clarify this point and to answer the question of why in zeolites sodium, which is the embodiment of basicity, manifests acidity of the cation comparable with that of weak Lewis sites of oxide surfaces, a special work was carried out (Ezdakova et al., 2022).

Calcium was chosen to explore the effect of the environment on the electron-accepting ability of cations since it can exist as an exchangeable cation in zeolites or be present at the surface of oxides or salts. To this end, the surface of calcium oxide was modified by chemisorption of CO_2, SO_2 and SO_3 and the strength of acidic sites of such surface was characterized by IR spectra of adsorbed CO and compared with that of initial CaO surface, of CaY zeolite and with the published data on calcium halides. The experimental data are completed with the results of quantum chemical calculations of cluster models of clean or modified oxide and models of zeolite fragment with calcium cation, as well as of CO interacting with a bare cation without any environment.

Figure 4. IR spectrum of CO adsorbed on CaO sample thermoevacuated at 700°C and modified by adsorption of H_2O with subsequent pumping at 20 (1) and 500°C (2), and adsorption of CO_2 (3), SO_2 (4) and SO_3 (5). 6 — spectrum of CO adsorbed on CaY zeolite. The spectra were recorded at —196°C and normalized in amplitude.

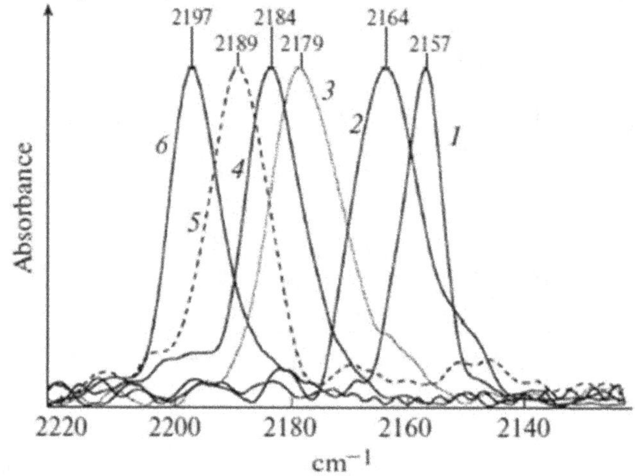

(from Ezdakova (Voronina), 2022)

The experimental data, presented in Figure 4, show that the conversion of oxygen ions surrounding the cation into larger carbonate, sulfite or sulfate anions, when negative charge becomes distributed over several atoms, results in the increase of adsorbed CO frequency. If we compare band positions of CO adsorbed on CaO (2156 cm^{-1}) with those for CaF_2 (2168 cm^{-1}), $CaCl_2$ (2185 cm^{-1}), $CaBr_2$ (2185 cm^{-1}) and CaI_2 (2190 cm^{-1}) reported in (Schmidt et al., 1984), we have to conclude that the greater the size and the lower the charge of anions surrounding the cation, the higher the electron-accepting ability. In zeolite, the negative charge distribution over the framework leads to a further decrease of the shielding effect of surrounding anions and the frequency of carbon monoxide adsorbed on the cation increases even more. Lower concentration of aluminum and cations in Y zeolites as compared with X is accompanied with lower negative charge of the framework and hence, higher electron-accepting ability of the former, reflected in slightly higher frequencies of CO adsorbed on Y or ZSM-5 as compared with X zeolites with the same cations.

The quantum-chemical calculations of cluster models (Ezdakova et al., 2022) reproduce the experimentally observed increase in the frequency of adsorbed CO in the series CaO < $CaCO_3$ < $CaSO_3$ < $CaSO_4$ < CaY zeolite and predict an even higher frequency of CO in complex with a bare Ca^{2+} cation, the experimental observation of which seems extremely problematic. This means that the electron-accepting ability or the strength of cationic acid sites of the salts of oxygen-containing acids should be higher as compared to oxides of the same metals In parallel with frequency growth, the calculated integrated absorbance of adsorbed CO decreases in fair agreement with the experimental results (Kondratieva et al., 2008) and electrostatic model. It should be noted that this accordance is a rule for the cations of non-transition metals.

CO adsorbed on Ag-ZSM-5, AgY, AgX (Hadjiivanov, 1998, Tarach et al., 2014) or Ag-mordenite (Tsyganenko, et al., 2025) absorb at 2195-2184 cm^{-1}, almost at the same frequencies as for CaY – at 2201–2197 cm^{-1} (Hadjiivanov et al., 2003; Tsyganenko, et al., 2004) however the absorption coefficient is extremely high, 6.98 ± 0.01 cm/μmol for AgZSM-5 and AgY (Tarach et al., 2014) or 8.0 cm/μmol for Ag-mordenite (Tsyganenko, et al., 2025), greater than 6.55 cm/μmol for gaseous CO (Kim, 1983). Even higher absorbance values were found for Ni-and Cu-containing zeolites, reaching for Cu-ZSM-5 16.50 cm/μmol (Góra-Marek et al., 2012; Tarach et al., 2014).

The band of CO adsorbed on CaY zeolite initially appears at 2197 cm^{-1}, then, with increasing coverage it starts to diminish while a band at 2190 cm^{-1} grows up to saturation (Tsyganenko, et al., 2004). Such behavior testifies for the adsorption of two molecules at the same cation. Spectra obtained for CaNaY by Hadjiivanov et al. (2003) are similar to ours, but accurate analysis of coverage dependence at -188°C led them to the conclusion that there could be up to three CO molecules bound to the same Ca^{2+} cation, with the bands at 2201, 2191 and 2185 cm^{-1} for $Ca^{2+}CO$, $Ca^{2+}(CO)_2$ and $Ca^{2+}(CO)_3$, respectively. Formation of three-, di- and monocarbonyls was established also at -188°C for zeolites SrY (bands at 2182.5, 2187.5 and 2193 cm^{-1}) and BaY (ca. 2172, 2177, 2181 cm^{-1}). Low-temperature CO adsorption on BeNaY results in formation of two types of Be–CO carbonyls (band at 2222 which shifts to 2212 cm^{-1} with the coverage increase and a weak band at 2198 cm^{-1}) and a coverage independent band at 2215 cm^{-1} for Mg–CO. No evidence for multicarbonyls was presented for Be- and Mg-zeolites (Hadjiivanov et al., 2003), unlike Na- (Hadjiivanov & Knözinger (1999a), Ni- (Hadjiivanov et al., 2002) and Cu- zeolites (Hadjiivanov & Knözinger, 2000), where the formation of dicarbonyl complexes was established. On NaY zeolite, low-temperature N_2 adsorption was also shown to result in the formation of geminal dinitrogen species (Hadjiivanov & Knözinger (1999b).

If the Si/Al ratio is not too high, two or more cations can be located in the same cavity of a zeolite and a molecule adsorbed on one of them can be affected by the presence of others. The case of CO adsorption on such dual and multiple cation sites was studied in a recent work (Nachtigall et al., 2012) that combined variable temperature infrared spectroscopy with periodic DFT calculations. Adsorption complexes formed on these cationic adsorption sites differ in structure and stability from those formed on single cations.

Figure 5 IR spectra of CO adsorbed on Na-FER with Si: Al=8:1. From top to bottom, the temperature goes from –66 to –9°C, and pressure from 0.19 to 1.17 mbar.

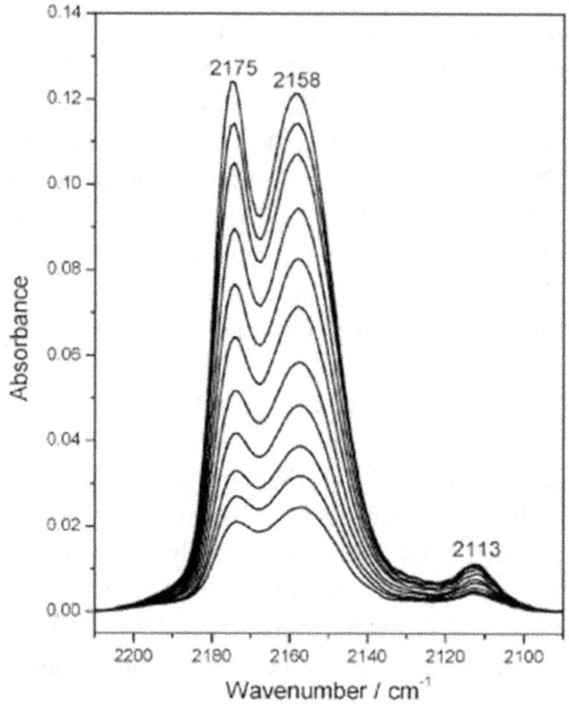

(from Nachtigall et al., 2012)

The spectra of carbon monoxide adsorbed on Na-FER with a Si/Al ratio of 8:1 are shown in Fig. 5 for a surface coverage range of 0.05 - 0.25 to avoid the formation of dicarbonyl species. One can see distinctive IR absorption bands at 2175, 2158 and 2113 cm^{-1}. The band at 2175 cm^{-1} corresponds to the C–O stretching mode of CO interacting through the carbon atom with a single Na^+ cation, while that at 2158 cm^{-1} should come from the C–O stretching mode of carbon monoxide bridging two nearby Na^+ cations, interacting with one of them through the C atom and with the other one through the O atom; that is, a CO molecule adsorbed on a dual cation site.

Note that the field of Na^+ cation facing the O atom of CO adsorbed on a dual cation site would partially counterbalance the frequency increase caused by the Na^+ cation facing the C atom. This gives a qualitative explanation for the smaller blue shift referred to free CO at 2143 cm^{-1} of the IR absorption band at 2158 cm^{-1} (dual site) as compared to 2175 cm^{-1} (single site). From the temperature dependences of integrated intensities of the bands at 2175 and 2158 cm^{-1}, the corresponding standard adsorption enthalpy ΔH^0 for the species that account for these bands were obtained, equal to 30 and 35 kJ/mol. The latter value is greater, as should be expected from higher thermal stability of the 2158 cm^{-1} band following from the figure. It was, thus, shown that, when dealing with alkaline zeolites, any C-bonded monocarbonyl showing a C–O stretching frequency significantly lower than expected for the corresponding species formed on a single cation should be suspected to denote a dual or multiple cation site; particularly when that carbonyl complex shows enhanced resistance to outgassing, or to the increase in temperature.

Another example of a dual cationic site in zeolite was observed in the diffuse-reflectance infrared spectra of dinitrogen adsorbed on the zeolite Cu-ZSM-5, where the formation of bridged Cu^+–N–N–Cu^+ species was confirmed by quantum chemical calculations of cluster models (Serykh, & Amirides, 2006). On oxide surfaces, dual sites arise as a result of bridged (type II) OH group desorption in the form of water after recombination with a proton of neighboring hydroxyl group. This process should lead for the surface of alumina to the appearance of dual coordinately unsaturated Al^{3+} cations (Tsyganenko & Mardilovich, 1996). In this case, the OH^- anion can be considered a test for such sites. It should be noted that charged particles have certain advantages as tests. In fact, CN^- ions formed as a result of HCN dissociation on basic zeolites (Tsyganenko, et al., 2010), do not attract to the negatively charged framework oxygen atoms, but interact exclusively with the cations. Just the same, NO^+ cations (Henriques et al., 2001) formed in exchange position in zeolites upon co-adsorption of NO and O_2 can be used as a probe for basic sites in zeolites.

LINKAGE ISOMERISM

Most spectroscopic studies of surface species concern relatively stable products of adsorption or chemical transformations that exist at low temperatures in the absence of reactive molecules as long as possible. If the sample temperature is high enough, the energetically excited or less favorable states of adsorbed molecules can be observed. When the amount of adsorbate is limited, temperature rising can lead to redistribution of molecules over surface sites so that their concentration on weaker sites increases. So, CO adsorbed on aerosil is mostly H-bonded to the silanol groups, and at temperatures below that of liquid nitrogen, the CO band at 2156 cm^{-1} is the most intense. However, at -196°C or higher, the band at about 2136 cm^{-1} due to molecules bound to siloxane bridges becomes more intense.

At high temperatures, vibrationally excited states of surface species, which account for so-called "hot bands" in the spectra, can be observed. Such a band of silanol groups was detected in the spectrum of silica at 800 °C (Peri, 1966) as a weak maximum at 3560 cm^{-1}. The estimate of the population of vibrationally excited state by the Boltzmann equation shows that the fraction of excited silanol groups is less than 1% even at such a temperature. Excitation of low-frequency vibrations usually results in a gradual shift of the maxima of absorption bands with temperature due to the anharmonicity of vibrations. Such shifts can be clearly seen in the spectra of isolated OH groups and especially of silanol groups perturbed by hydrogen bonding (Pavlov & Tsyganenko, 1994).

In the spectra of CO adsorbed on zeolites, weak bands shifted to low frequencies with respect to that of a free molecule are often found along with the usual bands of molecules bound to the cations, shifted to higher frequencies (HF band). Various hypotheses were advanced concerning the origin of these LF band. In the spectrum of Na zeolite, shown in Fig. 6, the position of LF bands is very close to that of ^{13}CO molecules, corresponding to the HF band. However, it was found (Otero Areán et al., 1998) that the intensities of the LF bands grow with temperature when that of the HF band decreases, which is completely unexpected for the isotope band. The intensity changes in the spectra are completely reversible, thus the LF bands cannot belong to some new products of activated CO adsorption.

Figure 6. a: FTIR spectra of CO (ca. 0.8 Torr) adsorbed on Na-Y at variable temperatures from –106 up to –40°C (1-7); b: The van't Hoff plots of the natural logarithm of the intensity ratio of LF and HF bands versus reciprocal temperature. Circles correspond to the CO/Na-Y, and squares to CO/Na-ZSM-5.

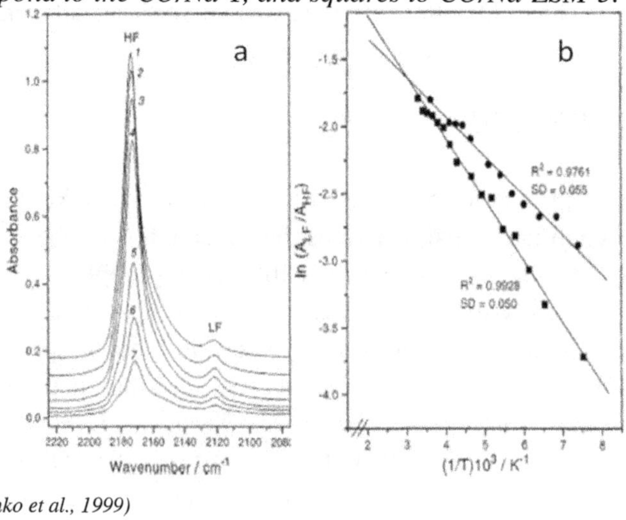

(from Tsyganenko et al., 1999)

It was shown that the logarithm of intensity ratio for the LF and HF band changes linearly with reciprocal temperature. Such dependence can be expected if the LF band corresponds to an excited state of a molecule with the excitation energy comparable to kT, as in the case of the above "hot" transitions from the vibrationally excited states. However, the shift of the LF band with respect to the HF band is too large, and the spectrally estimated excitation energy is nearly one order of magnitude lower than it could be expected for ν(CO) excitation.

The origin of the LF band was explained by the existence of isomeric CO complexes, in which the molecule is bound to a cation via the oxygen atom. The two isomeric forms, C- and O-bonded, are in thermodynamic equilibrium, and the ratio of their concentrations depends on the temperature and enthalpy of their interconversion ΔH^0 following the Vant-Hoff equation, which can be written for our case as

$$\ln(ALF/AHF) = -\Delta H^0/RT + \Delta S^0/R + \ln(\varepsilon LF/\varepsilon HF),$$

where the ratio of concentrations of substances participating in the equilibrium is expressed in terms of the integral intensities of the low-frequency and high-frequency absorption bands (ALF and AHF) and the corresponding extinction coefficients of surface compounds εLF and εHF.

From the slope of the Vant-Hoff plots for NaZSM-5 and NaY zeolites shown in fig. 6b, one obtained the values of CO-OC isomerization enthalpy ΔH^0. These slopes are different for the two straight lines and give ΔH^0 values of 3.8 and 2.4 kJ/mol for NaZSM-5 and NaY, respectively. Hence, CO interacts with the same Na^+ cation with different adsorption energies, which, besides the cation nature, depends on its environment, which is different in zeolites with various lattice structures.

The fact that two straight lines in fig. 6b do not pass through the origin can be due to the difference between εLF and εHF. The εLF/εHF ratio can be found from the position of intersection points of the two curves extrapolated to 1/T=0, if one estimates ΔS^0, the changes of entropy upon isomerization, taking into account the experimentally observed correlation between the entropy and enthalpy of adsorption for adsorbed CO. In this way, it was found that εLF/εHF is always above 1 and grows with the increase of ΔH^0.

For known values of ΔH^0, such as for CaY, another method of εLF estimation is possible. Having obtained the isomerization energy from the slope of Vant-Hoff plot we can find the relative population of the O-bonded species from the Boltzmann factor. Assuming that the concentration ratio of the C- and O-bonded complexes is equal to $\exp(\Delta H^\circ/RT)$, we have

$$\ln(AHF/ALF) + \ln(\varepsilon HF/\varepsilon LF) = \Delta H^\circ/RT.$$

Thus, having found $\Delta H°$ from the slope of the van't Hoff plot and having measured the intensity ratio of the two bands for a certain temperature T, one can readily determine the εLF/εHF ratio. The data of the paper (Ugliengo et al., 1999) allow the εLF/εHF ratio to be calculated for zeolite NaY and lead to the same value of 2.2 as the first method.

The similarity of results calculated using the above two methods shows that the values of εLF for SrY and CaY, which might seem unnaturally high, are quite realistic. Indeed, if the energy difference between the two states is about 10 kJ/mol, the concentration of O-bonded species estimated from the Boltzmann formula turns out to be so low that the LF bands would hardly be observed if their intensities were not enhanced by more than two orders of magnitude.

The existence of linkage isomerism of adsorbed CO was found for a broad series of zeolites with ZSM-5 or faujasite structures with Li, K, Rb, Cs, Ca and Sr cations, as well as for the H-forms of zeolites (Otero Areán et al., 2001) or silica (Storozhev et al., 2003) in which the hydroxyl groups form hydrogen bonds with CO molecules. Data on the spectra of adsorbed CO and ΔH^0exp values determined from the Vant-Hoff plots are summarized in (Tsyganenko et al., 2004). According to Hadjiivanov et al., (2003), low-temperature CO adsorption on CaY, SrY and BaY zeolites results in the formation of $Me^{2+}(CO)_3$ species which lose their CO ligands in a stepwise manner with decreasing equilibrium pressure and increasing temperature Then the bands of isocarbonyls Ca^{2+}–OC, Sr^{2+}–OC and Ba^{2+}–OC became visible at 2094.5, 2097 and 2100 cm^{-1}, respectively. At higher coverages, the bands of polycarbonyls with one C-bonded ligand could be seen. So, bands at 2107 cm^{-1} and 2104 cm^{-1} were assigned to the $Ba^{2+}(CO)_2(OC)$ and $Ba^{2+}(CO)(OC)$ complexes, respectively. The concentration of O-bonded species is too low to expect complexes like $Ba^{2+}(OC)_2$.

MODELING AND CALCULATIONS

CO is isoelectronic with nitrogen, and the two molecules have close physical properties. The difference in their structures is that instead of a positive charge of nuclei equally shared for N_2 between two nitrogen atoms, in CO, it is distributed in such a way as if a proton is taken from one nitrogen nucleus and added to another one. The electron density follows the positive charge and is shifted to oxygen, almost compensating the arising dipole momentum, so that the negative charge is still on carbon, but the total dipole is only 0.11D. The triple bond, very strong for nitrogen, is weakened due to electron displacement, and the frequency of stretching vibration is slightly lower. When CO forms a complex with a cation, it attaches to it via a carbon atom, and the field of the cation shifts the electron density back to carbon, partially compensating for the attraction of electrons to the oxygen nucleus. The electronic structure of the molecule becomes closer to that of N_2, and the frequency of CO increases, approaching that of N-N vibration. Moreover, the quantum mechanical calculation of CO placed in the electrostatic field directed along the molecular axis (Pacchioni et al., 1992), shows that at a certain moment νCO reaches its maximum, and what is important, the dipole momentum derivative by the coordinate that accounts for the intensity of IR band, decreases and vanishes changing its sign at this point. Thus, the external field of the cation compensates for the internal field of nuclei, and CO becomes close to N_2 not only in frequency but also in IR absorbance. The absorption coefficient decrease for CO adsorption on the cations was supported by the experiment (Kondratieva et al. 2008).

For moderate field strength, frequency shifts of adsorbed CO can be described by a simple electrostatic model as a result of the vibrational Stark effect. In (Storozhev et al. 2004), an attempt was made to explain the linkage isomerism by electrostatic model. The theory had to answer, why a dipole molecule forms linear complexes with a cation by both ends; why it does not when adsorbed on metal oxides or X zeolites; and how the formation of each isomeric state affects the frequency and absorbance of CO molecule.

Within the scope of this approximation, we consider a molecule as a polarizable dipole. For the linear M-CO and M-OC complexes, as well as for the side-on complex with a surface cation (M), the E_{ads} is respectively $(1/2)\alpha_{úú}E^2 \pm \mu_o E$ and $(1/2)\alpha_{\perp}E^2$. The contribution of the induced dipole to the total adsorption energy is always positive, while that of the permanent dipole of a molecule depends on orientation. For CO, with the dipole of 0.11 D, even for comparatively weak fields, as soon as $E>2\mu_o/\alpha_{úú}$, both C- and O- bonded adsorption should be energetically favorable. If we take into account the greater value of parallel polarizability as compared with the perpendicular one (2,33 and 1.8 A^3, respectively, for CO), for the field strength $E>2\mu_o/(\alpha_{úú}-\alpha_{\perp})$, the two states are divided by an energetic barrier, and the less favorable O-bonded state should become quasi-stable.

The ΔE value can be calculated as the energy of reorientation of the CO dipole in the field of a cation E whose strength can be estimated from the frequency shift upon adsorption

$$\Delta E = 2\mu_0 E = 2\mu_0\ \Delta\nu/Kst,$$

where for the CO molecule $\mu = 0.11$ D and Kst=4.29 x 10-9 cm^{-1}/Vm^{-1}. One can find a good agreement between the experimental ΔH^0 exp determined from the Vant- Hoff plots and values calculated using the electrostatic model. However, if we calculate the Eads considering a molecule as a polarizable dipole, the agreement with the experimental data is much poorer.

It should be noted that, besides the small dipole moment of CO, it has a large linear quadrupole momentum, as if both ends of the molecules are negatively charged, while the positive charge is in the middle. If we take into consideration cation interaction with the molecule quadrupole, the correspondence with the experiment becomes much better: the adsorption energy increases substantially, and the height of the barrier between the two states grows.

Quadrupole interaction also enables us to explain the manifestations of CO adsorption on basic sites. In fact, for the side-on adsorption on anions, the contribution of the quadrupole–anion interaction to the Eads is positive and comparable with the energy of adsorption on the cations. This is clearly seen if we compare the energy of side-on complexes, where the effective charge on oxygen is -1 or -1.5, with that of linear complexes of CO with Cs, Rb or K. For zeolites X or oxides with such a charge on oxygen atoms, the barrier between C- and O-bonded states can disappear if the transition between the two states occurs via the side-on complex with a neighboring oxygen ion. This could be the reason why CO-OC isomerism was not observed for metal oxides. For zeolites with large cations and high negative charge on the framework, such as for CsX, the energy of side-on CO interaction with oxygen becomes even greater than that of adsorption on the cations. This explains the appearance of a band with the frequency as low as 2124 cm^{-1} when CO is added to CsX (Tsyganenko et al., 2006).

The same kind of CO interaction with even more basic oxygen atoms of CaO or other basic oxides leads finally to the formation of the surface "carbonite" CO_2^{2-} ions (Babaeva et al., 1987) with further νCO lowering.

The presented experimental data are completed with the results of quantum chemical calculations using the density functional theory of cluster models of oxide, its products of interaction with different molecules, as well as the models of zeolite fragment with calcium cation and a free cation without any environment. In addition to the frequencies and absorbance of CO molecules bound to the cation through carbon, properties of molecules attached via oxygen were also investigated. (Ezdakova et al., 2022) As it was mentioned above, the calculations of cluster models well reproduce the experimentally observed increase in the frequency of adsorbed CO in the series CaO < CaO + CO_2 < CaO + SO_2 < CaO + SO_3 < CaY zeolite, as well as lower absorbance of CO adsorbed on stronger cationic sites. Calculations of O-bonded structures show that the effect of CO linkage isomerism with calcium cation, observed in experiments with CaY and CaZSM-5 zeolites, and not detected for adsorption on cations of oxide adsorbents, could be observed for the cations surrounded by polyatomic anions or hydroxyl groups. Formation of the O-bonded complexes should lead to the increase in IR absorbance. Our data (Kondratieva et al., 2008) confirm the existence of such dependence, although the experimental determination of the integral absorption coefficient meets serious difficulties, and the literature data are very contradictory (Bulanin et al., 2022). The results of quantum chemical calculation of the intensity of the CO absorption band in the spectra of studied complexes are generally consistent with the predictions of the electrostatic model.

The above analysis suggests that linkage isomerism can be anticipated for systems with a weak field of anions. Besides the compounds with polyanions, these can be metal halides where anions have lower charge and greater radii than in corresponding oxides. Data on the temperature dependence of band positions and intensities of CO adsorbed on evaporated films of NaCl as well as KBr persuade us that the weak low-frequency bands observed in the 2125–2110 cm^{-1} region of the spectra of CO adsorbed on NaCl or KBr films are due to the O-bonded CO molecules (Belykh et al., 2015). Not all of the surface cations are capable of isomerism. Two low-frequency maxima at ~2115 and 2124 cm^{-1} in the spectrum of NaCl (Fig. 7) correspond to the high-frequency CO bands at 2172 and 2160 cm^{-1}, respectively, which, in accordance with the earlier studies (Richardson et al., 1987), are due to the strongest cationic sites at the kinks or edges and steps of the rocksalt microcrystals. For the pair of bands at 2160 and 2124 cm^{-1}, the value of the isomerization enthalpy ΔHi^0 obtained from the van't Hoff plot was found to be the same as that estimated from the Boltzman's law and equal to 4.0 kJ/mol. Almost the same spectra were observed for KBr films, where the frequency shifts upon adsorption were smaller, and one HF band near 2152 cm^{-1} was observed, and the LF band could be clearly seen near 2124 cm^{-1}.

Isomerism is possible for systems with negative charge delocalized, as far as it could be, over a multiatomic framework. For zeolites, the latter is three-dimensional. To find isomerism in the systems with two-dimensional polyanions, IR spectra were studied at variable temperatures for CO adsorbed on two samples of clay minerals of the montmorillonite group, which can be considered as a layered analog of zeolites (Maevsky, Tsyganenko, 2012). Samples of commercial "blue clay", mostly of K- montmorillonite, and natural bentonite were studied.

The spectrum of CO adsorbed on blue clay exhibits a band at 2161 cm^{-1}, which moves to 2165.5 cm^{-1} with a decrease in coverage. On raising the temperature, a new band appears at 2125 cm^{-1}, whose intensity first grows and then diminishes so that the intensity ratio of the latter (LF band) to that of the high-frequency band at about 2165 cm^{-1} constantly increases. The estimated from the van't Hoff plot isomerization enthalpy was 3.8 ± 0.8 kJ mol^{-1}.

Figure 7 (a) Evolution of the IR spectrum of CO adsorbed on NaCl film with temperature increase from −184°C (top curve) to −140°C (bottom curve). (b) The same spectra with the expanded absorbance scale

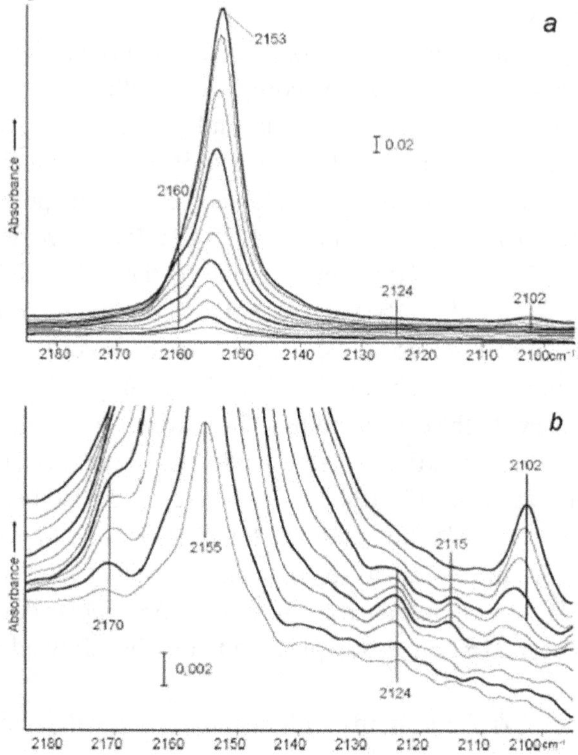

(Belykh et al., 2015).

In the spectrum of CO adsorbed on a bentonite sample at 77 K, besides the high frequency (HF) band at 2164 cm^{-1} of molecules bonded to the cations via C atom, a band at 2142 cm^{-1} with a shoulder at 2135 cm^{-1} can be seen. On heating the sample, the maximum of the HF band shifts to 2176 cm^{-1} and one more band appears at 2121 cm^{-1}. One can see the monotonous increase in the relative intensity of this band with temperature. The isomerization enthalpy estimated from the van't Hoff plot in this case was 5.6 ± 0.4 kJ mol^{-1}.

Thus, both for the blue clay and bentonite samples, spectra of CO at variable temperatures provide evidence for the linkage isomerism of molecules coordinately bound to the cations. Higher position of the HF band and greater isomerization enthalpy value for bentonite as compared with those of blue clay are naturally due to different cationic composition of the samples, mostly Ca for the former and K for the latter.

BASICITY TESTING

Basic sites of oxide surfaces manifest themselves in their reactions with the adsorbed acids. So, CO_2 reacts with oxygen ions to form surface carbonates CO_3^{2-}, SO_2 and SO_3 adsorption leads to formation of sulfites and sulfates. With basic OH groups, these molecules produce bicarbonate, bisulfite and bisulfates with their characteristic IR spectra. Measuring the quantities of adsorbed acids, one can estimate the number of basic sites capable of reacting with one or another acid. Studies on the basicity of zeolites were thoroughly reviewed (Barthomeuf, 1996; Lavalley, 1996; Bordiga et al., 2015; Busca, 2017). CO interaction with the most basic sites of oxides leads to the formation of "carbonite" CO_2^{2-} ions (Babaeva, & Tsyganenko, 1987).

It is not a case for zeolites, where CO interaction with Si–O–Al or siloxane bridges results in the formation of side-on complexes usually absorbing near 2136 cm^{-1}, but for the most basic sites of CsX the band assigned to such structure appeared at 2124 cm^{-1} (Tsyganenko et al., 2006). For silica, this band is often assigned to physisorption, however the heat of adsorption (11 kJ/mol) for this species is well above that of condensation (6.044 kJ/mol), the bathochromic frequency shift by 5–7 cm^{-1}, which is not observed for liquid or dissolved CO, and the clearly observed rotational wings of this band (Beebe et al., 1984) show that it is rather due to the side-on complexes with siloxanes retaining the rotational degree of freedom.

In X zeolites, the higher the Al content, the greater the negative charge of the framework, and the estimated energy of electrostatic attraction of CO to the framework oxygen atoms for the side-on complexes becomes greater than that in the linear complexes with the large cations such as Cs^+ or Ru^+ and it is not surprising that adsorption on oxygen atoms became the most stable.

From the variable temperature measurements, it was possible to study the entropy change on adsorption. It was found (Tsyganenko et al., 2006) that the relative entropy change $\Delta S^0/\Delta H^0$ for the 2136 cm^{-1} band of CO on silica is considerably (about 15%) lower than that for OH···CO complexes which account for the 2154 cm^{-1} band of adsorbed CO. This can be explained by the remaining rotational freedom or hindered rotation of molecules adsorbed in such a form, in agreement with the 'rotational' IR band contour. For the 2124 cm^{-1} band of CsX zeolite the relative entropy change was even lower, pointing to the remaining rotational mobility, or, maybe even the translational degree of freedom for this side-on surface complex.

According to Bertsch & Habgood, (1963) carbon dioxide on Li-, Na-, and KX-zeolites is both adsorbed in a linear configuration at a cation and also chemisorbed in bent configurations formed by interaction with surface oxide ions adjacent to exchangeable cations. One such species common to all three zeolites gives bands around 1700 and 1340 cm^{-1}. On NaX, a second species with bands at 1485 and 1425 cm^{-1} is more strongly held at low pressures. Spectra of carbon dioxide adsorbed on the Ca, Sr, and Ba forms of zeolite X showed no evidence of bent or carbonate-like structures (Ward & Habgood, 1966).

It is not easy, if possible, to find an acidic molecule whose interaction with the basic site would cause a frequency shift, which could be used as a universal measure of site basicity. Barthomeuf (1996) suggested for that pyrrole, whose N-H band shifted to lower wavenumbers on interaction with basic O atoms. Benzene was another molecule whose bending C-H vibrations are sensitive to the basicity of surface oxygen atoms (Dzwigaj et al.,1990). However, the complexity of spectra prevents its wide use for basicity testing. Paukshtis, & Yurchenko (1983) recommended to use for this purpose C-H (C-D) vibration of chloroform or $CDCl_3$, but on some surfaces these molecules dissociate modifying the properties of the surface. Acetylene and its derivatives were thought to be the best probe molecules among the tested compounds (Kustov, 1997). The formation of complexes between these molecules and basic centers results in a large low-frequency shift of the C-H stretching vibrations (up to 220 cm^{-1} as compared with the frequency in the gas phase).

Unlike these molecules, fluoroform CHF_3 is inert enough, and for all the studied systems reveals only reversible molecular adsorption. However, its adsorption often leads to the increase in CH frequency. The quantum chemical calculations (Hermansson, 2002) have shown that the dependence of XH stretching frequency on the strength of the electric field applied along the X-H bond has the shape of a curve with a maximum. While for most molecules, the field of negative charge placed from the side of the proton results in the XH frequency decrease, for fluoroform, the initial frequency position is on another side of the maximum, and the field should first cause νCH increase. Only after reaching its maximum value, νCH should start to drop with further increase of the field. Such frequency dependence is typical of the blue-shifting H-bond.

Figure 8. a- IR spectrum of NaX zeolite after adsorption of CHF_3 (1) and sequential desorption (2-6), registered at -196°C, spectrum of the sample before adsorption subtracted. b- νCH and the ratio of intgrated intensities of the stretching and bending CH vibrations of H-bonded fluoroform, plotted against δCH of the same species.

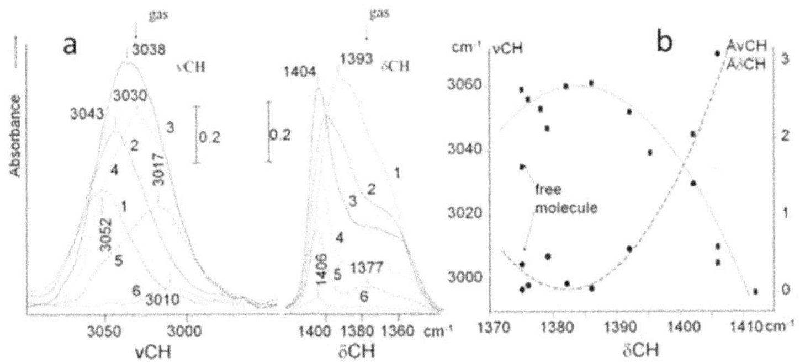

(Tsyganenko, 2014)

Fig. 8a shows coverage dependence of the spectrum of fluoroform adsorbed on NaX zeolite (Tsyganenko et al., 2014). After adding the excess of adsorbate, the bands appear at about 1392 and 3052 cm^{-1}. With the coverage decrease, besides the intensity changes, the maxima of these bands move up to 1406 and 3010 cm^{-1}, respectively. The position of the stretching CH band finally becomes well below that of a free fluoroform molecule, as expected for the strong H-bond with the most basic sites. One more important phenomenon is noticeable in fig 8. The intensity decrease of the δCH band observed on desorption is accompanied in the beginning (curves 1-3) by the growing intensity of the νCH band. More refined analysis shows that the ratio of integrated intensities of νCH and δCH bands steadily grows from about 0.5 up to 3 while νCH moves to lower wavenumbers. This effect is illustrated in fig 8b, where both νCH band position and the integrated intensity ratio AνCH/AδCH are shown as a function of the frequency of δCH mode, which increases with the increasing basicity of the remaining sites of of NaX zeolite. At the same plot, data on CHF_3 bound to basic sites of other oxide adsorbents are presented.

If we suppose that the integrated absorption coefficient of δCH vibration does not change too much with the H-bond formation, the measured values of AνCH/AδCH are proportional to the integrated absorption coefficient of the νCH mode, whose dependence on basic site strength, as seen from the figure, can be approximated by a curve with a minimum. The fact that the minimum of νCH band intensity is observed for the systems where the frequency has its maximum values is not by chance. In fact, according to the electrostatic model, the frequency of the X-H group changes as a result of the vibrational Stark effect in the electric field of the environment. The point of maximum frequency corresponds to the field that provides the highest force constant of the bond or the shortest X-H interatomic distance. The extremum of distance as a function of the electric field corresponds to the conditions where the vibrational polarizability and, hence, the absorbance is zero. The same antibate dependence of frequency and absorbance was found for adsorbed CO (Kondratieva et al., 2008).

The obtained results show that frequency shifts of adsorbed at low temperatures CH-containing molecules enable us to compare the basic properties of different adsorbents, providing semiquantitative information on the electron-donating ability of surface oxygen atoms. Stronger sites could be detected by CHF_3, while for less basic centers, more sensitive tests are needed, such as acetylene or HCN.

Another important consequence of the results presented in figure 8 is the evidence for the role of lateral interactions in adsorption properties of zeolites. In fact, removal of a part of adsorbate results in a serious change in the absorbance and vibrational frequencies and, apparently, the enthalpy of adsorption of the remaining molecules. This means that the attempts to study the heterogeneity of surface site strength by means of calorimetry or isotherm measurements, or other non-spectroscopic methods can lead to false conclusions when the seeming heterogeneity could be caused by the interaction between the adsorbed molecules.

LATERAL INTERACTIONS

First spectral studies of adsorption were concentrated on the changes in a molecule caused by its interaction with surface sites. However, soon it became clear that besides the influence of surface the spectra of adsorbed molecules are sensitive to the presence of other surrounding molecules. These lateral interactions in the adsorbed layers play an important role in adsorption, catalysis, and other processes at the surfaces.

The shift of band positions on surface coverage caused by lateral interactions was first observed for CO adsorbed on Pt by Eischens et al. (1956), who have applied the isotopic dilution method to study the nature of this phenomenon. Further development of these studies for CO adsorbed on metals was summarized by Hoffmann (1983). Later on, the same effect has been observed also for oxides (Denisenko, et., al.,1984) in the spectrum of CO adsorbed on ZnO at -196°C.

For zeolites, manifestations of the mutual influence of two CO molecules adsorbed on Ca^{2+} cation in NaCaY zeolite were noticed by Paukshtis et al. (1983). The same phenomenon was observed later for CO on CaY (Tsyganenko et al., 2004) and SrY (Garrone et al., (2003) zeolites, or for N_2 on NaY (Hadjiivanov & Knözinger, 1999b). In their spectra, the band of binary complex with a cation was shifted to lower wavenumbers relative to that of a single adsorbed molecule. Formation of geminal CO species on transition metal cations results in two distinct bands of symmetric and asymmetric CO stretching vibrations (Hadjiivanov & Knözinger, 2000, Hadjiivanov et al., 2002). In certain cases formation of complexes with even three ligands has been established for Ca- and Ni-zeolites (Hadjiivanov et al., 2001; 2002). Properties of these site-specified geminal complexes with strong ligand–ligand interaction extensively studied by Hadjiivanov et al., were summarized in the review (Hadjiivanov & Vayssilov, 2002).

To clarify the nature of the interaction between adsorbed CO molecules, the isotopic substitution method has been applied. Adsorption of ^{13}CO or of ^{12}CO–^{13}CO isotopic mixtures shows that the observed shifts of the bands caused by the addition of extra molecules in germinal CO species on CaY or Ag-mordenite are entirely static in character (Tsyganenko et al., 2004; 2025). Comparison of the observed and calculated values for the static and dynamic frequency shifts due to lateral interactions between CO molecules adsorbed on oxides has shown that the interactions are not direct but mediated by the solid, and the mechanism of static interaction includes relaxation, i.e., the displacement of surface atoms induced by adsorption (Tsyganenko, & Zverev, 1988). For the zeolites CO adsorption on cations modifies their positions, as shows DFT calculation (Qiu, 2012) changing the conditions for adsorption of another molecules. Interaction of CO_2 or fluoroform with the framework oxygen atoms, apparently, should also change interatomic distances and the electric field of neighboring atoms.

Different band positions of admixed and predominating CO isotopomers, observed for $Cu^{2+}(CO)_2$ (Hadjiivanov & Knözinger, 2000) or $Ni^{+}(CO)_2$ (Hadjiivanov & Knözinger, 2002; Belykh et al, 2019) species indicate a strong dynamic (resonance dipole-dipole) interaction between the ligands, which also accounts for a large distance between two components of stretching CO vibrations. Narrow bands of $Ni^{+}(CO)_2$ bands in the spectrum of Ni-USY zeolite enabled the authors to observe an interesting phenomenon called "isotopic isomerism" (Belykh et al, 2019). The bands of dicarbonyls of mixed isotopic composition $Ni^{+}(^{12}CO)(^{13}CO)$ turned to be split into two pairs of bands slightly different in their positions. This means that two ligands in these species somehow differ in their location. The quantum-mechanical calculation (Baranov et al., 2022) has shown that both the molecules in this structure are bound to Ni^{+} cation via carbon atom, but their positions inside a cavity of the framework are not identical.

Figure 9. The region of ν_3 vibration in FTIR spectra of SF_6 adsorbed on ZnO (1), a solid film of SF_6 on the cell windows (2) and adsorbed on silicalite-1 (3). The arrows show the position of the ν_3 band of SF_6 dissolved in liquid Ar at 939.3 cm^{-1}. The vertical dash lines represent the band positions in the calculated spectra of 2D and 1D model systems.

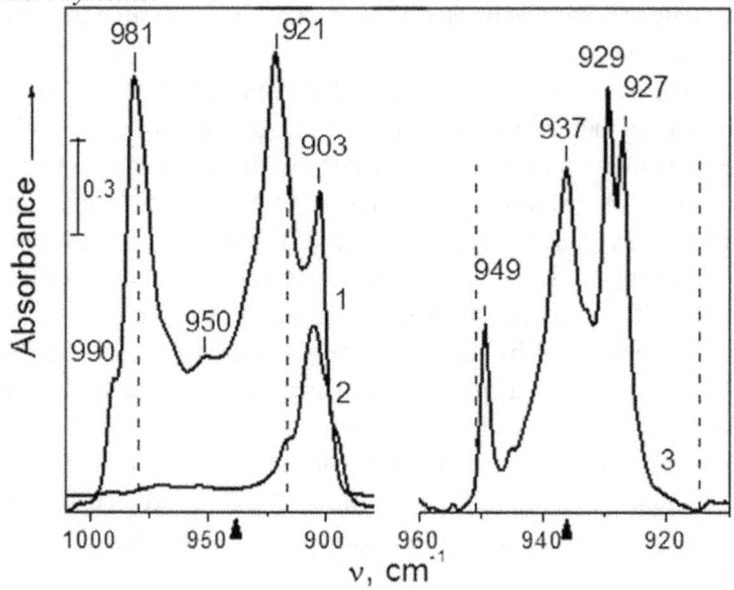

(from Dobrotvorskaia et al., 2017)

Dynamic interaction is particularly strong for vibrational transitions of high intensity, such as for the triply degenerate ν_3 mode of the octahedral sulfur hexafluoride molecule. Adsorption of SF_6 on ZnO and on silicalite-1 was investigated by a combination of IR spectroscopy and theoretical spectra calculations (Dobrotvorskaia et al., 2015). The theory says that spectra depend crucially on the mutual arrangement of interacting molecules, which can form one-dimensional (1D) chains, two-dimensional (2D) layers on flat faces of crystals, or three-dimensional (3D) phases of bulk solids. The interaction manifests itself in the spectra of 1D and 2D systems as the doublet band structure, whose components correspond to the transversal and longitudinal vibrations of the molecular dipole moment. According to the calculations, the maximal splitting between the two components should be 65 cm^{-1} for the dense layer on a plane surface and 36 cm^{-1} for 1D chains. The experiment gives even more significant difference. While for crystalline ZnO powder, it is 60–78 cm^{-1}, in fair agreement with the theory, for molecules confined in the channels of silicalite-1, it was only 20–22 cm^{-1} with the low-frequency band split into two sharp components (see Fig 9). Smaller distance between the components could mean that the intermolecular distance is determined not by the size of molecules, but by the geometry of the channel. The origin of the low-frequency band splitting is not clear.

The presented results show that IR spectra of adsorbed test molecules can provide information about discrete surface sites, the mutual influence of adsorbed species, and the geometry of pore structure inside the material.

ACKNOWLEDGEMENT

The work was supported by RSCF, grant 24-23-00606.

REFERENCES

Angell, C. L., & Schaffer, P. C. (1966). Infrared spectroscopic investigations of zeolites and adsorbed molecules. II. Adsorbed carbon monoxide. *Journal of Physical Chemistry*, *70*(5), 1413–1418. DOI: 10.1021/j100877a012

Angell, C. L., & Schaffer, P. C.Structural OH Groups. (1965). Infrared spectroscopic investigations of zeolites and adsorbed molecules. I. Structural OH groups. *Journal of Physical Chemistry*, *69*(10), 3463–3470. DOI: 10.1021/j100894a037

Babaeva, M. A., & Tsyganenko, A. A. (1987). Infrared spectroscopic evidence for the formation of carbonite CO_2^{2-} ions in CO interaction with basic oxide surfaces. *Reaction Kinetics and Catalysis Letters*, *34*(1), 9–14. DOI: 10.1007/BF02069193

Babaeva, M. A., Tsyganenko, A. A., & Filimonov, V. N. (1984). Infrared spectra of the adsorbed SO_2. *Kinetics and Catalysis*, *25*(4), 787–792.

Baranov, D. A., Krauklis, I. V., & Tsyganenko, A. A. (2022). Quantum chemical study of isotopic isomerism of CO adsorbed on NiY zeolite. *Optics and Spectroscopy*, *130*(5), 530–537.

Barthomeuf, D. (1996). Basic zeolites: Characterization and uses in adsorption and catalysis. *Catalysis Reviews. Science and Engineering*, *38*(3), 521–612. DOI: 10.1080/01614949608006465

Beebe, T. P., Gelin, P., & Yates, J. T.Jr. (1984). Infrared spectroscopic observations of surface bonding in physical adsorption: The physical adsorption of CO on SiO_2 surfaces. *Surface Science*, *148*(2-3), 526–550. DOI: 10.1016/0039-6028(84)90597-1

Belykh, R., Maevskaya, M., Krauklis, I., & Tsyganenko, A. (2015). Linkage isomerism of CO Adsorbed on alkali halides. *The Journal of Physical Chemistry A*, *119*(11), 2363–2370. Advance online publication. DOI: 10.1021/jp507394s PMID: 25222795

Belykh, R., Maugé, F., & Tsyganenko, A. (2019). Deep insight into CO adsorption on Ni-USY zeolite catalyst by FTIR spectroscopy. Evidence for isotopic isomerism. *Applied Catalysis A, General*, *583*, 117140. DOI: 10.1016/j.apcata.2019.117140

Bertsch, L., & Habgood, H. W. (1963). An Infrared Spectroscopic study of the Adsorption of water and Carbon Dioxide on Linde Molecular sieve X. *Journal of Physical Chemistry*, *67*(8), 1621–1628. DOI: 10.1021/j100802a013

Bordiga, S., Lamberti, C., Bonino, F., Travert, A., & Thibault-Starzyk, F. (2015). Probing zeolites by vibrational spectroscopies. *Chemical Society Reviews*, *44*(20), 7262–7341. DOI: 10.1039/C5CS00396B PMID: 26435467

Bulanin, K. M., Mikheleva, A. Yu., Shchepkin, D. N., & Rudakova, A. V. (2022). Determination of the Extinction Coefficient of Carbon Monoxide Adsorbed on Titanium Dioxide. *Optics and Spectroscopy*, *130*(4), 248–256. DOI: 10.1134/S0030400X2204004X

Busca, G. (2017). Acidity and basicity of zeolites: A fundamental approach. *Microporous and Mesoporous Materials*, *254*, 3–16. DOI: 10.1016/j.micromeso.2017.04.007

Busca, G., & Gervasini, A. (2020). Solid acids, surface acidity and heterogeneous acid catalysis. *Advances in Catalysis*, *67*, 1–90. DOI: 10.1016/bs.acat.2020.09.003

Chakarova, K., & Hadjiivanov, K. (2011). H-bonding of zeolite hydroxyls with weak bases: FTIR study of CO and N_2 adsorption on H-D-ZSM-5. *The Journal of Physical Chemistry. C, Nanomaterials and Interfaces*, *115*(11), 4806–4817. DOI: 10.1021/jp111961g

Chakarova, K., & Hadjiivanov, K. (2013). FTIR study of N_2 and CO adsorption on H-D-FER. *Microporous and Mesoporous Materials*, *177*, 59–65. DOI: 10.1016/j.micromeso.2013.04.022

Denisenko, L. A., Tsyganenko, A. A., & Filimonov, V. N. (1984). Infrared study of the interaction between adsorbed molecules in the Co/ZnO system. *Reaction Kinetics and Catalysis Letters*, *25*(1–2), 23–26. DOI: 10.1007/BF02076533

Dobrotvorskaia, A. N., Kolomiitsova, T. D., Petrov, S. N., Shchepkin, D. N., Smirnov, K. S., & Tsyganenko, A. A. (2015). Effect of resonance dipole-dipole interaction on spectra of adsorbed SF_6 molecules. *Spectrochimica Acta. Part A: Molecular and Biomolecular Spectroscopy*, *148*, 271–279. DOI: 10.1016/j.saa.2015.04.002 PMID: 25897721

Dobrotvorskaia, A. N., Pestsov, O. S., & Tsyganenko, A. A. (2017). Lateral Interaction between Molecules Adsorbed on the Surfaces of Non-Metals. *Topics in Catalysis*, *60*(19–20), 1506–1521. Advance online publication. DOI: 10.1007/s11244-017-0835-8

Dzwigaj, S., de Mallmann, A., & Barthomeuf, D. (1990). Adsorption of benzene and ethylbenzene on the acidic and basic sites of beta zeolite. *Journal of the Chemical Society, Faraday Transactions*, *86*(2), 431–435. DOI: 10.1039/ft9908600431

Eischens, R. P., Francis, S. A., & Pliskin, W. A. (1956). The effect of surface coverage on the spectra of chemisorbed CO. *Journal of Physical Chemistry*, *60*(2), 194–201. DOI: 10.1021/j150536a015

Ezdakova, K. V., Rudakova, A. V., Krauklis, I. V., Shergin, Y. V., & Tsyganenko, A. A. (2023). The Influence of the Local Environment on the Electron Accepting Ability of the Cation: Induced Lewis Acidity. *Optics and Spectroscopy*, *131*(10), 964–974.

Garrone, E., Bonelli, B., Tsyganenko, A. A., Rodríguez Delgado, M., Turnes Palomino, G., Manoilova, O. V., & Otero Areán, C. (2003). Spectroscopic and thermodynamic characterization of strontium carbonyls formed upon carbon monoxide adsorption on the zeolite Sr-Y. *The Journal of Physical Chemistry B*, *107*(11), 2537–2542. DOI: 10.1021/jp0217841

Góra-Marek, K., Glanowska, A., & Datka, J. (2012). Quantitative IR Studies of the Concentration of Different Nickel Sites in NiZSM-5 Zeolites. *Microporous and Mesoporous Materials*, *158*, 162–169. DOI: 10.1016/j.micromeso.2012.03.027

Hadjiivanov, K., Ivanova, E., & Knözinger, H. (2003). FTIR study of low-temperature CO adsorption on Y zeolite exchanged with Be2+, Mg2+, Ca2+, Sr2+ and Ba2+ cations. *Microporous and Mesoporous Materials*, *58*(3), 225–236. DOI: 10.1016/S1387-1811(02)00650-9

Hadjiivanov, K., & Knozinger, H. (1999a). FTIR study of the low-temperature adsorption and co-adsorption of CO and N_2 on NaY zeolite: Evidence of simultaneous coordination of two molecules to one Na^+ site. *Chemical Physics Letters*, *303*(5-6), 513–520. DOI: 10.1016/S0009-2614(99)00229-8

Hadjiivanov, K., & Knözinger, H. (1999b). FTIR spectroscopic evidence of formation of geminal dinitrogen species during the low-temperature N_2 adsorption on NaY zeolites. *Catalysis Letters*, *58*(1), 21–26. DOI: 10.1023/A:1019040825491

Hadjiivanov, K., & Knözinger, H. (2000). FTIR Study of Low-Temperature CO Adsorption on Cu-ZSM-5: Evidence of the Formation of $Cu^{2+}(CO)_2$ Species. *Journal of Catalysis*, *191*(2), 480–485. DOI: 10.1006/jcat.1999.2805

Hadjiivanov, K., Knözinger, H., Ivanova, E., & Dimitrov, L. (2001). FTIR study of low-temperature CO and $^{15}N_2$ adsorption on a CaNaY zeolite: Formation of site-specified and $Ca^{2+}(CO)_3$ complexes $Ca^{2+}(^{15}N_2)_3$. *Physical Chemistry Chemical Physics*, *3*, 2531–2536. DOI: 10.1039/b101782i

Hadjiivanov, K., Knözinger, H., & Mihaylov, M. (2002). FTIR study of CO adsorption on Ni-ZSM-5. *The Journal of Physical Chemistry B*, *106*(10), 2618–2624. DOI: 10.1021/jp0132782

Hadjiivanov, K. I. (1998). IR study of CO and NOx sorption on Ag-ZSM-5. *Microporous and Mesoporous Materials*, *24*(1–3), 41–49. DOI: 10.1016/S1387-1811(98)00147-4

Hadjiivanov, K. I., & Vayssilov, G. N. (2002). Characterization of oxide surfaces and zeolites by carbon monoxide as an IR probe molecule. *Advances in Catalysis*, *47*, 307–511. DOI: 10.1016/S0360-0564(02)47008-3

Henriques, C., Marie, O., Thibault-Starzyk, F., & Lavalley, J.-C. (2001). NO^+ ions as IR probes for the location of OH groups and Na^+ ions in main channels and side pockets of mordenite. *Microporous and Mesoporous Materials*, *50*(2-3), 167–171. DOI: 10.1016/S1387-1811(01)00444-9

Hermansson, K. (2002). Blue-Shifting Hydrogen Bonds. *The Journal of Physical Chemistry A*, *106*(18), 4695–4702. DOI: 10.1021/jp0143948

Hobza, P., & Havlas, Z. (2000). Blue-shifting hydrogen bonds. *Chemical Reviews*, *100*(11), 4253–4264. DOI: 10.1021/cr990050q PMID: 11749346

Kim, K. (1983). The integrated intensity of the carbon monoxide fundamental band. *Journal of Quantitative Spectroscopy & Radiative Transfer*, *30*(5), 413–416. DOI: 10.1016/0022-4073(83)90104-8

Kondratieva, E. V., Manoilova, O. V., & Tsyganenko, A. A. (2008). Integrated absorption coefficient of adsorbed CO. *Kinetics and Catalysis*, *49*(3), 451–456. DOI: 10.1134/S002315840803021X

Kustov, L. M. (1997). New trends in IR-spectroscopic characterization of acid and basic sites in zeolites and oxide catalysts. *Topics in Catalysis*, *1997*(4), 131–144. DOI: 10.1023/A:1019136121724

Kustov, L. M., Kazansky, V. B., Beran, S., Kubelková, L., & Jiru, P. (1987). Adsorption of carbon monoxide on ZSM-5 zeolites. Infrared spectroscopic study and quantum-chemical calculations. *Journal of Physical Chemistry*, *91*(20), 5247–5251. Advance online publication. DOI: 10.1021/j100304a023

Lavalley, J.-C. (1996). IR studies of the surface basicity of metal oxides and zeolites using adsorbed probe molecules. *Catalysis Today*, *27*(3-4), 377–401. DOI: 10.1016/0920-5861(95)00161-1

López-Hernández, I., García, C., Truttmann, V., Pollitt, S., Barrabés, N., Rupprechter, G., Rey, F., & Palomares, A. E. (2020). Evaluation of the silver species nature in Ag-ITQ2 zeolites by the CO oxidation reaction. *Catalysis Today*, *345*, 22–26. DOI: 10.1016/j.cattod.2019.12.001

Maevsky, A. V., & Tsyganenko, A. A. (2012). Linkage isomerism in layered silicates [Paper presentation]. *International Symposium on Zeolites and Mycroporous Crystals ZMPC2012, Hiroshima, Japan.* Abstracts, p. 124

Makarova, M. A., Al-Ghefaili, K. M., & Dwyer, J. (1994a). Brønsted acid strength in US-Y: FTIR study of CO adsorption. *Journal of the Chemical Society, Faraday Transactions*, *90*(2), 383–386. DOI: 10.1039/FT9949000383

Makarova, M. A., Zholobenko, V. L., Al-Ghefaili, K. M., Thompson, N. E., Dewing, J., & Dwyer, J. (1994b). Bronsted acid sites in zeolites. FTIR study of molecular hydrogen as a probe for acidity testing. *Journal of the Chemical Society, Faraday Transactions*, *90*(7), 1047–1054. DOI: 10.1039/FT9949001047

Mapes, J. E., & Eischens, R. P. (1954). The infrared spectra of ammonia chemisorbed on cracking catalysts. *Journal of Physical Chemistry*, *58*(12), 1059–1062. DOI: 10.1021/j150522a002

McDonald, R. S. (1958). Surface functionality of amorphous silica by infrared spectroscopy. *Journal of Physical Chemistry*, *62*(10), 1168–1178. DOI: 10.1021/j150568a004

Nachtigall, P., Delgado, M. R., Nachtigallova, D., & Areán, C. O. (2012). The nature of cationic adsorption sites in alkaline zeolites - Single, dual and multiple cation sites. In *Physical Chemistry Chemical Physics* (Vol. 14, Issue 5). DOI: 10.1039/C2CP23237E

Novikov, R. G., Konopnitskii, R. & Tsyganenko, A. A. (2018). Distortions in IR Spectra Related to Registration Conditions: II. The Influence of Scattering. *Optics and Spectroscopy 2018 124:5*, *124*(5), 655–659. DOI: 10.1134/S0030400X18050156

Oliviero, L., Vimont, A., Lavalley, J.-C., Sarria, F. R., Gaillard, M., & Maugé, F. (2005). 2,6-Dimethylpyridine as a probe of the strength of Brønsted acid sites: Study on zeolites. Application to alumina. *Physical Chemistry Chemical Physics*, *7*(8), 1861–1869. DOI: 10.1039/B500689A PMID: 19787950

Otero Areán, C., Manoilova, O. V., Tsyganenko, A. A., Turnes Palomino, G., Peñarroya Mentruit, M., Geobaldo, F. & Garrone, E. (2001). Thermodynamics of Hydrogen Bonding between CO and the Supercage Brønsted Acid Sites of the H-Y Zeolite – Studies from Variable Temperature IR Spectrometry. *European Journal of Inorganic Chemistry, 2001*(7). DOI: 10.1002/1099-0682(200107)2001:7<1739::AID-EJIC1739>3.3.CO;2-D

Otero Areán, C., Tsyganenko, A. A., Escalona Platero, E., Garrone, E., & Zecchina, A. (1998). Two Coordination Modes of CO in Zeolites: A Temperature-Dependent Equilibrium. *Angewandte Chemie International Edition, 37*(22), 3161–3163. DOI: 10.1002/(SICI)1521-3773(19981204)37:22<3161::AID-ANIE3161>3.0.CO;2-B PMID: 29711312

Otero Areán, C., Turnes Palomino, G., Tsyganenko, A. A., & Garrone, E. (2002). Quantum Chemical and FTIR Spectroscopic Studies on the Linkage Isomerism of Carbon Monoxide in Alkali-Metal-Exchanged Zeolites: A Review of Current Research. *International Journal of Molecular Sciences, 3*(7), 764–776. DOI: 10.3390/i3070764

Paukshtis, E. A., & Yurchenko, E. N. (1983). Study of the Acid–Base Properties of Heterogeneous Catalysts by Infrared Spectroscopy. *Russian Chemical Reviews, 52*(3), 242–258. DOI: 10.1070/RC1983v052n03ABEH002812

Pavlov, A. Y., & Tsyganenko, A. A. (1994). Band profile of hydroxyl groups in the infrared spectrum of hydrogen-bonded surface complexes: Ammonia on silicon dioxide. [English Translation of Optika i Spektroskopiya]. *Optics and Spectroscopy, 77*(1).

Peri, J. B. (1966). Infrared study of OH and NH_2 groups on the surface of a dry silica aerogel. *Journal of Physical Chemistry, 70*(9), 2937–2945. DOI: 10.1021/j100881a037

Qiu, Zh.-Zh. Yu., Yu, Y.-X., & Mi, J.-G. (2012). Adsorption of carbon monoxide on Ag(I)-ZSM-5 zeolite: An ab initio density functional theory study. *Applied Surface Science, 258*(24), 9629–9635. DOI: 10.1016/j.apsusc.2012.05.162

Richardson, H. H., Baumann, C., & Ewing, G. E. (1987). Infrared spectroscopy and thermodynamic measurements of CO on NaCl films. *Surface Science, 185*(1–2), 15–35. DOI: 10.1016/S0039-6028(87)80611-8

Rodionova, T. A., Tsyganenko, A. A., & Filimonov, V. N. (1982). *Infrared study of low-temperature CO adsorption on metal oxides. Adsorbtsiya i adsorbenty, 10, 33-42*. Russ.

Serykh, A. I., & Amirides, M. D. (2006). DRIFTS evidence for the formation of binuclear Cu(I)–dinitrogen complexes upon adsorption of N_2 on CuZSM-5. *Microporous and Mesoporous Materials*, *94*(1-3), 320–324. DOI: 10.1016/j.micromeso.2006.04.005

Smirnov, E. P., & Tsyganenko, A. A. (1977). CNDO study of the properties of the OH group. *Reaction Kinetics and Catalysis Letters*, *7*(4), 425–429. DOI: 10.1007/BF02073831

Storozhev, P. Y., Otero Otero Areán, C., Garrone, E., Ugliengo, P., Ermoshin, V. A., & Tsyganenko, A. A. (2003). FTIR spectroscopic and ab initio evidence for an amphipathic character of CO bonding with silanol groups. *Chemical Physics Letters*, *374*(5–6), 439–445. DOI: 10.1016/S0009-2614(03)00587-6

Storozhev, P. Y., Yanko, V. S., Tsyganenko, A. A., Palomino, G. T., Delgado, M. R., & Otero Areán, C. (2004). Isomeric states of polar molecules on ionic surfaces: Electrostatic model and FTIR studies. *Applied Surface Science*, *238*(1-4), 390–394. DOI: 10.1016/j.apsusc.2004.05.242

Storozheva, E. N., Sekushin, V. N., & Tsyganenko, A. A. (2006). FTIR spectroscopy evidence for the basicity induced by adsorption. *Catalysis Letters*, *107*(3–4), 185–188. Advance online publication. DOI: 10.1007/s10562-005-0008-4

Tarach, K., Góra-Marek, K., Chrzan, M., & Walas, S. (2014). Quantification of silver sites in zeolites: Carbon monoxide sorption monitored by IR spectroscopy. *The Journal of Physical Chemistry. C, Nanomaterials and Interfaces*, *118*(41), 23751–23760. Advance online publication. DOI: 10.1021/jp506820v

Tret'yakov, N. E., & Filimonov, V. N. (1972). An infrared spectroscopic study of the relative proton-donating power of the OH-groups of oxide surfaces. *Kinetics and Catalysis*, *13*(3), 735–737.

Tsyganenko, A. (2011). Low-temperature infrared spectroscopy of surface species. In Khriachtchev, L. (Ed.), *Physics and Chemistry at Low Temperatures* (pp. 267–296). Pan Stanford Publishing Pte., DOI: 10.1201/b11403-10

Tsyganenko, A. A. (1975). IR spectra and structure of the hydroxyl covering of oxides comparison with the spectra of hydroxides and silicates. *Journal of Structural Chemistry*, *16*(4), 536–540. DOI: 10.1007/BF00754034

Tsyganenko, A. A. (1980). A cell for the investigation of IR spectra of adsorbed molecules at liquid helium temperatures. *Приборы и техника эксперимента*, *1*, 265–267.

Tsyganenko, A. A. (2013). Variable temperature IR spectroscopy in the studies of oxide catalysts. *Topics in Catalysis, 56*(11), 905–913. DOI: 10.1007/s11244-013-0054-x

Tsyganenko, A. A., Chizhik, A. M., & Chizhik, A. I. (2010). A FTIR search for linkage isomerism of CN^- ions on oxides and zeolites. *Physical Chemistry Chemical Physics, 12*(24), 6387–6395. DOI: 10.1039/c003942j PMID: 20454717

Tsyganenko, A. A., Escalona Platero, E., Otero Areán, C., Garrone, E., & Zecchina, A. (1999). Variable temperature IR spectroscopic studies of CO adsorbed on Na-ZSM-5 and Na-Y zeolites. *Catalysis Letters, 61*(3-4), 187–192. DOI: 10.1023/A:1019089309446

Tsyganenko, A. A., & Filimonov, V. N. (1972). Infrared Spectra of Surface Hydroxyl Groups and Crystalline Structure of Oxides. *Spectroscopy Letters, 5*(12), 477–487. DOI: 10.1080/00387017208065418

Tsyganenko, A. A., & Filimonov, V. N. (1973). Infrared spectra of surface hydroxyl groups and crystalline structure of oxides. *Journal of Molecular Structure, 19*(2), 579–589. DOI: 10.1016/0022-2860(73)85136-1

Tsyganenko, A. A., Khomenya, A. V., & Filimonov, V. N. (1976). *Infrared spectroscopic study of the proton-donating ability of the hydroxyl groups of oxide surfaces. Adsorbtsiya i adsorbenty, 4, 86-91*. Russ.

Tsyganenko, A. A., Kondratieva, E. V., Yanko, V. S., & Storozhev, P. Y. (2006). FTIR study of CO adsorption on basic zeolites. *Journal of Materials Chemistry, 16*(24), 2358. Advance online publication. DOI: 10.1039/b514752b

Tsyganenko, A. A., & Mardilovich, P. P. (1996). Structure of alumina surfaces. *Journal of the Chemical Society, Faraday Transactions, 92*(23), 4843–4852. DOI: 10.1039/ft9969204843

Tsyganenko, A. A., Shergin, Y. V., Shelyapina, M. G., Satikova, E. A., & Petranovskii, V. (2025). (in press). FTIR Study of CO Adsorbed on Ag-mordenite. *Kinetics and Catalysis*. Advance online publication. DOI: 10.1134/S0023158425600142

Tsyganenko, A. A., Storozhev, P. Y., & Otero Areán, C. (2004). IR-spectroscopic study of the binding isomerism of adsorbed molecules. *Kinetics and Catalysis, 45*(4), 530–540. Advance online publication. DOI: 10.1023/B:KICA.0000038081.43384.56

Tsyganenko, A. A., Storozheva, E. N., & Manoilova, O. V. (2001). Manifestations of the acidity of adsorbed molecules in H-bonded complexes with silanol groups: Lewis acidity of ozone. *Catalysis Today, 70*(1–3), 59–71. DOI: 10.1016/S0920-5861(01)00407-2

Tsyganenko, A. A., & Trusov, E. A. (1978). Infrared study of the interaction between SO_3 and functional groups of SiO_2 surface. *Colloid Journal, 40*(5), 875–880.

Tsyganenko, A. A., Zakharov, N. V., & Murzin, P. D. (2014). CHF_3 as a probe for surface basicity. *Catalysis Today, 226*, 73–80. DOI: 10.1016/j.cattod.2013.11.049

Tsyganenko, A. A., & Zverev, S. M. (1988). Mechanism of lateral interactions between molecules adsorbed on oxide surfaces. *Reaction Kinetics and Catalysis Letters, 36*(2), 269–274. DOI: 10.1007/BF02063817

Ugliengo, P., Garrone, E., Ferrari, A. M., Zecchina, A., & Otero Areán, C. (1999). Quantum Chemical Calculations and Experimental Evidence for O-Bonding of Carbon Monoxide to Alkali Metal Cations in Zeolites. *The Journal of Physical Chemistry B, 103*(23), 4839–4846. DOI: 10.1021/jp982078w

Ward, J. W., & Habgood, H. W. (1966). The infrared spectra of carbon dioxide adsorbed on zeolite X. *Journal of Physical Chemistry, 70*(4), 1178–1182. DOI: 10.1021/j100876a034

Zaki, M. I., & Knözinger, H. (1987). Characterization of oxide surfaces by adsorption of carbon monoxide — A low temperature infrared spectroscopy study. *Spectrochimica Acta. Part A: Molecular Spectroscopy, 43*(12), 1455–1459. DOI: 10.1016/S0584-8539(87)80030-2

Zholobenko, V. L., Makarova, M. A., & Dwyer, J. (1993). Inhomogeneity of Brönsted acid sites in H-mordenite. *Journal of Physical Chemistry, 97*(22), 5962–5964. DOI: 10.1021/j100124a030

ADDITIONAL READING

Banwell, C. N. (1983). *Fundamentals of molecular spectroscopy*. McGraw-Hill Book Company.

Chalmers, J. M., & Griffiths, P. R. (2002). *Handbook of Vibrational Spectroscopy* (Vol. 1-5). Wiley.

Che, M., & Vedrine, J. C. (Eds.). (2012). Characterization of solid materials and heterogeneous catalysts. From structure to surface reactivity, v.1, 2. Wiley VCH Verlag & Co. KGaA. DOI: 10.1002/9783527645329

Davydov, A. A. (2003). *Molecular spectroscopy of oxide catalyst surfaces*. Wiley. DOI: 10.1002/0470867981

Grasselli, J. G., Snavely, M. K., & Bulkin, B. J. (1981). *Chemical applications of Raman spectroscopy*. J. Wiley & Sons.

Hair, M. L. (1967). *Infrared spectroscopy in surface chemistry*. Marcel Dekker, Inc.

Little, L. H. (1966). *Infrared spectra of adsorbed species*. Academic Press.

Niemantsverdriet, J. W. (2000). *Spectroscopy in Catalysis. An introduction* (2nd ed.). Wiley-VCH. DOI: 10.1002/9783527614127

Urban, M. W. (1993). *Vibrational spectroscopy of molecules and macromolecules on surfaces*. J.Wiley & Sons.

Van Bekkum, H., Flanigen, E. M., & Jansen, J. C. (Eds.). (1991) *Introduction to zeolite science and practice. (Studies in Surface Science and Catalysis. Vol. 58.)* Elsevier.

Willis, R. F. (Ed.). (1980). *Vibrational Spectroscopy of Adsorbates*. Springer. DOI: 10.1007/978-3-642-88644-7

KEY TERMS AND DEFINITIONS

Blue-shifting hydrogen bond: Unusual kind of H-bond when the frequency of the X-H vibration increases as a result of bonding.

Dynamic interaction (dipole coupling): The resonance dipole-dipole interaction between the adsorbed molecules leading to the splitting or shifts of the bands in their vibrational spectra.

Induced: Brønsted acidity: The ability of OH groups to protonate the adsorbed strong bases under the action of co-adsorbed gaseous acids.

Isotopic isomerism: A kind of linkage isomerism when two adsorbed species differ in the positions of isotopically labeled atoms.

Lateral interactions: Unlike adsorption interactions between the surface and adsorbed molecules, these are interactions between the molecules in the adsorbed layer.

Linkage isomerism of adsorbed species: A form of structural isomerism when certain adsorbed molecule form with the same site different species in which the molecule is bonded to the surface via different atoms.

Static interaction: A kind of lateral interactions which unlike the dynamic one does not depend on the vibrational frequencies of interacting molecules and for oxides can be enhanced by surface relaxation.

Chapter 4
Influence of Natural Zeolites on the Crystalline Properties of ZnO:
X-Ray Diffraction Technique and Its Application in Characterization

Karin M. Álvarez Gómez
https://orcid.org/0000-0002-0401-8078
Benemérita Universidad Autónoma de Puebla, Mexico

Miguel Ángel Hernández-Espinosa
Benemérita Universidad Autónoma de Puebla, Mexico

Luis Pérez-Cabrera
https://orcid.org/0000-0001-9896-3419
Universidad Nacional Autónoma de México, Mexico

Viridiana Evangelista-Hernández
https://orcid.org/0009-0009-4306-3638
Universidad Nacional Autónoma de México, Mexico

Vitalii Petranovskii
https://orcid.org/0000-0002-8794-0593
Universidad Nacional Autónoma de México, Mexico

ABSTRACT

ZnO-zeolite composites were synthesized and characterized by sol-gel method and

DOI: 10.4018/979-8-3373-0320-8.ch004

X-ray diffraction (XRD). Different proportions of clinoptilolite, mordenite, erionite and chabazite with ZnO were analyzed. The results show that the addition of zeolites affects the crystal structure, particle size (32.3-66.5 nm), dislocation density and microdeformations. Rietveld refinement confirmed variations in the lattice parameters. It is concluded that these materials can improve the dispersion and stability of ZnO, being promising for catalysis and gas purification.

1. INTRODUCTION

Semiconductors are materials that have a conductivity intermediate between a conductor and an insulator (Min et al. 2025). Conductivity is related to the forbidden band better known as bandgap. The forbidden band is the amount of energy needed for an electron to pass from the valence band to the conduction band (Fouad et al. 2017). It is worth mentioning that the size of the forbidden bands depends on the type of material, that is, for a conductor there is no bandwidth (the conduction and valence bands overlap) (Molinari et al. 2015; Ghorbani et al. 2019), for a semiconductor the bandwidth is given a value of 1 eV, while for an insulator the bandwidth value is 4 eV as shown in Figure 1 (Bird 2002; Karimi et al. 2021).

Figure 1. Band gap of different materials.

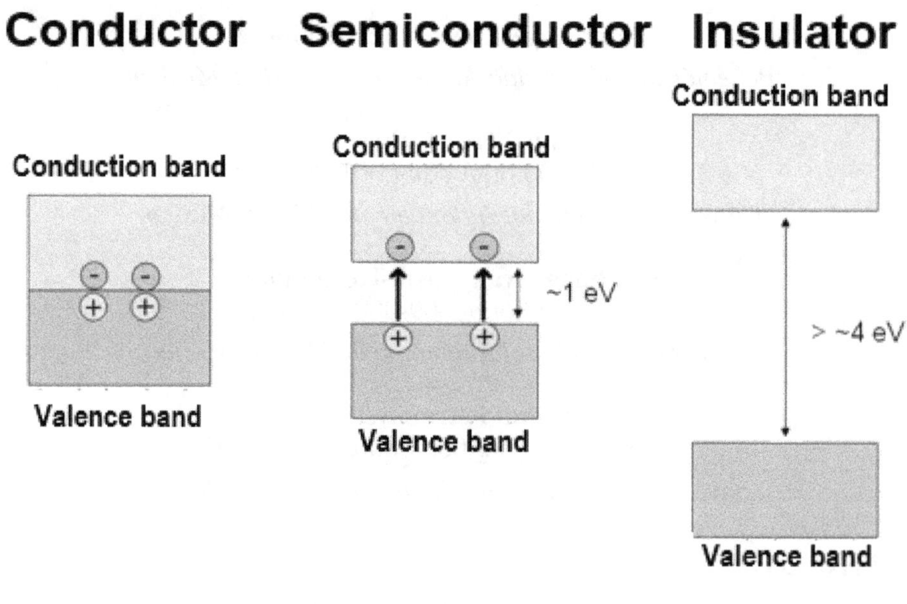

Their conductivity can be adjusted by temperature, light, electric field or doping. Therefore, they are essential for the manufacture of electronic devices such as transistors, diodes, solar cells, etc. (See Figure 2) (Min et al. 2025).

Figure 2. Electronic devices.

Semiconductors can be classified into composition semiconductors or semiconductors with the presence of impurities (See Figure 3). Composition semiconductors are in turn divided into elemental semiconductors and compound semiconductors. Elemental semiconductors are materials that are composed of a single element such as: Boron (Group III), Carbon, Silicon and Germanium (Group IV), Selenium and Telenide (Group VI). It is worth mentioning that compound semiconductors are divided into Inorganic and Organic. Within the inorganic compound semiconductors, we find (Yakout 2020):

- Binary semiconductors: BN, BP, BAs, AlN, AlP, AlAs, AlSb, GaN, GaP, GaAs, GaSb, InN, InP, InAs, InSb, (Grupos III-V), CdSe, CdS, CdTe, ZnO, ZnSe, ZnS, ZnTe (Grupo II-VI), CuCl (Grupos II-VI), CuCl (Grupos I-VII), Cu_2S, Cu_2O, CuO, Ag_2S (Grupos I-VI). PbSe, PbS, PbTe, SnS, SnS_2, SnTe, SnO_2, SnS (Grupos IV-VI), Bi_2Te_3, Bi_2S_3, As_2S_3, As_4S_4 (grupos V-VI), Cd_3P_2, Cd_3As_2, Zn_3P_2, ZnP_2, Zn_3As_2, Zn_3Sb_2 (Grupos II-V), TiO_2, GaSe, InSe, FeS_2 (Grupos III-VI), PbI_2 (Grupos IV-VII), MoS2 (Grupo VI), FeO, NiO, EuO, EuS (Grupos VIII-VI), $CrBr_3$ (VI-VII), Pt-Si (Grupos VIII-IV), BiI_3 (Grupos V-VII), HgI_2 (Grupos II-VII), TlBr (Grupos III-VII), $Si_{1-X}Ge_X$, $Si_{1-X}Sn$ (Grupo IV)

- Ternary semiconductors: $BaTiO_3$, $SrTiO_3$ (Grupos II-III-VI), $LiNbO_3$ (Grupos I-V-VI), GaMnAs (Grupos III-VII-V), PbMnTe (Grupos IV-VII-VI), $CuInSe_2$ (Grupos I-III-VI), $AgGaS_2$ (Grupos I-III-VI), $ZnSiP_2$ (Grupos II-IV-V), Cu_2SnS_3 (Grupos I-IV-VI), $Pb_{1-X}Sn_XTe$ (Grupos IV-VI), $Al_XGa_{1-X}As$, $In_XGa_{1-X}As$, $In_XGa_{1-X}P$, $Al_XIn_{1-X}As$, $Al_XGa_{1-X}Sb$, $Al_XIn_{1-X}Sb$, GaAsN, GaAsP, AlAsSb, GaAsSb, AlGaSb, AlGaP, AlGaN, InAsSb, InGaSb (Grupo III-V), CdZnTe, HgCdTe, HgZnTe, HgZnSe (grupos II-VI)
- Quaternary semiconductors: Cu_2ZnSnS_4 (Grupos I-II-IV-VI), InGaInP, AlGaAsP, InGaAsP, InGaAsSb, InAsSbP, AlInAsP, Al GaAsN, InGaAsN, InAlAsN, GaAsSbN (Grupos III-V), $(Zn_{1-X}Cd_X)_3(P_{1-Y}As_Y)_2$
- Quinary semiconductors: GaInNAsSb, GaIAsSbP (Grupos III-V).

In organic semiconductors we can find (Estrada Flores et al. 2022):

- Polymers: This type of semiconductor has very good mechanical properties because thin films with regular morphologies and good flexibility can be made. It is worth mentioning that due to the size of the conjugated chain, they are considered good and very stable semiconductors. However, they have low solubility, which makes it difficult to process during the manufacture of devices.
- Low molecular weight molecules: These semiconductors have a defined molecular structure and weight. It is worth mentioning that they can be deposited by evaporation due to their low melting point. They have greater solubility because they can be deposited on a substrate by centrifugal deposition. Although in this case, the molecules do not have good mechanical properties and tend to form crystalline aggregates, forming charge recombination traps.

Semiconductors with the presence of impurities (See Figure 4) can be classified as intrinsic (pure) or extrinsic (Balkan et al. 2021)

- Intrinsic: These are pure semiconductors because they are made up of only one type of atom (example: Silicon (Si), Germanium (Ge), Tin (Sn), Selenium (Se) and Tellurium (Te).
- Extrinsic: These are semiconductors to which impurities are added through a doping process, this is done to increase their conductivity. Extrinsic semiconductors can be divided into:
    - P-type: Certain types of atoms are added to increase the number of positive free charge carriers, better known as holes.

- N-type: These are obtained when atoms with five electrons in their last shell are added. With this increase, the number of negative free charge carriers, better known as electrons, increases.

Figure 3. Classification of semiconductors.

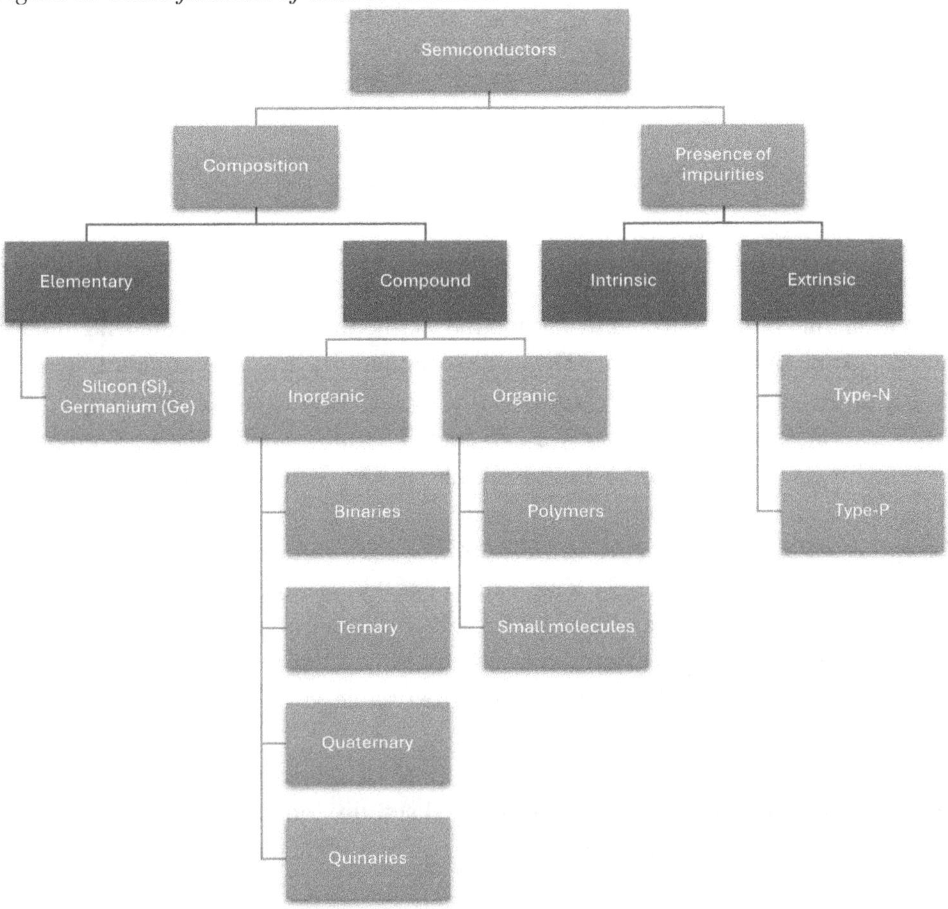

Figure 4. Intrinsic and extrinsic semiconductors.

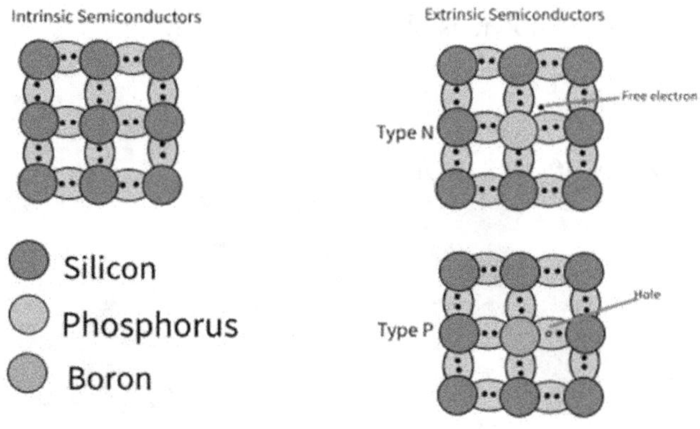

Within the binary inorganic compound semiconductors we can find binary metallic semiconductor oxides. These semiconductors are compounds that have oxygen molecules in their structure with a metal (Gao et al. 2018). Within this classification of semiconductors we can find: Titanium Oxide (TiO_2), Tin Oxide (SnO_2), Tungsten Oxide (WO_3), Indium Oxide (In_2O_3), Vanadium Pentoxide (V_2O_5), Ferric Oxide (Fe_2O_3), Ferrous Ferric Oxide (Fe_3O_4), Copper Oxide (CuO), Nickel Oxide (NiO), Cobalt Oxide (Co_3O_4), Chromic Oxide (Cr_2O_3) and Zinc Oxide (ZnO), among others (Uma et al. 2023). Zinc oxide (ZnO) is a semiconductor belonging to groups II-VI, has a wide band gap 3.37 eV, its exciton binding energy of 60 MeV, a high electron mobility, is easily fabricated at low temperature, efficient photon emissions, can be easily doped with n, low resistance, high conductivity, low cost and non-toxicity (Rasmidi et al. 2021).

Zinc oxide is attracting attention in the field of semiconductor devices for its versatile characteristics, such as high thermal stability and light transparency due to its wide band gap (3.37eV), and high exciton binding energy of 60 meV at room temperature (Lee et al. 2024)

Zinc oxide crystallizes in three forms: Rock salt (Cubic), Zincblende (Cubic) and Wurtzite (Hexagonal) (See Figure 5 and 6). The wurtzite phase has attracted much attention due to its higher thermodynamic stability and easy synthesis (Li et al. 2020; Shaba et al. 2021). However, ZnO does not possess a large surface area compared to zeolitic materials (Wahyuni et al. 2022).

Figure 5. Crystal structures of ZnO: a) Rock salt, b) Zinc Blende and c) Wurtzite.

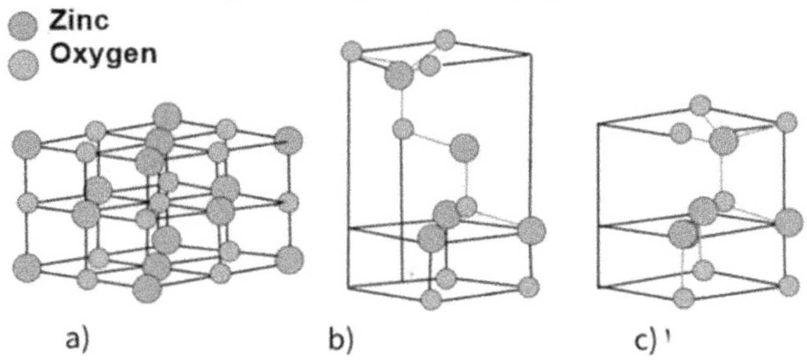

Figure 6. Three-dimensional crystalline structure.

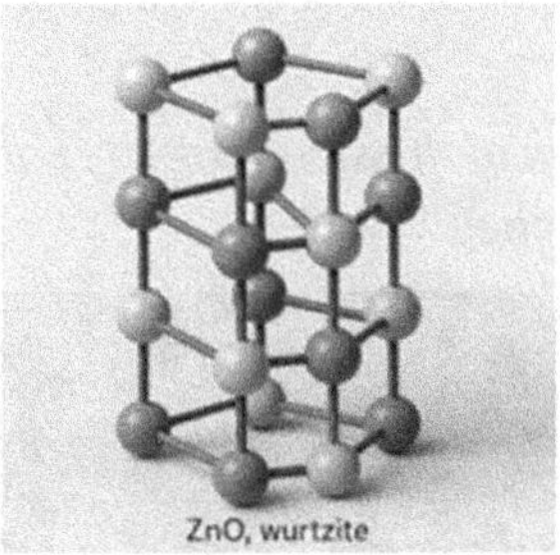

2. WHAT ARE ZEOLITES?

Zeolites are a type of crystalline hydrated aluminosilicate material with a regular pore structure which is occupied by water. Due to their unique three-dimensional pore structure, they can acquire exceptional adsorption properties (Pal 2017). Their structures reveal regular arrays of channels and cavities of approximately 3-15 A (Coudy et al. 2023). There are different topologies, which vary in pore diameter (small, medium or large pore systems with 1D, 2D or 3D channels, See Figure 7), and the presence of internal cages where compounds can react and become trapped (Vasconcelos et al. 2023). Its unique physical and chemical properties give it good adsorption, catalytic, shape selectivity and ion exchange characteristics (Li et al. 2021).

The International Zeolite Association (IZA) has reported 242 completely ordered and 11 partially disordered structures (Murrieta-Rico et al. 2022). There are synthetic and natural zeolites. Within the synthetic zeolites we can find zeolite A, zeolite X, zeolite Y and zeolite P. The formation of natural zeolites occurs from volcanic and sedimentary rocks and in this category we can find: erionite, chabasite, clinoptilolite and mordenite (See Figure 8) (Derbe et al. 2021).

Figure 7. 1D, 2D or 3D channels

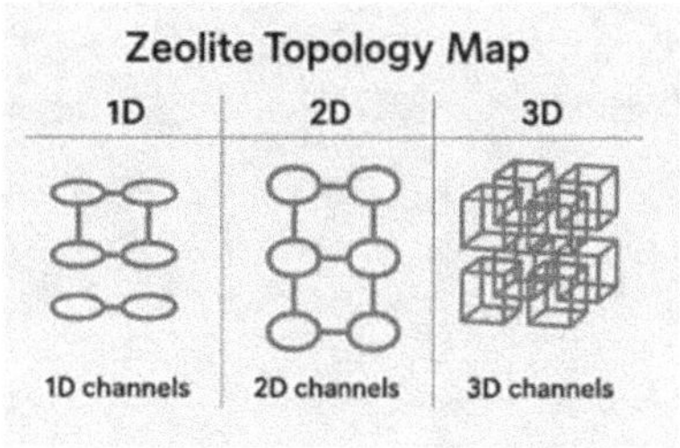

Figure 8. Schematic illustration of the structure of zeolites: a) erionite, b) chabasite, c) clinoptilolite and d) mordenite

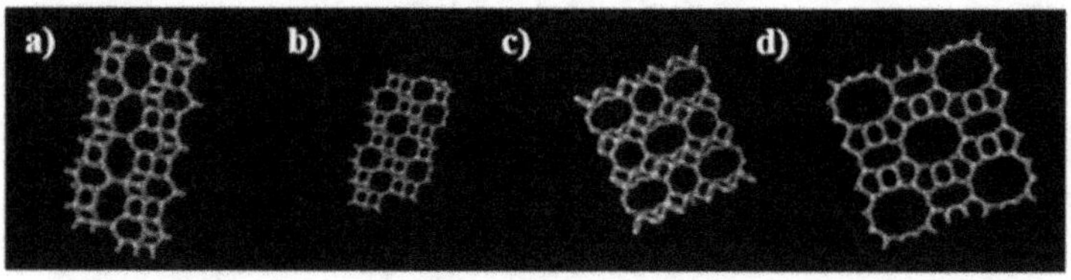

3. SOL-GEL A SYNTHESIS METHOD

3.1. Obtaining ZnO-Zeolite Composites

There is a great diversity of methods for the fabrication of nanomaterials among which we find: solid-state synthesis, microwave, coprecipitation, sonochemical, hydrothermal, thermal decomposition, sol-gel, among others (Mengyao et al. 2024; Zineb et al. 2023; Saba et al. 2024). Although all the methods are capable of producing large quantities of nanomaterials, the sol-gel method stands out as the most widely used and industrially applied. Due to its unique properties and advantages, it enables the production of high-quality, uniformly sized nanoparticles on an industrial scale (Bokov et al. 2021).

The sol-gel method is a liquid route technique for the synthesis of materials discovered in the mid-19th century (Murthi et al. 2025). This method has several advantages among them: it produces highly homogeneous compounds with high purity (99.99% purity), it is carried out at lower temperatures in a range of 70 to 320 ° C, materials with nanometric porosities are obtained, it has a narrow particle size distribution, as well as the achievement of uniform nanostructures, control of the chemical composition, applicability to large areas, easy operation, mild reaction conditions, controllable size, high crystallinity, good dispersion of the prepared materials, among others (Bokov et al. 2021; Zining et al. 2024; Xu et al. 2024; Ferreira et al. 2024). This method is used to obtain ceramic, glass and semiconductor oxide materials through obtaining a "sol" solution, followed by the gelation stage and then the elimination of the liquid phase. The steps involved in the Sol-Gel method are (see Figure 9 and 10) (Goñi et al. 2023; Bokov et al. 2021):

Precursors: In this first stage, the precursors are in the liquid phase.

Hydrolysis: In this stage, the reaction hydrolyzes the alkoxy chain of the alkoxides of the material present in the presence of water, with at least one OH group.

Sol: It consists of a colloidal suspension (particles that have diameters of a few nanometers).

Condensation: The OH groups condense, react with others or with an alkoxy.

Gel: Sol evolves through chemical reactions to form a high viscosity network.

Aging: After a certain period of rest, the evolution of the network occurs (due to the exit of liquid from its pores) through the reactions of the functional groups that remained in it.

Drying: The solvent is removed, leaving a porous structure.

Sintering: Heat is applied to densify the material and improve its mechanical and thermal properties.

Figure 9. Scheme sol-gel method.

Figure 10. Stages present in the sol-gel method.

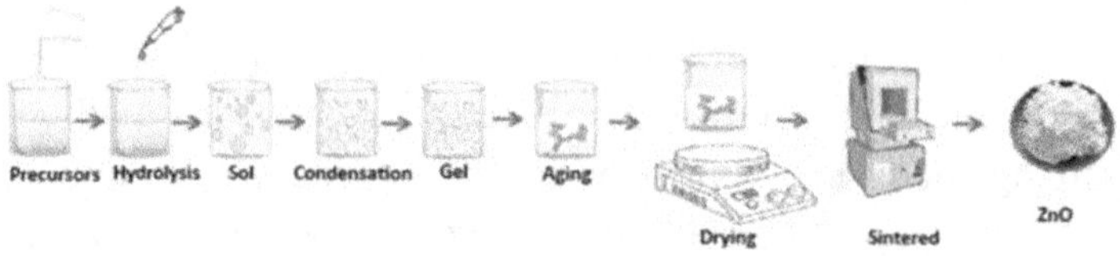

The preparation of ZnO-Zeolite composites is done as shown in Figure 11, starting as follows:

- Stage 1- Precursors: Sodium Hydroxide (NaOH) and Zinc Acetate Dihydrate ($Zn(O_2CCH_3)_2 (H_2O)_2$) are added and mixed. After mixing the precursors, the zeolites are added.
- Stage 2- Hydrolysis: It is carried out by adding water drop by drop.
- Stage 3- Sol: A colloidal solution begins to form.
- Stage 4- Condensation: The OH groups condense, react with others or with an alkoxy.
- Stage 5- Gel: Sol evolves through chemical reactions to form a high viscosity network.
- Stage 6- Aging: It is left to rest for 24 hours.
- Stage 7 - Drying: Dry at 100°C for approximately 6 hours.
- Stage 8 - Sintering: Place in the muffle at 500 °C for one hour.
- Stage 9 - Powder: The composites are obtained.

Figure 11 shows some of the stages mentioned above in the process of ZnO-Zeolite composites. Figure 11 a) shows how the precursors are deposited inside the beaker and mixed with the help of a magnetic stirrer. In Figure 11 b) the zeolite is added to the aforementioned mixture. While in Figure 11 c) the composite already made in the drying lid can be distinguished (it can be observed that it has a thermometer to maintain the temperature at 100°C) and finally, in Figure 11 d) the composites can be observed after having been sintered at 500 °C for approximately one hour. It is worth mentioning that the composites made are presented in Table 1.

Figure 11. Sol-gel synthesis of ZnO-Zeolite composites: a) Precursors, b) Aggregation of zeolites and hydrolysis, c) drying and d) sintering.

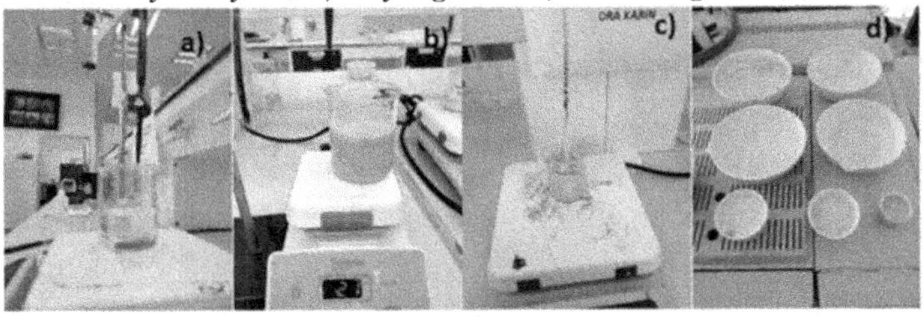

Table 1. Composer of ZnO-Zeolites.

Sample	Composition % by weight
ZnO	100% ZnO
ZnO-Clinoptilolite (75-25)	75% ZnO 25% Clinoptilolite
ZnO-Clinoptilolite (50-50)	50% ZnO 50% Clinoptilolite
ZnO-Clinoptilolite (25-75)	25% ZnO 75% Clinoptilolite
ZnO-Mordenite (75-25)	75% ZnO 25% Mordenite
ZnO-Mordenita (50-50)	50% ZnO 50% Mordenite
ZnO-Mordenita (25-75)	25% ZnO 75% Mordenite
ZnO-Erionite (75-25)	75% ZnO 25% Erionite
ZnO- Erionite (50-50)	50% ZnO 50% Erionite
ZnO- Erionite (25-75)	25% ZnO 75% Erionite
ZnO-Chabasite (75-25)	75% ZnO 25% Chabacite
ZnO- Chabasite (50-50)	50% ZnO 50% Chabacite
ZnO- Chabasite (25-75)	25% ZnO 75% Chabacite

4. STRUCTURAL CHARACTERIZATION BY X-RAY DIFFRACTION

4.1. What is X-ray Diffraction (XRD)?

X-ray diffraction (XRD) is a non-destructive characterization technique used to examine a wide range of materials (minerals, polymers, plastics, metals, semiconductors, ceramics, and solar cells, etc.) (Ali et al., 2022). It is used for nanoscale materials. This technique can provide the following information: phase identification, sample purity, and crystal size, among others (Holder et al., 2019).

X-ray diffraction is based on the constructive interference of monochromatic X-rays and a crystal. The X-rays are generated by a cathode ray tube, filtered to obtain monochromatic radiation, collimated to concentrate them, and then directed toward the sample as seen in Figure 12.

(Bunaciu et al., 2015). The sample consisting of atoms is organized into crystalline structures and microstructures. The periodicity of these atoms depends on external factors such as temperature, pressure and cooling rate during solidification. Materials can be classified as: amorphous, polycrystalline and monocrystalline (See Figure 13). Amorphous materials are known as isotropic because their atomic arrangements are not regular and have the same properties in all directions. While anisotropic materials are those whose atoms are arranged in a regular and repeated pattern, and their properties vary with the direction (Ameh, 2019).

Figure 12. Schematic diagram of a diffraction system.

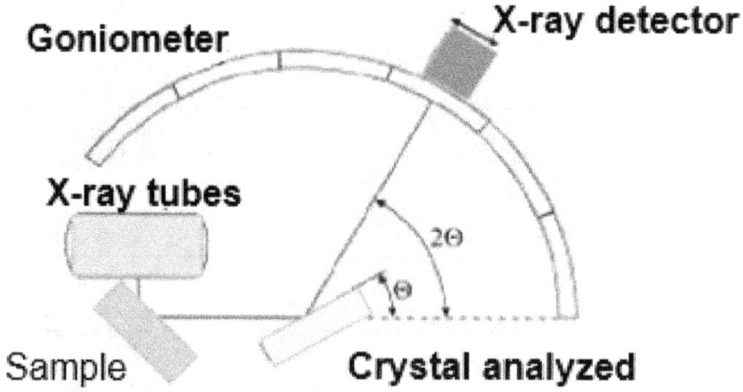

Figure 13. Materials: a) Monocrystalline or Crystalline, b) Polycrystalline and c) Amorphous.

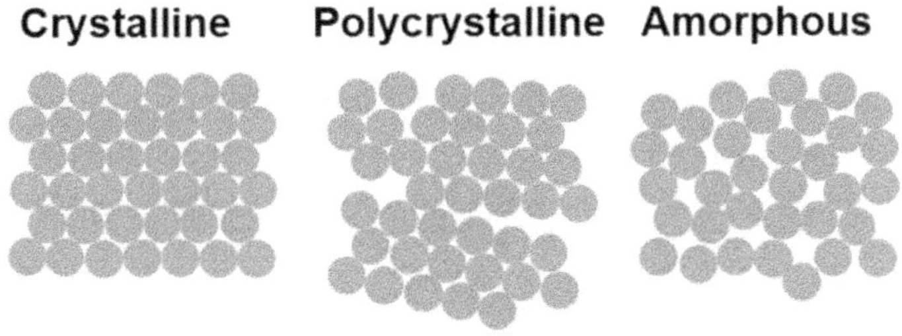

Interaction of the incident rays with the sample causes a constructive interference that results in a diffracted ray, provided that the conditions stipulated by Bragg's law are met:

$$2d_{hkl}\sin\theta = n\lambda$$

The diffraction of a crystalline sample can be explained and visualized using a simple concept: the specular reflection of the incident X-ray beam from a series of crystallographic planes. As previously stated, all planes with identical Miller index triplets are parallel to each other and equally spaced. Therefore, each plane in a given (hkl) set can be considered an independent scattering object. The arrangement is periodic in the direction perpendicular to these planes, and the repeating distance in this direction corresponds to the interplanar spacing, **dhkl** (Springer, 2009).

Diffraction phenomenon, generated by the interference of electromagnetic waves and particle beams with matter, is closely related to the properties of the crystalline phases present in the system. Figure 14 shows a schematic representation of the interference process between monochromatic electromagnetic waves scattered by the atoms of the crystal, as dictated by Bragg's law (Artioli, 2016).

Figure 14. Interference of electromagnetic waves of wavelength λ with crystallographic spacing planes d_{hkl}.

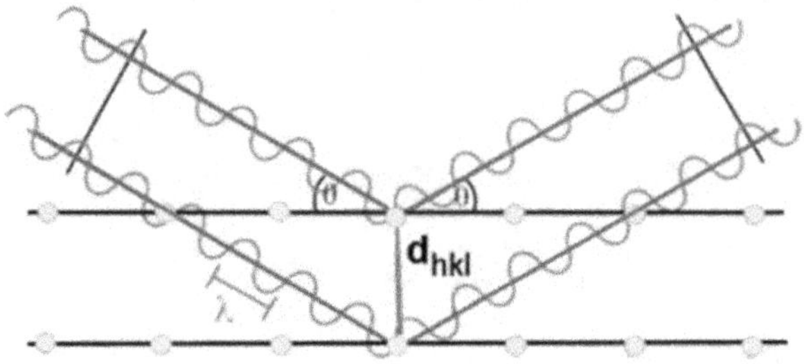

Diffractograms are graphs made in an X-ray study, used to view crystal forms in materials. They are shown as intensity (Y axis) versus diffraction angle (2θ, X axis) and have unique peaks that show the existence of specific crystal phases (Cullity et al., 2014).

The main features of a diffractogram are:

X axis corresponding to 2θ: Shows the dip angle, with units of degrees (°).

Y axis corresponding to Intensity: Shows the strength of the X-rays that are broken, linked to a pile of material that helps the signal.

Diffraction peaks: Each crystalline material generates a unique pattern of peaks at specific 2θ positions, according to the Bragg equation and the material that contains the sample (Klug & Alexander, 1974):

Types of diffractograms according to the material:

- Crystalline materials: They present well-defined and sharp peaks in specific positions (Cullity & Stock, 2014).
- Amorphous materials: They do not show defined peaks, but a broad diffraction band (Warren, 1990).
- Polycrystalline materials: They exhibit multiple peaks corresponding to different crystalline planes (Dinnebier & Billinge, 2008).

4.2. Determination of Crystalline Phases

The X-ray diffraction (XRD) technique is an extremely powerful tool that is applied to identify crystalline phases. The information it provides is essential to understand the properties and performance of a material, which is crucial in applications such as quality control, among others (Cullity & Stock). This technique can also provide more detailed information on crystallographic details, stress analysis and particle size measurements (Bertaut, 1964).

It is worth mentioning that knowing the chemical composition of materials is very important, but it does not provide a complete view. A common example is metals, which can be composed of different crystallographic phases that have radically different mechanical properties. In the same way, strength and durability is a function of the arrangement and type of particle in the crystalline structure (Liu et al., 2015). This information is crucial to optimize the performance of materials in their final applications (Sato et a., 2019). As mentioned in the previous section, the sample is irradiated by an X-ray beam, causing the material to scatter the incoming radiation. The scattered radiation reveals detailed information about the atomic arrangement of the material (Warren, 1969). It is important to mention that each crystalline material has its own unique diffraction pattern, and the pattern obtained from each sample can be compared to the thousands of patterns stored in international databases to identify the crystalline structure of the material (International Centre for Diffraction Data [ICDD], 2020).

It is estimated that more than 95% of solid materials are crystalline. It is worth mentioning that the International Centre for Diffraction Data (better known by its acronym in English as ICDD) contains around 50,000 databases of inorganic individual and crystalline components while 25,000 are organic. To determine their crystalline phases, the diffraction pattern is used together with the intensities of the crystallographic chips. This probes the arrangement of the atoms in each unit cell, the position of the atoms in each and the atomic spacing angles due to the comparative wavelength of the X-rays with the atomic size (Ameh, 2019).

Why is it important to identify the crystalline phases? It is necessary to understand that different materials can contain the same chemical composition but differ considerably in their crystalline structure. To better understand this, it is necessary to analyze the iron oxides Fe_2O_3 (hematite) and Fe_3O_4 (magnetite) (see Figure 15), which, although both have the same chemical composition of iron and oxygen, have different crystalline structures, which influences their properties and behavior (Gualtieri et al., 2003).

Figure 15. Crystalline structures: a) Hematite and b) Magnetite.

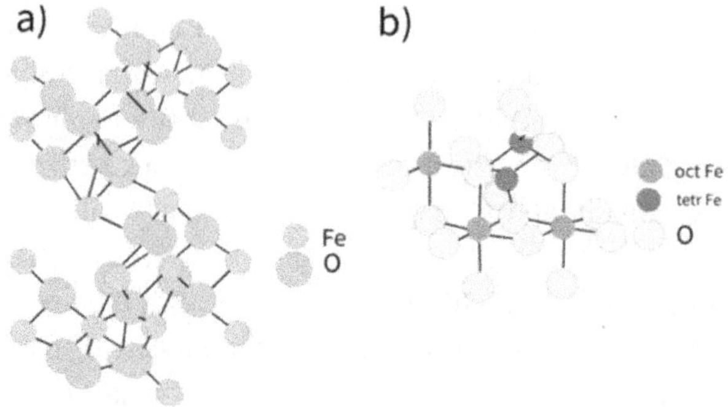

A PXRD pattern can be affected by peak overlap, challenges in background estimation, preferred orientation effects, and limited experimental resolution, all of which complicate structural resolution. More critically, it can be particularly difficult to carry out the initial key steps, such as indexing the pattern and determining the space group—especially when the compound contains more than one chemical phase. If the unit cell is poorly defined, structure determination may become unfeasible. These challenges persist despite advances in the availability and diversity of automated indexing software, both in terms of strategies and implemented methods (Coudy, 2003).

To understand this further, we will begin by analyzing the crystalline phases of the ZnO-Zeolite composites. The first thing to consider is the operating conditions of the equipment, which was a Bruker D8 diffractometer with the following measurement parameters: current intensity of 30 mA, voltage of 30 kV, step time of 1 second, increments of 0.07° and a measurement interval of $2\theta = 10°$ to 90°. The files must be saved with a .RAW 1 extension in order to be processed by the PANalytical HighScore Plus 3.0 program (See Figure 16).

Figure 16. Program used for indexing diffractograms.

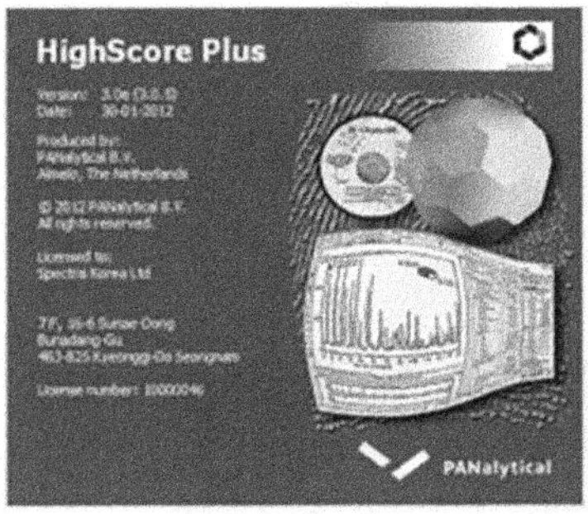

Program is opened and the corresponding file is loaded. In this case, the ZnO-Chabasite (25-75) compound has been selected, as shown in Figure 17. As can be seen, the diffractogram requires processing before indexing. To do this, the following steps will be initiated:

Figure 17. Diffractogram loaded into PANalytical's HighScore Plus program.

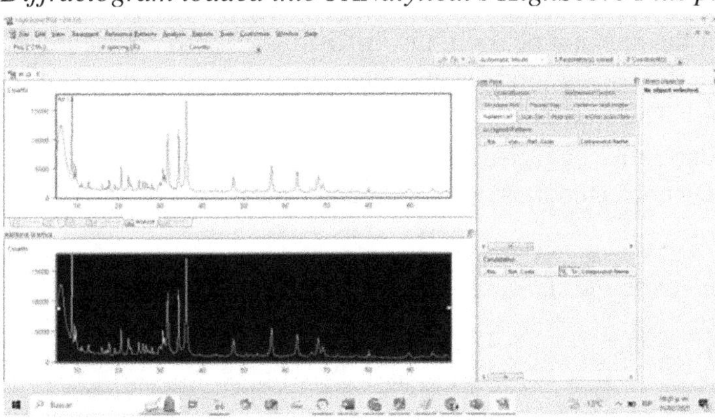

Step 1: Go to the Treatment section of the toolbar and click on Determine Background (See Figure 18). The "Determine Background" function is used to estimate and correct the background in X-ray diffraction (XRD) data. The background in a diffraction pattern is the base level of signal that does not correspond to the specific crystalline phases you are analyzing, but to unwanted signals such as noise, X-ray scattering in air, or signal from the sample itself with no significant contribution to the diffraction peaks.

Figure 18. Selecting the determine background option.

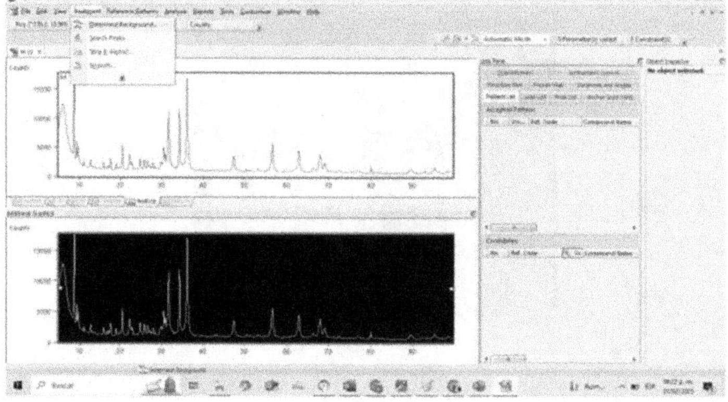

Step 2: A dialog box opens, and you choose the Subtract option, then the Replace option, and then the Close option (See Figure 19). The Subtract option subtracts the determined background line (which represents the estimated background) from the original data. That is, it removes the background contribution from the entire diffraction pattern. The purpose is to remove unwanted signals or “background noise” so that the relevant peaks in the diffraction pattern are clearer and easier to identify. After the background is subtracted, the Replace option is used to modify the data values. This option replaces the values of the original data points with the corrected values (after background subtraction). In other words, the processed data is saved as new corrected data, without the interference.

Figure 19. Selecting the subtract, replace and close option.

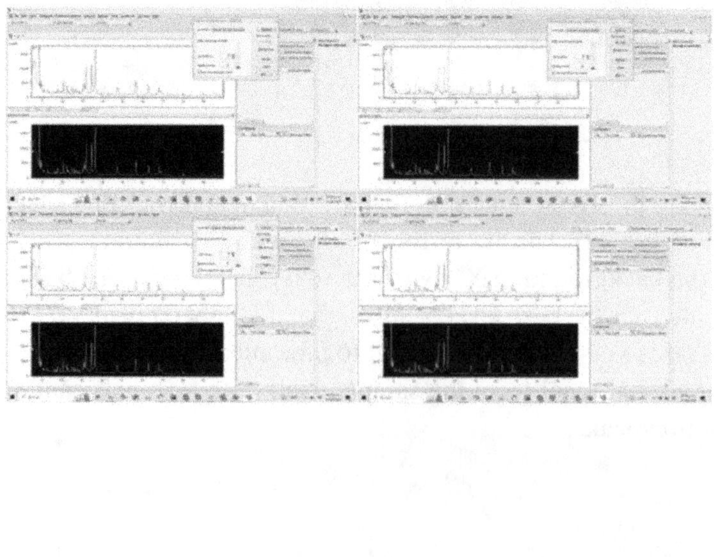

Step 3: To smooth a diffractogram, go to the toolbar and select Treatment again, go to the Smooth option and click Replace (See Figure 20). This option is used to smooth diffraction data, which can be useful for better visualization and analysis of diffraction patterns. However, in this case it can be seen that the smoothing generated by Smooth is not necessary because it affects the quality of the analysis and interpretation of the data. Therefore, Smooth is removed from the diffractogram (See Figure 21).

Figure 20. Selecting the smooth option.

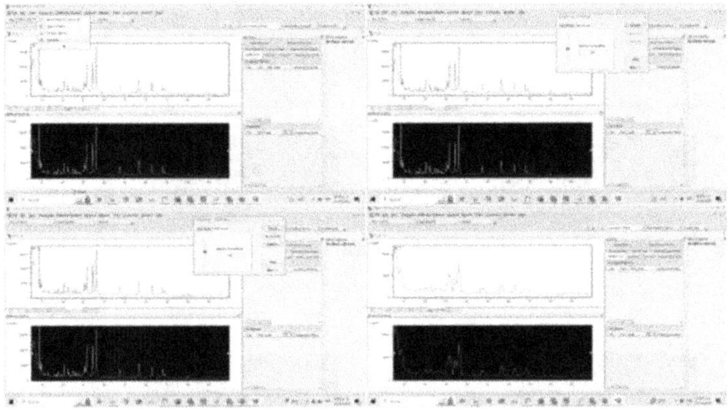

Figure 21. Diffractogram without smooth.

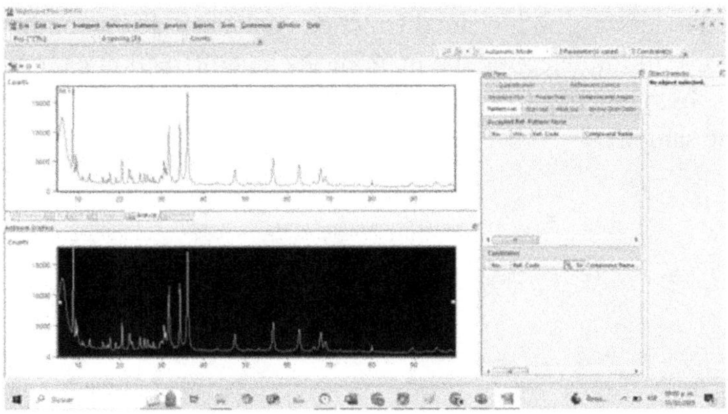

Step 4: In this step, the Strip K-alpha2 option is applied. This option is used to eliminate and correct the contribution of the k-alpha2 radiation from the elements present in the sample. This radiation is a type of characteristic radiation that can interfere with the diffraction pattern due to the X-ray fluorescence generated by the interaction of the X-rays with the elements of the sample. The first thing to do is to go to the toolbar in the Treatment section and then choose the Strip K-alpha2 option. Then, click on the Strip k-alpha2 button and click Replace (See Figure 22).

Figure 22. Selecting the Strip k-alpha2 option.

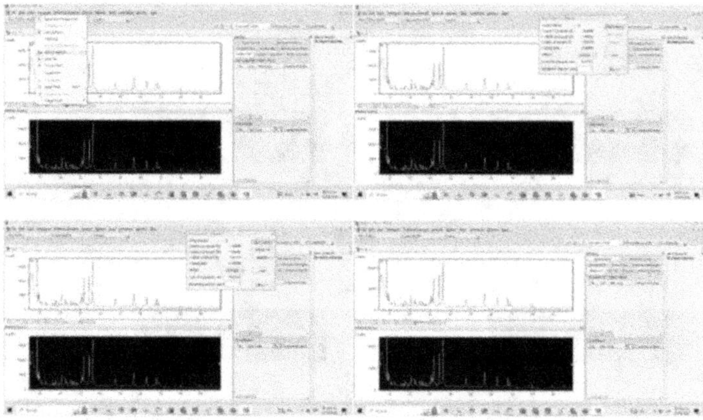

Step 5: Proceed to use the Search Peaks option. To do this, go back to the toolbar in the Treatment section, select Search Peaks and click OK (See Figure 23). This option is used to identify and locate diffraction peaks in a diffractogram. This tool is especially useful for detecting the main peaks in a diffraction pattern, which facilitates the subsequent identification of phases and the quantification of the phases present in the sample.

Figure 23. Selecting the search peaks option.

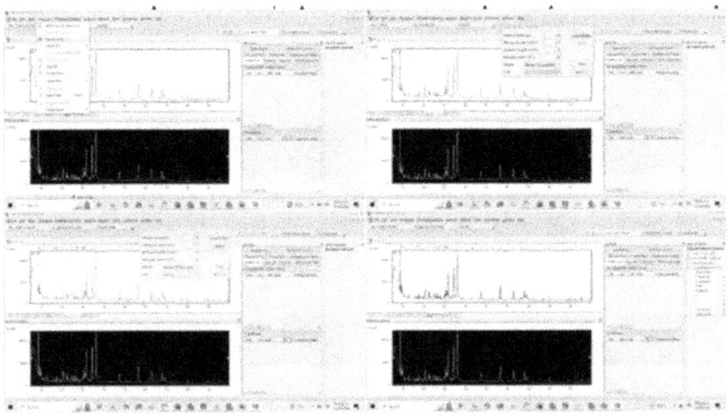

Step 6: To begin identifying the phases we must go to the toolbar and choose the option Reference Patterns/Retrieve Pattern by, there are two options to choose from: 1. Restrictions or 2. Reference code. The first option is used when you simply know the elements that the sample or compound contains, while option 2 is used if you have a crystallographic code. However, in this case it is known that the synthesis should have given ZnO with the zeolite that was added which was chabasite, therefore we proceed to option 1. Select the Periodic Table option and first select the elements belonging to the first compound in this case ZnO. This dialog box has four colors: Red that that element is not found, White that it is possible that that element is found, Blue that at least one of those elements is found and Green that it must be found. In this case, Zn and O must be present, so they are colored green, and therefore the others must not be present, so the color is red. After that, click OK. Then, click on load, load the possible crystallographic tokens, and click on the close button (See Figure 24). Proceed to identify which token corresponds to the ZnO peaks (See Figure 25), this is done by selecting and deselecting the loaded crystallographic tokens.

Figure 24. Identification of crystallographic phases.

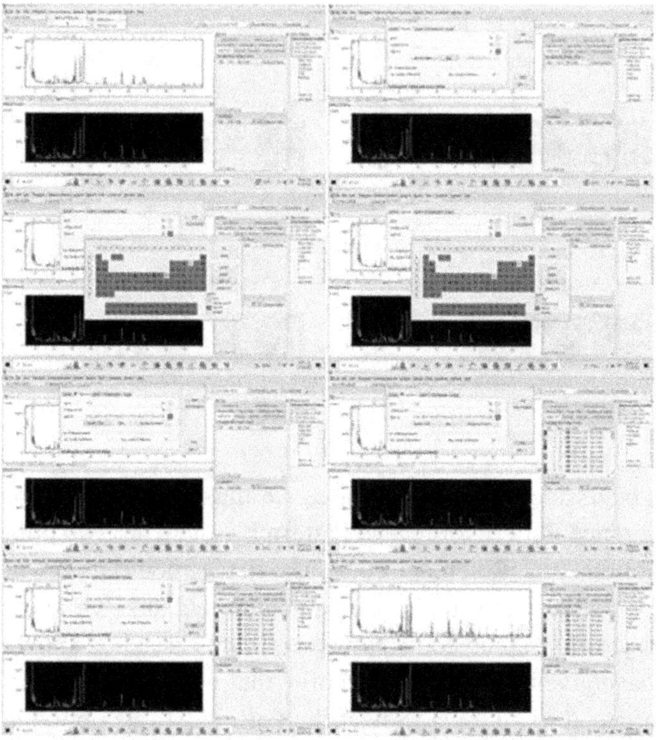

Figure 25. Indexing of ZnO with the ICDD crystallographic record 00-036-1451.

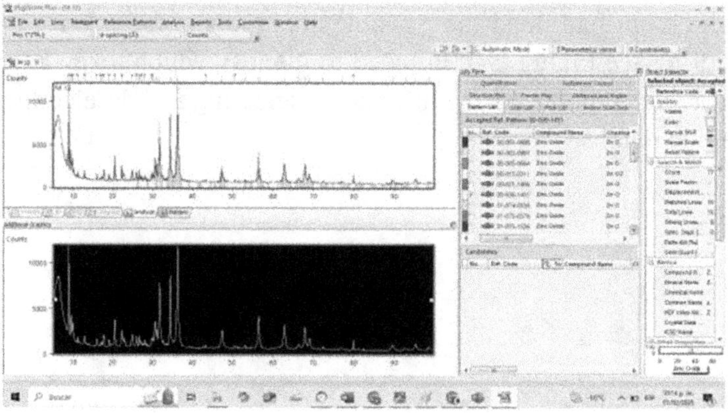

The same procedure is carried out for the case of cabasite zeolite, but in this case only the indexing is summarized in Figure 26.

Figure 26. Indexing of the chabazite with the ICDD crystallographic record 00-034-0137.

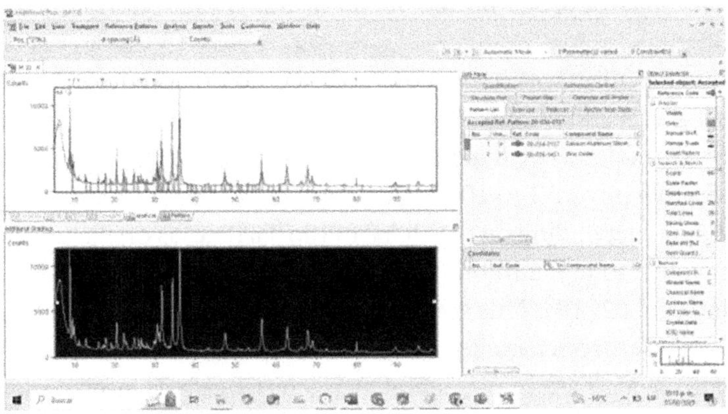

4.3. Debye-Scherrer Method

It is a method known as X-ray powder diffraction that provides qualitative and quantitative analysis information. It consists of emitting an X-ray beam on a crystalline material, generating diffraction patterns that reflect its structural characteristics. According to Hull, "Every crystalline substance gives a crystalline pattern; the same substance will always give the same crystalline pattern; and in a mixture of substances, each will produce its crystalline pattern independent of the others" (Khan et al., 2019).

This technique is used to measure crystal or particle size. The information obtained by determining this parameter is used to identify the nature of the materials, such as small crystals (soft), large crystals (hard), as well as diffusion and thermal behavior (Sivagami et al., 2022).

For the determination of particle size, the Scherrer technique is implemented, which ignores the contribution of deformation to peak broadening. It is worth mentioning that this method is very simple and is the most widely used for determining the size of crystals from a diffraction pattern. It can be implemented in both powders and thin films. The equation is as follows (Hassanzadeh-Tabrizi, 2023; Hargreaves, 2016):

$$D = \frac{K\lambda}{\beta cos\theta}$$

Where:

D is the particle size given in nm.

λ corresponds to the X-ray wavelength.

β corrected which is in radians.

θ is the Bragg angle in degrees.

K is a value that depends on the shape of the crystal, normally it is between 0.62 and 2.08.

The present equation was implemented for the determination of particle size as presented in Table 2 and Figure 27. This table shows that particle size ranges from 32.3 nm to 66.5 nm. It is worth mentioning that a nanometric particle is considered when its size range is from 1 to 100 nanometers (nm). That is, its diameter or length of any dimension (depending on the shape of the particle) is on the nanometer scale, which is equivalent to one millionth of a millimeter (1 nm = 10-9) (Vázquez, et al., 2020; Aravind et al., 2019). This size allows particles to have physical and chemical properties that differ from larger particles due to quantum effects and high surface-to-volume ratio, which are fundamental characteristics of nanomaterials (Sánchez et al., 2018). Particles in this size range can exhibit quantum behaviors, such as energy quantization, which is not observed in larger particles (Huang et al., 2020). Furthermore, due to the higher surface-to-volume ratio, nanosized particles are typically more reactive and have a higher adsorption capacity compared to larger-scale particles (Li et al., 2019). These characteristics make nanomaterials of interest in applications such as medicine, electronics, and energy, among others.

Table 2. Particle size.

Sample	D (nm)
ZnO	35.9
ZnO-Clinoptilolite (75-25)	39.4
ZnO-Clinoptilolite (50-50)	55
ZnO-Clinoptilolite (25-75)	58.8
ZnO-Mordenite (75-25)	40.2
ZnO-Mordenite (50-50)	32.3
ZnO-Mordenite (25-75)	35.5
ZnO-Erionite (75-25)	66.5
ZnO- Erionite (50-50)	41.1
ZnO- Erionite (25-75)	59.6
ZnO-Chabasite (75-25)	54.4
ZnO- Chabasite (50-50)	49.2
ZnO- Chabasite (25-75)	61.3

Figure 27. Particle size.

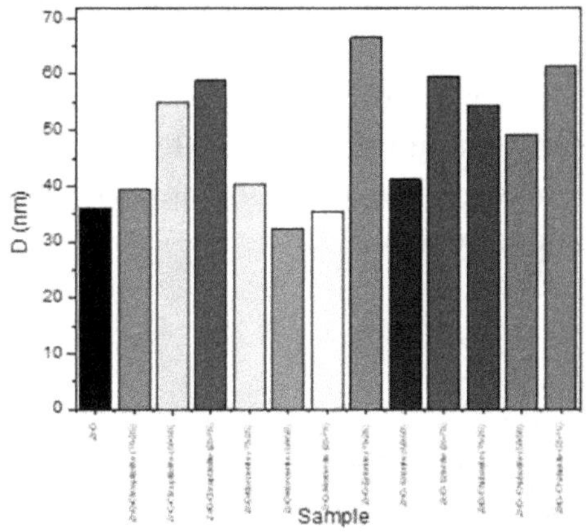

4.4. Dislocation Density

The ability of crystal lattices to be cut through dislocation movements is known as plastic deformation, this plastic deformation is given by the dislocation density. It is necessary to say that the interactions of the different deformation mechanisms are complex and depend on each other (Gallet et al., 2023).

Dislocations are responsible for most aspects of plastic deformation of materials (Sousa et al., 2015). Dislocation density is a parameter that can be measured by direct methods such as transmission electron microscopy (TEM), and also by indirect methods such as X-ray diffraction (XRD). It is worth mentioning that the aforementioned methods are complementary because direct methods can reveal microstructural information in an extremely small area of the sample, while indirect methods reveal average data over a relatively large area exposed to radiation (Shintani et al., 2011).

With the values obtained in section 4.5 (particle size), the value of the dislocation density of the ZnO-Zeolite composites can be determined. The dislocation density can be calculated with the following equation (Sutapa et al., 2018):

$$\delta = \frac{1}{D^2}$$

Where:
δ is the dislocation density
D is the particle size

It is necessary to mention that the value of the dislocation density denotes the number of dislocation lines per unit of crystal volume, which is the size of the crystalline defects that a crystal has. In conclusion, it will illustrate the degree of crystallinity of the nanoparticle profile (Sutapa et al., 2018). Figure 28 presents images of the dislocations that are presented, which are defined by the Burgués vector, which allows them to go from one point of the network to the one obtained after sliding the dislocation (Poblete., 2017).

Figure 28. Dislocations.

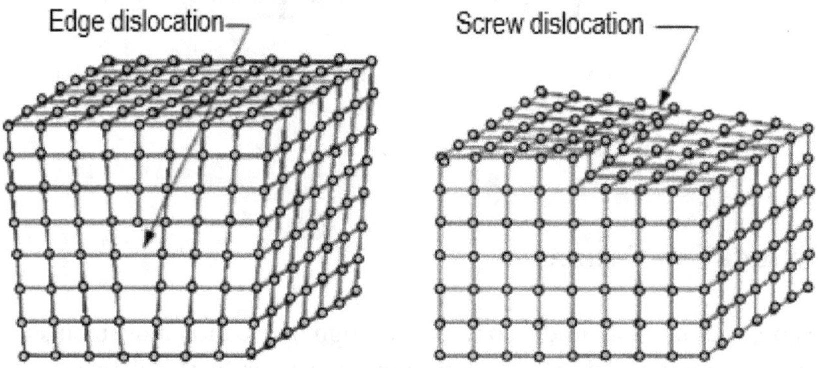

Table 3 and Figure 29 show the results of the dislocation density obtained from the different ZnO-Zeolite composites. It is worth mentioning that the one with the lowest value was the ZnO-Clinoptilolite (50-50) composite with an estimate of 5.0 X10^{-4} nm, while the one with the highest value was the ZnO-Erionite (50-50) composite with an estimate of 81.8 X10^{-4} nm.

It is worth mentioning that a small value of dislocation density, as is the case of the ZnO-Clinoptilolite (50-50) composite, indicates that the material has fewer dislocations per unit volume or area. This suggests that the material is relatively homogeneous and has fewer internal deformations. This means that with fewer dislocations, the mechanical properties such as resistance to deformation can be lower, since there are fewer internal barriers to the movement of the dislocations. This indicates that it is a more ductile material, that is, easier to deform without breaking, but at the same time less resistant.

On the other hand, in the case of the ZnO-Erionite (50-50) composite, which has the highest dislocation density value, it indicates that the material has a greater number of dislocations per unit of volume or area. This is characteristic of materials that have been deformed more or that have a more complex internal structure. It is worth mentioning that a high value of dislocation density generally translates into greater resistance to deformation of the material, because the dislocations act as a barrier that hinders the movement of other dislocations, increasing the mechanical resistance (this is known as dislocation hardening).

Table 3. Dislocation density of ZnO-Zeolite composites.

Sample	**δ nm² (X10⁻⁴)**
ZnO	8.4
ZnO-Clinoptilolita (75-25)	6.7
ZnO-Clinoptilolita (50-50)	5.0
ZnO-Clinoptilolita (25-75)	5.6
ZnO-Mordenita (75-25)	7.6
ZnO-Mordenita (50-50)	10.4
ZnO-Mordenita (25-75)	9.7
ZnO-Erionita (75-25)	62.2
ZnO- Erionita (50-50)	81.8
ZnO- Erionita (25-75)	55.3
ZnO-Chabasita (75-25)	55.3
ZnO- Chabasita (50-50)	65.0
ZnO- Chabasita (25-75)	55.3

Figure 29. Dislocation density of ZnO-Zeolite composites.

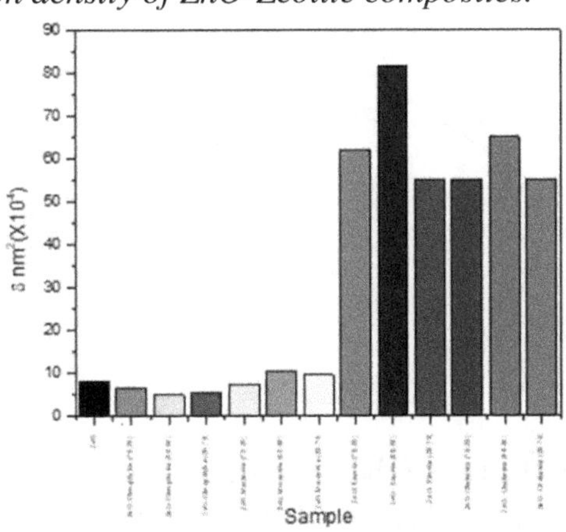

4.5. Microstrain or Microdeformations

Microdeformations obtained by X-ray diffraction are a technique that allows quantifying threading dislocations by measuring the radial microdeformation field surrounding these defects (Yon et al., 2020).

Expanding on the term microdeformations, we can mention that dilation and contraction are measurements that occur in the crystallographic network due to radial microdeformation, which come from edge, screw and mixed dislocations. These deformations extend in the normal plane to the dislocation line (Yon et al., 2020). In other words, the broadening of the diffraction lines in the X-ray diffraction pattern is caused by crystals of both small size and by distortions within the crystallites that behave as a mixture of crystallites with lattice parameters that vary within a range (See Figure 30) (Kimmel et al., 2013).

For the determination of microdeformation the following equation is used (Erdoğan et al., 2023):

$$\varepsilon = \frac{\beta}{4\tan\theta}$$

En donde:
β corregida que está en radianes.
θ es el ángulo de Bragg en grados.

Figure 30. Microdeformation.

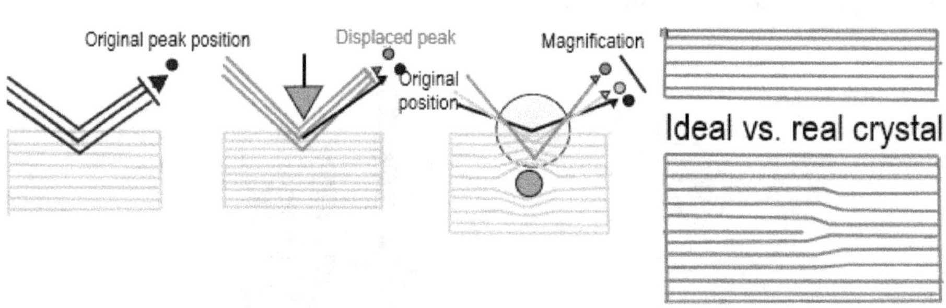

Table 4 and Figure 31 present the microdeformations of the ZnO-Zeolite composites. In this table it can be observed that the composite with the lowest value is ZnO-Clinoptilolite with a value of -813.2 X10^{-3}, while the composite with the highest value is ZnO-Mordenite with a value of 2250.0 X10^{-3}. It is worth mentioning that a large value in the microdeformation as in the case of the ZnO-Mordenite (75-25) composite with a value of 225.0 X10^{-3} indicates that the material presents a high degree of distortion or microtension in its crystalline structure. This may be due to several factors among which we can find internal tensions generated during the synthesis, presence of structural defects or perhaps the interaction with the environment such as thermal or pressure changes. These types of local stresses can affect both the mechanical and physical properties of the material (hardness, conductivity, and thermal stability, among others).

However, when we talk about negative values that can occur in microdeformation, it indicates compensation or relaxation in the crystalline structure. This means that the material is experiencing a slight compression instead of an expansion, which could be related to the compensation of internal stresses that were generated during the deformation process of the material. This may also reflect that the structure of a material is much more stable, indicating that the distortions are smaller, and the particles or crystalline units are more aligned or compact.

Table 4. Microdeformations of ZnO-Zeolite composites.

Sample	**ε (X10^{-3})**
ZnO	314.7
ZnO-Clinoptilolite (75-25)	-813.2
ZnO-Clinoptilolite (50-50)	1073.9
ZnO-Clinoptilolite (25-75)	-436.0
ZnO-Mordenite (75-25)	296.1
ZnO-Mordenite (50-50)	2250.0
ZnO-Mordenite (25-75)	1614.8
ZnO-Erionite (75-25)	1355.3
ZnO- Erionite (50-50)	1674.5
ZnO- Erionite (25-75)	458.9
ZnO-Chabasite (75-25)	951.6
ZnO- Chabasite (50-50)	1357.0
ZnO- Chabasite (25-75)	385.5

Figure 31. Microdeformations of ZnO-Zeolite composites.

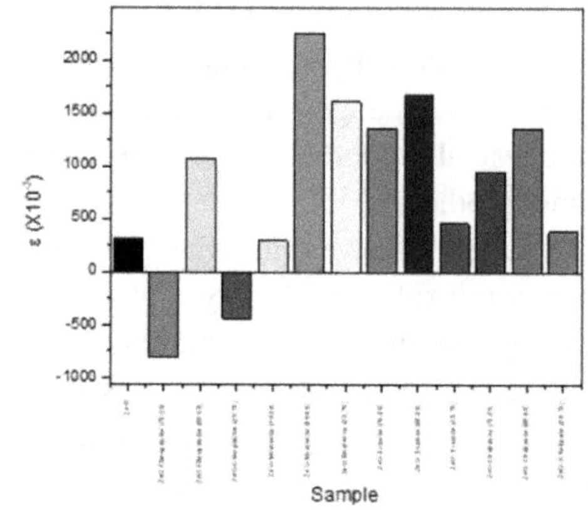

Figure 32 shows the diffractogram of the ZnO-Mordenite (50-50) composite, which was the composite that presented the highest microdeformation value. In this image it is enlarged by 2θ=10° up to 2θ=45° in order to see what happened. It can be seen in this image that the intensity and shape of the peaks belonging to the Mordenite zeolite are affected. This indicates that local deformations occur that affect the interplanar distances (this makes the peaks wider than expected for a material without defects). It is also observed that the diffraction peaks shift slightly or widen.

Figure 32. Intensities and displacement of planes.

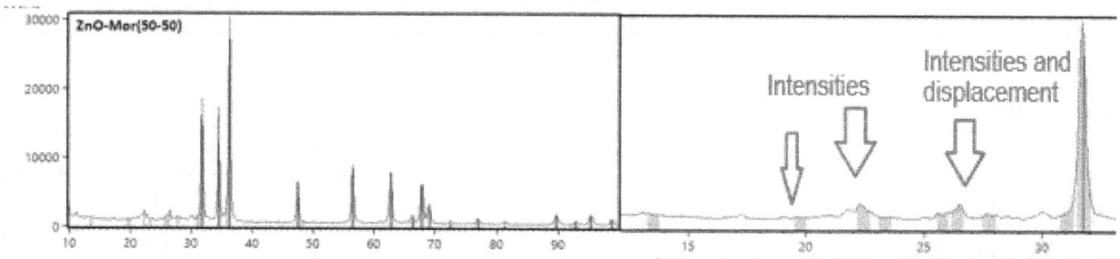

4.6. Lattice Parameters

4.6.1. What are Lattice Parameters?

Lattice parameters are the dimensions and angles that describe a unit cell (See Figure 33). It is worth mentioning that a unit cell is the smallest repetitive unit that defines the structure of a crystal (Cullity & Stock, 2001). Among the lattice parameters we can find:

- Lengths of the crystallographic axes represented by a, b and c which have units of measurement in nanometers or angstroms (West, 2014).
- Angles between the axes represented by α, β and γ which have units in degrees (Kittel, 2005).

Figure 33. Lattice parameters in a unit cell.

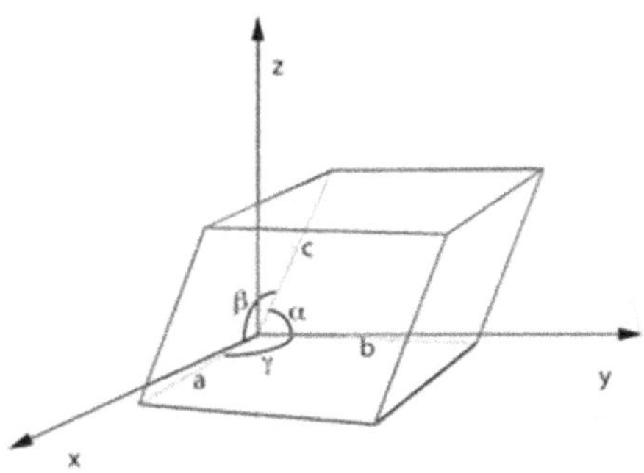

4.6.2. Crystalline Systems

Crystal systems are the different ways in which atoms, ions or molecules are organized in a crystalline solid (See Figure 35). They are classified into seven systems, each defined by the shape of its unit cell, which is the smallest repeating unit of the crystal structure (See Table 5) (Callister & Rethwisch, 2020).

Table 5. Crystal systems.

Crystal systems	Expression for d_{hkl}
Cubic	$\frac{1}{d_{hkl}^2} = \frac{h^2+k^2+l^2}{a^2}$
Tetragonal	$\frac{1}{d_{hkl}^2} = \frac{h^2+k^2}{a^2} + \frac{l^2}{c^2}$
Orthorhombic	$\frac{1}{d_{hkl}^2} = \frac{h^2}{a^2} + \frac{k^2}{b^2} + \frac{l^2}{c^2}$
Hexagonal	$\frac{1}{d_{hkl}^2} = \frac{4}{3}\left(\frac{h^2 + hk + k^2}{a^2}\right) + \frac{l^2}{c^2}$
Monoclinic	$\frac{1}{d_{hkl}^2} = \frac{1}{sin^2\beta}\left[\frac{h^2}{a^2} + \frac{k^2 sin^2\beta}{b^2} + \frac{l^2}{c^2} - \frac{2hl\,cos\beta}{ac}\right]$
Triclinic	$\frac{1}{d_{hkl}^2} = \frac{1}{V^2}[h^2b^2c^2sin^2\alpha + k^2a^2c^2sin^2\beta + l^2a^2b^2sin^2\gamma + 2hkabc^2(\cos\alpha\cos\beta - \cos\gamma) + 2kla^2(\cos\beta\cos\gamma - \cos\alpha) + 2hlab^2c(\cos\alpha\cos\gamma - \cos\beta)]$ where: $V = abc[1 - cos^2\alpha - cos^2\beta - cos^2\gamma + 2\cos\alpha\cos\beta\cos\gamma]^{\frac{1}{2}}$

4.6.3. How to Calculate Network Parameters?

The lattice parameters can be calculated by implementing various experimental techniques such as: X-ray diffraction (XRD) (Smith, 2010), Transmission Electron Microscopy (TEM) (Brown et al., 2015) or Atomic Force Microscopy (AFM) (Jones, 2018). Considering the X-ray diffraction technique we will have the following steps:

1. Calculation from the interplanar distance equation d_{hkl}.

For the determination of the interplanar distance of a cubic crystal system (see Table 5) with the lattice parameters, the following equation is used (Kumar et al., 2017):

$$d_{hkl} = \frac{a}{\sqrt{h^2 + k^2 + l^2}}$$

2. Using Bragg's Law

Bragg's law allows me to calculate dhkl from an X-ray diffraction pattern (Bragg, 1913):

$$n\lambda = 2d_{hkl}\sin\theta$$

Where:

λ is the wavelength of the X-ray

θ is the diffraction angle

n is the diffraction order

From the value of dhkl the values of a, b and c can be determined in the different crystal systems (Williams et al., 2009).

With the above data, the network parameters of the ZnO-Mordenite composites (See Figure 34) are determined using the values found in the crystallographic records (See Figures 35 and 36), which were obtained using the PANalytical HighScore Plus 3.0 program. As can be seen in Figure 34, the network parameter a with a value of 3.2498 corresponds to ZnO, while parameter b with a value of 20.248 corresponds to the Mordenite zeolite. It can be seen that the values calculated for the different planes that correspond to ZnO do not vary with respect to the theoretical value given in crystallographic record 00-036-1451, which indicates that neither compressive nor expansive stress is being generated. In the case of the Mordenite zeolite with its network parameter, it is observed that it has an expansive stress, this is because the calculated values of the parameter b in the different planes increase compared to the parameter b=20.248 indicating that the distances between the atoms expand (See Figure 37).

Figure 34. Crystal lattice parameters of Zn-Mordenite composites.

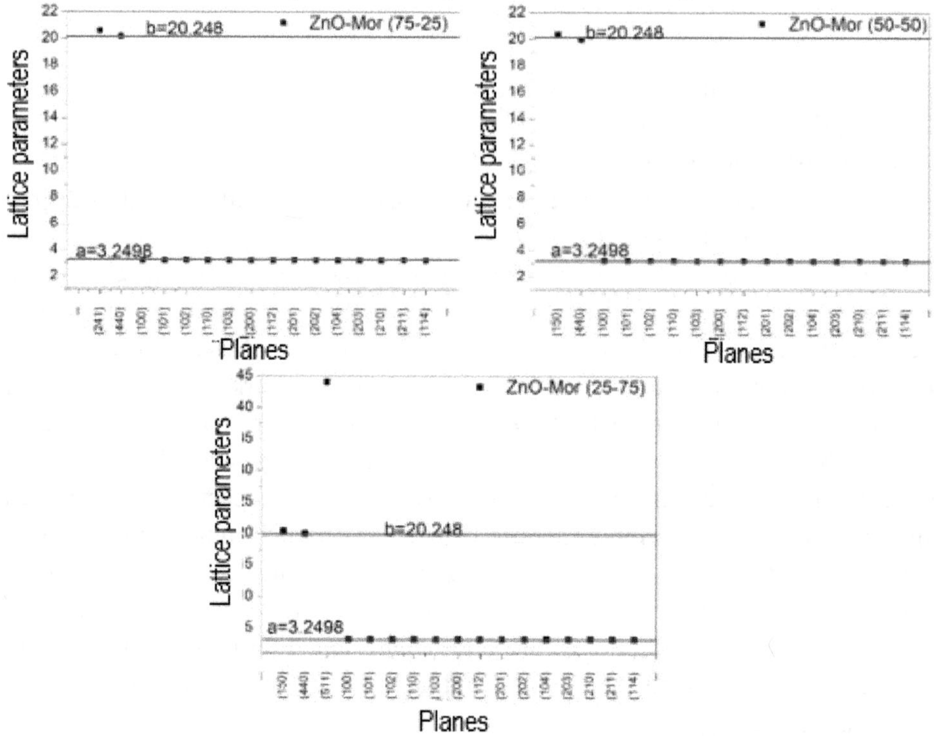

Figure 35. ZnO reference standard with reference code 00-036-1451.

Reference Pattern: 00-036-1451

Name and formula

Reference code:	00-036-1451
Mineral name:	Zincite, syn
Compound name:	Zinc Oxide
Common name:	zinc white
PDF index name:	Zinc Oxide
Empirical formula:	OZn
Chemical formula:	ZnO

Crystallographic parameters

Crystal system:	Hexagonal
Space group:	P63mc
Space group number:	186
a (Å):	3.2498
b (Å):	3.2498
c (Å):	5.2066
Alpha (°):	90.0000
Beta (°):	90.0000
Gamma (°):	120.0000
Volume of cell (10^6 pm^3):	47.62
Z:	2.00

Figure 36. Reference standard of the Mordenite zeolite with reference code 00-049-0924.

Reference Pattern: 00-049-0924

Name and formula

Reference code:	00-049-0924
Mineral name:	Mordenite, sy
Compound name:	Sodium Alum
PDF index name:	Sodium Alum
Zeolite name:	Zeolite Al-mo
Chemical formula:	$Na_2Al_2Si_{13.3}C$

Crystallographic parameters

Crystal system:	Orthorhombic
Space group:	Cmc21
Space group number:	36
a (Å):	18.0670
b (Å):	20.2840
c (Å):	7.4910
Alpha (°):	90.0000
Beta (°):	90.0000
Gamma (°):	90.0000
Volume of cell (10^6 pm^3):	2745.23
RIR:	-

Figure 37. Expansive and compressive stress.

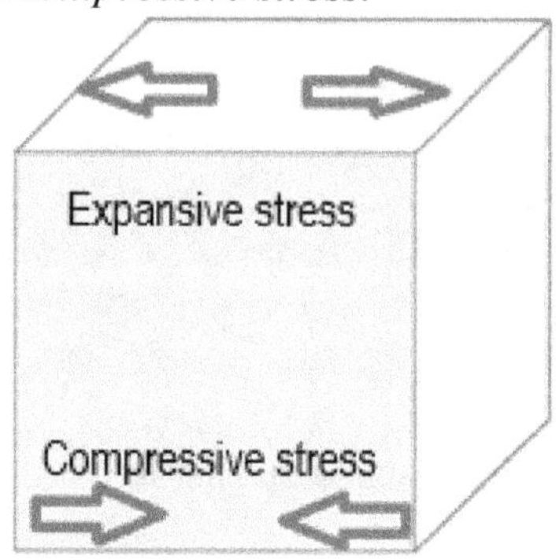

4.7. Rietveld Refinement of X-Ray Powder Diffraction.

This method was proposed by the scientist Hugo Rietveld in 1969 (Rietveld, 2021), it consists of adjusting the complete computer simulation spectrum to analyze neutron diffraction data which uses integral diffraction intensity data and diffraction spectrum information to obtain relevant information on the structure of the powder material (Ramkumar et al, 2018). To detail this method a little more, it is worth mentioning that it depends on the crystalline structure and the X-ray diffraction peaks through a least squares refinement process, which minimizes the differences between the observed and calculated patterns. The parameters that are refined for the samples in this method are those that control the intensity of the pattern (scale factors), the peak profile (width and shape), the peak position (zero point and unit cell), the background parameters and the thermal parameters of each atom (Asmi et al., 2014). To perform Rietveld refinement by X-ray diffraction, the PANalytical HighScore Plus 3.0 program is used. Where the two phases corresponding to the crystallographic cards 00-036-1451 of the ICDD for ZnO and 00-049-0924 of the ICDD for the Mordenite zeolite belonging to the ZnO-Mor (25-75) composite are already identified as seen in Figure 38. With the help of the aforementioned crystallographic cards (See Figure 39) the values of the network parameters of each of the phases are identified (For ZnO a=3.2498 , b=3.2498 , c= 5.2066 , α= 90°, β= 90°, γ=120 and V= 47.62 X106 pm3 and for the Mordenite zeolite a=18.670 , b=20.2840 , c= 7.4910 , α= 90°, β= 90°, γ=90° and V= 2745.23 X106 pm3. With this data provided, files with the .cif extension are obtained from the Crystallography Open Database (COD) website. It is worth mentioning that .cif files are standard formats used in crystallography to store and share data related to the structure of a crystal. On the COD website, the .cif formats corresponding to each of the aforementioned phases that match their network parameters are searched for (See Figure 40). The .cif files with the ID 9008877 corresponding to ZnO and the ID 1525216 corresponding to the Mordenite zeolite are downloaded. Once the .cif files have been downloaded, they are inserted into the HighScore Plus 3.0 program, which already contains the diffractogram of the loaded sample (ZnO-Mordenite (25-75)) as shown in Figure 41.

Once the two .cif files corresponding to the crystalline structures of ZnO and the mordenite zeolite have been inserted, the Rietveld refinement is performed as shown in Figure 42.

To ensure that the values obtained were determined correctly in the HighScore Plus 3.0 program, the most direct discrepancy index was taken into account, known as Rwp (weighted profile R-factor), which is derived directly from the square root of the minimized quantity, scaled by the weighted intensities:

$$R_{wp}^2 = \frac{\sum_i w_i \left(y_{c,i} - y_{O,i}\right)^2}{\sum_i w_i \left(y_{O,i}\right)^2}$$

Where $w = \frac{1}{\sigma^2\left[y_{O,i}\right]}$, relates to errors of a purely statistical nature arising from the smallest uncertainties in the fitting parameters. Considering that yC,i are the intensity values simulated from the model where C indicates that it is calculated from the model. By convention, intensity values are labeled yO,i. Where O indicates that these are observed values and i indicates that the intensity was measured at 2θ the value 2θi. To perform the Rietveld analysis, we must have an uncertainty estimate for yO,i which is called $\sigma\left[y_{O,i}\right]$. In the past, this was called the estimated standard deviation, but crystallographic convention now calls it the standard uncertainty term (Ozaki et al., 2020).

Another value given is Rexp (expected R-factor). To determine the Rexp, it should be considered to be the "best possible Rwp". Thus, the following equation remains: $R_{exp}^2 = \frac{N}{\sum_i w_i \left(y_{O,i}\right)^2}$ where N is used as a label for the number of data points (Ozaki et al., 2020).

The literature on single crystals often uses the term Goodness of Fit represented by the letter G, which is defined by G^2=χ^2. Recall that χ^2 or "Chi square" is a statistical concept that is represented as follows: $X^2 = \left(\frac{1}{N}\sum \frac{\left(y_{c,i} - y_{O,i}\right)^2}{\sigma\left[y_{O,i}\right]}\right)$. Note that χ^2 can also be determined from the factors Rwp and Rexp. Therefore, the equation is as follows $X^2 = \left(\frac{R_{wp}}{R_{exp}}\right)^2$ (Ozaki et al., 2020).

Figure 34 shows the values Rexp=2.232, Rwp=16.954 and G=57.662 of the first Rietveld refinement of the TiO2-Mordenite (25-75) composite. As the value of G is still high, the Rietveld refinement is carried out semi-automatically as shown in Figure 35.

The refinement results obtained for the real ZnO-Mordenite (25-75) sample were the following Rexp=30.698, Rwp=18.890 and G=0.3786. The statistical refinement indices (Rexp, Rwp and Godness) are considered satisfactory, since the general index of Goodness of fit = (Rwp/Rexp) was less than 5 (See Figure 36) [32]. This refinement helped to determine the percentage of phases found in the ZnO-Mordenite (25-75) composite powder. In the phase corresponding to ZnO with COD number 9008877 a value of 25.1% was found, while for the Zeolite Mordenite phase with COD number 1525216 a value of 74.9% was observed.

Table 6 presents a summary of the phase percentage of the different combinations of ZnO-Zeolite composites. It is observed that the pure ZnO sample exhibits 100% ZnO phase, without the presence of zeolite. As the zeolite is introduced in different proportions and types (Clinoptilolite, Mordenite, Erionite and Chabasite), the % of ZnO progressively decreases while the % of zeolite increases.

The combinations with the highest ZnO content correspond to ZnO-Chabasite (75-25) with 92% of ZnO phase, while those with a lower proportion of ZnO are ZnO-Clinoptilolite (25-75) and ZnO-Mordenite (25-75) with approximately 27.8% and 25.1% of ZnO, respectively.

These results indicate that the incorporation of zeolite significantly affects the crystalline phase of the material, with an inverse relationship between the percentage of ZnO and zeolite.

Figure 38. ZnO-Mordenite (75-25) composite.

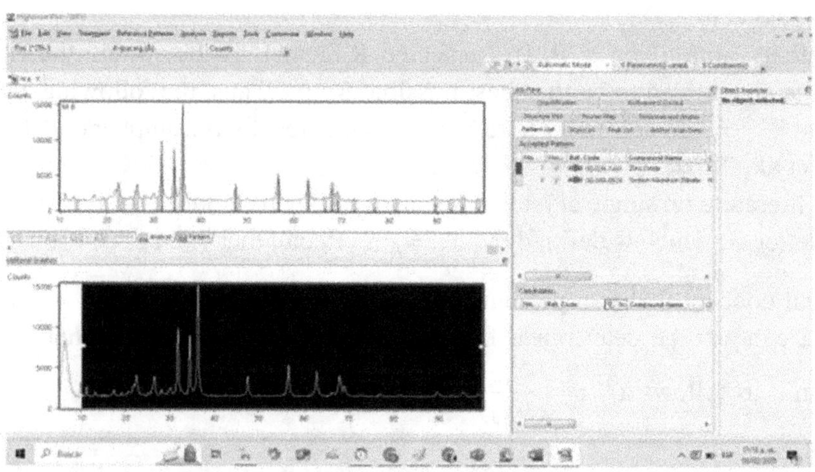

Figure 39. Lattice parameters of ZnO and Mordenite zeolite obtained from ICDD database.

Reference Pattern: 00-036-1451

Name and formula

Reference code:	00-036-1451
Mineral name	Zincite, syn
Compound name	Zinc Oxide
Common name	zinc white
PDF index name	Zinc Oxide
Empirical formula:	OZn
Chemical formula:	ZnO

Crystallographic parameters

Crystal system:	Hexagonal
Space group:	P63mc
Space group number:	186
a (Å):	3.2498
b (Å):	3.2498
c (Å):	5.2066
Alpha (°):	90.0000
Beta (°):	90.0000
Gamma (°):	120.0000
Volume of cell (10^6 pm^3):	47.62
Z:	2.00

Reference Pattern: 00-049-0924

Name and formula

Reference code:	00-049-0924
Mineral name	Mordenite, syn
Compound name	Sodium Aluminum Silicate
PDF index name	Sodium Aluminum Silicate
Zeolite name	Zeolite Al-mordenite
Chemical formula:	$Na_2Al_2Si_{13.3}O_{29.6+x}$

Crystallographic parameters

Crystal system:	Orthorhombic
Space group:	Cmc21
Space group number:	36
a (Å):	18.0670
b (Å):	20.2840
c (Å):	7.4910
Alpha (°):	90.0000
Beta (°):	90.0000
Gamma (°):	90.0000
Volume of cell (10^6 pm^3):	2745.23
RIR:	-

Figure 40. Identification of COD IDs to obtain .cif files.

COD ID ▲	Links	Formula ▲	Space group ▲	Cell parameters	Cell volume ▲	Bibliography
9008877	CIF	O Zn	P 63 m c	3.2495; 3.2495; 5.2069 90; 90; 120	47.615	Wyckoff, R. W. G. Second edition. Interscience Publishers, New York, New York Note: wurtzite structure *Crystal Structures*, **1963**, *1*, 85-237

COD ID ▲	Links	Formula ▲	Space group ▲	Cell parameters	Cell volume ▲	Bibliography
1525216	CIF	Al8.5 H7.9 Na0.6 O96 Si39.5	C m c m	18.178; 20.394; 7.488 90; 90; 90	2775.97	Schlenker, J.L.; Pluth, J.J.; Smith, J.V. Positions of cations and molecules in zeolites with the mordenite framework. IX. Dehydrated H-mordenite via acid exchange *Materials Research Bulletin*, **1979**, *14*, 849-856

Figure 41. Inserting .cif files.

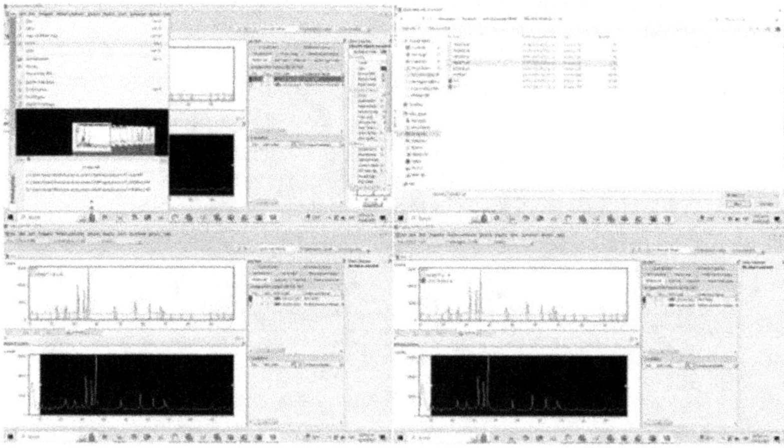

Figure 42. Rietveld refinement.

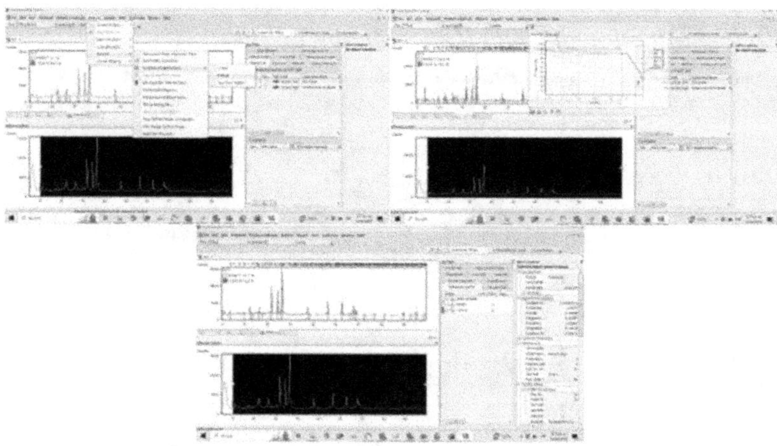

Figure 43. Semi-automatic selection.

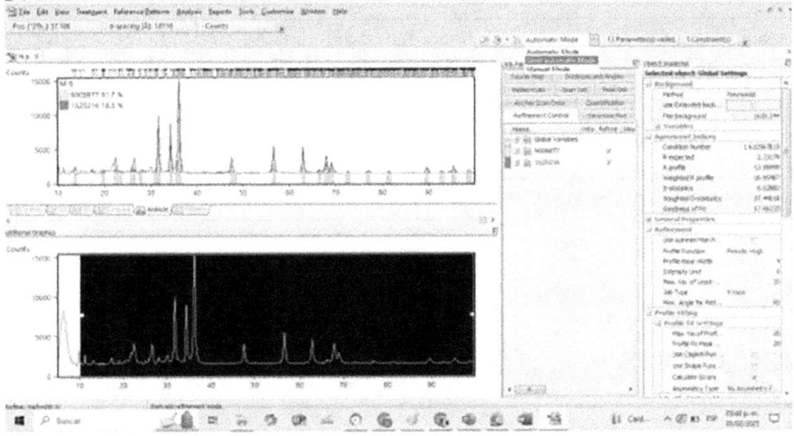

Figure 43. Semi-automatic selection.

Figure 44. Determination of the percentages of crystalline phases by Rietveld refinement.

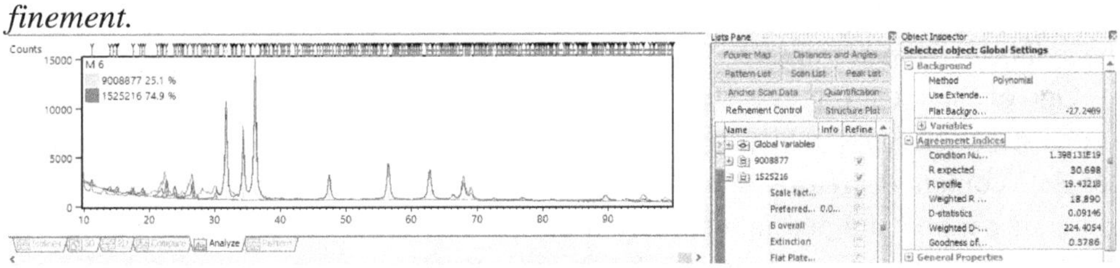

Table 6. Percentage of crystalline phases.

Sample	%ZnO	%Zeolite
ZnO	100	N/A
ZnO- Clinoptilolite (75-25)	83.0	17.0
ZnO- Clinoptilolite (50-50)	55.6	44.4
ZnO- Clinoptilolite (25-75)	27.8	72.2
ZnO- Mordenite (75-25)	72.8	27.2
ZnO- Mordenite (50-50)	26.6	73.4
ZnO- Mordenite (25-75)	25.1	74.9
ZnO- Erionite (75-25)	34.6	65.4

continued on following page

Table 6. Continued

Sample	%ZnO	%Zeolite
ZnO- Erionite (50-50)	55.3	44.7
ZnO- Erionite (25-75)	34.6	65.4
ZnO- Chabasite (75-25)	92.0	8.0
ZnO- Chabasite (50-50)	70.3	29.7
ZnO- Chabasite (25-75)	67.1	32.9

5. CONCLUSIONS

The present study on the synthesis and structural characterization of ZnO-Zeolite composites by X-ray diffraction has allowed us to understand in detail how the incorporation of different zeolites influences the crystalline phase of ZnO. Throughout the research, composites formed by ZnO in combination with different types of zeolites, including Clinoptilolite, Mordenite, Erionite and Chabasite, in various proportions, were analyzed. The results obtained have shown an inverse relationship between the amount of ZnO and the percentage of zeolitic phase present in the synthesized materials.

One of the most relevant findings is that the crystalline phase of pure ZnO remains stable at 100%, while, when incorporating zeolites in different proportions, the ZnO content decreases significantly. This suggests that zeolite plays a key role in the dispersion and possible reduction of the crystallinity of ZnO, depending on the amount and type of zeolite used. In particular, it was observed that Chabasite is the zeolite that maintains a higher proportion of ZnO in the crystalline phase, while Clinoptilolite and Mordenite favor a greater dispersion of ZnO within the zeolitic matrix.

From a structural point of view, this behavior can be explained due to the differences in the crystalline structure of the zeolites and their ability to interact with ZnO. Zeolite can act as a confinement medium, favoring the formation of nanocomposites with a better dispersion of ZnO. This effect is especially important for applications in which the active surface of ZnO is sought to be increased, such as in heterogeneous catalysis, gas sensors and optoelectronic devices.

Another important aspect to consider is that the combination of ZnO with different types of zeolites allows modulating its structural and functional properties. In this sense, the results obtained can contribute to the development of new materials with specific characteristics according to the desired application. For example, composites with a high ZnO content could be suitable for photocatalytic applications, while those with a higher proportion of zeolite could be used in gas adsorption and purification processes.

In addition to structural characterization, this study highlights the importance of X-ray diffraction as a fundamental tool to evaluate the phase composition in composite materials. The information obtained through this technique has allowed to accurately quantify the proportion of each phase present in the composites, thus facilitating the understanding of the influence of zeolites on the structure of ZnO.

In conclusion, the incorporation of zeolites in ZnO composites represents an effective strategy to modify their crystalline properties and expand their range of technological applications. The ability of zeolites to modulate the ZnO phase opens new opportunities in the design of advanced materials, especially those intended for environmental and energy conversion processes. In the future, it would be interesting to complement these findings with additional studies that include more advanced characterization techniques, such as Raman spectroscopy, scanning electron microscopy (SEM) and measurements of optical and electronic properties. In this way, the understanding of the interactions between ZnO and zeolites could be further deepened, optimizing their performance in specific applications.

REFERENCES

Ali, A., Chiang, Y. W., & Santos, R. M. (2022). X-ray diffraction techniques for mineral characterization: A review for engineers of the fundamentals, applications, and research directions. *Minerals (Basel)*, *12*(2), 205. DOI: 10.3390/min12020205

Ameh, E. S. (2019). A review of basic crystallography and X-ray diffraction applications. *International Journal of Advanced Manufacturing Technology*, *105*(7-8), 3289–3330. DOI: 10.1007/s00170-019-04508-1

Aravind, S.. (2019). Nanomaterials: Synthesis, properties, and applications. *Journal of Materials Science and Technology*, *35*(6), 1092–1104.

Artioli, G. (2016). X-ray diffraction (XRD). *Encyclopedia of Earth Sciences Series*, 1019–1025. https://doi.org/DOI: 10.1007/978-1-4020-4409-0_29

Asmi, D., & Low, I. M. (2014). Manufacture of graded ceramic matrix composites using infiltration techniques. In *Advances in Ceramic Matrix Composites* (pp. 109–140). Elsevier., DOI: 10.1533/9780857098825.1.109

Balkan, N., & Erol, A. (2021). Intrinsic and extrinsic semiconductors. In *Semiconductors for Optoelectronics* (pp. 37–78). Springer., DOI: 10.1007/978-3-319-44936-4_2

Bertaut, E. F. (1964). *La diffraction des rayons X*. Masson & Cie.

Bird, J. P. (2002). Semiconductors: An introduction. En *Encyclopedia of Materials: Science and Technology* (2nd ed., pp. 1-16). https://doi.org/DOI: 10.1016/B0-08-043152-6/01854-4

Bokov, D., Jalil, A. T., Chupradit, S., Suksatan, W., Ansari, M. J., Shewael, I. H., Valiev, G. H., & Kianfar, E. (2021). Nanomaterial by sol-gel method: Synthesis and application. *Advances in Materials Science and Engineering*, *5102014*(1), 5102014. Advance online publication. DOI: 10.1155/2021/5102014

Bokov, D., Jalil, A. T., Chupradit, S., Suksatan, W., Ansari, M. J., Shewael, I. H., Valiev, G. H., & Kianfar, E. (2021). Nanomaterial by sol-gel method: Synthesis and application. *Advances in Materials Science and Engineering*, *2021*(1), 5102014. Advance online publication. DOI: 10.1155/2021/5102014

Bragg, W. H. (1913). The diffraction of X-rays by crystals. *Proceedings - Royal Society. Mathematical, Physical and Engineering Sciences*, *88*(605), 428–438.

Brown, D., & Green, R. (2015). Transmission electron microscopy: Principles and applications. *Journal of Microscopy*, *158*(1), 55–65.

Bunaciu, A. A., Udriştioiu, E. G., & Aboul-Enein, H. Y. (2015). X-ray diffraction: Instrumentation and applications. *Critical Reviews in Analytical Chemistry*, *45*(4), 289–299. DOI: 10.1080/10408347.2014.949616 PMID: 25831472

Callister, W. D., & Rethwisch, D. G. (2020). *Materials Science and Engineering: An Introduction* (10th ed.). Wiley.

Cullity, B. D., & Stock, S. R. (2001). *Elements of X-ray Diffraction* (3rd ed.). Prentice Hall.

Cullity, B. D., & Stock, S. R. (2001). *Elements of X-ray Diffraction*. Prentice Hall.

Cullity, B. D., & Stock, S. R. (2014). *Elements of X-ray Diffraction* (3rd ed.). Pearson.

Cundy, C. S., & Cox, P. A. (2003). The hydrothermal synthesis of zeolites: History and development from the earliest days to the present time. *Chemical Reviews*, *103*(3), 663–701. DOI: 10.1021/cr020060i PMID: 12630849

Derbe, T., Temesgen, S., & Bitew, M. (2021). A short review on synthesis, characterization, and applications of zeolites. *Advances in Materials Science and Engineering*, *6637898*(1), 6637898. Advance online publication. DOI: 10.1155/2021/6637898

Erdoğan, N. N., & Başyiğit, A. B. (2023). An approach on determining micro-strain and crystallite size values of thermal spray barrier coated Inconel 601 super alloy. [TIJMET]. *The International Journal of Materials and Engineering Technology*, *6*(1), 21–25.

Estrada Flores, S. L., Garcias-Morales, C., Mirón Velázquez, R. S., Pérez Berumen, C. M., & De la Cruz Duran, F. N. (2022). *An overview of organic electronics*. Revistas UNAM., DOI: 10.22201/fq.18708404e.2022.2.79879

Fouad, M. M., Shihata, L. A., & Morgan, E. I. (2017). An integrated review of factors influencing the performance of photovoltaic panels. *Renewable and Sustainable Energy 23Reviews, 80*, 1499-1511. https://doi.org/DOI: 10.1016/j.rser.2017.05.141

Gallet, J., Perez, M., Guillou, R., Ernould, C., Le Bourlot, C., Langlois, C., Beausir, B., Bouzy, E., Chaise, T., & Cazottes, S. (2023). Experimental measurement of dislocation density in metallic materials: A quantitative comparison between measurement techniques (XRD, R-ECCI, HR-EBSD, TEM). *Materials Characterization*, *199*, 112842. DOI: 10.1016/j.matchar.2023.112842

Gao, X., & Zhang, T. (2018). An overview: Facet-dependent metal oxide semiconductor gas sensors. *Sensors and Actuators. B, Chemical*, *277*, 604–633. DOI: 10.1016/j.snb.2018.08.129

Ghorbani, M. M., & Taherian, R. (2019). Methods of measuring electrical properties of material. En *Electrical Conductivity in Polymer-Based Composites: Experiments, Modelling, and Applications* (pp. 365-394). Plastics Design Library. https://doi.org/ DOI: 10.1016/B978-0-12-812541-0.00012-4

Goñi Echave, I., & García Arnáez, I. (2023). *Incorporación de simvastatina a recubrimientos sol-gel para implantes dentales* [Tesis de licenciatura, Universidad del País Vasco]. https://addi.ehu.es/bitstream/handle/10810/62777/TFG_IreneToledo.pdf?sequence=2&isAllowed=y

Gualtieri, A. F., Lanci, L., & Tommasini, S. (2003). Crystallographic structure of natural magnetite. *Journal of Applied Crystallography*, *36*(3), 738–744.

Hargreaves, J. S. J. (2016). Some considerations related to the use of the Scherrer equation in powder X-ray diffraction as applied to heterogeneous catalysts. *Catalysis. Structure & Reactivity*, *2*(1-4), 33–37. DOI: 10.1080/2055074X.2016.1252548

Hassanzadeh-Tabrizi, S. A. (2023). Precise calculation of crystallite size of nanomaterials: A review. *Journal of Alloys and Compounds*, *968*, 171914. DOI: 10.1016/j.jallcom.2023.171914

Holder, C. F., & Schaak, R. E. (2019). Tutorial on powder X-ray diffraction for characterizing nanoscale materials. *ACS Nano*, *13*(7), 7359–7365. DOI: 10.1021/acsnano.9b05157 PMID: 31336433

Huang, Z.. (2020). Quantum effects in nanomaterials: Insights and challenges. *Journal of Nanoscience*, *25*(4), 237–250.

International Centre for Diffraction Data (ICDD). (2020). *PDF-4+ 2020*. ICDD.

Jones, A. (2018). Atomic force microscopy: Techniques and applications. *Surface Science Reports*, *71*(3), 123–136.

Karimi, M., Mehrabadi, Z., Farsadrooh, M., Bafkary, R., Derikvandi, H., Hayati, P., & Mohammadi, K. (2021). Metal–organic framework. En *Interface Science and Technology, 33*, 279-387. https://doi.org/DOI: 10.1016/B978-0-12-818805-7.00010-2

Khan, H., Yerramilli, A. S., D'Oliveira, A., Alford, T. L., Boffito, D. C., & Patience, G. S. (2020). Experimental methods in chemical engineering: X-ray diffraction spectroscopy—XRD. *Canadian Journal of Chemical Engineering*, *98*(6), 1255–1266. Advance online publication. DOI: 10.1002/cjce.23747

Kimmel, G., & Dayan, D. (2013). X-ray diffraction characterization of microstrain in some uranium alloys. *Powder Diffraction*, *13*(2), 89–95. DOI: 10.1017/S088571560000991X

Kittel, C. (2005). *Introduction to Solid State Physics*. Wiley.

Klug, H. P., & Alexander, L. E. (1974). *X-ray Diffraction Procedures for Polycrystalline and Amorphous Materials* (2nd ed.). Wiley.

Kumar, P., Sharma, S., & Patel, V. (2017). Calculation of interplanar distance in cubic crystals using X-ray diffraction. *Journal of Crystal Growth*, *461*, 38–42.

Lee, D.-Y., Ko, W.-S., Kim, K.-N., Byun, J.-H., Kim, E.-G., Kwon, S.-Y., & Lee, G.-W. (2024). Analysis of metal and zinc oxide semiconductor interface resistance using transmission line method. *Solid-State Electronics*, *216*, 108916. DOI: 10.1016/j.sse.2024.108916

Li, T., Cai, H., Li, C., Liu, X., & Huang, F. (2020). Rocksalt-zincblende–wurtzite mixed-phase ZnO crystals with high activity as photocatalysts for visible-light-driven water splitting. *Frontiers in Chemistry*, *8*, 351. DOI: 10.3389/fchem.2020.00351 PMID: 32411673

Li, W.. (2019). Nanoparticles with enhanced surface area for catalysis and energy storage. *Journal of Nanotechnology*, *22*(3), 194–210.

Li, Y., Zhu, G., Wang, Y., Chai, Y., & Liu, C. (2021). Preparation, mechanism and applications of oriented MFI zeolite membranes: A review. *Microporous and Mesoporous Materials*, *312*, 110790. DOI: 10.1016/j.micromeso.2020.110790

Liu, J., Guo, H., & Wang, F. (2015). Influence of crystallography on the properties of cementitious materials. *Cement and Concrete Research*, *70*, 36–44.

Mengyao, L., Changging, F., Youliang, C., Xin, Z., Hanzhi, H., Jiarui, Z., & Yue, Z. (2024). A facile pathway to fabricate VO2 (M) nanoparticles via sol-gel method for flexible thermochromic films with efficient infrared stealth. *Vacuum*, *221*, 112885. DOI: 10.1016/j.vacuum.2023.112885

Min, G., Wei, W., Fan, Q., Wan, T., Ye, M., & Yun, S. (2025). High-throughput exploration of stable semiconductors using deep learning and density functional theory. *Materials Science in Semiconductor Processing*, *188*, 109150. DOI: 10.1016/j.mssp.2024.109150

Molinari, R., Argurio, P., & Lavorato, C. (2015). Photocatalytic hydrogenation of organic compounds in membrane reactors. In *Membrane Reactors for Energy Applications and Basic Chemical Production* (pp. 605–639). Woodhead Publishing., DOI: 10.1016/B978-1-78242-223-5.00020-0

Murrieta-Rico, F. N., Antúnez-García, J., Yocupicio-Gaxiola, R. I., Galván, D. H., González, J. C., & Petranovskii, V. (2022). Zeolites as initial structures for the preparation of functional materials. *Journal of Applied Research and Technology, 20*(1), 92–116. DOI: 10.22201/icat.24486736e.2022.20.1.1561

Murthi, M., & Destino, J. F. (2025). Sol-gel technologies for additive manufacturing glass materials. *Journal of Sol-Gel Science and Technology*. Advance online publication. DOI: 10.1007/s10971-025-06664-1

Ozaki, Y., Suzuki, Y., Hawai, T., Saito, K., Onishi, M., & Ono, K. (2020).. . *Computational Materials, 6*, 1. DOI: 10.1038/s41524-019-0267-z

Pal, P. (2017). Nanotechnology in water treatment. En *Industrial Water Treatment Process Technology* (pp. 513-536). https://doi.org/DOI: 10.1016/B978-0-12-810391-3.00007-2

Pecharsky, V. K., & Zavalij, P. Y. (2009). Fundamentals of diffraction. Fundamentals of Powder Diffraction and Structural Characterization of Materials, 133-149.

Poblete Palacios, G. E. (2017). *Estudio de la influencia de la densidad de dislocaciones en la resistencia mecánica del cobre* [Tesis de ingeniería civil mecánico, Universidad de Chile]. Facultad de Ciencias Físicas y Matemáticas, Departamento de Ingeniería Mecánica.

Ramkumar, T., Selvakumar, M., Vasanthsankar, R., Sathishkumar, A. S., Narayanasamy, P., & Girija, G. (2018). Rietveld refinement of powder X-ray diffraction, microstructural and mechanical studies of magnesium matrix composites processed by high energy ball milling. *Journal of Magnesium and Alloys, 6*(4), 390–398. DOI: 10.1016/j.jma.2018.08.002

Rasmidi, R., Duinong, M., & Chee, F. P. (2021). Radiation damage effects on zinc oxide (ZnO) based semiconductor devices – a review. *Radiation Physics and Chemistry, 184*, 109455. DOI: 10.1016/j.radphyschem.2021.109455

Rietveld, H. M. (2021). The Rietveld refinement method: Half of a century anniversary. *Crystal Growth & Design, 21*(8), 4821–4822. DOI: 10.1021/acs.cgd.1c00854

Saba, M., Esmaeil, M., & Saeed, H. (2024). The effect of apple pectin agent on the structural, magnetic, and optical properties of cobalt ferrite nanoparticles synthesized by the auto-combustion sol-gel method. *Materials Chemistry and Physics, 315*, 129015. DOI: 10.1016/j.matchemphys.2024.129015

Sanchez, C.. (2018). Nanomaterials for drug delivery and diagnostics. *Advanced Drug Delivery Reviews, 56*(6), 705–747.

Sato, T., Tanaka, T., & Harada, H. (2019). Structural analysis of materials for performance optimization. *Materials Science and Engineering R Reports*, *138*, 47–56.

Shaba, E. Y., Jacob, J. O., Tijani, J. O., & Suleiman, M. A. T. (2021). A critical review of synthesis parameters affecting the properties of zinc oxide nanoparticle and its application in wastewater treatment. *Applied Water Science*, *11*(2), 48. DOI: 10.1007/s13201-021-01370-z

Shintani, T., & Murata, Y. (2011). Evaluation of the dislocation density and dislocation character in cold rolled Type 304 steel determined by profile analysis of X-ray diffraction. *Acta Materialia*, *59*(11), 4314–4322. DOI: 10.1016/j.actamat.2011.03.055

Sivagami, M., & Asharani, I. V. (2022). Phyto-mediated Ni/NiO NPs and their catalytic applications: A short review. *Inorganic Chemistry Communications*, *145*, 110054. DOI: 10.1016/j.inoche.2022.110054

Smith, D. (2010). X-ray diffraction methods for materials characterization. *Materials Science and Engineering*, *210*(2), 104–109.

Sousa, T. G., Diniz, S. B., Pinto, A. L., & Brandão, L. P. (2015). Dislocation density by X-ray diffraction in α brass deformed by rolling and ECAE. *Materials Research*, *18*(Suppl 2), 246–249. DOI: 10.1590/1516-1439.369214

Sutapa, I. W., Wahid, A. W., Taba, P., & Nafie, N. L. (2018). Dislocation, crystallite size distribution, and lattice strain of magnesium oxide nanoparticles. *Journal of Physics: Conference Series*, *979*, 012021. DOI: 10.1088/1742-6596/979/1/012021

Uma, S., & Shobana, M. K. (2023). Metal oxide semiconductor gas sensors in clinical diagnosis and environmental monitoring. *Sensors and Actuators. A, Physical*, *349*, 114044. DOI: 10.1016/j.sna.2022.114044

Vasconcelos, A. A., Len, T., de Nazaré de Oliveira, A., da Costa, A. A. F., da Silva Souza, A. R., da Costa, C. E. F., Luque, R., da Rocha Filho, G. N., Noronha, R. C. R., & do Nascimento, L. A. S. (2023). Zeolites: A theoretical and practical approach with uses in (bio)chemical processes. *Applied Sciences (Basel, Switzerland)*, *13*(3), 1897. DOI: 10.3390/app13031897

Vázquez, M.. (2020). *Introduction to nanomaterials and nanotechnology: Structure, properties, and applications*. Elsevier.

Wahyuni, E. T., Diantariani, N. P., Kartini, I., & Kuncaka, A. (2022). Enhancement of the photostability and visible photoactivity of ZnO photocatalyst used for reduction of Cr (VI) ions. *Results in Engineering*, *13*, 100351. DOI: 10.1016/j.rineng.2022.100351

Warren, B. E. (1969). *X-ray Diffraction*. Dover Publications.

West, A. R. (2014). *Solid State Chemistry and its Applications*. Wiley.

Williams, D. B., & Carter, C. B. (2009). *Transmission Electron Microscopy: A Textbook for Materials Science*. Springer Science & Business Media. DOI: 10.1007/978-0-387-76501-3

Xu, Z., Wang, W., Li, Y., Meng, S., Liu, Y., Cao, Y., You, J., Zhang, B., & Zang, J. (2024). The influence of solvent-temperature parameters of sodium titanium phosphate anode materials prepared by sol-gel method. *Materials Chemistry and Physics*, *315*, 128946. DOI: 10.1016/j.matchemphys.2024.128946

Yakout, S. M. (2020). Spintronics: Future technology for new data storage and communication devices. *Journal of Superconductivity and Novel Magnetism*, *33*(9), 2557–2580. DOI: 10.1007/s10948-020-05545-8

Yon, V., Rochat, N., Charles, M., Nolot, E., & Gergaud, P. (2020). X-ray diffraction micro-strain analysis for extraction of threading dislocation density of GaN films grown on silicon, sapphire and SiC substrates. [b]. *Physica Status Solidi. B, Basic Research*, *257*(4), 1900579. Advance online publication. DOI: 10.1002/pssb.201900579

Zineb, C. B., Amina, K., Sanaa, E. K., Mohammed, R. R., Amel, B., Khaldoun, B., & Abderrahim, C. B. (2023). Enhancing structural properties of simple perovskite materials based on zirconate and titanate prepared by sol-gel method. *Materials Chemistry and Physics*, *310*, 128460. DOI: 10.1016/j.matchemphys.2023.128460

Zining, Y., Anlong, F., Kangyu, Z., Dmitry, V. K., Qi, G., Jiaojiao, Y., & Lisha, L. (2024). Great enhancement of ferroelectric properties of Al2O3-modified BiFeO3 thin films obtained by sol-gel method. *Journal of the European Ceramic Society*, *44*(1), 224–232. DOI: 10.1016/j.jeurceramsoc.2023.08.058

Chapter 5
Ion Exchange and Zeolites:
Sustainable Approaches to Water Pollution Mitigation

Amita Somya
https://orcid.org/0000-0003-2855-899X
Amity University, India

Amit Prakash Varshney
Independent Researcher, India

ABSTRACT

Water pollution is a critical global issue requiring sustainable treatment methods. Ion exchange and zeolite-based processes are effective solutions due to their efficiency in pollutant removal and alignment with sustainability principles. This chapter explores their role in mitigating water pollution. Ion exchange is widely used for removing heavy metals, nitrates, and harmful ions due to its high selectivity and regenerability. Zeolites, naturally occurring aluminosilicate minerals, have a porous structure, large surface area, and high cation exchange capacity, making them ideal for water treatment. The present chapter examines ion exchange mechanisms and zeolite properties that enhance pollution control. It also highlights advancements in zeolite modification for improved contaminant removal and discusses environmental benefits like reduced chemical usage and lower energy consumption. Finally, future prospects of these technologies in addressing emerging pollutants are explored, emphasizing their role in sustainable water management.

DOI: 10.4018/979-8-3373-0320-8.ch005

1. INTRODUCTION

1.1 Introduction to Water Pollution and Challenges for its Treatment

Rapid urbanization, industrialization, and agricultural growth have made water contamination (Qasem et al., 2021) a serious environmental and public health concern. Water quality (Somya, 2024) is severely deteriorated by contaminants such organic pollutants, heavy metals, phosphates, nitrates, and microbiological diseases, endangering aquatic ecosystems as well as human health. In order to reduce pollution and guarantee safe water supplies, the growing demand for clean water calls on the creation of sustainable, effective, and affordable treatment techniques.

Numerous man-made and natural causes contribute to water pollution, which has a major negative influence on aquatic ecosystems and human health. One of the main causes is industrial discharge, which occurs when manufacturers release wastewater that has been partially or completely treated and contains synthetic colours, hazardous organic compounds, and heavy metals including lead, cadmium, and mercury (Somya, 2023). These contaminants build up in bodies of water, endangering aquatic life as well as people who drink tainted water. Due to nutrient contamination caused by the overuse of chemical fertilizers, pesticides, and herbicides, agricultural runoff is another significant source. Water bodies with high nitrate and phosphate concentrations experience eutrophication, which reduces oxygen levels and disturbs aquatic biodiversity.

Additionally, the discharge of organic materials, germs, medications, and personal care items from sewage and residential wastewater contaminates water. The issue is made worse by poorly maintained sewage systems, which permit untreated wastewater to reach rivers and groundwater supplies, particularly in developing nations. Hazardous substances are also introduced into water sources via landfills and leachates. Heavy metals, plastics, and many other hazardous materials are carried into groundwater by rainwater seeping through landfill debris, rendering it unfit for human use.

Furthermore, by changing oxygen levels and influencing aquatic species' ability to survive, thermal pollution from power plants and industrial facilities that release hot water into lakes and rivers disturb aquatic ecosystems. Transportation-related oil spills and offshore drilling-related marine pollution are major hazards to marine life, polluting large regions and harming the environment over time. The necessity for sustainable water treatment methods, such as ion exchange and zeolite-based filtration (Somya et al., 2025), to reduce contamination and provide clean water for future generations is underscored by the combination of these pollution sources.

The effects of water pollution are severe and widespread, impacting aquatic ecosystems, human health, and the world economy. Since exposure to tainted water can result in deadly illnesses including cholera, dysentery, and hepatitis, human health is among the most affected. The body may store heavy metals like lead and mercury, which can harm organs, cause neurological diseases, and affect a child's development. Nitrates in drinking water can also cause methemoglobinemia, sometimes known as "blue baby syndrome," which lowers the amount of oxygen that is transported in babies' blood.

Water contamination has equally significant effects on the ecosystem. Agricultural runoff that contains too many nutrients causes eutrophication, a condition in which algae blooms lower the oxygen content of water bodies, smothering aquatic life and upsetting food networks. As toxic chemicals and microplastics build up in marine life, they can infiltrate the food chain and have an impact on biodiversity. In addition, industrial discharges and oil spills release persistent organic pollutants (POPs), which can permanently harm ecosystems for decades.

Water pollution reduces access to clean water, particularly in poor nations, by raising the cost of water treatment and purification. It also causes large financial losses for sectors like tourism, agriculture, and fisheries that depend on clean water. In addition to upsetting ecosystems, the loss in fish populations brought on by pollution also jeopardises the livelihoods of people who depend on fishing. Additionally, because governments and people must spend more to cure waterborne infections, dirty water sources raise healthcare costs.

Effective, economical, and ecologically friendly sustainable water treatment technologies (Carolin et al., 2017; Kumar et al., 2020) have been developed as a result of the increasing demand for clean water and the negative impacts of pollution. Conventional techniques such as chemical precipitation, chlorination, and coagulation are efficient, but they can produce hazardous byproducts, demand a lot of energy, or cause secondary contamination. Sustainable technology, on the other hand, aims to reduce waste, energy use, and environmental damage. Reverse osmosis and ultrafiltration are two membrane filtration methods that purify water to a high standard, although they can be costly because of membrane fouling and upkeep. An environmentally acceptable method of eliminating both organic and inorganic pollutants is adsorption employing biochar, activated carbon, and nanomaterials (Somya et al., 2025; Rosaline et al., 2024; Latifian et al., 2023; Nidheesh et al., 2024). Furthermore, although their effectiveness varies depending on the environmental circumstances, bioremediation and phytoremediation - which use microorganisms and plants to break down pollutants - offer a sustainable and ecological option.

Ion exchange and zeolite-based filtration are two of these developments that have drawn a lot of interest because of their excellent selectivity, reusability, and capacity to eliminate a variety of pollutants, such as ammonium ions, nitrates, and heavy metals. While zeolites provide superior adsorption and catalytic qualities due to their microporous structure, ion exchange resins function by specifically substituting hazardous ions with non-toxic ones. Through the regeneration and reuse of materials (Katsou et al., 2011), these technologies not only guarantee efficient pollution removal but also support a circular economy. The future of sustainable water treatment lies in the fusion of hybrid systems, which combine zeolites or ion exchange with nanomaterials, smart sensors, and AI-powered monitoring (Nishu and Kumar, 2023). Water purification may become more effective, affordable, and accessible by using such cutting-edge techniques, which will aid in addressing the problems of pollution and water shortages worldwide.

VOSviewer is a widely utilized software for constructing and examining bibliometric maps in various formats. It is frequently employed in research for co-occurrence analysis (Qin et al., 2022). Figure 1 illustrates a term co-occurrence analysis conducted using VOSviewer (version 1.6.20). In this representation, the node size signifies the frequency of a keyword's appearance, while the closeness between terms indicates the strength of their relationship. Terms positioned nearer to each other exhibit stronger associations. Within the red cluster, prominent terms such as "Adsorption" and "zeolite" demonstrate a high level of research focus. Meanwhile, the green cluster encompasses potential pollution sources, including "Effluents" and "water," among others. Brighter nodes highlight research areas currently receiving considerable attention in this field.

This chapter presents the principles, structures, and applications of ion exchange and zeolite-based materials in the context of reducing water pollution. Through the description of their structural and functional principles, the chapter demonstrates how these materials work at the molecular level to eliminate contaminants from a range of water sources, including heavy metals, hardness ions, dyes, and novel organic pollutants. Ion exchange resins, natural and synthetic zeolites have been highlighted for their increasing importance in contemporary water treatment due to their high ion exchange capacity, selectivity, reusability, and adaptability to various industrial and environmental conditions.

1.2. Ion Exchange: A Promising Water Purification Technique

Ion exchange (Alexandratos, 2008) is a reversible process that substitutes desired ions from a solid-phase exchanger - usually an ion exchange resin or a natural substance like zeolites—for undesired ions in a solution. Ions with comparable charges vie for exchange sites according to variables including ionic charge, hydrated radius, and selectivity coefficient in this process, which is controlled by electrostatic attraction.

Figure 1. Co-occurrence overlay visualization of keywords

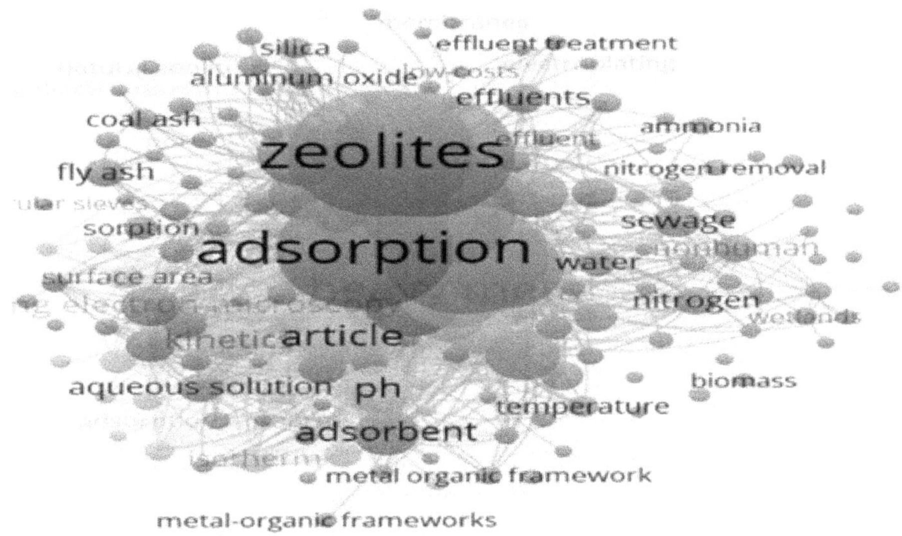

Anion exchange resins have positively charged groups (like quaternary ammonium) that eliminate anions like NO_3^-, SO_4^{2-}, and Cl^-, whereas cation exchange resins have negatively charged functional groups (like sulfonates) that draw in and replace positively charged contaminants like Pb^{2+}, Ca^{2+}, and Fe^{3+}. Ion exchange is a flexible technique for treating industrial effluents, eliminating heavy metals from wastewater, and cleaning drinking water. Its effectiveness is dependent on multiple factors, including temperature, pH, resin type, and ion concentration. An important benefit of ion exchange is its regenerability, which ensures sustainability and cost-effectiveness by restoring depleted resins with the use of appropriate regenerants (such as NaOH for anion resins and NaCl for cation resins).

Ion exchange is a sustainable and very successful water treatment technology that has a number of benefits over traditional techniques. Its great selectivity, which enables the targeted removal of particular contaminants including ammonium (NH_4^+), nitrates (NO_3^-), and heavy metals (Pb^{2+}, Cd^{2+}, and Cr^{3+}) even at low concentrations, is one of its most important advantages. Because of its selectivity, the method effectively eliminates harmful compounds from the water without compromising other vital ions (Somya et al., 2015). Renewability, or the ability to reuse ion exchange resins repeatedly by appropriate regeneration with the right chemicals (e.g., NaCl for cation exchange and NaOH for anion exchange), is another significant benefit. Compared to alternative treatment techniques like chemical precipitation, this property lowers waste formation and makes ion exchange more economical.

Furthermore, ion exchange generates less sludge than precipitation or coagulation, which require the disposal of substantial volumes of solid waste. Additionally, the method is flexible and scalable, which makes it appropriate for both large-scale industrial wastewater treatment facilities and small-scale residential water filters. Furthermore, ion exchange systems can be used with other treatment techniques including membrane filtration, adsorption, and sophisticated oxidation processes to improve overall performance. They also function well in a variety of environments, including those with fluctuating pH and temperature. In order to further reduce its negative effects on the environment, ion exchange technology can be paired with biodegradable or bio-based resins. Because of these benefits, ion exchange is a recommended option for wastewater cleanup and sustainable water purification.

Reusing and regenerating ion exchange materials is one of the main benefits of ion exchange technology, which makes the procedure sustainable and economical. Ion exchange resins or zeolites lose their effectiveness over time due to saturation with undesirable ions. A regeneration procedure is utilized to restore their functioning, in which the absorbed pollutants are replaced with new exchangeable ions using a regenerant solution (such as sodium chloride (NaCl), hydrochloric acid (HCl), sulphuric acid (H_2SO_4), or sodium hydroxide (NaOH)). Heavy metals and other cations are frequently displaced by sodium (Na^+) or hydrogen (H^+) ions in cation exchange resins.

Oppositely, the ability of anion exchange resins to eliminate anionic pollutants like nitrates and sulphates is also restored by regenerating them with hydroxide (OH^-) or chloride (Cl^-) ions. The lifespan of ion exchange materials can be greatly increased and operating costs can be decreased by repeating the regeneration process. Furthermore, developments in green regeneration methods that make use of hybrid nanomaterials and eco-friendly treatments reduce chemical waste and their negative effects on the environment. In order to support a circular economy strategy, recovered metal ions and other contaminants from the regeneration process can also be retrieved and used again in industries. By lowering waste and resource consumption, appropriate regeneration techniques not only preserve the effectiveness of ion exchange systems but also support sustainable water treatment methods.

1.3. Zeolites: The Natural Water Purifiers

Zeolites are microporous aluminosilicate minerals (Salahudeen, 2022) that can be found naturally or artificially. They are distinguished by their distinctive three-dimensional crystalline structure, which consists of a network of interconnected pores and channels. These materials are very successful for gas separation, water purification, and catalysis because they are made of silicon (Si), aluminium (Al), and oxygen (O) atoms that form a rigid framework that can trap and exchange ions.

The two main categories of zeolites employed for water treatment are natural and synthetic. Figure 2 depicts categorization of various zeolite types along with their benefits (Kumari et al., 2024). Each type has certain qualities that enable it to effectively remove pollutants from water. Natural zeolites, including clinoptilolite, mordenite, chabazite, and phillipsite, are inexpensive, widely accessible, and highly effective at eliminating ammonium (NH_4^+) and heavy metals (such as Pb^{2+}, Cd^{2+}, and Cu^{2+}) from wastewater. Because of its exceptional cation exchange capability and durability in a variety of environmental situations, clinoptilolite is the most widely used of them. Because of its tiny pore structure, mordenite works very well at adsorbing organic contaminants and smaller ions.

Synthetic zeolites (Derbe et al., 2021; Cheng et al., 2016), on the other hand, like Zeolite A, Zeolite X, Zeolite Y, and Zeolite P, are designed to have improved ion exchange capabilities, uniform pore diameters, and higher purity. Because of its great affinity for calcium and magnesium ions, zeolite A is frequently employed in industrial wastewater treatment and softening applications. Zeolite X and Zeolite Y (Sellaoui et al., 2021) are members of the faujasite family and are very good at adsorbing heavy metals and organic pollutants because of their greater pore diameters. In wastewater treatment plants, zeolite P is frequently used to remove phosphate and ammonium (Abukhadra et al., 2020). Both synthetic and natural zeolite varieties are essential for effective and sustainable water filtration systems, and the selection of zeolite is influenced by variables like temperature, pH, regeneration capacity, and pollutant type.

Figure 2. Categorization of various zeolite types along with their benefits

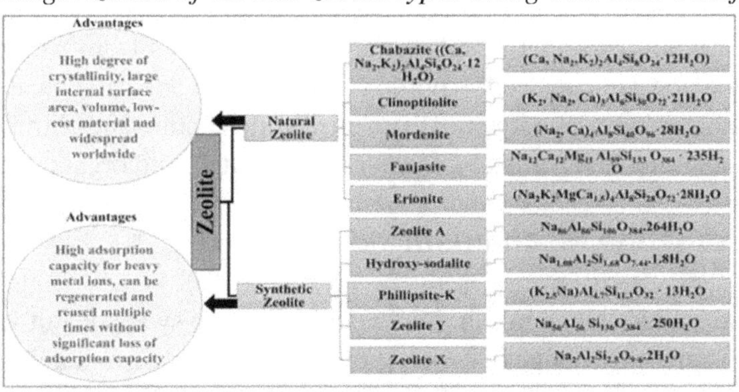

Zeolites can be produced (Khaleque et al., 2020; Nguyen et al., 2023; Latifian et al., 2023) in a variety of ways that replicate their natural production, but under carefully monitored circumstances, to obtain desired characteristics including particular pore diameters, crystallinity, and ion exchange capacity. The most used process is hydrothermal synthesis, which involves heating a mixture of silica (SiO_2) and alumina (Al_2O_3) sources plus a mineralizing agent (such NaOH or KOH) to moderate temperatures (80–200°C) in an autoclave. Zeolite crystal nucleation and growth are facilitated by this mechanism over a period of hours or days. Another method for creating extremely pure and homogeneous zeolites is sol-gel synthesis, which generates customized pore architectures by regulating the gelation of precursor solutions. Because microwave-assisted synthesis can drastically cut down on synthesis time without sacrificing efficiency or crystallinity, it has drawn interest. This technique uses microwave radiation to speed up the kinetics of the reaction, which causes the crystal to develop more quickly.

Another cutting-edge method is template-assisted synthesis, which produces mesoporous or hierarchical zeolites with improved adsorption and catalytic qualities by using organic templates (such as surfactants or structure-directing agents). Furthermore, an environmentally beneficial substitute that reduces waste production is ionothermal synthesis, which makes use of ionic liquids rather than traditional solvents. The type of zeolite needed, its use in water treatment, and financial concerns all influence the synthesis method selection; research is still being done to optimize these techniques for sustainable, large-scale production.

2. MECHANISMS OF REMOVAL OF CONTAMINANTS USING ZEOLITES AND ION EXCHANGE

Zeolites and ion exchange resins remove contaminants from water through a combination of ion exchange, adsorption, size exclusion, and catalytic degradation. These mechanisms work synergistically to efficiently capture and eliminate various pollutants, including heavy metals, ammonium, nitrates, and organic compounds.

2.1. Ion Exchange Mechanism

Ion exchange, in which undesirable ions in water are swapped out for less hazardous ions from the exchange material, is the main method of contaminant removal in both zeolites and ion exchange resins (Joseph et al., 2020; Somya et al., 2024). Zeolites attract and exchange cations like Pb^{2+}, Cd^{2+}, NH_4^+, and Ca^{2+} with Na^+, K^+, or H^+ from their structure because of their negatively charged aluminosilicate framework, as revealed in Figure 3. Functional groups like quaternary ammonium

($-NR_4^+$) for anion exchange and sulfonic acid ($-SO_3H$) for cation exchange are also present in synthetic ion exchange resins. Ionic charge, hydrated radius, and exchanger material affinity are some of the variables that affect ion exchange selectivity. The technique is very effective in eliminating hazardous metals (Somya, 2024) from industrial effluent because higher-charged ions (such as $Pb^{2+} > Na^+$) have a tendency to attach more firmly.

2.2. Adsorption Mechanism

Additionally, zeolites and ion exchange resins have adsorption capabilities (Jiang et al., 2020; Khan et al., 2021; Luo et al., 2024), signifying that contaminants adhere to their active sites and large surface area pores. Because zeolites have micro- and mesoporous structures, both organic and inorganic contaminants can be trapped in them. Within the porous framework, π-π interactions and Van der Waals forces can be used to eliminate hydrophobic organic contaminants, including pesticides and medications. Figure 4 depicts common adsorption mechanisms of copious pollutants on zeolites.

Surfactants play a significant role in modifying the adsorption mechanism (Somya et al., 2018) on ion-exchanger and zeolite surfaces by altering surface charge, hydrophobicity, and ion-exchange capacity. Depending on their nature—cationic, anionic, or nonionic—surfactants can either enhance or inhibit pollutant adsorption (Somya, 2019). Cationic surfactants, for instance, can increase the positive charge on zeolite surfaces, improving the adsorption of anionic contaminants through electrostatic attraction. Conversely, anionic surfactants may compete with negatively charged pollutants, reducing adsorption efficiency. Nonionic surfactants can form micellar structures on the surface, impacting the accessibility of active sites. Overall, surfactants influence pollutant interactions with zeolite, affecting selectivity and adsorption capacity in water treatment and remediation processes. Surfactant-modified ion exchangers and zeolites (SMZs) enhance surface charge attraction, significantly improving the adsorption of different types of pollutants (Somya, 2023). By modifying the surface with cationic surfactants, these materials alter their hydrophobicity and electrostatic interactions, promoting stronger binding of negatively charged contaminants. The formation of surfactant bilayers or monolayers on the zeolite surface creates additional adsorption sites, increasing selectivity and capacity for pollutant removal. This modification expands the applicability of ion exchangers and zeolites in wastewater treatment, enabling the efficient removal of various organic and inorganic pollutants through ion exchange and surface adsorption mechanisms (Somya et al., 2017; Somya and Maheria, 2020; Somya and Singh, 2021).

2.3. Size Exclusion and Molecular Sieving

Zeolites act as molecular sieves, rejecting bigger pollutants while allowing only molecules or ions smaller than their pore size to get through (Wang et al., 2021; Liu et al., 2020; Ghojavand et al., 2022). This method of selective filtration works especially well at eliminating organic contaminants, ammonium, and heavy metals from wastewater. Synthetic zeolites are very customizable for water filtration applications because their pore size can be designed to target particular impurities.

Figure 3. Ion exchange mechanism for the removal of heavy metals from wastewater using zeolites

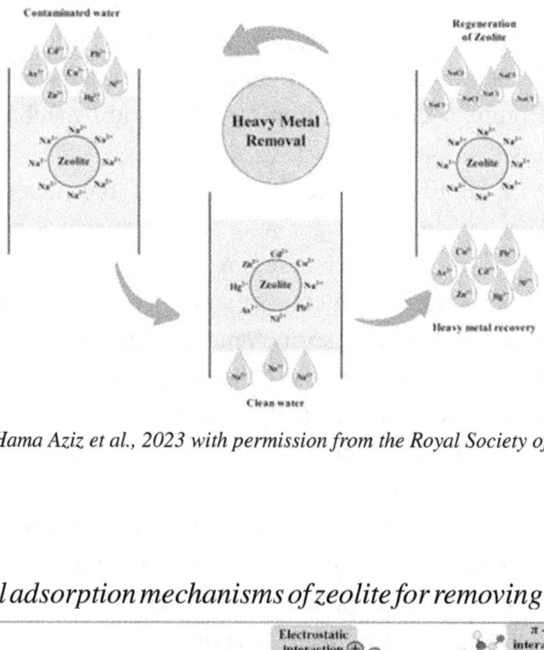

(Reproduced from Hama Aziz et al., 2023 with permission from the Royal Society of Chemistry)

Figure 4. Potential adsorption mechanisms of zeolite for removing various pollutants

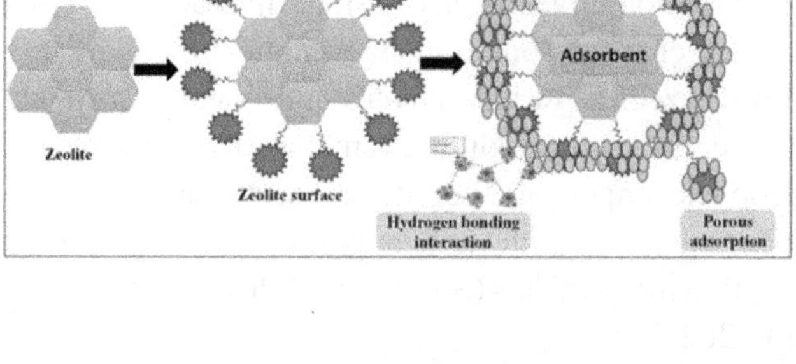

2.4. Catalytic Degradation and Redox Reactions

Because of their catalytic qualities, some zeolites and modified ion exchange resins can break down contaminants through oxidation-reduction (redox) reactions (Kumar et al., 2023). To break down organic pollutants like colours, medications, and pesticides into innocuous byproducts, for instance, zeolites coated with metal oxides (such as Fe, Mn, or Ti) function as catalysts in advanced oxidation processes (AOPs) (Andana et al., 2021; He et al., 2024). When paired with nanomaterials such as TiO_2, certain zeolites augment photocatalytic degradation, enabling the elimination of persistent contaminants when exposed to UV or visible light.

3. APPLICATIONS OF ION EXCHANGE AND ZEOLITES IN WATER POLLUTION MITIGATION

Ion exchange and zeolites play a crucial role in addressing water pollution by efficiently removing contaminants such as heavy metals, nitrates, ammonium, organic pollutants, and industrial waste compounds (Somya, 2020; Turan and Ergun, 2009). Their high selectivity, regenerability, and eco-friendly nature make them ideal for various water treatment applications. The key applications include-

3.1. Removal of Heavy Metals from Industrial Wastewater

Because mining, electroplating, battery manufacturing, and textile production release substantial levels of hazardous heavy metals into aquatic environments, these industrial processes are major contributors to water pollution. Lead (Pb^{2+}), cadmium (Cd^{2+}), chromium (Cr^{3+}/Cr^{6+}), copper (Cu^{2+}), zinc (Zn^{2+}), nickel (Ni^{2+}), and mercury (Hg^{2+}) are among the metals that are extremely persistent, non-biodegradable, and poisonous, even at low densities. Through acid mine drainage, where sulphide minerals oxidize and dissolve metals into nearby water bodies, mining operations discharge heavy metals into the environment. High amounts of chromium, nickel, and cadmium are contributed by the electroplating and metal finishing industries and are frequently utilized in coatings that resist corrosion. If not adequately managed, wastewater from battery production processes that are high in lead, mercury, and cadmium can seep into groundwater. The textile industry releases copper and chromium into wastewater streams, especially when they use metal-based dyes and mordants.

Ion exchange resins and zeolites are frequently employed to reduce this pollution because of their great selectivity, effectiveness, and regenerability. In order to reduce metal concentrations in wastewater, ion exchange resins, which are made of synthetic polymer matrices with functional groups like sulfonic acid (-SO_3H), carboxyl (-COOH), and quaternary ammonium (-NR_4^+), selectively replace toxic metal ions with safer ones like sodium (Na^+) or hydrogen (H^+). The negatively charged aluminosilicate structure of zeolites, especially clinoptilolite, chabazite, and mordenite (Grifasi et al., 2024; Ghojavand et al., 2024; Wakchaure et al., 2025), promotes cation exchange and enables them to efficiently absorb and hold onto metal ions. These materials are very effective at eliminating dangerous metal pollutants because of their excellent adsorption capabilities and high cation exchange capacity (CEC).

Industries can reduce their ecological footprint and adhere to strict environmental standards by combining zeolite-based treatment systems with ion exchange. By drastically lowering the levels of heavy metals in released water, these technologies shield human health, groundwater supplies, and aquatic life from harmful exposure. Furthermore, developments in functionalized zeolites and modified resins, such as those covered with nanoparticles and bio-based materials, improve the sustainability and removal efficiency of these materials.

3.2. Ammonium and Nitrate Removal from Municipal Wastewater

The main sources of high nitrate (NO_3^-) and ammonium (NH_4^+) contamination are industrial discharges, household wastewater, and agricultural runoff. These pollutants have grown to be a serious environmental concern. Increased nitrogen levels in surface and groundwater are caused by the overuse of fertilizers, animal dung, and sewage effluents. Ammonium is frequently found in wastewater from septic systems, sewage treatment plants, and industrial effluents, whereas nitrates from synthetic fertilizers seep into aquifers in agricultural areas. These nitrogen compounds cause eutrophication, which results in excessive algae growth, oxygen depletion, and the decline of aquatic life, when they enter aquatic habitats. Furthermore, excessive quantities of nitrate in drinking water can cause methemoglobinemia (blue baby syndrome) in babies and other health problems, making it very dangerous.

Anion and cation exchange resins, together with natural zeolites, are commonly employed in water treatment systems to remove ammonium (Gagliano et al., 2022; Sarioglu, 2025) in order to combat this pollution. In order to prevent nitrate buildup in drinking water, anion exchange resins efficiently remove nitrate (NO_3^-) ions by substituting them with chloride (Cl^-) ions. These resins are especially useful for purifying municipal water supplies and groundwater in regions where agricultural runoff is present. Similar to this, cation exchange resins are made to absorb ammonium (NH_4^+) ions and replace them with hydrogen (H^+) or sodium (Na^+) ions. This greatly lowers the amount of ammonium in wastewater.

Natural zeolites, like clinoptilolite, are very useful in wastewater treatment facilities, aquaculture systems, and industrial water recycling procedures because of their high cation exchange capacity (CEC) and significant affinity for ammonium ions. Furthermore, nitrate removal selectivity is improved by modified zeolites, such as surfactant-modified and nano-enhanced zeolites (El Hanache et al., 2019; Schick et al., 2011), which makes them a viable substitute for traditional biological treatments like denitrification. Zeolite-based and ion exchange techniques provide low-maintenance, effective, and environmentally friendly nitrogen removal alternatives to biological treatments, which call for certain microbe conditions and energy-intensive procedures. By incorporating these materials into water treatment systems, sustainable water management is promoted by reducing the dangers to public health, groundwater contamination, and eutrophication.

3.3. Softening of Hard Water

High calcium (Ca^{2+}) and magnesium (Mg^{2+}) ion concentrations in hard water lead to scaling in water lines, boilers, water heaters, and home appliances. Scaling is the result of these dissolved minerals precipitating and forming deposits, which lowers water flow, increases energy use, and shortens the life of industrial equipment and plumbing systems. Additionally, hard water makes soaps and detergents less efficient, which results in overuse of soap and residue accumulation on textiles and surfaces.

Ion exchange water softeners are frequently employed in residential, commercial, and municipal water treatment systems to help address these problems. The sodium-based cation exchange resins found in these softeners, which are usually composed of sulfonated polystyrene beads, exchange sodium (Na^+) ions for calcium (Ca^{2+}) and magnesium (Mg^{2+}) ions in the water in a selective manner. Hard water is successfully softened as it flows through the resin bed, which absorbs hardness ions and releases sodium. A strong sodium chloride ($NaCl$) brine solution is needed for regeneration when the resin gets saturated with calcium and magnesium over time. This restores the resin's exchange capacity.

Zeolites, especially synthetic Zeolite A, are employed extensively in water softening (Cinar & Beler-Baykal, 2005; Wang & Peng, 2010) in addition to ion exchange resins because of their high cation exchange capacity (CEC) and potent affinity for hardness ions. Zeolites preserve water quality and inhibit the formation of scale by specifically adsorbing Ca^{2+} and Mg^{2+}. Because of their great durability, thermal stability, and capacity to operate under a range of pH settings, they are frequently utilized in laundry detergent, boiler feed water treatment, and commercial water purification systems (Onyutha et al., 2024). Zeolites are also reusable and regenerable, which makes them an economical and sustainable substitute for conventional ion exchange systems.

Using ion exchange and zeolite-based water softening techniques greatly increases heating system efficiency, lowers maintenance costs, and prolongs the life of industrial and domestic equipment. The efficacy of modified and nano-enhanced zeolites is continuously being improved, providing long-term solutions for the treatment of hard water in commercial as well as residential settings.

3.4. Removal of Radioactive Contaminants

One of the most important problems in nuclear waste management and environmental cleanup is the safe containment and disposal of radioactive pollutants. Because of their lengthy half-lives, high mobility, and serious health concerns, radioactive isotopes including cesium-137 (Cs^{+}), strontium-90 (Sr^{2+}), and uranium (UO_2^{2+}) that are frequently found in nuclear power plant effluents, weapon development sites, and radioactive accidents offer serious threats. These radionuclides have the potential to contaminate the environment widely and bioaccumulate in living things if they are not adequately confined. They can also penetrate surface and groundwater.

Zeolites and ion exchange resins are widely employed to absorb and immobilize radioactive materials (Jiménez-Reyes et al., 2021; Ambrosino et al., 2024) in order to address this problem and stop them from spreading throughout the environment. Because of their high cation exchange capacity (CEC), molecular sieving capabilities, and structural stability under radiation, natural zeolites—clinoptilolite in particular—show a remarkable affinity for caesium and strontium. These minerals efficiently trap radioactive ions inside their porous structure by preferentially exchanging their native ions (such Na or K) for radioactive ions. Furthermore, nuclear power reactors, reprocessing facilities, and decontamination systems use synthetic zeolites with improved selectivity to extract particular radionuclides from liquid waste streams and cooling water.

Likewise, in nuclear wastewater treatment, uranium, plutonium, and other transuranic elements are captured using ion exchange resins, such as chelating resins and strong acid cation resins. To ensure effective separation and long-term confinement, these resins have functional groups that combine with radioactive metal ions to create stable complexes. Particularly useful in post-Fukushima decontamination operations, selective resins, such those based on ammonium molybdophosphate (AMP), are particularly good at binding caesium ions.

Because of their thermal and chemical resilience, which enables them to hold onto trapped radionuclides even in harsh environments, zeolites and ion exchange resins are important tools for treating radioactive waste. After becoming saturated, these materials can be protected from further leaching and disposed of safely in geological repositories by encasing them in glass or cement matrices through a process called solidification and vitrification.

The effectiveness of radioactive decontamination is increased by integrating zeolite-based barriers (Zhang et al., 2022), ion exchange filtering systems, and hybrid remediation techniques; as a result, these materials are essential for radioactive waste management, atmospheric restoration, and disaster response activities. In order to mitigate radioactive pollution in a safer and more efficient manner, ongoing research in functionalized ion exchange materials and nano-engineered zeolites seeks to further increase selectivity, durability, and scalability.

3.5. Treatment of Textile and Dye Industry Wastewater

Large volumes of synthetic dyes, organic pollutants, and heavy metals are released into water bodies by the textile and dye industries, making them one of the worst causes of water pollution. In addition to hazardous substances including surfactants, heavy metals (Cr^{6+}, Pb^{2+}, Cd^{2+}, and Cu^{2+}), and toxic residues from dyeing, bleaching, and finishing operations, these effluent streams frequently contain reactive, acidic, and basic dyes. Although dyes can be removed using well-known photocatalytic degradation methods (Al-Zahrani et al., 2023), many synthetic dyes are designed to be highly durable, non-biodegradable, and resistant to oxidation and light. As a result, they are frequently difficult to remove using standard treatment techniques like biodegradation and basic filtering. When dye-laden effluents are released without treatment, they cause serious water pollution, lower oxygen levels, bioaccumulation, and possible health risks for both people and wildlife.

In order to address this issue, ion exchange resins and modified zeolites have become very successful materials for wastewater treatment that remove dyes (Wang and Zhu, 2006; Rakanović et al., 2022). Zeolites have improved adsorption capacity for both cationic (basic dyes) and anionic (acid, direct, and reactive dyes) pollutants, especially those treated with cationic surfactants or metal oxides. Through the processes of ion exchange, electrostatic attraction, and molecular sieving, these materials absorb dye molecules and lower the wastewater's overall colour intensity, COD, and BOD levels (Gadore et al., 2024). Zeolite Y and Zeolite A are examples of synthetic zeolites that are particularly helpful for treating dye effluents because of their high porosity, huge surface area, and capacity for regeneration and reuse.

In addition, ion exchange resins—both cation and anion exchange varieties—are essential for the removal of metal and colour. The positively charged dyes and metal ions are eliminated by cation exchange resins, whereas negatively charged dye molecules are efficiently adsorbed by anion exchange molecules. Furthermore, chelation-based ion exchange resins are frequently used to remove harmful heavy metals like chromium (Cr^{6+}) from tanning industry effluents. These metals are known to be highly hazardous, to cause cancer, and to have the potential to contaminate groundwater. These resins ensure safer discharge into the environment by specifically exchanging poisonous metal ions with less dangerous ones, such Na^+ or H^+.

Zeolites and ion exchange resins are integrated into textile wastewater treatment (Chang et al., 2002; Giraldo-Loaiza et al., 2024) facilities to improve pollutant removal efficiency and to provide reusable, economical, and environmentally friendly treatment options. Furthermore, the adsorption and regeneration characteristics of functionalized and nanostructured zeolites have been further enhanced by developments, making them a viable substitute for sophisticated oxidation and conventional chemical coagulation procedures. Businesses can greatly lower environmental pollution, adhere to wastewater discharge standards, and support sustainable water management by implementing zeolite-based and ion exchange-based filtering systems.

3.6. Removal of Organic Contaminants and Pharmaceuticals

A significant environmental and public health concern nowadays is the existence of emerging contaminants (ECs) in water sources. These pollutants, which come from industrial discharge, hospital waste, residential wastewater, and agricultural runoff, include hormones, pesticides, pharmaceutical residues, endocrine-disrupting chemicals (EDCs), and personal care products (PCPs). They are frequently difficult to fully remove using traditional wastewater treatment techniques such sedimentation, filtration, and biological degradation because of their high persistence, low biodegradability, and probable toxicity. The buildup of these contaminants in aquatic environments can result in antibiotic resistance, wildlife hormone abnormalities, and long-term health hazards for people due to contaminated drinking water.

To solve this problem, ion exchange resins with hydrophobic functional groups and modified-surface zeolites have shown great efficacy in adsorbing and eliminating ECs from water (Somya et al., 2023). By covering natural or synthetic zeolites with cationic surfactants (such as hexadecyltrimethylammonium bromide, or HDTMA), surfactant-modified zeolites (SMZs) improve the adsorption of negatively charged organic contaminants such pesticides, antibiotics, and pharmaceutical residues (Li and Bowman, 1998; Araneta et al, 2024; Mittal et al., 2024). Because of their altered hydrophobic surface, SMZs are able to capture and hold onto organic molecules that would normally evade standard treatment. These contaminants are especially well-captured by Zeolite Y, Zeolite Beta, and mordenite because of their large surface area, adjustable pore structure, and potent attraction for organic substances.

Likewise, ion exchange resins—particularly those containing hydrophobic or functionalized groups (such as quaternary ammonium, sulfonic acid, or chelating ligands)—are frequently employed to extract pesticides, hormones, antibiotics, and other persistent organic contaminants from water. Hydrophobic adsorption, ion exchange, and electrostatic interactions are the ways in which these resins bind organic pollutants. Remainders of personal care products and hydrophobic EDCs are captured by non-polar resins, whereas strong anion exchange resins are very efficient against negatively charged insecticides and medications.

When zeolite-based and ion exchange-based filtration systems are used in water treatment facilities, the removal efficiency of ECs is greatly increased, lowering the dangers to human health and the environment. Furthermore, by dissolving or immobilizing persistent contaminants, hybrid techniques that combine zeolite adsorption, ion exchange, and advanced oxidation processes (AOPs) enhance water purification even more. The efficiency and sustainability of new pollutant removal technologies are being improved by research into nanostructured zeolites, molecularly imprinted ion exchange resins, and bio-based adsorbents. Ensuring safer drinking water, limiting the buildup of EDCs in aquatic environments, and advancing sustainable water management techniques globally all depend on the implementation of these creative solutions.

3.7. Desalination and Brackish Water Treatment

With desalination, fresh water from brackish and saltwater sources is produced, making it a vital solution to the world's water shortage. Scaling ions, dissolved organics, and suspended particles, however, frequently reduce the effectiveness of desalination operations, especially reverse osmosis (RO) and capacitive deionization (CDI). Through the removal of troublesome ions, the prevention of fouling, and the reduction of energy consumption, ion exchange resins and zeolites are essential for improving the efficiency of these systems.

Calcium (Ca^{2+}), magnesium (Mg^{2+}), iron (Fe^{3+}), and silica are examples of scaling ions that can precipitate on membrane surfaces during RO desalination. This can result in decreased water flux, higher energy needs, and membrane breakdown. In order to replace these scale-forming ions with sodium (Na^{+}) or hydrogen (H^{+}) ions that do not produce scaling, ion exchange resins—especially potent acid cation resins—are utilized in the pre-treatment phase. Likewise, anion exchange resins aid in the removal of bicarbonate (HCO_3^{-}) and sulphate (SO_4^{2-}) ions, reducing the possibility of scaling and extending membrane life. Desalination plants can improve total water recovery rates, cut maintenance costs, and require less chemical dosing by integrating ion exchange into RO feedwater treatment.

In desalination technologies, zeolites (Dong et al., 2015; Abukhadra et al., 2022; Rasyid et al., 2021) are also being used more and more because of their excellent ion exchange capacity, molecular sieving capabilities, and adsorption effectiveness. By acting as ion-capturing materials, zeolites enhance charge storage and selectivity for the removal of salt from brackish water in capacitive deionization (CDI), a new low-energy desalination technique. Through the specific elimination of divalent and monovalent salts, the reduction of electrode degradation, and the increase in ion adsorption rates, nano-engineered and modified zeolites improve CDI performance. These developments make CDI a viable substitute for energy-efficient, low-cost desalination in areas with saline water supplies.

Zeolites are employed in hybrid desalination systems in addition to membrane pre-treatment and CDI. In these systems, they serve as adsorbents for trace pollutants including fluoride, boron, and arsenic that are challenging to eliminate using traditional RO. Furthermore, studies are looking into how to improve salt rejection and lower energy usage in membrane distillation and forward osmosis systems by integrating zeolite-based membranes.

By lowering chemical waste and energy requirements, the use of ion exchange resins and zeolites in desalination not only increases operational efficacy and cost-effectiveness but also supports sustainable water management. Improvements in functionalized resins, nano-zeolites, and bio-inspired materials will further strengthen desalination technology's role in supplying dependable, clean water supplies around the world.

4. FINAL THOUGHTS

4.1 Challenges

Despite their success in reducing water pollution, ion exchange resins and zeolites still face a number of environmental, financial, and technical obstacles that must be overcome for broad use and long-term viability-

- When ion exchange resins need to be periodically regenerated with chemical solutions (such as NaCl, HCl, or NaOH), secondary waste brine is produced, which must be properly disposed of or treated.
- Zeolites, especially natural ones, need to be replaced or regenerated frequently due to their limited ion exchange capability, which can cause waste buildup and disposal problems.
- Without further adjustments, conventional ion exchange resins might not effectively remove new contaminants including medications, endocrine-disrupting chemicals (EDCs), and microplastics.
- Zeolites frequently show preferential adsorption, which means that in complicated wastewater matrices with competing ions, their effectiveness may be diminished.

- Ion exchange resins and zeolites may experience scaling, biofouling, and organic fouling during desalination and wastewater treatment, which lowers their efficacy and raises maintenance expenses.
- Frequent cleaning and chemical treatments are necessary to clear clogged pores caused by oil, grease, and microbial growth.
- Synthetic and modified zeolites (such as nanostructured, surfactant-modified, and functionalized versions) are costly due to the need for specialized processing, but natural zeolites are plentiful and reasonably priced.
- Similarly, the application of sophisticated ion exchange resins with customized functional groups for selective pollutant removal is limited in developing regions due to their high synthesis and operating costs.
- Petroleum-based polymers, which make up a large portion of ion exchange resins, have the potential to break down over time, generating microplastics and adding to the plastic pollution problem.
- To provide sustainable water treatment solutions, it is still difficult to develop biodegradable or bio-based polymers.

4.2 Future Prospects

With continuous research aimed at improving their effectiveness, sustainability, and affordability, ion exchange and zeolite-based technologies in water treatment appear to have a bright future. To increase the selectivity and adsorption capability for new contaminants such as pharmaceuticals, endocrine-disrupting chemicals (EDCs), and microplastics, advanced functionalized resins and nano-engineered zeolites are being created. Enhanced pollutant removal and reduced energy usage are anticipated with the incorporation of hybrid treatment systems that combine ion exchange with membrane filtration, advanced oxidation processes (AOPs), and capacitive deionization (CDI). To lessen the environmental impact of chemical regenerants, efforts are also being made to use green regeneration techniques as electrochemical and CO_2-based regeneration. Furthermore, natural polymers like chitosan, cellulose, and alginate are being used to create biodegradable and bio-based ion exchange resins, which are becoming more sustainable than petroleum-based resins and lowering the issue of plastic waste. There is also growing interest in the possibility of waste valorization, such as recycling discarded zeolites and resins for construction materials, soil remediation, or catalytic uses. Furthermore, real-time water monitoring and adaptive treatment systems may undergo a revolution thanks to the development of smart materials like sensor-integrated, programmable, self-cleaning resins and zeolites. These developments will be essential to developing sustainable, decentralized, and energy-efficient water purification systems as the world's water problems worsen, guaranteeing safe and clean water for household and commercial usage.

4.3 Conclusion

Ion exchange and zeolite-based technologies have proven to be highly effective in mitigating water pollution by removing heavy metals, nitrates, ammonium, hardness ions, and emerging contaminants from wastewater and drinking water sources. Their high selectivity, regeneration capability, and adaptability make them indispensable for industries, municipal water treatment, and environmental remediation. However, challenges such as scaling, regeneration waste, high production costs, and selective removal efficiency must be addressed to ensure their long-term sustainability. Advancements in nano-engineered materials, hybrid treatment approaches, biodegradable resins, and green regeneration methods are paving the way for more efficient and eco-friendly solutions. As water scarcity and pollution continue to threaten global ecosystems and human health, further research and innovation in ion exchange and zeolite-based water treatment will be essential in achieving sustainable, cost-effective, and scalable clean water solutions for the future. In order to ensure safe, clean water access worldwide, this chapter not only highlights the current capabilities of ion exchange and zeolites in water filtration but also offers the groundwork for their potential as long-term, affordable, and scalable solutions. The goal has been to provide the reader a thorough grasp of the concepts and potential of these materials to encourage more research and development of sustainable water treatment methods.

ACKNOWLEDGEMENT

The author is grateful and thankful to the Chancellor, Vice-Chancellor, Director, Amity University, Bengaluru for the above project work. Thanks are also due to Miss. Anushi Varshney and Mr. Anirudh for supporting to compile this work.

REFERENCES

Abukhadra, M. R., Ali, S. M., Nasr, E. A., Mahmoud, H. A. A., & Awwad, E. M. (2020). Effective Sequestration of Phosphate and Ammonium Ions by the Bentonite/Zeolite Na–P Composite as a Simple Technique to Control the Eutrophication Phenomenon: Realistic Studies. In ACS Omega (Vol. 5, Issue 24, pp. 14656–14668). American Chemical Society (ACS). https://doi.org/DOI: 10.1021/acsomega.0c01399

Abukhadra, M. R., Eid, M. H., El-Sherbeeny, A. M., Abd Elgawad, A. E. E., & Shim, J.-J. (2022). Effective desalination of brackish groundwater using zeolitized diatomite/kaolinite geopolymer as low-cost inorganic membrane; Siwa Oasis in Egypt as a realistic case study. In *Journal of Contaminant Hydrology* (Vol. 244, p. 103923). Elsevier BV., DOI: 10.1016/j.jconhyd.2021.103923

Al-Zahrani, S. A., Patil, M. B., Mathad, S. N., Patil, A. Y., Al Otaibi, A., Masood, N., Mansour, D., Khan, A., Gupta, V., Topare, N. S., Somya, A., & Ayyar, M. (2023). Photocatalytic Azo Dye Degradation Using Graphite Carbon Nitride Photocatalyst and UV-A Irradiation. In Crystals (Vol. 13, Issue 4, p. 577). MDPI AG. https://doi.org/DOI: 10.3390/cryst13040577

Alexandratos, S. D. (2008). Ion-Exchange Resins: A Retrospective from Industrial and Engineering Chemistry Research. In Industrial & Engineering Chemistry Research (Vol. 48, Issue 1, pp. 388–398). American Chemical Society (ACS). https://doi.org/DOI: 10.1021/ie801242v

Ambrosino, F., Gargiulo, N., Peruta, G. D., Gravino, C., Gagliardo, G., Pisciotta, F., Esposito, S., La Verde, G., Caputo, D., & Pugliese, M. (2024). Study and characterization of zeolites for the removal of artificial radionuclides in wastewater samples from nuclear power plants. In *Journal of Hazardous Materials Advances* (Vol. 16, p. 100458). Elsevier BV., DOI: 10.1016/j.hazadv.2024.100458

Andana, T., Rappé, K. G., Gao, F., Szanyi, J., Pereira-Hernandez, X., & Wang, Y. (2021). Recent advances in hybrid metal oxide–zeolite catalysts for low-temperature selective catalytic reduction of NOx by ammonia. In Applied Catalysis B: Environmental (Vol. 291, p. 120054). Elsevier BV. https://doi.org/DOI: 10.1016/j.apcatb.2021.120054

Araneta, M. J. P., Gaw, E. A. P., Seagan, C. G. V., Dolina, M. N. E., Bonto, A. P., Casalme, E. C. G., Salvador, A. A., & Nuñez, J. A. P. (2024). Enhanced Removal of Acetylsalicylic Acid Using Surfactant-Modified Natural Zeolites. In Trends in Sciences (Vol. 22, Issue 2, p. 8908). College of Graduate Studies, Walailak University. https://doi.org/DOI: 10.48048/tis.2025.8908

Carolin, C. F., Kumar, P. S., Saravanan, A., Joshiba, G. J., & Naushad, Mu. (2017). Efficient techniques for the removal of toxic heavy metals from aquatic environment: A review. In Journal of Environmental Chemical Engineering (Vol. 5, Issue 3, pp. 2782–2799). Elsevier BV. https://doi.org/DOI: 10.1016/j.jece.2017.05.029

Chang, W.-S., Hong, S.-W., & Park, J. (2002). Effect of zeolite media for the treatment of textile wastewater in a biological aerated filter. In Process Biochemistry (Vol. 37, Issue 7, pp. 693–698). Elsevier BV. https://doi.org/DOI: 10.1016/S0032-9592(01)00258-8

Cheng, W. P., Gao, W., Cui, X., Ma, J. H., & Li, R. F. (2016). Phenol adsorption equilibrium and kinetics on zeolite X/activated carbon composite. In Journal of the Taiwan Institute of Chemical Engineers (Vol. 62, pp. 192–198). Elsevier BV. https://doi.org/DOI: 10.1016/j.jtice.2016.02.004

Cinar, S., & Beler-Baykal, B. (2005). Ion exchange with natural zeolites: an alternative for water softening? In Water Science and Technology (Vol. 51, Issue 11, pp. 71–77). IWA Publishing. https://doi.org/DOI: 10.2166/wst.2005.0392

Derbe, T., Temesgen, S., & Bitew, M. (2021). A Short Review on Synthesis, Characterization, and Applications of Zeolites. In M. Fedel (Ed.), Advances in Materials Science and Engineering (Vol. 2021, Issue 1). Wiley. https://doi.org/DOI: 10.1155/2021/6637898

Dong, H., Zhao, L., Zhang, L., Chen, H., Gao, C., & Winston Ho, W. S. (2015). High-flux reverse osmosis membranes incorporated with NaY zeolite nanoparticles for brackish water desalination. In *Journal of Membrane Science* (Vol. 476, pp. 373–383). Elsevier BV., DOI: 10.1016/j.memsci.2014.11.054

El Hanache, L., Lebeau, B., Nouali, H., Toufaily, J., Hamieh, T., & Daou, T. J. (2019). Performance of surfactant-modified *BEA-type zeolite nanosponges for the removal of nitrate in contaminated water: Effect of the external surface. In *Journal of Hazardous Materials* (Vol. 364, pp. 206–217). Elsevier BV., DOI: 10.1016/j.jhazmat.2018.10.015

Gadore, V., Mishra, S. R., Yadav, N., Yadav, G., & Ahmaruzzaman, Md. (2024). Advances in zeolite-based materials for dye removal: Current trends and future prospects. In *Inorganic Chemistry Communications* (Vol. 166, p. 112606). Elsevier BV., DOI: 10.1016/j.inoche.2024.112606

Gagliano, E., Sgroi, M., Falciglia, P. P., Belviso, C., Cavalcante, F., Lettino, A., Vagliasindi, F. G. A., & Roccaro, P. (2022). Removal of ammonium from wastewater by zeolite synthetized from volcanic ash: Batch and column tests. In Journal of Environmental Chemical Engineering (Vol. 10, Issue 3, p. 107539). Elsevier BV. https://doi.org/DOI: 10.1016/j.jece.2022.107539

Ghojavand, S., Coasne, B., Clatworthy, E. B., Guillet-Nicolas, R., Bazin, P., Desmurs, M., Jacobo Aguilera, L., Ruaux, V., & Mintova, S. (2022). Alkali Metal Cations Influence the CO2 Adsorption Capacity of Nanosized Chabazite: Modeling vs Experiment. In ACS Applied Nano Materials (Vol. 5, Issue 4, pp. 5578–5588). American Chemical Society (ACS). https://doi.org/DOI: 10.1021/acsanm.2c00537

Giraldo-Loaiza, C., Salazar-Loaiza, A. M., Sandoval-Barrera, M. A., Macías-Quiroga, I. F., Ocampo-Serna, D. M., & Sanabria-González, N. R. (2024). Integration of Ion Exchange—AOP—Biological System for the Treatment of Real Textile Wastewater. In ChemEngineering (Vol. 8, Issue 4, p. 76). MDPI AG. https://doi.org/DOI: 10.3390/chemengineering8040076

Grifasi, N., Ziantoni, B., Fino, D., & Piumetti, M. (2024). Fundamental properties and sustainable applications of the natural zeolite clinoptilolite. In *Environmental Science and Pollution Research*. Springer Science and Business Media LLC., DOI: 10.1007/s11356-024-33656-5

Hama Aziz, K. H., Mustafa, F. S., Omer, K. M., Hama, S., Hamarawf, R. F., & Rahman, K. O. (2023). Heavy metal pollution in the aquatic environment: efficient and low-cost removal approaches to eliminate their toxicity: a review. In RSC Advances (Vol. 13, Issue 26, pp. 17595–17610). Royal Society of Chemistry (RSC). https://doi.org/DOI: 10.1039/D3RA00723E

He, J., Deng, J., Lan, T., Liu, X., Shen, Y., Han, L., Wang, J., & Zhang, D. (2024). Strong metal oxide-zeolite interactions during selective catalytic reduction of nitrogen oxides. In *Journal of Hazardous Materials* (Vol. 465, p. 133164). Elsevier BV., DOI: 10.1016/j.jhazmat.2023.133164

Jiang, N., Shang, R., Heijman, S. G. J., & Rietveld, L. C. (2020). Adsorption of triclosan, trichlorophenol and phenol by high-silica zeolites: Adsorption efficiencies and mechanisms. In *Separation and Purification Technology* (Vol. 235, p. 116152). Elsevier BV., DOI: 10.1016/j.seppur.2019.116152

Jiménez-Reyes, M., Almazán-Sánchez, P. T., & Solache-Ríos, M. (2021). Radioactive waste treatments by using zeolites. A short review. In *Journal of Environmental Radioactivity* (Vol. 233, p. 106610). Elsevier BV., DOI: 10.1016/j.jenvrad.2021.106610

Joseph, J., Radhakrishnan, R. C., Johnson, J. K., Joy, S. P., & Thomas, J. (2020). Ion-exchange mediated removal of cationic dye-stuffs from water using ammonium phosphomolybdate. In *Materials Chemistry and Physics* (Vol. 242, p. 122488). Elsevier BV., DOI: 10.1016/j.matchemphys.2019.122488

Katsou, E., Malamis, S., Tzanoudaki, M., Haralambous, K. J., & Loizidou, M. (2011). Regeneration of natural zeolite polluted by lead and zinc in wastewater treatment systems. In Journal of Hazardous Materials (Vol. 189, Issue 3, pp. 773–786). Elsevier BV. https://doi.org/DOI: 10.1016/j.jhazmat.2010.12.061

Khaleque, A., Alam, M. M., Hoque, M., Mondal, S., Haider, J. B., Xu, B., Johir, M. A. H., Karmakar, A. K., Zhou, J. L., Ahmed, M. B., & Moni, M. A. (2020). Zeolite synthesis from low-cost materials and environmental applications: A review. In *Environmental Advances* (Vol. 2, p. 100019). Elsevier BV., DOI: 10.1016/j.envadv.2020.100019

Khan, S., Idrees, M., & Bilal, M. (2021). Revealing and elucidating chemical speciation mechanisms for lead and nickel adsorption on zeolite in aqueous solutions. In Colloids and Surfaces A: Physicochemical and Engineering Aspects (Vol. 623, p. 126711). Elsevier BV. https://doi.org/DOI: 10.1016/j.colsurfa.2021.126711

Kumar, M., Ambika, S., Hassani, A., & Nidheesh, P. V. (2023). Waste to catalyst: Role of agricultural waste in water and wastewater treatment. In *Science of The Total Environment* (Vol. 858, p. 159762). Elsevier BV., DOI: 10.1016/j.scitotenv.2022.159762

Kumar, M., Xiong, X., Sun, Y., Yu, I. K. M., Tsang, D. C. W., Hou, D., Gupta, J., Bhaskar, T., & Pandey, A. (2020). Critical Review on Biochar-Supported Catalysts for Pollutant Degradation and Sustainable Biorefinery. In Advanced Sustainable Systems (Vol. 4, Issue 10). Wiley. https://doi.org/DOI: 10.1002/adsu.201900149

Kumari, S., Chowdhry, J., Kumar, M., & Chandra Garg, M. (2024). Zeolites in wastewater treatment: A comprehensive review on scientometric analysis, adsorption mechanisms, and future prospects. In *Environmental Research* (Vol. 260, p. 119782). Elsevier BV., DOI: 10.1016/j.envres.2024.119782

Latifian, B., Taghizade Firozjaee, T., & Abdi, J. (2023). Synthesis of biochar-iron oxide magnetic nanocomposite for efficient and rapid removal of Pb(II) heavy metal from water. In Water Practice & Technology (Vol. 19, Issue 1, pp. 181–199). IWA Publishing. https://doi.org/DOI: 10.2166/wpt.2023.223

Li, Z., & Bowman, R. S. (1998). Sorption of Perchloroethylene by Surfactant-Modified Zeolite as Controlled by Surfactant Loading. In Environmental Science & Technology (Vol. 32, Issue 15, pp. 2278–2282). American Chemical Society (ACS). https://doi.org/DOI: 10.1021/es971118r

Liu, J., Huang, Z., Sun, J., Zou, Y., & Gong, B. (2020). Enhancing the removal performance of Cd(II) from aqueous solutions by NaA zeolite through doped thiourea reduced GO which is trapped within zeolite crystals. In *Journal of Alloys and Compounds* (Vol. 815, p. 152514). Elsevier BV., DOI: 10.1016/j.jallcom.2019.152514

Luo, Q., Wei, J., Guo, Z., & Song, Y. (2024). Adsorption and immobilization of phosphorus from water and sediments using a lanthanum-modified natural zeolite: Performance, mechanism and effect. In *Separation and Purification Technology* (Vol. 329, p. 125187). Elsevier BV., DOI: 10.1016/j.seppur.2023.125187

Mittal, A., Nagpal, M., Sharma, D., Das, D., & Chahar, J. (2024). A Comprehensive Review of Natural and Surfactant- Modified Zeolites and Their Applications. *Letters in Applied NanoBioScience.*, *13*(4), 188. DOI: 10.33263/LIANBS134.188

Nguyen, D.-K., Dinh, V.-P., Dang, N. T., Khan, D. T., Hung, N. T., & Thi Tran, N. H. (2023). Effects of aging and hydrothermal treatment on the crystallization of ZSM-5 zeolite synthesis from bentonite. In RSC Advances (Vol. 13, Issue 30, pp. 20565–20574). Royal Society of Chemistry (RSC). https://doi.org/DOI: 10.1039/D3RA02552G

Nidheesh, P. V., Kumar, M., Venkateshwaran, G., Ambika, S., & Bhaskar, S. Vinay, & Ghosh, P. (2024). Conversion of locally available materials to biochar and activated carbon for drinking water treatment. In Chemosphere (Vol. 353, p. 141566). Elsevier BV. https://doi.org/DOI: 10.1016/j.chemosphere.2024.141566

Kumar, S. (2023). Smart and innovative nanotechnology applications for water purification. *Hybrid Advances*, *3*, 100044.

Onyutha, C., Okello, E., Atukwase, R., Nduhukiire, P., Ecodu, M., & Kwiringira, J. N. (2024). Improving household water treatment: using zeolite to remove lead, fluoride and arsenic following optimized turbidity reduction in slow sand filtration. In Sustainable Environment Research (Vol. 34, Issue 1). Springer Science and Business Media LLC. https://doi.org/DOI: 10.1186/s42834-024-00209-x

Qasem, N. A. A., Mohammed, R. H., & Lawal, D. U. (2021). Removal of heavy metal ions from wastewater: a comprehensive and critical review. In npj Clean Water (Vol. 4, Issue 1). Springer Science and Business Media LLC. https://doi.org/DOI: 10.1038/s41545-021-00127-0

Qin, F., Li, J., Zhang, C., Zeng, G., Huang, D., Tan, X., Qin, D., Tan, H., 2022. Biochar in the 21st century: a data-driven visualization of collaboration, frontier identification, and future trend. Sci. Total Environ. 818, 151774 https://doi.org/. scitotenv.2021.151774.DOI: 10.1016/j

Rakanović, M., Vukojević, A., Savanović, M. M., Armaković, S., Pelemiš, S., Živić, F., Sladojević, S., & Armaković, S. J. (2022). Zeolites as Adsorbents and Photocatalysts for Removal of Dyes from the Aqueous Environment. In Molecules (Vol. 27, Issue 19, p. 6582). MDPI AG. https://doi.org/DOI: 10.3390/molecules27196582

Rasyid, F., & Masri, M. (2019). Desalination of sea water using activated zeolite by cloride acid in tropical area based on column ion exchange method. *AMINA*, *1*(3), 114–119.

Rosaline, D. R., Daphne, V., Srividya, G., Nivetha, P., Jarvin, M., Ashok Kumar, S., Manikandan, A., Inbanathan, S. S. R., Dinesh, A., Thanrasu, K., Raja, K. K., Almessiere, M. A., Slimani, Y., Baykal, A., Khan, A., Asiri, A. M., Topare, N. S., & Somya, A. (2024). Magnetic nanocomposites for biomedical and environmental applications. In *Magnetic Nanoparticles and Polymer Nanocomposites* (pp. 225–287). Elsevier., DOI: 10.1016/B978-0-323-85748-2.00011-6

Salahudeen, N. (2022). A Review on Zeolite: Application, Synthesis and Effect of Synthesis Parameters on Product Properties. In Chemistry Africa (Vol. 5, Issue 6, pp. 1889–1906). Springer Science and Business Media LLC. https://doi.org/DOI: 10.1007/s42250-022-00471-9

Sarioglu, M. (2005). Removal of ammonium from municipal wastewater using natural Turkish (Dogantepe) zeolite. In Separation and Purification Technology (Vol. 41, Issue 1, pp. 1–11). Elsevier BV. https://doi.org/DOI: 10.1016/j.seppur.2004.03.008

Schick, J., Caullet, P., Paillaud, J.-L., Patarin, J., & Mangold-Callarec, C. (2011). Nitrate sorption from water on a Surfactant-Modified Zeolite. Fixed-bed column experiments. In Microporous and Mesoporous Materials (Vol. 142, Issues 2–3, pp. 549–556). Elsevier BV. https://doi.org/DOI: 10.1016/j.micromeso.2010.12.039

Sellaoui, L., Hessou, E. P., Badawi, M., Netto, M. S., Dotto, G. L., Silva, L. F. O., Tielens, F., Ifthikar, J., Bonilla-Petriciolet, A., & Chen, Z. (2021). Trapping of Ag+, Cu2+, and Co2+ by faujasite zeolite Y: New interpretations of the adsorption mechanism via DFT and statistical modeling investigation. In *Chemical Engineering Journal* (Vol. 420, p. 127712). Elsevier BV., DOI: 10.1016/j.cej.2020.127712

Somya, A. (2019). Surfactant based hybrid ion exchangers. *Research Journal of Chemistry and Environment*, *23*(3), 96.

Somya, A. (2020). Hybrid Ion Exchangers. In Hybrid Nanomaterials - Flexible Electronics Materials. IntechOpen. https://doi.org/DOI: 10.5772/intechopen.92116

Somya, A. (2023). Metal Phosphates: Their Role as Ion Exchangers in Water Purification. In Metal Phosphates and Phosphonates (pp. 341–356). Springer International Publishing. https://doi.org/DOI: 10.1007/978-3-031-27062-8_19

Somya, A. (2023). N-Cetyl-N N. N-tri methyl ammonium bromide based tin (IV) phosphate, a new surfactant based hybrid ion exchanger: Synthesis, ion exchange and physico-chemical characterization. *Research Journal of Chemistry and Environment*, *27*(6), 18–23. DOI: 10.25303/2706rjce018023

Somya, A. (2024). An investigation study on water quality of few lakes of Bengaluru. In *Industry 4.0 with Modern Technology* (pp. 200–204). CRC Press., DOI: 10.1201/9781003450924-37

Somya, A. (2024). Hybrid cation exchangers: A boon for analytical chemistry. Journal of Applied Optics, 263–270. Retrieved from https://appliedopticsjournal.net/index.php/JAO/article/view/111

Somya, A., Khan, A., Hussein, M. A., Al-Romaizan, A. N., Malash, A. M., Topare, N. S., & Asiri, A. M. (2023). Biological and environmental studies on pectin-zirconium (IV) phosphate, a biopolymer doped hybrid cation exchanger. In *International Journal of Biological Macromolecules* (Vol. 250, p. 125803). Elsevier BV., DOI: 10.1016/j.ijbiomac.2023.125803

Somya, A., Khan, A., & Varshney, A. P. (2025). Nanocomposites and Their Role in Wastewater Treatment. In Singh, L. (Ed.), *Design, Fabrication, and Significance of Advanced Nanostructured Materials* (pp. 75–100). IGI Global Scientific Publishing., DOI: 10.4018/979-8-3693-5320-2.ch004

Somya, A., Kumar, S., & K., S. (2024). An exploration of AFLC-14-doped ceric phosphate, a novel liquid crystal hybrid ion exchanger: synthesis, ion exchange studies and analytical insights. In Materials Research Innovations (pp. 1–9). Informa UK Limited. https://doi.org/DOI: 10.1080/14328917.2024.2396655

Somya, A., & Maheria, K. (2020). Studies on polyoxyethylene octyl phenyl ether supported thorium(IV) phosphates: a new cation exchange material. In Bulletin of Materials Science (Vol. 43, Issue 1). Springer Science and Business Media LLC. https://doi.org/DOI: 10.1007/s12034-020-02153-z

Somya, A., Rafiquee, M. Z. A., & Upadhyaya, V. (2018). Role of surfactants in the adsorption studies of n-butyl acetate-cerium (IV) phosphate for some alkaline earths and heavy metal ions. *Research Journal of Chemistry and Environment*, *22*(6), 45.

Somya, A., Rafiquee, M. Z. A., Varshney, K. G., & Al-Lohedan, H. A. (2015). Synthesis, characterization and analytical applications of N-dodecyl pyridinium chloride-cerium (IV) phosphate fibrous ion exchanger: selective for Hg(II) and its binary separations. In Desalination and Water Treatment (Vol. 56, Issue 3, pp. 638–645). Elsevier BV. https://doi.org/DOI: 10.1080/19443994.2014.940388

Somya, A., & Singh, M. (2022). Studies on Sodium Lauryl Sulphate-supported Thorium (IV) Phosphate: A New Surfactant-supported Cation Exchange Resin, Useful in Water Purification. In Current Analytical Chemistry (Vol. 18, Issue 3, pp. 383–390). Bentham Science Publishers Ltd. https://doi.org/DOI: 10.2174/157 3411017666210804102748

Somya, A., & Upadhyaya, V. (2017). Synthesis, Characterization and Adsorption behaviour of N-dodecyl pyridinium chloride based Tin (IV) phosphate: A new surfactant based hybrid ion exchanger. *Research Journal of Chemistry and Environment*, *21*(1), 18.

Somya, A., Varshney, A. P., & Thakur, A. (2025). Hybrid Ion Exchange Materials Useful in Water Purification. In Amrousse, R., Bachar, A., Mabrouk, A., & Azat, S. (Eds.), *Innovative Materials for Environmental and Aerospace Applications* (pp. 363–388). IGI Global Scientific Publishing., DOI: 10.4018/979-8-3373-0669-8.ch010

Turan, N. G., & Ergun, O. N. (2009). Removal of Cu(II) from leachate using natural zeolite as a landfill liner material. In Journal of Hazardous Materials (Vol. 167, Issues 1–3, pp. 696–700). Elsevier BV. https://doi.org/DOI: 10.1016/j.jhazmat.2009.01.047

Wakchaure, S. G., Dabhade, G. B., Borhade, A. V., Malpure, A. A., Palde, S., Shewale, Y., & Yannam, R. (2025). Mordenite's mineralogical study and application as an efficient adsorbent for heavy metal detection from waste water. In Journal of the Indian Chemical Society (Vol. 102, Issue 1, p. 101502). Elsevier BV. https://doi.org/DOI: 10.1016/j.jics.2024.101502

Wang, S., & Peng, Y. (2010). Natural zeolites as effective adsorbents in water and wastewater treatment. In Chemical Engineering Journal (Vol. 156, Issue 1, pp. 11–24). Elsevier BV. https://doi.org/DOI: 10.1016/j.cej.2009.10.029

Wang, S., & Zhu, Z. (2006). Characterisation and environmental application of an Australian natural zeolite for basic dye removal from aqueous solution. In Journal of Hazardous Materials (Vol. 136, Issue 3, pp. 946–952). Elsevier BV. https://doi.org/DOI: 10.1016/j.jhazmat.2006.01.038

Wang, X., Yan, N., Xie, M., Liu, P., Bai, P., Su, H., Wang, B., Wang, Y., Li, L., Cheng, T., Guo, P., Yan, W., & Yu, J. (2021). The inorganic cation-tailored "trapdoor" effect of silicoaluminophosphate zeolite for highly selective CO2 separation. In Chemical Science (Vol. 12, Issue 25, pp. 8803–8810). Royal Society of Chemistry (RSC). https://doi.org/DOI: 10.1039/D1SC00619C

Zhang, Y., Cao, B., Yin, H., Meng, L., Jin, W., Wang, F., Xu, J., & Al-Tabbaa, A. (2022). Application of zeolites in permeable reactive barriers (PRBs) for in-situ groundwater remediation: A critical review. In *Chemosphere* (Vol. 308, p. 136290). Elsevier BV., DOI: 10.1016/j.chemosphere.2022.136290

Chapter 6
Comparison of ZSM-5 and MgO-Al_2O_3 Nanosheet Supports and Their Performance in Hydrodesulfurization

Luis Perez-Cabrera
https://orcid.org/0000-0001-9896-3419
Centro de Nanociencias y Nanotecnología Universidad Nacional Autónoma de México, Mexico

Victor Alfredo Reyes-Villegas
Centro de Investigación Científica y de Educación Superior de Ensenada, Mexico

Jesus Isaías De León-Ramírez
https://orcid.org/0000-0003-1393-610X
Centro de Investigación Científica y de Educación Superior de Ensenada, Mexico

Joel Antúnez-García
Universidad Autónoma de Baja California, Mexico

Rosario Isidro Yocupicio-Gaxiola
Instituto Tecnológico Superior de Guasave, Mexico

Miguel Angel Hernández-Espinosa
Benemerita Universidade Autónoma de Puebla, Mexico

Viridiana Evangelista-Hernández
https://orcid.org/0009-0009-4306-3638
Universidad Autónoma de Baja California, Mexico

Armando Reyes-Serrato
https://orcid.org/0000-0001-6741-8813
Universidad Nacional Autónoma de México, Mexico

Sergio Fuentes-Moyado
Universidad Nacional Autónoma de México, Mexico

DOI: 10.4018/979-8-3373-0320-8.ch006

ABSTRACT

This chapter developed pure and iron-modified ZSM-5 by soft template synthesis—CTABr functions as a layered director. ZSM-5 was post-pillaring by TEOS to create interlamellar mesopores. Pillared ZSM-5 and iron-modified ZSM-5 were efficient catalytic supports for NiMoS used in the HDS of DBT reaction, which is significant for clean fuel production. From both NiMoS-supported, ZSM-5 and MgO-Al_2O_3, ZSM-5 exhibited better HDS performance. Al_2O_3-MgO oxides were obtained from hydrotalcite with different Al/(Al+Mg) ratios. Characterization of the catalysts by XRD, UV-Vis, and TEM indicated that a good dispersion of the sulfide phase was obtained due to the specific acid-base properties of the supports. The most active catalyst in the reaction of desulfurization of dibenzothiophene was the one presenting the highest acidity along the MgAl-x series, which was alumina due to improved dispersion of the nickel-promoted MoS_2 crystallites (low stacking and shorter lengths). All catalysts on Al_2O_3-MgO binary oxides exhibited better selectivity to the hydrogenation pathway than the alumina catalyst.

1. INTRODUCTION

Developing novel porous-ordered catalyst supports for the hydrodesulfurization (HDS) process is an important challenge considering the growing global demand for diesel and gasoline and stringent fuel quality regulations. It is necessary to develop transition metal sulfide catalysts (TMSs) to produce ultra-low sulfur diesel ULSD and gasolines (EN 590:2009, S content; 10 ppm) (Stanislaus, Marafi, & Rana, 2010; S. Zhang, Liu, Zu, Wang, & Li, 2025). Low-sulfur fuels are urgently necessary for a healthier and more sustainable future.

From 2000 to the present, a concerted global scientific effort has borne fruit with robust catalysts developed and commercialized, like Nebula in 2004. Nebula catalyst is a well-ordered porous material composed primarily of nickel, molybdenum, and tungsten. However, its use requires more stringent hydrogen consumption, pressure, and operating temperature conditions (Kerby, Degnan, Marler, & Beck, 2005). (S. Zhang et al., 2025).

The global market for refinery catalysts has seen significant growth, from \$2.32 billion in 2001 to nearly \$2.65 billion in 2005, with an annual growth rate of 3.6% (Silvy, 2004). The estimated demand for oil refinery catalysts approached \$6 billion in 2013, with an annual growth rate of 5.7% (de Jong, 2009). In 2014, the global catalyst revenue was \$17.2 billion, especially since the cost of refinery catalysts is less than 0.1% of the revenue generated by the products in which they are used (Hagen, 2015).

Concerning the influence of the porous support in HDS reactions, new supports include nanostructured and mesostructured silica (Zepeda, 2025), zeolites micro-mesoporous zeolites (Na, Park, Seo, & Ryoo, 2011), and others (Suresh et al., 2017; Zepeda et al., 2008).

MFI, also known as ZMS-5 or silicalite-1, has an orthorhombic crystal structure. MFI micropores are characterized by two intersecting 10-ring channels (5.3 x 5.6 Å, 5.1 x 5.5 Å). These channels typically limit the diffusion of catalytic products, resulting in the stagnation of macromolecular products in the pores, eventually blocking them and deactivating zeolite. However, incorporating meso- and macropores has broadened its applications (Fu, Chang, Shao, & Li, 2017), including adsorption processes, catalysis, thin films, and membranes. It is also a model system in zeolite crystal growth studies (Zhou, Jiang, Chen, Wang, & Hou, 2021). Zeolites, particularly ZSM-5, can convert methanol into C9-C10 chain olefins and adjust their size under reaction conditions, opening up exciting possibilities for converting larger molecules like benzene and naphthalene (Timoshev et al., 2024). Its main application is as a catalytic cracking catalyst in petroleum refineries, which handle more than 20 million barrels of oil daily (Samanta & Das, 2021). ZSM-5 zeolite presents application in fine chemistry, advanced medical technology, as catalysts to convert methane to liquid hydrocarbons (Anggoro & Amin, 2006), methanol to gasoline (Y.-H. Chen, Han, Zhang, & Cui, 2020; Fu et al., 2017), catalytic cracking in petroleum refining (Tian et al., 2024), benzene alkylation (producing ethylbenzene and 2-phenylpropane). In particular, iron-modified ZSM-5 zeolites are applied as catalysts in selective transformation reactions such as oxidation of benzene to produce phenol (Koekkoek, Xin, Yang, Li, & Hensen, 2011), selective catalytic oxidation of ammonia to dinitrogen (Borcuch et al., 2020), oxidation of methane to methanol, transformation of ethanol to produce hydrogen (Suriya et al., 2024), and selective dehydrogenation of hydrocarbons to olefines (Tian et al., 2024).

However, developing new methods for preparing silica-based materials with ordered mesopores (called hierarchical) has favored the incorporation of mesopores in ZSM-5 as well (in addition to micropores) and thus has broadened its applications. Mesopores are created by incorporating organic soft templates, been known as quaternary amines such as TPABr (Hernández, Petranovskii, & Quiroz, 2018) (Demikhova et al., 2023) or CTABr (Shelyapina et al., 2020), into the synthesis process. CTABr induce stacked sheet growth and the formation of materials with the desired intercrystalline mesoporosity. A strategy that makes a difference in the development of zeolites is to obtain zeolites in the form of nanometer-sized lamellae; Ryoo et al. 2009 obtained MFI zeolite in the form of lamellae; such lamellae can be arranged with pillars between the lamellae using SiO_2 pillars and lamellar thickness of 2 nm (Na et al., 2011).

Figure 1 (left) shows that using zeolites such as ZSM-5 with nanosheet morphology and tunable pore size properties holds great potential for industrial applications. **Figure 1** also shows the use of hydrocarbon chains to enhance mesopore pillars intercalated between nanosheet zeolites.

Figure 1. Advantages and catalytic performance of nanosheet/pillared ZSM-5 in various potential industrial reactions.

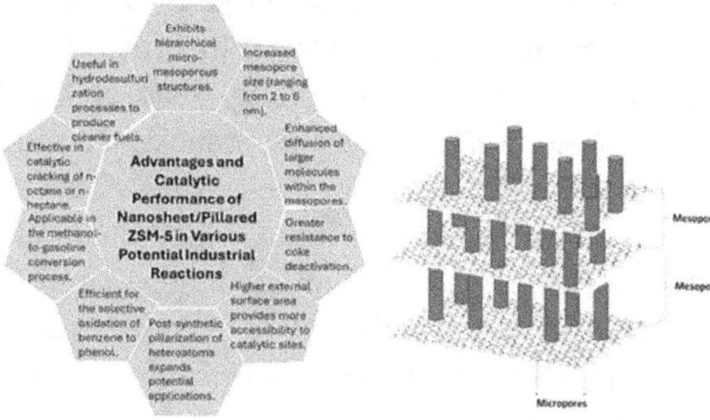

The role of TMSs in the HDS process is crucial, as they facilitate the conversion of organosulfur compounds into hydrogen sulfide and clean hydrocarbon fuels . Alkyldibenzothiophene-type molecules (such as 4,6 DBT) found in heavy crudes pose an important challenge to ultra-deep desulfurization. Their methyl groups hinder the interaction of sulfur with the catalyst surface (Zepeda, 2025). The HID pathway becomes dominant (Egorova & Prins, 2004; Stanislaus et al., 2010) in line with theoretical calculations indicate that the activation barrier for C-S bond scission increases as one moves from thiophene to DBT and almost doubles for 4,6-DMDBT relative to thiophene (Prabhu et al., 2025).

Mesoporous materials, such as ZSM-5 with silica pillars, could lead to interesting developments in their application as TMS supports to catalyze the transformation of 4,6-DMDBT (Tanimu & Alhooshani, 2019) through isomerization (Breysse, Geantet, Afanasiev, Blanchard, & Vrinat, 2008). Enhancing porosity by introducing silica pillars into the ZSM-5 lamellar support increases the BET area and reduces the overall acidity of the catalyst (Fermoso et al., 2016). Then, it is crucial to moderate such acidity to increase the catalyst's lifetime by enhancing its tolerance to coke poisoning due to the accumulation of hydrocarbon molecules (Breysse et al., 2008). (Alabdullah et al., 2020). (Hartmann, Thommes, & Schwieger, 2021).

In the past, we reported the synthesis of CoMo catalysts for hydrodesulfurizing dibenzothiophene reaction as a function of morphological effect and heteroatom substitution on KIT-6 supports (Suresh et al., 2017). The interconnected pores of

KIT-6 seem to play a crucial role in preparing active catalysts, facilitating the dispersion of these nano-sized CoMoS species.The present work primarily evaluates the catalysts' ability to remove sulfur from dibenzothiophene. However, it does not delve into studying other key factors for transforming molecules more resistant to such removal, such as alkyl-dibenzothiophenes. It does not analyze their capacity for hydrogenation or isomerization. The potential for future, more in-depth studies with more refractory molecules is promising and could lead to significant advancements in the field. This aspect is important in our research and should be carefully considered in future studies.This work compares the ZSM-5 pillared nanosheet materials with other typical nanosheet materials known as hydrotalcite. We compare ZSM-5 with Mg-Al hydrotalcite double-layered compounds. Nanosheet materials could be a promising alternative to promote the formation of molybdenum sulfide nanoparticles with high dispersion and exposed reactive sites, paving the way for future advancements in catalyst development.

Previous results on layered compounds MgAl mixed oxides showed an important correlation about the optimal selectivity/activity ratio at Al(Al+Mg) molar ratio (x=0.25) (Pérez, & Linares, 2010; Linares et al., 2015; R. Zhao, Yin, Zhao, & Liu, 2003). In a different research, Wu et al. reported that CoMo/MgO-Al_2O_3 synthesized from calcined hydrotalcite with x=0.20 molar ratio, have a higher activity for the HDS of DBT than CoMo/Al_2O_3 and CoMo/MgO respectively (Li Bao Wu, 2010; Wu, Jiao, Wang, Chen, & Cao, 2009). Guevara-Lara et al. suggest that NiMo sulfides supported in binary oxides (MgO with lower Al content ~5%) exhibited higher activity for the HDS of DBT than NiMo/Al_2O_3 (Guevara-Lara et al., 2010). They also reported that acid-basic support such as Al_2O_3-MgO increase the hydrogenolysis pathway and this effect is correlated with the presence of polymolybdates structures in the precursor oxide, which governs the dispersion of the Ni-Mo-S active phases (Guevara-Lara et al., 2010). López-Cruz et al. encounter that the increased hydrogenation activity of NiMoS nanoslabs supported on acid-basic materials (with MgO-Al_2O_3 oxide) is due to an electronic enrichment induced by the support through an electron transfer phenomenon that favors the formation of S vacancy sites on the MoS_2 active sites, where reactive molecules (methylnaphthalene) are adsorbed and activated(López-Cruz, Guzman, Cao, Martínez, & Corma, 2021).

This research provides a comprehensive comparison of pure versus iron-modified ZSM-5 mesoporous materials for the hydrodesulfurization of the dibenzothiophene reaction. The study also compares catalysts with Fe modifier, Ni, and Co promoters supported by lamellar silica-pillared zeolites -ZSM5. This comparison provides valuable insights into the performance of different catalysts, enhancing the knowledge of hydrotreatment catalysts.

Additionally, we prepared five nickel-promoted MoS_2 sulfides supported on binary traditional Al_2O_3-MgO oxides obtained from calcined layered double hydrotalcite with different Al/(Al+Mg) ratios. Characterization of the catalysts by XRD, UV-Vis, and TEM indicated that a good dispersion of the sulfide phase was obtained due to the specific acid-base properties of the supports. The most active catalyst in the reaction of desulfurization of dibenzothiophene was the one presenting the highest acidity along the MgAl-x series, which was pure alumina due to improved dispersion of the nickel-promoted MoS_2 crystallites (low stacking and shorter lengths). All catalysts on Al_2O_3-MgO binary oxides exhibited better selectivity to the hydrogenation pathway than the alumina catalyst.

2. EXPERIMENTAL PROCEDURE

2.1. Supports and Catalysts Preparation

For the hierarchical zeolite ZSM-5 synthesis, we used the structure-directing agents TPABr and PEG-20000, which induce the formation of the microporous structure of the MFI zeolite and cetyltrimethylammonium bromide (CTAB), which favors the growth in the form of stacked sheets, as well as the formation of inter-crystalline mesopores as previously report with some modifications (Shelyapina et al., 2020; Yocupicio-Gaxiola, Petranovskii, Antúnez-García, & Fuentes Moyado, 2019). In a typical experiment, we prepared an aqueous solution by dissolving 6.2 g of CTABr, 0.52 g of PEG20000, 2.66 g of TPABr, and 0.47 g of NaOH in 36.3 mL of deionized water. Then, 21.46 g of sodium silicate (25% w/w SiO_2 and 10.6% w/w Na_2O) was added to the previous solution and mixed with continuous stirring for 30 min. A third solution of sodium aluminate, prepared by dissolving 0.48 g of $NaAlO_2$ in 26.6 g of H_2O, is then added drop by drop. Finally, precipitation and gelation are obtained by adding, with vigorous stirring, an acid solution of 26 g of a 10% by-weight solution of H_2SO_4. The resulting gel was then hydrothermally treated in a Teflon-lined stainless-steel autoclave in an oven at 150°C for 97 hours. The product was separated by filtration and washed with deionized water to pH 9-10, and the obtained fresh lamellar ZSM-5 zeolite was dried at 120 °C for 24 hours.

In the second stage, pillarization of the layered ZSM-5 was performed. To do this, we used the top-down method to perform hydrolysis of TEOS in the presence or absence of iron nanoparticles. In the pillarization method without iron, the non-calcined layered zeolite (as synthesized) was mixed with TEOS (1:5 w/w) and stirred for 12 hours at room temperature. The product was separated by filtration and washed with 1 ml of TEOS per gram of sample after the sample was dried at

70 ^{0}C for 12 hours. The TEOS aqueous hydrolysis solution was prepared with 10 ml of water per gram of sample by continuous stirring for 12 hours at 90 ^{0}C.

Similarly, the iron charge and simultaneous TEOS pillaring process occurred in another flask, where fresh nanosheet zeolite was modified by reaction with TEOS (1:5 w/w) for 12 hours. After that, TEOS polymerizes by acid hydrolysis in the presence of low-force acids (citric acid and acetic acid). However, the acid pH of the hydrolysis solution containing iron nanoparticles has decreased to pH=2. It was prepared as follows: in a volume of 10 mL of water, 0.15 g of citric acid and 0.85 mL of acetic acid were dissolved, then the desired 3% of FeO as the amount of iron acetate ($Fe(CO_2CH_3)_2$, Aldrich 95%) was dissolved, and the resulting solution was mixed with the TEOS-modified zeolite.

The dried samples labeled as ZSM-5 as synthesized with or without iron modification were calcined with a heat ramp of 1 °C / min to 550 °C and held at this temperature for 4 hours without airflow to obtain catalysts in their oxide state. The calcined zeolite materials (pure and iron modified) were labeled as ZSM-5-P or 3%Fe-ZSM-5-P, with P referred to as the SiO_2 pillared material.

A series of Mg-Al hydrotalcite materials with Al/(Al+Mg) molar ratios of 0.20, 0.25 and 0.33 was synthesized by a modified co-precipitation method (Cavani, Trifirò, & Vaccari, 1991). For comparison, MgO and γ-Al_2O_3 materials were prepared as we reported previously (Pérez-Cabrera et al., 2019). The Mg-Al calcined hydrotalcite materials (mixed oxides) were labeled as MgAl-x, with x=Al/(Al+Mg) referred to the Al content (MgAl-0, MgAl-0.20, MgAl-0.25, MgAl-0.33, and MgAl-1).

Figure 2 shows the experimental methodology for the catalyst's obtention.

Figure 2. An experimental procedure was performed to obtain Ni-MoS_2-supported nanoparticles on nanosheet ZSM-5 materials.

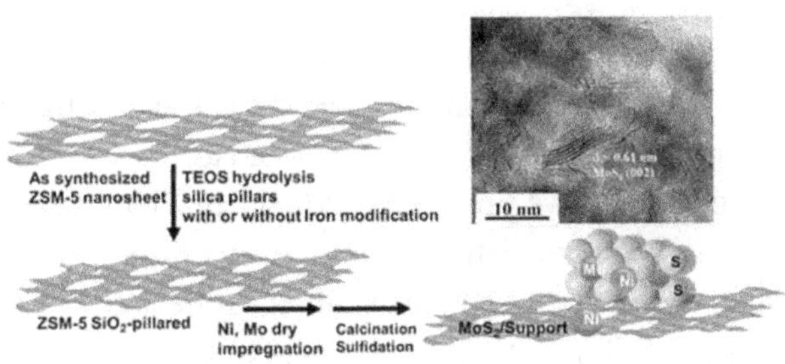

Transition metal sulfides (Ni-MoS_2) were similarly obtained onto ZSM-5 or Mg-Al layered supports.

The catalysts were prepared by incipient impregnation: the precursors of metals were ammonium heptamolybdate, and Ni acetate (Sigma–Aldrich, >99%). Metals were loaded via the co-impregnation method; citric acid was added as a chelant agent to the impregnating solutions. The molar ratio of nickel acetate: ammonium molybdate: citric acid relation was 1:2.4:1.8. The metal content was MoO_3 = 13%, NiO = 3% per gram of support. Double promoter CoNiMo/ZSM-5-P catalyst was prepared with 3% of the corresponding total promoter content (0.95 CoO+ 0.05 NiO). The impregnated materials were dried at 80 °C, and then calcined at 500 °C under air atmosphere during four hours. After calcination, the catalysts were sulfided under a gas atmosphere 15% H_2S/H_2 with a 4 Lh-1 volumetric flux and a heating rate of 10 °C min-1 until 400 °C. They were kep at 400 °C for 1 hour. After sulfidation the NiMo/ZSM-5 and NiMo/MgAl-x sulfided catalysts were carefully preserved under inert gas atmosphere, transferred to the DBT-hexadecane mixture and finally introduced into the stainless-steel autoclave reactor.

The chemical composition of the catalysts was determined by inductively coupled plasma optical emission spectroscopy (ICP-OES). An ICP-OES Vista, MPX Varian instrument, equipped with CCD detectors was used. The determination of the chemical composition of the samples was performed by disaggregating each catalyst into 100 mL of a solution, which contained one ml of each concentrated acid HF, HNO_3 and HCl. After digestion and dissolution in distilled water, the composition was analyzed by ICP spectroscopy with a coupled plasma emission spectrophotometer. The considered elements were Ni, Mo, Al, and Mg.

The structural characterization of the supports and catalysts was obtained from the corresponding powder X-Ray Diffraction (XRD) patterns obtained in an X-ray diffractometer equipment, Philips X'pert, with CuKα radiation (1.542 Å), using 45 kV voltage and a current of 40 mA.

The textural properties were characterized by analysis of N2 physisorption in a Tristar-II 3020 Micromeritics analyzer. Samples were treated in vacuum at 300 °C for four h using a degassing system; FloVac-2 Quantachrome Instruments. The textural properties were calculated from the data of adsorption-desorption isotherms, measured in the range of relative pressures (P/P0) from 10-5 to 0.995. The surface area, pore volume, and pore size distribution for each of the solids were determined by the classical approaches of Brunauer-Emmett-Teller (BET) (Sing, 2014) and Barret-Joyner-Halenda (BJH) methods.

The metallic coordination symmetry of the catalysts in the oxide state was analyzed by UV-Vis spectroscopy on an Avantes AvaSpec-2048 spectrometer, using an optical fiber. The spectra were obtained at room temperature in the wavelength range of 190-900 nm.

The MoS_2 dispersion and morphology features of the supports and catalysts were studied by transmission electron microscopy (TEM) using a JEOL 2200 FS microscope. Catalysts were dispersed in isopropanol using ultrasound, and one drop was added to a grid carbon-copper for observation. Statistics of average stacking and length of layers of MoS_2 nanoparticles were performed for each catalyst.

The activity tests were performed in a batch reactor (Parr 4848). The reaction mixture consists of dissolving DBT ~0.2267 g in 100 ml of n-hexadecane and by adding ~0.20±0.01 g of crushed and sieved catalyst, (80-100 mesh). The particle diameter was controlled, to avoid diffusion limitations: diameter average particle ~0.165 μm, P = 5.5 MPa, T = 320 °C and 700 rpm of stirring speed. Gas chromatography equipment (Agilent 7890A with flame ionization detector and HP-5 Agilent Al_2O_3 capillary column model 19091J-413 of 30mx320μmx0.25μm) have analyzed aliquots of reaction. The catalytic activities (mol DBT transformed per second and a gram of catalyst) were calculated from the initial reaction rate. After the hydrodesulfurization reaction tests, the spent catalysts were washed with isopropyl alcohol, dried and collected in a beaker within an inactive helium atmosphere, and stored for XRD, physisorption or TEM studies.

3. RESULTS

3.1. Characterization of the Catalysts At The Oxide State

3.1.1 Composition Analysis

The surface composition for ZSM-5 samples was obtained by EDX and XRF spectroscopies. The composition data of Si and Al are listed in **Table 1**. The Si/Al experimental ratios were 34.77 and 39.53, corresponding to ZSM-5-P and Fe-ZSM-5-P, and are double respect the nominal composition (Si/Al=19); the nominal value was consistent with previous EDX Si/Al=16 reported earlier (Shelyapina et al., 2020). This double value Si/Al was also observed for the NiMo/ZSM-5-P and NiMo/Fe ZSM-5-P materials. The difference in the surface EDX concerning the nominal Si/Al ratio could be related to a decreased aluminum surface species.

Table 1. Composition analysis of the NiMo/ZSM-5-P catalysts in the oxide state.

ZSM-5 material	x^a EDX	EDX Si% Al%		x^b XRF	Ni/(Ni+Mo) EDX wt. %
ZSM-5-P	34.77	41.38	1.19	25.41	--
3%Fe ZSM-5-P	39.53	40.72	1.03	45.56	--
NiMo/ZSM-5-P	35.15	37.97	1.08	--	0.27
NiMo/Fe ZSM-5-P	38.75	36.82	0.95	--	0.23
CoNiMo/ZSM-5-P	17.25	37.62	2.18	--	0.24

[a, b] x = Si/Al surface EDX and XRF compositions.

EDX indicated that the FeO, NiO, and MoO_3 composition of the mixed oxides was 1.60%, 2.46%, and 15:35 MoO_3% for FeNiMo/ZSM-5 material. CoNiMo sample (with nominal composition 0.95 CoO+ 0.05 NiO) exhibited CoO 1.84%, NiO 1.44%, and 11.80% of MoO_3 content. In change, the bimetallic NiMo/ZSM-5 catalyst exhibited a composition of 2.71 NiO% and 14.03% MoO_3 mixed oxides.

The promoter molar ratio Ni/(Ni+Mo) for the NiMo/ZSM-5-P was 0.27, and NiMo/Fe ZSM-5-P catalyst was 0.23, then those samples are positively correlated with the optimal promoter ratio of Ni/(Ni+Mo) = 0.30, which agrees with the nominal intended content. CoNiMo/ZSM-5-P catalyst exhibited a good correlation between the nominal and experimental Si/Al values but exhibited Co+Ni/(Co+Ni+Mo) = 0.16 promoter content, which is shifted more than 50% from the optimal molar ratio reported as Co/(Co+Mo)=0.47(Zepeda, 2025).

In the other hand, the bulk composition for NiMo/MgAl-x samples was obtained by ICP. The ICP data of Mg and Al are listed in **Table 2**. The Al/(Al+Mg) experimental ratios were 0.22, 0.20 and 0.35 corresponding to MgAl-0.20, MgAl-0.25, and MgAl-0.33, respectively, and are consistent with the nominal composition. The promoter content for all catalysts presented a good correlation with the optimal promoter ratio of Ni/(Ni+Mo) = 0.30 which is in accord with the nominal content.

Table 2. Composition analysis of the NiMo/MgAl-x catalysts in the oxide state.

NiMo/MgAl-x $x^a_{nominal}$	x^b XPS	ICP Al% Mg%		x^c ICP	Ni/(Ni+Mo) ICP At. %
0.00	0	0	26.64	0.00	0.29
0.20	0.28	4.59	16.48	0.22	0.30
0.25	0.30	6.3	17.97	0.26	0.31
0.33	0.40	9.41	17.59	0.35	0.31
1.00	1	30.88	0	1.00	0.29

[a, b, c] x = Al/(Al+Mg) nominal, surface and bulk compositions.

The surface composition for MgAl-x samples was obtained by XPS. The XPS data of Mg and Al are included in **Table 2**. The Al/(Al+Mg) experimental ratios were 0.28, 0.30 and 0.40 corresponding to MgAl-0.20, MgAl-0.25, and MgAl-0.33 respectively. These values indicate a surface enrichment by aluminum; the MgAl-0.20 and MgAl-0.33 oxides presented the highest Al-surface enrichment.

3.1.1 Textural properties

Textural properties (S_{BET}, V_{BJH}, and D_{BJH}) of bimetallic and trimetallic catalysts at the oxide state are shown in **Table 3**. In pure 3%Fe ZSM-5-P support, the highest surface area was obtained. Still, some textural properties, such as pore volume and size, decreased directly due to the Fe content concerning the ZSM-5-P without Fe modification, then pH decreased in the TEOS hydrolysis solution decreased the mesopore size and pore volume due to the iron addition during the SiO_2 pillars formation. As expected, after the impregnation of active transition metals and calcination steps, catalysts in the oxide state exhibited a decrease in the specific surface area. NiMo/Fe ZSM-5-P and CoNiMo/ZSM-5-P decreased by 35% in surface area for pure Fe ZSM-5-P and ZSM-5-P supports, respectively. In change the lower (18%) surface area decrease (after Ni-Mo loading) was obtained for NiMo/ZSM-5-P catalyst, which indicates that the last catalyst shows better Ni-Mo nanoparticles dispersion.

Table 3. Textural properties of the catalysts in the oxide state.

Catalyst	$^aS_{BET}$ ^bP_s ^cP_v		
ZSM-5-P	744	3.1	0.61
3%Fe ZSM-5-P	1040	2.8	0.44
NiMo/ZSM-5-P	611	2.8	0.36
NiMo/Fe ZSM-5-P	680	2.5	0.39
CoNiMo/ZSM-5-P	483	3.0	0.41

a, b, c Surface area $_{BET}$ (m^2 g^{-1}), pore size $_{BJH}$ (nm) and pore volume $_{BJH}$ (cm^3 g^{-1}).

Figure 3 shows N_2 adsorption-desorption isotherms-77K of ZSM-5 related materials. The hysteresis loops shown by ZSM-5 zeolites are of Type H3 or H4, attributed to intercrystallite adsorption within aggregates. It can be observed that iron modification shifts N_2-physisorption at low pressures (in the range of 10-3 $\leq$ p/p0 $\leq$ 0.5), then the higher surface area was obtained in Fe-ZSM-5 concerning pure ZSM-5-P. For pure pillared ZSM-5P zeolite and their corresponding-supported catalyst, broad mesopore distribution between 2-4 nm was found with maxima centered at 3.5, 3-3.5, and 3.9 nm for pure ZSM-5-P, NiMo, and CoNiMo, catalysts. Meanwhile, for the iron-modified FeZSM-5 and NiMo/Fe-ZSM-5 materials, the iron introduction and decrease of pH of the hydrolysis reaction, the more extensive the intensity of small mesopores of 2.3 nm was obtained, and poor quantity of pores of 4 nm was also observed. That porosity decrease could affect the molecular diffusion of the NiMo/Fe-ZSM-5 catalyst.

Figure 3. N_2 physisorption isotherms and pore-size distributions of pure and iron modified ZSM-5 supports (A and B) and their corresponding NiMo/ZSM-5 catalyst (C and D).

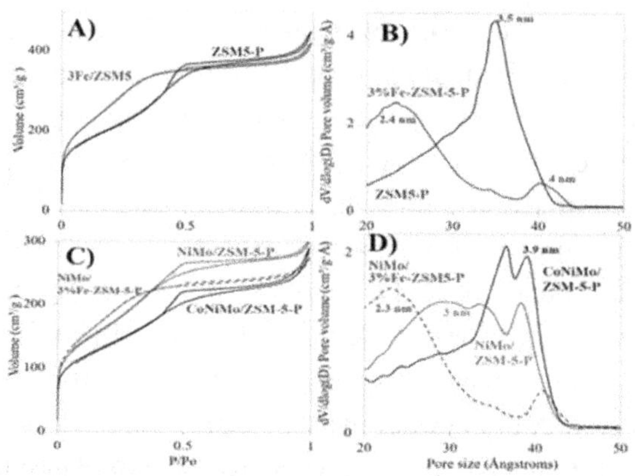

Table 4 reports the textural properties of the NiMo/MgO-Al_2O_3 catalysts at the oxide state. All materials exhibit a significant decrease of the specific surface area after impregnation, due to the metallic charge of Ni and Mo active metals. The pore volume decreased directly as a function of the Al content in the oxide catalyst precursors with the Al/(Al+Mg) ratio. In the case of the NiMo/MgAl-0-33 material, there is a lower decrease of volume than in NiMoMgAl/0.20 and NiMoMgAl/0.25 which is related to the observed shifts in the pore-size distribution toward macropores at high x-composition.

Table 4. Textural properties of the catalysts in the oxide and sulfide states.

Sample MgAl-x	**Support** [a]S_{BET} [b]P_s [c]P_v			**NiMo oxide** S_{BET} P_s P_v			**NiMo sulfide** S_{BET} P_s P_v		
0.00	200	17.0	1.14	114	9.1	0.28	65	4.5	0.14
0.20	133	22.3	0.84	63	20	0.34	130	7.8	0.33
0.25	127	12.8	0.51	73	17	0.36	39	24	0.25

continued on following page

Table 4. Continued

Sample MgAl-x	Support $^{a}S_{BET}$ $^{b}P_s$ $^{c}P_v$			NiMo oxide S_{BET} P_s P_v			NiMo sulfide S_{BET} P_s P_v		
0.33	119	17.9	0.24	79	16	0.22	23	35	0.21
1.00	262	5.4	0.46	59	9.4	0.12	74	9.9	0.21

[a,b,c] Surface area $_{BET}$ (m^2 g^{-1}), pore size $_{BJH}$ (nm) and pore volume $_{BJH}$ (cm^3 g^{-1}).

3.1.2 UV-Vis spectroscopy

Figure 4 presents the UV-Vis spectra obtained for the NiMo/ZSM-5-P series. The CoNiMo/ZSM-5-P catalyst exhibited low-intensity signals between 200 and 400 nm. Those signals are assigned to charge-transfer transitions (O^{2-} to Mo^{6+}) species in octahedral and tetrahedral coordination. In the CoNiMo catalyst, the blue-shifted signal position decreases. Then, octahedral Mo^{6+} may be decomposed into small tetrahedral molybdate particles.

Figure 4. UV-Vis spectra from the NiMo/ZSM-5-P catalyst in their oxide states.

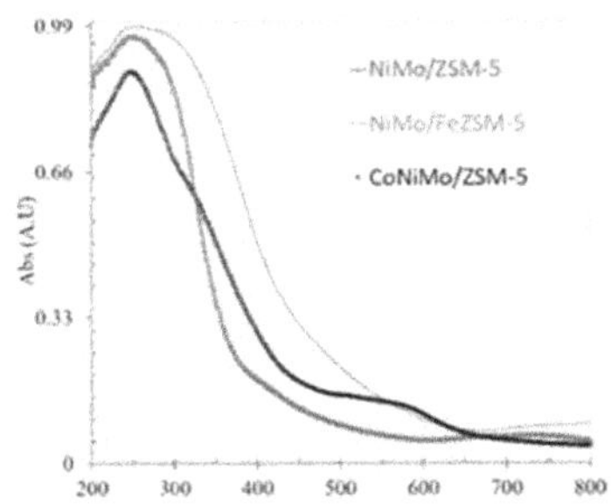

This effect on the coordination is also evident in the NiMo/FeZSM-5 sample (iron support modified), which displayed broader signals than the NiMo/ZSM-5 catalyst. The NiMo/ZSM-5-P catalyst exhibited a band with a maximum of approximately 255 nm. At around 400 nm, the band that appears as a broad shoulder for the other NiMo/FeZSM-5 and CoNiMo/ZSM-5 materials confirms the presence of octahedral Ni^{2+}. Concerning the Ni coordination, the Uv-Vis spectra of NiMo/FeZSM-5 and CoNiMo/ZSM-5 materials in the visible region showed a plateau; this matter could be assigned to octahedral Ni^{2+} species. This plateau has been attributed to inactive nickel compounds, which gradually diminish as observed. The last signal could also be related to molybdates as a β-$NiMoO_4$, where the Mo^{6+} presents tetrahedral coordination and Ni^{2+} stills in octahedral coordination.

The coordination of Ni and Mo oxides $MgO-Al_2O_3$ supported was analyzed by UV-Vis spectroscopy. Results in **Figure 5** exhibit Mo signals in the range 200-400 nm which correspond to the charge transfer oxygen to metal ($O^{2-} \rightarrow Mo^{6+}$). In this region two bands are observed, one as a large shoulder centered at 320 nm present for NiMoMgAl-1, the other in the 220 to 260 nm region for the catalysts supported on mixed oxides, NiMo/MgAl-0.20, NiMo/MgAl-0.25, and NiMo/MgAl-0.33 materials, whose maximum signal is located about 240 nm. The observed bands centered at 240 nm and 320 nm have been assigned to isolated tetrahedral MoO_4^{2-} and octahedral Mo^{6+}, respectively (Duan et al., 2007; Wu et al., 2009; Zepeda et al., 2008). Thus, the difference in position of these two signals is related with the metallic coordination of Mo atoms. The bands of NiMo/MgAl-0.20, 0.25 and 0.33 to low wavelengths indicate a decrease of the number of octahedral Mo in favor of tetrahedral Mo species. (Klimova, Solís Casados, & Ramírez, 1998). For the NiMo/MgAl-0 catalyst a broad shoulder around 320 nm is observed corresponding to Mo in octahedral coordination. The signal at 320 nm also has been related to the agglomeration of polymolybdates with octahedral coordination (Y. Zhang, Han, Long, & Nie, 2016). (Pawelec, Halachev, Olivas, & Zepeda, 2008).

Figure 5. UV-Vis spectra from the NiMo/MgAl-x catalyst in their oxide states. Catalysts supported on alumina and MgO (a), catalysts supported on mixed oxides (b).

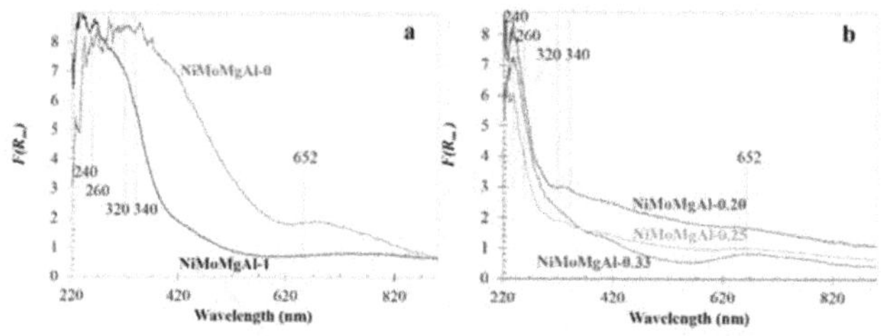

For the nickel coordination two bands at 660 nm and 750 nm were observed as shown in NiMo/MgAl-0.33 and NiMo/MgAl-0 samples. These bands correspond to the Ni^{2+} cation in tetrahedral and octahedral coordination, [Ni^{2+} $4O^{2-}$] and [Ni^{2+} $6O^{4-}$], respectively (Guevara-Lara et al., 2010). Thus, in the oxide precursors, the tetrahedral nickel species decreased on NiMo/MgAl-0.25 and NiMo/MgAl-0.20 and more octahedral nickel resulted.

3.1.3 X-ray diffraction powder analysis

Figure 6 shows the X-ray diffraction pattern of fresh ZSM-5 zeolites (ZSM-5 as synthesized) and silica pillared (ZSM-5-P). For these two materials, reflections were recorded at low angles, 2θ~2-3°(interplanar distance d=4.1nm), confirming the acquisition of the mesoporous-ordered structure characteristic of laminar zeolites intercalated with organic or silicon pillars, whose typical signal is associated with the interplane distance (100), located at 2θ~2°. This type of diffraction peak corresponds to a lamellar phase, that is stacked along the c-direction and has a 2D morphology for the MFI material, as discussed in previous work (Shelyapina et al., 2020). It is observed that the ZSM-5 iron-modified support is characterized by having the reflection centered at the angle 2θ~2.5° compared to the XRD pattern of the material with pure SiO_2 pillars (ZSM-5-P). Similarly, Figure 6D shows that zeolites impregnated with 3% promoter and 12% MoO_3 showed low intensity, but broad reflections shifted slightly to the right by 2θ ~ 2.7° (interplanar distance d=3.3 nm).

Figure 6. XRD patterns at low angle and high angle of pure and iron modified ZSM-5 supports (A and B) and their corresponding NiMo/ZSM-5 catalyst (C and D).

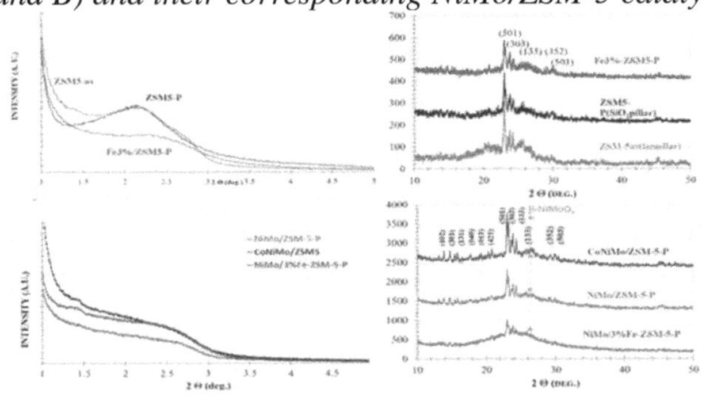

The decrease in intensity is greater for the NiMo/ZSM-5-P catalyst than for the iron- or cobalt-containing catalysts, and the observed angle shift of 2θ ~ 2.7° (d=3.3 nm) occurs after the addition of the cobalt and molybdenum oxides. This shift can be correlated with a decrease in the intercrystalline d(100) distance, which affects the mesostructure associated with intercrystalline mesopores. This suggests that MoO_3 and NiO crystallites may be located in the intercrystalline mesopores, so NiO and MoO_3 crystallites are expected to be larger in the CoNiMo/ZSM-5 and NiMo/FeZSM-5-P materials than in the NiMo/ZSM-5-P catalyst, with lower intensity in the low angle diffraction pattern.

It should be noted that in the wide-angle XRD pattern, none of the materials showed XRD peaks corresponding to aggregated iron oxide species, indicating the presence of well-dispersed and/or zeolite-incorporated Fe species in all these samples whose size is below the detection limit of the instrument.

In these patterns, a small reflection is observed, the intensity of which is most prominent at $2\theta = 26.7$ for the CoNiMo/ZSM-5-P catalyst. This signal can be attributed to the formation of the β-$CoMoO_4$ phase (JCPDS-21-0868) when the bimetallic Ni(Co)-Mo complex decomposes by calcination (Zepeda, 2025). Although the analogous nickel β-$NiMoO_4$ oxide also shows the main diffraction signal, it is very small at $2\theta = 26.6$ (JCPDS-45-0142). The diffraction lines of MoO_3, NiO, and Co_2O_3 are not observed, suggesting that their particles are very small and highly dispersed on the support. It is also possible that these oxides have been incorporated into the molybdate structure to form bimetallic mixed oxides (Brito, Barbosa, Albornoz, Severino, & Laine, 1994).

Figure 7 shows the XRD patterns of catalysts in the oxide state. The catalysts supported on MgO and mixed oxides present characteristic peaks of the periclase structure (JCPDS No. 45-0946) with reflections at 2θ= 36.5, 42.9°, 62.7°. The catalyst supported on alumina presents only small signals corresponding to γ-Al_2O_3 at 2θ=32°, 39°, 46° and 67° degrees corresponding to the JCPDS No. 47-1308 card. The catalysts NiMo/MgAl-0.20, NiMo/MgAl-0.25 and NiMo/MgAl-0.33 present a primary peak located at 2θ = 26.60°, and small peaks at 2θ=19.2°, and 23.2° that is characteristic of the β-$NiMoO_4$ phase (JCPDS No. 45-0142). Therefore, the diffraction results suggest that Ni incorporates into the structure of molybdates to form bimetallic mixed oxides in the case of catalysts supported on the mixed MgAl-x oxides. For the NiMo/MgAl-1 catalyst the NiMo oxides do not appear in the diffraction pattern suggesting that they are very dispersed.

Figure 7. XRD pattern of oxide (a) and sulfide (c) catalysts supported on alumina and MgO. XRD pattern of oxide (b) and sulfide (d) catalysts supported on MgAl-x oxides.

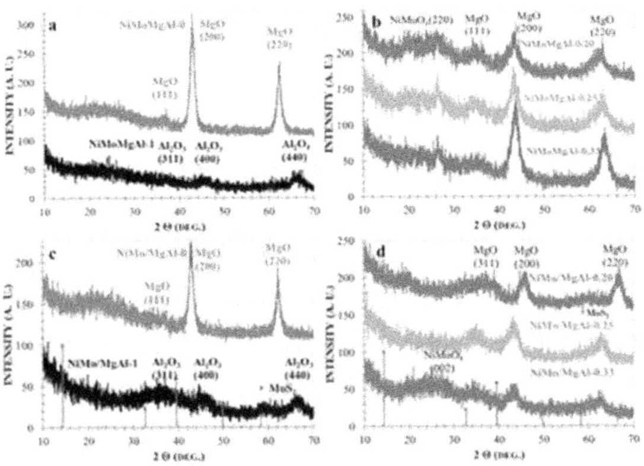

3.2. Characterization of the Catalysts At The Sulfide State

3.2.1 Textural Properties

Table 4 reports the textural properties of the sulfide catalysts. The surface areas decrease considerably compared with the areas of the supports indicating that blocking of pores is occurring due to impregnation. For NiMo/MgAl-0 (NiMo/MgO) the surface area dropped from the oxide to the sulfide state by 43%. Contrarily, for the NiMoMgAl/0.20 the sulfidation recovers the previous surface area of the support (130 m^2/g). In the case of the NiMoMgAl/0.25 and NiMoMgAl/0.33 sulfides the surface area decrease can be explained by a phenomenon observed by Klimova and Solís Casados et al., 1998, showing that the MgO periclase dissolves to $Mg(OH)_2$ during the process of impregnation, however after calcination and sulfidation steps the periclase MgO structure is restored (Klimova et al., 1998). For the NiMo/MgAl-1 catalyst (NiMo/Al_2O_3) the sulfidation increases the area due to the formation of nickel-promoted MoS_2 slabs.

3.2.2 XRD of the NiMo/MgAl-x Sulfides

The XRD patterns of the sulfided catalysts (**Figure 7** c-d). present diffraction peaks corresponding to typical MgO periclase or alumina. The catalysts NiMo/MgAl-0 and NiMo/MgAl-0.33 showed the crystalline β-$NiMoO_4$ phase located at $2\theta = 26.60°$, indicating that there is a strong interaction of Ni and Mo oxide phases forming irreducible species. In the other catalysts did not appear Ni or Mo oxide or sulfide species, indicating that a good dispersion was obtained.

The **Figure 7**-c shows the diffraction pattern of NiMo/MgAl-1 catalyst which present planes of the Al_2O_3 support, including those located at $2\theta = 37.60°$ and 45.86°. The last diffraction was associated with the increased interaction of sulfides species with the acidic support.

3.2.3 TEM Analysis of the NiMo/ZSM-5, NiMo/MgAl-x Sulfides

The TEM microscopy image in **Figure 8** shown in the gray box on the left shows the ordering of the mesostructure due to the intercalation of inorganic silica pillars (SiO_2). In the inset (the result of filtering the image using digital micrograph software), distances of 4.11 nm are observed, corresponding to the interplanar distance (001) also observed in the low-angle XRD pattern and associated with intercrystallite mesopores. In the central micrograph, iron nanoparticles of 3-4 nm size are observed deposited on ZSM-5. Micrographs show MoS_2 nanoparticles as black slabs with different lengths and stacks number 2-4 for the NiMo/FeZSM-5 catalyst. We measured interplanar distances of about 0.61 nm, which correspond to the (002) planes in a MoS_2-HCP structure.

Figure 8. TEM micrographs of iron modified ZSM-5 supports (A and B) and their corresponding NiMo/Fe-ZSM-5-P catalyst (C and D).

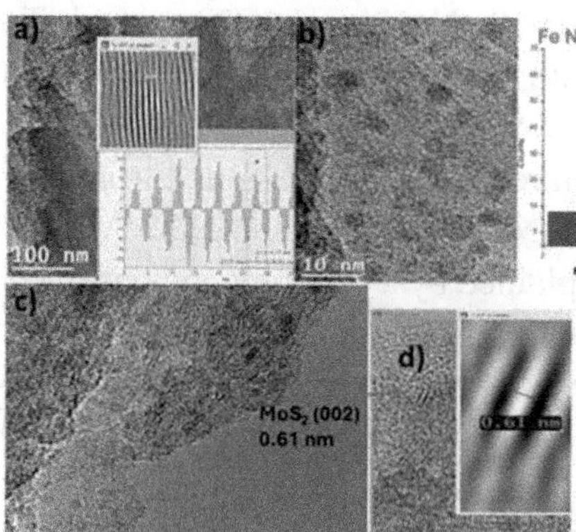

In **Figure 9** we observed the size and number of layers of these nanoparticles change depending on the catalyst. The Figure 9a show MgO-supported catalysts (NiMo/MgAl-0) with MoS_2 slabs decorating the borders of MgO crystallites. It also exhibits curved or flat nanoslabs randomly oriented decorating MgO nanorods with an edge-on orientation respect to the (200) planes (d=0.21 nm) (Cinibulk, Kooyman, Vit, & Zdražil, 2003).

Figure 9. TEM images of NiMoAlMg-(x) sulfides: 0 (a), 0.20 (b), 0.25 (c), 0.33 (d), and 1 (e).

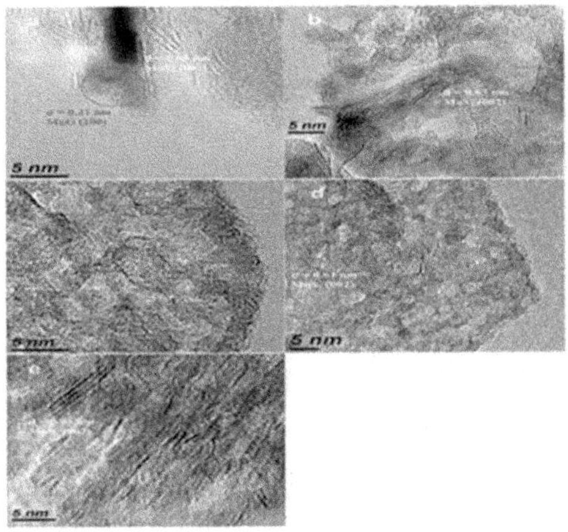

The catalysts NiMo/MgAl-0.20, NiMo/MgAl-0.25 and NiMo/MgAl-0.33 exhibit different morphologies of the MgAl-x oxide structure (**Figure 9** b-d). We observe cauliflower-like morphology made of MgO nanoparticles with semi-hexagonal arrays (bricks near to 5 nm in size). These NiMo/MgAl-0.33 and NiMo/MgAl-0.25 exhibit MoS_2 crystallites of small slab-length in the porous-zones (**Figure 9** c-d). For the catalyst NiMo/MgAl-1 (**Figure 10** a-c), the HR-TEM analysis and fast-Fourier transform (FFT) image show stacked slabs of MoS_2 randomly oriented on the (311) direction of the Al_2O_3-support. It has been suggested that the γ-Al_2O_3, unlike other supports, stabilizes better the MoS_2 phase (Gutiérrez et al., 2014).

Figure 10. TEM image of sulfided NiMo/MgAl-1 (a), filtering of the FFT image to focus on planes of the support, of the selected square marked (b), and contrast distribution profile (c).

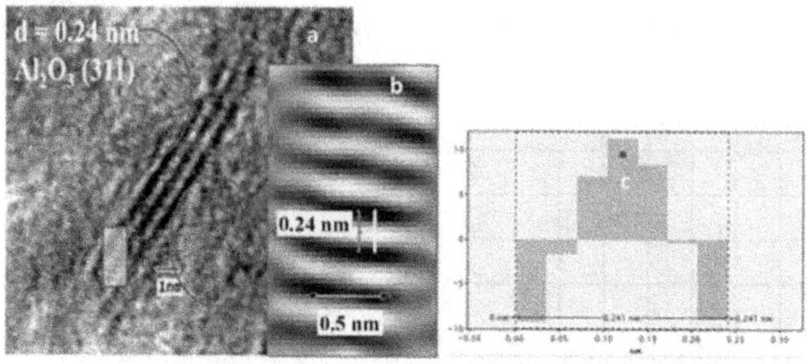

Selected area diffraction patterns (SAED) were obtained for NiMo/MgAl-0 and NiMo/MgAl-0.20 and NiMo/MgAl-1. The first two catalysts showed diffraction rings from the MgO periclase (**Figure 11** a-b) with intense reflections corresponding to (111), (200), (220), (222), (400), (420) and (422) planes. For the third catalyst (**Figure 11**-c) the SAED pattern presented a ring localized at $2\theta = 49.8$ with $d_{(hkl)} = 0.18$ nm corresponding to the plane (105) of MoS_2.

Figure 11. SAED patterns of NiMo/MgAl-x sulfides: 0 (a), 0.2 (b), and 1 (c).

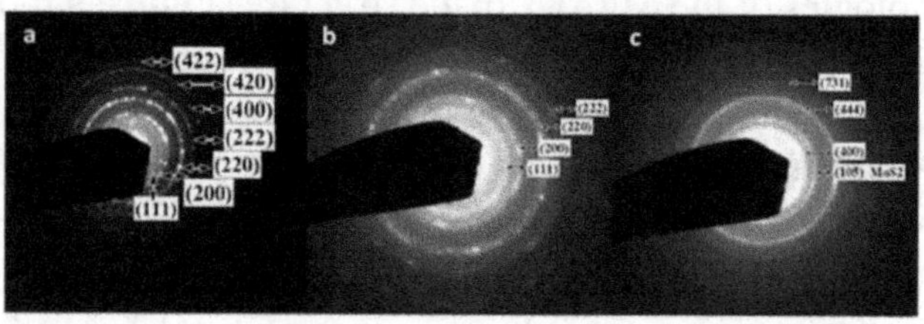

The statistical analysis of average length and stacking of molybdenum sulfide nanoparticles of the catalysts is given in **Table 5**. These data were analyzed statistically from 300 MoS_2 crystallites for each sample (see MoS_2 slab length and stacking number in **Figure 12**). The relative order for the decrease of slab length of the catalysts is: MgAl-0.0 (3.42) <MgAl-0.33 (3.40) <MgAl-0.20 (3.17) <MgAl-0.25 (2.88) < MgAl-1.0 (2.32). The trend of slab length and stacking degree parameters follow the same trend the acidity of the supports previously determined by DFT calculations (Pérez-Cabrera et al., 2019) for the Mulliken atomic charges: MgAl-0.0 <MgAl-0.33 <MgAl-0.20 <MgAl-0.25 <MgAl-1.0. This result suggest that acidity of the support improves the dispersion of MoS_2 nanoparticles on the surface. In the case of mixed oxides, the NiMo/MgAl-0.25 catalyst exhibited shorter slabs

and lower stacking than the other catalysts suggesting also a favorable electronic contribution at the Al content. The effect of acidity of the supports is due to formation of bonds with the sulfide active phase promoting a better dispersion (W. Chen et al., 2013; Hensen, de Beer, van Veen, & van Santen, 2002; Yocupicio, Díaz de León, Zepeda, & Fuentes, 2017).

Table 5. Dimensions of MoS_2 crystallites of the catalysts at the sulfided state.

Catalyst	L[a] (nm)	N[b]
NiMo/MgAl-0	3.42	2.60
NiMo/MgAl-0.20	3.17	2.58
NiMo/MgAl-0.25	2.88	2.26
NiMo/MgAl-0.33	3.40	2.42
NiMo/MgAl-1	2.32	2.17
Industrial	-	-

[a] average slab length, [b] average slab stacking.

Figure 12. Dispersion parameters of NiMo/MgAl-x sulfides. Slab length (a) and stacking number (b).

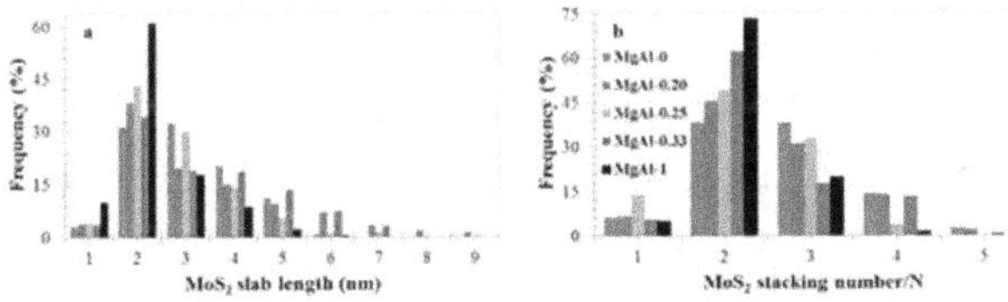

3.2.4 Hydrodesulfurization activity of NiMo/ZSM-5-P and NiMo/MgAl-x sulfides

The HDS of DBT was used as reference reaction test, to evaluate the catalytic features of NiMo/ZSM-5-P and NiMo/MgAl-x sulfides. This reaction occurs by two parallel and independent reaction paths (see **Figure 13**) . (i) The hydrogenation (HYD) of aromatic rings yields first the tetra and hexahydrodibenzothiophene intermediates (T-HDBT or H-HDBT), then desulfurization to cyclohexylbenzene (CHB) and H_2S products, and; (ii) the direct desulfurization (DDS) pathway (also referred as hydrogenolysis), involves the cleavage of the C-S bond, yielding phenyl-benzene (BP) and H_2S products. Furtherly BP can be slowly hydrogenated leading to cyclohexylcyclohexane (CCH).

Figure 13. Initial reaction rate of NiMo/ZSM-5 sulfides for the hydrodesulfurization reaction of dibenzothiophene (DBT).

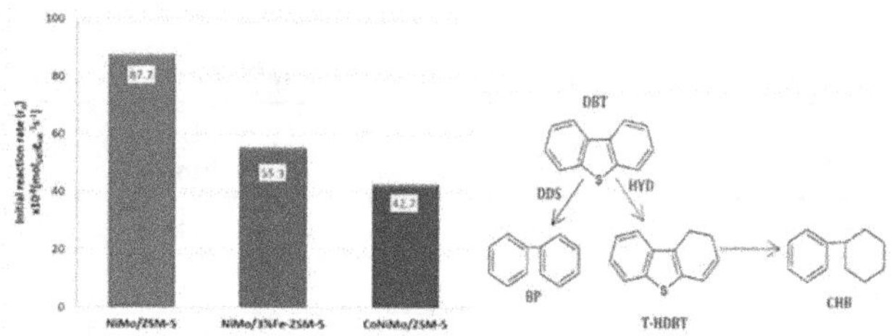

In **Figure 13** the catalysts supported on ZSM-5 presented the following order of catalytic activity: CoNiMo < NiMo/FeZSM-5 < NiMo/ZSM-5. The exact order from lowest to highest activity was observed by varying the nickel content and promoter/promoter + metal molar ratio by X-ray emission spectroscopy (EDX). The most active catalyst of the series was NiMo/ZSM-5-P (with a -rDBT rate of 87.7 x 10($^{-8}$) (mol DBT g($^{-1}$) s($^{-1}$)) and a HYD/DDS selectivity of 12.4). An increased selectivity towards direct desulfurization (HYD/DDS = 5.7 and 7.1) was obtained for the CoNiMo/ZSM-5 and NiMo/FeZSM-5 catalysts.

In **Figure 14**, from the catalytic activity data for MgO-Al_2O_3 samples, the HDS of DBT activity order was: NiMo/MgAl-0 < NiMo/MgAl-0.33 < NiMo/MgAl-0.20 < NiMo/MgAl-0.25 < NiMo/MgAl-1. Therefore, as the Al content increased the activity increased up to x = 1 which corresponds to pure alumina. This NiMo/MgAl-1 catalyst exhibited similar activity to an industrial NiMo/Al_2O_3 catalyst ($-r_{DBT}$= $60x10^{-8}$ (mol_{DBT} $g^{-1}s^{-1}$) with a HYD/DDS selectivity = 0.15 at 30% of conversion).

Figure 14. Initial reaction rate (a) and selectivity (b) of NiMo/MgAl-x sulfides for the hydrodesulfurization reaction of dibenzothiophene (DBT).

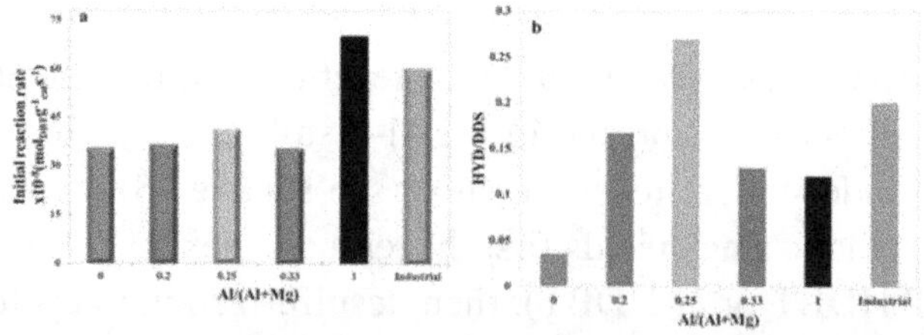

The selectivity data at 30% of conversion (see **Table 6**) show that the NiMo/MgAl-0 catalyst show very good selectivity to the DDS path (96.5%). Otherwise, the NiMo catalysts supported on binary MgAl-x oxides showed an effect of increased acidity on the HYD selectivity.

Table 6. Selectivity and initial reaction rate of all NiMo/MgAl catalysts.

Catalyst	HYD	DDS	[a]HYD/DDS	Initial reaction rate $x10^{-8}$ [mol_{DBT} $(g_{cat}s)^{-1}$]
NiMo/MgAl-0	3.45	96.55	0.036	35.2
NiMo/MgAl-0.20	14.29	85.71	0.167	36.3
NiMo/MgAl-0.25	21.21	78.79	0.269	41.1
NiMo/MgAl-0.33	11.42	88.58	0.129	35.3
NiMo/MgAl-1	10.71	89.29	0.120	69.9
[b] Industrial	16.67	83.33	0.200	60.0

[a] HYD/DDS is derived from selectivity of products (TH-DBT+CHB)/BP, at 30% of conversion.

[b] conventional $NiMo/Al_2O_3$.

Then the HYD selectivity for the $NiMoS_2$ catalysts present the order MgAl-0.0 (3.45) <MgAl 1.0 (10.71) <MgAl-0.33 (11.42) ~MgAl-0.20 (14.29) <MgAl-0.25 (21.21).

4. DISCUSSION

According to EDX compositional data, the directing agent CTABr contributes to a twofold increase in surface silicon. Given that the amount of CTAB has been doubled compared to earlier work by our group (Egorova & Prins, 2004; Shelyapina et al., 2020), we anticipate a proportional increase in the number of surfactant cations interacting primarily with the silanol groups of the zeolitic network. Moreover, we expect a larger population of pillars formed through TEOS hydrolysis, which are intercalated between the ZSM-5-P sheets.

Conversely, the incorporation of iron nanoparticles into the support during the TEOS hydrolysis process has reduced mesopore size from 4 nm to 2 nm, a change attributed to its catalytic effect on the hydrolysis reaction. The acid hydrolysis of TEOS has been enhanced by using weak inorganic acids (citric and acetic acid) to stabilize iron species, thereby modifying SiO_2 crystal growth. Consequently, the decreased pore size may impact the dispersion and reduction of polymolybdates, along with the diffusion and reactivity of DBT in the well, particularly kept for the catalyst with iron-modified support. Additionally, UV-visible and X-ray data indicated the formation of cobalt molybdate on the CoNiMo/ZSM-5 catalyst, which also displayed a lower surface area, thus explaining the reduced reaction rate in comparison to the NiMo/ZSM-5 catalyst.

From theoretical DFT results, we observed that isomorphic substitution of Mg by Al in MgO causes the basicity drop experimentally observed. The Al electronic contribution at the Fermi level is enhanced at the composition x=Al/(Al+Mg) = 0.25, showing the highest acidity.

The support acidity can induce an electronic deficiency on sulfide phase, then a direct correlation is expected between the electronic deficiency and the catalytic activity of the catalysts (W. Chen et al., 2013). In line with this, different experimental studies have indicated that MgO as support decreases the interaction support-active metal (Kaluža, Gulková, Vít, & Zdražil, 2015; Klicpera & Zdražil, 2002; Semeykina et al., 2016). Among the most active catalyst supported on binary oxides, the composition x=0.25, corresponds to the maximum participation of the acidity provided by the Al atoms calculated by theoretical DFT approximation.

According to our UV-Vis results, the non-agglomerated octahedral coordination of molybdenum exhibited by the most acidic support (γ-Al_2O_3) is more suitable to produce better Mo dispersion. The blue shift observed on NiMo/MgAl-0.20, 0.25 and 0.33 materials was assigned to the formation of tetrahedral molybdenum at 220-250 nm that could correspond to the nickel-molybdate spinel in agreement with our results of XRD (Figure 7-b) (Calderón-Magdaleno, Mendoza-Nieto, & Klimova, 2014).

The formation of dispersed nickel promoted MoS_2 nanoslabs on the NiMo/MgAl-0.25 sulfide catalyst is associated with the interaction of Ni and Mo species in the oxide state with mixed MgAl-x acidic support that decreases the crystallite size (Figure 7-d). The β-$NiMoO_4$ phase exhibited by the NiMoMgAl-0.33 catalyst is due to segregation of Ni promoter and goes in decrement of the HDS of DBT activity (González-Cortés, Xiao, Lin, & Green, 2006).

According to Topsoe (Topsøe et al., 1995) the Ni role resulted in a decrease of the Mo-S bond strength. Hence a correlation could be proposed between the support acidity, and the reducibility of nickel molybdates.

From TEM analysis of NiMo/MgAl-1 catalyst, the decrease of MoS_2 particle size was associated to promoter localization at the edge of MoS_2 particles, to produce edge-decoration of MoS_2 slabs. These results are in line with previous and theoretical works which explained the smallest slabs observed by a promoter edge-decoration effect of MoS_2 slabs on Al_2O_3 (Costa, Arrouvel, Breysse, Toulhoat, & Raybaud, 2007).

In the case of the mixed oxides series, the NiMo/MgAl-0.25 catalyst showed the highest activity and shortest slab length (Figure 12). As the Al content increased, the activity goes through an optimum value at x=0.25 (Figure 14). This result coincides with the maximal electronic participation of Al at the Fermi level in the mixed oxides predicted by the DFT method at x=0.25. Thus, DFT calculations approximation is found to fit well with support acidity.

The HDS activity results as a function of the MoS_2 dispersion is related to the slab length, and stacking parameters suggests that the promoter gives a better synergetic effect on acid supports than in basic-supports. Results of the general HDS of DBT activity order for all catalysts was: NiMo/MgAl-0 < NiMo/MgAl-0.33 < NiMo/MgAl-0.20 < NiMo/MgAl-0.25 < NiMo/MgAl-1 indicating that the HDS activity is influenced by the slab length (Figure 12).

Previous results on MgAl mixed oxides showed an important correlation about the optimal selectivity/activity ratio at Al(Al+Mg) molar ratio (x=0.25) (Linares et al., 2015; R. Zhao et al., 2003). In a different research, Wu et al. reported that CoMo/MgO-Al_2O_3 synthesized from calcined hydrotalcite with x=0.20 molar ratio, have a higher activity for the HDS of DBT than CoMo/Al_2O_3 and CoMo/MgO respectively (Li Bao Wu, 2010; Wu et al., 2009). Guevara-Lara et al. suggest that NiMo sulfides supported in binary oxides (MgO with lower Al content ~5%) exhibited higher activity for the HDS of DBT than NiMo/Al_2O_3 (Guevara-Lara et al., 2010). They also reported that acid-basic support such as Al_2O_3-MgO increase the hydrogenolysis pathway and this effect is correlated with the presence of polymolybdates structures in the precursor oxide, which governs the dispersion of the Ni-Mo-S active phases (Guevara-Lara et al., 2010).

Our results of catalytic activity as a function of the Al/(Al+Mg) ratio of the support are in good correspondence with those firstly described works. The Al role in MgAl mixed oxides was to strength the acidity, which in turns had an increase on HYD route and catalytic activity. Nevertheless, the Al_2O_3 support exhibited better acidic properties to disperse MoS_2 particles, and consequently, it enhances catalytic activities. The pure-MgO presents sulfides with less dispersion (large slabs and higher stacking number of MoS_2), and the DDS route is predominant. The DDS route occurs at the corner sites of MoS_2 or in S edge-vacancy in the catalyst (Novais Antunes et al., 2015), thus, the NiMo/MgAl-0 catalyst did not show any yield to tetrahydro-dibenzothiophene (see **Table 6**). Therefore, the interaction of nickel-promoted MoS_2 in MgO must favor the perpendicular adsorption of DBT at the edges of MoS_2, and the direct desulfurization pathway. In the case of NiMo/MgAl-1 and NiMo/MgAl-0.25 sulfides, the rim adsorption appears where are involved the carbon atoms in the aromatic rings, leading to further hydrogenation (Daage & Chianelli, 1994).

Reports of Hinnemann by theoretical DFT approach show that MoS_2 edges interact with oxygen atoms of the Al_2O_3 support due to Mo-O-Al linkages (Hinnemann, Nørskov, & Topsøe, 2005). The support interaction decreases the Mo-S bond and then restrict the hydrogenation route on the Al_2O_3 support. Although, it was found that the slab length exhibited more impact on HDS activity than on the HYD selectivity, the addition of Ni promoter enhances the catalytic activity of NiMo/MgAl-1 in HDS but leads to a slight decrease in the hydrogenation selectivity.

5. CONCLUSION

As UV-Vis and XRD analyses demonstrated, the Fe-modified NiMo/ZSM-5 and CoNiMo/ZSM-5 catalysts displayed $CoMoO_4$ and $NiMoO_4$ phase spinels within our mixed oxide samples. Consequently, $Co(Ni)MoO_4$ may reduce the coordination of the active metal, leading to lower dispersion of the (Co)NiMoS active sites. The highest selectivity toward DBT hydrogenation was noted among the bimetallic Ni-Mo zeolites. In contrast, the Co-Ni-Mo/ZSM-5 and Ni-Mo/FeZSM-5 samples exhibited decreased activity, although all three catalysts achieved DBT conversions of up to 50 percent.

The iron-content catalyst demonstrated medium reactivity within the Ni-MoS2/ZSM-5 series, while all catalysts supported by ZSM-5 showed enhanced reaction rates compared to the MgO-containing catalyst. Therefore, ZSM-5 pillared compounds and TEOS acid hydrolysis present promising strategies to facilitate effective diffusion and sulfur removal from refractory DBT model molecules. Continued improvement of the iron deposition method is essential to achieve a better synergistic effect from the iron to the MoS_2 nanoparticles and enhance the hydrodesulfurization process.

We report Al_2O_3-MgO modified materials derived from hydrotalcites to achieve a suitable strategy in the preparation of NiMo/MgAl-x catalysts for the reaction of HDS of DBT. As reported previously, by changing the Al content a decrease on the basic-properties of MgO was found and an effect on the selectivity of NiMo/MgAl-x catalysts was revealed (Pérez-Cabrera et al., 2019).

Characterization of the catalysts by XRD, UV-Vis, and TEM indicated that a good dispersion of the sulfide phase was obtained on these AlMg-X supports because of the specific acid-base properties as a function of the Al/ (Al +Mg) ratio.

The statistical analysis of the TEM micrographs showed that the Ni-Mo sulfides are present in the characteristic form of stacked slabs randomly oriented on the surface of the supports. Additionally, the micrograph analysis showed Ni-Mo sulfides nanoparticles with a different dispersion on each support. For catalysts supported on the more acidic Al_2O_3-MgO binary oxides, a better sulfide dispersion is observed than in pure MgO-supported sulfide.

The most active catalyst in the reaction of desulfurization of dibenzothiophene was the NiMo/MgAl-1 (NiMo/ Al_2O_3) presenting the highest acidity along the MgAl-x oxides due to an improved dispersion of the nickel-promoted MoS_2 crystallites (low stacking and shorter lengths). All catalysts on Al_2O_3-MgO binary oxides presented only a slight change on activity; however, they exhibited a better selectivity to the hydrogenation pathway compared to the alumina catalyst.

The mixed oxides employment for HDS of DBT offers significant enhancement toward hydrogenation selectivity reaching the maximum for Al/(Al+Mg) molar ratio of 0.25. The NiMo/MgAl-0 catalyst is mostly selective toward partial hydrogenation of the aromatic rings.

ACKNOWLEDGEMENTS

We acknowledge the valuable assistance of F. Ruiz, E. Aparicio, D. Domínguez, E. Flores, A. Mízquez, A. Rodríguez, P. Casillas, and J. Mendoza. Pérez Cabrera L. is also indebted to the Consejo Nacional de Ciencia y Tecnología (postdoctoral research CAB-CONAHCYT I1200/320/2022) and the SENER-CONACyT-Hydrocarbons project (number 117373) for the financial support granted. Also, thanks to DGAPA-PAPIIT-UNAM-IG101623.

Also, thanks to the supercomputing time provided by UNAM through the LANCAD-UNAM-DGTIC-423 and 084 projects, and the support of the zeolite research department at ICUAP.

REFERENCES

Alabdullah, M. A., Gomez, A. R., Vittenet, J., Bendjeriou-Sedjerari, A., Xu, W., Abba, I. A., & Gascon, J. (2020). A Viewpoint on the Refinery of the Future: Catalyst and Process Challenges. *ACS Catalysis*, *10*(15), 8131–8140. DOI: 10.1021/acscatal.0c02209

Anggoro, D. D., & Amin, N. A. S. (2006). Methane to Liquid Hydrocarbons over Tungsten-ZSM-5 and Tungsten Loaded Cu/ZSM-5 Catalysts. *Journal of Natural Gas Chemistry*, *15*(4), 340–347. DOI: 10.1016/S1003-9953(07)60016-4

Antunes, F. P. N., Vaiss, V. S., Tavares, S. R., Chiaro, S. S., Souza, W. F., & Leitao, A. A. (2015). Vacancy formation in MoS2 supported on MgO: Electronic and energetic analysis. Chemical Physics Letters, 626, 55-62.

Borcuch, A., Rutkowska, M., Marzec, A., Kowalczyk, A., Michalik, M., Moreno, J. M., & Chmielarz, L. (2020). Selective ammonia oxidation over ZSM-5 zeolite: Impact of catalyst's support porosity and type of deposited iron species. *Catalysis Today*, *348*, 223–229. DOI: 10.1016/j.cattod.2019.08.054

Breysse, M., Geantet, C., Afanasiev, P., Blanchard, J., & Vrinat, M. (2008). Recent studies on the preparation, activation and design of active phases and supports of hydrotreating catalysts. *Catalysis Today*, *130*(1), 3–13. DOI: 10.1016/j.cattod.2007.08.018

Brito, J. L., Barbosa, A. L., Albornoz, A., Severino, F., & Laine, J. (1994). Nickel molybdate as precursor of HDS catalysts: Effect of phase composition. *Catalysis Letters*, *26*(3), 329–337. DOI: 10.1007/BF00810606

Calderón-Magdaleno, M. Á., Mendoza-Nieto, J. A., & Klimova, T. E. (2014). Effect of the amount of citric acid used in the preparation of NiMo/SBA-15 catalysts on their performance in HDS of dibenzothiophene-type compounds. *Catalysis Today*, *220-222*(Supplement C), 78–88. DOI: 10.1016/j.cattod.2013.06.002

Cavani, F., Trifirò, F., & Vaccari, A. (1991). Hydrotalcite-type anionic clays: Preparation, properties and applications. *Catalysis Today*, *11*(2), 173–301. DOI: 10.1016/0920-5861(91)80068-K

Chen, W., Maugé, F., van Gestel, J., Nie, H., Li, D., & Long, X. (2013). Effect of modification of the alumina acidity on the properties of supported Mo and CoMo sulfide catalysts. *Journal of Catalysis*, *304*, 47–62. DOI: 10.1016/j.jcat.2013.03.004

Chen, Y.-H., Han, D.-M., Zhang, Q., & Cui, H.-X. (2020). In-situ synthesis of hierarchical lamellar ZSM-5 zeolite with enhanced MTP catalytic performance by a facile seed-assisted method. *Journal of Porous Materials*, *27*(5), 1265–1275. DOI: 10.1007/s10934-020-00898-w

Cinibulk, J., Kooyman, P. J., Vit, Z., & Zdražil, M. (2003). Magnesia-Supported Mo, CoMo and NiMo Sulfide Catalysts Prepared by Nonaqueous Impregnation: Parallel HDS/HDN of Thiophene and Pyridine and TEM Microstructure. *Catalysis Letters*, *89*(1), 147–152. DOI: 10.1023/A:1024748232539

Costa, D., Arrouvel, C., Breysse, M., Toulhoat, H., & Raybaud, P. (2007). Edge wetting effects of γ-Al2O3 and anatase-TiO2 supports by MoS2 and CoMoS active phases: A DFT study. *Journal of Catalysis*, *246*(2), 325–343. DOI: 10.1016/j.jcat.2006.12.007

Daage, M., & Chianelli, R. R. (1994). Structure-Function Relations in Molybdenum Sulfide Catalysts: The "Rim-Edge" Model. *Journal of Catalysis*, *149*(2), 414–427. DOI: 10.1006/jcat.1994.1308

de Jong, K. P. (2009). General Aspects *Synthesis of Solid Catalysts* (pp. 1-11).

Demikhova, N. R., Rubtsova, M. I., Kireev, G. A., Cherednichenko, K. A., Vinokurov, V. A., & Glotov, A. P. (2023). Micro-mesoporous catalysts based on ZSM-5 zeolite synthesized from natural clay nanotubes: Preparation and application in the isomerization of C-8 aromatic fraction. *Chemical Engineering Journal*, *453*, 139581. DOI: 10.1016/j.cej.2022.139581

Duan, A., Wan, G., Zhao, Z., Xu, C., Zheng, Y., Zhang, Y., Dou, T., Bao, X., & Chung, K. (2007). Characterization and activity of Mo supported catalysts for diesel deep hydrodesulphurization. *Catalysis Today*, *119*(1), 13–18. DOI: 10.1016/j.cattod.2006.08.049

Hagen, J. (2015). Economic Importance of Catalysts. In J. Hagen (Ed.) *Industrial Catalysis*. Wiley. doi: 9783527684625.ch17

Egorova, M., & Prins, R. (2004). Hydrodesulfurization of dibenzothiophene and 4,6-dimethyldibenzothiophene over sulfided NiMo/γ-Al2O3, CoMo/γ-Al2O3, and Mo/γ-Al2O3 catalysts. *Journal of Catalysis*, *225*(2), 417–427. DOI: 10.1016/j.jcat.2004.05.002

Fermoso, J., Hernando, H., Jana, P., Moreno, I., Přech, J., Ochoa-Hernández, C., Pizarro, P., Coronado, J. M., Čejka, J., & Serrano, D. P. (2016). Lamellar and pillared ZSM-5 zeolites modified with MgO and ZnO for catalytic fast-pyrolysis of eucalyptus woodchips. *Catalysis Today*, *277*, 171–181. DOI: 10.1016/j.cattod.2015.12.009

Fu, T., Chang, J., Shao, J., & Li, Z. (2017). Fabrication of a nano-sized ZSM-5 zeolite with intercrystalline mesopores for conversion of methanol to gasoline. *Journal of Energy Chemistry*, *26*(1), 139–146. DOI: 10.1016/j.jechem.2016.09.011

González-Cortés, S. L., Xiao, T.-C., Lin, T.-W., & Green, M. L. H. (2006). Influence of double promotion on HDS catalysts prepared by urea-matrix combustion synthesis. *Applied Catalysis A, General*, *302*(2), 264–273. DOI: 10.1016/j.apcata.2006.01.019

Guevara-Lara, A., Cruz-Pérez, A. E., Contreras-Valdez, Z., Mogica-Betancourt, J., Alvarez-Hernández, A., & Vrinat, M. (2010). Effect of Ni promoter in the oxide precursors of MoS2/MgO–Al2O3 catalysts tested in dibenzothiophene hydrodesulphurization. *Catalysis Today*, *149*(3), 288–294. DOI: 10.1016/j.cattod.2009.09.014

Gutiérrez, O. Y., Singh, S., Schachtl, E., Kim, J., Kondratieva, E., Hein, J., & Lercher, J. A. (2014). Effects of the Support on the Performance and Promotion of (Ni)MoS2 Catalysts for Simultaneous Hydrodenitrogenation and Hydrodesulfurization. *ACS Catalysis*, *4*(5), 1487–1499. DOI: 10.1021/cs500034d

Hartmann, M., Thommes, M., & Schwieger, W. (2021). Hierarchically-Ordered Zeolites: A Critical Assessment. *Advanced Materials Interfaces*, *8*(4), 2001841. DOI: 10.1002/admi.202001841

Hensen, E. J. M., de Beer, V. H. J., van Veen, J. A. R., & van Santen, R. A. (2002). A Refinement on the Notion of Type I and II (Co)MoS Phases in Hydrotreating Catalysts. *Catalysis Letters*, *84*(1), 59–67. DOI: 10.1023/A:1021024617582

Hernández, M. A., Petranovskii, V., & Quiroz, K. (2018). Estimation of Nanoporosity of ZSM-5 Zeolites as Hierarchical Materials. In Rashed, M. N., & Palanisamy, P. N. (Eds.), *Zeolites and Their Applications*. IntechOpen. DOI: 10.5772/intechopen.73624

Hinnemann, B., Nørskov, J. K., & Topsøe, H. (2005). A Density Functional Study of the Chemical Differences between Type I and Type II MoS2-Based Structures in Hydrotreating Catalysts. *The Journal of Physical Chemistry B, 109*(6), 2245–2253. DOI: 10.1021/jp048842y PMID: 16851217

Kaluža, L., Gulková, D., Vít, Z., & Zdražil, M. (2015). High-activity MgO-supported CoMo hydrodesulfurization catalysts prepared by non-aqueous impregnation. *Applied Catalysis B: Environmental, 162*(Supplement C), 430–436. DOI: 10.1016/j.apcatb.2014.07.016

Kerby, M. C., Degnan, T. F.Jr, Marler, D. O., & Beck, J. S. (2005). Advanced catalyst technology and applications for high quality fuels and lubricants. *Catalysis Today, 104*(1), 55–63. DOI: 10.1016/j.cattod.2005.03.028

Klicpera, T., & Zdražil, M. (2002). Preparation of High-Activity MgO-Supported Co–Mo and Ni–Mo Sulfide Hydrodesulfurization Catalysts. *Journal of Catalysis, 206*(2), 314–320. DOI: 10.1006/jcat.2001.3488

Klimova, T., Solís Casados, D., & Ramírez, J. (1998). New selective Mo and NiMo HDS catalysts supported on Al2O3–MgO(x) mixed oxides. *Catalysis Today, 43*(1), 135–146. DOI: 10.1016/S0920-5861(98)00142-4

Koekkoek, A. J. J., Xin, H., Yang, Q., Li, C., & Hensen, E. J. M. (2011). Hierarchically structured Fe/ZSM-5 as catalysts for the oxidation of benzene to phenol. *Microporous and Mesoporous Materials, 145*(1), 172–181. DOI: 10.1016/j.micromeso.2011.05.013

Linares, C. F., Vásquez, M., Castillo, R., Bretto, P., Solano, R., & Rincón, A. (2015). Applications of CoMo/calcined quaternary hydrotalcites for hydrotreatment reactions. *Fuel Processing Technology, 132*(Supplement C), 105–110. DOI: 10.1016/j.fuproc.2014.12.043

López-Cruz, C., Guzman, J., Cao, G., Martínez, C., & Corma, A. (2021). Modifying the catalytic properties of hydrotreating NiMo–S phases by changing the electrodonor capacity of the support. *Catalysis Today, 382*, 130–141. DOI: 10.1016/j.cattod.2021.08.002

Na, K., Park, W., Seo, Y., & Ryoo, R. (2011). Disordered Assembly of MFI Zeolite Nanosheets with a Large Volume of Intersheet Mesopores. *Chemistry of Materials, 23*(5), 1273–1279. DOI: 10.1021/cm103245m

Pawelec, B., Halachev, T., Olivas, A., & Zepeda, T. A. (2008). Impact of preparation method and support modification on the activity of mesoporous hydrotreating CoMo catalysts. *Applied Catalysis A, General*, *348*(1), 30–41. DOI: 10.1016/j.apcata.2008.06.014

Pérez-Cabrera, L., Díaz de León, J. N., Antúnez-García, J., Galván, D. H., Alonso-Núñez, G., & Fuentes-Moyado, S. (2019). Isomorphic substitution of Mg2+ by Al3+ on MgO: Effects on basicity, textural properties and microstructure. *Revista Mexicana de Ingeniería Química*, *18*(1), 9. DOI: 10.24275/uam/izt/dcbi/revmexingquim/2019v18n1/Perez

Prabhu, M. K., Louwen, J. N., Vogt, E. T. C., & Groot, I. M. N. (2025). Catalytic Hydrodesulfurization of Thiophene, Dibenzothiophene and 4,6-Dimethyldibenzothiophene on a CoMoS Catalyst. *Catalysis Letters*, *155*(3), 113. DOI: 10.1007/s10562-025-04953-6

Samanta, C., & Das, R. K. (2021). C3-Based Petrochemicals: Recent Advances in Processes and Catalysts. In Pant, K. K., Gupta, S. K., & Ahmad, E. (Eds.), *Catalysis for Clean Energy and Environmental Sustainability: Petrochemicals and Refining Processes* - (Vol. 2, pp. 149–204). Springer International Publishing. DOI: 10.1007/978-3-030-65021-6_5

Semeykina, V. S., Parkhomchuk, E. V., Polukhin, A. V., Parunin, P. D., Lysikov, A. I., Ayupov, A. B., & Glazneva, T. S. (2016). CoMoNi Catalyst Texture and Surface Properties in Heavy Oil Processing. Part II: Macroporous Sepiolite-Like Mineral. *Industrial & Engineering Chemistry Research*, *55*(34), 9129–9139. DOI: 10.1021/acs.iecr.6b01208

Shelyapina, M. G., Yocupicio-Gaxiola, R. I., Zhelezniak, I. V., Chislov, M. V., Antúnez-García, J., Murrieta-Rico, F. N., & Fuentes-Moyado, S. (2020). Local Structures of Two-Dimensional Zeolites—Mordenite and ZSM-5—Probed by Multinuclear NMR. *Molecules (Basel, Switzerland)*, *25*(20), 4678. DOI: 10.3390/molecules25204678 PMID: 33066351

Silvy, R. P. (2004). Future trends in the refining catalyst market. *Applied Catalysis A, General*, *261*(2), 247–252. DOI: 10.1016/j.apcata.2003.11.019

Sing, K. S. W. (2014). *7 - Assessment of Surface Area by Gas Adsorption, Adsorption by Powders and Porous Solids* (2nd ed.). Academic Press.

Stanislaus, A., Marafi, A., & Rana, M. S. (2010). Recent advances in the science and technology of ultra low sulfur diesel (ULSD) production. *Catalysis Today*, *153*(1), 1–68. DOI: 10.1016/j.cattod.2010.05.011

Suresh, C., Pérez-Cabrera, L., Díaz de León, J. N., Zepeda, T. A., Alonso-Núñez, G., & Fuentes-Moyado, S. (2017). Highly active CoMo/Al (10) KIT-6 catalysts for HDS of DBT: Role of structure and aluminum heteroatom in the support matrix. *Catalysis Today*, *296*(Supplement C), 214–218. DOI: 10.1016/j.cattod.2017.04.048

Suriya, P., Xu, S., Ding, S., Chansai, S., Jiao, Y., Hurd, J., Lee, D., Zhang, Y., Hardacre, C., Reubroycharoen, P., & Fan, X. (2024). Ethanol steam reforming over Ni/ZSM-5 nanosheet for hydrogen production. *Chinese Journal of Chemical Engineering*, *67*, 247–256. DOI: 10.1016/j.cjche.2023.12.006

Tanimu, A., & Alhooshani, K. (2019). Advanced Hydrodesulfurization Catalysts: A Review of Design and Synthesis. *Energy & Fuels*, *33*(4), 2810–2838. DOI: 10.1021/acs.energyfuels.9b00354

Tian, Y., He, X., Chen, X., Qiao, C., Wang, H., Diao, Z., & Liu, G. (2024). Transition metal modified hierarchical ZSM-5 nanosheet for catalytic cracking of n-pentane to light olefins. *Fuel*, *363*, 130902. DOI: 10.1016/j.fuel.2024.130902

Timoshev, V., Haufe, L. A., Busse, O., Hamedi, H., Seifert, M., & Weigand, J. J. (2024). Recycling of Spent FCC Catalysts: Conversion of Leached Residues to Zeolite ZSM-5. *ChemSusChem*, *17*(14), e202301642. DOI: 10.1002/cssc.202301642 PMID: 38462539

Topsøe, H., Clausen, B. S., & Massoth, F. E. (1996). Hydrotreating Catalysis. In Anderson, J. R., & Boudart, M. (Eds.), *Catalysis: Science and Technology* (pp. 1–269). Springer Berlin Heidelberg. DOI: 10.1007/978-3-642-61040-0_1

Topsøe, H., Clausen, B. S., Topsøe, N. Y., Nørskov, J. K., Ovesen, C. V., & Jacobsen, C. J. H. (1995). The Bond Energy Model for Hydrotreating Reactions: Theoretical and Experimental Aspects. *Bulletin des Sociétés Chimiques Belges*, *104*(4-5), 283–291. DOI: 10.1002/bscb.19951040415

Wu, L., Jiao, D., Wang, J., Chen, L., & Cao, F. (2009). The role of MgO in the formation of surface active phases of CoMo/Al2O3–MgO catalysts for hydrodesulfurization of dibenzothiophene. *Catalysis Communications*, *11*(4), 302–305. DOI: 10.1016/j.catcom.2009.10.021

Wu, L. B., Jiao, D. M., Chen, L. F., Wang, J. A., & Cao, F. H. (2010). Comparative studies of the CoMo/MgO, CoMo/Al2O3 and CoMo/MgO-MgAl2O4 catalysts prepared by a urea-matrix combustion method. Advanced Materials Research, 132, 45-54.

Yocupicio, R. I., Díaz de León, J. N., Zepeda, T. A., & Fuentes, S. (2017). Study of CoMo catalysts supported on hierarchical mesoporous zeolites for hydrodesulfurization of dibenzothiophene. *Revista Mexicana de Ingeniería Química*, *16*(2), 503–520.

Yocupicio-Gaxiola, R. I., Petranovskii, V., Antúnez-García, J., & Fuentes Moyado, S. (2019). One-pot synthesis of lamellar mordenite and ZSM-5 zeolites and subsequent pillaring by amorphous SiO2. *Applied Nanoscience*, *9*(4), 557–565. DOI: 10.1007/s13204-018-0935-1

Zepeda, T. A. (2025). Ultrasonic-assistance influence on the impregnation method for high metal loading CoMo/Ti-HMS catalysts. *Applied Catalysis A, General*, *693*, 120135. DOI: 10.1016/j.apcata.2025.120135

Zepeda, T. A., Pawelec, B., Fierro, J. L. G., Olivas, A., Fuentes, S., & Halachev, T. (2008). Effect of Al and Ti content in HMS material on the catalytic activity of NiMo and CoMo hydrotreating catalysts in the HDS of DBT. *Microporous and Mesoporous Materials*, *111*(1), 157–170. DOI: 10.1016/j.micromeso.2007.07.025

Zhang, S., Liu, J., Zu, L., Wang, B., & Li, K. (2025). Analysis of real-world diesel sulphur in diesel engines in China's key regions. *Transportation Research Part D, Transport and Environment*, *139*, 104562. DOI: 10.1016/j.trd.2024.104562

Zhang, Y., Han, W., Long, X., & Nie, H. (2016). Redispersion effects of citric acid on CoMo/γ-Al2O3 hydrodesulfurization catalysts. *Catalysis Communications*, *82*(Supplement C), 20–23. DOI: 10.1016/j.catcom.2016.04.012

Zhao, R., Yin, C., Zhao, H., & Liu, C. (2003). Synthesis, characterization, and application of hydotalcites in hydrodesulfurization of FCC gasoline. *Fuel Processing Technology*, *81*(3), 201–209. DOI: 10.1016/S0378-3820(03)00012-2

Zhou, Z., Jiang, R., Chen, X., Wang, X., & Hou, H. (2021). One-step synthesis of hierarchical lamellar H-ZSM-5 zeolite and catalytic performance of methanol to olefin. *Journal of Solid State Chemistry*, *298*, 122132. DOI: 10.1016/j.jssc.2021.122132

Chapter 7 Clinoptilolite Perspectives as a Zeolite for Removing Pollutants From Water: A Review

Manuel Alberto Flores-Hidalgo
https://orcid.org/0000-0001-5406-0795
Universidad Juarez del Estado de Durango, Mexico

Diana Barraza-Jiménez
Universidad Juárez del Estado de Durango, Mexico

Heimi V. Ríos-Arámbula
Universidad Juárez del Estado de Durango, Mexico

Hugo I. Flores-Hidalgo
TecNM, Jiménez Institute of Technology, Mexico

Sandra I. Torres-Herrera
Universidad Juárez del Estado de Durango, Mexico

ABSTRACT

A review on clinoptilolite, a natural zeolite, which is widely used in water remediation due to its porous structure and ion exchange properties. It effectively removes pollutants like heavy metals, ammonium, nitrates, and organic compounds from water sources through selective adsorption, offering a cost-effective, eco-friendly solution for treating industrial wastewater, groundwater, and drinking water. Its efficacy varies with factors such as origin and modification, requiring tailored optimization. Research employs advanced characterization (e.g., electron microscopy,

DOI: 10.4018/979-8-3373-0320-8.ch007

X-ray diffraction, etc.) and computational simulations to enhance understanding of its pollutant removal mechanisms. With its versatility, affordability, and environmental benefits, clinoptilolite is a promising tool for sustainable water management. This review synthesizes its strengths, limitations, and future potential, emphasizing the need for continued innovation to address global water quality challenges.

1. INTRODUCTION

Water pollution remains a pressing issue, endangering human health and ecosystems worldwide despite mitigation efforts (Pérez-Botella et al., 2022; Li et al., 2021). Persistent pollutants, including HM and EPs, require innovative remediation strategies (Abdelwahab et al., 2023). Clinoptilolite, a naturally abundant zeolite, stands out for its high cation exchange capacity, large SA, and selective adsorption, effectively targeting pollutants like arsenic, fluoride, and pharmaceutical residues (Helmi et al., 2024; Restiawaty et al., 2024). This review explores clinoptilolite's role in water remediation, detailing its structure, properties, and pollutant removal mechanisms (Velarde et al., 2023; Hu et al., 2022). We evaluate its efficacy in applications such as industrial wastewater treatment, groundwater purification, and desalination, and consider its broader benefits in soil remediation and medical uses (Reid et al., 2021; Javaid et al., 2024). The manuscript also covers preparation techniques, advanced characterization, and computational simulations that enhance its performance (Jin et al., 2023). With its versatility, low cost, and environmental friendliness, clinoptilolite offers a promising solution for global water quality challenges, though its effectiveness requires optimization based on specific conditions (Perego et al., 2021; Sun et al., 2022). This work serves as a comprehensive resource, highlighting clinoptilolite's strengths, limitations, and future potential in sustainable water management.

2. CLINOPTILOLITE STRUCTURE AND MORPHOLOGY

More than 200 years ago, the discovery of a new porous mineral called zeolite revolutionized various applications over time. Zeolite is a hydrated aluminosilicate that can be either natural or synthetic. Natural zeolites are of volcanic origin and are formed by the action of alkaline or sea water on volcanic glasses in marine sediments. There are more than 40 types of natural zeolites, among which clinoptilolite is the most notable due to its three-dimensional microporous structure, which gives it unique properties (Bahmanzadegan et al., 2025; Ullah et al., 2020). Synthetic zeolites are generated through preparation in laboratories using specialized equipment, and there are around 150 types manufactured from natural or synthetic precursors by hydrothermal processes (Bahmanzadegan et al., 2025).

2.1 Natural Clinoptilolite

Clinoptilolite, a natural zeolite, occurs in volcanic and sedimentary formations, notably in tuffs from the devitrification of siliceous volcanic crystals or within cavities of rhyolitic, andesitic, or basaltic rocks (Xu et al., 2025). Its ion exchange (IE) and adsorption properties emerged in the 20th century, though its mineral nature puzzled scientists until Schaller, between 1923 and 1932, identified it as a monoclinic dimorph of ptiloptilolite, naming it clinoptilolite (Xu et al., 2025). Chemically, it's a "hydrated aluminosilicate of potassium, calcium, and sodium" with the formula $K_2\ Na_2\ Ca\ \left(Al_6 Si_{30} O_{72}\right) \bullet H_2O$ per unit cell. Rich in silicon and aluminum, it features exchangeable cations—sodium, potassium, and calcium—forming channels and cavities that house cations and water molecules. These structural traits enable its exceptional adsorption and IE capabilities, making it valuable for pollutant removal and enhancing water and soil quality (Rodríguez-Iznaga et al., 2022; Grass et al., 2021). Major producers of natural clinoptilolite today include the United States (Nevada, Oregon, California), China, Russia (Siberia), Japan, Hungary, Slovakia, Bulgaria, Mexico (Sonora, Chihuahua), Cuba, and Turkey (Velarde et al., 2023; Mortazavi et al., 2021).

2.2 Clinoptilolite Species

Clinoptilolite has different species or variations that are the result of the geological conditions and elements present during the formation of the mineral. However, they are still considered a single mineral species, clinoptilolite, because they have the same crystalline structure and similar properties. (Shekarchi et al., 2023; Grifasi et al., 2024). Among the main species are: Clinoptilolite-Na, K, Ca, Mg, NH_4. Each one has different characteristics and properties since they come from different geological environments, therefore they will have applications that will differentiate them, observed in Table 1.

Table 1. Species and characteristics of clinoptilolite found in Mexico and around the world.

Species/ variation	Chemical formula	Dominant cation	Place of origin	Application	Ref.
Clinoptilolite-Na	$(Na,K)_6(Si_{30}Al_6)O_{72}\cdot 24H_2O$	Sodium (Na^+)	Oaxaca, Mexico	Water purification and HM removal.	(Velarde et al., 2023)
Clinoptilolite-K	$(K,Na)_6(Si_{30}Al_6)O_{72}\cdot 24H_2O$	Potassium (K^+)	Sonora, Mexico	Soil amendment and nutrient retention.	(De la Rosa-Gómez, et al., 2007)
Clinoptilolite-Ca	$(Ca,Na,K)_6(Si_{30}Al_6)O_{72}\cdot 24H_2O$	Calcium (Ca^{2+})	Chihuahua, Mexico	Wastewater treatment and ammonia removal.	(Velarde et al., 2023)
Clinoptilolite-NH_4	$(NH_4,Na,K)_6(Si_{30}Al_6)O_{72}\cdot 24H_2O$	Ammonium (NH_4^+)	Hidalgo, Mexico	Ammonia adsorption in wastewater treatment.	(De Pablo et al., 2014)
Clinoptilolite-Ca	$(Ca,Na,K)_6(Si_{30}Al_6)O_{72}\cdot 24H_2O$	Calcium (Ca^{2+})	Zacatecas, Mexico	Environmental remediation and soil conditioning.	(De Pablo et al., 2014)
Clinoptilolite-K	$(K,Na)_6(Si_{30}Al_6)O_{72}\cdot 24H_2O$	Potassium (K^+)	Morelia, Mexico	Agricultural use, water and soil decontamination.	(Ostroumov et al., 2011)
Clinoptilolite-Ca	$(Ca,Na,K)_6(Si_{30}Al_6)O_{72}\cdot 24H_2O$	Calcium (Ca^{2+})	Turkey	Wastewater and ammonia treatment.	(Sarioglu et al., 2004)
Clinoptilolite-Mg	$(Mg,Na,K)_6(Si_{30}Al_6)O_{72}\cdot 24H_2O$	Magnesium(Mg^{2+})	Greece	Environmental remediation and soil conditioning.	(Velarde et al., 2023)
Clinoptilolite-Na	$(Na,K)_6(Si_{30}Al_6)O_{72}\cdot 24H_2O$	Sodium (Na^+)	Bulgaria	Used for gas separation and environmental protection.	(Velarde et al., 2023)
Clinoptilolite-K	$(K,Na)_6(Si_{30}Al_6)O_{72}\cdot 24H_2O$	Potassium (K^+)	China	Used in agriculture and industrial catalyst.	(Velarde et al., 2023)
Clinoptilolite-Ca	$(Ca,Na,K)_6(Si_{30}Al_6)O_{72}\cdot 24H_2O$	Calcium (Ca^{2+})	Iran	HM removal and water purification.	(Velarde et al., 2023)
Clinoptilolite-Mg	$(Mg,Na,K)_6(Si_{30}Al_6)O_{72}\cdot 24H_2O$	Magnesium(Mg^{2+})	Australia	Environmental remediation and as a soil conditioner.	(Velarde et al., 2023)
Clinoptilolite-NH_4	$(NH_4,Na,K)_6(Si_{30}Al_6)O_{72}\cdot 24H_2O$	Ammonium (NH_4^+)	Hungary	Ammonia removal and agricultural applications.	(Velarde et al., 2023)

2.3 Clinoptilolite Properties

Clinoptilolite is a versatile material with unique properties ranging from adsorption of pollutants to its use in medicine (Wahono et al., 2019). Due to the defined pore size, internal cavities and presence of active centers, clinoptilolite has a high IE capacity, mechanical strength, resistance to corrosive media, good porosity, among others; which makes it valuable in various industrial, environmental and health applications (Dikić et al., 2019; Yang S et al., 2024). As observed in Figure 1, this zeolite has properties with desirable indices for water treatment, such as a surface area (SA) of 27 m2/g, a pore size of 7.0, a density of 2.2-2.4 g/cm^3, with a high adsorption capacity for HM (95.4% Fe, 96.0% Cu and 85.1% Cr) and with good thermal stability that is essential for its characterization. (Grifasi et al., 2024)

Figure 1. Classification of clinoptilolite properties.

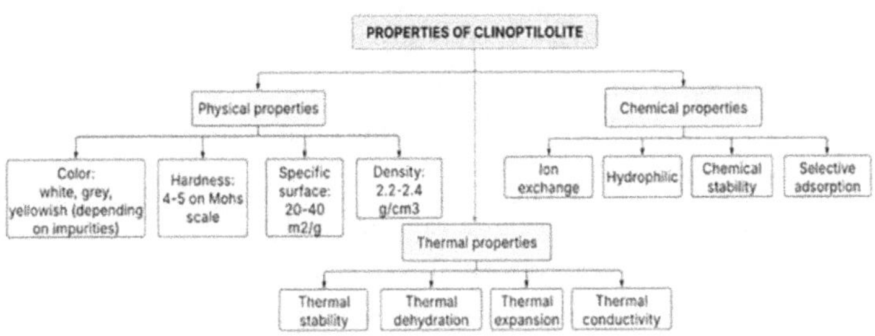

3. PREPARATION METHODS

The purification process refers to the process of removing impurities and other unwanted minerals that may be present in the sample, thus improving the properties of the clinoptilolite. (Grifasi et al., 2024; Biblioteca et al., 2023). Conditioning is a process designed to optimize the physical and surface properties of the clinoptilolite and thus improve its performance. Purification seeks to improve the chemical and mineralogical quality of the clinoptilolite, while conditioning focuses on optimizing its presentation and functionality for practical use (Rapado et al., 2023). A study was conducted comparing natural clinoptilolite and clinoptilolite conditioned with Na in order to change its structure and remove a greater amount of Ni^{2+}, resulting in a removal efficiency of natural clinoptilolite of 73.2%, while after conditioning it reaches 96.6%. (Biblioteca et al., 2023).

3.1 Purification Methods.

3.1.1 Density Flotation

This is a widely used method to separate materials based on their density differences, and is carried out when particles with a lower density than that of the liquid in which they are immersed float, while denser particles sink (Rabia et al., 2019; Kashi et al., 2023). In one study, this process was carried out when HM such as Pb, Ba and Zn were raised to clarify the solution. The best recoveries could be obtained in the particle size range of 20 to 50 μm, with removal of up to 97%. (Ahmed et al., 2022)

3.1.2 Ion Exchange

It is a process that consists of replacing ions in the clinoptilolite with specific ions that can then be removed, leaving the purified clinoptilolite, and is repeated several times to ensure high purity in the separation of the clinoptilolite. Its purpose is to improve the ionic composition to improve selectivity for certain ions (Li et al., 2020; Kabuba et al., 2020). Depending on the type of functional group, ion exchangers can be synthetic or natural and can be divided into several types that are strongly acidic, strongly basic, weakly acidic or weakly basic (Moradi et al., 2018). In a study, the IE process for the removal of Ni (II) and Co (II) from wastewater was analyzed using clinoptilolite modified with succinic acid and ethylenediamine tetraacetic acid, resulting in a maximum removal percentage of 92.80% for Ni (II) and 33.67% for Co (II) (Kabuba et al., 2020).

3.1.3 Magnetization

This is a technique that consists of coating its particles or impregnating its pores with magnetic materials in its structure or surface to improve the capacity of the clinoptilolite and facilitate its recovery by magnetic methods (Badeenezhad et al., 2019; Foroutan et al., 2021). The general process begins with a solution containing iron salts, such as ferric chloride or ferrous sulfate, then the clinoptilolite is immersed for a certain time allowing the iron ions to penetrate its porous structure, allowing it to be separated from non-magnetic impurities (Badeenezhad et al., 2019). In a study, a magnetization of the iron oxide-coated clinoptilolite zeolite was performed to remove methylene blue from dye solutions with pH 7 giving an elimination effectiveness of 98.3%. [50] (Badeenezhad et al., 2019)

3.2 Conditioning Methods.

3.2.1 Sieving or Classification by Particle Size

This is a simple method in which sieves with specific meshes are used to select particles of the desired size, eliminating fine material (dust or clays), large fragments (non-zeolitic rocks, sand) or other undesirable components (Znak et al., 2021). These processes allow obtaining a high quality and purity product, optimizing its performance in various industrial and environmental applications (Bukka et al., 2024; Manzoli et al., 2021). As can be seen in Figure 2, the natural clinoptilolite arrived from the mine with a particle size of 0.5 cm, so it was subjected to a size reduction with the Laboratory Sieve Shaker model RX-29 equipment with which a particle size of less than 38 μm was achieved. (Montes-Luna et al., 2015).

Figure 2. Clinoptilolite classified by particle size. (a) natural clinoptilolite with a particle size <0.5 cm (b) clinoptilolite particle size less than 38 μm.

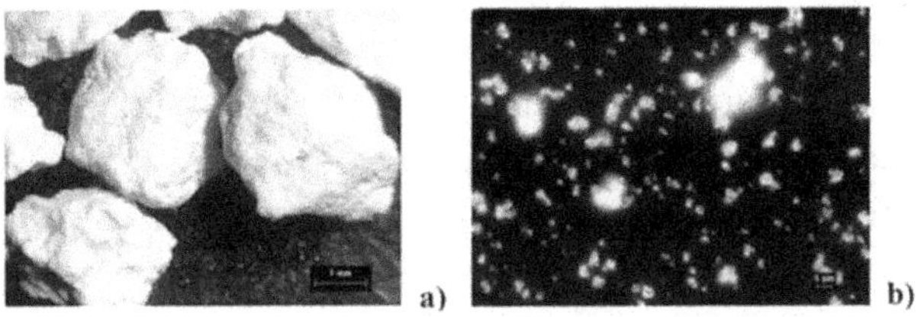

(Montes-Luna et al., 2015).

3.2.2 Drying

Drying It is a crucial stage in the conditioning of minerals to maximize their adsorption capacity. This process removes the water adsorbed on the surface, extracts the water present inside the pores and channels of the crystalline structure and optimizes the capacity of the mineral (Ottoboni et al., 2020). Some of the most common methods are: hot air drying, freeze drying, microwave drying and infrared drying (Shoinkhorova et al., 2019). In a study it was proven that drying is a very important process for the modification of natural clinoptilolite and thus have better properties. After washing, filtering and adding iron chloride, the sample was dried in the oven at 70 °C for 24 h, obtaining the iron-modified clinoptilolite. (Nami et al., 2023)

3.2.3 Compaction

This is an important process to improve the physical properties of the material, which involves transforming the clinoptilolite into dense or compact forms (granules, pellets or blocks) without altering its chemical and adsorption properties (Bauer et al., 2024). Some of the most common methods include: mechanical compaction (roller presses or extruders), bond compaction, extrusion compaction, sinter compaction and block compaction (Ogorek et al., 2024). In a study, clinoptilolite/Na-PAA modified compacted clay was tested for water resistance by static compaction and press molding to make the sample better serve as an effective and economical barrier for preventing seepage and blocking gases in landfills. (Sun et al., 2023)

4. SYNTHESIS AND CHARACTERIZATION OF CLINOPTILOLITE

Characterization is a set of analytical techniques that allow the selection of the appropriate zeolite, its chemical composition, crystalline structure, physical properties and the development of new zeolites with improved properties (Gattucci et al., 2025). Synthesis is a process that involves the artificial creation of this mineral from basic chemical components. The objective is to obtain a material with specific characteristics for specific applications (Jin et al., 2023; Güngör et al., 2021). Some of the most commonly used clinoptilolites in water treatment are: manganese oxide coated clinoptilolite for the removal of Mn(II) with a removal effectiveness of 95.51% (Jin et al., 2023)., clinoptilolite treated with HCl, HNO_3 and $C_2H_2O_4$ to have up to 96% hydrogen adsorption (Akyalcin et al., 2024) and clinoptilolite modified with (Fe) salts for the removal of arsenic with an effectiveness of 97%.

4.1 Synthesis Methods.

4.1.1 Hydrothermal Synthesis

It is one of the most common methods for obtaining synthetic clinoptilolite. This process takes advantage of extreme temperature and pressure conditions in the presence of water to dissolve and recrystallize the precursor materials (Derbe et al., 2021; Gul et al., 2021). In its general process, the precursors are first selected, then mixed with mineralizing agents, then this mixture is introduced into an autoclave, cooled slowly and filtered to separate the crystals and finally, the crystals are dried and characterized by different techniques (XRD, SEM, IR, etc.) (Deneyer et al., 2020). In studies carried out, TiO_2-supported clinoptilolite was synthesized by hydrothermal crystallization for the removal of crystalline violet dyes from aqueous media with an effectiveness of 95%. (Ullah et al., 2020)

4.1.2 Sol-Gel Synthesis

It is a technique used to synthesize materials from chemical precursors in liquid phase. It allows to control the structure and morphology of the material at nanometric level (Kianfar et al., 2019; Carstens et al., 2019). The process begins when the inorganic precursors are dissolved in an organic solvent so that they are hydrolyzed and condensed forming a sol. This sol evolves into a gel. The gel undergoes an aging process and dries, eliminating the solvent and obtaining a xerogel. The xerogel is calcined to eliminate the organic groups and crystallize the clinoptilolite. (Ibrahim et al., 2023). In a study, a sol-gel synthesis was carried out to prepare a new FeO(OH) clinoptilolite adsorbent effective for the removal of antimony with 96% removal effectiveness. (Chmielewská et al., 2017)

4.1.3 Impregnation

This is a technique that consists of immersing the clinoptilolite in a concentrated solution of a chemical substance (metals, metal oxides, organic compounds and acids). The purpose of this technique is to improve the adsorption capacity, add catalytic properties, modify the selectivity and increase the antimicrobial activity (Mehrabadi et al., 2018; Kusrini et al., 2018). The process begins with the cleaning and purification of the clinoptilolite, then the impregnation solution is prepared to immerse the clinoptilolite and be able to carry out this process, and finally the material is dried at moderate temperatures to eliminate the solvent. (Ugrina et al., 2023). In a study, the removal of sulfide in an aqueous solution using clinoptilolite impregnated with transition metal ions was analyzed, obtaining an effectiveness of 86.5%. (Karimi et al., 2017)

4.2 Characterization Methods

4.2.1 Electron Microscopy

This is a technique that allows obtaining images at the nanometric scale and analyzing their morphology, particle size, elemental composition, structure and crystalline defects. Within this characterization method there are various techniques, of which two stand out: SEM, which is responsible for analyzing the shape and size of the particles (Zhang et al., 2020; Favvas et al., 2016). TEM is responsible for studying the crystalline structure at the atomic level. (Pabiś-Mazgaj et al., 2021). As seen in Figure 3, in a study natural clinoptilolite was analyzed for the adsorption of an industrial dye using the SEM technique, which revealed a specific particle and pore size. (Shankar et al., 2023).

Figure 3. Natural clinoptilolite subjected to SEM technique, with a particle size of 13 nm (Fig. a) and pores ranging from 125 nm to 410 nm (Fig. b).

(Shankar et al., 2023).

4.2.2 X-Ray Diffraction

It is a technique used to confirm the presence of clinoptilolite, identify impurities, crystal lattice parameters, crystal size, and evaluate crystallinity (Loiola et al., 2022; Islam et al., 2023). X-rays are incident on a crystalline sample and a resulting diffraction pattern is created for each material containing information about the distance between atomic planes, the symmetry of the unit cell, and the chemical composition (Wang et al., 2025). In a study, natural clinoptilolite was analyzed for adsorbing industrial dye where XRD technique was used. As shown in Figure 4, the narrow and intense peaks at $2\theta = 20.850°$, $26.652°$, which were assigned to SiO_2, there are also peaks at $2\theta = 9.483°$, $11.159°$, which were assigned to Ca-clinoptilolite. This indicates that the clinoptilolite used in this study is composed of quartz sand and calcium-type clinoptilolite (Jin et al., 2023).

Figure 4. Clinoptilolite by XRD technique, resulting in a composition of quartz sand and calcium-type clinoptilolite

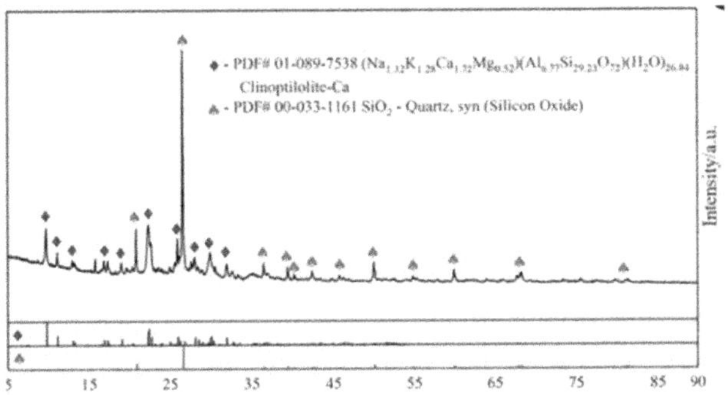

(Jin et al., 2023).

4.2.3 BET Characterization

This technique is based on the physical adsorption of a gas (usually nitrogen) on the surface of a solid to obtain an isotherm and thus be able to calculate the pore volume, the specific surface, the adsorption capacity and the catalytic properties of the material. In a study, natural clinoptilolite was analyzed to adsorb an industrial dye using the BET technique. (Shimizu et al., 2020; Aili et al., 2020). As can be seen in Figure 5, from the N_2 adsorption isotherm there is a curve between relative pressures of 0.4 to 1, so this type of adsorption occurs by the extension of the pore width depending on the adsorption system and the temperature. (Shankar et al., 2023).

Figure 5. Clinoptilolite using the BET technique, producing an N_2 adsorption isotherm.

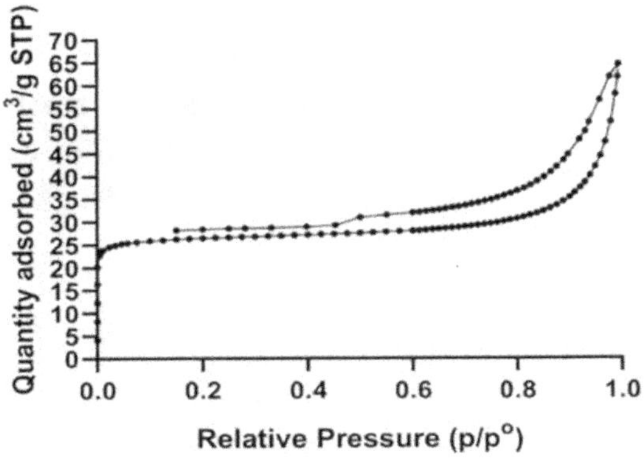

(Shankar et al., 2023).

5. COMPUTATIONAL SIMULATION OF CLINOPTILOLITE

Computer simulation is a tool to study and predict the behavior of materials at the molecular and macroscopic level. It allows exploring atomic properties and behaviors that are difficult or impossible to observe experimentally, such as the visualization and quantification of their structure, adsorption capacity at the molecular level and optimizing the material with specific properties for particular applications (Harami et al., 2020; Abatal et al., 2019). Combining methods such as DFT, molecular dynamics and Monte Carlo or programs such as ANSYS and machine learning (ML) allow analyzing their structure, properties and performance in specific applications, optimizing their functionality and reducing experimental costs. (Dryaz et al., 2021; Lappala et al., 2024)

5.1 Density Functional Theory

It is based on two fundamental theorems: the Hohenberg-Kohn theorem, which states that the ground state energy is determined by the electron density, and the Kohn-Sham theorem, which introduces a system of independent particles that generates the same electron density as the real system (Farooq et al., 2024; Fischer et al., 2023). It is a tool used to obtain detailed information about: electron distribution, energy states, charge density, interatomic distances, interaction of molecules, adsorption energy and adsorption geometry (Luo et al., 2025). In a study, the DFT based simulation of nitro-organic pollutants 4-nitrophenol, 2-nitrophenol, 9-nitroanthracene interacting with water in the imidazole framework of zeolite (ZIF-8) was analyzed. According to Figure 6, all peaks in the simulated DFT spectrum are in good agreement with the experimental XRD measurement, in parentheses the calculated DFTs are: (001) 7.5o [7.4o], (200) 10.5o [10.4o], (211) 12.8o[12.8o], (220). (Adjal et al., 2024).

Figure 6. Experimental XRD pattern (black) and calculated DFT (blue) of ZIF-8.

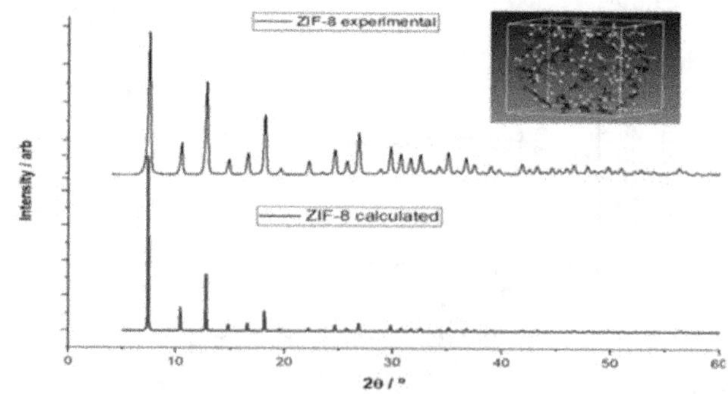

(Adjal et al., 2024).

5.2 Multi-physics Simulations using ANSYS

It is one of the most widely used finite element simulation software in the industry, offering tools and capabilities to perform multiphysics analysis (Chwatal et al., 2024; Morgan et al., 2020). Clinoptilolite can be simulated with ANSYS because it has a complex and difficult-to-model structure, different processes are optimized (adsorption, IE and catalysis) and it allows reducing the number of physical experiments, which translates into savings in time and resources. (Kwon et al., 2023; Faisal et al., 2023). In a study, the mechanical properties of high-strength concrete using natural zeolite were analyzed, where a deformation diagram was made for the compressive strength of the concrete in ANSYS (Gowram et al., 2024)

5.3 Machine Learning

An artificial intelligence tool that allows analyzing experimental data and simulations to build predictive models that can be used in various applications such as porosity analysis, adsorption energies, synthetic data generation and dynamic process simulation (Mahesh et al., 2020; Li et al., 2023). As shown in Figure 7, in one study, descriptors for a total of 590,448 zeolite structures from the PCOD2 database and 195 zeolite structures from IZA were calculated using ML to provide general characteristics, elastic properties, and pore volume enhancement. (Evans et al., 2017).

Figure 7. Summary of the descriptors used in the ML model, classified into local, structural, and porosity-related descriptors.

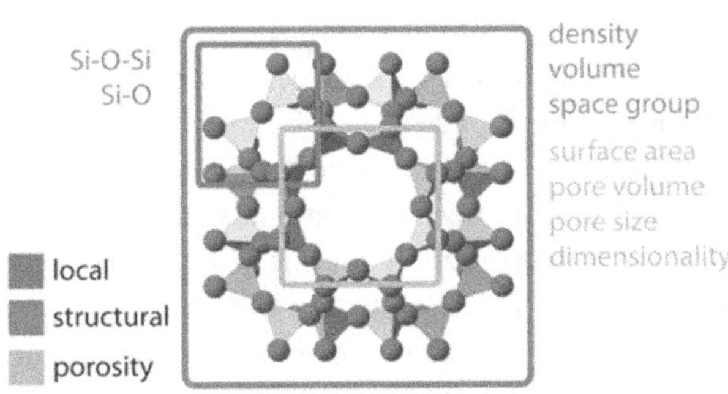

(Evans et al., 2017).

6. CLINOPTILOLITE FOR THE REMOVAL OF POLLUTANTS FROM WATER

Clinoptilolite has unique adsorption and IE properties and its high selectivity towards various pollutants, which makes it have various applications in different areas, one of which is the removal of pollutants in water (Grifasi et al., 2024; Nami et al., 2023). Its effectiveness, versatility and relative abundance make it an attractive material for the treatment and purification of water in different contexts, thanks to its adsorption, IE and filtration properties (Grifasi et al., 2024). The natural zeolite clinoptilolite is used in the purification of water from industrial effluents, in environmental protection and in aquaristics to improve water quality. It is ideal for treating water in ponds, biological pools and aquariums and reduces ammonium and phosphates, with the benefits that this brings to the health of fish and to prevent the proliferation of algae in the pond. (Grifasi et al., 2024). Clinoptilolite is used in wastewater treatment systems to remove pollutants such as oils, greases, dyes and chemicals through adsorption and IE processes. The general process begins with pretreatment of wastewater, then it comes into contact with clinoptilolite, then it is absorbed, thus removing pollutants where they are trapped in the porous structure of the clinoptilolite (Grifasi et al., 2024; Sahu et al., 2024).

6.1 Clinoptilolite for the Removal of Arsenic from Water

Arsenic is a highly toxic chemical element found naturally in the Earth's crust. When groundwater comes into contact with rocks and soils containing arsenic, arsenic can dissolve and pollutants water sources (Liu et al., 2019). This HM, in its forms arsenite (As^{3+}) and arsenate (As^{5+}), is highly toxic, so its efficient removal is crucial to ensure the safety of drinking water (Nguyen et al., 2019). Clinoptilolite has been shown to be a highly effective adsorbent for the removal of arsenic from water. The mechanism of arsenic adsorption by this zeolite begins when arsenic ions are exchanged with cations at the zeolite sites, then arsenic molecules can be adsorbed on the clinoptilolite surface and finally arsenic can form complexes with the functional groups present on the zeolite surface (Bilici Baskan et al., 2021; Khan et al., 2021). For clinoptilolite to be effective in water treatment, it depends greatly on its place of origin, whether it is natural or modified, its chemical composition, or the conditions under which it is treated.

In Table 2, it can be observed that modified clinoptilolite has a higher As removal efficiency than natural clinoptilolite, since when this material is modified, its properties and adsorption characteristics are improved.

Table 2. Adsorption effect of arsenic by different adsorbent materials.

Material	Place of origin	Adsorbent amount (g)	Contact time (h)	T (°C)	pH	Mechanism Removal	Efficiency (%)	Ref.
Zeolite-reduced graphene oxide (ZrGO)	Merck Laboratory	0.1	1.5	25	8	Adsorption	97	(Soni et al., 2018)
Natural clinoptilolite with sulfuric acid	Afrazand Company, Iran	480	4	-	8	Adsorption	96.7	(Shokrolahzadeh et al., 2017)
Iron oxide-modified clinoptilolite	New Mexico	1	67	Environment	6.5	Adsorption	93	(Camacho et al., 2011)
Natural clinoptilolite-Ca zeolite	New Mexico	1	48	Environment	6.5	Adsorption	92	(Camacho et al., 2011)
Magnesium dioxide-modified clinoptilolite	Vietnam	10	4	Environment	7	Adsorption	50-60	(Nguyen et al., 2019)
Natural magnetic rock	Yunlin, Taiwan	0.002	2	25	8.3	Adsorption	95-99	(Maity et al., 2020)
Iron oxide filters	Sigma Aldrich Laboratory, Merck.	0.02	0.5	30	7	Adsorption	87-95	(Casentini et al., 2019)
Titanium dioxide (TiO_2)	Mississippi, USA	1.0	1-2	35	6-8	Adsorption	75-90	(Mukherjee et al., 2021)
Skye sand	Victoria, Australia	1.0	24	22	7.0	Adsorption	45	(Khan et al., 2021)
Granular ferric hydroxide sand (GFH)	Osnabrueck, Germany.	1.0	24	22	7.0	Adsorption	40	(Khan et al., 2021)
Red sand NT	Alice Springs, Australia	1.0	24	22	7.0	Adsorption	13	(Khan et al., 2021)

6.2 Clinoptilolite for the Removal of Fluor from Water

Fluoride is found naturally in some water sources, but can also be released into the environment by industrial activities. However, excess fluoride in drinking water can cause health problems such as dental and skeletal fluorosis. Clinoptilolite has proven to be an effective adsorbent for removing excess fluoride from drinking water. The main removal mechanisms include IE, physical and chemical adsorption, and electrostatic attraction (Chen et al., 2025). This zeolite in its modified form is a more effective and economical material to improve fluoride adsorption. Among the most common modifications are modifications with metal oxides, acid activation, cation exchange and nanocomposites. (Ramasamy et al., 2025). As shown in Figure 8, there are different processes used in water defluorination, but the most common is adsorption. A study was carried out where, through an adsorption process, clinoptilolite treated with acid had an 87% fluoride removal efficiency in an aqueous solution, with pH = 3 and a time of 2 h. (Mahmoudi et al., 2019)

Figure 8. Diagram of processes used in defluorination.

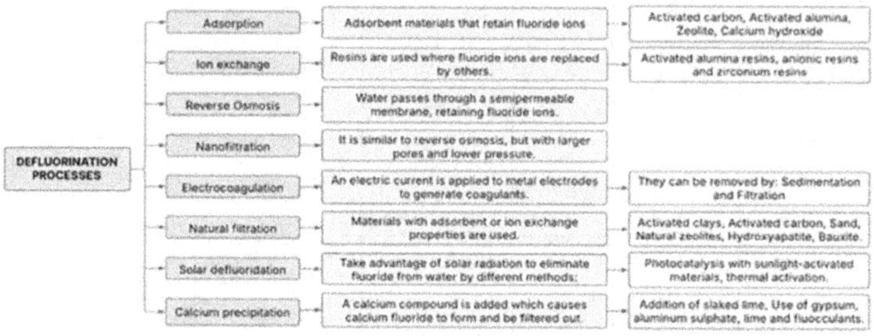

(Hossein et al., 2024)

6.3 Clinoptilolite for the Removal of Heavy Metals from Water

Heavy metals (HM) are chemical elements with a relatively high density and toxic properties even at small concentrations, so their presence in water represents a significant risk to human health and the environment (Kuldeyev et al., 2023). These metals are usually lead (Pb), cadmium (Cd), copper (Cu), zinc (Zn), nickel (Ni), and mercury (Hg). In addition, other pollutants such as free ions, soluble salts, ions associated with ligands or ions bound to particulate matter may exist in an aqueous solution. (Sun et al., 2022). Water toxicity by hazardous substances such as HM is a critical problem worldwide, requiring that all harmful effluents be cured to reduce their harmful effects. Many industries, for example, battery manufacturing, metal plating, mining, ceramics, glass, etc., are the main factors that increase the concentration of HM in the environment and are responsible for HM pollution (Hu et al., 2022; Biblioteca et al., 2023). Most of these HM are toxic in nature and some are also carcinogenic. They can cause liver and kidney damage, cancer, diarrhea, nausea, bleeding, headache, abdominal pain, vomiting, liver and respiratory problems. Therefore, a limit should be set on the discharge of HM into the environment to avoid serious damage to human and aquatic life. (Hu et al., 2022)

Clinoptilolite is a material of great relevance in the field of water treatment due to its ability to remove a wide variety of pollutants, including HM such as lead, zinc, cadmium, nickel and dyes from an aqueous medium. When water contaminated with HM comes into contact with clinoptilolite, the most common process for the removal of pollutants occurs, IE, which is when HM ions are attracted to the zeolite and replace the cations present in its structure. Other mechanisms also occur, such as physical and chemical adsorption and electrostatic attraction. (Grifasi et al., 2024; Biblioteca et al., 2023)

The choice of the best treatment technique depends on the final purpose and the type of effluent. In this sense, low-cost, robust and more efficient techniques are highly beneficial for treating wastewater without endangering human health or damaging the environment. Adsorption and IE are some of the low-cost and efficient techniques for removing pollution (Galletti et al., 2020). The basic process of HM removal begins when the clinoptilolite particles come into contact with the polluted solution, the metal ions are attracted to the clinoptilolite due to electrostatic forces until they reach the adsorption sites by means of different mechanisms (IE, complexation, physical or chemical adsorption), and then reach an equilibrium when the amount of metal ions adsorbed on the surface of the clinoptilolite is maximum and constant. (Liu et al., 2019)

As shown in Table 3, there are various adsorbent materials for the removal of HM in water, but the most notable is natural clinoptilolite since it has high availability, good cost and selectivity for HM, or modified clinoptilolite (with HCI, H_2SO_4, Ag, Fe, HDTMA, CTBA) since they offer better performance in specific applications. (Abdelwahab et al., 2023)

Table 3. Removal of heavy metals by means of different adsorbent materials

Adsorbent material	Adsorbed HM	Adsorption capacity (mg/g)	Cost	Environmental impact	Optimal conditions	Ref.
Zeolites	Pb, Cd, Cu, Zn, Cr, Ni	30–120	Moderate	Low	pH 6–7, 25°C, 2–3 contact hours	(Babel et al., 2003)
Natural Clinoptilolite	Pb, Cd, Cu, Zn, Cr, Ni	10–50	Moderate	Low	pH 7, room temperature, 1-2 hour contact	(Wang et al., 2010)
Modified Clinoptilolite	Pb, Cd, Cu, Zn, Cr, Ni	50–150	Moderate	Low	pH 5-8, 20-50°C, 1 hour contact	(Senila et al., 2024)
Activated Carbon	Pb, Cd, Cu, Zn, Hg, Cr, Ni	50–200	High	Moderate	pH 5–6, 25–30°C, 2–4 hours of contact	(Babel et al., 2003)
Clay Minerals	Pb, Cd, Cu, Zn, Cr, Ni	10–100	Very Low	Low	pH 5–6, 25°C, 3–5 contact hours	(Babel et al., 2003)

continued on following page

Table 3. Continued

Adsorbent material	Adsorbed HM	Adsorption capacity (mg/g)	Cost	Environmental impact	Optimal conditions	Ref.
Chitosan	Pb, Cd, Cu, Zn, Hg, Cr	50–300	Moderate	Low	pH 5–6, 25°C, 1–2 hours of contact	(Babel et al., 2003)
Metal oxides	As, Pb, Cd, Cu, Zn, Cr	50–200	Moderate	Moderate	pH 6–7, 25°C, 1–2 contact hours	(Babel et al., 2003)
Graphene Oxide	Pb, Cd, Cu, Zn, Hg, Cr	100–500	Very High	Moderate	pH 5–6, 25°C, 1–2 hours of contact	(Babel et al., 2003)

6.4 Clinoptilolite for the Removal of Emerging Pollutants from Water

These are chemicals that have recently been detected in water. The sources of emerging pollutants (EPs are diverse and constantly evolving, the most notable being pharmaceutical and personal care products (antibiotics, analgesics, cosmetics, etc.), industrial chemicals (pesticides, flame retardants, etc.), nanomaterials, and microplastics (Gogoi et al., 2018; Thakuret al., 2024). These pollutants represent a growing concern due to their presence in water sources and their potential adverse effects on human health and the environment. Clinoptilolite has emerged as a promising adsorbent for the removal of EPs from water. Its unique properties and low cost make it an attractive material for application in various water treatment processes (Sossou et al., 2020). The most common process for removing EPs using clinoptilolite is adsorption. This begins when water containing the EPs comes into contact with the clinoptilolite and is attracted to its surface by intermolecular forces (IE or physical adsorption). The pollutants are subsequently retained in the porous structure of the zeolite, preventing their return to the water. Finally, clinoptilolite can be regenerated by washing with chemical or thermal solutions to remove adsorbed pollutants and restore it to its original state. (Zhou et al., 2024; Rathi et al., 2021)

As shown in Table 4, numerous EPs are present in wastewater, surface water, groundwater, and drinking water. The most notable clinoptilolites for the removal of these pollutants are natural clinoptilolite, NZVI-impregnated clinoptilolite, and HCl-activated clinoptilolite (Sossou et al., 2020). In one study, cetylpyridinium chloride-modified clinoptilolite was tested for the removal of diclofenac sodium and ketoprofen, with removal efficiencies of 58% and 69%, respectively. (Smiljanić et al., 2020)

Table 4. Different emerging pollutants and their characteristics.

ECs	Type of pollutant	Potential Sources	Concentrations in various environmental samples				Ref.
			Sewage (μg/L)	Surface water (μg/L)	Groundwater (μg/L)	Drinking water (μg/L)	
Analgesics and anti-inflammatories	Pharmacist	Wastewater treatment plants, agricultural runoff, improper disposal of medicines	60	5 μg/L	-	0.12	(Gogoi et al., 2018)
Lipid regulators	Pharmacist	Human excrements, hospital discharges, pharmaceutical industry.	5	0.2 μg/L	7.5	0.17	(Gogoi et al., 2018)
Blockers	Pharmacist	Use of cardiovascular treatments, excretion through urine and feces.	9	2 μg/L	-	0.27	(Gogoi et al., 2018)
Antibiotics	Pharmacist	Human and veterinary use, pharmaceutical industry, food and animal treatments.	6	1.9	0.2	-	(Gogoi et al., 2018)
Antiepileptic drugs	Pharmacist	Medical use, human excretion, and improper disposal.	22	1.8	1.1	0.05	(Gogoi et al., 2018)
Estrone (E1)	Hormonal descriptors	Hormone excretion, household wastewater, hormones in animal growth.	< 0.1–19	< 0.1–17	-	0.20–0.60	(Gogoi et al., 2018)
Bisphenol A	Hormonal descriptors	Plastics industry, recycling and combustion, use in medical and electronic products.	4.8–258	0.5–250	3–1410	0.50–44	(Gogoi et al., 2018)
Pesticides	Agrochemicals	Agricultural runoff, urban use, industrial discharges	-	223.4	-	-	(Schwarzenbach et al., 2006)

continued on following page

Table 4. Continued

ECs	Type of pollutant	Potential Sources	Concentrations in various environmental samples				Ref.
			Sewage (µg/L)	Surface water (µg/L)	Groundwater (µg/L)	Drinking water (µg/L)	
Nanoparticles	Chemists	Industrial Processes, Consumer Products, Wastewater Treatment Plants	0.1-100	0.01-10	0.001-1	$<$0.001-0.1	(Klaine et al., 2008)
Plastic micros	Physicists	Plastic waste, synthetic textiles, industrial effluents	1000-100000 particles/L	0.01-100 particles/L	0.01-10 particles/L	0.01-10 particles/L	(Andrady et al., 2011)

6.5 Removal of Other Pollutants from Water

In addition to the types of pollutants already mentioned, clinoptilolite is capable of removing a variety of pollutants, both organic and inorganic, such as:

Ammonia: Ammonia is a colorless gas composed of nitrogen and hydrogen, with a very strong and penetrating characteristic odor. It is found naturally in the environment in small quantities, but in high concentrations it can be very harmful to aquatic ecosystems and human health. (Shaban et al., 2017). In a study, ammonia removal was carried out using clinoptilolite treated with NaCl and an efficiency greater than 69% was achieved.(Kannan et al., 2021)

Colorants and dyes: Colorants and dyes are chemical substances that are used to give color to textile products and foods, however, when these compounds are released into the environment, they are a serious problem (Kochubei et al., 2020). In a study, the application of natural clinoptilolite for the removal of basic textile dyes from water was investigated with an effectiveness of 85%. (Amin et al., 2023)

Volatile organic compounds (VOCs): They are chemical substances that evaporate easily at room temperature, passing from the liquid or solid state to the gaseous state, representing a threat to human health and the environment (Rochard et al., 2021). In a study, VOC removal was performed in aqueous solutions using HDTMA-modified clinoptilolite and an efficiency greater than 79% was achieved. (Wołowiec et al., 2017).

Figure 9 shows the most commonly used pollutants and their characteristics at each point mentioned above. The clinoptilolites that stand out the most in ammonia removal are: natural, activated with NaCl and activated with Fe. The clinoptilolites most commonly used in the removal of colorants and dyes are: natural and modified with HDTMA. The clinoptilolites most commonly used in VOC removal are: impregnated with activated carbon and modified with HDTMA.

Figure 9. Elimination of different pollutants and their characteristics by means of clinoptilolite.

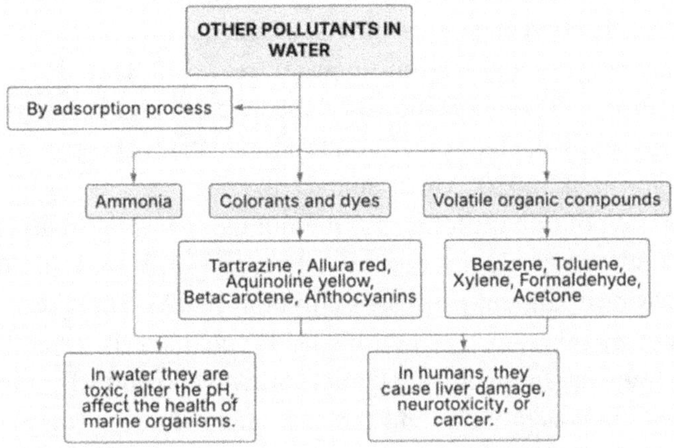

6.6 Clinoptilolite for Water Filtration

Water filters are an effective, economical and environmentally friendly technology for removing pollutants from water. Adaptable to both domestic and industrial uses, they represent a versatile solution to address water quality challenges in global and local contexts (Taghavi et al., 2021; Cescon et al., 2020). Clinoptilolite water filters work by a process known as adsorption. Pollutant molecules present in the water are attracted to the surface of the clinoptilolite particles, where they are retained (Taghavi et al., 2021; Grifasi et al., 2024). This process has been favored because clinoptilolite provides a large contact surface for the adsorption of pollutants, they can be exchanged with the ions present in the structure, and through Van der Waals forces attract the pollutant molecules to the surface of the clinoptilolite. (Taghavi et al., 2021)

The basic components of filters are filter beds, structural supports for the filter bed, and inlet and outlet systems for water circulation. Filters come in different sizes depending on their application: domestic (small cartridges for home filters), industrial (large-capacity tanks for wastewater or industrial treatment), and portable (compact filters for use in camps or emergencies) (Yu et al., 2019). For specific pollutants, clinoptilolite can be chemically modified with surfactants and metal impregnations or combined with other filter media such as activated carbon, sand, or IE resins to achieve more complete purification. (Xiang et al., 2021; Lima et al., 2021)

6.7 Clinoptilolite for Membrane in Water Treatment

Membranes for water treatment are semipermeable barriers that allow certain substances to pass through and retain others, acting as filters at the molecular level. They have a wide range of applications in water treatment, including: production of drinking water from salt water, removal of specific pollutants and water recovery, use in the pharmaceutical, electronics and food industries and elimination of water hardness. (Ezugbe et al., 2020; Yusuf et al., 2020)

The combination of clinoptilolite and membranes in water treatment allows the advantages of both materials to be taken advantage of, improving the efficiency and effectiveness of treatment processes (Jin et al., 2023, Hallajiqomi). There are different types of membranes, each with specific characteristics and applications: microfiltration (they retain particles and microorganisms), ultrafiltration (they retain larger molecules, such as proteins and viruses), nanofiltration (they retain divalent ions and low molecular weight organic molecules), RO (this is the most common since it retains most solutes)(Estrada-Cabrera et al., 2021). Clinoptilolite is placed before the membrane, acting as a coarse filter that retains large particles, organic matter and some ions. Clinoptilolite is also used after the membrane to polish the treated water, removing any pollutants that have managed to pass through the membrane. In some cases, clinoptilolite can be incorporated directly into the membrane matrix, modifying its properties and improving its performance. (Hamad et al., 2024)

As can be seen in Figure 10, membranes generally have four main processes in water treatment, where the most commonly used clinoptilolites are: natural, impregnated with Ag and treated with PVDF (Estrada-Cabrera et al., 2021). In a study, membranes containing PVDF/HDTMA modified clinoptilolite were manufactured for the treatment of effluents containing dyes, using the RO method with 98.5% effectiveness. (Hosseinifard et al., 2020)

Figure 10. Processes using membranes.

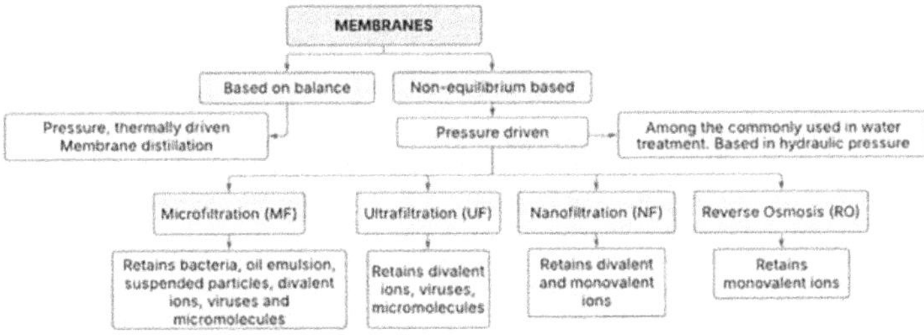

(Ezugbe et al., 2020; Estrada-Cabrera et al., 2021)

6.8 Clinoptilolite for Reverse Osmosis

It is a water purification process that uses a semipermeable membrane to remove dissolved salts, minerals and other impurities. It is an effective process used to produce high quality drinking water from various water sources such as seawater, groundwater and well water(Field et al., 2021; Almansouri et al., 2024). Clinoptilolite is used in some RO systems primarily as a filter media in the pretreatment stage. It has the ability to adsorb pollutants which helps protect the RO membrane from fouling and prolongs its life and reduces maintenance costs (Anis et al., 2019).

The process generally begins when raw water is subjected to a pretreatment to eliminate pollutants that may damage the membrane (filtration process with sand, clinoptilolite, activated carbon, etc.), then the membrane allows the passage of water but blocks the passage of salts, minerals and other impurities and at the end of the process it can be subjected to a post-treatment to adjust its pH, add beneficial minerals or disinfect it (Safarpour et al., 2017; Otitoju et al., 2018). It is used in a wide variety of applications, such as: in plants to produce drinking water on a large scale, in various industries (food, pharmaceutical, electronics), at home to purify water for domestic consumption (Otitoju et al., 2018). As can be seen in Table 5, there are different clinoptilolites, both natural and modified. For example, in one study, a RO membrane was manufactured using plasma-treated clinoptilolite, where it achieved 88% effectiveness. (Safarpour et al., 2017)

Table 5. Advantages and disadvantages of different types of clinoptilolites used in reverse osmosis.

Type of clinoptilolite	Advantages	Disadvantages	Ref.
Natural Clinoptilolite	- Pretreatment and clogging for RO systems. - Removes ammonia, HM and suspended solids.	- Limited capacity for certain pollutants. - Requires regeneration.	(Wang et al., 2010)
Acid-Activated Clinoptilolite	- Increased cation adsorption capacity. - Improves membrane life by reducing clogging.	- Acid treatment increases cost. - It can release ions (e.g., Al^{3+}) into the water.	(Wang et al., 2015)
Heat-treated clinoptilolite	- Improved porosity and adsorption efficiency. - Effective in removing organic pollutants.	- Energy-intensive process: It can reduce IE capacity for some ions.	(Akkoca et al., 2013)
Surfactant treated clinoptilolite	- Effective in removing organic pollutants and anions. - Improves RO performance by reducing organic clogging.	- Higher cost due to modification. - Surfactants can leach into the water.	(Xie et al., 2013)
Plasma-treated clinoptilolite	- Reduce RO membrane stress, while the dry plasma process avoids chemical waste.	-Overtreatment carries the risk of structural damage. -Energy use and cost could limit its viability.	Safarpour et al., 2017
Natural Clinoptilolite	- Clinoptilolite effectively removes ammonia, which can damage RO membranes.	- Requires regeneration with brine or other solutions. - Limited capacity for high concentrations of ammonia.	(Flores et al., 2002)

7. APPLICATIONS OF CLINOPTILOLITE IN THE REMOVAL OF ENVIRONMENTAL POLLUTANTS

Clinoptilolite zeolite has been highlighted as an effective and sustainable material for the removal of pollutants in water due to its unique crystalline structure, high IE capacity and adsorbent properties. (Grifasi et al., 2024; El-Azim et al., 2018). Because these types of materials are crystalline, they have narrow pore size distributions, and have a wide variety of applications and uses in various areas due to their unique adsorption, IE properties and porous structure (Reid et al., 2021b). It has a very defined crystalline structure and unique adsorption, IE properties and its porous structure that make it useful in a variety of applications such as in the manufacture of urea sensors, deodorant production, agricultural fertilizers and HM removal. (Abal et al., 2021).

7.1 Removal of Pollutants in Soil (Agriculture And Gardening)

Pollution in agriculture can originate from several sources, including: HM (lead, cadmium and zinc), excess nutrients (ammonium and nitrates), pesticides and herbicides, radioactive substances (cesium and strontium) (Grifasi et al., 2024; Cataldo et al., 2021). These pollutants reduce soil fertility, contaminate water sources and affect the quality of agricultural products. Clinoptilolite is used in agriculture as an efficient material to remove pollutants from water, soil and fertilizers due to its IE and absorption properties (Velarde et al., 2023). It improves water and nutrient retention in the soil, increasing nutrient availability to plants and improving soil structure (Rodríguez-Iznaga et al., 2022; Kadium et al., 2025). Its ability to adsorb and retain a wide variety of pollutants, together with its low cost and non-toxic nature, makes it a valuable tool for environmental protection (Reid et al., 2021; Leal et al., 2017). Table 6 shows some pollutants found in soil and agriculture that can be removed by natural or modified clinoptilolite, carried out by adsorption or IE mechanisms.

Table 6. Clinoptilolite applied for the removal of different pollutants in soil and agriculture.

Pollutant	Material used	Elimination mechanism	Removal Efficiency (%)	Application	Ref.
Mg, Mn, Ni, Cu, Zn, Pb	Tuffs rich in clinoptilolite	IE, and adsorption	Mg (60), Mn (47), Ni (15), Cu (84), Zn (50), Pb (100)	They can be applied in water remediation and contaminated soils.	(Shi et al., 2023)
Nitrogen	Natural Zeolite	IE	60	It is applied in wastewater treatment, filtration systems, soil remediation.	(Kukowska et al., 2024)
Hydrocarbons	Zeolite clinoptilolite-Na	Adsorption	65	Acts as an adsorbent helping to clean contaminated soils and water	(Shi et al., 2023)
Ammonia	N-loaded clinoptilolite	Adsorption	39	High water retention and nutrient supplementation, which benefit crop growth.	(Yang S et al., 2024)

continued on following page

Table 6. Continued

Pollutant	Material used	Elimination mechanism	Removal Efficiency (%)	Application	Ref.
4-acetamidoantipyrine	Clinoptilolite	Adsorption	99.3	In the treatment of wastewater containing pharmaceutical residues.	(Cataldo et al., 2021)
Acrylic acid	Clinoptilolite treated phosphate.	Adsorption	51	Solution to reduce water consumption, especially in agricultural applications.	(Shi et al., 2023)
Ammonium (NH_4^+)	Clinoptilolite (natural zeolite)	IE	56	Use in fertilizers for slow nitrogen release, reducing nutrient runoff.	(He et al., 2015)
Radionuclides	Clay minerals	Adsorption	52	Remediation of soils due to nuclear accidents or radioactive waste.	(Borai et al., 2009)

7.2 Clinoptilolite uses in Poultry and Cattle Farming

Poultry farming faces challenges related to the accumulation of pollutants in production systems that affect bird health, product quality and the environment (Biesek et al., 2021). Clinoptilolite can be added to poultry feed, where it improves digestive health by adsorbing toxic substances and ammonia produced by fungi that contaminate feed, it also reduces metabolic stress, promotes better growth, reduces unpleasant odors, slows the spread of respiratory diseases and improves the safety of poultry products (Pavlak et al., 2023; Elsherbeni et al., 2024). Clinoptilolite represents a valuable tool in poultry farming for the removal of pollutants and the improvement of environmental and sanitary conditions in production systems. Its integration into sustainable management strategies allows not only to increase poultry productivity, but also to mitigate negative impacts on the environment (Saeed et al., 2019). A study was conducted in which ninety-one-day-old Ross 308 broiler chickens were divided into three groups of 30 each. The same diet was used in all groups, but groups II and III received a 2% and 3% clinoptilolite additive. The groups that received this zeolite showed increased weight, improved antigen presentation, and no clinical symptoms of disease were observed. (Jarosz et al., 2017)

7.3 Removal of Pollutants in Air and Gases

Clinoptilolite has proven to be a highly effective adsorbent for the removal of a wide variety of air and gas pollutants. This type of zeolite is sometimes routinely added to small air filters for the absorption of certain gases and reduction of allergy problems (Grifasi et al., 2024). It is used in purification systems to remove pollutants including: SO_2, NOx, HCl, VOCs, HM and greenhouse gases (Pour et al., 2024; Zhang et al., 2016). One of the main causes of climate change is the excessive presence in the atmosphere of greenhouse gases that persist for a hundred years after their emission. In recent decades, the natural zeolite clinoptilolite has been the subject of intense research for its application as a sustainable and low-cost sorbent for the purification of gaseous streams from different pollutants (Grifasi et al., 2024; Xu et al., 2023). In a study, three different clinoptilolites were tested as sorbents in the pressure swing adsorption (PSA) process for CO_2 removal and found that the clinoptilolite with the highest sodium content was up to 2 times more efficient. (Grifasi et al., 2024)

7.4 Clinoptilolite Uses in Medical Applications

Clinoptilolite is used as dietary supplements, toxin and HM absorbers as it has a negative chemical charge which allows it to attract and trap HM and positively charged toxic chemicals, therefore, it is an important detoxifier (Hovhannisyan et al., 2021; Souza et al., 2021). Several scientific studies have demonstrated the beneficial effects in rheumatism, kidney infections, treatment of dermatological diseases, elimination of harmful residues left at the end of chemotherapy and radiotherapy and recently, for the high dissolution of directly adsorbed anticancer drugs, mainly letrozole (LTZ) (Hao et al., 2021; Alqarni et al., 2024). The most important medical applications of clinoptilolite are shown in Figure 11. In one study, clinoptilolite conditioned with NaCl in the form of micronized powder was analyzed to reduce nickel from the human gastrointestinal tract, with an elimination efficiency of 98%. (Biblioteca et al., 2023)

Figure 11. Main medical applications.

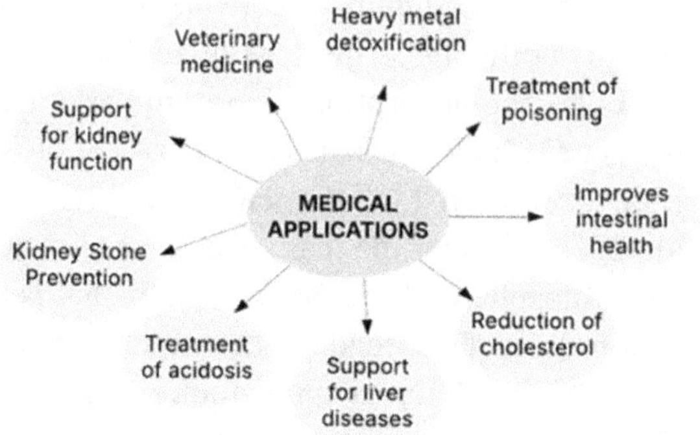

(Grifasi et al., 2024; Mastinu et al., 2019; Pavelić et al., 2018)

8. CONCLUSIONS

Water pollution threatens human health and ecosystems, driving the need for sustainable remediation strategies. Clinoptilolite, a naturally occurring zeolite, emerges as a versatile solution due to its high cation exchange capacity, expansive SA, and selective adsorption properties. This review confirms its effectiveness in removing various pollutants, including HM, arsenic, fluoride, and ECs, through IE and adsorption mechanisms. Preparation techniques like purification, conditioning, and modification (e.g., IE or metal impregnation) significantly enhance its performance, as evidenced by studies achieving up to 97% removal efficiency for specific pollutants. Advanced characterization tools, such as electron microscopy and XRD, elucidate its structural properties, while computational simulations offer molecular-level insights into optimization strategies. However, clinoptilolite's effectiveness varies with factors like origin and application conditions, necessitating tailored approaches. Future research should prioritize novel modifications, synergistic material combinations, and industrial-scale applications to overcome these challenges. By addressing these areas, clinoptilolite can significantly advance sustainable water management, providing a robust framework for tackling global water quality issues.

ACKNOWLEDGEMENTS

Thanks go to the Scientific Computing Laboratory at FCQ-UJED for computational resources. Thanks go to the Academic Group UJED-CA-129 for valuable discussions.

Credits Authorship and Contribution Statement: DBJ, MAFH: Writing—original draft, Investigation, Formal analysis, Data curation, Conceptualization, Formal analysis, Writing—review & editing. DBJ, MAFH, SITH, HIFH: Project administration, Funding acquisition. HVRA, HIFH: Writing—review & editing, Project administration, Formal analysis, Conceptualization. All authors have read and agreed to the published version of the manuscript.

Conflict of Interest: The authors declare no conflict of interest.

Data Availability: The Supporting Information is available upon request by direct request to the authors.

REFERENCES

Abal, . (2021). Molecular fluid flow in MoS2 nanoporous membranes and hydrodynamics interactions. *The Journal of Chemical Physics*, *154*(18), 184704. Advance online publication. DOI: 10.1063/5.0039963 PMID: 33832278

Abatal, , Ruiz-Salvador, A. R., & Hernández, N. C. (2019). A DFT-based simulated annealing method for the optimization of global energy in zeolite framework systems: Application to natrolite, chabazite and clinoptilolite. *Microporous and Mesoporous Materials*, *294*, 109885. Advance online publication. DOI: 10.1016/j.micromeso.2019.109885

Abdelwahab, , & Thabet, W. M. (2023). Natural zeolites and zeolite composites for heavy metal removal from contaminated water and their applications in aquaculture systems: A review. *Egyptian Journal of Aquatic Research*, *49*(4), 431–443. DOI: 10.1016/j.ejar.2023.11.004

Adjal, , Guechtouli, N., Timón, V., Boussassi, R., Hammoutène, D., & Senent, M. L. (2024). DFT-based, Monte Carlo and Grand Canonical Monte Carlo simulations of nitro-organic pollutants 4-nitrophenol, 2-nitrophenol, 9-nitroanthracene and nitrogen trifluoride interacting with water in zeolite imidazole framework (ZIF-8). *Computational & Theoretical Chemistry*, *1243*, 114995. Advance online publication. DOI: 10.1016/j.comptc.2024.114995

Ahmed, , Mavukkandy, M. O., Giwa, A., Elektorowicz, M., Katsou, E., Khelifi, O., Naddeo, V., & Hasan, S. W. (2022). Recent developments in hazardous pollutants removal from wastewater and water reuse within a circular economy. *NPJ Clean Water, 12*(1), 12. Advance online publication. DOI: 10.1038/s41545-022-00154-5

Aili, , & Maruyama, I. (2020). Review of several experimental methods for characterization of micro- and nano-scale pores in cement-based material. *International Journal of Concrete Structures and Materials, 14*(1), 53. Advance online publication. DOI: 10.1186/s40069-020-00431-y

Akkoca, , Yılgın, M., Ural, M., Akçin, H., & Mergen, A. (2013). Hydrothermal and thermal treatment of the natural zeolite clinoptilolite from Bigadiç, Turkey: An experimental study. *Geochemistry International, 51*(4), 315–323. DOI: 10.1134/S0016702913040022

Akyalcin, , Akyalcin, L., & Ertugrul, E. (2024). Modification of natural clinoptilolite zeolite to enhance its hydrogen adsorption capacity. *Research on Chemical Intermediates, 1473*(3), 1455–1473. Advance online publication. DOI: 10.1007/s11164-023-05212-2

Almansouri, , Edokali, M., & Abu Seman, M. N. (2024). Flat sheet thin film composite forward osmosis membranes for water treatment and purification: A review on modification techniques and concepts. *Desalination and Water Treatment, 320*, 100815. Advance online publication. DOI: 10.1016/j.dwt.2024.100815

Alqarni, M., Ashour, A. A., Shafie, A., Alqarni, A., Felemban, M. F., Shukr, B. S., & Algahtani, F. S. (2024). Intelligence computational analysis of letrozole solubility in supercritical solvent via machine learning models. *Scientific Reports, 14*(1), 21677.

Amin, G., Konstantinovic, S., Jordanov, I., & Djordjevic, D. (2023). Adsorption study of textile dye Basic Red 46 on clinoptilolite. Iran. J. Chem. Chem. Eng. Research Article Vol, 42(11).

Andrady, A. L. (2011). Microplastics in the marine environment. *Marine Pollution Bulletin, 62*(8), 1596–1605.

Anis, , Hashaikeh, R., & Hilal, N. (2019). Flux and salt rejection enhancement of polyvinyl(alcohol) reverse osmosis membranes using nano-zeolite. *Desalination, 463*, 18–27. DOI: 10.1016/j.desal.2019.114104

Babel, S., & Kurniawan, T. A. (2003). Low-cost adsorbents for heavy metals uptake from contaminated water: A review. *Journal of Hazardous Materials, 97*(1-3), 219–243.

Badeenezhad, , & Azhdarpoor, A. (2019a). Efficiency of the activated carbon and clinoptilolite particles coated with iron oxide magnetic nanoparticles in removal of methylene blue. *Desalination and Water Treatment*, *154*, 347–355. DOI: 10.5004/dwt.2019.24039

Badeenezhad, , Azhdarpoor, A., Bahrami, S., & Yousefinejad, S. (2019b). Removal of methylene blue dye from aqueous solutions by natural clinoptilolite and clinoptilolite modified by iron oxide nanoparticles. *Molecular Simulation*, *45*(7), 564–571. DOI: 10.1080/08927022.2018.1564077

Bahmanzadegan, , & Ghaemi, A. (2025). A comprehensive review on novel zeolite adsorbents for environmental pollutant. *Journal of Hazardous Materials Advances*, *100617*, 100617. Advance online publication. DOI: 10.1016/j.hazadv.2025.100617

Baskan, B.. (2021). Removal of arsenate using graphene oxide-iron modified clinoptilolite-based composites. *Journal of Analytical Science and Technology*, *12*(1), 14. Advance online publication. DOI: 10.1186/s40543-021-00274-6

Bauer, , Broome, S. T., Kibikas, W. M., & Wilson, J. E. (2024). Mechanical and poroelastic behavior of porous tuff under drained and undrained conditions. *Bulletin of Engineering Geology and the Environment*, *83*(10), 415. Advance online publication. DOI: 10.1007/s10064-024-03916-2

Biblioteca, I., Sambucci, M., & Valente, M. (2023). Zeolite-Clinoptilolite conditioning for improved heavy metals ions removal: A preliminary assessment. *Ceramics International*, *49*(23), 39649–39656.

Biesek, , Banaszak, M., & Adamski, M. (2021). Ducks' growth, meat quality, bone strength, and jejunum strength depend on zeolite in feed and long-term factors. *Animals (Basel)*, *11*(4), 1015. Advance online publication. DOI: 10.3390/ani11041015 PMID: 33916880

Borai, E. H., Harjula, R., malinen, L., & Paajanen, A. (2009). Efficient removal of cesium from low-level radioactive liquid waste using natural and impregnated zeolite minerals. *Journal of Hazardous Materials*, *172*(1), 416–422. DOI: 10.1016/j.jhazmat.2009.07.033 PMID: 19656622

Bukka, , & Sarin, P. (2024). Effects of particle size reduction on the pore structure and accessibility in natural porous materials. *Energy & Fuels*, *38*(11), 9876. DOI: 10.1021/acs.energyfuels.4c00194

Camacho, , Parra, R. R., & Deng, S. (2011). Arsenic removal from groundwater by MnO2-modified natural clinoptilolite zeolite: Effects of pH and initial feed concentration. *Journal of Hazardous Materials*, *189*(1–2), 286–293. DOI: 10.1016/j.jhazmat.2011.02.035 PMID: 21398033

Carstens, , Splith, C., & Enke, D. (2019). Sol-gel synthesis of α-Al2O3 with enhanced porosity via dicarboxylic acid templating. *Scientific Reports*, *9*(1), 19982. Advance online publication. DOI: 10.1038/s41598-019-56294-1 PMID: 31882695

Casentini, , Gallo, M., & Baldi, F. (2019). Arsenate and arsenite removal from contaminated water by iron oxides nanoparticles formed inside a bacterial exopolysaccharide. *Journal of Environmental Chemical Engineering*, *7*(1), 102908. Advance online publication. DOI: 10.1016/j.jece.2019.102908

Cataldo, , Salvi, L., Paoli, F., Fucile, M., Masciandaro, G., Manzi, D., Masini, C. M., & Mattii, G. B. (2021). Application of zeolites in agriculture and other potential uses: A review. *Agronomy (Basel)*, *11*(8), 1547. Advance online publication. DOI: 10.3390/agronomy11081547

Cescon, , & Jiang, J.-Q. (2020). Filtration process and alternative filter media material in water treatment. *Water (Basel)*, *12*(12), 3377. Advance online publication. DOI: 10.3390/w12123377

Chen, , Wang, X., Li, Q., Chen, J., Zhu, L., Wang, L., & Zhang, L. (2025). Soil fluoride enrichment process and the possible adaptation prevention principle in coal-burning fluorosis. *Scientific Reports*, *15*(1), 84381. DOI: 10.1038/s41598-024-84381-5 PMID: 39762348

Chmielewská, E., Tylus, W., Drábik, M., Majzlan, J., Kravčak, J., Williams, C., Čaplovičová, M., & Čaplovič, Ľ. (2017). Structure investigation of nano-FeO(OH) modified clinoptilolite tuff for antimony removal. *Microporous and Mesoporous Materials*, *248*, 222–228. DOI: 10.1016/j.micromeso.2017.04.022

Chwatal, , Parizek, H., & Waldhauser, W. (2024). Simulation-based analysis of the atmospheric pressure plasma and its temperature distribution using Ansys Mechanical. *BHM. Berg- und Hüttenmännische Monatshefte*, *169*(8), 351–356. DOI: 10.1007/s00501-024-01477-1

De la Rosa-Gómez, I., Olguín, M. T., & Alcántara, D. (2008). Antibacterial behavior of silver-modified clinoptilolite–heulandite rich tuff on coliform microorganisms from wastewater in a column system. *Journal of Environmental Management*, *88*(4), 853–863.

de Pablo, L., Doval, M., La Iglesia, A., & Soriano, J. (2014). CaK-clinoptilolite, KNa-chabazite, KNa-heulandite, KNa-erionite and Na-phillipsite from tuffaceous rocks, Province of the Mesa Central, Mexico. *Revista Mexicana de Ciencias Geológicas*, *31*(1), 116–126.

Deneyer, , Ke, Q., Devos, J., & Dusselier, M. (2020). Zeolite synthesis under nonconventional conditions: Reagents, reactors, and modi operandi. *Chemistry of Materials*, *32*(12), 4884–4919. Advance online publication. DOI: 10.1021/acs.chemmater.9b04741

Derbe, , Temesgen, S., & Bitew, M. (2021). A short review on synthesis, characterization, and applications of zeolites. *Advances in Materials Science and Engineering*, *2021*(1), 6637898. Advance online publication. DOI: 10.1155/2021/6637898

Dikić, , Hrenović, J., Durn, G., Kovačić, A., & Rajić, N. (2019). Antibacterial activity of metal-containing clinoptilolite in natural seawater. *Desalination and Water Treatment*, *169*, 135–142. DOI: 10.5004/dwt.2019.24712

Dryaz, , Shaban, M., AlMohamadi, H., Al-Ola, K. A. A., Hamd, A., Soliman, N. K., & Ahmed, S. A. (2021). Design, characterization, and adsorption properties of Padina gymnospora/zeolite nanocomposite for Congo red dye removal from wastewater. *Scientific Reports*, *11*(1), 21062. Advance online publication. DOI: 10.1038/s41598-021-00025-y PMID: 34702834

El-Azim, , & Mourad, F. (2018). Removal of heavy metals Cd (II), Fe (III) and Ni (II), from aqueous solutions by natural (clinoptilolite) zeolites and application to industrial wastewater. *Asian Journal of Environment & Ecology*, *7*(1), 1–13. DOI: 10.9734/AJEE/2018/41004

Elsherbeni, , Youssef, I. M., Hamouda, R. E., Kamal, M., El-Gendi, G. M., El-Garhi, O. H., Alfassam, H. E., Rudayni, H. A., Allam, A. A., Moustafa, M., Alshaharn, M. O., & El Kholy, M. S. (2024). Performance and economic efficiency of laying hens in response to adding zeolite to feed and litter. *Poultry Science*, *103*(6), 103799. DOI: 10.1016/j.psj.2024.103799 PMID: 38759566

Estrada-Cabrera, E., Torres-Ferrer, L. R., Luna-Barcenas, G., & Ramirez-Bon, R. (2021). Cellulose dialysis membrane containing raw clinoptilolite enhances the removal of Rhodamine 6G from aqueous solutions. *Microporous and Mesoporous Materials*, *321*, 111113. Advance online publication. DOI: 10.1016/j.micromeso.2021.111113

Evans, , & Coudert, F.-X. (2017). Predicting the mechanical properties of zeolite frameworks by machine learning. *Chemistry of Materials*, *29*(18), 7833–7839. Advance online publication. DOI: 10.1021/acs.chemmater.7b02532

Ezugbe, . (2020). Membrane technologies in wastewater treatment: A review. *Membranes*, *10*(5), 89. Advance online publication. DOI: 10.3390/membranes10050089 PMID: 32365810

Faisal, F., Rasul, M. G., Jahirul, M. I., & Chowdhury, A. A. (2023). Waste plastics pyrolytic oil is a source of diesel fuel: A recent review on diesel engine performance, emissions, and combustion characteristics. *The Science of the Total Environment*, *886*, 163756. Advance online publication. DOI: 10.1016/j.scitotenv.2023.163756 PMID: 37146806

Farooq, , Muneer, M., Shahid, A., Rehman, M. A., Ullah, K., Murtaza, G., Iqbal, R., Iqbal, J., & Rahimi, M. (2024). Synthesis and characterization of fluorenone derivatives with electrical properties explored using density functional theory. *Scientific Reports*, *14*(1), 80477. Advance online publication. DOI: 10.1038/s41598-024-80477-0 PMID: 39578658

Favvas, , Tsanaktsidis, C. G., Sapalidis, A. A., Tzilantonis, G. T., Papageorgiou, S. K., & Mitropoulos, A. C. (2016). Clinoptilolite, a natural zeolite material: Structural characterization and performance evaluation on its dehydration properties of hydrocarbon-based fuels. *Microporous and Mesoporous Materials*, *225*, 385–391. Advance online publication. DOI: 10.1016/j.micromeso.2016.01.021

Field, , She, Q., Siddiqui, F. A., & Fane, A. G. (2021). Reverse osmosis and forward osmosis fouling: A comparison. *Discover Chemical Engineering*, *1*(1), 6. Advance online publication. DOI: 10.1007/s43938-021-00006-7

Fischer, . (2023). Density functional theory study of hydrophobic zeolites for the removal of triclosan from aqueous solution. *Environmental Science. Advances*, *2*(5), 678–689. DOI: 10.1039/D3VA00078H

Flores, et al. (2002). Ammonia removal in clarified wastewater by the ion exchange process with zeolites.

Foroutan, R., Peighambardoust, S. J., Hemmati, S., Khatooni, H., & Ramavandi, B. (2021). Preparation of clinoptilolite/starch/CoFe2O4 magnetic nanocomposite powder and its elimination properties for cationic dyes from water and wastewater. *International Journal of Biological Macromolecules*, *189*, 432–442.

Foroutan, , Peighambardoust, S. J., Hemmati, S., Khatooni, H., & Ramavandi, B. (2021). Preparation of clinoptilolite magnetic nanocomposite powder and its elimination properties for cationic dyes from water and wastewater. *International Journal of Biological Macromolecules*, *189*, 432–442. Advance online publication. DOI: 10.1016/j.ijbiomac.2021.08.144 PMID: 34450143

Galletti, . (2020). Zn2 and Cd2 removal from wastewater using clinoptilolite as adsorbent. *Environmental Sciences (Ruse)*, *27*(31), 38791–38802. DOI: 10.1007/s11356-020-08483-z PMID: 32212074

Gattucci, . (2025). Tannic acid-assisted green functionalization of clinoptilolite A step-by- characterization. *Ceramics International*. Advance online publication. DOI: 10.1016/j.ceramint.2025.01.151

Gogoi, , Mazumder, P., Tyagi, V. K., Tushara Chaminda, G. G., An, A. K., & Kumar, M. (2018). Occurrence and fate of emerging contaminants in water environment: A review. *Groundwater for Sustainable Development*, *6*, 169–180. DOI: 10.1016/j.gsd.2017.12.009

Gowram, , Beulah, M., Ibrahim Mohamed, M. M., Ibrahim, S. A. M., Puneeth, V., Ben Moussa, S., & Nasr, S. (2024). Experimental and finite element studies on the mechanical properties of high-strength concrete using natural zeolite and additives. *Alexandria Engineering Journal*, *109*, 989–993. DOI: 10.1016/j.aej.2024.09.094

Grass, , Pals, U., Inayat, A., Schwieger, W., Hartmann, M., & Dathe, W. (2021). Uptake and release characteristics of serotonin hydrochloride by natural Cuban zeolite containing clinoptilolite and mordenite. *Scientific Reports*, *11*(1), 14277. Advance online publication. DOI: 10.1038/s41598-021-93487-z PMID: 34253745

Grifasi, , Ziantoni, B., Fino, D., & Piumetti, M. (2024). Fundamental properties and sustainable applications of the natural zeolite clinoptilolite. *Environmental Science and Pollution Research International*, *31*(32), 44567–44580. DOI: 10.1007/s11356-024-33656-5 PMID: 38780851

Gul, , Ullah, R., Sun, J., Munir, T., & Bai, S. (2021). The fabrication of TiO2-supported clinoptilolite via F− contained hydrothermal etching and a resultant highly energetic {001} facet for the enhancement of its photocatalytic activity. *RSC Advances*, *11*(29), 17849–17859. DOI: 10.1039/D1RA02269E PMID: 35480182

Güngör, , & Özen, S. (2021). Development and characterization of clinoptilolite-, mordenite-, and analcime-based geopolymers: A comparative study. *Case Studies in Construction Materials*, *15*, e00576. Advance online publication. DOI: 10.1016/j.cscm.2021.e00576

Hallajiqomi, , Mehdipourghazi, M., & Varaminian, F. (2019). Degradation of tetracycline using photocatalytic membrane reactor with nanocomposite Ag doped clinoptilolite zeolite photocatalyst. *Desalination and Water Treatment*, *139*, 111–122. DOI: 10.5004/dwt.2019.23306

Hamad, , Abdallah, H., Aly, S. T., Abobeah, R., & Amin, S. K. (2024). Fabrication and assessment of performance of clay based ceramic membranes impregnated with CNTs in dye removal. *Scientific Reports*, *77015*(1), 26728. Advance online publication. DOI: 10.1038/s41598-024-77015-3 PMID: 39496693

Hao, , Stavljenić Milašin, I., Batu Eken, Z., Mravak-Stipetic, M., Pavelić, K., & Ozer, F. (2021). Effects of zeolite as a drug delivery system on cancer therapy: A systematic review. *Molecules (Basel, Switzerland)*, *26*(20), 6196. Advance online publication. DOI: 10.3390/molecules26206196 PMID: 34684777

Harami, . (2020). Molecular simulation and computational modeling of gas separation through polycarbonate/p-nitroaniline/zeolite 4A mixed matrix membranes. *Industrial & Engineering Chemistry Research*, *59*(38), 16772–16785. Advance online publication. DOI: 10.1021/acs.iecr.0c02827

He, . (2015). Removal of heavy metal ions from aqueous solution by zeolite synthesized from fly ash. *Environmental Science and Pollution Research International*, *4576*. Advance online publication. DOI: 10.1007/s11356-015-5422-6 PMID: 26446735

Helmi, , Hemmati, A., & Ghaemi, A. (2024). Clinoptilolite and MCM-41 impregnated with chitosan as a green novel adsorbent for CO2 capture. *Case Studies in Chemical and Environmental Engineering*, *9*, 100764. Advance online publication. DOI: 10.1016/j.cscee.2024.100764

Hossein, , Rwiza, M. J., Nyanza, E. C., Bakari, R., Ripanda, A., Nkrumah, S., Selemani, J. R., & Machunda, R. L. (2024). Fluoride contamination a silent global water crisis: A case of Africa. *Scientific African*, *25*, e02485. Advance online publication. DOI: 10.1016/j.sciaf.2024.e02485

Hosseinifard, , Aroon, M. A., & Dahrazma, B. (2020). Application of PVDF/HDTMA-modified clinoptilolite nanocomposite membranes in removal of reactive dye from aqueous solution. *Separation and Purification Technology*, *251*, 117294. Advance online publication. DOI: 10.1016/j.seppur.2020.117294

Hovhannisyan, , Siposova, K., Musatov, A., & Chen, S.-J. (2021). Development of multifunctional nanocomposites for controlled drug delivery and hyperthermia. *Scientific Reports*, *11*(1), 5528. Advance online publication. DOI: 10.1038/s41598-021-84927-x PMID: 33750868

Hu, , Han, J., Wei, Y., & Liu, Z. (2022). Dynamic evolution of zeolite framework and metal-zeolite interface. *ACS Catalysis*, *12*(9), 5060–5076. DOI: 10.1021/acscatal.2c01233

Ibrahim, F. A., & El-Desoky, M. M. (2023). Erbium-doped (Pb0.9-Er0.01-Zr0.09) and (Pb0.9-Er0.01-Zr0.045-Ti0.045) nano-crystalline films and nano-rods ceramic synthesis by sol–gel technique for energy-storage application. *Scientific Reports*, *13*(1), 4192. Advance online publication. DOI: 10.1038/s41598-023-31279-3 PMID: 36918632

Islam, , & Mohr, B. J. (2023). Performance of clinoptilolite zeolite after milling as a pretreatment on hydration kinetics, shrinkage, and alkali-silica reaction of cementitious materials. *Cement*, *12*, 100069. Advance online publication. DOI: 10.1016/j.cement.2023.100069

Jarosz, et al. (2017). The effect of feed supplementation with Zakarpacki zeolite (clinoptilolite) on percentages of T and B lymphocytes. 2095–2100. DOI: 10.3382/ps/pex030

Javaid, , Munir, N., Abideen, Z., Siddiqui, Z. S., & Yong, J. W. H. (2024). The role of natural and synthetic zeolites as soil amendments for mitigating the harmful impacts of abiotic stresses to improve agricultural resilience. *Plant Stress*, *12*, 100627. Advance online publication. DOI: 10.1016/j.stress.2024.100627

Jin, , Fu, J., Yu, P., & Luo, D. (2023). Characterization and properties of manganese oxide film coated clinoptilolite as filter material in fixed-bed columns for removal of Mn(II) from aqueous solution. *Scientific Reports*, *13*(1), 17440. Advance online publication. DOI: 10.1038/s41598-023-44611-8 PMID: 37838822

Kabuba, , & Banza, M. (2020). Ion-exchange process for the removal of Ni (II) and Co (II) from wastewater using modified clinoptilolite: Modeling by response surface methodology and artificial neural network. *Results in Engineering*, *100189*, 100189. Advance online publication. DOI: 10.1016/j.rineng.2020.100189

Kadium, A. H., Sharif, J. A., & Akhgar, B. N. (2025). Evaluating microstructure and surface changes of clinoptilolite after mechanical activation. *Colloids and Surfaces. A, Physicochemical and Engineering Aspects*, *686*, 136523. Advance online publication. DOI: 10.1016/j.colsurfa.2025.136523

Kannan, . (2021). Ammonia adsorption and recovery from swine wastewater permeate using naturally occurring clinoptilolite. *Journal of Water Process Engineering*, *43*, 102234. Advance online publication. DOI: 10.1016/j.jwpe.2021.102234

Karimi, , & Azadmehr, A. (2017). Removal of sulfide from aqueous solution with transition metals ion impregnated clinoptilolite: Characteristics, equilibrium, kinetic and thermodynamic approach. *Desalination and Water Treatment*, *74*, 296–307. DOI: 10.5004/dwt.2017.20608

Kashi, . (2023). Electrocoagulation/flotation process for removing copper from an aqueous environment. *Scientific Reports, 13*(1), 13334. Advance online publication. DOI: 10.1038/s41598-023-40512-y PMID: 37587185

Khan, . (2021). *Experimental studies on arsenic removal efficiencies through adsorption using different natural adsorbents.* Water & Soil Pollution., DOI: 10.1007/s11270-020-04977-1

Kianfar, . (2019). Nanozeolites: Synthesized, properties, applications. *Journal of Sol-Gel Science and Technology, 91*(2), 415–429. DOI: 10.1007/s10971-019-05012-4

Klaine, , Alvarez, P. J. J., Batley, G. E., Fernandes, T. F., Handy, R. D., Lyon, D. Y., Mahendra, S., McLaughlin, M. J., & Lead, J. R. (2008). Nanomaterials in the environment: Behavior, fate, bioavailability, and effects. *Environmental Toxicology and Chemistry, 27*(9), 1825–1851. DOI: 10.1897/08-090.1 PMID: 19086204

Kochubei, , Yaholnyk, S., Bets, M., & Malovanyy, M. (2020). Use of activated clinoptilolite for direct dye-contained wastewater treatment. *Chemistry & Chemical Technology, 14*(3), 323–329. DOI: 10.23939/chcht14.03.386

Kukowska, . (2024). Management of the soil environment using biochar and zeolite in various combinations: Impact on soil condition and economical aspects. *Journal of Soils and Sediments, 24*(10), 3456–3468. DOI: 10.1007/s11368-024-03927-2

Kuldeyev, , Seitzhanova, M., Tanirbergenova, S., Tazhu, K., Doszhanov, E., Mansurov, Z., Azat, S., Nurlybaev, R., & Berndtsson, R. (2023). Modifying natural zeolites to improve heavy metal adsorption. *Water (Basel), 15*(12), 2215. Advance online publication. DOI: 10.3390/w15122215

Kusrini, et al. (2018). The role of praseodymium oxide-impregnated clinoptilolite zeolite catalyst to increase octane number in gasoline. E3S Web of Conferences, 67, Article 03051. DOI: 10.1051/e3sconf/20186703051

Kwon, , Yang, D. S., Koo, M. S., Ji, S. M., Jeong, J., Oh, S., Kuk, S. K., Heo, H., Ham, D. J., Kim, M., Choi, H., Lee, J.-M., Shur, J.-W., Lee, W.-J., Bin, C.-O., Timofeev, N., Wu, H., Wang, L., Lee, T., & Lee, H. C. (2023). Long-lifetime water-washable ceramic catalyst filter for air purification. *Nature Communications, 14*(1), 36050. Advance online publication. DOI: 10.1038/s41467-023-36050-w PMID: 36792592

Lappala, . (2024). The next revolution in computational simulations: Harnessing AI and quantum computing in molecular dynamics. *Current Opinion in Structural Biology, 89*, 102919. Advance online publication. DOI: 10.1016/j.sbi.2024.102919 PMID: 39306949

Leal, M., Martínez-Hernández, V., Meffe, R., Lillo, J., & de Bustamante, I. (2017). Clinoptilolite and palygorskite as sorbents of neutral emerging organic contaminants in treated wastewater: Sorption-desorption studies. *Chemosphere*, *175*, 534–542. DOI: 10.1016/j.chemosphere.2017.02.057 PMID: 28254718

Li, , Han, H., Evangelou, N., Wichrowski, N. J., Lu, P., Xu, W., Hwang, S.-J., Zhao, W., Song, C., Guo, X., Bhan, A., Kevrekidis, I. G., & Tsapatsis, M. (2023). Machine learning-assisted crystal engineering of a zeolite. *Nature Communications*, *14*(1), 3152. Advance online publication. DOI: 10.1038/s41467-023-38738-5 PMID: 37258522

Li, , Ullah, R., Jiao, J., Sun, J., & Bai, S. (2020). Ion exchange of cations from different groups with ammonium-modified clinoptilolite and selectivity for methane and nitrogen. *Materials Chemistry and Physics*, *256*, 123760. Advance online publication. DOI: 10.1016/j.matchemphys.2020.123760

Li, , & Yu, J. (2021a). Emerging applications of zeolites in catalysis, separation and host–guest assembly. *Nature Reviews. Materials*, *6*(12), 1156–1174. DOI: 10.1038/s41578-021-00347-3

Lima, , Silva, Y. F., & Lima, P. L. T. (2021). Iron removal efficiency in irrigation water by zeolite added to sand media filters. *Desalination and Water Treatment*, *220*, 134–141. DOI: 10.5004/dwt.2021.27024

Liu, , Kim, K.-H., Kumar, V., & Kim, S. (2019b). A review of functional sorbents for adsorptive removal of arsenic ions in aqueous systems. *Journal of Hazardous Materials*, *393*, 121815. Advance online publication. DOI: 10.1016/j.jhazmat.2019.121815 PMID: 31831285

Liu, , Tian, R., Ding, W., He, Y., & Li, H. (2019a). Adsorption selectivity of heavy metals by Na-clinoptilolite in aqueous solutions. *Adsorption*, *25*(4), 747–755. DOI: 10.1007/s10450-019-00081-x

Loiola, , Bessa, R. A., Oliveira, C. P., Freitas, A. D. L., Soares, S. A., Bohn, F., & Pergher, S. B. C. (2022). Magnetic zeolite composites: Classification, synthesis routes, and technological applications. *Journal of Magnetism and Magnetic Materials*, *560*, 169651. Advance online publication. DOI: 10.1016/j.jmmm.2022.169651

Luo, et al. (2025). Atomic-scale insights into topotactic transformations in an extra-large-pore zeolite using time-resolved 3D electron diffraction. 145–153. DOI: 10.1038/s44160-024-00715-1

Mahesh, . (2020). Machine learning algorithms - a review. [IJSR]. *International Journal of Science and Research (Raipur, India)*, *9*(1), 381–386. DOI: 10.21275/ART20203995

Mahmoudi, , Nasseri, S., Mahvi, A. H., Dargahi, A., Khubestani, M. S., & Salari, M. (2019). Fluoride removal from aqueous solution by acid-treated clinoptilolite: Isotherm and kinetic study. *Desalination and Water Treatment*, *141*, 246–253. DOI: 10.5004/dwt.2019.23625

Maity, , Chen, C.-Y., Bhattacharya, P., Sharma, R. K., Ahmad, A., Patnaik, S., & Bundschuh, J. (2020). Advanced application of nano-technological and biological processes as well as mitigation options for arsenic removal. *Journal of Hazardous Materials*, *405*, 123885. Advance online publication. DOI: 10.1016/j.jhazmat.2020.123885 PMID: 33183836

Manzoli, et al. (2021). New insights in the production of simulated moon agglutinates: The use of natural zeolite-bearing rocks. ACS, 5(10), 2878–2886. DOI: 10.1021/acsearthspacechem.1c00118

Mastinu, , Kumar, A., Maccarinelli, G., Bonini, S. A., Premoli, M., Aria, F., Gianoncelli, A., & Memo, M. (2019). Zeolite clinoptilolite: Therapeutic virtues of an ancient mineral. *Molecules (Basel, Switzerland)*, *24*(8), 1517. Advance online publication. DOI: 10.3390/molecules24081517 PMID: 30999685

Mehrabadi, , & Faghihian, H. (2018). Elimination of highly consumed herbicide; 2,4-dichlorophenoxyacetic acid from aqueous solution by TiO2 impregnated clinoptilolite, study of degradation pathway. *Spectrochimica Acta. Part A: Molecular and Biomolecular Spectroscopy*, *204*, 676–684. DOI: 10.1016/j.saa.2018.06.047 PMID: 29936222

Montes-Luna, . (2015). Characterization of natural and Ca2+-modified clinoptilolite by different physical-chemical methods for its possible application in gas separation processes. *Superficies y Vacío*, *28*(1). https://www.scielo.org.mx/scielo.php?script=sci_arttext&pid=S1665-35212015000100002

Moradi, , Karimzadeh, R., & Moosavi, E. S. (2018). Modified and ion exchanged clinoptilolite for the adsorptive removal of sulfur compounds in a model fuel. *Fuel*, *217*, 467–477. DOI: 10.1016/j.fuel.2017.12.095

Morgan, , & Jacobs, R. (2020). Opportunities and challenges for machine learning in materials science. *Annual Review of Materials Research*, *50*(1), 71–103. DOI: 10.1146/annurev-matsci-070218-010015

Mortazavi, , Bahadori, M., Marandi, A., Tangestaninejad, S., Moghadam, M., Mirkhani, V., & Mohammadpoor-Baltork, I. (2021). Enhancement of CO2 adsorption on natural zeolite, modified clinoptilolite with cations, amines and ionic liquids. *Sustainable Chemistry and Pharmacy*, *22*, 100495. Advance online publication. DOI: 10.1016/j.scp.2021.100495

Mukherjee, . (2021). Iron oxide-based nanomaterials for arsenic removal: A review. *Journal of Hazardous Materials*, *403*, 123558. Advance online publication. DOI: 10.1016/j.jhazmat.2020.123558

Nami, . (2023a). Nitrate removal performance of different granular adsorbents using a novel Fe-exchanged nanoporous clinoptilolite. *Industrial & Engineering Chemistry Research*, *62*(8), 3659–3671. DOI: 10.1021/acs.iecr.2c03308

Nguyen, , Loganathan, P., Nguyen, T. V., & Vigneswaran, S. (2019). Removing arsenic from water with an original and modified natural manganese oxide ore: Batch kinetic and equilibrium adsorption studies. *Environmental Science and Pollution Research International*, *27*(5), 5490–5502. DOI: 10.1007/s11356-019-07284-3 PMID: 31853842

Ogorek, . (2024). Soil compaction sensing mechanisms and root responses. *Trends in Plant Science*, *29*(12), 1345–1356. DOI: 10.1016/j.tplants.2024.10.014

Ostrooumov, , Cappelletti, P., & de'Gennaro, R. (2011). Mineralogical study of zeolite from New Mexican deposits (Cuitzeo area, Michoacan, Mexico). *Applied Clay Science*, *55*, 156–160. DOI: 10.1016/j.clay.2011.09.011

Otitoju, T. A., Saari, R. A., & Ahmad, A. L. (2018). Progress in the modification of reverse osmosis membranes for enhanced performance. *Journal of Industrial and Engineering Chemistry*, *67*, 52–71. Advance online publication. DOI: 10.1016/j.jiec.2018.07.010

Ottoboni, , Coleman, S. J., Steven, C., Siddique, M., Fraissinet, M., Joannes, M., Laux, A., Barton, A., Firth, P., Price, C. J., & Mulheran, P. A. (2020). Understanding API static drying with hot gas flow: Design and test of a drying rig prototype and drying modeling development. *Organic Process Research & Development*, *24*(11), 2505–2520. DOI: 10.1021/acs.oprd.0c00035 PMID: 33250628

Pabiś-Mazgaj, , Gawenda, T., Pichniarczyk, P., & Stempkowska, A. (2021). Mineral composition and structural characterization of the clinoptilolite powders obtained from zeolite-rich tuffs. *Minerals (Basel)*, *1030*(10), 1030. Advance online publication. DOI: 10.3390/min11101030

Pavelić, . (2018). Critical review on zeolite clinoptilolite safety and medical applications in vivo. *Frontiers in Pharmacology*, *9*, 1350. Advance online publication. DOI: 10.3389/fphar.2018.01350 PMID: 30538633

Pavlak, , Kaufmann, C., Eyng, C., Carvalho, P. L. O., Pozza, P. C., Vieites, F. M., Rohloff Junior, N., Avila, A. S., Polese, C., & Nunes, R. V. (2023). Zeolite and corn with different compositions in broiler chickens feeding. *Poultry Science*, *102*(4), 102494. Advance online publication. DOI: 10.1016/j.psj.2023.102494 PMID: 36745958

Perego, , de Angelis, A., Pollesel, P., & Millini, R. (2021). Zeolite-based catalysis for phenol production. *Industrial & Engineering Chemistry Research*, *60*(18), 6379–6402. DOI: 10.1021/acs.iecr.0c05886

Pérez-Botella, , Valencia, S., & Rey, F. (2022). Zeolites in adsorption processes: State of the art and future prospects. *Chemical Reviews*, *17647*(24), 17647–17695. Advance online publication. DOI: 10.1021/acs.chemrev.2c00140 PMID: 36260918

Pour, . (2024). Rapid SO2 gas removal using MgO/AC/CaCO3/zeolite nanocomposite at room temperature. *Case Studies in Chemical and Environmental Engineering*, *10*, 101042. Advance online publication. DOI: 10.1016/j.cscee.2024.101042

Rabia, , Ould Hamou, M., Kasperkiewicz, K., Brożek, J., & Augustyniak, M. (2019). Adhesion abilities and biosorption of Cd and Mg by microorganisms - first step for beneficiation of phosphate ore. *Scientific Reports*, *9*(1), 12929. Advance online publication. DOI: 10.1038/s41598-019-49406-4 PMID: 31506570

Ramasamy, , Senthil Kumar, R., Dharmaraj, R., & Kannan, M. (2025). Fluoride removal from contaminated water using natural bio-adsorbents: A numerical and experimental investigation. *Desalination and Water Treatment*, *324*, 100984. Advance online publication. DOI: 10.1016/j.dwt.2024.100984

Rapado, , Garcés, D., Faba, L., & Ordóñez, S. (2024). Zeolite-assisted acid hydrolysis of cellulose: Optimization of reaction conditions and chemical pretreatments for enhancing HMF yields. *Industrial Crops and Products*, *212*, 118309. Advance online publication. DOI: 10.1016/j.indcrop.2024.118309

Rathi, , & Kumar, P. S. (2021). Application of adsorption process for effective removal of emerging contaminants from water and wastewater. *Environmental Pollution*, *116995*, 116995. Advance online publication. DOI: 10.1016/j.envpol.2021.116995 PMID: 33789220

Reid, , Klebe, S., van Zandwijk, N., & George, A. M. (2021). Asbestos and zeolites: From A to Z via a common ion. *Chemical Research in Toxicology*, *34*(4), 936–951. DOI: 10.1021/acs.chemrestox.0c00286 PMID: 33749247

Restiawaty, , Gozali, V. A., Wibisono, T. A. S. E., & Budhi, Y. W. (2024). Utilizing modified clinoptilolite for the adsorption of heavy metal ions in acid mine drainage. *Case Studies in Chemical and Environmental Engineering*, *9*, 100706. Advance online publication. DOI: 10.1016/j.cscee.2024.100706

Rochard, , Olivet, L., Tannous, M., Poupin, C., Siffert, S., & Cousin, R. (2021). Recent advances in the catalytic treatment of volatile organic compounds: A review based on the mixture effect. *Catalysts*, *11*(10), 1218. Advance online publication. DOI: 10.3390/catal11101218

Rodríguez-Iznaga, , Shelyapina, M. G., & Petranovskii, V. (2022). Ion exchange in natural clinoptilolite: Aspects related to its structure and applications. *Minerals (Basel)*, *12*(12), 1628. Advance online publication. DOI: 10.3390/min12121628

Saeed, , Abbas, G., Alagawany, M., Kamboh, A. A., Abd El-Hack, M. E., Khafaga, A. F., & Chao, S. (2019). Heat stress management in poultry farms: A comprehensive overview. *Journal of Thermal Biology*, *425*, 414–425. Advance online publication. DOI: 10.1016/j.jtherbio.2019.07.025 PMID: 31466781

Safarpour, , Vatanpour, V., Khataee, A., Zarrabi, H., Gholami, P., & Yekavalangi, M. E. (2017). High flux and fouling resistant reverse osmosis membrane modified with plasma treated natural zeolite. *Desalination*, *411*, 89–100. DOI: 10.1016/j.desal.2017.02.012

Sahu, , Pervez, S., Karbhal, I., Tamrakar, A., Mishra, A., Verma, S. R., Deb, M. K., Ghosh, K. K., Pervez, Y. F., Shrivas, K., & Satnami, M. L. (2024). Applications of different adsorbent materials for the removal of organic and inorganic contaminants from water and wastewater – A review. *Desalination and Water Treatment*, *317*, 100253. Advance online publication. DOI: 10.1016/j.dwt.2024.100253

Sarioglu, M. (2004). Removal of ammonium from municipal wastewater using natural Turkish (Dogantepe) zeolite, 41(1), 1–11. DOI: 10.1016/j.seppur.2004.03.008

Schwarzenbach, , Escher, B. I., Fenner, K., Hofstetter, T. B., Johnson, C. A., von Gunten, U., & Wehrli, B. (2006). The challenge of micropollutants in aquatic systems. *Science*, *313*(5790), 1072–1077. DOI: 10.1126/science.1127291 PMID: 16931750

Senila, , & Cadar, O. (2024). Modification of natural zeolites and their applications for heavy metal removal from polluted environments: Challenges, recent advances, and perspectives. *Heliyon, 10*(3), e25303. Advance online publication. DOI: 10.1016/j.heliyon.2024.e25303 PMID: 38352776

Shaban, , AbuKhadra, M. R., Nasief, F. M., & Abd El-Salam, H. M. (2017). Removal of ammonia from aqueous solutions, ground water, and wastewater using mechanically activated clinoptilolite and synthetic zeolite-A: Kinetic and equilibrium studies. *Water, Air, and Soil Pollution, 228*(1), 20. Advance online publication. DOI: 10.1007/s11270-017-3643-7

Shankar, . (2023). Adsorption kinetics, equilibrium and thermodynamics of a textile dye V5BN by a natural nanocomplex material. *Energy Nexus, 100197*, 100197. Advance online publication. DOI: 10.1016/j.nexus.2023.100197

Shekarchi, , Ahmadi, B., Azarhomayun, F., Shafei, B., & Kioumarsi, M. (2023). Natural zeolite as a supplementary cementitious material – A holistic review of main properties and applications. *Construction & Building Materials, 409*, 133766. Advance online publication. DOI: 10.1016/j.conbuildmat.2023.133766

Shi, et al. (2023). Adsorption properties of natural zeolite supported by chitosan on Cd(II) in micropolluted irrigation water. ACS, 12. DOI: 10.1021/acsomega.3c06041

Shimizu, , & Matubayasi, N. (2020). Fluctuation adsorption theory: Quantifying adsorbate–adsorbate interaction and interfacial phase transition from an isotherm. *Physical Chemistry Chemical Physics, 22*(48), 28304–28316. DOI: 10.1039/D0CP05122E PMID: 33295900

Shoinkhorova, , Dikhtiarenko, A., Ramirez, A., Dutta Chowdhury, A., Caglayan, M., Vittenet, J., Bendjeriou-Sedjerari, A., Ali, O. S., Morales-Osorio, I., Xu, W., & Gascon, J. (2019). Shaping of ZSM-5-based catalysts via spray drying: Effect on methanol-to-olefins performance. *ACS Applied Materials & Interfaces, 11*(47), 44133–44143. Advance online publication. DOI: 10.1021/acsami.9b14082 PMID: 31612697

Shokrolahzadeh, . (2017). Modification of nano clinoptilolite zeolite using sulfuric acid and its application toward removal of arsenic from water sample. *Journal of Nanoanalysis, 4*(1), 48–58. DOI: 10.22034/jna.2017.01.006

Smiljanić, , de Gennaro, B., Izzo, F., Langella, A., Daković, A., Germinario, C., Rottinghaus, G. E., Spasojević, M., & Mercurio, M. (2020). Removal of emerging contaminants from water by zeolite-rich composites: A first approach aiming at diclofenac and ketoprofen. *Microporous and Mesoporous Materials, 298*, 110057. Advance online publication. DOI: 10.1016/j.micromeso.2020.110057

Soni, et al. (2018). Synthesis of fly ash based zeolite-reduced graphene oxide composite and its evaluation as adsorbent for arsenic removal. DOI: 10.1016/j.chemosphere.2018.11.203

Sossou, , Prasad, S. B., Agbotsou, E. K., & Saidou Souley, H. (2024). Evaluation of the performance of magnetic zeolite nanocomposites in removing various water contaminants as heavy metals, organic pollutants, and emerging contaminants. *Article, 100075*, 100075. Advance online publication. DOI: 10.1016/j.nxnano.2024.100075

Souza, , Borrego-Sánchez, A., Rigoti, E., Sainz-Díaz, C. I., Viseras, C., & Pergher, S. B. C. (2021). Experimental and molecular modelling study of beta zeolite as drug delivery system. *Microporous and Mesoporous Materials, 111152*, 111152. Advance online publication. DOI: 10.1016/j.micromeso.2021.111152

Sun, , Abbass, R., Ghoroqi, M., Patra, I., Dwijendra, N. K. A., Uktamov, K. F., & Jasem, H. (2022). [Retracted] Optimization of dyes and toxic metals removal from environmental water samples by clinoptilolite zeolite using response surface methodology approach. *Scientific Reports, 12*(1), 13218. Advance online publication. DOI: 10.1038/s41598-022-17636-8 PMID: 35918466

Sun, , Ding, Z.-H., Bi, Y.-Z., & Pu, S. (2023). Experimental investigation on water/gas resistance and strength properties of clinoptilolite/Na-PAA amended compacted clay cover. Geophysics for Geo-Energy and Geo-Resources. *Geomechanics and Geophysics for Geo-energy and Geo-resources, 66*(1), 66. Advance online publication. DOI: 10.1007/s40948-023-00591-7

Thakur, , Ojha, A., Kansal, S. K., Gupta, N. K., Swart, H. C., Cho, J., Kuznetsov, A., Sun, S., & Prakash, J. (2024). Advances in powder nano-photocatalysts as pollutant removal and as emerging contaminants in water: Analysis of pros and cons on health and environment. *Advanced Powder Materials, 100233*(6), 100233. Advance online publication. DOI: 10.1016/j.apmate.2024.100233

Ugrina, M., Gabrilo, I., & Jurić, A. (2023). Modeling, simulation, optimization, and experimental verification of mercury removal onto natural and sulfur-impregnated zeolite clinoptilolite—Assessment of feasibility for remediation of mercury-contaminated soil. *Processes (Basel, Switzerland), 11*(2), 606. Advance online publication. DOI: 10.3390/pr11020606

Ullah, , Sun, J., Gul, A., & Bai, S. (2020b). One-step hydrothermal synthesis of TiO2-supported clinoptilolite: An integrated photocatalytic adsorbent for removal of crystal violet dye from aqueous media. Journal of Environmental Chemical Engineering. *Journal of Environmental Chemical Engineering, 103852*(4), 103852. Advance online publication. DOI: 10.1016/j.jece.2020.103852

Ullah, , Sun, J., Gul, A., Munir, T., & Wu, X. (2020a). Evaluations of physico-chemical properties of TiO2/clinoptilolite synthesized via three methods on photo-catalytic degradation of crystal violet. *Chinese Journal of Chemical Engineering*, *33*, 181–189. DOI: 10.1016/j.cjche.2020.09.045

Velarde, , Nabavi, M. S., Escalera, E., Antti, M.-L., & Akhtar, F. (2023). Adsorption of heavy metals on natural zeolites: A review. *Chemosphere*, *328*, 138508. Advance online publication. DOI: 10.1016/j.chemosphere.2023.138508 PMID: 36972873

Vitale, , Barbato, C., Crispo, A., Habetswallner, F., De Martino, B. M., Riccardi, F., Maione, A., Eisenwagen, S., Vitale, G., & Cartenì, G. (2020). ZeOxaNMulti trial: A randomized, double-blinded, placebo-controlled trial of oral PMA-zeolite to prevent chemotherapy-induced side effects, in particular, peripheral neuropathy. *Molecules (Basel, Switzerland)*, *2297*(10), 2297. Advance online publication. DOI: 10.3390/molecules25102297 PMID: 32414185

Wahono, , Suwanto, A., Prasetyo, D. J., Hernawan, , Jatmiko, T. H., & Vasilev, K. (2019). Plasma activation on natural mordenite-clinoptilolite zeolite for water vapor adsorption enhancement. *Applied Surface Science*, *483*, 940–946. Advance online publication. DOI: 10.1016/j.apsusc.2019.04.033

Wang, , & Nguyen, A. V. (2015). Characterisation of electrokinetic properties of clinoptilolite before and after activation by sulphuric acid for treating CSG water. *Microporous and Mesoporous Materials*, *220*, 175–182. DOI: 10.1016/j.micromeso.2015.09.003

Wang, , & Peng, Y. (2010). Natural zeolites as effective adsorbents in water and wastewater treatment. *Chemical Engineering Journal*, *156*(1), 11–24. Advance online publication. DOI: 10.1016/j.cej.2009.10.029

Wang, , Zhang, T., & Wang, X. (2025). Removal mechanism of lead by the composite of clinoptilolite, biochar, and Ca(OH)2 and its application in acid wastewater. *Environmental Technology & Innovation*, *37*, 104024. Advance online publication. DOI: 10.1016/j.eti.2025.104024

Wołowiec, , Muir, B., Zięba, K., Bajda, T., Kowalik, M., & Franus, W. (2017). Experimental study on the removal of VOCs and PAHs by zeolites and surfactant-modified zeolites. *Energy & Fuels*, *31*(8), 8812. DOI: 10.1021/acs.energyfuels.7b01124

Xiang, , Han, Y., Jiang, C., Li, M., Wei, L., Fu, J., & Zhu, L. (2021). Composite biologically active filter (BAF) with zeolite, granular activated carbon, and suspended biological carrier for treating algae-laden raw water. Journal of Water Process Engineering. *Journal of Water Process Engineering*, *102188*, 102188. Advance online publication. DOI: 10.1016/j.jwpe.2021.102188

Xie, . (2013). *Adsorption of organic pollutants by surfactant modified zeolite as controlled by surfactant chain length*. Microporous., DOI: 10.1016/j.micromeso.2013.05.027

Xu, , Huang, Z., Ye, J., & Zheng, J. (2023). Hourly emissions of air pollutants and greenhouse gases from open biomass burning in China during. *Scientific Data*, *10*(1), 629. Advance online publication. DOI: 10.1038/s41597-023-02541-0 PMID: 37717027

Xu, , Liu, H., Zhang, J., Xu, M., Zhao, X., & Wang, J. (2025). Functionalized zeolite regulates bone metabolic microenvironment. *Materials Today. Bio*, *101558*, 101558. Advance online publication. DOI: 10.1016/j.mtbio.2025.101558 PMID: 40034985

Yang, . (2024). Water-saving irrigation combined with N-loaded clinoptilolite enhances nutrient yield, and water productivity by improving rice root characteristics: A combined PCA-SEM analysis. *Agricultural Water Management*, *301*, 109203. Advance online publication. DOI: 10.1016/j.agwat.2024.109203

Yu, , Rengasamy, K., Lim, K.-Y., Tan, L.-S., Tarawneh, M. A., Zulkoffli, Z. B., & Se Yong, E. N. (2019). Comparison of activated carbon and zeolites filtering efficiency in freshwater. Journal of Environmental Chemical Engineering. *Journal of Environmental Chemical Engineering*, *103223*(4), 103223. Advance online publication. DOI: 10.1016/j.jece.2019.103223

Yusuf, , Sodiq, A., Giwa, A., Eke, J., Pikuda, O., De Luca, G., Di Salvo, J. L., & Chakraborty, S. (2020). A review of emerging trends in membrane science and technology. *Journal of Cleaner Production*, *266*, 121867. Advance online publication. DOI: 10.1016/j.jclepro.2020.121867

Zhang, . (2020). *Electron microscopy studies of local structural modulations in zeolite crystals* (International Edition). Angewandte Chemie., DOI: 10.1002/anie.202007490

Zhang, , Liu, N., Lei, Z., & Chen, B. (2016). Selective transformation of various nitrogen exhaust gases toward N2 over zeolite catalysts. *Chemical Reviews*, *117*(10), 6838–6880. DOI: 10.1021/acs.chemrev.5b00474 PMID: 26889565

Zhou, , Shen, Y., Zhao, S., Chen, Y., Gao, S., Liu, W., & Wei, D. (2020). Hydrothermal synthesis of novel ternary hierarchical MoS2/TiO2/clinoptilolite nanocomposites with remarkably enhanced visible light response towards xanthates. *Applied Surface Science*, *542*, 148578. Advance online publication. DOI: 10.1016/j.apsusc.2020.148578

Znak, Z., Zin, O., Mashtaler, A., Korniy, S., Sukhatskiy, Y., Gogate, P. R., Mnykh, R., & Thanekar, P. (2021). Improved modification of clinoptilolite with silver using ultrasonic radiation. *Ultrasonics Sonochemistry*, *73*, 105496. DOI: 10.1016/j.ultsonch.2021.105496 PMID: 33636551

ADDITIONAL READING

Yadav, M., et al. (2020). Current Status of Heavy Metal Contaminants and Their Removal/Recovery Techniques. 41–64. doi:DOI: 10.1021/bk-2020-1352.ch003

Chmielewská, E. (2019). Natural zeolite, Alternative adsorbent in purification or post-treatment of waters. Elsevier. 87–112. doi:DOI: 10.1016/b978-0-12-814617-0.00012-8

Milić, J., Daković, et al. (2014). Modified Natural Zeolites-Functional Characterization and Biomedical Application. *Scrivener Publishing, LLC*, 359–403. DOI: 10.1002/9781118774205.ch10

Krstić, V. (2021). Role of zeolite adsorbent in water treatment. Elsevier Inc. 417–481. doi:DOI: 10.1016/b978-0-12-821496-1.00024-6

Soetaredjo, F. E.. (2018). Removal of Hazardous Contaminants from Water or Wastewater Using Polymer Nanocomposites Materials. *Scrivener Publishing, LLC*, 103–139. DOI: 10.1002/9781119323655.ch4

Ejder-Korucu, . (2015). Removal of Organic Dyes from Industrial Effluents: An Overview of Physical and Biotechnological Applications. *Scrivener Publishing, LLC*, 1–34. DOI: 10.1002/9781118721001.ch1

Younas, F.. (2023). *Environmental Applications of Natural and Surface-Modified Zeolite. Advances in Material Research and Technology*. Springer., DOI: 10.1007/978-981-99-2544-5_17

Samanta, N.S., et al. (2024). Waste-Based Zeolites and their Advanced Composites for Wastewater and Environmental Remediation Applications. Springer. 451–490. Doi: org/DOI: 10.1007/978-3-031-42731-2_14

Ray, S. S.. (2025). *Water Purification. Springer Series in Materials Science* (Vol. 350). Springer., DOI: 10.1007/978-3-031-85304-3_7

KEY TERMS AND DEFINITIONS

Water Remediation: Refers to restoring water quality and damaged properties after water damage.

Heavy Metals (HM): Heavy metals in water include arsenic, cadmium, chromium, copper, lead, mercury, and nickel.

Emerging Pollutants (EP): Include pharmaceuticals, pesticides, personal care products, and microplastics. These substances are found in surface water, groundwater, and wastewater.

Reverse Osmosis: It is a water purification technique that uses a semipermeable membrane to remove pollutants.

Soil Remediation: Refers to restoring soil from pollutants damage.

Porous Structure: Term used to define porosity, relates with pore size, pore size distribution, and pore morphology of a porous medium.

Cation Exchange Capacity (CEC): It is a measure of how much a soil can hold positively charged ions, or cations.

Surface Area (SA): In a porous material refers to the total area of its pore space relative to the volume of the material.

Thermal Stability: It is a material's ability to resist changes in its structure or properties when heated.

Hydrothermal Synthesis: It is a technique for creating crystals and nanomaterials using high-pressure, high-temperature water solutions

Sol-Gel Synthesis: It is a wet chemical process that uses a liquid precursor to create solid materials.

Impregnation: Involves filling the pores of a material with a substance to change its properties.

Electron Microscopy: Uses a beam of electrons as a source of illumination.

X-Ray Diffraction (XRD): Used to determine the atomic and molecular structure of a crystal, in which the crystal structure causes an incident X-ray beam to be diffracted in specific directions.

BET: It is a technique that measures the surface area of solid, porous materials by gas adsorption, covered.

Density Functional Theory (DFT): It is an atomistic simulation method to compute a wide variety of properties of almost any kind of atomic system.

Chapter 8
Are NEREA® Zeolitic Nanostructured Materials Equal to NPK Fertilizers Mixed With Natural Zeolites?

Gerardo Rodríguez Fuentes
https://orcid.org/0000-0001-5846-2863
Instituto de Ciencia y Tecnología de Materiales, Cuba

Inocente Rodríguez-Iznaga
https://orcid.org/0000-0002-0729-1096
Instituto de Ciencia y Tecnología de Materiales, Cuba

ABSTRACT

The most recommended use of natural zeolite in agriculture is the addition to the soil as amending agent or in mixture with fertilizers, but the results are not observed in short time, and its reproducibility is not entirely positive. A NPK chemical fertilizer requires water to be solubilized and make nutrients available for plants. If zeolite particles are present can retain part of these nutrients, other part remains in the soil increasing the pollution of the environment. New nanostructured zeolitic materials named NEREA® have changed the paradigm of natural zeolites in agriculture. The distinction is to introduce steadily and homogeneously all the nutrients that plants require into the zeolite particles, before incorporating them into the soil. Only the plants take up the nutrients modifying the architecture and morphology of their roots to involve NEREA particles. If the mechanical mixing of zeolite mineral with NPK fertilizers has represented an important contribution, NEREA products solve

DOI: 10.4018/979-8-3373-0320-8.ch008

its shortcomings, representing an innovation for agriculture and zeolite science.

INTRODUCTION

Professor Frederick A. Mumpton first documented in *Zeo-Agriculture: Use of Natural Zeolites in Agriculture & Aquaculture* (Mumpton, F.A., 1984); and then in T*he magic rock: Uses of natural zeolites in agriculture and industry* (Mumpton, F.A., 1999) that the oldest use of zeolites in agriculture is in Japan, as a soil improver and additive to fertilizers, dating back several centuries (Minato, H., 1968). While in other countries its use was reported during the second half of the 20th century, with similar objectives of improving the properties of soils, increasing the efficiency of fertilizers, and consequently increase crop yields. Mumpton established the uses of natural zeolites in agriculture: 1) Soil improver, 2) Additive to chemical fertilizers, 3) Treatment of animal excreta for use as compost, and 4) Obtaining artificial soils or substrates for growing plants, *zeoponic* indicating that natural zeolite is used as a substrate in cultivation of plants, employing its properties.

Ming, D.W and Allen, E.R. (2001) established that the correctly documented uses of natural zeolites in agriculture were: 1) Substrates for plants in both greenhouses and potting in the houses, 2) Fertilizers and soil conditioners on golf courses, 3) Slow release fertilizers in agricultural applications, 4) Fertilizers, insecticides, and slow release herbicides in areas where environmental problems are important (sandy soils near surface waters or aquifers), 5) Remedial agents for soils with environmental problems of contamination by heavy metals and radionuclides. From this review up to 2024, numerous works have been reported that use zeolitic minerals directly in agricultural technologies or modified for the elaboration of products for that purpose. The modification are aimed at the *highest possible incorporation of the macronutrients* N, P and K to the mineral, for their subsequent slow or controlled release to plant nutrition which can be grouped into: 1) Mixture with chemical fertilizers that can contain up to 50% of the zeolitic mineral, 2) Mixture with organic matter to be used as a fertilizer, which extends to the treatment of the excreta of production animals with zeolitic mineral, 3) Mix with minerals that provide nutrients such as apatite, calcium phosphate carbonate, etc. 4) Modification of the previously ground zeolitic mineral with different particle sizes, applying heat, treatments with acids or surfactants, or a combination of these treatments, and the subsequent incorporation of the nutrient by ion exchange or adsorption of the fertilizer, 5) The combination of the above procedures to obtain a fertilizer material that provides nutrients and organic matter, and 6) The elaboration of substrates for the cultivation of plants without soils called *zeoponic*, which is the most complex of the procedures for the use of natural zeolites in agriculture since its objective is to obtain an artificial soil.

Natural zeolites are part of the mineral composition of the soils close to zeolite occurrence and deposits, or due to the transformation of volcanic materials into zeolites (Ming, D.W. and Boettinger, J.L. 2001). The types of zeolites that formed in soils include analcime, phillipsite, mordenite, clinoptilolite, chabazite, and natrolite. In Cuba, where 106 occurrences of zeolitic rocks have been documented by A. Brito and D. Coutín (2017), almost perfect match can be observed of the occurrences, and the best soils used for agriculture production.

Three primary procedures for obtaining substrates using natural zeolitic rocks, recognized by specialists are: 1) BALKANIN substrate: US 4337078 patent granted to Georgi S. Petrov et al. This procedure is based on a mixture of zeolite, vermiculite, and peat, treated with solid-state ammonium and phosphorus fertilizers. It only uses water to irrigate the plants that grow in it. The three basic components have particles with different sizes, densities, and nutrient absorption capacities, which gives the substrate little homogeneity, and limits the uptake of nutrients in the same way by plants. The Balkanin was successfully tested on one of the MIR manned cosmic flights. 2) Substrate NASA. Since the late 1980s, the United States National Aeronautics Administration (NASA) developed Project Biosphere 2 to support the life of astronauts in space. The project comprised the development of a zeoponic substrate for plant growth that was developed by Ming, D.W. and Allen E.R. (1999). The substrate is a mixture of several solid components: clinoptilolite zeolite exchanged with NH_4^+, zeolite exchanged with K^+, and natural and synthetic apatite as sources of P. A fertilization of slow release due to the dissolution of natural apatite that contributes Ca and PO_4; synthetic apatite is hydroxyapatite and contributes Mg, S and micronutrients previously inserted in its structure. The substrate only requires water for the irrigation of the plants. This product is marketed in the United States with the ZeoPro™ brand. Its production includes several slow and costly production processes, which contribute residual pollutants. 3) NEREA substrates. During the years 1984 to 1989, G. Rodríguez-Fuentes and L.A. Rivero-González developed five procedures for the modification of the zeolite mineral with particle size -1+4mm directly in the hydroponic cultivation facilities of Cuba, using their irrigation systems, made possible the incorporation of the nutrients not contained in the zeolitic mineral and the balance of those present; all the macro and micro nutrients contained in the "loading" solution were taken completely during the successive irrigations over several hours, without generating residual pollutants. The zeolitic substrate obtained was called NEREA, acronym for Nutrients In Rock and Water Use (in Spanish). The procedures are described in *The Commercial Zeoponic in Cuba: Handbook for Engineers* limited publication only for the personnel of the zeoponic facilities of the Ministry of Agriculture of Cuba (MINAG) (1990). During their use they only require water for the irrigation of the plants, when the nutrients are depleted by the plants, they can be reincorporated formulating a new nutrient solution balancing their contents with the rest in the substrate. These substrates are distinguished from Balkanin and NASA, in 1) They only contain zeolitic mineral as basic raw material, 2) All the macro and micronutrients necessary for the development of the plants are incorporated into the mineral particles, 3) They are homogeneous substrates, 4) The substrates preserve the nutrients until the plants extract them, thus minimizing losses due to leaching, and 5) They can be recharged when nutrients are depleted.

NEREA substrates were positively validated for their use in plant cultivation by the specialists of the Cuban Agriculture (Pérez, J.E. et al. 1991). The main deficiencies reported were: 1) The large volumes of nutrient dissolution necessary to obtain them in the zeoponic facilities, 2) The long treatment time of the zeolitic mineral, and 3) The loss of nutrient dissolution in zeoponic facilities with an open irrigation system.

Different variants of elaboration of NEREA substrates using Cuban zeolitic minerals have been development to be implemented industrially: **LITONITA®** and **CELTONITA®** used the original procedure developed by G. Rodríguez Fuentes and L. A. Rivero González, however, it was modified to be based on the immersion of the zeolitic mineral in a solution of NPK fertilizer for 24 - 48 hours, releasing liquid waste with a high load of nutrients not adsorbed by the mineral. Celtonita is a trademark of the Celta Brasil Group. The technical sheet recommends incorporating 15% Celtonita in the traditional substrate used in seedbeds and nurseries. As fertilizer dose of 500 to 1000 kg ha^{-1}, replace the bottom NPK fertilizer during sowing, but requires the application of the complementary nitrogen fertilizer recommended for the crop (Febles, J. A. et al. 2014).

NEREA's technology and products have undergone constant development over time. This has been materialized through research and development studies, product diversification, new high-value products, and substantial improvements to the technology, all of which have been accompanied with a protection by industrial secret, patents, and trademarks. This work is aimed to analyze the innovative potentiality of the NEREA's technology and products for agriculture and zeolite science compared with the use of zeolitic raw materials from Cuban and other country in the formulation of fertilizer treatments with NPK or urea formulas.

STATE OF THE ART OF PRODUCTC FOR THE CULTIVATION OF PLANTS USING ZEOLITIC MINERALS

Substrates. Several procedures for obtaining substrates based on modified zeolitic rocks have been reported. Overall, they are inventions based on the three basic procedures mentioned above. CN1017395 B. **Zeolite cultivating substrate for plant**. In this invention acid treatments of the mineral, heat treatment for drying and mix with a solid chemical fertilizer that does not directly incorporate the nutrients into the zeolite. US20080116141 A1. **Complete plant growth medium.** This invention modifies the zeolitic mineral particles with an iron oxide compound to increase the nutrient adsorption capacity, the product is mixed with other materials to make substrates. US5106405 A. JP63-232589. **Horticultural medium consisting essentially of natural zeolite particles**. In this invention, the natural zeolite is mixed with the other components that provide the nutrients. CN103553832 A1, CN1096994395 A1, CN107827568 A1, CN107827569 A1, CN107721659 A1, CN108147904 A1. They are some of the many inventions patented in China that describe substrates and fertilizers for growing plants, which are mixtures of natural zeolite with natural organic materials, fertilizers, waste from agricultural, food and medicinal productions. A comprehensive review was published by T.G. Andronikashvili et al. (2009) where zeolite containing substrates are described as a *new way from plant-growing*. T.G. Andronikashvili et al. (2011) describe a substrate based on a mixture of brown carbon or lignite, natural zeolite modified with ammonium, and with potassium. The bioproduction of the plants grown in the substrate was 2.9 times higher than that obtained in the soil.

Fertilizers based on natural zeolites. This practice is based on the studies on the use of **natural zeolites** in plant cultivation that demonstrated the benefit in the physical and chemical properties of soils, specifically in the dynamics of nitrogen (Vilcek, J. et al.2013) (Torma, S., et al. 2014). The authors observed in the mixture soil + zeolite that "more than 90% of the cations were fixed in the zeolite in the first few minutes while the ammonia cations reacted fastest." One month after zeolite application

> *a decreased ammonium content was apparent in the variants with zeolite in comparison with control without zeolite owing to the specific fixation of* NH_4^+ *cations in the zeolite lattice. Three months later, there was an increase in ammonium content in the variants with zeolite of 24% to 59% in comparison with the control variant owing to gradual ammonium release from the zeolite lattice. Six months later, the nitrification process in the soil was less intense under the influence of applied zeolite. The nitrate nitrogen contents in the soil decreased by 66% to 78% in comparison with the control; therefore, the amount of nitrate leaching from the soil horizons to the groundwater was less. In summary, zeolite can be a slow releasing nitrogen fertilizer.*

Same results have also been observed when using natural zeolite previously mixed with NPK fertilizers or directly at the time of application in the field or added to the soil before or after the NPK application. This practice determines a time for plants to benefit from the nutrients added to the soil. 1) Solubilization of the salts that contain the nutrients (NH_4NO_3, NH_4HPO_4, KCl, KPO_4) that make up NPK mixed fertilizer, by the action of irrigation or rainwater, after its addition to the soil, 2) Part of the nutrients are taken up by plant roots, part by zeolite particles, part by the organic and mineral components of the soil, and most are leached into groundwater, 3) The processes of incorporation the nutrients into the zeolite particles are spontaneous and are determined by the soil conditions, and their proximity to the fertilizer components, 4) The times for each process to occur are long, days and weeks, and 5) The homogeneity of the processes is scarce.

Fertilizers based on the modification of zeolitic rocks. The ways of modifying the zeolite mineral use thermal treatments, acids, organic and inorganic chemical compounds, and a combination of these treatments, all with the aim of increasing the adsorption capacity of the zeolite to retain plant nutrients. US5695542 A. **Method of preparing a slow release fertilizer**: This invention procedure uses the zeolite in a physical mixture with fertilizers preserved in a gelatinous substance. The product obtained must be solubilized by rain or irrigation water that acts on the gelatinous substance. CA2998756. **Synthetic agricultural product** is obtained starting from silicic materials such as montmorillonite, kaolinite, smectite and zeolite, mixed with the solution of fertilizers, and organic compounds. Energy is applied to effectively achieve nutrient incorporation. O2013119108 A1. **Controlled release fertilizer composite comprising zeolite**: it is a dry compound of modified clinoptilolite and mordenite type zeolites, activated carbon, molasses, nitrogen fertilizer, Fe, Mn, Zn nutrients and a binder, for the controlled release of nutrients for the cultivation of plants. US20040099027 A1. **Manufacturing method for zeolite-containing fertilizer**: it is slow release fertilizer containing natural zeolite, residual from the refining of beet sugar, and calcium carbonate, mixed with water. The dry mix must be loaded with commercial fertilizers. US20130327103 A1. **Use of zeolites in supplying micronutrients**: The micronutrients Cu, Mn, Zn, Mo, Co or Fe, and the meso-elements Ca or Mg, are incorporated into the zeolite in 1.5 to 15% of the mass of the zeolite. WO2013110106 A1. **Method for producing mineral foliar fertilizer**: a mixture of clinoptilolite, dolomite and calcite materials ground to obtain particles sizes <40 μm.

Nanoporous zeolite-based N fertilizer. Manikandan, A. and Subramanian, K. S. (2014) reported a laboratory study to increase the efficient use of nitrogen using a natural zeolite (Z) or a zeolite with nanopores (NZ) as an adsorbent substrate to which urea is impregnated. The commercial urea and the adsorbent in the ratio were mixed or fused in the ratio of 1:1 to 1:10 using simple liquid immersion with the hydrothermal technique. The 1:1 ratio registered the highest total N content of 18.5 zeourea (ZU) and 28% nano-zeourea (NZU), respectively. The study suggests that a fertilizer based on a zeolite with nanopores can be used as a strategy to improve the efficient use of nitrogen. Zhaohui Li. (2003), and Bansiwal, A.K. et al. (2006) described procedures for obtaining fertilizers with the use of surfactants (such as hexadecyltrimethylammonium bromide) to modify the surface of the crystals and zeolite particles, which increases the adsorption capacity of anions (nitrates and phosphates). The material obtained containing significant amounts of the nutrients N and P, and releasing them slowly, the cost of the procedure and the toxicity are discussed.

Patent Analytics Report: Agricultural Nanomaterials an Australian Government IP Australia study, analyzed the impact of nanomaterials on agriculture through the lens of intellectual property, assessing the scale and intensity of patent activity as a reflection of the innovation in this technology. 128 innovations related to nanomaterials for agriculture were identified. The main contributions to this field come from the United States, China, and Germany. The study defines zeolites among the main raw materials for the development of nanomaterials for agriculture. It states that, when considering agriculture, nanomaterials and zeolites should be divided into two categories: **soil improvers** and **protective agents**. Between 2012 and 2021, the Chinese company Shikefeng Chemical Industry Co Ltd producer and distributor of fertilizers obtained the family of fertilizers patents for different crops (WO2014/ 101265 A1/ 101266 A1/ 101267 A1/ 101268 A1/ 101269 A1/ 101270 A1) characterized in the preparation of a solid and granulated compound of chloride or sulfate-based fertilizers, ammonium and potassium, which are coated first with stearic acid and then with modified zeolite powder, and classified as slow release fertilizers. The patent family WO2005/117581 A1 owned by Envirofocus of New Zealand, describes a treatment agent containing a zeolite core coated with diatomite, which are useful to control the release of fertilizers and nutrients to plants.

Zeolites as carriers of nano-fertilizers, include synthetic zeolites and hierarchical ones. Sharma, V. et al. (2022) reviewed

> *A multitude of engineered nano-zeolites have promising applications in the field of agriculture as nutrient delivery molecules and soil conditioners. However, there are a lot of stones unturned towards the exploitation of their full potential due to physico-chemical properties which will revolutionize the field of agronomy and for the welfare of global food security.* **Nano-zeolites** *are the best candidates for increasing crop yield and reducing agriculture inputs costs. There is a need to examine the synthesis approach to prepare nano-zeolites that are eco-friendly, cost-effective, and straightforward. Use of inorganic structure directing agents (SDAs) in hydrothermal treatment is the best applied approach for achieving homogenous nano-sized zeolites. Their application as a carrier molecule will solve issues associated with contemporary conventional techniques and may confer enhanced agri-potential by preserving the physiological fitness of plants and soil. Their ion exchange properties and adsorption capabilities can be fully utilized in agronomy. Comprehensive tools need to be developed for the assessment of nano-zeolites and their framework for improving knowledge on toxicity and risk assessments for sustainable food chains and ecosystem functioning.*

The authors highlight the possibility of synthesizing zeolites with the desired properties to carry nutrients. The use of zeolites as carriers for nano-fertilizers can be understood as a contribution to precision agriculture, reducing the fertilizer content used in plant nutrition. However, they increase the cost of the fertilizer production process and the risks of using materials with particle sizes that increase the toxicity of the product not only for plants but also for animals and humans who will consume the agricultural products.

In summary, several common points can be outlined for the aforesaid substrates and fertilizers made using modified zeolitic rocks: 1) They do not consider that zeolitic minerals are a natural part of soils, 2) They only make partial use of the zeolitic material properties, 3) Inventions using modified natural zeolites incorporate one or two essential plant nutrients into the particles, 4) They are based on the concept of incorporating as much of the nutrient as possible into the product to achieve higher yields, 5) Modifications of the zeolitic mineral are carried out with treatments which make the process more expensive, emitting residual pollutants, and 6) Nano zeolite or zeolite containing nanofertilizer reject that zeolites are nanostructured materials.

NEREA® TECHNOLOGY AND PRODUCTS.

"Nature operates in the shortest way possible" –Aristotle

Since its beginnings, 1985, the NEREA® technology has been innovated to solve the deficiencies of the original procedures for obtaining the NEREA® formulations (Rodríguez-Fuentes et al. 2023), regulating the characteristics of the product obtained in relation to containing only the amount necessary for the nutrition of the plants of the macronutrients N, P, K, also Ca, Mg, Fe and Si, and of the micronutrients Mn, Zn, Cu, B, optimizing the production time of the product, establishing quality control parameters: pH, conductivity, humidity, size of particles, storage conditions, specific rules for its use as a substrate, edaphic and foliar fertilizer. The new procedure eliminates the emission of residues of aqueous solutions of the fertilizers used in the modification of the zeolitic raw material. The products thus obtained do not constitute a risk to the environment during their use since the chemical elements contained in their structure are only extracted by the plants, and their leaching by irrigation water or rain is very limited. These fertilizer products are incorporated into of the soil, increasing its storage capacity for nutrients and water, balancing the pH and the composition of particle sizes. The products provide silicon in an assimilable way. The innovation use the zeolitic minerals as fundamental raw materials, designing the process and the products: 1) The selective extraction of the zeolitic mineral in the deposit as a fundamental raw material, 2) Processing the mineral includes crushing, drying, grinding to obtaining particles with sizes -

1.0+5.0 mm, preserved of humidity, 3) Produce them using a solution that contains all the essential macro and micronutrients for plants, 4) The treatment is carried out using optimized times and temperatures, 5) The incorporation of the nutrients species (ions and molecules) is based on zeolitic material properties, mainly multi-element ion exchange,-and adsorption, which is called "*curing-drying*" process, newly innovated to allow greater adaptability of the technology, occurring directly in the final container (Molina-Amat, J.C. et al. 2024), 6) Some products are processed using controlled mechanochemistry, and 7) Each product has particles size adjusted after modification.

NEREA® substrate allows the germination of the seeds, the development of the plants, their flowering and fruiting, optimally, only using water for irrigation, and the nutrients can be restored in the same amount extracted by the plants during their development. **NEREA fertilizer** is incorporated into the soil as a natural or friendly supplement, so the nutrient content does not exceed the limits considered as pollutants. In the NEREA® substrate and fertilizer, the **plants take the nutrients in the amounts they need**. What remains is preserved in the product. **NEREA® Plus foliar fertilizer** allows the nutrition of the plants in their aerial parts, without the need to use other foliar fertilizers. It provides Si as nutrients and stimulator. It protects the flowers, leaves, and fruits from the attack of insects by creating a physical barrier with the particles smaller than 0.091 mm and protects the plants from the presence of fungi and bacteria. NEREA® technology avoids including or supporting nanofertilizer, because NEREA® particles contain nanostructured crystals of zeolites, where nutrient species are incorporated.

Why do selective mineral extraction? Rodríguez-Fuentes, G. and Rodríguez-Iznaga I. (2009) proved that

> *representative samples from two natural clinoptilolite deposits has shown that the morphology of the crystal, produces a significant influence on the ion exchange process of the zeolite. A sample formed from ore mined at different depths exhibits variability in chemical composition, crystal morphology and tuff characteristics. Homogenization can only be performed chemically to stabilize the composition of exchangeable cations. The sample formed from ore mined to a given depth only exhibits variability in the composition of mineral phases, which can be compensated by fluidized bed separation. The chemical composition is little variable, and the crystal morphology is the same. The determined kinetic parameters of the ammonium exchange at room temperature and 100º C are different. The activation energy of the exchange process is 14.1 KJmol^{-1} when the ammonium was exchanged in the sample representative of the mineral from a drill hole (0 - 30 meters), while the energy of the same process is about 3 KJmol^{-1} for a sample representative of the weathered horizon (0 - 7 meters). The*

sample formed with the mineral extracted to a certain depth is the closest to the optimum for the design and development of a nanostructured zeolitic material.

That is the reason the selection of the zeolitic mineral as raw material for the NEREA® products manufacture is the keystone, and must meet the characteristics of table 1:

Table 1. Characteristics of zeolitic mineral for the manufacture of NEREA® products.

Type of zeolite:	**Clinoptilolite CLI, heulandite HEU and mordenite MOR.**
Minimum total zeolite content:	60%.
Maximum total content of non-zeolitic phases (calcite, clays, quartz, feldspar, iron oxides):	40%.
Minimum total ion exchange capacity:	120 milliequivalent per 100 grams of mineral.
Water absorption capacity:	16.3 – 23.50%
Humidity:	5.0 – 16.5%
Specific gravity:	2.0 - 2.37gcm^{-3}
Volumetric weight:	1.35 - 2.06 gcm^{-3}
Porosity:	13.0 - 40.0%

Chemical composition of zeolitic mineral [% wt.]

	SiO_2	Al_2O_3	Fe_2O_3	FeO	MgO	CaO	Na_2O	K_2O	lost by ignition
Max	72.4	13.7	5.3	2.7	7.2	7.5	3.5	1.7	12.1
Min	58.3	8.6	0.9	0.2	0.5	1.9	1.0	0.4	8.5

Content of toxic elements [g t^{-1}]

F	Pb	As	Cd	Hg
<200	< 10	< 3	< 2	< 5

How are the nutrients inserted in the NEREA® particles? The kinetic study of the obtaining of the NEREA III formula allowed to establish the incorporation of nutrients into the clinoptilolite (NH_4^+, K^+ and P) and the exit of some of its natural cations (Na^+ and Mg^{2+}) as shown in Figure 1 left. In the case of calcium, a fundamental nutrient, its exchange of the clinoptilolite was stopped by incorporating $CaCl_2$ into the multielement solution, otherwise the entry of nutrients would have occurred by exchange of this natural cation. P is adsorbed as phosphate (Rodríguez-Fuentes, G. 2015). The use of nutrients by plants, for example, tomato, was studied in detail, and the consumption of these nutrients could be established throughout the growth of the plant (Figure 1 right). As can be seen, the highest consumption was of ammonium and phosphorus, as expected. These results allowed the initial nutrient load levels to be established, and the "recharge" at the end of each productive cycle of tomato. These studies made it possible to understand that the incorporation of nutrients in ionic form into the particles of the zeolitic raw material is verified simultaneously and in a controlled manner by the nanostructures of the CLI, HEU and MOR zeolites present. However, the production and marketing of complete formulas of NPK fertilizers, the so-called Triple 15, 17, 18, 19, 20 and 22, which include micronutrients Zn, Cu, B, Mo, S, etc. in which all nutrients are supplied in the form of highly water-soluble molecules, determined a modification in the process of obtaining NEREA® products, and a more in-depth study to answer the question How are the nutrients inserted in the NEREA® particles? Rodríguez-Fuentes, G. et al. (2025) answered:

> *each particle of a NEREA® product supports the same nutrient contents and in a similar ratio, balanced with the contents of the mineral's natural nutrients, including Si and Al. The STEM-EDX micrographics displayed the elemental mapping for N, P and K distributed on NEREA® particles, validating that there is a homogeneous distribution of the three nutrients on the zeolitic support and that NEREA products are not a physical mixture of natural zeolite particles and fertilizers.*

Figure 1. Kinetic curves of exchanged cations and adsorbed phosphorus on zeolitic mineral from Tasajeras deposit when producing NEREA material (left). Consumption of nutrients from NEREA substrate by tomato crops (right).

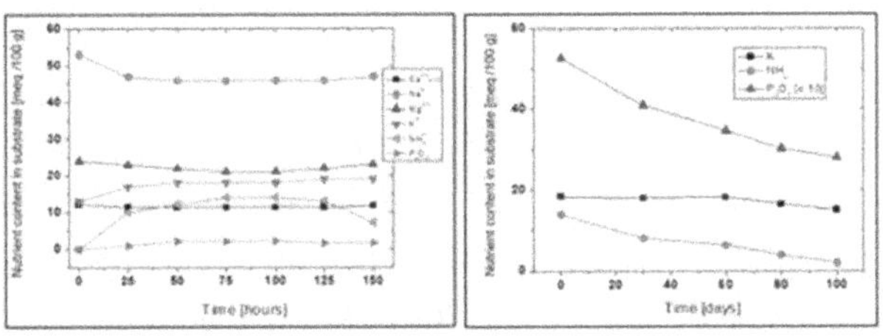

E.Y. de la Nuez et al. (2024) studied

> *the development of zeolitic materials enriched in phosphorus and nitrogen, with a pronounced content for this last element, obtained from natural clinoptilolite (CLI) treated with ammonium hydrogenphosphate (DAP) and urea. The interaction of phosphorus and nitrogen species retained on CLI was analyzed using different research methods. FTIR spectra showed the presence of NH_4^+, PO_4^{3-} and urea on CLI. The retention of ammonium involves both ion-exchange in clinoptilolite, and adsorption. Phosphate is retained by adsorption, and it can be incorporated into an electrical double layer with accompanying cations including NH_4^+. The FTIR bands of the amino group of urea appear to be shifted relative to their location in the urea spectrum. This is associated with weaken-*

ing of N–H bond from amino group due to its interaction with the clinoptilolite framework via hydrogen bonding. The urea is arranged on the zeolite surface, which is consistent with SEM observations and elemental (N and P) mapping by STEM. As a result, the surface area and micropore volume of these materials, treated with DAP and urea, decrease with respect to those modified with DAP alone. Accordingly, the release studies showed that the subsequent delivery of phosphorus species occurs in a delayed mode. These materials have great potential to be used as slow-release nutrient fertilizers for crop plants. They have a significant amount of nitrogen available on their surface, allowing plants to easily assimilate it early in their growing cycle. Zeolitic matrix leads to obtain materials at low-cost for massive applications, efficient use of nutrients and environmental benefits.

NEREA® Fertilizer (Edaphic) and NEREA® Plus (Foliar) Fertilizers *vs* NPK Fertilizer + Natural Zeolite

Agustina T. E. et al. (2018) considered that the incorporation of natural zeolites into so-called slow-release fertilizers has represented a setback. They cited Rizkya, I. et al. (2018) who

characterized and utilized nitrogen, phosphorus, and potassium (NPK) slow release fertilizer by using natural zeolite in Indonesia. Zeolite was utilized as a mixture of Slow Release Fertilizer (SRF). Based on the experimental results, it was found that addition of zeolite in the single (N-P-K) or multi-nutrients (NPK) fertilizer reduced the average rate of solubility of nutrients in the leaching medium. The results of this study showed that (N-P-K) + zeolite fertilizer has the lowest average rate of solubility than the other samples, it was proved by its average rate of solubility in soil, domestic water, and demineralized water medium, i.e. 9.63, 23.97, and 14.68 ppm/h, respectively.

Figure 2 presents the comparison between the mechanism of action of the NPK fertilizer and NEREA® fertilizer/substrate products. NPK fertilizer needs water to solubilize nutrients from the particles of each component. The roots of the plant take the nutrients from the dissolution. NEREA® particle containing all the nutrients inside, absorbs water and the roots of the plants wrapped the particle to extract the nutrients.

Figure 2. Comparison of the mechanism of action of the NPK fertilizer and NEREA® fertilizer/substrate products

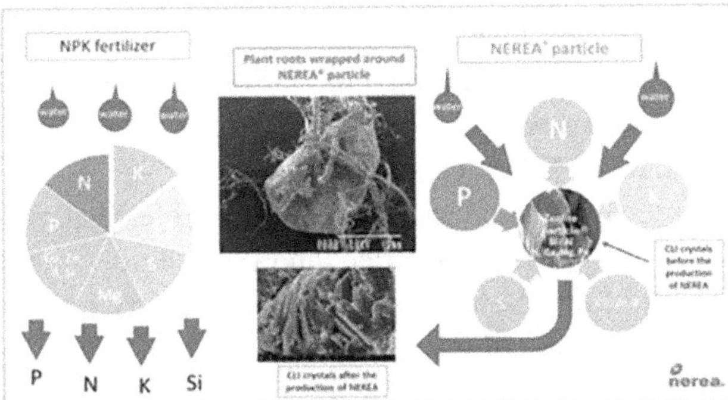

In order to establish a fair comparison between NEREA® fertilizer and NEREA® Plus foliar fertilizer products, and the NPK fertilizer + natural zeolite formula blends, we will focus on five essential crops: beans, rice, potato, tomato, and sugarcane, due to their importance as food crops and profitability for farmers. The comparative tables by crop present the authors of the study and the publications from which the data were taken. The treatment doses used, NPK doses per treatment, crop yield, ratio of kg of product obtained to kg of NPK used, and the amount of NPK saved for each treatment are compared.

Tomato *(Solanum Lycopersicum L.)* Crop

The first report of a NEREA® product was made by Rivero González, L.A. and Rodríguez-Fuentes (1988) as a substrate to solve the problems of hydroponic cultivation technology in Cuba that used silica or basalt rock and fertigation with a Steiner solution. The benefit of only using water for irrigation with NEREA substrate resulted in significant savings in water volume and the systematic use of fertilizers to keep nutrition of the plants. All the nutrients necessary for a 100-day tomato cultivation campaign were incorporated into the zeolitic mineral at the beginning. Later, Espino, C. and Soca Nuñez, M. (1991) verified that similar yields could be obtained from tomato plants of the Tegula variety grown in soil and in natural zeolite with fertigation, and in the NEREA IV substrate with the nutrients initially incorporated, irrigated only with water (Table 2).

In 2006, Febles González, J.A. (2006) reported a 20-year study comparing the results of growing tomato variety 84-84 in Havana's organoponic and zeoponic gardens. In organoponic gardens, the applied procedure was to add organic matter to the original NEREA III substrates, or to mix organic matter with natural zeolite. In the study, Febles compares the yields of the two variants of organoponic substrates with NEREA III. Table 2 shows that the average yield of tomato grown in NEREA III was 193.13 t ha^{-1}, much higher than the yields in the organoponic variants, which were 103.64 t ha^{-1} and 124.55 t ha^{-1}.

Ponce Cruz, L. et al. (2011), presented the results of the use of the NEREA III substrate in the industrial production of cucumber and tomato at the Victoria de Girón company, Matanzas, Cuba. The authors observed significant savings in the use of fertilizers to prepare the nutrient solution for soil fertigation and only water in crops grown in the NEREA III substrate, and the use of pesticides. The yields of the Giulia F1 variety tomato crop grown in soil with fertigation was 70.0 t ha^{-1} and in NEREA III 100.0 t ha^{-1} (Table 2). The values of efficiency **kg tomato/kg nutrient used** ratio do not differ much, they were 12.5 for crop grown in soil and 18.3 in NEREA.

In Zeoecological Fertilization Rivero-González, L.A. and Rivero-Robaina, E.L. (2016), compared the yields of several crops using NPK formula 8-9-17 and NEREA® Fertilizer (NPK 2% wt.). In the case of tomato crop, the authors show the yields when grown in soil at open-air and in a protected cultivation house. The standard use of NPK formula 8-9-17 for tomato in soil is 800 kgha^{-1} for a yield of 80.0 t ha^{-1}. The authors obtained a similar yield with the application of 5 t $ha^{-1}$of NEREA® Fertilizer. The observed efficiency **kg tomato/kg nutrient** was 100.0 for cultivation in soil and 800.0 for fertilization with NEREA®, with a saving of 700.0 kg of NPK.

Soca Nuñez, M. and Lorente, E. (2015) evaluated tomato cultivation with 1,000.0 kg of urea fertilization (1/3 urea at sowing + 2/3 at 30 days) and urea with 15% Zeolite at sowing. Yields were 34.80 and 45.50 t ha^{-1} respectively, while the corresponding efficiency was 34.8 and 53.5, with a saving of 150 kg of urea. López Paz (2019) evaluated three NPK formulas with the addition (mechanical mix) of natural zeolite: formula 8-9-14, formula 6-7-10 + 25% natural zeolite and formula 4-4.5-7 + 50% natural zeolite. The respective yields were similar 29.81, 33.73 and 28.23 t ha^{-1}, however, the efficiency showed that the mixture of NPK and zeolite 152.0 and 165.0 was higher than that observed (109.0) when fertilizing only with NPK.

Finally, Terry Alfonso, E. et al. (2024) compared the fertilization of tomato with NPK formula 9-13-17, 1,000.0 kg ha^{-1} with the use of Agromenas®, a product that contains natural zeolite (10%) and NPK fertilizer (278.4 kg ha^{-1}) mixed with other minerals. The yields reported by the authors are 30.63 and 32.60 t ha^{-1}, while the efficiency was 30.63 and 117.0, respectively.

All these reports confirm that the use of natural zeolite in tomato fertilizer formulations is more efficient and saves nutrients. Significantly, NEREA®Fertilizer-2% yields the highest efficiency values kg of tomato/kg nutrient used, for the same yield as using a complete NPK formula.

Table 2. Comparison of treatments in tomato cultivation.

Treatment / author		Yield (kg/ plant)		
Espino, C. and Soca Nuñez, M. (1991)				
Soil + fertirrigation KNO_3+$CaNO_3$+NH_4PO_4		4.17	-	
Zeolite natural + fertirrigation		4.10	-	
NEREA IV + water		4.29	-	
Febles González, J.A. (2006)	**Dose NPK** (kgha^{-1})	**Yield** (t ha^{-1})	**kg tomato / kg nutrient**	**NPK saved** (kgha^{-1})
Organoponic (40% soil + 60% organic matter) and water		103.64	-	
Organoponic + zeolite + water		124.55	-	
Zeoponic NEREA III + water	16.0	193.13	-	
Ponce Cruz, L. et al. (2011)				
Fertirrigation in soil	5614.0	70	12.5	
NEREA III + water	5477.0	100	18.3	
Rivero-González and Rivero-Robaina (2016)				
NPK Formula 8-9-17	800.0	80.0	100.0	-
NEREA fertilizer 2% (open air)	40.0	18.0	450.0	760
NEREA fertilizer 2% (greenhouse)	100.0	80.0	800.0	700
Soca Nuñez, M. and Lorente, E. (2015)				
Application 1/3 Urea sowing + 2/3 a los 30 días	1000.0	34.80	34.8	
Urea con 15% de Zeolite in sowing	850.0	45.50	53.5	150
López Paz, C. (2019)				
Control (without fertilization)	-	14.62	-	
Formula 8-9-14	273.5	29.81	109.0	
Formula 6-7-10 + 25% natural zeolite.	221.9	33.73	152.0	52
Formula 4-4.5-7 + 50% natural zeolite.	171.1	28.23	165.0	102.4
Terry Alfonso, E. et al. (2024)				
Formula 9-13-17	1000.0	30.63	30.63	-
Agromenas®	278.4	32.60	117.0	721.6

Common Beans *(Phaseolus Vulgaris L.)* Crop

Table 3 compares the results obtained by several authors in common bean cultivation using different fertilization treatments that evaluate the inclusion of natural zeolite and the use of NEREA® products. Rancho Chinobampo in its Report and Validation of Results for Zeolitic (2008), compares the yields of common beans Jamapa variety, fertilized with NPK triple 17 and with NEREA® Chinobampo (2%), these were 600.0 and 3,200.0 kg.ha^{-1} respectively, 5.34 times higher with the NEREA® product. The efficiency in the use of nutrients with NEREA® Chinobampo Fertilizer was 320.

A study on the use of natural zeolite mixed with chemical fertilizer in the cultivation of common beans was conducted by Diaz Coronel, G.T. et al. (2009). In their experimental model, they did not replace fertilizer with natural zeolite, but rather added zeolite to the same amount of fertilizer, which is understood as a dilution of the fertilizer or extension of the volume of fertilizer used. The authors demonstrated that by adding natural zeolite to NPK, crop yields are higher than those of the control, but they do not report fertilizer savings since that was not their objective. The values of **kg bean/kg NPK** did not vary significantly between the study groups.

Rivero-Gonzalez, L.A. and Rivero-Robaina, E.L. (2016), to answer the best combination to achieve high yields with less fertilizer use, studied the treatment doses of NPK formula 8-9-17 + natural zeolite, and NEREA® Fertilizer prepared with 2% wt. of the same NPK to achieve a yield of 1,800.0 kg ha^{-1}. The authors demonstrated that the greatest NPK savings were achieved with the use of NEREA® Fertilizer made with 20 kg of the same fertilizer formula 8-9-17 and the same natural zeolite from the Tasajeras deposit. Table 4 presents the comparison of fertilization treatments of black beans *Phaseolus vulgaris* in red Ferralitic soil. The fertilizers were NPK formula 8-9-17, NEREA® Fertilizer, and mixes of natural zeolite and NPK formula. A 500.0 kgha^{-1} dose of NPK 8-9-17 produced a mean yield of 1,800.0 kg ha^{-1} for Cuban black beans. A 500.0 kg ha^{-1} dose of the mix with 25% natural zeolite and 75% NPK 8-9-17 proved best for the same yield. This mixture saved 125.0 kg of NPK. However, a 1,000.0 kg ha^{-1} dose of NEREA® Fertilizer containing 2.0% NPK (equivalent to 20 kg) produced the same yield, saving 480.0 kg of NPK. The **kg beans/kg nutrient** ratios were higher when NEREA® fertilizer was used, even though the same yield of 1,800.0 kgha^{-1} was achieved. Similarly, fertilizer savings were 480.0 kg when using NEREA® fertilizer, while the optimal mixture of NPK + 25% zeolite only saved 125.0 kg. The ratio of kg beans/kg nutrients was 90 for NEREA® fertilizer 2.0 t ha^{-1} compared to 14.4 for the NPK+25% zeolite mixture, which means that with fewer nutrients they obtained the same mass of beans more efficiently.

Ozbahcea, A. et al. (2015) studied

the effects of different zeolite doses (Z_0:0, Z_3:30, Z_6:60, Z_9:90, and Z_{12}:120 t ha^{-1}) and water levels (I_{50}: 0.50, I_{75}: 0.75, and I_{100}:1.00) on yield, quality, and nutrient absorption of common bean were evaluated under arid conditions in Karapinar Konya, Turkey, in 2011 and 2012. The experiment was carried out in a split-plot design with three replications. Fifty kg N, 40.0 kg P, 50.0 kg K, 30.0 kg Fe, and 30.0 kg Zn per hectare (200.0 kgha^{-1}) were applied to the experimental area. One-third of N and all of P and K were applied to the soil before the sowing. The remaining fertilizer was applied by a drip irrigation system during the growing period. The results showed that in each treatment different doses of zeolite together with changes in the rate of irrigation levels effected yield and yield components. Water use efficiency was also affected by zeolite and irrigation treatments. The highest yields, 4,777.0 and 4,114.0 kg ha^{-1}, respectively, were obtained from the Z_9I_{100} and $Z_{12}I_{100}$ applications. These nutrient levels increased with raising zeolite doses.

The study confirmed that applying large amount of natural zeolite to the soil affects crop yield and nutrient efficiency.

Possible soil damage with prolonged use of NEREA® fertilizer or high doses described by Rodríguez-Fuentes, G. et al., (2025), different doses of the NEREA® Fertilizer product were evaluated over 5 years on a compacted red Ferralitic soil, which showed that in accumulated doses up to 100.0 t ha^{-1}.

The study shows that the nutrient content has remained within normal parameters. Over time, it can enrich the soil already containing NEREA® Fertilizer with full formula fertilizers, given the possible degradation of potassium, since we are incorporating the other nutrients with the applications of NEREA® in sufficient quantities for several years. In the case of pH in the soil samples where a total content of 100.0 t ha^{-1} of NEREA® Fertilizer was applied in 5 years, it has risen by one unit. In the case of acid soils, of which more than 40% exist in Cuba, the continuous use of zeolite would be of great importance. The sodium content increased from 0.06 to 0.52 mol kg^{-1} after five years of introducing a dose 20.0 t ha^{-1} of NEREA® every year to accumulate 100.0 tons. However, the plants did not show a lack of nutrition and damage produced due to increased sodium.

Under extreme conditions of fertilizer and pesticide scarcity, plus a high infestation of thrips *Megalurothrips usitatus*, Meneses, P. et al. (2022) conducted a study of NEREA® Plus as the sole fertilizer in common bean crops, varieties Buenaventura, CUL 156, and CUFIG-48, in three provinces in central Cuba: Villa Clara, Sancti Spiritus and Ciego de Ávila, under production conditions on a total of 90 ha. In this study, there was no absolute control, only control groups in each province to which base fertilization was applied with 250.0 kg ha^{-1} of the complete NPK formula 9-13-17, and two foliar applications of the stimulant Fitomas® E. Without base fertilization to the soil, only NEREA® Plus was applied as a foliar fertilizer at a dose of 4.0 kg ha^{-1} dispersed in water, in the growth phases: 1st. vegetative phase stage V4 (third trifoliate leaf), 2nd pre-flowering stage R5, 3rd flowering stage R6, and 4th pod formation stage R7.

The average yield of the control groups was 650.0 kg ha^{-1}, while the average yield of the groups that only fertilized the plants with NEREA® Plus foliar was 1,630.0 kg ha^{-1}. Although this yield is lower than the 1,800.0kg ha^{-1}, the result showed that the plants are well nourished with the NEREA® Plus foliar product. Another result is that the product protects the aerial parts of the plants from predation by thrips insects, which did not occur in the control groups, where the plants were not protected by the stimulant Fitomas® E, and the insects destroyed the flowers and pods, determining the low yield. The values of **kg bean/kg nutrients** were extreme, while in the control groups it was 2.6, where NEREA® Plus was applied it was 636.7. The nutrient content in each application of 4.0 kg of NEREA® Plus is 0.64 kg. The 4 applications contained only 2.56 kg ha^{-1}, which directly reached the leaves, flowers, and pods. Figure 3 summarizes the results of this study. The average number of pods per plant was 16, with 6 beans per pod, much higher than the 6.5 pods and 4 beans per plant in the control groups. The exclusive foliar use of NEREA® Plus allowed for a saving of 247.0 kg ha^{-1} of nutrients and an additional effect of preserving the plants from insect attack. With these results, the authors estimated that 1,520 tons of NEREA® Plus could fertilize the 95,000 ha of common beans that must be grown to meet the demand of the Cuban population. The savings of the NPK formula 9-13-17 would be 23,500 tons.

Figure 3. Comparison of common bean crop at production scale fertilized with NPK and NEREA® Plus foliar fertilizer.

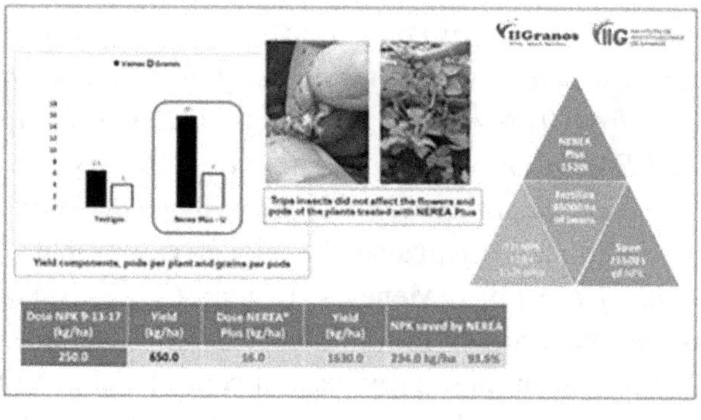

Laffita Suárez, M. and Martínez Valdés, M.E. (2022), evaluated the zeolitic products NEREA® Fertilizer and NEREA® Plus in the cultivation of common bean variety BAT 304 in the years 2021 - 2023. The results of the two campaigns are presented in Table 3. In these studies, in addition to the absolute control without fertilization, the entire technological package recommended by the authorities of the Cuban Agriculture for the cultivation of common beans was used: Production control (TP) fertilization at the bottom of the furrow with complete NPK formula 8-23-4 (300.0 kg ha^{-1}) + urea (200.0 kg ha^{-1}) at 15 days of germination (DAG) + Bayfolan® foliar fertilizer applied (N 11.47 + K_2O 6.0 + P_2O_5 8.0% wt.) (3.0 L ha^{-1}) in stages V4 and R7. A second group with TP + EcoMic® mycorrhiza to the seed + urea + NEREA® Plus 4.0 kg ha^{-1} in each stage V4, R5, R6 and R7 as in the study by Meneses et al. The third group was the combination of NEREA fertilizer 2.0 t ha^{-1} as base fertilization + EcoMic® in seed + NEREA® Plus-U 4.0 kg ha^{-1} in V4, R5, R6 and R7. Here the results were different, the absence of thrips did not disturb the control group to reach a yield of 1,000.0 kg ha^{-1}, however the second group where NPK base fertilizer was combined with urea and NEREA® Plus the yield was 800.0 kg ha^{-1}. The group that combined the two products, NEREA® soil fertilizer and Plus foliar fertilizer, with the use of EcoMic®, obtained a yield of 1,900.0 kg ha^{-1}, almost double that of the control group, and 1,100.0 kg more than the second group, where excess nutrients affected plant development.

In the 2023 campaign, authors Laffita Suárez, M. and Martínez Valdés, M.E. (2024) introduced a study group to evaluate whether the NEREA® Plus product could be used as a base soil fertilizer. The results showed that the use of NEREA® Plus as the sole fertilizer + EcoMic® mycorrhizal in seed achieved the highest yield (2,175.0 kg ha^{-1}) and the production control (1,702.0 kg ha^{-1}), while the combination of NEREA® and EcoMic® products achieved 2,055.0 kg ha^{-1}. However, when using NEREA® Plus as a soil and foliar fertilizer, the yield fell to 1,396.0 kg ha^{-1}, very close to the yield of the absolute control without fertilization (1,514.0 kg ha^{-1}), and below the production control. In both studies, the **kg bean/kg nutrient** ratios for the trials with NEREA® products were higher than those of the control groups, and the NPK savings were also higher. The results confirmed that NEREA® products contain the nutrients necessary for optimal bean plant development.

Table 3. Comparison of common bean fertilization using natural zeolite and NEREA® products by different authors.

Treatment / author	*Treatment (kg ha^{-1})*	*NPK (kg)*	*Yield (kg ha^{-1})*	*kg beans/ kg nutrient*	*NPK saved (kg ha^{-1})*
RanchoChinobampo (2008).					
NPK Fertilizer Triple 17	500.0	500.0	600.0	1.2	-
NEREA Chinobampo	500.0	10.0	3200.0	320.0	490.0

continued on following page

Table 3. Continued

Treatment / author	*Treatment (kg ha^{-1})*	*NPK (kg)*	*Yield (kg ha^{-1})*	*kg beans/ kg nutrient*	*NPK saved (kg ha^{-1})*
G. T.Díaz Coronel, et al. (2009)					
NPK Fertilizer 80 $kgha^{-1}$ de N (urea) + 40 $kgha^{-1}$ de P (super phosphate triple)	120.0	120.0	994.0	8.28	-
NPK + urea + 25% zeolite - C	150.0	120.0	1374.15	11.45	-
NPK + urea + 50% zeolite - C	180.0	120.0	1195.76	9.96	-
NPK + urea + 75% zeolite - C	210.0	120.0	1309.18	10.9	-
Rivero-González, L.A. and Rivero-Robaina, E.L. (2016)					
Control (without fertilization)	0	.	60.0	-	-
NPK Fertilizer 8-9-17	500.0	500.0	1800.0	3.6	-
NEREA® Fertilizer (2% NPK 8-9-17)	500.0	10.0	1000.0	100.0	490.0
NEREA® Fertilizer (2% NPK8-9-17)	1000.0	20.0	1800.0	90.0	480.0
Natural Zeolite 75%+ (NPK 8-9-17) 25%	500.0	125.0	1600.0	4.2	375.0
Natural Zeolite 50%+(NPK 8-9-17) 50%	500.0	250.0	1600.0	6.4	250.0
Natural Zeolite 25% + (NPK 8-9-17) 75%	500.0	375.0	1800.0	14.4	125.0
A. Ozbahce et al. (2015)					
Z_0: control NPK: (N 50/P 40/K 50/ Fe 30/ Zn 30)+ 0 zeolite	200.0	200.0	3773.0	18.8	-
Z_3: NPK + zeolite 30 t ha^{-1}	30,200.0	200.0	3638.0	18.2	-
Z_6: NPK + zeolite 60 t ha^{-1}	60,200.0	200.0	4297.0	21.4	-
Z_9: NPK + zeolite 90 t ha^{-1}	90,200.0	200.0	4777.0	23.8	-
Z_{12}: NPK + zeolite 120 t ha^{-1}	120,200.0	200.0	4690.0	23.5	-
Menses, P. et al. (2022)					
NPK 9-13-17 + 2 foliar applications Fitomas® E + (N6/ $P_2O_5$2.7/K_2O5.5)	250.0	250.0	650.0	2.6	-
NEREA® Plus foliar 4 applications 4 kg h^{-1} V4, R5, R6, R7	16.0	2.56	1630.0	636.7	247.0
Laffita Suárez, M. and Martínez Valdés, M.E. (2022).					
Production control (TP) NPK bottom furrow complete formula 8-23-4 (300 $kgha^{-1}$) + 15 DAG urea 200 $kgha^{-1}$ + Bayfolán® V4, R7	500.0	500.0	1000.0	2.0	-
TP + EcoMic® in seed + 15 DAG Urea 200 $kgha^{-1}$+ NEREA® Plus (V4, R5, R6 y R7).	516	502.6	800	1.6	-
NEREA® fertilizer 2 tha^{-1} bottom furrow before sowing + EcoMic® micorrizal in seed + NEREA® Plus (V4, R5, R6, R7).	2016	42.6	1900	44.6	457.4
Laffita Suárez, M. and Martínez Valdés, M.E. (2024)					
Control (without fertilization)	-	-	1396.0	-	
Production control (TP) NPK bottom furrow complete formula 8-23-4 (300 kg ha^{-1}) + Azofert® rhizobium in seed + Urea in V2 (60 kg ha^{-1}) V4 (50 kg ha^{-1}) + Fitomas® V3 and R5.	410.0	410.0	1702.0	4.15	-

continued on following page

Table 3. Continued

Treatment / author	*Treatment (kg ha^{-1})*	*NPK (kg)*	*Yield (kg ha^{-1})*	*kg beans/ kg nutrient*	*NPK saved (kg ha^{-1})*
EcoMic® in seed + NEREA® Plus-U 4 kgha^{-1} (V4, R5, R6, R7).	16.0	2.46	2175.0	884.0	397.64
NEREA® fertilizer (2%) 2 tha^{-1} + EcoMic®in seed + NEREA® Plus-U 4 kgha^{-1} (V4, R5, R6, R7).	2016.0	42.46	2055.0	48.4	357.54
NEREA® Plus-U edaphic (100 kgha^{-1}) + EcoMic® in seed + NEREA® Plus-U 4 kgha^{-1} (V4, R5, R6, R7).	116.0	18.46	1514.0	82.0	381.54

So far, the comparison is largely in favor of NEREA® products. The question of how it is possible that with so few nutrients in NEREA® particles, bean plants nutrition is good, and yields are equal or higher, is answered by the fact of the synergy with the soil microbiota, supplying the necessary nutrients for proper development. In a subsequent study conducted by Arias Cruz, C.L. and Martín Alonso, G.M. (2023) under controlled conditions in pots and a protected cultivation house, they used a completely randomized design with four replications and a factorial arrangement. The factors studied were: 1) NEREA® edaphic fertilizer and NEREA® Plus foliar fertilizer; 2) inoculation with rhizobia (*Rhizobium tropici*, strain CIAT 899, and 3) the arbuscular mycorrhizal fungus (AMF) *Glomus cubense*, strain INCAM-4 (DAOM241198), at a concentration of 45 spores/g of inoculant. The rhizobia and mycorrhiza were applied to the seeds. Base fertilization with NEREA® Fertilizer was carried out at the time of sowing, at a rate of 4.0 g per plant, equivalent to 1.0 t ha^{-1}. Foliar application of NEREA® Plus was carried out using a 1.0 L sprayer, starting 10 days after germination, every 10 days, until completing 5 applications, at a dose of 10.0 g L^{-1}. With absolute control without any treatment. In this study, the only nutrient contribution was made by applying NEREA products. In Table 4, the results indicate that the combination of the NEREA® products + AMF + Rhizobia nourishes the plants well, providing the highest values of N, P, K and chlorophyll to the aerial parts, and the highest yield.

Table 4. Nitrogen, Phosphorus, Potassium, and chlorophyll (SPAD) content in aerial parts of common bean plants and yields observed by Arias Cruz, C.L. and Martín Alonso, G.M. (2023)

Treatments	**% N**		**% P**		**% K**		**SPAD**		**Yield (g/pot)**	
Control	**3,47**	**d**	**0,67**	**b**	**1,73**	**e**	**36,60**		**1.8**	c
Rizobios	3,86	b	0,60	d	1,83	d	35,85		2.6	bc
HMA	3,54	c	0,66	bc	1,82	d	36,30		2.65	bc

continued on following page

Table 4. Continued

Treatments	% N		% P		% K		SPAD		Yield (g/pot)	
Rizobios + HMA	3,19	g	0,64	bc	1,57	f	32,05		3.2	abc
NEREA®	3,40	f	0,64	bc	2,13	b	31,48		2.9	abc
NEREA® + Rizobios	3,44	e	0,63	c	2,05	c	33,00		3.45	ab
NEREA® + HMA	3,56	c	0,66	bc	1,75	e	35,50		3.3	ab
NEREA® + Rizobios + HMA	**4,34**	**a**	**0,72**	**a**	**2,46**	**a**	**38,48**		**3.6**	**a**
ES	0,01	**	0,01	**	0,01	**	2,41	n.s.	0.26	

**Identical letters in the same column do not differ significantly according to Duncan's test for $p \leq 0.5$.
Averages of two years evaluated, n = 8

Rice *(Oryza Sativa)* Crop

Studies on the use of natural zeolites in rice cultivation have focused on the direct application of the mineral to the soil to evaluate the effects of zeolite on physiological characteristics and grain quality in rice under alternate wetting and drying irrigation (Sun, Y. et al. (2023), and, secondarily, in mixtures with NPK fertilizers and urea. This trend must be based on the vast expanses of rice cultivation to meet the demands of billions of people, the depletion of soils due to intensive use, and the large volumes of water required. In Table 5 is presented the comparison of the results reported by different authors in the use of NPK fertilizer + natural zeolites and the utilization of NEREA® products in the rice crops.

Kavoosi, M. (2007) investigated

the feasibility of using a natural clinoptilolite to enhance rice grain yield, nitrogen (N) recovery and N use efficiency, an experiment was conducted in a coarse-textured rice field in the Guilan province of Iran. Three levels of 8, 16, or 24 tons of zeolite ha^{-1} were incorporated into the soil with and without 60.0 $kgNha^{-1}$ as urea, and their effects on the grain and straw yields, available K, apparent N recovery, N use efficiency, and some soil properties were determined. Results indicated a significant positive effect of treatments on the grain yield, straw, and tiller numbers of rice. All levels of zeolite with N showed a significant increase in grain yield compared with 60.0 $kgNha^{-1}$ and control treatment. Zeolite application also increased soil-available potassium and its uptake by rice straw significantly. The lowest apparent N recovery (40%) was obtained from application of 60.0 $kgNha^{-1}$, whereas application of 8 or 16 tons of zeolite ha^{-1} increased apparent N recovery up to 65%. Nitrogen use efficiency was also improved.

Qi Wu et al. (2016), evaluate the coupling effects of zeolite (Z_0, 0 tha^{-1}; $Z_{0.9}$ and $Z_{0.22}$ 10 tha^{-1} in different diameter of 0.45-0.9 mm and 0.17-0.22 mm) and Nitrogen (N_0, $N_{52.5}$, N_{105}, $N_{157.5}$ $kg.ha^{-1}$) application on N use, leaf area index (LAI) and rice grain yield in 2014 and 2015. Results showed that both $Z_{0.9}$ and $Z_{0.22}$ at N_{105} and $N_{157.5}$ application rates could greatly enhance the biomass, dry weight of root, LAI, and N uptake. However, at initial tillering stage of rice, Z decreased N content in plant for its high absorption capacity and hence lowered biomass and N uptake. The highest value of N recovery efficiency (NRE) and N use efficiency (NUE) was obtained from $N_{105}Z_{0.22}$. Z input contributed to higher NRE and NUE. At $N_{157.5}$ application rate, $Z_{0.9}$ and $Z_{0.22}$ significantly increased rice grain yield by 14.6% and 15.3%, respectively. However, Z mattered little to rice grain yield at N_0 and $N_{52.5}$. There was no significant difference in rice grain yield between $Z_{0.22}$ and $Z_{0.9}$. $Z_{0.22}$ and $Z_{0.9}$ clearly increased the 1000-grain weight. N application significantly lowered unfilled grain percentage.

Weiyu Hu et al. (2023) studied the Zeolite as a Tool to Recycle Nitrogen and Phosphorus in Paddy Fields under Straw Returning Conditions.

Excess nitrogen (N) caused by straw returning to paddy fields undergoing flooding irrigation deteriorates the water quality. The purpose of this research was to use both simulated field and pot experiments to explore a new approach using zeolite to recycle this excess N. The results from simulated field experiments in stagnant water showed N adsorption with different zeolite applications (25, 50, 75, 100, 125, and 150 g L^{-1}). Pot experiments revealed how straw and reused zeolite applications affected the concentrations of ammonia N (NH_4^+-N), nitrate N (NO_3^--N), total N (TN),and total phosphorus (TP) in the surface water and soil layers of the paddy field. Zeolite showed a strong ability to adsorb NH_4^+-N in wastewater, even in a simulated drainage ditch (100 g L^{-1} zeolite adsorbed 74% NH_4^+-N). The zeolite recycled from the drainage ditch was still able to reduce N concentration caused by straw decomposition in the surface water. Zeolite adsorption reduced the peak values of NH_4^+-N, TN, and TP by 30%, 19%, and 5%, respectively. Based on these findings and conventional field designs, the use of 20 t ha^{-1} zeolite in the field is effective for recycling N and P. This research provides a sustainable development method to mitigate the water quality deterioration caused by straw returning to the field.

Chen, T. et al (2017) studied The Influence of Zeolite Amendment on Yield Performance, Quality Characteristics, and Nitrogen Use Efficiency of Paddy Rice:

A 2-yrfield experiment using a strip-plot design was conducted to evaluate the impact of Z (zeolite) amendment on yield performance, quality characteristics, and nitrogen (N) use efficiency of paddy rice. Japonica rice (cv. Gangyu 6) was cultivated in a silty loam soil with N application (0, 52.5,105.0, and 157.5 kg N ha^{-1}) as whole plots and Z amendment (0, 5, 10, and 15 t Z ha^{-1}) as strip lots. Compared with the paddy field without Z amendment, there was 14.2 to 35.8% higher potential postharvest residual soil mineral N and20.1 to 44.6% higher exchangeable potassium when the upper 30 cm of the soil was amended with

5 to 15 t Z ha^{-1}. Application of 5 to 15 t Z ha^{-1} increased applied N fertilizer use efficiency by 39.0 to 64.4% and N recovery efficiency by 20.7 to 85.2%, respectively. Milling, appearance, and eating quality traits were not affected by Z amendment, whereas rice protein content was increased by Z addition. The enhanced yield performance due to Z was mainly caused by improved essential plant-available nutrients. In addition, these positive residual activities were maintained for at least an additional year after initial application in the low land rice production system.

In 2015, Villarreal, L.A. et al (2015) evaluated

the effect of zeolite on the efficiency of nitrogen fertilizer in rice crop variety ID-IAP FL 106-11 with seeding density of 110.0 kg ha^{-1} in a medium-fertility vertic Fluvisol. This study consisted of five treatments with four replications, following a completely randomized block design using N rate of 80.0 kg ha^{-1} mixed with different percentages of zeolite (0; 15; 25; 35 and45%). The isotopic technique 15N 3% excess to determine the nitrogen fertilizer use efficiency (EUNF) was used. Top grain yields were obtained with 80.0 kg N + 45% zeolite/ha and 80.0 kg N + 15% zeolite/ha. A 7% EUNF increase was achieved with the addition of natural zeolite.

They concluded that "small doses of 12.0 kg/ha (15%) of zeolite can improve EUNF; after subsequent experiments, it is recommended to reduce the amount of N applied in soils with similar soil and climatic conditions suitable for growing rice." However, the **ratio kg rice/kg nutrient only increases** from the control without zeolite 44.3 to 48.4 when using 45% zeolite, denoting the relevance of the mineral in the efficiency of nutrient supply to plants.

Soca Nuñez, M. and Constanza Daza, M. (2018) studied the effect of zeolite on the yields of rice Oryza sativa L. where two application methods (spreading and transplanting) were evaluated with four treatments (T control, Urea, Urea + Zeolite 10%, and U+ Z15%) and their effect on crop yield.

Zeolite application in rice achieved 60% RVNA and improved yields with no differences found between application methods. Its application also improved grain weight per ear and the dry weight of corn grain. Zeolite improved the efficiency of nitrogen fertilization. The kg rice/kg nutrient values increased from 50.0 in the control to 59.1 and 60.9 with the substitution of 10% and 15% urea for zeolite, respectively, while yield only increased by 8% for 10% zeolite and 6.6% for 15%.

Martínez Grillo, J. F. et al. (2022, 2023) evaluated the individual and combined use of NEREA® Fertilizer and NEREA® Plus products in rice cultivation. The results obtained allowed them to establish that the combined use provided the best results expressed in yield, efficient use of nutrients, and their savings. The observed yield in the grain with 14% moisture with the combined use of NEREAs was 3.55 t ha^{-1}, while in the reference control with fertilization according to the recommended technical instructions it was 3.48 t ha^{-1} (Table 5). In the spring 2023 campaign, they repeated the experiment on the IACuba 31 rice variety, well recognized in the country for that season, and achieved similar results in both freshly harvested and dry grain, "even though weather conditions affected yields." The authors concluded that

> *the application of the zeolitic products NEREA® Fertilizer (edaphic) and NEREA® Plus (foliar)improves growth and yield indicators in rice crops. The most pronounced response was obtained when combining both products T5 formulations, achieving up to 82% compliance with the fertilization recommendation according to the Technical Instructions. The results achieved demonstrate that NEREA® applications in rice crops are positive.*

In rice cultivation, the reports on the use of NPK fertilizers mixed with natural zeolite use very different doses with different results. However, in the case of the use of NEREA® products combined in the two reports by Martínez Grillo, J. F. et al. (2022, 2023), there is a logical need to reduce nutrient use.

Table 5. Comparison of rice fertilization using NPK + natural zeolite and NEREA® products

Treatment / Author	**Dose NPK (kg ha^{-1})**	**Yield (t ha-1)**		**kg rice/ kg nutrient**	**NPK saved (kgha^{-1})**
Villarreal, J.E.et al. (2015)					
N 80 kg.ha^{-1}+ 0 Zeolite	80.0	3.55		44.3	-
15	80.0	3.69		46.2	-
25	80.0	3.57		44.7	-
35	80.0	3.36		42.0	-
45	80.0	3.87		48.4	-
Soca Nuñez, M. and Constanza Daza, M. (2018)					
Control without fertilization		4.0 b			
Urea	120.0	6.0 a		50	
Urea + zeolite (10%)	110.0	6.5 a		59.1	10.0
Urea + zeolite(15%)	105.0	6.4 a		60.9	15.0
Martínez Grillo, J-F. et al. (2022)		**Yield** (t ha^{-1})	**Yield 14% humidity** (t ha^{-1})		
Control without fertilization		1.96	2,03	-	-

continued on following page

Table 5. Continued

Treatment / Author	Dose NPK (kg ha^{-1})	Yield (t ha-1)		kg rice/ kg nutrient	NPK saved (kgha^{-1})
Technical Instructions with fertilizer doses (N 150 kg ha^{-1}, P_2O_5 70 kg ha^{-1} y K_2O 70 kg ha^{-1}) for a performance of 6 t ha^{-1}.	290.0	6.49	3,48	22.4	-
NEREA® Fertilizer (2%) in background application 2 tha^{-1}	40.0	5.13	2,87	128.3	250
NEREA® Plus 3 kg ha^{-1}in active tillering, cotton point and paniculation	1.44	4.55	2,89	3159.7	288.56
NEREA® Fertilizer (2%) background 2 t ha^{-1} + NEREA® Plus foliar fertilizer 3 kg ha^{-1}in cotton knitting and paniculation	40.96	5.97	3.55	145.8	249.04
Martínez Grillo, J-F. et al. (2023)		**Yield test 2023** (t ha^{-1})	**Mean yield two tests** (t ha^{-1})		
Control without fertilization	-	2.67	2.35		
Technical Instructions with fertilizer doses (N 150, P_2O_5 70, K_2O 70 kg ha^{-1}) for a performance of 6 t ha^{-1}.	290.0	3,36	3.42	11.6	
NEREA® fertilizer (2%) background 2 tha^{-1}.	40.0	2.97	2.92	74.3	250.0
NEREA® Plus: foliar fertilizer, 3 kg ha^{-1}in active tillering, cotton point and paniculation.	1.44	3.01	2.95	2090.0	288.56
NEREA®: fertilizer (2%) background 2 t ha^{-1} + NEREA® Plus: Foliar fertilizer 3 kgha^{-1} in cotton point and paniculation.	40.96	2.09	2.82	51.1	249.04
Saborit Reyes, R. et al. 2021	**Dose NPK** (kgha^{-1})	**Yield** (t ha^{-1})		**kg rice/ kg nutrient**	**NPK saved** (kgha^{-1})
T-1: Control without fertilization.		1.7	-		
T-2: Technical Instructions NPK: N 100+P_2O_5 90+K_2O 90 kg ha^{-1}.	280.0	4.3	-	15.4	-
T-3: Nerea Plus (4 applications of 4.0 kg ha^{-1}) + 50% NPK	142.6	5.4	-	37.9	138.0
T-4: Nerea Plus + 100% NPK	282.6	5.7	-	20.2	-
T-5: Nerea Plus + Fitomas-E 2.0 Lha^{-1} + 50% NPK	142.6	6.9	-	48.4	138.0
T-6: Nerea Plus + Fitomas-E 2.0 Lha^{-1} + 100% NPK	282.6	6.7	-	23.7	-
T-7: Nerea Plus + Bayfolan 3.0 Lha^{-1} + 50% NPK	142.6	6.8	-	47.7	138.0

continued on following page

Table 5. Continued

Treatment / Author	Dose NPK (kg ha^{-1})	Yield (t ha-1)		kg rice/ kg nutrient	NPK saved (kgha^{-1})
T-8: Nerea Plus + Bayfolan 3.0 Lha^{-1} + 100% NPK	282.6	7.4	-	26.2	-
T-9: Nerea Plus + Fitomas-E 2 Lha^{-1} + Bayfolan 3 Lha^{-1} + 50% NPK	142.6	7.6	-	53.3	138.0
T-10: Nerea Plus +Fitomas-E 2 Lha^{-1} +Bayfolan 3 Lha^{-1} +100% NPK	282.6	7.9	-	27.9	-

How will NEREA products perform combined with traditional fertilization?Saborit Reyes, R. et al. (2021) took the experiences of Martínez Grillo, J.F. et al. (2021-3), and designed an experiment considering that farmers do not completely modify their agricultural procedures. They evaluated possible combinations of the NEREA® Plus foliar fertilizer product applied in the four stages of vegetative development of the rice crop: 1st application at the beginning of tillering. (15 - 20 days after germination (DAG), 2nd application in active tillering (30 - 35 DAG), 3rd application at primordium change (40 - 45 DAG), 4th application at the beginning of panicle (Spike Formation) (65–70 DAG); combined with NPK bottom fertilization (N 100.0 kg ha^{-1} P_2O_5 90.0 kg ha^{-1}, K_2O 90.0 kg ha^{-1}), Fitomas®-E biostimulant, and the Bayfolan® foliar fertilizer (Table 6). The authors concluded:

> *A positive response was observed in the applications of NEREA® Plus, combined with the biostimulant, foliar and NPK, on the agricultural yields of the rice crop. The most accentuated responses were achieved in treatments 8, 9 and 10, with yields of 7.4, 7.6 and 7.9 t ha^{-1} respectively, with no significant difference between them, applied at different times of vegetative development of the rice crop. In the economic analysis, where treatment 9, which obtained a yield of 7.6 t ha^{-1}, was compared with the reference control (NPK, according to Technical Instructions) 4.3 t ha^{-1}, showing a relative gain of 2253.46 CUP per hectare in favor of the first case.*

Treatment 9, 50% of NPK was used and a yield of 7.6 t ha^{-1} was obtained, 43% higher than the reference control, with a better efficient use of nutrients kg rice/ kg nutrient of 53.3.These studies demonstrate that the use of NEREA® products in combined fertilization schemes of rice provides better production results and efficient use of nutrients, compared to the use of a mixture of NPK fertilizer and natural zeolite (Table 5).

Potato (*Solanum Tuberosum* L.) Crop

Ozbahcea, A. et al. (2018) examined the

effect of application of different doses of the natural zeolite clinoptilolite and different irrigation levels on yield and quality of potatoes (Solanum tuberosum L.), cv. Agria, and the nutrient contents of the soil under water deficit stress in Konya-Karapınar, Turkey. The study was established in a split-plot design with three replications. The main factor was the zeolite dose (Z_0:0, Z_3:30, Z_6:60, Z_9:90 and Z_{12}:120 t ha^{-1}) and the sub factor the irrigation level (I_{50}: 0.50, I_{75}: 0.75 and I_{100}:1.00). InI_{100} treatment, irrigation was applied to fill 0–60-cm soil-depth until field capacity. In other treatments (I_{75}-I_{50}), it was given up to 75 and 50% of water applied to I_{100} treatment at 6-day intervals. For potato tuber yield, several quality characteristics and some nutrient element content in the soil, significant interaction occurred between zeolite doses and irrigation level ($P < 0.01$). The highest crop yields, 33.9–39.1 and 33.5–34.3 kg ha^{-1}, respectively, were obtained from Z_6I_{100} and Z_6I_{75} applications in both years. There were no significant differences between these two treatments and several other treatments (Z_3I_{100}, Z_9I_{100} and $Z_{12}I_{100}$). Some physical and chemical contents (the cation exchange capacity, exchangeable sodium percentage and total P, K, Ca, Mg, Zn and Mn contents) of the experiment soil were affected by zeolite treatments. These results showed that certain zeolite doses with optimum irrigation can be helpful for potato grown in water-deficit stress conditions because of positive effects on some soil physical and chemical properties and crop quality.

Rivero González, L.A. et al. (2018) conducted a study in potato growing under production conditions in a red Ferralitic soil. The control group was NPK fertilization formula 9-13-17 dose of 1,340.0 kg ha^{-1} established for potato cultivation in Cuba, a formula that contains 25% of natural zeolite particles +1-4 mm from the Tasajeras deposit. Two NEREA® Fertilizer products were produced using the same zeolitic mineral but with different nutrients content. One was produced using 2% wt. of DAP fertilizer, and the other 2.5% wt. of the fertilizer NPK formula 9-13-17. The evaluation was made through an experimental design of random blocks, with 7 treatments and 4 repetitions with the aim of evaluating various possibilities with the use of zeolite in fertilization. Each plot had an area of 20 m^2 and the production was evaluated in two m^2 of each plot in the potato field. The potato variety CIMEGA trial demonstrated that increasing from 2% to 2.5% of NPK inside NEREA® Fertilizer, raised the yield to nearly that obtained in the control fertilized with 1340.0 kgha^{-1} of NPK formula 9-13-17 and 25% natural (Table 6). The use of NEREA® Fertilizer saved 1,290.0 and 1,300.0 kg of NPK accounting for 96% and 97%, respectively, of the 1,340.0 kg of the formula 9-13-17. The increase of nutrients in NEREA® Fertilizer in only 0.5% wt. allowed to almost reach the yield when using chemical fertilization NPK 9-13-17 with 25% natural zeolite The **kg potato/kg nutrient** values were 31.0 for NPK + 25% natural zeolite, 668.0 for NEREA® Fertilizer (2%), and 599.0 for NEREA® Fertilizer (2.5%).

Al-Hassan Jamal Abdul Razzaq et al. (2023) evaluated the use of 6 and 12 t ha^{-1} of natural zeolite combined with biostimulator and NPK fertilizers in potato crop.

> *The first factor is adding natural zeolite to the soil in three levels: the first level without addition, the second level 6 t ha^{-1}, and the third level 12 t ha^{-1}. The second factor: the addition of Biohealth, the first level without addition, the second level, the addition of 5.0 kg ha^{-1}. The third factor adds mineral fertilizers, the first level without addition, the second level adds 50% of the fertilizer recommendation, and the third level adds 75%. The fertilizer recommendation is 300.0 kg ha^{-1} N, 100.0 kg ha^{-1} P, and 300.0 kg ha^{-1} K. The addition of zeolite significantly affected the achievement of the highest value in treatment Z2 (27.07 Mg ha^{-1}) compared to the control treatment (Z0) 23.27 (Mg ha^{-1}).The total yield increased with the increased levels of adding mineral fertilizer. The treatment (F2), which is the second level of mineral fertilizer, recorded 29.02 (Mgha^{-1}) compared to the control treatment (F0), 20.39 (Mgha^{-1}), an increase of 42.32%. The effect of the biinteraction between the bio-enhancer and the adding zeolite (ZxB) was significant in increasing the total yield rate, where the highest value was recorded in treatment Z2B1 (29.42 Mgha^{-1}) compared to the control treatment Z0B0 (19.52 Mgha^{-1}), with an increase rate of (50.71%). The highest treatment (Z2F2) was recorded with a value of 32.04 (Mgha^{-1}) compared to the control treatment (Z0F0), 19.04(Mgha^{-1}), with an increase of 68.27%. The triple interaction of the study factors (F x B x Z) achieved a significant effect on this trait, where the treatment (Z2B1F2) excelled in achieving the best average of the total yield, which amounted to 33.36 (Mgha^{-1}) compared to the control treatment, which achieved the lowest value in the total yield. It reached (17.47 Mgha^{-1}), with an increase of (90.95%).*

The doses of 6 and 12 t ha^{-1} of zeolite are large even when doses of NPK are the 50 and 75% of the recommended one for the crop in the country, while the **kg potato/ kg nutrient** values comparable with those achieved by other authors (Table 6).

L.A. Rivero González et al. (2023) conducted a study using NEREA® Fertilizer and NEREA® Plus foliar fertilizer in potato variety Rudol cultivated in bags of 20.0 kg at open air, using watering can. The best results were confirmed with the combined use of NEREA® products, however the yields were lower than the ones observed with var Cimega in 2018 at production scale. Besides, the values of **kg potato/kg nutrient** are comparable. In 2025, Fundación UH and NEREA Team conducted a test at production scale on 14 ha of potato crop in Pinar del Rio province, Cuba, using 300 kg of NEREA® Fertilizer (6%) (18 kg of NPK) before sowing, combined with the foliar stimulator BIOBRAS® 16. The yield obtained was 16.5

t ha^{-1}. This yield was 2 tons higher than the potato average yield of the province in 2024 campaign. The **kg potato/kg nutrient** was 916.0.

Table 6. Comparison of potato crops in soils fertilized with NPK formula + natural zeolite vs NEREA® fertilizer

Treatment / Author	Dose fertilizer (kgha^{-1})	Dose NPK (kgha^{-1})	Yield (t ha^{-1})	kg potato/ kg nutrient	NPK saved (kgha^{-1})
A.Ozbahcea et al. (2018)					
Z_0:Control NPK: (160 kg ha^{-1}ammoniumsulphate, 21% N), phosphorus (100 kg ha^{-1} 46 DAP, 46% P_2O_5) and potassium (100 kg ha^{-1} KNO_3, 50% K_2O) + 0 zeolite	360.0	360.0	18.7-22.4	62.22	-
Z_3: NPK + zeolite 30 t ha^{-1}	390.0	360.0	30.1 – 36.5	101.38	-
Z_6: NPK + zeolite 60 t ha^{-1} I_{75}	420.0	360.0	33.5 - 34.3	95.28	-
Z_6: NPK + zeolite 60 t ha$^{-1}I_{100}$	420.0	360.0	33.9 - 39.1	108.61	-
Z_9: NPK + zeolite 90 t ha^{-1}	450.0	360.0	30.9 - 38.7	107.5	-
Z_{12}: NPK + zeolite 120 t ha^{-1}	480.0	360.0	30.7 –29.1	80.83	-
Rivero González, L.A. et al. (2018)					
Control (withoutfertilization)			17.00		
Control NPK 9-13-17 + 25% zeolite	1340.0	1000.0	31.10	31.0	335
NEREA® Fertilizer (2.0% DAP)	2000.0	40.0	26.72	668.0	1300.0
NEREA® Fertilizer (2.5% NPK)	2000.0	50.0	29.95	599.0	1290.0
Díaz Álvarez, H.J.et al. (2019)					
Control (without fertilization)			18.43		
Complete formula 8-9-14 *	1 490.0	1 640.0	24.31	46.0	-
Complete formula 6-7-10 + 25% natural zeolite *	1 490.0	1 267.5	29.46	61.0	372.5
Complete formula 4-4.5-7+ 50% natural zeolite *	1 490.0	895.0	25.06	95	745.0
*additional dose of 150 kg.ha-1 of urea.					
Al-Hassan J. Abdul Razzaq, et al. (2023)					
F0 - Control (without fertilization)			17.47		
Z1 6 t ha^{-1}			20.58		

continued on following page

Table 6. Continued

Treatment / Author	Dose fertilizer (kgha^{-1})	Dose NPK (kgha^{-1})	Yield (t ha^{-1})	kg potato/ kg nutrient	NPK saved (kgha^{-1})
Z2 12 t ha^{-1}			27.07		
FNPK 300 kg ha^{-1} N, 100 kg ha^{-1} P, and 300 kg ha^{-1} K	700.0				
F1 50%: 150 kg ha^{-1} N, 50 kg ha^{-1} P, and 150 kg ha^{-1} K	350.0	350.0	27.54	78.68	
F2 75%: 225 kg ha^{-1} N, 75 kg ha^{-1} P, and 225 kg ha^{-1} K	525.0	525.0	29.02	55.27	
Z0F0	-		19.04		
Z0F1	350.0	350.0	24.39	69.68	-
Z0F2	350.0	350.0	26.38	75.37	-
Z1F0	6,000.0	-	20.08		-
Z1F1	6,350.0	350.0	24.11	68.88	-
Z1F2	6,525.0	525.0	28.62	54.51	-
Z2F1	12,350.0	350.0	27.09	77.4	-
Z2F2	12,525.0	525.0	32.04	61.03	-
L.A.Rivero González et al. (2023)					
Control (without fertilization)	-		17.00	-	-
NEREA Plus foliar 11 applications	44.0	7.04	22.40	3182.0	
NEREA Fertilizer 2 tha^{-1} + NEREA Plus Foliar 11 applications	2,044.0	27.04	25.80	954.0	
NEREA Plus to soil 0.1 tha^{-1} + NEREA Plus foliar 11 applications	144.0	71.04	20.00	282.0	
NEREA Plus foliar 8 applications	32.0	5.12	20.20	3945.0	
NEREA Fertilizer 2 tha^{-1} + NEREA Plus foliar 8 applications	2,032.0	25.12	22.20	884.0	
NEREA Plus to soil 0.1 tha^{-1} + NEREA Plus foliar 8 applications	132.0	69.12	21.50	311.0	

Sugarcane (*Saccharum Officinarum)* Crop

The first reports of the use of natural zeolite in sugarcane cultivation in Cuba date back to the 1990s. M. López et al. (1991) evaluated the incorporation of natural zeolite into the soil at doses of 3.0 t ha^{-1} and 6.0 t ha^{-1} prior to fertilization with NPK formula 25-35-120. Although yields increased compared to those observed in the control (80.13 t ha^{-1}), 82.55 and 87.12 t ha^{-1} respectively, the authors did not consider them significant. Bouzo, L. et al. (1991) also evaluated the use of natural zeolite alone at 4.5 and 15.0 t ha^{-1} with mixed results. However, R. Crespo (1998) repeated the previous studies and did observe a positive response of sugarcane cultivation with the progressive addition of only natural zeolite to the soil, with 15.0 t ha^{-1} he observed a yield of 92.00 t ha^{-1}, remaining as an agricultural practice for sugarcane in Cuba (Table 7).

Suprapron, J. et al. (2002) evaluated the effect of doses of 125.0 and 250.0 kg ha^{-1} of natural zeolite incorporated into the soil, and together with triple NPK triple 15 fertilizer doses of 312.5 and 625 kg ha^{-1}. Although the zeolite doses are low compared to the results of other authors, in this study an increase in sugarcane crop yield was observed. The control without fertilization provided a yield of 51.96 t ha^{-1} and the addition of 250.0 kg ha^{-1} zeolite to the 625.0 kg ha^{-1} of NPK triple 15 yielded 90.14 t ha^{-1}, a significant increase.

Rancho Chinobampo (2008), reported on a validation study of the NEREA Chinobampo zeolitic fertilizer, formula 14-05-15 (with 25% natural zeolite from the San Luis deposit) at a dose of 400.0 kg ha^{-1}, compare with the traditional sugarcane NPK formula 18-05-21 at a dose of 500 kg ha^{-1} as a control. The study was conducted in the sugarcane areas of the Tres Valles Sugar Mill, part of the PIASA Group, Veracruz, Mexico. Yields were observed at 100.0 t ha^{-1} for the control fertilizer and 135.0 t ha^{-1} for the areas fertilized with NEREA Chinobampo. Sugar production increased from 20.14 to 22.33 tons per hectare.

Méndez Días, N. and Soca Núñez, M. (2016) evaluated the NPK formula 25-05-24 in sugar cane cultivation with 25% substitution of natural zeolite from the Tasajeras deposit, processed (+1-3mm). The dose used was 500.0 kg ha^{-1}. The authors report that the yield observed in the control was 100.8 t ha^{-1} while the yield of the cane fertilized with the incorporation of 25% zeolite was 113.20 t ha^{-1}. Similarly, the extracted sugar increased from 19.02 to 20.93 t sugar ha^{-1}.

Cairo Cairo, P. et al. (2017) evaluated

the effect of zeolite and its combination with chemical fertilizer and organic fertilizers on soil quality and sugarcane yield in the north coast of Cuba, Villa Clara province. The organic fertilizer (sugarcane filter cake-SFC), and natural mineral (zeolite-Z) and chemical fertilization (nitrogen-phosphorus-potassium-NPK) were applied as treatments. The experiment was laid out in a randomized complete block design with nine treatments and four replicates. The soil samples were taken at a depth of 0-20 cm to determine soil properties. The sugarcane yield components such as cane yield (t ha^{-1}) and sucrose yield (t ha^{-1}) were evaluated. The best results on cane yield were obtained with the treatment Z 7.5 t ha^{-1} + SFC 22.5 t ha^{-1}, which represent a relative increase of 200% vs control (without fertilization). Relationships between soil properties and principal component *analysis confirmed the good response of zeolite application on soil quality and sugarcane yield.*

The authors observed the need to use zeolite in quantities greater than a few kilograms, up to 7.5 and 15.0 t ha^{-1} combined with NPK or SFC to achieve higher agricultural yields and the sugar extracted from the crop.

De Antoni Migliorati, M. et al. (2024) answered the question: *Can zeolites be used in sugarcane cropping systems to help reduce nitrogen losses into the Australian Great Barrier Reef?*

The aim of this study was to determine the effectiveness of natural (NatZeo) and acid-treated (AcidZeo) zeolites in increasing the nitrogen (N) recovery of sugarcane grown under conditions highly conducive for N losses. This glasshouse trial replicated the pedoclimatic conditions typical of the Australian sugarcane industry in the Wet Tropics to evaluate the capacity and economic feasibility of using NatZeo and AcidZeo to reduce leaching and increase plant and soil N recovery in a highly permeable sandy-loam soil. Nitrogen fertilizer was applied at two rates (84.0 and 120.0 kg N ha^{-1}) as 15N-enriched urea. Compared with the unamended treatments, AcidZeo led to significantly higher average fertilizer N recoveries in both the soil (+24%) and plant (+54%), which resulted in significantly lower overall N losses (−22%). However, the significantly higher capacity of AcidZeo in retaining ions limited the availability of potassium and sulphur to the plants and led to a 4% yield reduction. The use of NatZeo led to lower abatements of N losses (−6%) but increased N recoveries in the plant (+22%). NatZeo did not result in as much cation immobilization, which, combined with the increased plant/soil N recovery, resulted in the highest yield of the trial (+8% compared with the unamended treatments). The economic analysis indicated that only NatZeo could potentially increase farmer's income when applied in 20 cm wide strips around the fertilizer band. Conclusions: This study highlights that zeolites have potential to improve the environmental sustainability and profitability of sugarcane cropping systems.

Table 7. Comparison of sugarcane treated with NEREA fertilizer vs NPK + natural zeolite by different authors.

Treatment / Author	**Dose fertilizer (kg ha^{-1})**	**Dose NPK (kg)**	**Yield (t ha^{-1})**	**Pol (t sugar ha^{-1})**
Crespo, R. (1998)				
Natural zeolite	3,000.0	-	75.90	12.42
Natural zeolite	6,000.0	-	80.10	12.72

continued on following page

Table 7. Continued

Treatment / Author	Dose fertilizer (kg ha^{-1})	Dose NPK (kg)	Yield (t ha^{-1})	Pol (t sugar ha^{-1})
Natural zeolite	7,500.0	-	81.50	14.93
Natural zeolite	15,000.0	-	92.00	14.59
RanchoChinobampo (2008).				
Control Formula 18-05-21	500.0	500.0	100.0	20.14
Zeolitic Fertilizer NEREA Chinobampo 14-05-15 (25% zeolite)	400.0	300.0	135.0	22.33
Méndez Días, N. and Soca Núñez, M. (2016)				
Control Formula 25-05-24	600.0	600.0	100.8	19.02
Formula 25-05-24 + 25% natural zeolite	600.0	450.0	113.20	20.93
Cairo Cairo, P. et al. (2017)				
Control	-	-	86.37	14.99
Z 7.5 t ha^{-1}	7,500.0	-	89.41	15.31
Z 15 t ha^{-1}	15,000.0	-	114.40	19.48
Z 7.5 t ha^{-1}+ N (100.0 kg ha^{-1})	7,600.0	100.0	154.84	26.14
Z 15 t ha^{-1}+ N (100 0kg ha^{-1})	15,100.0	100.0	142.30	23.84
Z 7.5 t ha^{-1}+SFC (22.5 t ha^{-1})	30,000.0		162.00	24.42
Z 3 t ha^{-1}+SFC (18 t ha^{-1})	21,000.0		137.66	22.95
Z 7.5 t ha^{-1}+NPK (100-60-200 kg ha^{-1})	7,860.0	360.0	135.79	22.19
Z 15 t ha^{-1}+NPK (100-60-200 kg ha^{-1})	15,360.0	360.0	106.93	17.29
Rivero González, L.A. et al. (2018)				
Control 1: Triple 18 Commercial Formula	181.44	181.44	134,37	19,50
FC Triple 18 (181.44 kg ha^{-1}) + 250 kg zeolite (58%)	431.44	181.44	130.09	20,39
NEREA fertilizer (9% NPK)	2,181.44	181.44	140.47	20,34
Control 2: FC Triple 18 (136.08 kg ha^{-1} (75% control 1)	136.08	136.08	127.16	18.51
FC 136.08 kg ha^{-1} + 250 kg zeolite (65%)	386.08	136.08	122.43	18.14
NEREA fertilizer (12% NPK)	1,136.08	136.08	138.30	20.20
NEREA fertilizer (7% NPK)	2,136.08	136.08	132.15	19.64

Rivero González, L.A. et al (2018) used natural zeolite from the San Andrés deposit, Holguín, Cuba to make the NEREA® fertilizer product with different doses of nutrients. The study was carried out in areas of the Ingenio Tolulá, Guatemala, and the use of NEREA® Fertilizer products was compared with the commercial NPK formula triple 18, with fertilizations made with NPK + zeolite. The most significant results are presented in Table 7. The highest yield 140.0 t ha^{-1} with a

Pol 20.34 t sugar ha^{-1} was observed when using 2.0 tons of NEREA® Fertilizer with 9% NPK incorporated, while control 1 had a yield of 134.37 t ha^{-1} and a Pol of 19.50 t sugar ha^{-1}. In control 2 with a fertilization of 136.08 kg ha^{-1} equivalent to 75% of that of witness 1, a yield of 127.16 t ha^{-1} and Pol of 18.51 t sugar ha^{-1} were observed, however, when using one ton of NEREA® Fertilizer with 12% NPK incorporated the observed yield was 138.30 t ha^{-1}, the Pol of 20.20 t sugar ha^{-1}. The yields observed with the NEREA® Fertilizer product show that agricultural and industrial yields can be achieved by optimizing the amount of nutrients incorporated into the zeolite particles according to NEREA® technology. In the Figure 4 it can be seen for the two NPK fertilizer contents triple 18, 181.44 and 136.08 kg ha^{-1}, as they are incorporated into larger masses of zeolite with particle class size +1-3 mm, the yield of sugarcane increased, except when incorporating 136.08 kg in 2,000.0 kg of zeolite, which decreased.

Figure 4. Yield of sugarcane fertilized with NEREA fertilizer formulated with different natural zeolite contents.

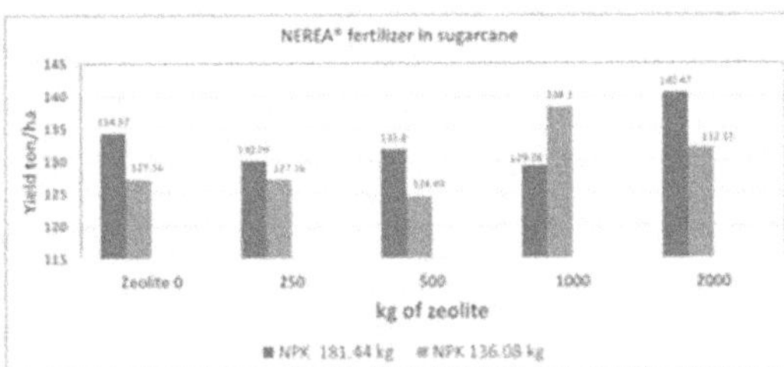

It can be observed in the Figure 5 that the highest agricultural (t ha^{-1}) and industrial Pol (t sugar ha^{-1}) yields were obtained with NPK contents (% wt.) of 9 and 12%, higher than those observed with commercial fertilization using NPK (18-18-18) at different doses.

Figure 5. Sugarcane yield and sugar produced per hectare fertilized with NEREA fertilizer with different NPK contents.

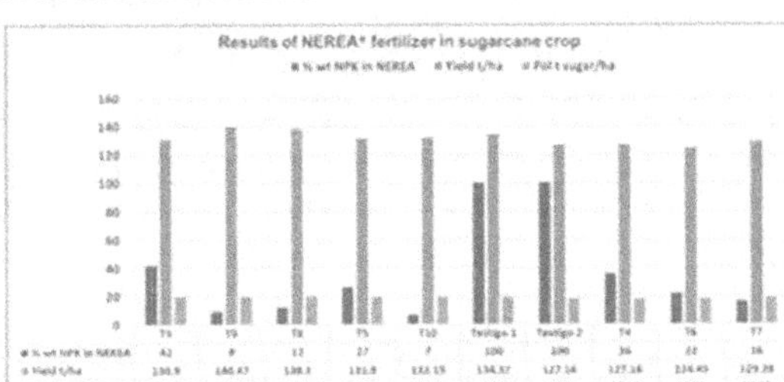

The positive response of sugarcane to the presence of zeolite in the soil can be explained by the high silicon requirements for its development. Savant, N. K. et al. (1999) expressed

Silicon (Si) is one of the most abundant elements found in the earth's crust but is mostly inert and only slightly soluble. Agriculture activity tends to remove large quantities of Si from soil. Sugarcane is known to absorb more Si than any other mineral nutrient, accumulating approximately 380 kg ha^{-1} of Si, in a 12-month old crop. Sugarcane (plant growth and development) responses to silicon fertilization have been documented in some areas of the world, and applications on commercial fields are routine in certain areas. Some studies indicate that sugarcane yield responses to silicon may be associated with induced resistance to biotic and abiotic stresses, such as disease and pest resistance, Al, Mn and Fe toxicity alleviation, increased P availability, reduced lodging, improved leaf and stalk erectness, freeze resistance, and improvement in plant water economy.

Filgueiras Dutra, A. et al. (2023) reported that

Silicon (Si) fertilization is widely recognized to improve the development of crops, especially in tropical soils and cultivation under dryland management. Si stoichiometry favors the efficient use of carbon (C), nitrogen (N), and phosphorus (P) in sugarcane plants. The most prominent impacts were found in the leaf Si-sprayed RB92579 cultivar, with a significant increase of 7.0% (11 Mg ha^{-1}) in stalk yield, 9.0% (12 Mg ha^{-1}) in total recoverable sugar, and 20% (4 Mg ha^{-1}) in sugar yield compared to the Si-without control."

Farias da Silva, J. L. et al. (2023) reported that

Adopting a Si supply strategy can amplify the sugarcane response. Thus, this study aimed to verify whether Si supply in the pre-sprouted seedling (PSS) formation phase would have an effect after field transplanting similar to Si supply only in the field phase (via foliar spraying or fertigation). Si supplied increasing photosynthetic pigment content and the antioxidative defense system, and by increasing the crop's productivity and sustainability.

In the specific case of NEREA® products, it has been shown that Si is more available to plants than when using only natural zeolite. Figure 6 shows the silicon content in corn leaves—a Gramineae that requires high levels of Si for its development—grown in compact red Ferralitic soil, fertilized with increasing doses of NEREA fertilizer (2%) up to 100%, resulting in total soil replacement. We have used the corn model because "in general, plants belonging to family Gramineae accumulate much more silicon than that by other species belonging to other families" (Amin M.et al2015).

Figure 6. Silicon content in corn leaf fertilized with NEREA® (2%)

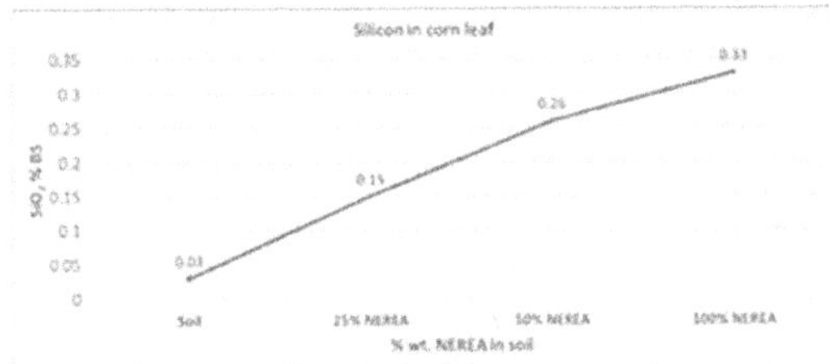

However, R.A. Hussien (2020) reported that a

> *partial replacement of synthetic chemical fertilizers by naturally occurring alternatives is environmentally recommended. Feldspars (F), quartz silica (S), and zeolites (Z) are silicon (Si)-rich minerals that may be utilized as Si fertilizers. This study aims to assess the agronomic efficiency of the mentioned minerals as Si fertilizers and to estimate Si-use efficiency sandy soil.*

A field experiment was carried out in which F, S, and Z were mixed with surface soil in an application rate 500.0 mg kg^{-1} soil with and without potassium humate (K-H, 2 mg kg^{-1} soil). Treatments were distributed in a complete randomized block design with three replicates including control before cultivation of soybean (Glycine max L.). Yield (kg ha^{-1}) of soybean, available Si (mg kg^{-1}) in soil and uptake (mg kg^{-1}) of N, P, K, Cu, Fe, Mn, Zn, and Si by soybean seeds and straw were estimated. The most significant increase was 67.87% followed 38.69% was recorded for the S and S K-H treatments, compared to the control. "Silica was more efficient agronomically than feldspar and zeolite. Absorption of more biocompatible Si-organo species may depend on Si source."

G. Conversa et al. (2024) have reported that

> *the foliar application of zeolites has been proven to positively affect plant physiology suggesting that it could improve plant growth, however, this topic has been little studied. Two field trials were carried out to study the potential beneficial effect of natural zeolite particle films on processing tomato crops. Natural zeolites i) clinoptinolite + mordenite (CubZeo) and ii) micronized clinoptinolite (BigZeo) were sprayed onto the plants six times. Si was observed at the highest level in fruits and stems of BigZeo-treated plants, followed by CubZeo and Control ones. The accumulation of Si could explain the best crop responses elicited by BigZeo, whereas the observed responses did not appear to involve changes in the photosynthetic machinery. The authors could not explain the observed results and proposed comprehensive investigations to define the zeolite particle characteristics related to physiological and nutritional changes in the plants. Rivero-Robaina, E. et al. (2024) compare the foliar application*

of NEREA® Plus and CubZeo (the same dose and particle size -0.040mm) in potato and maize concept study in German field soil, observing a higher yield of the crops treated with NEREA® Plus, confirming its values for plant nutrition and Si supplier stimulator.

How NEREA products containing small quantity of nutrients can nourish the plants? Rodríguez-Fuentes, G. et al. (2025) answered

more than a simple physical and chemical mechanism of NEREA® particles to supply nutrients to plants, the processes involved are several: 1) The direct action of plant roots in the uptake of nutrients supported by NEREA® particles through extraction and ion exchange processes mediated by the transport of water and hydronium ions, 2) The increase in the number, size and change in the morphology of the roots to involve the NEREA® particles, 3) The Biota-NEREA® particles-plants-soil relationship. The participation of soil microorganisms with their transformation of N and P in the modification of the nutrients contained in the soil and in the NEREA® particles and their delivery to the roots of the plants, 4) The availability of all nutrients in each of the NEREA® particles, 5) The greater availability of high silicon content compared to natural zeolite, and 6) Oxygenation of the roots.

Environmental benefits of NEREA products were reported by Rodríguez-Fuentes et al (2025). No residuals are generated during production, nutrients are not leached in the soil, and they are made up of more than 90% zeolitic mineral naturally appears in soils.

NEREA® technology generates a significant reduction in the emission of Greenhouse Gases (GHG) due to increased crop yields, decreased consumption of fertilizers, fuel for their production and application, reduced impact of the nitrogen cycle. According to a study carried out by International Zeolite Corp (2022) using Canadian agricultural standards, and the results of the validation of NEREA® fertilizer in potato cultivation in Cuba, 2018, the reduction emission of 575.68 kg CO_2e per hectare per crop cycle, a significant contribution to reducing the Carbon Footprint and Climate Change.

How much NPK can be saved using NEREA® products? In table 8Rodríguez-Fuentes, G. et al. (2025) answered using the conversion of raw materials and zeolitic raw material NZ to produce 1 ton of NEREA® products, and the dose to be used in three representative crops of Cuban agriculture (tobacco, potato, common beans) to obtain similar performance to the dose of NPK fertilizers that is regularly used, which allows determining nutrient savings about 96 to 99%.

Table 8. Conversion of raw materials and zeolitic raw material NZ to produce 1 ton of NEREA products, application dose to obtain same yield with regular NPK fertilizer, and NPK saved.

NEREA® product	Raw materials for 1 t NEREA® (kg)	Dose NEREA® (kg ha^{-1})	Crop	Dose NPK fertilizer (kg ha^{-1})	Yield	NPK saved	
						(kg ha^{-1})	(%)
Substrate	20 NPK T-18 980 NZ	1,000.0	Tobacco nursery	NPK (5-11-5+2.5) 2,250.0	Seed germinate in 27 days vs 40 days with regular NPK	2,230.0	99.1
Fertilizer	25 NPK T-18 975 NZ	2,000.0	Potato	NPK (9-13-17) 1,340.0	30 t ha^{-1}	1,290.0	96.3
Plus	60 NPK T-18 100 urea 840 NZ	16.0	Common beans	NPK (8-23-14) 300.0 Urea 200.0 Bayfolan® 3 L	1.9 t ha^{-1}	497.0	99.4

The studies funded by International Zeolite Corp in Canada and conducted by Vineland Research & Innovation Centre (Yin, Q. and McAughey, S. 2022), and the Agriculture & Environmental Technologies Innovation Centre, of Niagara College(-Schulze, D. et al. 2021), George, C. et al. 2022) confirmed the results achieved by Cuban researchers. In this sense, several evidences were summarized: 1) Greater number of germinated seeds, less lost seeds, 2) Rapid growth of seedlings, 3) Similar or higher agricultural yields of hydroponic crops with NPK fertirrigation when using NEREA® substrate (2% NPK) in doses of 10 to 25% and irrigation with water, 4) Provision of all nutrients in NEREA® particles, 5) Efficient development of plants in contact with NEREA® particles even with low nutrient content, 6) Plants selectively extract the nutrients they need. The presence of Sodium in NEREA®, its selective uptake by plants without evidence of contamination or deterioration has also been studied, 7) Great development of the roots of plants in contact with NEREA®, seeking more NEREA® particles, 8) The roots surround the NEREA® particles, 9) White roots indicative of good health and oxygenation, and 10) Preference of the plant roots for NEREA® particles than other substrates.

NEREA® technology transfer to Cuban companies. In 1988, the technology of zeoponic cultures based on obtaining NEREA® substrate *in situ* was transferred to the Ministry of Agriculture of Cuba (MINAG), from the creators to provide the engineers the transformation process from hydroponics to zeoponic technology. Thirty years later, the innovating NEREA® technology was assumed by MINAG, to "produce and introduce into the Cuban agriculture the NEREA® products, to bring fertilization to all crops, reduce fertilization, nutrients losses, and aquifer contamination." In 2019, Foundation of University of Havana (UH) signed a merger production agreement with Geominera Oriente, to construct the first NEREA®

Experimental Plant (PEN) (Holguín) to produce, register, and marketing the NEREA® products: Substrate (RCF 019/20), Fertilizer (RCF 018/20) and Plus foliar fertilizer (RCF 017/20). The Foundation UH have also signed technology license and transfer agreements with other six Cuban enterprises for the industrial production of NEREA® products. The value of NEREA® products commercialized until January 2025 is over 24.5 million of CUP (Cuban pesos).

Can NEREA® technology be used with any natural zeolite raw material? In 2006, the NEREA® technology was licensed and transferred to the Mexican company Grupo Bampoahi S.A. de C.V. to produce, validate, register, and marketing NEREA® substrate and fertilizer. The zeolitic raw material from San Luis Potosi deposit was used. The quality of NEREA® Chinobampo products was like those produced with the zeolitic raw material from Cuban deposits. In 2024, samples of natural zeolites from USA deposits Ida Ore, Beer River, and KMI, were used to produce NEREA substrate following patent CU 20210017 A7. They were compared with NEREA® substrate produced by the PEN of Tasajeras and San Andrés. The comparison of Fourier Transform Infrared spectra of the products showed no statistically significant differences with a 95% confidence level. The quality of NEREA® substrates was also tested *in vivo* in greenhouse conditions with arugula seeds sown and lettuce seedling transplanted. The tests used 100% of NEREA® substrates and fresh water for irrigation. The tests demonstrated that the samples produced with the USA zeolite ores worked equally to Cuban zeolites.

NEREA® technology and products have received several awards. Science and Technology Provincial Forum sponsored by the National Association of Private Farmers, 2018; the Province of Ontario Honors International Zeolite Corp. as a Recipient of the Fertilizer Accelerating Solutions and Technological Challenge, and awarded by the Government Award of the Government of Canada-Funded i.d.e.a. Fund™, 2022; the CITMA Award for Technological Innovation, Havana 2022; Distinguish Speakers Awards Conference on Nanomaterials & Machine Learning by BIC^{2}UK, 2023. Recently, the minister of Science, Technology and Environment (CITMA) of Cuba granted the two top awards presented by the Cuban Government for distinguished scientific and technological accomplishments, that is, the CITMA 2023 National Award for Technological Innovation, and the CITMA 2023 Special Award for Highest Economic Impact. In 2024, the Rector of University of Havana and the Minister of High Education of Cuba recognized NEREA® technology and products with the awards for Technological Innovation of Highest Economic Impact, and Food Production.

CONCLUSION AND FUTURE RESEARCH DIRECTIONS

"Everything should be kept as simple as possible, but not simpler." –Ockham´s razor principle interpreted by Albert Einstein.

NEREA® zeolitic nanostructured materials are not likewise to NPK fertilizer blended with natural zeolites. They are unique materials. NEREA® technology solves most of the deficiencies of the NPK fertilizer mixed with natural zeolites practice. All the spontaneous, complex, and random process to dissolve NPK fertilizer and absorption of the nutrients by zeolite particles in the soil, were controlled and optimized industrially without residual emission before the incorporation to the soil. A) Nutrients are homogeneously distributed in each NEREA® particle. B) Each NEREA® particle contains the same nutrients in the quantities supplied. C) The use of NEREA® significantly reduces the action time of natural zeolite particles mixed with chemical fertilizers. D) There is no loss of the nutrients contained in the NEREA® particles due to leaching, due to irrigation and rainwater. E) The greater availability of high Silicon content compared to natural zeolite because the modification process. F) The dose of NPK fertilizer regularly employed in crops like tomato, rice, beans, potato, and sugar cane can be significantly reduced adopting NEREA® products. H) The efficiency of the nutrients applied to plants growth using NEREA® products is higher than the use of NPK + natural zeolite. The numbers of **kg crop/kg nutrients** are largest for NEREA® products. I) The mechanism of action of NEREA® products is multifactorial, involving the physical chemistry of the zeolitic support, the Biota, the soil, and the plants.

The new development based on natural zeolites must start from a well-founded knowledge of the characteristics of the zeolitic mineral used, including its natural nutritional potential for crop plants. Also, there are many questions to be answered regarding the challenges and prospects for innovation: 1) Are natural zeolites the cornerstone of agricultural fertilizers? 2) Which of the 67 types of natural zeolites reported and accepted by the International Zeolite Association can be used as soil improvers and in fertilizer formulations? 3) What is the limit for including zeolites in a soil that naturally contains or does not contain the mineral, and consequently, what is the potential harm or benefit? 4) Once natural zeolites are present in the soil, can they continue to do the same job of incorporating nutrients and releasing them slowly? This leads to study of the method and amounts of nutrient inclusion using NPK and organic fertilizers. Similarly, these studies must be conducted for NEREA® products.

ACKNOWLEDGEMENTS

The authors want to recognize the support of the University of Havana, and its Foundation of Innovation and Development. The effort of the colleagues from the Institute of Grains Research, and the authorities of the Ministry of Agriculture of Cuba. The colleagues of the enterprise Geominera Holguín where the first NEREA Experimental Plant was built and still work efficiently. Special thanks to the members of the NEREA-NatZEng Team of University of Havana: L. A. Rivero-González, E. L. Rivero-Robaina, E.Y de la Nuez-Pantoja, B. Concepción-Rosabal, J.C. Molina-Amat, C. Lariot-Sánchez and A. Caisés-Ávalos.

REFERENCES

Agustina, T. E., Rizkya, I., Utamaa, M. E. W., & Amalb, M. I. (2018). Characterization and Utilization of Zeolite for NPK Slow Release Fertilizer. *International Journal of Engineering TRANSACTIONS A: Basics*, *31*(4), 622–628.

Amin, M., Ahmad, R., Ali, A., Aslam, M., & Lee, D. J. (2015). Silicon Fertilization Improves the Maize (Zea mays L.) Performance under Limited Moisture Supply. *Cereal Research Communications*. Advance online publication. DOI: 10.1556/0806.43.2015.035

Andronikashvili, T. G., Zautashvili, M., Eprikashvili, L., Pirtskhalava, N., & Dzagani, M. (2009). Zeolite containing substrates - A new way from plant-growing in plant production. *Annals of Agrarian Science*, *7*(4), 14–45.

Andronikashvili, T. G., Zautashvili, M., Eprikashvili, L., Pirtskhalava, N., & Dzagania, M. (2011). *New Substrate of Prolonging Action on the Basis of Natural Zeolite and Brown Coal for Growing Agricultural Crops* Bull. *Georg. Nat. Acad. Sci.*, *5*, 2.

Arias Cruz, C.L. and Martín Alonso, G.M. (2023) *personal report.*

Bansiwal, A. K., Rayalu, S. S., Labhasetwar, N. K., Juwarkar, A. A., & Devotta, S. (2006). Surfactant-Modified Zeolite as a Slow Release Fertilizer for Phosphorus. *Journal of Agricultural and Food Chemistry*, *54*(13), 4773–4779. DOI: 10.1021/jf060034b PMID: 16787027

Bouzo, L., López, M., & Acosta, J. A. (1991) *Estudio de formas de aplicación de zeolita en caña de azúcar en suelos de textura ligera. Resultados de caña planta.* Memoirs 3rd Int. Conf. Occur. Prop. Util. Nat. Zeolites, Part II, 104-105

Brito, A., & Coutín, D. (2017). *Las zeolititas de Cuba: Results of 45 years dedicated to its study* Inst. Geol. Paleont., Geological Survey of Cuba Cairo Cairo, P., Machado de Armas, J., Torres Artiles, P., Diaz Martin, B., Jiménez Carrazana, R. and Rodríguez López, O. (2017) *Effects of zeolite and organic fertilizers on soil quality and yield of sugarcane.Australian Journal of Crop Science, 11*(06), 733–738. DOI: 10.21475/ajcs.17.11.06.p501

Chen, T., Xia, G., Wu, Q., Zheng, J., Jin, Y., Sun, D., Wang, S., & Chi, D. (2017). The Influence of Zeolite Amendment on Yield Performance, Quality Characteristics, and Nitrogen Use Efficiency of Paddy Rice. *Crop Science*, *57*(5), 1–11. DOI: 10.2135/cropsci2016.04.0228

Chinobampo, R. (2008). *Report and Validation of Results for Zeolitic products to INIFAP*, state of Morelos, Mexico

Conversa, G., Pacifico, S., La Rotonda, P., Lazzizera, C., Bonasia, A., & Elia, A. (2024). Foliar application of natural zeolites affects the growth and productivity of processing tomato. *European Journal of Agronomy*, *154*, 127100. DOI: 10.1016/j.eja.2024.127100

Crespo, R. (1998), Producción *cañera más eficiente y ecológica.* Revista Cañaveral, Vol. 4, No. 1, Jan-Mar, MINAZ, p. 27.

De Antoni Migliorati, M., Leo, S., Millar, G. J., Bell, M. J., Strazzabosco, A., & Grace, P. R. (2024). Can zeolites be used in sugarcane cropping systems to help reduce nitrogen losses into the Australian Great Barrier Reef? *Plant and Soil*, *504*(1-2), 575–592. DOI: 10.1007/s11104-024-06643-5

De la Nuez, E. Y., Rodríguez Iznaga, I., Rodríguez Fuentes, G., Petranovskii, V., Martínez García, A., Calvino Gámez, J. J., Goma Jiménez, D., Cauqui, M. A., Rivero González, L. A., & Collazo García, O. (2024). Natural Clinoptilolite Materials Enriched in Nitrogen and Phosphorous for Agricultural Purposes: A Comprehensive Study on Their Development. *Journal of Inorganic and Organometallic Polymers and Materials*, *34*(6), 2522–2542. Advance online publication. DOI: 10.1007/s10904-023-02991-4

Díaz Coronel, G. T., Sánchez Mora, F. D., Llerena Ramos, L. T., & Vásconez, G. H. (2009). Use of natural zeolites in the fertilization and production of beans (Phaseolus vulgaris L.) in the Quevedo area. *Science and Technology*, *3*, 1–6.

Espino, C., & Soca, M. (1991) Influencia de la Zeolita natural y modificada en tomate cultivado en invernadero. Memoirs 3rd Int. Conf. Occur. Prop. Util. Nat. Zeolites, Part II, 65-67.

Farias da Silva, J. L., de Mello Prado, R., Alves, T. L., Lata-Tenesaca, L. F., & Bomfim Soares, M. (2023). *New strategy for silicon supply through fertigation in sugarcane integrating the pre-*sprouted *seedling phase and field cultivation.Scientific Reports*, *13*(1), 1230. DOI: 10.1038/s41598-022-27323-3 PMID: 36681705

Febles (2006) Zeoponic, the antecedent of Organoponic. Report to Ministry of Agriculture of Cuba

Febles, J. A., Borsatto, F., Ferreira, J. P. & Ferreira, F. (2014) Celtonita: an environmentally friendly product from natural clinoptilolite for high technology agriculture. Zeolite 2014 Book of Abstract - 9th Int. Conf. Natural Zeolites, Belgrade, Serbia, 8 - 13 June 2014, 67-68

Filgueiras Dutra, A., Lima Leite, M. R., de Faria Melo, C. C., Silva Amaral, D., Farias da Silva, J. L., de Mello Prado, R., de Cássia Piccolo, M., de Souza Miranda, R., Barbosa da Silva Júnior, G., dos Santos Apollo Sousa, T. K., Mendes, L. W., Ferreira Araújo, A. S., Zuffo, A. M., & de Alcântara Neto, F. (2023). Soil and foliar Si fertilization alters elemental stoichiometry and increases yield of sugarcane cultivars. *Scientific Reports*, *13*(1), 16040. . DOI: 10.1038/s41598-023-43351-z PMID: 37749306

Hu, W., Li, J., Jiao, X., & Jiang, H. (2023). *Jiao*, X. and Jiang, H. (2023). *Zeolite as a Tool to Recycle Nitrogen and Phosphorus in Paddy Fields under Straw Returning Conditions.Agronomy (Basel)*, *13*(2), 327. DOI: 10.3390/agronomy13020327

International *Zeolite Corp (2022)* Greenhouse Gas Impact: Analysis & Estimation of NEREA on CO_2e reduction.

Jarosz, R., Szerement, J., Gondek, K., & Mierzwa-Hersztek, M. (2022). *The use of zeolites as an addition to fertilizers – A review.Catena*, *213*(June), 10612. DOI: 10.1016/j.catena.2022.106125

Kavoosi, M. (2007). *Effect of zeolite application on rice yield, nitrogen recovery, and nitrogen use efficiency.* Communications in soil science and Plat Analysis Vo. 38, Issue 1-2, 69-76

Laffita Suárez, M. and Martínez Valdés, M. E. (2022). *Results of the Evaluation of NEREA® zeolitic products in the common beans (Phaseolus vulgaris L.) var BAT 304.Campaign* 2021. Report Grain Research Institute, MINAG

Laffita Suárez, M. and Martínez Valdés, M.E. (2024). *Results of the Evaluation of NEREA® products in the common beans (Phaseolus vulgaris L.) var BAT 304. Campaign* 2023. Report Institute of Grains Researchs, MINAG

Li, Z. (2003). Use of surfactant-modified zeolite as fertilizer carriers to control nitrate release. *Microporous and Mesoporous Materials*, *61*(1), 181–188. DOI: 10.1016/S1387-1811(03)00366-4

López, M., Bouza, L., & Acosta, A. (1991) Estudio de la interacción de la zeolita con diferentes dosis de nitrógeno en el cultivo de la caña de azúcar. I: resultados de caña planta var C266-70. Memoirs 3rd Int. Conf. Occur. Prop. Util. Nat. Zeolites, Part II, 102-103

López Paz, C. (2019). Fertilizantes completos con zeolita natural. Una alternativa sostenible en el cultivo del tomate. Tesis de Diploma. Facultad de Ciencias Agropecuarias Universidad de Matanzas, Cuba.

Manikandan, A., & Subramanian, K. S. (2014). Evaluation of Zeolite Based Nitrogen Nano-fertilizers on Maize Growth, Yield and Quality on Inceptisols and Alfisols. *African Journal of Agricultural Research*, *9*(2), 276–284.

Martínez Grillo, J. F., Martínez Valdés, M. and Rodríguez Reyes, M. (2022). *Results of the Evaluation of NEREA® zeolitic products in rice cultivation, 2021 winter campaign. Rice var. Selection-1.* Report Grain Research Institute, MINAG.

Martínez Grillo, J. F., Martínez Valdés, M. and Rodríguez Reyes, M. (2024). *Results of the Evaluation of NEREA® zeolitic products in rice cultivation, 2023 spring campaign. Rice var. IACuba 31.* Report Grain Research Institute, MINAG.

Méndez Días, N. y Soca Núñez, M. (2016) *Evaluación agroindustrial del impacto de la zeolita en el cultivo de la caña de azúcar.* Revista electrónica de la Agencia de Medio Ambiente Año 16, No.31, jul-dic.

Meneses, P., Méndez Parra, W., Hernández Gómez, A. and Bouza Magdaleno, Y. (2022). *Report on the results of the study of Nerea Plus foliar fertilizer in common bean cultivation.* Territorial Grain Research Station "Sur del Jíbaro", Grain Research Institute, MINAG.

Minato, H. (1968). Characteristics and Uses of Natural Zeolites. *Koatsugasu*, *5*, 536–547.

Ming, D. W., & Allen, E. R. (2001) *Use of Natural Zeolites in Agronomy, Horticulture and Environmental Soil Remediation* Chapter 18. Nat. Zeol. Occur. Prop., Appl., Mineralogical Society of America. Geochemical Society ISSN 1529-6466

Ming, D. W., & Boettinger, J. L. (2001). Zeolites in Soil Environments, Chapter 11 of Natural Zeolites: Occurrence, Properties, Applications (Mineralogical Society of America, Geochemical Society ISSN 1529-6466

Ming, D.W. and Allen E.R. (1999). Zeoponic Substrates for Space Applications: Advances in the Use of Natural Zeolites for Plant Growth. Misailidis, P., Macásek, F., Pinnavaia, T.J., Colella, C. (eds) Natural Microporous Materials in Environmental Technology. NATO Science Series book series NSSE, vol 362, 157-176. Springer, Dordrecht. https://doi.org/doi:10.1007/978-94-011-4499-5_11

Molina-Amat, J. C., Rodríguez-Iznaga, I., Rodríguez-Fuentes, G., de la Nuez-Pantoja, E. Y., & Caisés Ávalos, A. (2024, September-December). Industrial production of NEREA® zeolitic products for the agricultural industry: Alternatives for the curing-drying process. *Tecnología Química*, *44*(3), 529–543.

Mumpton, F. A. (1984). *Zeo-Agriculture: Use of Natural Zeolites in Agriculture & Aquaculture*. Westview.

Mumpton, F. A. (1999). *The magic rock: Uses of natural zeolites in agriculture and industry* colloquium "Geology, Mineralogy and Human Wellbeing" organized by the National Academy of Sciences of the United States of America on November 8-9, 1998, *Proc. Nat.* Science Academy of USA, Vol. 96, Issue 7, 3463-3470. DOI: 10.1073/pnas.96.7.3463

Ozbahcea, A., Fuat Tarib, A., Gönülalc, E., Simseklic, N., & Pademd, H. (2015). The effect of zeolitic zeolite applications on yield components and nutrient uptake of common bean under water stress. *Archives of Agronomy and Soil Science*, *61*(5), 615–626. DOI: 10.1080/03650340.2014.946021

Ozbahcea, A., Tarib, A. F., Gönülalc, E., & Simseklic, N. (2018). *Zeolite for Enhancing Yield and* Quality *of Potatoes Cultivated Under Water Deficit Conditions.Potato Research*, *61*(3), 247–259. DOI: 10.1007/s11540-018-9372-5

Pérez, J. E., Rivero, L., & Arozarena, N. (1991). *Estudio del contenido de nutrientes en sustratos NEREA*. Memoirs 3rd Int. Conf. Occur. Prop. Util. Nat. Zeolites, Part II, 3-7.

Ponce Cruz, L., Blanco Concepción, N., & García González, A. (2011). *Cultivos sin suelo empleando Zeolita en* Jagüey *Grande Matanzas.1st Workshop Use Industrial Minerals in Agricultural Branch*, Villa Clara, Cuba

Rivero-Gonzalez, L. A., & Rivero-Robaina, E. L. (2016). *Zeoecological Fertilization*, Proceedings of the 11th Meeting on Organic and Sustainable Agriculture. ACTAF, 24 – 26 mayo 2016, Matanzas, Cuba.

Rivero-González, L. A., Rivero-Robaina, E. L., Rodríguez-Iznaga, I., & Rodríguez-Fuentes, G. (2023). *Empleo de los productos NEREA® plus como fertilizante edáfico y foliar en el cultivo de papa var Rudol.* Report Institute of Material Science and Technology (IMRE), University of Havana.

Rivero González, L. A., & Rodríguez-Fuentes, G. (1988). *Cuban experience with the use of natural zeolite substrates in soilless culture. Proceeding of Seventh International Congress on Soilless Culture*, The Netherlands, International Society for Soilless Culture. 405-416

Rivero-González, L. A., Rodríguez-Fuentes, G., Rivero-Robaina, E. L., & Rodríguez-Iznaga, I. (2018). *Report Validation of the effect of Cuban natural zeolite on sugarcane cultivation in Ingenio Tolulá S.A. Suchitepéquez.*

Rivero González, L.A., Rodríguez Fuentes, G. and Rodríguez Iznaga, I. (2018). *Report on the field test of NEREA® zeolitic fertilizers in potato crop*, April 2018. NatZEng-IMRE

Rivero-Robaina, E.L., Coll, Y., Castillo C. R., Hansen, J., Köhler, S., Muskolus, A., Rodríguez-Iznaga. I. and Rodríguez-Fuentes, G. (2024). *NEREA® Plus-U and BIOBRAS® 16 as foliar fertilizer / biostimulant for potato and maize grown in a German field soil: study of concept.* Universidad de La Habana (IMRE, CEPN), Fundación Universitaria de Innovación y Desarrollo, Humboldt-Universität zu Berlin, IASP, WIWEX GmbH.

Rodríguez-Fuentes, G. (2015). Natural Zeolite Engineering: Development of New Materials and Utilization Technologies. [Associazione Italiana Zeoliti]. *Extended Abstract in Bollettino AIZ*, (45), 13–31.

Rodríguez-Fuentes, G., Rivero González, L. A., Rodríguez-Iznaga, I., & Rivero Robaina, E. L. (2023). *Procedimiento para la obtención de sustratos y fertilizantes zeolíticos de liberación controlada y métodos de tratamiento de las plantas.* Patent CU20210017 A7. Patent application in Canada H8327186CA

Rodríguez-Fuentes, G., & Rodríguez-Iznaga, I. (2009). Caracterización del mineral zeolítico para el desarrollo de materiales nanoestructurados. Rev. Cub. Física vol. 26, No. 1ª, 55-60

Rodríguez-Fuentes, G., Rodríguez-Iznaga, I., Lariot-Sánchez, C. A., Concepción Rosabal, B., Molina-Amat, J. C., Rivero-González, L. A., Rivero-Robaina, E. L., De la Nuez-Pantoja, E. Y., & Caisés Ávalos, A. (2025). *Cuba Sedimentary Zeolite Deposits. Second report. Natural Zeolite Engineering Innovation: Industrial production of NEREA® materials for Agriculture*.https://www.iza-online.org/natural/Insights/Cuba_Insights

Saborit Reyes, R., Meneses Dartayet, P., & Cañizares Sierra, A. (2021). *Effect of the foliating Nerea* Plus - *U, on the rice's cultivation performances.Rev. Infociencia, 25*(3), 36–48.

Savant, N. K., Korndorfer, G. H., Datnoff, L. E., & Snyder, G. H. (1999). Silicon nutrition and sugarcane production: A review. *Journal of Plant Nutrition, 22*(12), 1853–1903. DOI: 10.1080/01904169909365761

Schulze, D., George, C., Cathline, K., & Germani Torrens, G. (2021). *Evaluation of the performance of zeolite-based NEREA® as a media amendment for greenhouse grown basil in Canada.* Agriculture & Environmental Technologies Innovation Centre, Niagara College.

Sharma, V., Javed, B., Byrne, H., Curtin, J., & Tian, F. (2022). *Zeolites as Carriers of Nano-Fertilizers: From Structures and Principles to Prospects and Challenges* Appl. *Nano, 2022*(3), 163–187. DOI: 10.3390/applnano3030013

Soca Nuñez, M., & Constanza Daza, M. (2018). Zeolite and its effect on nitrogen efficiency in rice and corn Use of the zeolite in cultivations of economic importance. *INFOMIN, 10*(1), 46–56.

Soca Nuñez, M., & Lorente, E. (2015). Efecto de las zeolitas naturales y activadas en la producción de tomate. *Agrotecnia de Cuba Volumen, 39*(1), 67–78.

Sun, Y., Xie, J., Hou, H., Li, M., Wang, Y., & Wang, X. (2023). Effects of Zeolite on physiological characteristics and grain quality in rice under alternate wetting and drying irrigation. *Water (Basel), 15*(13), 2406. DOI: 10.3390/w15132406

Suprapron, J., Pitayakon, L., Kamalapa, W., & Touchamon, P. (2002). *Effect of zeolite and chemical fertilizer on the change of physical and chemical properties of Lat Ya soil series for sugar cane.* 17th WCSS, 14-21 august, Thailand. 1897.

Torma, S., Vilcek, J., Adamisin, P., & Hronec, O. (2014). Influence of natural zeolite on the dynamics of nitrogen in the soil. *Turkish Journal of Agriculture and Forestry.* Advance online publication. DOI: 10.3906/tar-1311-13

Vilcek, J., Torma, S., Adamisin, P., & Hronec, O. (2013). Nitrogen sorption and its release in the soil after zeolite application. *Bulgarian Journal of Agricultural Science, 19*, 228–234.

Villarreal-Nuñez, L. A., Barahona-Amores, O. A., & Castillo-Ortiz, O. A. (2015). Effect of zeolite on the nitrogen fertilizer efficiency in rice crop. *Agronomía Mesoamericana, 26*(2), 315–321. DOI: 10.15517/am.v26i2.19324

Wu, Q. (2016). *Xia*, G., Chen, T, Wang, X., Chi, D. and Sun, D. (2016). *Nitrogen use and rice yield formation response to zeolite and nitrogen coupling effects: Enhancement in nitrogen use efficiency.Journal of Soil Science and Plant Nutrition, 16*(4), 999–1009.

Yin, Q., & McAughey, S. (2022). *IZC_NEREA Validation. Greenhouse basil, lettuce, gerbera production and chrysanthemum rooting*. Vineland Research & Innovation Centre.

Chapter 9
Post-Synthesis Surfactant-Induced Mesoporosity in Zeolite Crystals:
State-of-the-Art, Open Questions and Perspectives

Joaquín Pérez-Pariente
https://orcid.org/0000-0001-5322-9528
Instituto de Catálisis y Petroleoquímica, Spain

Manuel Sánchez-Sánchez
Instituto de Catálisis y Petroleoquímica, Spain

Julia Aguilar Pliego
UAM-Azcapotzalco, Mexico

ABSTRACT

In this chapter, the state-of-the-art on the generation of intracrystalline mesopores in zeolite crystals by post-synthesis treatments with cationic surfactants in alkaline medium is critically reviewed. So far, five 12R large pore FAU, BEA, MAZ, LTL and MOR, and one medium pore 10R MFI zeolites have been subjected to this treatment. The main aspects of the process and the chemical and structural factors of the parent zeolites that trigger the mesostructuring process have been thoroughly examined. Particular attention has also been paid to discuss the influence of the treatment on the total acidity and acid strength of the resulting materials. Finally, some general considerations have been elaborated to serve as guidelines for future work, with the

DOI: 10.4018/979-8-3373-0320-8.ch009

aim of helping to develop the most suitable materials for specific catalytic processes.

1. INTRODUCTION

1.1. Background

Zeolites are crystalline porous materials formed by tetrahedral TO_4 units in which each oxygen atom of the vertices is shared by a neighbour tetrahedral unit, in such a way that the three-dimensional arrangement of interconnected tetrahedra forming the framework leaves cavities of various sizes and shapes inside the crystals. The tetrahedral central T atom is usually Si or Al in classic zeolites, but it could also be P, Ge, Ga, B, Sn or Ti, in various composition. For T = Al (or other M^{3+} ion), the overall negative charge of the TO_4 unit must be counterbalanced by a positive charge. If this charge is a proton, the zeolite has strong Brønsted acidity. Most of the industrial uses of zeolites relies upon their acidity, which enables them to catalyse a number of important reactions of carbon-based molecules, such as cracking, isomerization, hydrogen transfer or polymerization. Together with their acidity, the other property that endows zeolite their unique catalytic performance, is the existence of molecular-size cavities and channels where reactions can take place. These cavities are accessible to incoming molecules through rings or windows whose effective free diameter increases with the number of tetrahedra forming the ring, from ~0.35 nm for 8-membered (8R) rings, until ~1.5 nm for materials containing extra-large pores (> 12R).

The accessibility of the reacting molecules to the interior of the crystals where most of the active site are located (few are present on the external surface), and the size of the product molecules that are able to diffuse out, are therefore limited by the pore openings characteristic of each zeolite structure. Hence, what constitutes the identity mark of zeolitic catalysts imposes at the same time size limitations regarding the molecules that can be processed or formed.

Several strategies have been used to overcome the diffusional constraints intrinsic to zeolites. For a given structure, they consist in the generation of intracrystalline porosity (mesoporosity in most cases), or in the reduction of crystal size. The latter approach has two main drawbacks. First, for molecules whose size is well beyond that of the pore openings, the bulk of the crystals remains unreachable, and only the active sites located at the very external surface would be able to catalyze the target reaction. Therefore, most of the crystal remains unused, and the confinement effect so important in zeolite catalysis is largely lost. Second, the thermal and hydrothermal stability of zeolites, as it happens to many other solids, decreases with crystal size, and this makes nanosized zeolites in general too sensitive to the severe thermal treatments required to restore their activity once become poisoned by side-products during reaction.

This leaves the generation of intracrystalline (meso)porosity as the most suitable mean to alleviate diffusional constraints in real zeolite catalysts for increasing their efficiency. For this purpose, in broad sense two pathways have been followed: addition of mesopore-forming (porogen) agents or particular zeolite precursors to the zeolite synthesis mixture, or post-synthesis treatments of pre-synthesized zeolites. There is a large *corpus* of scientific literature on both approaches (Chen et al., 2020; Ivanova & Knyazeva, 2013; Kerstens et al., 2020; Li et al., 2014; Mallette et al., 2022; Pérez-Ramírez et al., 2008; Schwieger et al., 2016; Serrano et al., 2013; Verboekend & Perez-Ramirez, 2011), but we will focus on the latter in this chapter.

Post-synthesis approaches basically consist of chemical or thermal treatments (often both are used in sequential manner) that remove part of the framework tetrahedral T atoms to leave cavities of varied and irregular morphologies and sizes, usually in the meso and macropore range (> 2 nm) inside the eroded crystals. Depending on the treatment, part of the T atoms once forming part of the framework would eventually remain as extraframework species housed at the micropore cavities, which can be removed by additional chemical treatments if needed. An early and relevant example of catalyst engineering through this approach is the dealumination of the large-pore zeolite faujasite. For this purpose, the zeolite is submitted to cycles of ion exchange/steaming to increase the Si/Al ratio in order to improve its stability during the severe conditions ruling during regeneration of exhausted cracking catalysts. In this process, the removal of part of the framework Al is accompanied by the formation of additional porosity that permeate the crystals, which plays a positive role in the overall catalyst performance.

Among the chemical treatments, the selective removal of part of the framework silicon atoms (known as desilication) by contacting the zeolite with basic solutions (generally sodium hydroxide) has been widely used to generate intracrystalline mesoporosity, as it was first shown for ZSM-5 by Ogura et al. (Ogura et al., 2000). However, the generation of mesopores also produces unwanted side effects, namely a notable decrease of microporosity while a large fraction of the initial zeolite dissolved in the caustic solution, and a reduction of crystal size. However, a better control of the hydrolysis of the Si-O-Si bonds can be reached if tetraalkylammonium cations (such as tetrapropylammonium, TPA^+), so-called "pore-growth moderators", are added to the NaOH solution (Pérez-Ramírez et al., 2009).

Following the approaches involving chemical attack of the zeolite crystals, it is desirable to gain control on the size and morphology of the newly formed mesopores, while avoiding at the same time the undesired effects of desilication, which would allow a better tuning of the catalyst's performance in the processing of the target bulky feedstock. A breakthrough contribution to tackle this objective has been reported by Ying and García-Martínez (Ying & García-Martínez , 2005), which originally consists of the hydrothermal treatment of a dealuminated faujasite with a solution of a cationic surfactant (generally hexadecyltrimethylammonium, CTA) in alkaline medium (generally aqueous ammonia). Under these conditions, the surfactant is able to form micelles, whose size is typically within the mesopore size, (~ 4 nm), inside the crystals. Upon calcination of the micelle-containing material, the surfactant is removed leaving behind mesopores very uniform in size that permeates the entire crystals. This process is named "mesostructuring" or "surfactant-templating" (Mendoza-Castro et al., 2021), and has been first developed for zeolite FAU. This approach has been extended by other research groups to other 12R zeolites, such as zeolite L and mordenite (Al-Ani et al., 2019), and more recently by us to omega (Vidal de la Fuente et al., 2024) and high-silica zeolite beta (Suárez et al., 2019). However, attempts on the 10R zeolite ZSM-5 have apparently not been so successful (Al-Ani et al., 2019).

In nearly coincidence with these studies, Ivanova et al. (Ivanova et al., 2004; Ivanova & Knyazeva, 2013) reported a very similar CTA-based treatment initially focused mainly on mordenite, which also leads to the formation of intracrystalline mesopores together with some ordered mesophase (MCM-41), whose contribution is a function of the specific treatment conditions. This process is named "zeolite recrystallization."

The main difference between zeolite "mesostructuring" and "recrystallization" is due to the fact that in the second method the zeolite is first treated for a short time with a basic solution, generally a NaOH solution, before the surfactant is added and the mixture hydrothermally heated at 100 ºC. The idea behind this approach is to produce a sufficiently intense chemical attack of the zeolite structure (breaking of Si-O-T (T = Si or Al) bonds) to facilitate the generation of intracrystalline mesoporosity upon contacting the degraded zeolite with the surfactant. This first step of chemical attack is skipped in the "surfactant-templating" method, where the zeolite is put directly in contact with the basic solution which already contains the surfactant. In this case, the cleavage of the Si-O-Si bonds occurs under the protective shield of the CTA^+ cations which in addition would be available on-site to generate mesopores. This is not a minor difference, as it will be discussed later. Nevertheless, this chapter will concentrate on a critical review of the first method, the so-called zeolite mesostructuring, for two reasons because: (i) it is a direct method, and (ii) more importantly, it in principle produces materials that fulfil the 'dream' of achieving zeolites with controlled intracrystalline mesoporosity.

1.2. Purpose

Based on these grounds, the purpose of this chapter is to critically review the already existing knowledge on the generation of mesoporosity in zeolite crystals by post-synthesis treatments with surfactants following the mesostructuring general strategy above underlined. Attention will be paid along the chapter to several factors that, in our view, most affect the effectiveness of the treatment in creating mesopores and the catalytic performance of the resulting materials. The final aim of this analysis is to establish, as far as this is possible based on the data so far available, a general pattern on both the synthetic aspects and the acidity/catalytic activity that could eventually serve as a roadmap to the preparation and use of current and other hierarchical materials.

This chapter deals separately with large-pore (12R) and medium-pore (10R) zeolites, since both the experimental protocols and the effects in terms of generated intercrystalline mesoporosity are quite different. Special emphasis will be given to zeolite faujasite due to its industrial interest and because the methodology has been mainly developed with this zeolite. Particular aspects to be critically analysed and discussed are the following:

(a) Concerning the effective generation of mesopores, the conditions that have to be fulfilled by the starting zeolite materials have yet to be clearly determined. In particular, the necessity to create defects in the structure that facilitate the access of surfactant molecules to the crystals; the Si/Al of the starting material, and the nature of the structure. The method has so far been effective under certain conditions for large-pore 12R zeolites, and we will analyse possible reasons for this experimental finding, in particular the accessibility of the surfactant to the interior of the crystals as a function of the pore openings, as well as the impact of the framework density and the related intrinsic zeolite stability.
(b) In the cases where mesopores are successfully formed, the question arises on their degree of pore interconnectivity, which is most relevant for catalysis. For instance, spherical micelle-derived cavities spotted all across the crystals do not necessarily produced alleviation of diffusional constraints unless they are interconnected.
(c) Reports on the acidity of hierarchical materials will be analysed, as this is a critical issue for their catalytic application. In our view, this topic has not yet been in-deep discussed.
(d) Thermal and hydrothermal stability of hierarchical materials is also a matter of concern for catalytic applications, and it should be systematically evaluated. This aspect will be also discussed.

Based on these considerations, a section on perspectives for future work will close the chapter.

2. MESOSTRUCTURED 12R ZEOLITES

2.1. Ultrastable Faujasite (USY)

According to the procedure described by García-Martínez and coworkers (García-Martínez et al., 2012), a commercial ultrastable faujasite USY material (Zeolyst CBV 720, average Si/Al = 15, $Si/Al_{Framework}$ = 35.3, calculated from the Fichtner-Schmittler equation (Fichtner-Schmittler et al., 1990) that relates unit cell with Al framework in faujasite), was stirred for 20 minutes with a solution of cetyltrimethylammonium bromide (CTAB) in NH_3 (aq) at mild pH (9-11) to avoid desilication, and the mixture heated statically at 150 °C under autogenous pressure for a given time, usually 10 h, which can be extended to 48 h. In a typical synthesis, 1.00 g of zeolite, 0.7 g of CTAB and 64 mL of a 0.37 M solution of NH_3 (aq) were used. After the hydrothermal treatment, the solid was filtered, washed and dried. To remove the occluded surfactant, the dried solid was first heated in a nitrogen flow at 550 °C for 2 h and then in air for 2 h at the same temperature.

It has been repeatedly claimed by García-Martínez's group that the recovery of solid (after calcination) was close to 100%, and no significant leaching of Si and Al was found in the filtrate after the hydrothermal treatment. Furthermore, the Si/Al ratio of the mesoporous zeolite was 16, practically the same as that of CBV 720 (Sachse et al., 2017), and a similar value is also reported by Chal et al. (Chal et al., 2010).

The XRD pattern of the solids products reveals that the intensity of the peaks corresponding to the faujasite crystals is lower than that of the pristine, initial CBV 720 sample, while low-angle (100), (110) and (200) peaks characteristics of a hexagonal ordering of mesopores usually corresponding to MCM-41 type materials are clearly detected (García-Martínez et al., 2012) (Figure 1). Moreover, the nitrogen-adsorption/desorption isotherm shows a stepped, well-defined increase of volume adsorbed at the p/p_0 characteristic of ordered mesoporous materials synthesized from CTA surfactant, with a narrow pore-size distribution centered at ~3.8 nm (Figure 2). Indeed, there is a linear inverse correlation between micropore and mesopore volume for samples prepared at different degrees of severity.

Figure 1. X-ray diffraction pattern of mesostructured CBV 720 showing the (100), (110) and (200) peaks (asterisked) due to the hexagonal ordering (a_0=44.7 Å, determined by $a_0=2d_{100}/\sqrt{3}$) of the mesopores (2θ < 5°) and peaks that are characteristic of zeolite Y (2θ > 5°).

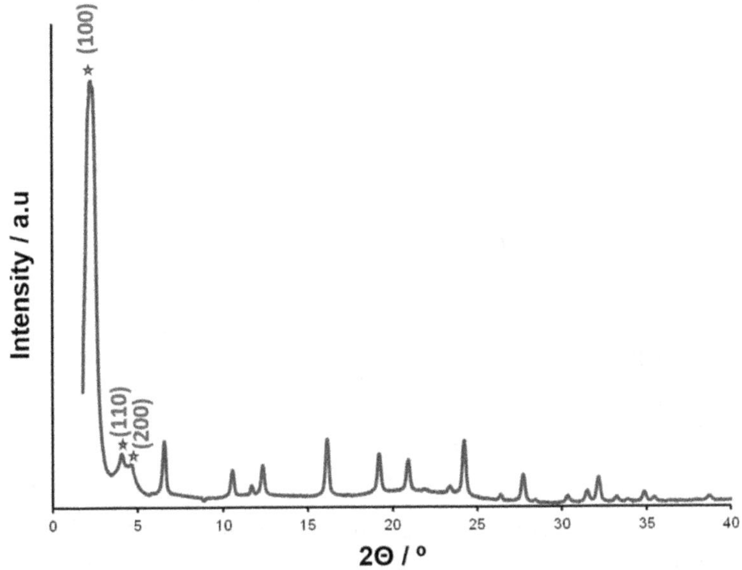

Adapted from (García-Martínez et al., 2012) with permission from the Royal Society of Chemistry, Copyright 2012.

Figure 2. N_2 adsorption–desorption isotherms at -196 °C for CBV 720 (black circle) and the mesostructured CBV 720 (blue square). The corresponding BJH pore size distribution of mesostructured CBV 720 is shown in the inset.

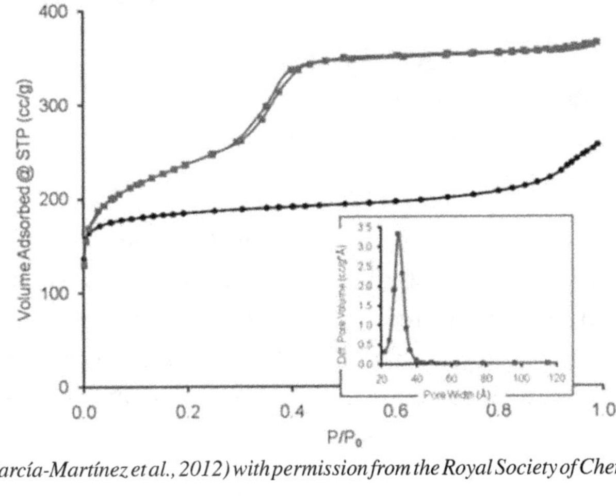

Adapted from (García-Martínez et al., 2012) with permission from the Royal Society of Chemistry, Copyright 2012.

All these data would have suggested at first glance that the method leads to the formation of MCM-41 at the expenses of silicon and aluminium species that would have dissolved from the zeolite during the hydrothermal alkali treatment in the presence of CTA. However, a thorough study carried out along the years by García-Martínez and co-workers mainly using advanced microscopy techniques have indeed succeeded in showing that intracrystalline mesopores permeate the entire zeolite crystals (Figure 3).

Figure 3. On the right, FE-SEM images of the untreated NaY zeolite (top) and the mesostructured Y-zeolite crystals (bottom). Adapted from (Li et al., 2014). On the right: TEM micrographs of a zeolite exposed to (top) desilication and (middle) surfactant-templating.

Figure adapted from (Sachse et al., 2017) with permission from the American Chemical Society, Copyright 2017.

Moreover, the mesopores are interconnected forming a 3D network that also connected them to the zeolite crystalline micropores (Mendoza-Castro et al., 2021). An early confirmation of these findings was published by Chal et al. (Chal et al., 2010), by replacing ammonia by a TMAOH solution. A pore-to-pore correlation distance of 45.8 Å was determined from XRD, while a narrow mesopore size distribution centered at 4.3 nm was found.

All these studies clearly show that the low-angle XRD reflections correspond to an ordered array of intracrystalline mesopores, and not to a MCM-41-like mesophase that might have eventually been mixed with the zeolite, and such mesophase has never been reported to be present in these preparations. Indeed, it has been claimed that the low-angle pattern is consistent with small domains, in the range 100-150 nm, of hexagonally ordered mesopores, but also with certain disorder in the pore-to-pore distance (Mendoza-Castro et al., 2021). Moreover, it has been shown that, for the mesostructuring to occur, the polar head of the surfactant must be able to penetrate along the micropore system of the zeolite attracted by the negatively charged defect SiO^- sites resulting from the base attack of the framework (García-Martínez et al., 2012) (Figure 4). However, it should be considered that these sites would be neutralized by NH_4^+, in such a way that an ammonium/CTA^+ cation exchange would better describe the process at molecular scale. For this reason, the method does not work for surfactants with excessively big polar heads (Sachse & García-Martínez, 2017).

Figure 4. Schematic of the speculated zeolite mesopore formation process: (a) original zeolite Y, (b) Si–O–Si bond opening/reconstruction in basic media, (c) crystal rearrangement to accommodate the surfactant micelles, and (d) removal of the template to expose the mesoporosity introduced.

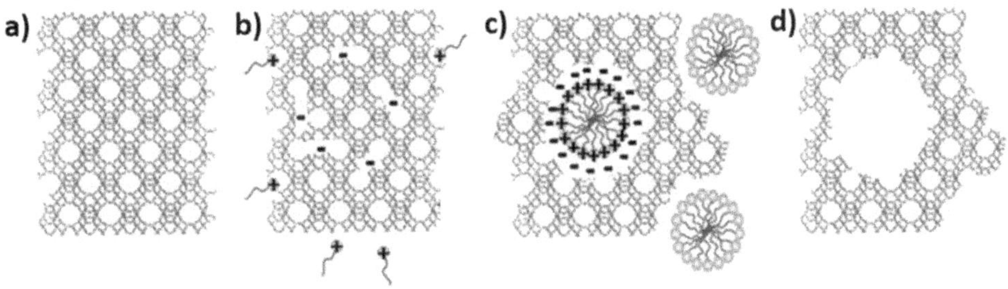

Adapted from (García-Martínez et al., 2012) with permission from the Royal Society of Chemistry, Copyright 2012.

While the basic features of the generation and nature of intracrystalline 3D mesoporous system in these zeolite materials is well-established, some aspects would still need further clarification. From the pore-to-pore distances and the average pore size of the mesopores, it can be deduced that the pore wall thickness of the ordered hexagonal arrangement of pores should be about 0.3 nm (Chal et al., 2010) or 0.8 nm (Linares et al., 2016) (Figure 5). These values are not compatible with the unit cell size of CBV 720, about 2.43 nm. In other words, the atoms located in the walls will not contribute to the overall intensity of the XRD spectra of the faujasite. A decrease of the XRD pattern intensity of the faujasite has been experimentally observed in the mesostructured materials, but if these domains of well-ordered mesopores would encompass the whole crystals, no faujasite peaks would whatsoever be detected, which is not the case. Clearly, there must be a heterogeneous distribution of pores across the crystals, and disordered pores with long average pore-to-pore distances, several nanometres apart each other, should co-exist together with well-ordered mesopore domains, but only the latter would contribute to the experimental low-angle XRD pattern.

Figure 5. Structural simulated model of the surfactant-templated USY zeolite from the experimental and simulated XRD patterns.

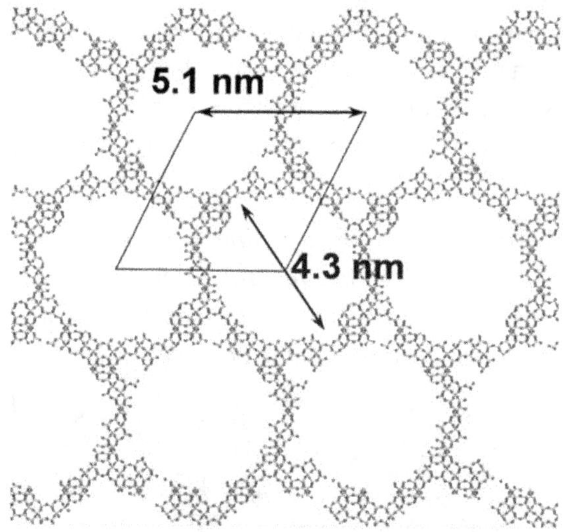

Adapted from (Linares et al., 2016) with permission from the American Chemical Society.

A second aspect concerns the claimed nearly 100% yield of solid recovered after the process. Sachse et al. (Sachse et al., 2017, p. 4300) attempted to give an answer to that question: "One would wonder how porosity is created in the zeolites without any significant material leaching. One possibility would be that the crystal expands to accommodate the additional porosity introduced by the surfactant." By analyzing the skeletal density determined by He pycnometry, they conclude that the zeolite "expands", and that

> *this crystal expansion involves short-scale breaking and reconstruction of the zeolite framework to accommodate the mesoporosity in its interior. This process does not involve any significant silica leaching, which would produce material loss, i.e., desilication (if no surfactant is present) or the formation of amorphous mesoporous phases, i.e., dissolution and reassembly (if surfactant is present).*

As mentioned before, recovery yields were close to 100% in all cases, and only intracrystalline mesoporosity was observed.

But the so-called eventual "reconstruction" of the zeolite actually means that the zeolite grows by using the T atoms removed from the crystal regions where the mesopores are formed. This will produce an increase of the crystal size, which has never been reported to occur, not to mention that the conditions prevalent during surfactant treatment (150 °C, absence of sodium cations) are not really favourable for faujasite crystal growth. Indeed, the eventual size increase of the zeolite crystals during mesoporization should be much more clearly detected when starting from FAU samples which do not contain already mesopores, which will be discussed in the next section. However, neither in this case an eventual increase of the crystal size has ever been reported. On the other hand, the conclusion on the eventual zeolite expansion has been derived from He pycnometry. However, the data collected in (Sachse et al., 2017, Table 1, p. 4298) shows that the skeletal density actually decreases from 2.34 g/cm^3 for CBV 720 to 2.27 g/cm^3 for the more severely treated sample, which cannot lead to a constant skeletal volume.

2.1.1. Acidity

The evolution of the total acidity and acid strength determined by pyridine adsorption/desorption has been reported by (Sachse et al., 2017) for mesoporous faujasite derived from CBV 720 as a function of the duration of the hydrothermal treatment, from 3 h to 48 h, in the presence of CTA. The total acidity decreases as a function of treatment time, together with the microporosity and crystallinity. Moreover, the density of strong acid sites follows the same decreasing trend, while the (weak + very weak) sites increase, and the Brønsted/Lewis sites ratio decreases. These results lead to interesting conclusions. First, that the strong acid sites are associated to the zeolite micropores that remain in the sample after CTA treatment. Second, that weak and very weak acid sites are associated to the aluminium present in the mesopore domains. And, third, the observed decrease of the total acidity (while the Si/Al remains the same as that of the starting sample) and the B/L ratio are indicative of low thermal stability of the acid sites located at the mesopores, and part of the associated aluminium atoms is dislodged forming extraframework species (EFAL) during the calcination step required to remove the surfactant.

These conclusions have been confirmed by Chal et al. in his reproduction of the procedure (Chal et al., 2010). They have reported that the total acidity determined by NH_3-TPD of the parent CBV 720 is quite similar to the CTA-treated sample. However, the concentration of strong acids sites decreases and that of weak acid sites increased for the mesoporous sample.

Nevertheless, in a very recent article, García-Martínez and co-workers maintain that the surfactant generation of intracrystalline mesoporosity in faujasite CBV 720 does not significantly change the whole acidity and the strong/weak acid sites distribution (Singh et al., 2024, Figure 3b), although the NH_3-TPD profiles are clearly different (Singh et al., 2024, Figure 3a), as well as the amount of strong acid sites, which suggest that the mesostructured zeolite Y contains less and weaker acidity than the pristine zeolite.

The weak acid strength of the acid sites associated to the intracrystalline mesopores would have important consequences for catalysis, because the reported gain in conversion for bulky molecules would be much higher if the acid strength characteristic of the zeolite structure would have been preserved during the treatment. Moreover, it also suggests that the catalytic benefits of the mesostructured faujasite materials in the processing of bulky reactants would be mainly contributed by the enhanced accessibility to the zeolitic acid sites. On the other hand, the method would have much less effect for reactions demanding strong acid sites.

2.2. Low Si/Al faujasite

The method for faujasite mesostructuring above outlined only works for high-silica materials, such as CBV 720. For Al-rich faujasite (Si/Al = 2.5 – 3), it has been stated that the samples must be previously dealuminated with an acid solution to produce the Si-O^- defects required to interact with the CTA^+ cations to initiate the mesostructuring process (Dight et al., 2010; García-Martínez et al., 2012; García-Martínez & Johnson, 2008; García-Martínez & Jonhson, 2010). As an example, the material Zeolyst CBV 300 (NH_4^+-form, Si/ Al = 2.5, a_0 = 24.70 Å), is pretreated with citric acid for 1 h before it is contacted with the basic solution containing CTA. The acid treatment not only breaks some Al-O bonds, but also removes a substantial amount of framework Al, as the unit-cell value a_0 decreases to 24.61 Å (Si/Al ~3.5). Moreover, the mesostructured material does not show XRD lines at low angle, which indicates that the mesopores are disordered (García-Martínez et al., 2012, p. 992). It has also been reported by the same team that no pre-existing mesoporosity in the parent zeolite is required for mesostructuring to occur. This has been shown by using NaY (Zeolyst CBV 100, Si/Al = 2.6). However, the sample must be previously dealuminated by acid citric treatment to Si/Al = 3.5 (Sachse et al., 2017).

Al-Ani et al. have also reported the generation of intracrystalline mesoporosity from CBV 100, but in this case the sample was dispersed in water and the pH adjusted to 5.5-5.7 by adding few drops of nitric acid before the citric acid treatment (Al-Ani et al., 2019). In addition, the ammonia solution was replaced by a NaOH solution to dissolve the CTA. It was found that the crystallinity and microporosity decreases by increasing the citric acid/zeolite ratio, i.e., by increasing the dealumination, while the mesoporosity follows the opposite trend. In this case, a peak at 2-theta 2.5° is observed, indicating the presence of ordered domains of mesopores. The total acidity determined by pyridine adsorption and desorption at 150 °C and the B/L ratio decrease, which agrees with the occurrence of dealumination.

The Al-Ani procedure (Al-Ani et al., 2019) has been more recently followed by Siakpebru et al. also starting from CBV 100, with similar results, although no information is given on the eventual presence of low-angle peaks in the X-ray diffraction pattern (Siakpebru et al., 2025). The total acidity determined by NH_3-TPD of the mesostructured samples is nearly halved by the CTA treatment and further calcination. Moreover, NH_3-TPD shows that the peak corresponding to weak acid sites shifts to lower temperature, and less acid sites of medium strength as compared with the parent HY (which was obtained by calcination of the ammonium-form CBV 300, Si/Al = 2.5) are present. FTIR studies on pyridine adsorption/desorption at 150 °C shows higher density of both Brønsted and Lewis sites than in the parent material, with an increase of the L/B ratio.

2.3. Zeolite L

Al-Ani and coworkers have reported the treatment of zeolite L, a large-pore material having unidirectional 12R channels, with CTA for the generation of hierarchical or mesoporous zeolite (Al-Ani et al., 2019). For this purpose, the NaK-LTL zeolite (Tosoh, HSZ-500KOA, Si/Al = 3.2) was first partially dealuminated with a solution of sulphuric acid (0.2 mol L^{-1} or 0.4 mol L^{-1}), and then treated with a CTA solution in sodium hydroxide at 80–100 °C overnight. The resulting solid was ion-exchanged with ammonium nitrate and then calcined (first in N_2 and then in air) to obtain the H-LTL form. The two CTA-samples resulting from the treatment have Si/Al ratio of 4.4 and 5.8, determined by EDX, evidencing the dealumination of the parent zeolite. Unfortunately, the Si/Al ratio of the acid-treated samples before they were contacted with the CTA solution is not given in the paper.

Two intense peaks are detected at low-angle in the XRD pattern of the CTA-material treated with the less concentrated sulphuric acid (Figure 6 left), corresponding to *d*-spacing of 4.39 nm and 2.19 nm. These values led to a pore-to-pore distance of 5.1 nm, assuming a hexagonal arrangement of pores. Interestingly, this is the same distance reported for mesostructured faujasite (Mendoza-Castro et al., 2021). Moreover, the crystallinity of the sample decreases as the acid treatment gets more severe, dropping to 59% for the most concentrated acid solution (Figure 6 middle).

Figure 6. On the left, low-angle XRD pattern of the zeolite L after being hydrothermally treated at 80 °C with the surfactant CTA in NaOH solution (0.1 g NaOH/3 g zeolite L) (sample ML-1). The starred peaks indicate long-range ordering in the mesostructured samples. On the middle, wide-angle XRD pattern of the parent zeolite L, sample ML-1 and sample ML-2 (hydrothermally treated at 100 °C in NaOH solutions with 0.2 g NaOH/3g zeolite L). On the right, N_2 adsorption/desorption isotherms at -196 °C of the parent zeolite L and the sample ML-1.

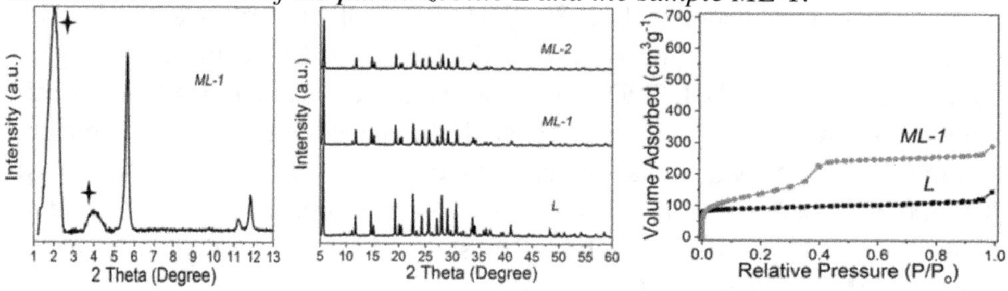

Adapted from (Al-Ani et al., 2019) with permission from the Royal Society of Chemistry, Copyright 2019.

The N_2 adsorption-desorption isotherm shows a marked increase of the amount of adsorbed nitrogen at p/p_0 ~0.35 (Figure 6 right), characteristic of mesoporous materials with a narrow pore size distribution, although, unfortunately, no information on the average pore size is given. However, the presence of a hexagonal arrangement of pores can be seen in the HRTEM images of the treated sample, with pore size in the range 3-5 nm (Figure 7).

Figure 7. Bright field TEM and HRTEM images of modified zeolite ML-1 viewing along two orthogonal directions (a) and (b). Higher magnification TEM image of a single rectangle L particle (c) and enlargement image of the edge (marked with a white rectangle) are given as inserts. The aligned pores (bright contrast) on the surface of the L crystal are observed from both directions.

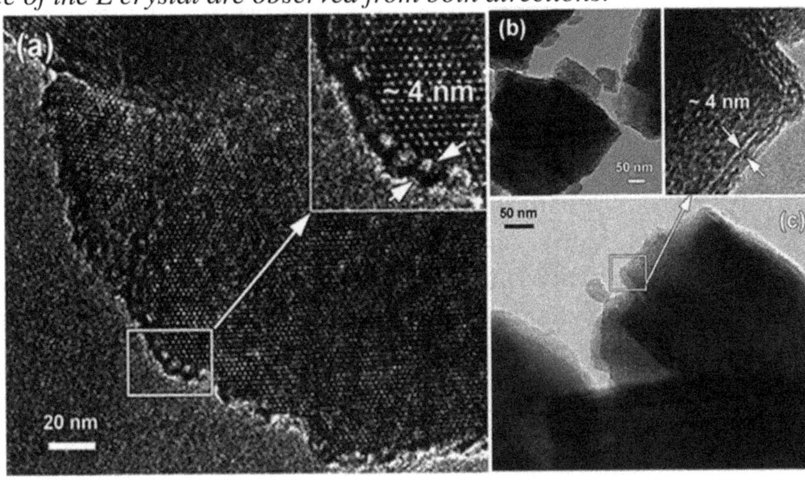

Adapted from (Al-Ani et al., 2019) with permission from the Royal Society of Chemistry, Copyright 2019.

It has been claimed the formation of a "core-shell type structure consisting of a microporous core and a mesoporous shell" (Al-Ani et al., 2019). These are clearly visible in the TEM image of Figure 7, which agrees with the low-angle diffraction peaks observed in the XRD pattern. However, it can also be seen in addition the presence of small particles on the external surface of the crystals that seem to be constituted only by a mesophase. The mechanism of the formation of the core-shell mesostructure has been further investigated by using time-resolved in situ synchrotron SAXS and WAXS (Zholobenko et al., 2025). Indeed, from the pore-to-pore distance and the pore size determined by HRTEM, the wall thickness of the mesophase should be ~1 nm at most, which is smaller than the zeolite unit cell. This fact rises again reasonable doubts on the actual presence of zeolitic T atoms in the mesopores walls.

2.3.1. Acidity

The total Brønsted and Lewis acidity of the H-LTL and one of the mesoporous samples (the one dealuminated with the less concentrated sulphuric acid solution) were determined by FTIR of pyridine adsorption and further desorption at 150 ºC, but, unfortunately, the acid strength distribution was not measured. As expected, the Brønsted acidity decreases, in agreement with the lower Al content of the sample, but there is a large increase of the Lewis acidity (the L/B ratio increases from 0.32

to 0.82), which suggests that the calcination of the CTA-sample leads to its partial dealumination, even if it has higher Si/Al ratio than the H-form, which should have made it more resistant to thermal dealumination. As it has been noticed in this review, this increase of the L/B ratio has been also systematically found in other mesostructured zeolites, including FAU. This evidences that the thermal stability of the Al sites in the mesopores is lower than those located at the zeolite micropores, and hence suggests that the chemical/structural environment of these Al atoms must be substantially different from the one they have in the zeolite framework. However, no signal attributed to octahedral extraframework Al species is found in the ^{27}Al MAS NMR spectra, and the broadening of the NMR signal of tetrahedral Al is interpreted by Al-Ani et al. as indicative of an increasing disorder of the environment of the framework Al atoms (Al-Ani et al., 2019).

The increase of the activity of the CTA-sample in the dealkylation of 1,3,5-tri-isoproylbenzene (13% conversion *vs* 1% for the H-LTL-form) is attributed to the enhancing of the accessibility of this bulky reagent to the acid sites.

2.4. Zeolite Omega

The mesostructuring of the zeolite omega (framework type code MAZ), the synthetic counterpart of the natural zeolite mazzite, has been recently explored by Vidal de la Fuente et al. (Vidal de la Fuente et al., 2024). Its structure contains 4R and 6R, but at difference of zeolites FAU and LTL, 5R are also present (the Secondary Building Units (SBU) are 4-2 and 5-1), which introduces an additional degree of structural complexity at molecular scale. The starting zeolite was synthesized by hydrothermal treatment of a gel containing a mixture of TMA^+/Na^+ cations, and its average Si/Al was 3.1 (determined by EDX). The purpose of the work was to determine the influence of the framework Si/Al and the extraframework Al species in the generation of intracrystalline mesoporosity, how this process affects the total acidity and acid strength of the resulting materials, and the way in which the textural and chemical changes influence the catalytic performance of the resulting materials in the isomerization/disproportionation of m-xylene.

The as-made sample was calcined and ion-exchanged with a solution of NH_4Cl and calcined again in air to obtain the acid form of the zeolite (H-form, Si/Al = 5.7, determined by NMR). The H-form was dealuminated by steaming (100% steam) at 620 °C under shallow bed conditions. Portions of the dealuminated sample were treated with HCl to remove EFAL species, leading to a dealuminated sample with framework Si/Al = 7.4 (as determined by NMR). The mesostructuring solution consisted in all cases of CTAB dissolved in NH_3 (aq), using different conditions and T in the range 80 –100 °C.

It was found that the H-form zeolite remained unaffected by the CTA treatment. However, this treatment led to intracrystalline mesoporosity in the steamed sample provided most of the EFAL species are previously removed by acid leaching. This behaviour suggests that a framework Si/Al threshold value seems to exist to generate mesoporosity, in the Si/Al range 5.7–7.4, but however it should also be considered that the steaming treatment produces a large number of structural defects that can facilitate the process, in the same way described above for low Si/Al faujasite. This study also reveals the necessity for the CTA to diffuse inside the intracrystalline zeolite channels to generate mesopores, because if the zeolite channels are plugged with EFAL species (the case of the dealuminated sample before acid treatment), the process is ineffective.

The CTA-treatment of the dealuminated/acid treated sample led to the generation of mesopores with a very narrow distribution of sizes centred at 4.0 nm (Figure 8 left), and to the decrease of the zeolite micropores, being their relative volume a function of the treatment conditions. No low-angle peaks were observed in the XRD pattern of the mesostructured samples (Figure 8 right), which of course excludes the presence of a segregated MCM-41-like phase mixed up with the zeolite in the solid product recovered after the CTA treatment, pointing instead to a disordered arrangement of mesopores.

Figure 8. Pore size distribution (left) and low-angle XRD (right) of some omega samples. The sample H-form is the pristine acid omega sample; the sample steamed+HCl is the pristine sample after undergoing a steam treatment followed by a treatment in HCl. The mesostructured omega samples were denoted CTA-xh-yC, where x and y refer to the time (in hours) and temperature (in °C) of the CTA/NH_4OH treatment, respectively, whereas the '0.7' in the sample CTA0.7-6h-80C indicates that the CTA content (in g) in the corresponding treatment was double than in the rest of samples (for which, 0.35 g was used), and '32mL' in the sample CTA-6h-80C-32mL indicates than half of ammonia solution (64 mL of 0.37 M was the volume and concentration for the rest of samples).

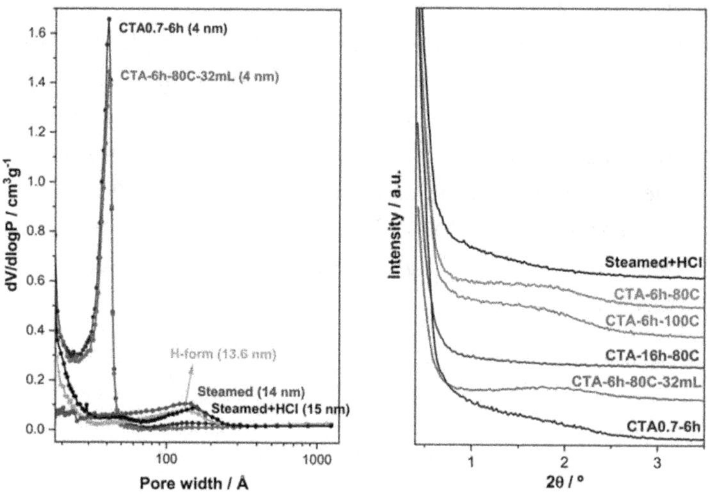

Taken from (Vidal de la Fuente et al., 2024).

The fact that the CTA treatment of H-FAU of Si/Al = 3.5 generates structural mesopores while it does not so for H-MAZ of Si/Al = 5.7, suggests that, in addition to the framework Al content and the presence of SiO^- defects, the zeolite structure also influences its reactivity against the basic CTA solution. The higher stability of MAZ to the treatment as compared with FAU could be related to the higher framework density (FD) of the former, 16.3 *vs* 12.7, as the stability of zeolites generally increases with the FD (Henson et al., 1994; Piccione et al., 2000). We will back later in this review to this point of paramount importance for a proper understanding of the process.

2.4.1. Acidity

The acidity and acid strength of the several samples were measured by FTIR spectroscopy of pyridine adsorption and desorption between 150 °C and 400 °C at intervals of 50 °C. The steaming/acid leaching of the H-form decreases the total acidity (acid centers retaining pyridine at a desorption T of 150 °C) in agreement with the removal of Al framework atoms, but increased the fraction of the strong aid sites, those retaining pyridine at 300 °C, which represent nearly 90% of the total acid sites for this sample. However, it has been found that the total acidity further decreased for all the CTA-samples, as well as the fraction of strong acid sites, which drops from 90% to the range 60-70% for all treated samples. These results show that the process to produce intracrystalline mesopores strongly reduces at the same time the average acid strength. As the acid strength of the acid sites located in the zeolite micropores should not be affected by the treatment, this leads to think that the acid strength of those located at the mesopore surface should be much weaker than those sitting at the zeolite micropores. The distribution of acid sites in zeolitic micropores and in the mesopores has been represented in Figure 9.

Figure 9. Scheme of mesoporosity development by treatment of the zeolite omega with CTA solutions in basic medium. The active (mainly strong) sites in micropores are marked with red dots, whereas the active (mainly weak) are marked with orange dots.

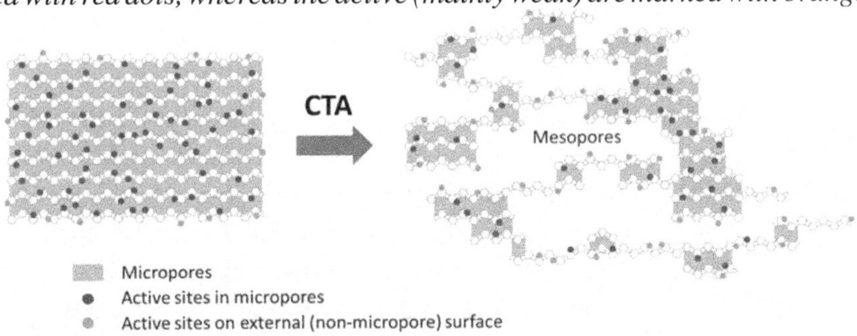

Taken from (Vidal de la Fuente et al., 2024).

The activity of these set of samples has been tested in the isomerization/disproportionation (I/D) of m-xylene. It has been shown elsewhere that both reactions are sensitive to the size of the zeolite channels, particularly the disproportionation, that involves a bimolecular transition state that cannot be easily accommodate in the omega channels, leading to an I/D ratio of 3-5 at low conversion (Martens et al., 1988).

The benefits of the generation of mesopores in the omega crystal in the m-xylene reaction is shown in the Figure 10, where the catalysts activity has been plotted as a function of the concentration of strong acid sites (those retaining pyridine at 300 °C).

Figure 10. Conversion in the isomerization/disproportionation of meta-xylene vs concentration of Brønsted sites that retain pyridine at 300 ºC in the omega zeolite catalysts. For nomenclature of the samples, see Figure 8 caption. Blue dashed line is the linear correlation between all plotted points, excluding H-form.

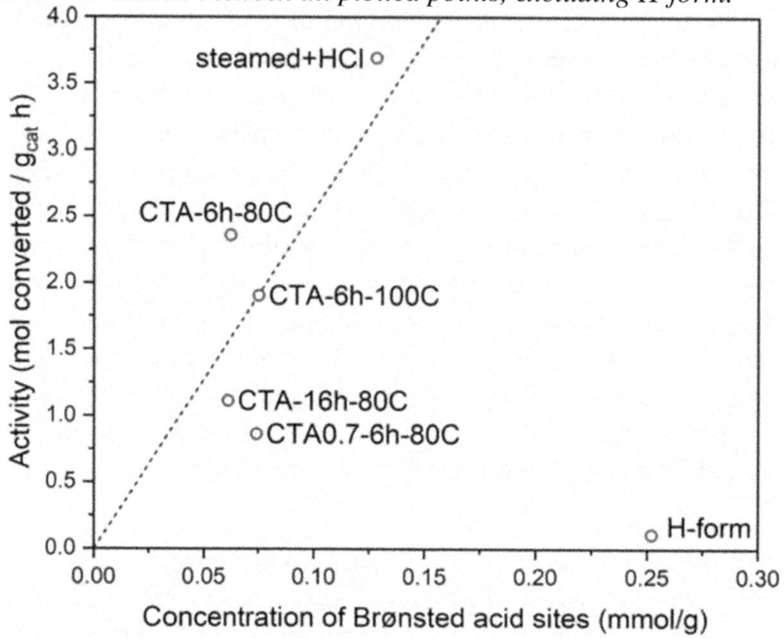

Taken from (Vidal de la Fuente et al., 2024).

The steamed/acid leached sample displays the highest activity, although it contains half the strong acid sites than the H-form. The CTA-samples are less active, in agreement with their lower content of strong acid sites, but also show noticeable differences in activity that have been attributed to differences in the mesopore architecture resulting from the different treatment conditions. The observed strong enhancement of activity both in the steamed sample and in the CTA-ones is due to the large improvement of the diffusion process due to the presence of mesopores. It could eventually be expected that this relief of diffusional constraints in the CTA-crystals should improve the formation of the bulky disproportionation products. However, the I/D ratio of the CTA-samples is insensitive to the presence of mesopores (Figure 11), and it is indeed higher than that of faujasite catalysts of different Si/Al ratio. These results led to conclude that the acid sites located at the mesopores surface have a very minor contribution to the total activity of the catalysts in this particular reaction, owing to their weak acid strength. It would be interesting to check further this conclusion by testing the activity of these catalysts in two type of reactions: first, the transformation of molecules too bulky to penetrate in the zeolite micropore channels. In this case, only the mesopore acid sites would be involved. Second, the transformation of acidity-demanding reaction of small molecules, such as the cracking of C6-C7 paraffins, requiring strong acid sites. If the conclusions derived from this study are correct, then the activity of the CTA-samples in the last reaction would be basically insensitive to the presence of mesopores and would correlate instead with the concentration of strong acid sites.

Figure 11. Variation of I/D (isomerization/disproportionation) given by samples of zeolite omega as a function of conversion. Pristine sample H-form (orange circles); Dealuminated sample steamed+HCl (black triangles), and mesostructured omega samples CTA samples (green squares).

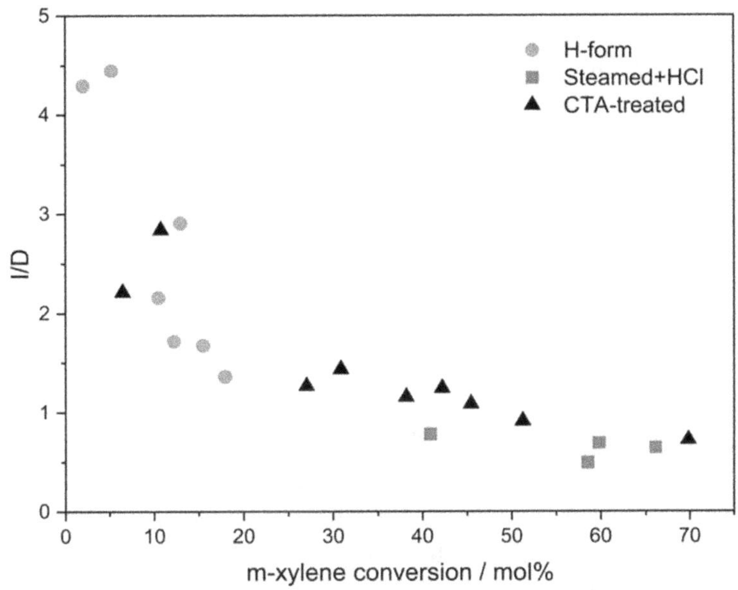

Taken from (Vidal de la Fuente et al., 2024).

2.5. Zeolite Mordenite

Mordenite is one of most industrially demanded zeolites. The catalytic benefits of mesoporosity within mordenite crystals were made clear even before the surfactant-based methodology were developed (Tromp et al., 2000); mesoporosity in that mordenite was simply generated by dealuminating a H-form mordenite with 1 M HCl for 1 hour at room temperature. However, in spite of the evident commercial interest of mordenite (Narayanan et al., 2020) and that surfactant-templated mesoporous MOR was one of the three zeolites explicitly included in the original patent (Ying & García-Martínez , 2005), the approach above described for other 12R zeolites has only been reported for this zeolite in just one synthesis work, using CTA as surfactant and TMAOH as base (Al-Ani et al., 2019). The positive effect of the intracrystalline mesoporosity of these materials on their catalytic performance in esterification and aldol condensation reactions was made clear in a different work (Al-Ani et al., 2020). Both studies did not only focus on the zeolite MOR, but also on some other 12R zeolites (FAU, BEA and LTL) and on the 10R zeolite MFI. Remarkably, even in that systematic work, the experimental treatments for generating CTA-templated mesoporosity is quite different for each zeolite (excepting for MOR and BEA, for which the methodology was the same). It suggests that one post-synthesis surfactant-template protocol cannot be universally applied for different zeolites but it surely requires notable optimization effort to find the right conditions (Si/Al ratio, nature of the base, media basicity, etc) for each studied zeolite. In the case of the parent MOR, which was CBV 21A from Zeolyst (Si/Al ratio = 10), no previous dealumination treatment was required, unlike the low Si//Al zeolites FAU (2.6) an LTL (3.1), which were pre-treated with acids.

Amongst all addressed zeolites in that work, mordenite is probably the one that better responds to the CTA/basic treatment paying attention to the characterization results show in Figure 12. Thus, the low-angle XRD pattern (Figure 12a) shows two very well-defined 2θ peaks corresponding to *d* spacing of 3.95 and 4.3 nm, which presumably correspond to (100) and overlapped (110)/(200) reflections, respectively, typical of mesoporous hexagonal arrangement. The introduction of so ordered mesoporosity entails a substantial decrease in the peaks' intensity of the wide-angle XRD pattern (Figure 12b). Moreover, the N_2 adsorption/desorption isotherms shown in Figure 12c make clear that the surfactant treatment induces notable mesoporosity of narrow size distribution. Finally, the TEM images of mordenite after being treated with ammonia solution containing the surfactant CTA evidence: (i) the very high mesopore volume of the particles, (ii) the ordered spatial arrangement of the mesopores, and (iii) that the size of the mesopores is perfectly compatible with having been directed by micelles of the CTA surfactant (Figure 12d).

Figure 12. a) Low-angle XRD pattern of the zeolite mordenite after being hydrothermally treated at 150 °C for 5 h under microwave radiation with the surfactant CTA in 0.13 M TMAOH solution (sample MMOR-1). The starred peaks indicate long-range ordering in the mesostructured samples. B) High-angle XRD pattern of the parent zeolite MOR, sample MMOR-1 and sample MMOR-2 (hydrothermally treated at 150 °C for 15 h under microwave radiation in 0.5 M TMAOH. C) N_2 adsorption/desorption isotherms at -196 °C of the parent zeolite mordenite and the sample MMOR-1. d) TEM images of a CTA-treated mordenite sample.

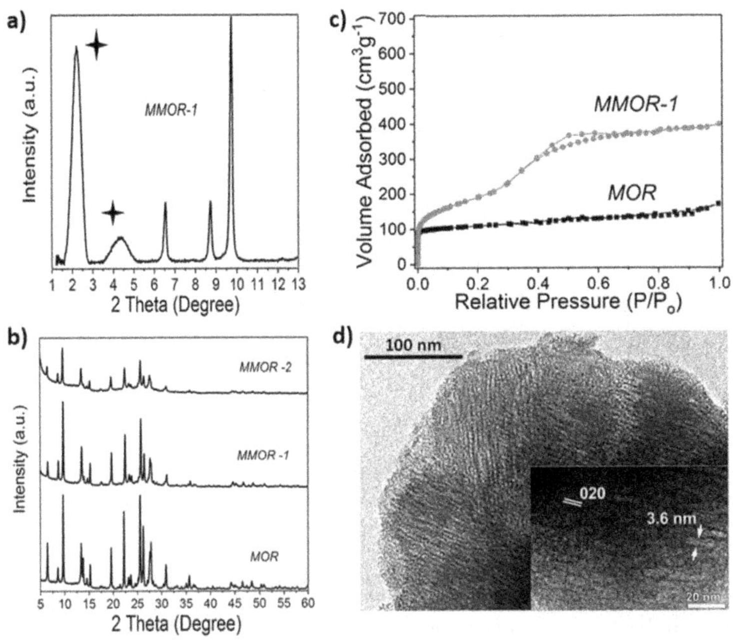

Adapted from (Al-Ani et al., 2019) with permission from the Royal Society of Chemistry, Copyright 2019.

2.5.1. Acidity

Figure 13 shows the FTIR spectra of the original mordenite (black line) and a mesostructured mordenite (red line) (Al-Ani et al., 2019). The basic treatment in the presence of the surfactant CTA produces a significant decrease in the number of Brønsted acid sites and a parallel increase of the number of Lewis acid sites. Accordingly, the intensity of the characteristic OH band in Si-OH-Al environments at 3610 cm^{-1} significantly decreases. It could be interpreted assuming a scenario similar to that proposed for omega zeolite (Figure 9) and it could also be related to the substantial decrease of crystallinity (Figure 12b) and microporosity (Figure 12c). In the same terms, acidity in mesostructured faujasite has been widely discussed in section 2.1. For us, it is surprising that the authors affirm that this behaviour is not found for the rest of investigated zeolites in that work (FAU, BEA, LTL and MFI), although no similar experimental FTIR results are shown for these zeolites. No explanation or hypothesis is given to the different behaviour of mordenite.

Figure 13. FTIR spectra of pyridine adsorbed on the parent MOR zeolite (a) and modified sample MMOR-1 (b).

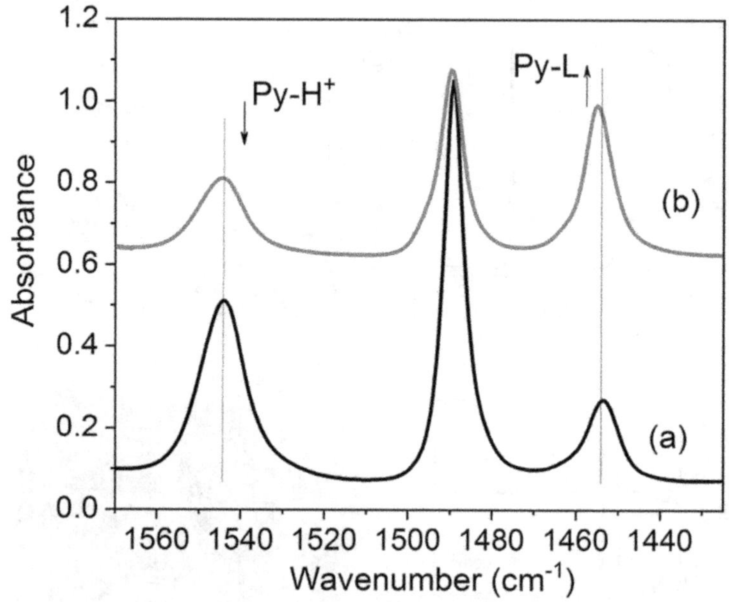

Adapted from (Al-Ani et al., 2019) with permission from the Royal Society of Chemistry, Copyright 2019.

Despite the acidity decrease in mordenite treated with surfactant CTA, it became more active in the dealkylation of the bulky molecule 1,3,5-tri-isoproylbenzene (9% conversion) than the pristine mordenite CBV 21A (2%), what highlights the relevant role of accessibility of the reactants to the active sites.

2.6. Zeolite Beta

The first two papers (to the best of our knowledge) on the mesostructuring of the large-pore, high-silica zeolite beta were published in 2019, both accepted for publication in February of that year, and reporting independent investigations on this subject. In the first one, Suárez et al. (Suárez et al., 2019) reported that one gram of a commercial sample of zeolite Beta (Zeolyst, CP81PC, Si/Al = 19) was stirred in 0.37 M solutions of NaOH (6.3 mL) or NH_3 (aq) (64 mL), to which 0.5 or

0.7 g of CTAB were respectively added. After 20 minutes of stirring, the samples were heated at 80 °C for the former case and 150 °C for the latter, during 6 and 12 h in each case, with no stirring. The CTA was removed by calcination at 550 °C first in N_2 and then in air. The NaOH-samples were ion-exchanged with an ammonium chloride solution and then calcined in air to obtain the acid form. The amount of zeolite recovered after calcination/ion exchange was in the range 60-80%. The synthesis procedure is quite similar to that previously reported in several publications by J. García-Martínez and co-workers for zeolite faujasite (García-Martínez et al., 2012), and its purpose was to determine the influence of the basicity and treatment conditions on the mesostructuring of this high-silica zeolite, as well as on the acidity and catalytic activity of the resulting materials in the isomerization/ disproportionation of m-xylene.

The zeolite was very resistant to the treatment, and only in the case of NaOH the crystallinity slightly decreases up to 83%. Moreover, the Si/Al ratio drops to 17 for the NaOH sample treated for 6 h, suggesting that some desilication took place, while it was 21-22 for the other three samples. In addition, no diffraction peaks were observed in the low-angle XRD region of the four samples (Figure 14).

Figure 14. Low-angle XRD patterns of the surfactant-treated beta samples.

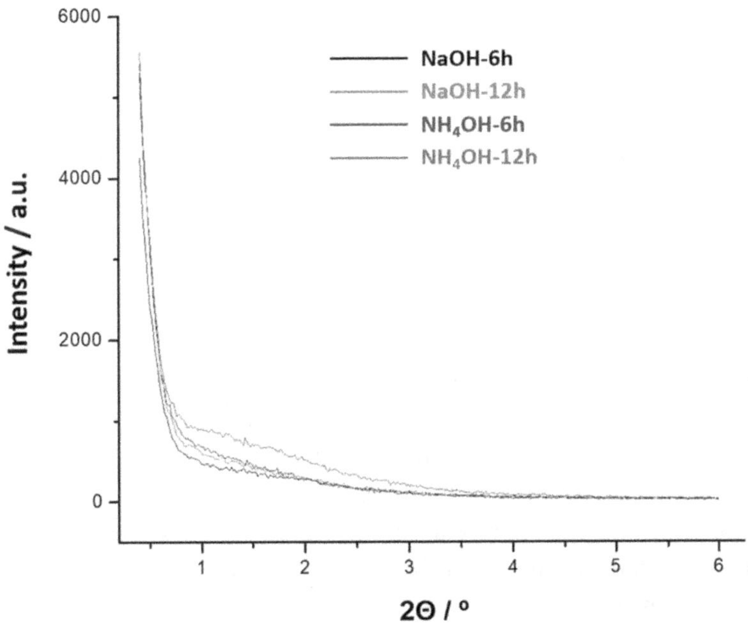

Taken from (Suárez et al., 2019).

Following what has been above discussed on the origin of the low-angle diffraction peaks, the absence of the latter evidences that no MCM-41-type materials are formed as a consequence of the treatment. Moreover, the N_2 adsorption/desorption isotherms show a stepped increase in the amount of nitrogen adsorbed in the p/p_0 range 0.2-0.4, which is much more pronounced for the NaOH-samples, denoting the presence of abundant mesoporosity (Figure 15).

Figure 15. N_2 adsorption/desorption isotherms at -196 °C of the parent and CTA-containing beta zeolite materials.

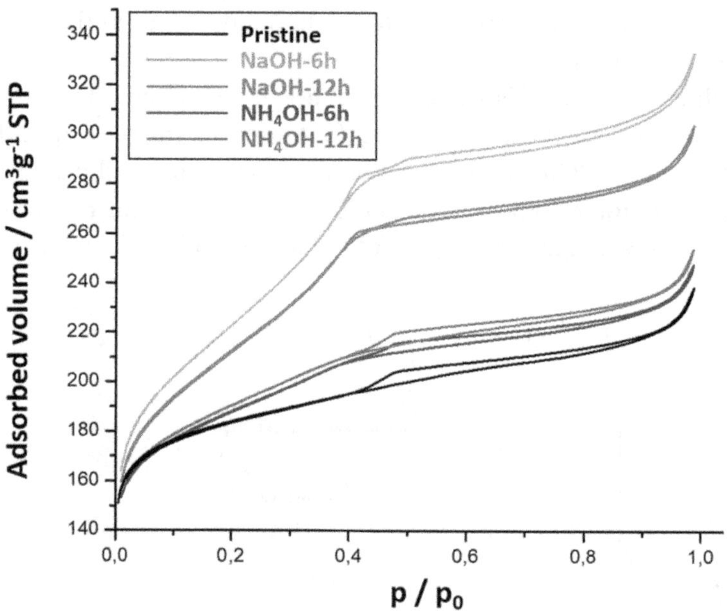

Taken from (Suárez et al., 2019).

Indeed, the pore size distribution is very narrow in all the samplės, centered at 2.5-3 nm (inset of Figure 16), and in the range 3-4 nm from the NLDFT method. Mesopore are visible in the TEM images (Figure 16), where they do not show any particular geometrical arrangement, which agrees with the lack of low-angle diffraction peaks. A large mesopore volume is formed at the expenses of micropores, particularly in the NaOH samples (Table 1). TGA show that the latter samples contain higher amount of CTA (in the range 25% - 26%) than the ammonia-samples, about 20 wt%. Hence, there is a nice positive correlation between the amount of surfactant occluded in the samples and the mesopore volume. These features of the mesostructuring of the zeolite seems to be governed by the pH of the two treating solutions, which was higher for the NaOH-samples (pH = 13.5) that show the largest mesopores volume than for the ammonia-samples (pH = 11.4).

Table 1. Textural properties of the parent and CTA-treated beta samples. Data taken from (Suárez et al., 2019).

Sample	S_{BET} [a] (m^2/g)	V_{micro} [b] (cm^3/g)	S_{meso} [b] (m^2/g)	V_{pore} [c] (cm^3/g)
Parent	707	0.22	142	0.14
NaOH-6h	793	0.12	499	0.38
NaOH-12h	765	0.14	427	0.31
NH_4OH-6h	696	0.18	245	0.19
NH_4OH-12h	707	0.18	245	0.20

[a] BET surface area. [b] Micropore t-plot. [c] V_{total} (P/Po=0,997)- V_{micro}.

Figure 16. Left: N_2 adsorption/desorption isotherms of the commercial beta sample, plotted on the background of the TEM image. Right: N_2 adsorption/desorption isotherm of the sample treated for 6 h with the CTA/NaOH solution plotted on the background of the TEM image of this sample. Some of the mesopores are indicated by yellow arrows. Inset: pore size distribution of the parent and CTA-samples.

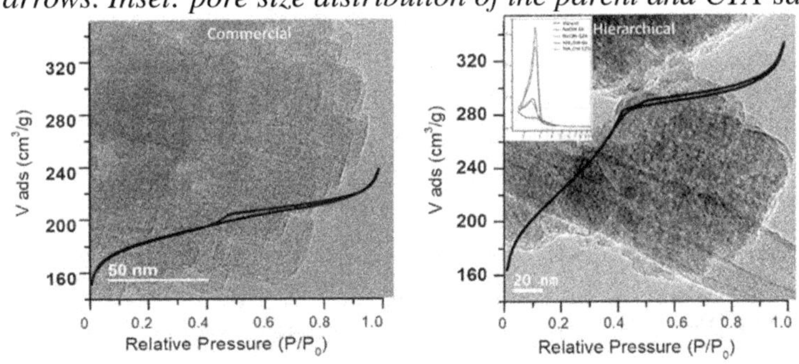

Adapted from (Suárez et al., 2019).

2.6.1. Acidity

The total acidity of the CTA-samples (Table 2), determined from the amount of pyridine retained at a desorption temperature of 250 ºC, increased as compared with the parent sample, which suggests that the Si/Al decreases as a consequence of partial desilication of the structure during the CTA treatment, in agreement with the results of the chemical analysis of the NaOH/CTA, although the Si/Al ratio of the NH_3(aq)/CTA material remained similar to that of the parent zeolite. It can be noticed that the ratio of strong acid + very strong acid sites (those retaining pyridine at 350 °C) to the total acid sites remained unchanged (0.62 - 0.64) after the CTA treatment, and only the ratio of strongest acid sites (at 400 °C of desorption T)/total acids sites decreases, particularly in the NaOH/CTA sample. This evidences that some rearrangement of the framework Al takes place during the mesostructuring of the material as a function of the specific CTA-treatment conditions, but, overall, the acid strength of the associated Brønsted sites is still very high. This behaviour is in contrast with that reported for zeolites FAU and MAZ, in which, as above discussed, a marked decrease of the average acid strength upon CTA treatment takes place. Moreover, it is also interesting to notice that the mesostructuring procedure applied to this specimen of zeolite Beta actually leads to lower L/B ratio, opposite to what has been found for FAU and LTL.

Table 2. Concentration of pyridine adsorbed on the Brønsted and Lewis acid centers at different desorption temperatures in beta samples. Data taken from (Suárez et al., 2019).

Sample	Brønsted[a] (μmol/g)				Lewis[b] (μmol/g)			
	250°C	300°C	350°C	400°C	250°C	300°C	350°C	400°C
Parent	151	129	94	72	127	105	99	103
NaOH-6h	166	133	107	51	133	104	93	104
NH_4OH-6h	209	174	134	76	108	89	87	101

[a] Determined by integration of the band located at 1545 cm^{-1}

[b] Determined by integration of the band located at 1450 cm^{-1}

The catalytic activity of the samples in the isomerization/disproportionation of m-xylene increase with the acid strength, in the sequence NaOH-6h < parent < NH_4OH-6h. The P/O ratio is not affected by the treatment, but the I/D ratio decreases from 3.3 for the parent material to 2.8. Considering that the overall selectivity is an average of the selectivity contributed by the acid sites located at both the micropores and the mesopores, this decrease of the I/D evidences a significant influence of the mesopore acid sites, due to the alleviation of the size-constraints for the formation of the bulky trimethylbenzenes. Moreover, as this reaction is acidity-demanding, it can be concluded that strong acid sites are located in both the micropores and the mesopores of the mesostructured beta materials, while, as it has been above discussed, the acid strength of the acid sites located in the mesopores of mesostructured FAU and MAZ would be comparatively lower.

2.6.2. Nanozeolite Beta with Si/Al = 25 and 50

This methodology has been also applied to nanocrystalline zeolite Beta (average crystal size = 50- 80 nm) with Si/Al = 25 and 50 synthesized from Na-free gels containing tetraethyl ammonium hydroxide (Chaida-Chenni et al., 2021). For the CTA-treated samples, the crystallinity is largely unaffected by using an ammonia solution, but it decreases if NaOH is used, being the effect more pronounced for the sample with lowest Al content. A very broad signal of very low intensity is observed in the XRD pattern of the two ammonia/CTA samples at $2\theta = 1.7°$. The analysis of the N_2 isotherms of the two CTA/ammonia materials prepared from sample with Si/Al = 25 indicates the presence of mesopores whose average size is centered at 3.0 nm. In this case, the micropore volume drops from 0.19 cm^3g^{-1} of the parent material to 0.15 $cm^3.g^{-1}$, while the mesopore volume raises to 0.43 cm^3g^{-1}. The effect of the treatment for the sample with the highest Si/Al ratio is less pronounced than that described for the sample richer in Al. Overall, it can be concluded that this two nanozeolites depleted in Al are less prone to mesostructuring than the commercial sample with Si/Al = 19 ratio above described.

2.6.3. Zeolite Beta treated with TMAOH/CTA solutions

In the second paper, Al-Ani and coworkers reported the treatment of the same commercial zeolite beta with CTA dissolved in solutions of tetramethyl ammonium hydroxide (TMAOH) (Al-Ani et al., 2019). Following this procedure, 2 grams of the zeolite (previously calcined in air) were stirred in 50 mL of TMAOH solutions at two different concentrations, 0.15 mol L^{-1} or 0.5 mol L^{-1}. Then, CTAB was added, the mixture stirred for 1 h and then heated statically at 150 °C for 5–18 h by using microwave heating. The resulting materials were calcined first in N_2 and then in O_2.

The two CTA-materials did not show low-angle reflections in the XRD pattern, while their crystallinity decreases as the TMAOH concentration in the treating solution increases. The same decreasing pattern is followed by the Si/Al (Table 3), which evidences a noticeable desilication of the samples. As shown above, slight desilication also occurs if the parent zeolite is treated in NaOH/CTA solution, but the effect is much more pronounced in this case.

Table 3. Properties of the beta samples. Data taken from (Al-Ani et al., 2019).

Sample	Si/Al[a]	Cryst (%)	S_{BET} (m^2 g^{-1})	V_{micro} (cm^3 g^{-1})	V_{meso} (cm^3 g^{-1})	C_B[b]	C_L	C_L/C_B
BEA	19	98	695	0.26	0.06	409	105	0.26
BEA-1	15	81	740	0.20	0.12	397	406	1.02
BEA-2[c]	10	70	810	0.16	0.25	-	-	-

[a] Determined by SEM;

[b] C_B = Brønsted acid sites, C_L = Lewis acid sites, in μmol g^{-1};

[c] [TMAOH] was 0.15 mol L^{-1} and 0.5 mol L^{-1} for BEA-1 and BEA-2, respectively.

The N_2 isotherm of mesostructured sample MBEA-1 does not show the characteristic stepped increase of nitrogen adsorption at $p/p_0 \sim 0.3 - 0.4$ (Figure 17) that was observed in the samples above discussed treated with CTA solutions in either NaOH or ammonia, while the nitrogen uptake at $p/p_0 > 0.5$ indicates the presence of large mesopores with a broad size distribution. In agreement with this, HRTEM images show pores in the range 5 to 8 nm.

Figure 17. N_2 isotherms of the parent zeolite beta and material BEA-1.

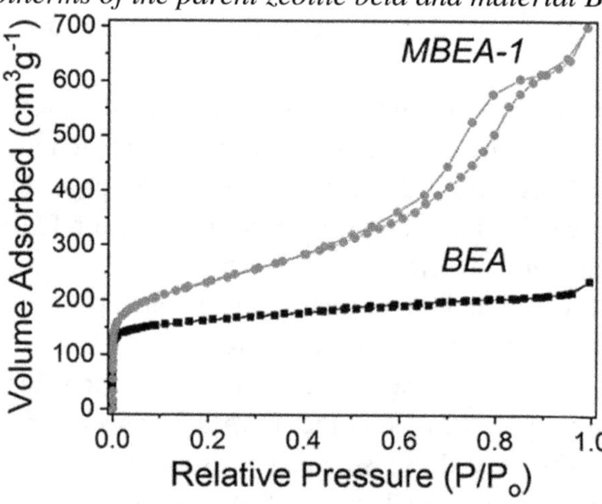

Adapted from (Al-Ani et al., 2019) with permission from the Royal Society of Chemistry, Copyright 2019.

These results suggest that the mesoporosity present in the CTA samples is essentially due to desilication, and not to true mesostructuring. This conclusion suggests that the generation of surfactant-driving porosity seems to be governed by the cation present in the alkaline solution, being Na^+ and NH_4^+ more favourable than TMA^+ to trigger the process. It has been argued that the incorporation of the surfactant is due to the electrostatic interaction that take place between the negatively charged Si-O$^-$ framework defects produced by the breaking of Si-O-Si bonds due to OH^- attack, and the positively charge polar head of the CTA surfactant (Mendoza-Castro et al., 2021). However, it should be considered that the inorganic cations present in the alkaline solution (Na^+ or NH_4^+) would first compensate the negative charge of the defect sites, and would progressively be replaced by CTA^+ cations. This ion-exchange process seems to be less favourable if TMA^+ is the countercation. However, in the experimental procedure above outlined it is stated that the parent zeolite was first treated with the TMAOH solution and then the surfactant was added, but no information is given on how long did the first treatment lasted. It could well have led to partial desilication of the sample and to the decrease of the Si/Al ratio, making the subsequent CTA treatment much less effective.

2.6.4 Acidity

The acidity of the samples was evaluated by FTIR of pyridine adsorption and further desorption at 150 ºC, which allow to determine the total content of the Brønsted and Lewis sites but unfortunately not the acid strength distribution. Sample MBEA-1 has higher Al content than the parent zeolite, but lower total content of Brønsted sites. This suggests that a substantial fraction of the Al present in the material should correspond to EFAL species, which is experimentally evidenced by the four-fold increase of the concentration of Lewis acid sites as compared with the parent zeolite, and the intense ^{27}Al signal corresponding to octahedral aluminium visible in the MAS NMR spectrum of this sample. It can be concluded that the CTA/TMAOH treatment followed by calcination leads to a severe dealumination of the parent zeolite.

These two procedures illustrate the determinant influence that the experimental conditions applied to the parent zeolite with the purpose of generating intracrystalline mesoporosity have on the properties of the final material. These conditions must be properly adjusted to produce the desired effect.

3. MESOSTRUCTURED 10R ZEOLITE: ZSM-5

The methodology above described for 12R zeolites has also been applied to the zeolite ZSM-5 (MFI-type structure), that contains a bidirectional system of intersecting 10R channels, with a very limited success. Al-Ani et al. treated the parent ZSM-5 (Zeolyst CBV 8014, Si/Al = 40) with a NaOH/TPAOH solution, and then CTAB is added and the mixture heated at 150 ºC for 9 or 16 h (Al-Ani et al., 2019). The authors stated that they have followed the desilication method of ZSM-5 reported by Pérez-Ramírez et al. (Pérez-Ramírez et al., 2009). It was found that the Si/Al ratio of the resulting materials decreased to 32 and 35 as a function of the treatment time, the crystallinity also decreases to 85%, and no low-angle reflections were found in the XRD patters. The N_2 isotherm (Figure 18) shows a small increase of the nitrogen uptake at relative pressure $p/p_0 > 0.6$, and there a slight decrease of the micropore volume, and a modest increase of the mesopore volume from 0.03 $cm^3 g^{-1}$ to 0.15 $cm^3 g^{-1}$ for the 16h-treated sample.

Figure 18. N_2 adsorption/desorption isotherms at -196 °C of the parent zeolite ZSM-5 and a mesostructured zeolite ZSM-5.

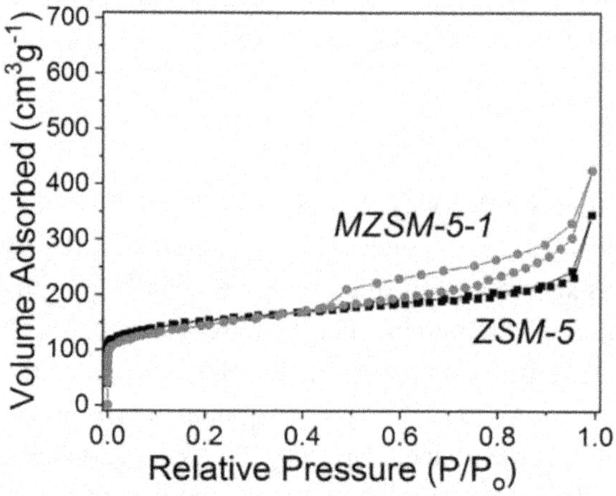

Adapted from (Al-Ani et al., 2019) with permission from the Royal Society of Chemistry, Copyright 2019.

Moreover, the total Brønsted acidity (pyridine retained at 150 °C) is lower for the 9h-treated sample as compared with the parent one, while, on the contrary, the total Lewis acidity increased.

Al-Ani et al. concluded that the mesostructuring process is difficult in the 10R, medium-pore MFI structure (Al-Ani et al., 2019), and attributed this behaviour to the impossibility for the CTA surfactant cations to penetrate inside the ZSM-5 channels due to their small size of the 10R pores, ~ 0.5 nm. However, it has been found by us that a calcined MFI zeolite (Zeolyst CBV 3024E, Si/Al = 15) treated with a CTA solution in aqueous ammonia at 100 °C show a TG loss of 11 wt% centered at 400 °C (Figure 19), attributed to the desorption/combustion of CTA. This weight loss nearly corresponds to the one commonly observed in the TGA profile of ZSM-5 synthesized from tetrapropylammonium cations (TPA), evidencing that CTA fills the channels as TPA does in the as-synthesized material. In other words, CTA is indeed able to penetrate the 10R channels of MFI.

Figure 19. TGA profile of MFI (Zeolyst CBV 3024E, Si/Al = 15) treated with a CTAB solution in aqueous ammonia at 100 °C.

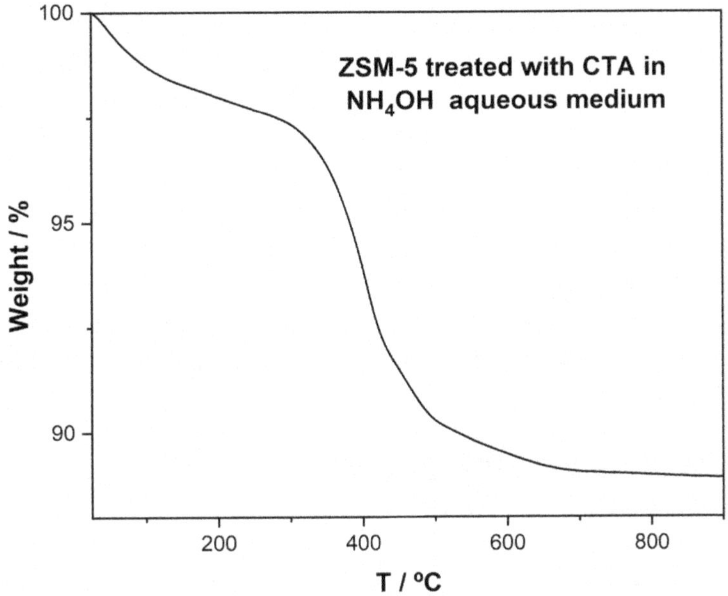

The method followed by Al-Ani et al. produces partial desilication of the MFI structure (Al-Ani et al., 2019), although they do not inform in the paper for how long the MFI zeolite was treated with the NaOH/TPAOH solution before CTA was added. This approach is quite similar to the procedure used for the zeolite recrystallization method mainly developed by Ivanova et al. above discussed (Ivanova & Knyazeva, 2013).

Following a similar protocol, Talebian-Kakalieh and Tarigi have reported the treatment of a synthetic MFI zeolite (SiO_2/Al_2O_3 = 65) with a NaOH solution for desilication, followed by acid leaching for partial dealumination (Talebian-Kiakalaieh & Tarighi, 2020). After this, the desilicated/dealuminated zeolite was treated with a solution of CTA in 0.3 M NH_3 (aq) at 150 °C for 12 h. The Si/Al ratio of the resulting material is 43.6. A strong reduction in crystallinity was observed, and the mesopore volume increased from 0.06 $cm^3\ g^{-1}$ in the parent MFI to 0.47 $cm^3\ g^{-1}$ in the CTA-containing material. However, the N_2 adsorption/desorption isotherm closely resembles that of desilicated MFI materials prepared in the absence of surfactant, and no increase of the nitrogen uptake at relative pressure p/p_0 of 0.35 is found. The authors attributed the pores with d < 6 nm to the desilication/dealumination process, and those with d > 6 nm to the effect of surfactant. The acidity determined by NH_3-TPD decreases, in agreement with the lower Al content of the treated sample.

More recently, the same procedure has been applied to MFI with higher Al content (Zeolyst, CBV 3024E, ammonium form, Si/Al = 15), which was calcined prior the chemical treatments (Martins et al., 2024). The zeolite treatment was the same as that above reported, and different concentrations of CTA were used to determined their influence on the process. The crystallinity decreased to 50%, and the mesopore volume corresponding to pores < 10 nm increased, but this mesoporosity was already present in the desilicated/dealuminated material prior CTA treatment. To summarize, for this specific MFI zeolite CTA had a very little influence on the developing of mesoporosity. The differences observed with the results reported by Talebian-Kakalieh and Tarigi (Talebian-Kiakalaieh & Tarighi, 2020) could be due to the lower Al content of the latter. In any case, none of these works report the development of additional mesoporosity assisted by the CTA surfactant, being the one observed due to the basic/acid desilication/dealumination pre-treatment.

4. FRAMEWORK DENSITY (FD) AND ZEOLITE MESOSTRUCTURING

Following the several examples of zeolites that have so far been submitted to mesostructuring processes by using post-synthesis treatments with surfactants (CTA in most cases) in alkaline medium, two main conclusions can be obtained. First, for each zeolite structure the generation of surfactant-driven intracrystalline mesoporosity only takes place if the framework Al content is below a threshold limit. Second, this Al content is different for different structures. Both factors seem to be related to the stability of the zeolite framework. The intrinsic stability of zeolite structures is determined by their framework density (FD), defined as the number of TO_4 tetrahedra per 1000 Å^3. It has been shown that the relative stability of pure silica molecular sieves with respect to quartz correlates linearly with FD, being most stable the denser phases (Henson et al., 1994; Piccione et al., 2000). This would account for the failed efforts so far reported for mesostructuring the high FD zeolite MFI while they succeeded for the low FD zeolite FAU. In other words, the overall energy gain of the process to transform FAU into mesostructured FAU is not enough to transform the more stable MFI into mesostructured-MFI.

Concerning the Al content, FAU, LTL, MAZ and MOR are high-Al content materials that require the presence of Al in the framework to be stable in aqueous alkaline medium in which they are usually synthesized. Therefore, these structures become less stable as their framework Al content decreases. BEA and MFI are high-silica materials, but BEA is able to accommodate more Al in its structure than MFI. In this way, the mesostructuring process will be in general facilitated as the FD and the Al decreases. Conversely, it will become increasingly difficult for denser structures. However, it can also be considered that Al-rich MFI zeolite is less stable than high Si/Al materials, and all-silica MFI can be easily synthesized.

Figure 20. Relationship between Framework Density (FD) and Al content (expressed as Al/(Al + Si)) of mesostructured zeolites. Zeolites able to become mesostructured under surfactant treatment are represented as open black circles and they are found below the black line (grey region). Full black circles represent the zeolites that do not become mesostructured after surfactant treatment.

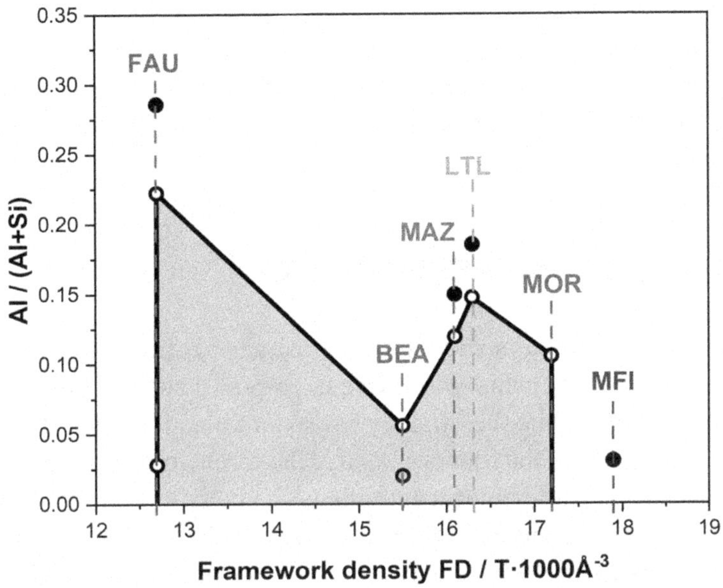

Figure 20 depicts the experimental FD and Al content of the zeolite materials discussed in this work. The grey area corresponds to materials for which the mesostructuring process has been successfully achieved, and defines the "stability field" of the mesostructured materials in terms of the FD and Al content of the parent zeolites. As far as new results on the response of other zeolite structures to the CTA-treatment will be available, the limits of the stability field outlined in the figure will be better defined. However, it can be already used in a predictive manner with respect to the nature (FD) and Al content of the zeolites that would be more sensitive to the surfactant-assisted mesostructuring process.

Another aspect that deserves consideration for outlining a broad picture of the mesostructuring process is the influence of the structural $Si\text{-}O^-$ defects present in the parent materials. The dealumination required to activate zeolites with high Al content as well as the steaming of FAU, or the desilication/dealumination of MFI, certainly increase the number of defects per unit cell. On the other hand, typical high-silica zeolites synthesized at high pH usually contains silanol defects, and BEA is probably the one with the largest number of defects among the pure silica molecular sieves (Piccione et al., 2000, p. 10006). It has been reported that silanols defects have a light destabilizing effect on zeolite stability, which has been calculated to be of an upper bound of 2.4 kJ/mol for zeolite BEA (Piccione et al., 2000, p. 10006). It has been calculated that the enthalpies of transition for the reaction quartz → molecular sieves at 298 K is in the range 6.8 – 14.4 kJ/mol. Therefore, the contribution of the silanol defects to the overall energy of any crystalline molecular sieves would be about 30% of the total range of energy that separate the most from the less stable structures. This energy of 2.4 kJ/mol could be in most cases much higher than the difference in energy between two structures with different FD. Therefore, the number of defects is indeed a factor to consider in the overall mesostructuring process of a given zeolite material.

5. CONCLUSIONS AND PERSPECTIVES

The generation of intracrystalline mesoporosity in zeolites assisted by cationic surfactants has so far been achieved for five large-pore 12R structures, FAU, LTL, MAZ, BEA and MOR, while this strategy has so far failed for 10R MFI. Moreover, for Al- rich zeolites, part of the framework Al of the parent material must be removed to trigger the process. These empirical findings can be rationalized by considering the stability of the corresponding zeolite framework, which is basically dependent on the framework density (FD) and the Al content. The combination of both parameters outlines a "field of stability" of the mesostructured zeolites, which in addition can eventually be modulated by the contribution of the concentration of silanol Si-OH defects groups present in the parent material. Following this picture of the overall process, the ability of zeolite crystals to accommodate surfactant-induced mesopores decreases as the FD and Al content increases. Moreover, for the zeolite MFI, the one with the highest FD so far reported that has been submitted to the process, no evidence of mesostructuring has been found even for materials with very low Al content. It is expected that future work can add new zeolites that will contribute to improve the limits of the field of stability.

The arrangement of the intracrystalline mesopores varies from one structure to another, and even for the same structure it is affected by the Al content, as it has been shown for FAU. For some materials, a well-ordered assembly of mesopores is found, while no such ordered mesostructuring occurs for others. Moreover, there is evidence that, at least for FAU, a heterogeneous distribution of pores results from the surfactant treatment, and regions containing a well-ordered arrange of mesopores coexist together with disordered mesoporosity.

While the general picture of the generation of mesopores has been reasonable described, based mostly on the comprehensive and systematic work on mesostructured FAU carried out by García Martínez's team, many important issues remain yet to be properly investigated. One of them concerns the eventual dissolution of part of the parent zeolite during the process. This is an important aspect, that has not only practical implications, but can also contribute to improve the thermodynamic understanding of the process. If part of the zeolite is dissolved in the liquid phase, then the leached silicate and aluminate species should be taken into consideration to properly described the energetic of the mesostructuring process. It has been repeatedly claimed that no appreciable dissolution takes place for FAU, while for BEA yields in the range 60 – 80% have been reported. Unfortunately, this information is not available for the other zeolites.

The main driving force to research on mesostructured zeolites has been the expected improvement that the mesopores would have on the performance of the materials in acid-catalyzed reactions as compared with the parent zeolites. While with almost no exception the hypothesized improvement is actually observed, in varies degrees, for different reactions, it is to notice that only in very few cases the nature and acid strength of the acid sites of the mesostructured materials is reported. For some of them, the total acidity determined by NH_3 TPD or pyridine desorption at 150 °C is given, but only in very few cases the acid strength distribution is examined. From the small number of examples in which this information is provided, it can be concluded that the overall acid strength of the mesozeolite is lower than that of the parent material. The high-silica zeolite BEA is less affected, but even in this case the concentration of strongest Brønsted acid sites decreases after the mesostructuring treatment. These experimental observations lead us to conclude that the acid strength of the acid sites associated to Al located at the mesopores is clearly lower than that of sites corresponding to Al sitting at the micropores. However, the behaviour of zeolite BEA suggests that the degradation of the acid strength upon the surfactant treatment is structure-sensitive. BEA is a typical high-silica material possessing abundant 5R, and it could well be that its structural features largely preserve for the Al atoms located at mesopores the chemical environment they had in the pristine material. Further insights into this aspect would be provided by the preparation of additional high-silica, low FD mesostructured materials.

If, on average, the strength of the acid sites located at the mesopores decreases, then the improvement of catalytic activity often reported for mesostructured materials would be mainly due to a strong enhancement of diffusion. In this way, the benefits on diffusion would counterbalance the decrease of activity due to lower acid strength. It would be advisable to determine the acid strength of acid sites located at the micropores and at the mesopore surface by using a combination of probes with different molecular size, one small able to check all acid sites, as pyridine, and other bigger, that would interact only with the acid sites located at the large mesopores. A similar strategy should be implemented in the study of their catalytic activity. For instance, the comparison of the cracking of reagents with different molecular size, a small one, generally n-heptane, and a bigger one, gas-oil or 1,3,5-triisopropylbenzene, has been used in the past to determine the relative influence of mesopores or the external surface of FAU catalysts on the overall cracking activity of the catalysts (Aguiar et al., 1995; Arribas et al., 1987). We believe that this approach, the systematic use of a combination of acid-demanding reactions of small and large molecules, would shed light on aspects of the catalytic behaviour of mesostructured zeolite materials that are still poorly understood. Moreover, the presence of accessible acid sites of mild acid strength located at the mesopore surface can be desirable for reactions of bulky substrates that not only do not required strong sites, but for which such sites should be avoided as much as possible. As one among many examples, hydroisomerization of long-chain hydrocarbons meets such condition, mild acid sites work well while undesired strong acid sites lead to unwanted cracking. A scientific understanding of the chemical phenomena involved in the mesostructuring process of the zeolite crystals that leads to solid guidelines would be much better than playing around with different reactions to determine which one show a positive effect.

ACKNOWLEDGMENTS

The authors acknowledge funding from MCIN/AEI grant code: PID2022-136321OB-C21/AEI/10.13039/501100011033/FEDER, EU, as well as the project: PID2022-138481NB-I00.

REFERENCES

Aguiar, E. F. S., Murta Valle, M. L., Silva, M. P., & Silva, D. F. (1995). Influence of external surface area of rare-earth containing Y zeolites on the cracking of 1,3,5-triisopropylbenzene. *Zeolites, 15*(7), 620–623. https://doi.org/https://doi.org/10.1016/0144-2449(95)00069-I

Al-Ani, A., Freitas, C., & Zholobenko, V. (2020). Nanostructured large-pore zeolite: The enhanced accessibility of active sites and its effect on the catalytic performance. *Microporous and Mesoporous Materials, 293*, 109805. https://doi.org/https://doi.org/10.1016/j.micromeso.2019.109805. DOI: 10.1016/j.micromeso.2019.109805

Al-Ani, A., Haslam, J. J. C., Mordvinova, N. E., Lebedev, O. I., Vicente, A., Fernandez, C., & Zholobenko, V. (2019). Synthesis of nanostructured catalysts by surfactantlating of large-pore zeolites. *Nanoscale Advances, 1*(5), 2029–2039. DOI: 10.1039/C9NA00004F PMID: 36134217

Arribas, J., Corma, A., Fornés, V., & Melo, F. (1987). Influence of framework aluminum gradients on the catalytic activity of Y zeolites: Cracking of gas-oil on Y zeolites dealuminated by different procedures. *Journal of Catalysis, 108*(1), 135–142. https://doi.org/https://doi.org/10.1016/0021-9517(87)90160-6. DOI: 10.1016/0021-9517(87)90160-6

Chaida-Chenni, F. Z., Grande Casas, M., Belhadj, F., Hamacha, R., & Pérez-Pariente, J. (2021). No Title. *Post-Synthesis Treatment of High-Silica Nanozeolite Beta with Cationic Surfactants for Enhancing Mesoporosity.*, 2.

Chal, R., Cacciaguerra, T., van Donk, S., & Gérardin, C. (2010). Pseudomorphic synthesis of mesoporous zeolite Y crystals. *Chemical Communications, 46*(41), 7840–7842. DOI: 10.1039/c0cc02073g PMID: 20714567

Chen, L.-H., Sun, M.-H., Wang, Z., Yang, W., Xie, Z., & Su, B.-L. (2020). Hierarchically Structured Zeolites: From Design to Application. *Chemical Reviews, 120*(20), 11194–11294. DOI: 10.1021/acs.chemrev.0c00016 PMID: 32915551

Dight, L., García-Martínez, J., Valla, J., & Johnson, M. (2010). *No Title* (Patent No. US 2010/0190632.).

Fichtner-Schmittler, H., Lohse, U., Miessner, H., & Maneck, H.-E. (1990). *Correlation between Unit-Cell Parameter, Skeletal Stretching Vibrations and Molar Fraction of Aluminium of Faujasite Type Zeolites for Si:Al = 1.1 - 1,000. 271O*(1), 69–80. https://doi.org/doi:10.1515/zpch-1990-27110

García-Martínez, J., & Johnson, M. (2008). (Patent No. US 2008/0138274.).

García-Martínez, J., Johnson, M., Valla, J., Li, K., & Ying, J. Y. (2012). Mesostructured zeolite Y—High hydrothermal stability and superior FCC catalytic performance. *Catalysis Science & Technology*, *2*(5), 987–994. DOI: 10.1039/c2cy00309k

García-Martínez , J., & Jonhson, M. (2010). (Patent No. US 2010/0021321).

Henson, N. J., Cheetham, A. K., & Gale, J. D. (1994). Theoretical Calculations on Silica Frameworks and Their Correlation with Experiment. *Chemistry of Materials*, *6*(10), 1647–1650. DOI: 10.1021/cm00046a015

Ivanova, I. I., & Knyazeva, E. E. (2013). Micro–mesoporous materials obtained by zeolite recrystallization: Synthesis, characterization and catalytic applications. *Chemical Society Reviews*, *42*(9), 3671–3688. DOI: 10.1039/C2CS35341E PMID: 23223641

Ivanova, I. I., Kuznetsov, A. S., Yuschenko, V. V., & Knyazeva, E. E. (2004). *Design of composite micro/mesoporous molecular sieve catalysts. 76*(9), 1647–1657. https://doi.org/doi:10.1351/pac200476091647

Kerstens, D., Smeyers, B., Van Waeyenberg, J., Zhang, Q., Yu, J., & Sels, B. F. (2020). State of the Art and Perspectives of Hierarchical Zeolites: Practical Overview of Synthesis Methods and Use in Catalysis. *Advanced Materials*, *32*(44), 2004690. https://doi.org/https://doi.org/10.1002/adma.202004690. DOI: 10.1002/adma.202004690 PMID: 32969083

Li, K., Valla, J., & García-Martínez , J. (2014). Realizing the Commercial Potential of Hierarchical Zeolites: New Opportunities in Catalytic Cracking. *ChemCatChem*, *6*(1), 46–66. https://doi.org/https://doi.org/10.1002/cctc.201300345. DOI: 10.1002/cctc.201300345

Linares, N., Sachse, A., Serrano, E., Grau-Atienza, A., De Oliveira Jardim, E., Silvestre-Albero, J., Cordeiro, M. A. L., Fauth, F., Beobide, G., Castillo, O., & García-Martínez, J. (2016). In Situ Time-Resolved Observation of the Development of Intracrystalline Mesoporosity in USY Zeolite. *Chemistry of Materials*, *28*(24), 8971–8979. DOI: 10.1021/acs.chemmater.6b03688

Mallette, A. J., Seo, S., & Rimer, J. D. (2022). Synthesis strategies and design principles for nanosized and hierarchical zeolites. *Nature Synthesis*, *1*(7), 521–534. DOI: 10.1038/s44160-022-00091-8

Martens, J. A., Perez-Pariente, J., Sastre, E., Corma, A., & Jacobs, P. A. (1988). Isomerization and disproportionation of m-xylene: Selectivities Induced by the Void Structure of the Zeolite Framework. *Applied Catalysis*, *45*(1), 85–101. https://doi.org/https://doi.org/10.1016/S0166-9834(00)82395-7. DOI: 10.1016/S0166-9834(00)82395-7

Martins, A., Amaro, B., Santos, M. S. C. S., Nunes, N., Elvas-Leitão, R., & Carvalho, A. P. (2024). Hierarchical Zeolites Prepared Using a Surfactant-Mediated Strategy: ZSM-5 vs. Y as Catalysts for Friedel–Crafts Acylation Reaction. In *Molecules* (Vol. 29, Issue 2). DOI: 10.3390/molecules29020517

Mendoza-Castro, M. J., Serrano, E., Linares, N., & García-Martínez, J. (2021). Surfactant-Templated Zeolites: From Thermodynamics to Direct Observation. *Advanced Materials Interfaces*, *8*(4), 2001388. https://doi.org/https://doi.org/10.1002/admi.202001388. DOI: 10.1002/admi.202001388

Narayanan, S., Tamizhdurai, P., Mangesh, V. L., Ragupathi, C., Santhana krishnan, P., & Ramesh, A. (2020). Recent advances in the synthesis and applications of mordenite zeolite - review. *RSC Advances*, *11*(1), 250–267. DOI: 10.1039/D0RA09434J PMID: 35423021

Ogura, M., Shinomiya, S.-Y., Tateno, J., Nara, Y., Kikuchi, E., & Matsukata, M. (2000). Formation of uniform mesopores in ZSM-5 zeolite through treatment in alkaline solution. *Chemistry Letters*, *8*(8), 882–883. DOI: 10.1246/cl.2000.882

Pérez-Ramírez, J., Christensen, C. H., Egeblad, K., Christensen, C. H., & Groen, J. C. (2008). Hierarchical zeolites: Enhanced utilisation of microporous crystals in catalysis by advances in materials design. *Chemical Society Reviews*, *37*(11), 2530–2542. DOI: 10.1039/b809030k PMID: 18949124

Pérez-Ramírez, J., Verboekend, D., Bonilla, A., & Abelló, S. (2009). Zeolite Catalysts with Tunable Hierarchy Factor by Pore-Growth Moderators. *Advanced Functional Materials*, *19*(24), 3972–3979. https://doi.org/https://doi.org/10.1002/adfm.200901394. DOI: 10.1002/adfm.200901394

Piccione, P. M., Laberty, C., Yang, S., Camblor, M. A., Navrotsky, A., & Davis, M. E. (2000). Thermochemistry of Pure-Silica Zeolites. *The Journal of Physical Chemistry B*, *104*(43), 10001–10011. DOI: 10.1021/jp002148a

Sachse, A., & García-Martínez, J. (2017). Surfactant-Templating of Zeolites: From Design to Application. *Chemistry of Materials*, *29*(9), 3827–3853. DOI: 10.1021/acs.chemmater.7b00599

Sachse, A., Grau-Atienza, A., Jardim, E. O., Linares, N., Thommes, M., & García-Martínez, J. (2017). Development of Intracrystalline Mesoporosity in Zeolites through Surfactant-Templating. *Crystal Growth & Design*, *17*(8), 4289–4305. DOI: 10.1021/acs.cgd.7b00619

Schwieger, W., Machoke, A. G., Weissenberger, T., Inayat, A., Selvam, T., Klumpp, M., & Inayat, A. (2016). Hierarchy concepts: Classification and preparation strategies for zeolite containing materials with hierarchical porosity. *Chemical Society Reviews*, *45*(12), 3353–3376. DOI: 10.1039/C5CS00599J PMID: 26477329

Serrano, D. P., Escola, J. M., & Pizarro, P. (2013). Synthesis strategies in the search for hierarchical zeolites. *Chemical Society Reviews*, *42*(9), 4004–4035. DOI: 10.1039/C2CS35330J PMID: 23138888

Siakpebru, O. K., Gurrala, L., Uchagawkar, A., Norris, O., Bartlett, J., & Morais, A. R. C. (2025). Application of Hierarchical H–Y Zeolites in Hydroxyalkylation–Alkylation of 2-Methylfuran with Furfural to Produce Jet Fuel Precursors. *Industrial & Engineering Chemistry Research*, *64*(11), 5830–5840. Advance online publication. DOI: 10.1021/acs.iecr.4c03317

Singh, S., Martínez-Ortigosa, J., Ortuño, N., Polshettiwar, V., & García-Martínez, J. (2024). Enhanced efficiency in plastic waste upcycling: The role of mesoporosity and acidity in zeolites. *Chemical Science (Cambridge)*, *15*(48), 20240–20250. DOI: 10.1039/D4SC05121A PMID: 39568899

Suárez, N., Pérez-Pariente, J., Mondragón, F., & Moreno, A. (2019). Generation of hierarchical porosity in beta zeolite by post-synthesis treatment with the cetyltrimethylammonium cationic surfactant under alkaline conditions. *Microporous and Mesoporous Materials*, *280*, 144–150. https://doi.org/https://doi.org/10.1016/j.micromeso.2019.02.001. DOI: 10.1016/j.micromeso.2019.02.001

Talebian-Kiakalaieh, A., & Tarighi, S. (2020). Synthesis of hierarchical Y and ZSM-5 zeolites using post-treatment approach to maximize catalytic cracking performance. *Journal of Industrial and Engineering Chemistry*, *88*, 167–177. https://doi.org/https://doi.org/10.1016/j.jiec.2020.04.009. DOI: 10.1016/j.jiec.2020.04.009

Tromp, M., van Bokhoven, J. A., Garriga Oostenbrink, M. T., Bitter, J. H., de Jong, K. P., & Koningsberger, D. C. (2000). Influence of the Generation of Mesopores on the Hydroisomerization Activity and Selectivity of n-Hexane over Pt/Mordenite. *Journal of Catalysis*, *190*(2), 209–214. https://doi.org/https://doi.org/10.1006/jcat.1999.2778. DOI: 10.1006/jcat.1999.2778

Verboekend, D., & Perez-Ramirez, J. (2011). Design of hierarchical zeolite catalysts by desilication. *Catalysis Science & Technology*, *1*(6), 879–890. DOI: 10.1039/c1cy00150g

Vidal de la Fuente, A., Mekki, A., Boukoussa, B., Grande-Casas, M., Sastre, E., Márquez-Álvarez, C., Sánchez-Sánchez, M., & Pérez-Pariente, J. (2024). Generation of mesoporosity in crystals of zeolite omega by post-synthesis treatments with surfactants. *Catalysis Today*, *425*, 114364. https://doi.org/https://doi.org/10.1016/j.cattod.2023.114364. DOI: 10.1016/j.cattod.2023.114364

Ying, J. Y., & García-Martínez , J. (2005). (Patent No. US 2005/0239634).

Zholobenko, V., Hollamby, M., Al-Ani, A., Lebedev, O. I., Smith, A. J., & Snow, T. (2025). A nanoscale investigation of the formation of mesostructured zeolites FAU and LTL. *Microporous and Mesoporous Materials*, *381*, 113363. https://doi.org/https://doi.org/10.1016/j.micromeso.2024.113363. DOI: 10.1016/j.micromeso.2024.113363

ADDITIONAL READING

Brinker, C. J., & Scherer, G. W. (1990). *Sol-gel science*. Academic Press.

Čejka, J., Pérez-Pariente, J., & Roth, W. J. (2008). *Zeolites: from model materials to industrial catalysts*. Transworld Research Network.

García-Martínez, J., & Kunhau, L. (Eds.). (2015). *Mesoporous zeolites. Preparation, characterization and applications*. Wiley-VCH. DOI: 10.1002/9783527673957

Martínez, C., & Pérez-Pariente, J. (Eds.). (2011). *Zeolites and ordered porous solids: fundamentals and applications*. Editorial Universitat Politècnica de València.

Pérez-Pariente, J., & Sánchez-Sánchez, M. (Eds.). (2018). *Structure and reactivity of metals in zeolites*. Springer. DOI: 10.1007/978-3-319-98905-1

Thommes, M., & Schlumberger, C. (2021). Characterization of nanoporous materials. *Annual Review of Chemical and Biomolecular Engineering*, *12*(1), 137–162. DOI: 10.1146/annurev-chembioeng-061720-081242 PMID: 33770464

Van Bekkum, H., Flanigen, E. M., Jacobs, P. A., & Jansen, J. C. (Eds.). (2001). *Introduction to zeolite science and practice* (2nd ed.). Elsevier.

Van Bekkum, H., & Kouwenhoven, H. W. (2012). *Zeolite manual for the organic chemist*. Mijnbestseller.

Wright, P. A. (2008). *Microporous framework solids*. RSC Publishing.

KEY TERMS AND DEFINITIONS

Acidity in zeolites: Ability of these microporous aluminosilicate materials to donate protons (Brønsted acidity) or to accept electrons (Lewis acidity), in such a way that they can act as heterogeneous catalysts in different acid-demanding organic reactions.

Adsorption/desorption isotherms: Characterization technique to evaluate the textural properties of solid materials by registering their adsorption capacity (and subsequent desorption) of an inert gas at increasing pressure values (decreasing values in the case of desorption branch) for a given temperature, which is maintained constant along all experiment. N_2 and Ar at their boiling points (77 and 87 K, respectively) are the most commonly used inert gases.

CTA: Acronym for CetylTrimethylAmmonium (also known as Hexadecyltrimethylammonium), which is a monovalent surfactant cation usually commercialized as bromide salt (that is why it is also abbreviated as CTAB), and the most widely used porogen for ordered mesoporosity in zeolites.

Mesoporosity: Nanoporosity with pore diameter dimensions between 2 and 50 nm.

Microporous material: Material containing pores with typical sizes less than 2 nm in diameter.

Porogen: Chemical agent capable of generating or inducing porosity in a material.

Surfactant: Chemical substance that, when added to a liquid, reduces its surface tension; in aqueous solution, they create self-assembled molecular clusters called micelles. Surfactants are commonly used as porogens in synthesis of silica-based materials.

Zeolites: Crystalline porous materials formed by tetrahedral TO_4 units in which each oxygen atom of the vertices is shared by a neighbour tetrahedral unit, in such a way that the three-dimensional arrangement of interconnected tetrahedra forming the framework leaves cavities or pores of various sizes and shapes inside the crystals.

Chapter 10
Recent Advances in Sonochemistry for the Synthesis and Modification of FeY Zeolite Catalyst

Victor Alfredo Reyes
http://orcid.org/0000-0001-5358-4837
Centro de Investigación Científica y de Educación Superior de Ensenada, Mexico

Jesus Isaias De Leon Ramirez
Centro de Nanociencias y Nanotecnologia UNAM, Mexico

Luis Perez-Cabrera
http://orcid.org/0000-0001-9896-3419
Centro de Nanociencias y Nanotecnologia UNAM, Mexico

Arturo Estolano-Cobian
Facultad de Ciencias Quimicas e Ingenieria, Universidad Autonoma de Baja California, Mexico

Jose Roman Chavez-Mendez
Facultad de Ciencias de la Salud, Universidad Autonoma de Baja California, Mexico

Vitalii Petranovskii
http://orcid.org/0000-0002-8794-0593
Centro de Nanociencias y Nanotecnologia UNAM, Mexico

ABSTRACT

Sonochemistry is recently used to improve the zeolitization process and post-synthetic treatments of potential zeolite-based catalysts. This work explores the use of ultrasound to synthesize a faujasite zeolite with a Y-type structure and the reported interconversions associated with FAU. Reviewing the effects of sono-assisted treatments of Y zeolite for the formation of mesopores and modification with alkaline metals, alkaline earth metals, and metals. Based on this specified literature review,

DOI: 10.4018/979-8-3373-0320-8.ch010

preliminary work was conducted, employing ultrasound irradiation to synthesize a selective catalyst for oxidizing benzene to phenol (OBP). The sono-assisted synthesized zeolite Y was sonochemically ion-exchanged and modified in an ammonia buffer solution with dissolved Fe. The catalyst activity for the OBP was measured by UV-Vis spectroscopy and GC/MS, with no significant difference between the techniques, and an enhanced selectivity. This preliminary study highlights the importance of the sonochemical method for potentially designing a catalyst for the benzene-to-phenol reaction.

1 INTRODUCTION

The application of sonocrystallization in zeolites has many benefits in the zeolitization process, such as an increase in crystal growth rate, decreased induction periods and metastable zone width, improving particle size distribution (PSD) with more homogeneous morphology and phases, crystallinity with potential for crystal size tuning, and more control over the nucleation process (Askari et al., 2013). Making ultrasonic treatments another physical radiation method to accelerate zeolite synthesis (Liang et al., 2023). The known mechanisms of this method is by disrupting the depolymerization–polymerization equilibrium, increasing the concentration of soluble species needed to form zeolites. The main effect of ultrasound comes from the increase of the nucleation rates due to the cavitation bubbles, which act as nuclei (Askari et al., 2013). In these cavitation bubbles, water and oxygen molecules are excited and dissociated to form reactive species, such as $H^{\bullet}$, $HO^{\bullet}$ and $O^{\bullet}$(Liang et al., 2023). Hydroxyl radicals accelerate the nucleation and growth of zeolites, especially in the nucleation stage. From theoretical studies comparing the energy of the Si-O-Si bond cleavage in quartzite, $HO^{\bullet}$ only needs 16.7 kJ/mol of energy, while under traditional OH^- ion conditions, 79.1 kJ/mol is required. Hence, these theoretical calculations show that $HO^{\bullet}$ facilitates the fracture of Si-O-Si bonds compared to OH^-. Therefore, hydroxyl radicals accelerate the synthesis of zeolites mainly by catalyzing the fracture and formation of Si-O-Si bonds(Liang et al., 2023).

The advantages of ultrasonic treatments for the synthesis of zeolites are their simplicity, rapid reactions, lack of complex facilities, high crystal growth rate, suitable particle size distribution and morphology, and high control over the nucleation process (Khaleque et al., 2020). For these reasons, much attention has been focused on synthesizing zeolites for major environmental issues, demanding low-cost synthesis processes and exploring different methods and raw materials. As presented in a review published in 2020, several raw materials (blast furnace slag, municipal solid waste, paper sludge, lithium slag, fly ash, rice husk ash and kaolin) have been used by different methods. As for ultrasonic-assisted methods, various raw materials have successfully functioned as precursors for zeolitization, maintaining the attractive tunable properties embedded in sonochemistry being temperature, time, and the molar ratio of reactants the main parameters to determine the type and characteristics of the zeolite formed (Khaleque et al., 2020). To our knowledge, different zeolites (FAU, GIS, ANA, MFI, CHA, MWW, ERI/OFF, ERI and EMT) have been synthesized by sonochemistry from the classical pure reagents aluminosilicate solutions, seeded or mineral material, coal fly ash, and waste by-products as precursors (Table 1). For the following articles, the zeolitization process is divided into three main stages needed for the zeolite synthesis that have been explored and exposed from the sono-assisted methods: I) Ageing, which is the induction period involving the dissolution and condensation of the precursors; II)Nucleation: formation of nuclei and polymerization, and finally III) crystal growth: addition of atoms or molecules to the nuclei or aggregation of nanoparticles (Bai et al., 2022; Deneyer et al., 2020; Grand et al., 2016). Within each precursor type, the order of appearance follows the decision tree constructed from the trained XGBoost model (Muraoka et al., 2019).

Table 1. Influence of ultrasound on the zeolitization process of zeolite Y and possible interconversions of FAU zeolites

COD	Precursor	Zeolite	Stage	Effect of Sonication	Ref.
FAU	C*	Y	Aging	Higher crystallinity, surface area, and smaller crystal size.	(Reinoso et al., 2018)
FAU	CFA	X	Aging/Nucleation	Not evaluated	(Boycheva et al., 2020)
FAU	CFA	X	Aging/Nucleation	Increased zeolitization efficiency.	(Czuma et al., 2017)
FAU	CFA	X	Nucleation	Reduced crystallization time and improved crystallinity.	(Belviso et al., 2011)
FAU	CFA	X	Nucleation	Reduced crystallization time and improved crystallinity.	(Belviso et al., 2013)
FAU	Fertilizer	X	Nucleation	Control over morphology and quantity of different zeolites formed.	(Vaičiukynienė et al., 2021)
FAU	Seed	X	Crystallization	Increased reaction kinetics and reduced crystal agglomeration.	(Ramirez Mendoza et al., 2020)
GIS	Mineral	P	Aging/Nucleation	Accelerates zeolitization and fine-tuning of material properties.	(Behin et al., 2016)
GIS	Fertilizer	P	Nucleation	Control over the quantity and morphology of the zeolites formed.	(Vaičiukynienė et al., 2021)
GIS	CFA	P	Nucleation/Crystallization	Improves crystallinity and reduces crystallization time.	(Aldahri et al., 2016)
GIS	C*	P	Nucleation/Crystallization	Increased Si dissolution and promotes zeolitization in one step.	(Pal et al., 2013)
ANA	C*	ANA	Aging/Nucleation	Improves nucleation rate with morphology fine-tuning potential.	(Azizi & Yousefpour, 2010)
CHA	C*	CHA	Aging/Nucleation	Reduces crystallization time and crystal size, improving crystallinity.	(Yin et al., 2019)

continued on following page

Table 1. Continued

COD	Precursor	Zeolite	Stage	Effect of Sonication	Ref.
CHA	Carbon	SSZ-13	Aging/Nucleation	Time optimization and Si/Al ratio, reducing crystallization time.	(Han et al., 2019)
CHA	C*	SSZ-13	Aging/Nucleation	Fine-tuning of properties and synthesis time by controlling power.	(Mu et al., 2017)
CHA	C*	SAPO-34	Nucleation	Less synthesis time.	(Askari & Halladj, 2012)
MFI	C*	ZSM-5	Aging	9-fold increase in reusability.	(Huang et al., 2016)
MFI	C*	BZSM-5	Aging	Improve the nucleation rate and shorten the nucleation period.	(Franko et al., 2010)
MFI	Rice Husks	ZSM-5	Aging/Nucleation	Smaller crystal size and ability to fine-tune catalytic properties.	(Khoshbin & Karimzadeh, 2017)
MFI	C*	Silicalite	Nucleation/Crystallization	Control over particle size as a function of power.	(Gürbüz et al., 2012)
ERI/ OFF	C*	T	Nucleation	Reduces crystallization time and increases crystal uniformity.	(Jusoh et al., 2017)
	C*	T	Nucleation	Reduces crystallization time and improves material properties.	(R. Wang et al., 2018)
ERI	Seed	ERI	Nucleation	Improved crystallization speed and crystallinity.	(Y. Liu et al., 2020)
MWW	C*	MCM-22	Aging/Nucleation	Reduces crystallization time and improved control in synthesis.	(B. Wang et al., 2008)
EMT	Rice Husks	EMT	Nucleation/Crystallization	Decreased crystallization time and improved crystallinity.	(Ng et al., 2015)

This table provides a comprehensive overview of studies investigating the impact of ultrasonication at different stages of zeolite synthesis for various types of zeolite. C*=The conventional method involves the dissolution of pure chemical reagents in an alkaline medium. CFA= coal fly ash.

2 SONO-ASSISTED ZEOLITIZATION

As for ultrasound-assisted methods, different raw materials have successfully functioned as precursors to zeolitization, maintaining the attractive tunable properties embedded in sonochemistry, which are temperature, time and molar ratio of the reactants as the main parameters to determine the type and characteristics of the zeolite form (Khaleque et al., 2020). The zeolitization process has been studied by sonochemical methods divided into three main stages: I) Aging, an induction period that involves the dissolution and condensation of the precursors; II) Nucleation, formation of nuclei and polymerization, and finally, III) Crystal growth, the addition of atoms or molecules to nuclei or aggregation of nanoparticles (Bai et al., 2022; Deneyer et al., 2020; Grand et al., 2016).

FAU structure Y zeolite has been synthesized without any organic structure directing agents (SDAs) at low temperatures, and ultrasound-assisted aging has been applied to find a change in the properties of the zeolite. When homogenizing the precursor gel for 1 h under sonication (47 kHz, 130W), crystal and particle size were reduced while the bulk Si/Al increased compared to the conventional method. These sonoassisted prepared samples showed high phase purity, crystallinity, and elevated surface area (SBET ~ 950 m^2 g^{-1}) with hierarchical mesopores. The changes were attributed to the ultrasound-induced increase in nuclei formation (Reinoso et al., 2018).

On the other hand, most of the studies have obtained zeolite X when crystallizing FAU structure rather than zeolite Y. When using coal fly ash (CFA) as the precursor, all studies conducted a pre-fusion treatment of the raw material before ultrasonic irradiation. Once the zeolitization started in the crystallization stage, the obtained zeolite X was as nanocrystals. In this procedure, performing the synthesis in an intermittent 30-second On and Off sonication for an overall 45 min is insufficient to crystallize zeolite X. However, when subjecting the mixture to crystallization (90 °C for 6 hours), the yield of zeolite X was approximately three times higher than in the conventional synthesis. With zeolite P and SOD co-crystallizing only in the ultrasonically synthesized sample (Czuma et al., 2017). Another study demonstrated that 15 min of sonication of the fused fly ash before crystallization (90 °C for 4 hours) was sufficient to obtain zeolite X nanocrystals with a relatively good crystallinity and surface area (486 m^2 g^{-1}) with potential for adsorption applications. Nevertheless, although in a small quantity, the phase SOD was also crystallized (Boycheva et al., 2020). This highlights that assisting the fly-ash mother liquor with ultrasound irradiation also benefits the ultrasound-induced increase in nuclei formation detected from the conventional pure chemical precursor zeolitization process. Nevertheless, comparing both precursors, the FAU crystals from the pure chemicals generate a Y-type structure, while the fused fly ash is X-type. It is worth mentioning that when using fly ash as the precursor, co-crystallization is promoted due to impurities and zeolitization conditions, as explored next.

A set of experiments with high insight into these concerns have been performed by Belviso et al., where the synthesis of zeolite X by the conventional hydrothermal method was compared to the sono-assisted hydrothermal process (1 h continuous, 240 W; 35 kHz), where the following factors are considered: major chemical constituents (SiO_2, Al_2O_3, Fe_2O_3, TiO, MnO, MgO, CaO, Na_2O, K_2O and P_2O_5), trace elements (As, Cd, Co, Cr, Cs, Cu, Pb, Tl, Zn and Ce), crystallization temperature (20, 30, 35, 40, 45 and 60 °C), sonication time (3, 6, 9, 38, 40 and 45 min) and water source (distilled water and seawater). When applying ultrasound, the pre-treated mother liquor of fused CFA crystalizes in zeolite X below 40 °C, regardless of the impurities. However, elevating the crystallization temperature promotes the co-crystallization of different zeolite structures, minimizing the effect of ultrasonic irradiation. Nevertheless, the role of ultrasound in the pre-treatment increases the nucleation rate by shifting the depolymerization-polymerization equilibria towards the amorphous phase, thus producing an enriched medium of Si and Al species. Given this effect, zeolitization is not favored under ultrasonic irradiation at temperatures equal to or above room temperature, given that cavitation bubbles act as nuclei, disrupting the existing nuclei. Concluding that impurities drive co-crystallization only when crystallizing above 40 ºC and that the zeolitization conditions strongly depend on the pre-formed nuclei, not the existing crystals (Belviso et al., 2011).

When replicating the experiments with seawater, a more heterogeneous zeolite crystallization is obtained, with zeolite X still in higher proportion. The ions present in the seawater promoted the crystallization at lower temperatures, however increasing co-crystalization and shifting the distribution of crystallized zeolites from A and X to SOD, ZK-5 and X. When treating the CFA with ultrasound, the effect of the seawater ions was observed at even lower temperatures where increasing crystallization temperature reduces co-crystallization. Confirming that impurities in the fly ash affect the crystallization of zeolite X at low temperatures (< 40 ºC) and that treating with ultrasound enhances crystallization (Belviso et al., 2013).

Another Si by-product used to synthesize FAU is from a fertilizer plant, which is obtained from the processing of Al fluoride. The effect of pretreatment sonication time (0-20 min) before the hydrothermal crystallization (3-24 h at 95 ºC) modified the phase distribution between zeolite X and P. Compared to the non-sonicated samples, the sono-assisted zeolitization improved with a higher yield of crystalline material and smaller particle size. In addition, the sonicated precursors transformed to a zeolite P and X mixture, revealing that the crystallization of X competed in the transition of amorphous to zeolite P. Hinting that the zeolitization conditions favored zeolite P. Given that sonication positively influenced the conversion of the pseudo-spherical forms constituted by small plates of P (primary crystallization form of zeolite P) to clear crystal edges morphology (well-crystallized P zeolite). Furthermore, it was determined that sono-assisted hydrothermal treatment affected a reduction in the crystal size and reduced the crystallization time from 24 h to 12 h (Vaičiukynienė et al., 2021). If considering the above studies, the most obvious modification for the crystallization of FAU would be to test the crystallization at lower temperatures. Nevertheless, these low-cost raw materials, such as fertilizer silica by-products, are potentially valuable for producing FAU zeolites.

Focusing on the effect of ultrasound on the synthesis of FAU, in a seeded process, crystallization (80 ºC for 3 h) can occur in the presence of ultrasound. In this experiment, the crystallization rate is accelerated when only 30 min of ultrasonic irradiation is applied within the first 60 minutes of the crystallization stage. This change in the reaction kinetics was attributed to a mass transfer effect caused by bubble cavitation, enhancing the solubility of silicate/aluminate species. Although crystal size was unaffected by ultrasound irradiation, the particle size distribution did show a higher tendency to smaller particles with increasing ultrasound irradiation.

Ultrasound only accelerates mass transfer in a mother liquor with pre-formed FAU nuclei, and this acceleration is limited. In addition, when applying ultrasound (20-24 kHz) to crystallized FAU particles at low temperatures in the mother liquor, the particle size distribution shifts to the size of individual crystals (monocrystalline particles). Hence, ultrasonic irradiation reduces the degree of agglomeration of the final powder by disintegrating agglomerates and crystal clusters without affecting the morphology of individual zeolite crystals. This ultrasonic irradiation time deagglomerating procedure did not impact crystallinity and was positively correlated with the surface area of the zeolite (Ramirez Mendoza et al., 2020).

Applying ultrasound to synthesize FAU zeolites only impacts the nuclei formation stage. In addition, this literature suggests that FAU is preferred for crystallizing in an ultrasonic pre-treated mother liquor at low temperatures, regardless of the present impurities.

Zeolite P has a GIS-type framework, a framework that is known to crystallize at elevated temperatures near 100 °C. This framework competes with the crystallization of FAU, where only temperature guides the crystallization of GIS or FAU at the required conditions. In addition, by comparing their relative Gibbs free energy, GIS (lower energy) would be more stable than FAU, hence the preference of zeolite P over zeolite X or even Y. Given this interconversion route of clinoptilolite to zeolite P, there should also be a route for the conversion of clinoptilolite to zeolite X or Y. When applying ultrasound (20 kHz, 15 W, 3 h at 90 °C) to an alkaline suspension of clinoptilolite, this slowly dissolves clinoptilolite, enriching the alkaline solution with Si and Al species, which, when the amount is accurate, re-crystallizes into GIS. This demonstrates that GIS can also crystallize in ultrasound, supporting the above evidence that the sonochemical effects are reduced at higher temperatures, and physical effects dominate. In addition, increasing sonication time did not affect the crystal's shape; hence, ultrasonic irradiation accelerates the re-crystallization process of converting natural clinoptilolite to synthetic NaP zeolite (Behin et al., 2016). Since ultrasonic irradiation creates acoustic cavitation, which cracks the surface of the clinoptilolite particles and increases the concentration of soluble aluminosilicate species, crystallization and crystal growth of zeolite P occurs faster. Lowering the crystallization temperature might allow the crystallization of FAU; however, the crystallization time might be longer.

Another study treated CFA at different ultrasonic times for insight into the effect on the crystallinity of zeolite P, reducing crystallization time from 24 to 4 hours. However, ultrasonic irradiation treatment alone was insufficient to crystallize zeolite P. Showing the importance of pre-formed nuclei for the zeolitization process. For zeolite P to crystallize, the CFA must be first heated (1 h at 100 °C) and then subjected to ultrasonic irradiation (Aldahri et al., 2016). This experiment suggests that the zeolitization of FAU can be achieved by pre-treating CFA with ultrasound at relatively low temperatures (100 °C) in contrast to the 500 °C of the pre-fusion treatment.

When a milky white solution with the gel composition ratios for the synthesis of FAU is aged for 2 days and then crystallized by the hydrothermal method, NaY is formed. However, if this solution is not aged and sonicated for at least 3 h, zeolite P crystallizes, and further hydrothermal treatments do not impact crystallinity. The study claims that no significant change in the crystalline properties is observed by shifting sonication time and power. Finding that P can be synthesized by a sonochemical method in 3 h at room temperature, which is sufficient for aging, nucleation, and crystallization. (Pal et al., 2013).

If not subjected to the right conditions, a potential solution for the synthesis of FAU would crystallize another structure. However, as seen from this analysis, if sonicated for a short time, the effect of temperature on the mother liquor for the zeolitization shifts to lower temperatures, preferring the crystallization of FAU and further temperature increase would compete with GIS and, if increased even more ANA.

For example, when crystallizing at 160 °C a pre-treated mother liquor for 30-90 min, the predominant phase was ANA, with GIS as the competing phase. It is worth mentioning that applying ultrasound produced altered morphologies of the particles.

The synthesis of zeolite ANA was accompanied by additional ultrasonic-assisted aging of the reaction mixture in an ultrasonic bath for 30-90 min at 25 °C before crystallization at 160 °C. This sonochemical treatment was compared to static, stirring-assisted, microwave-assisted, and without aging. It was found that the aging of the reaction mixture caused a reduction of the induction period and the duration of crystallization, as well as directing the synthesis for ANA relative to zeolite P. Where static and ultrasonic-assisted aging can further shorten the crystallization time of the ANA phase. Moreover, ultrasonic-assisted aging leads to altered morphologies of the particles (Azizi & Yousefpour, 2010).

However, presenting low levels of Na or high levels of K in the mother liquors allowed the zeolitization of CHA (Askari & Halladj, 2012; Han et al., 2019; Mu et al., 2017; Yin et al., 2019), MFI (Franko et al., 2010; Gürbüz et al., 2012; Huang et al., 2016; Khoshbin & Karimzadeh, 2017), ERI/OFF (R. Wang et al., 2018) (Jusoh et al., 2017)and ERI (Y. Liu et al., 2020). This shows the importance of Na for the synthesis of FAU, suggesting that if FAU is wanted as the resulting zeolite, elevating the amount of Na might be the solution. Other frameworks that can be interconverted to FAU are MWW (B. Wang et al., 2008) and EMT (Ng et al., 2015), zeolites that can be grown in the presence of an organic structure-directing agent. If this OSDA is not provided, the equilibrium might shift to the FAU structure.

However, EMT is a somewhat unique case, given that it is constructed from the same SBU as FAU, changing only the orientation of the stacking layers of the sodalite cage. For FAU, the inversion center in each layer is altered, giving an ABCABC stacking order; in contrast, the stacking layers in EMT are ABAB. If EMT can crystallize under ultrasonic irradiation at room temperature (25 °C) when sufficient time is given, FAU should not be restricted to crystallize under these same conditions. As evidenced above, FAU has been crystallized in the presence of ultrasound, even at elevated temperatures when seeds are added. Suggesting that evaluating FAU crystallization under ultrasonic irradiation without seeds at lower temperatures is a potential study for gaining insight into the crystallization of zeolite Y.

Ultrasound energy accelerates Al/Si dissolution and strengthens the solid particle/gel phase interface bonds. This increases the condensation process, thus speeding up the zeolite crystallization process. Aging of the source of Al or Si, either from pure chemical precursors or CFA using ultrasonic energy, has been shown to reduce the crystallization time. Solution aged with ultrasound has the shortest crystallization time in the second step, followed by mechanically stirred and static solutions. The mixing abilities of the ultrasonic bath are considered to provide the soluble silica required to produce the zeolite. Reduction in induction time can also be associated with ultrasonic aging. It is known that the amount of agitation produced due to the cavitation caused by the US is higher than the agitation produced through mechanical stirring. Furthermore, the cavitation bubbles caused by the ultrasound can provide nucleation sites for the species, resulting in secondary crystallization; high-speed collisions caused by the ultrasonic field in the mixture could be another source of secondary nucleation, reducing induction time and thus decreasing the crystallization time. Ultrasonic irradiation throughout aging and conventional hydrothermal synthesis crystallization was studied, and it was found that longer aging time corresponded to a shortening of the nucleation period and the eradication of the secondary amorphous phase. Experiments conducted in ultrasonic baths have also been reported to decrease the crystallization time between conventional water baths and ultrasonic baths (Bukhari et al., 2015).

The application of ultrasound power to the mother liquor is a popular pretreatment for zeolite synthesis, and it offers a simple way of accelerating the crystallization process and fine-tuning the properties of crystalline zeolites. The reduced synthesis time is mainly attributed to the enhanced crystallization kinetics caused by the fragmentation of seeds or nuclei. Compared with the conventional methods, the sonochemical method modifies the properties of the zeolites (Zhou et al., 2022).

FAU zeolite with Y-structure can be synthesized without any organic structure directing agent (OSDAs) at low temperatures and applying ultrasound-assisted aging, finding a change in the properties of the zeolite, decreasing crystal and particle size and increasing the Si/Al ratio, crystallinity and surface area with hierarchical mesopores (Reinoso et al., 2018). These and many more benefits have been obtained by applying ultrasound in zeolitization processes (Table 1) of other zeolites that can co-crystallize or interconvert into FAU. One of the most significant advantages of producing smaller crystals of zeolite is the increase in the number of acid sites on its surface, greater pore volume, and enhanced stability, improving the diffusion of reagents and products, which allows it to have more significant activity in a large distribution of applications, such as in the FCC process, being able to generate lighter gasoline and diesel of better quality (Ahmedzeki et al., 2016).

3 POST SYNTHESIS TREATMENT

The first study on ultrasound-assisted modification of zeolite Y was reported by (Hosseini et al., 2015), who investigated an ultrasound-assisted dealumination of zeolite Y in an ethanol-acetylacetone solution as a chelating agent (Table 2). Achieving a fast (30 min) and comfortable dealumination under simple conditions and combining conventional and ultrasonic methods. This process allows the removal of aluminum or silica from the structure, producing a migration and reorganization of structural atoms, generating a silica-rich zeolite (Zang et al., 2023). Positioning dealumination and desilication by increasing the solid and solvent interaction (Hosseini et al., 2015).

Table 2. Post-synthesis treatment under irradiation with ultrasound for modifying faujasite with different elements.

Modification	Zeolite	Modified Zeolite	Sonication conditions	Ref.
Mesopore Formation	Y	Dealumination	28 kHz, 600 W, 30 min	(Hosseini et al., 2015)
	Y	Mesoporous Y	60 Hz, 5 min, 65°C	(R. Zhang et al., 2020)
	Y	Desilication	20 kHz, 60 W, 30 min, 80 ºC	(Kuterasiński et al., 2020)
	Y	Hierarchical Y	20 kHz, 200 W, 20, 40 y 60 min, 30 °C	(Oruji et al., 2018)
Alkaline and Alkaline Earth Metals	NaY	K/Na-Y	28 kHz, 100 W, 10 min, RT	(Rakmae et al., 2016)
	NaX	Li-X	20 kHz, 500 W, 0-300 min, 70 ºC	(Erten-Kaya & Cakicioglu-Ozkan, 2012)
	NaX	Ca-X	20-35 kHz, 125-320 W, 0-300 min, 70 ºC	(Çakıcıoğlu-Özkan & Erten-Kaya, 2017)
	NaX	Ce-X	20-35 kHz, 125-320 W, 0-300 min, 70 ºC	
	NaX	Ce-X	20-35 kHz, 125-320 W, 0-300 min, 70 ºC	

continued on following page

Table 2. Continued

Modification	Zeolite	Modified Zeolite	Sonication conditions	Ref.
Metals	Y	La NPs/Y	200 W, 20 kHz, 30 min, RT	(Oruji et al., 2019)
	Y	Pt NPs/Y	150 W, 1 h, RT	(Dehghani & Tadjarodi, 2020)
	Y	$NiFe_2O_4$/Pd NPs/Y	150 W, 1 h	(Dehghani et al., 2019)
	Y	Pd NPs/Y	150 W, 1 h	(Tadjarodi et al., 2018)
	Y	Ni/Co NPs@Y	20 kHz, 90 W, 45 min	(Sharifi et al., 2015)
	Y	Fe@Y	5 o 30 min, 60 °C, pH=3-9	(Reyes Villegas, De León Ramirez, Perez-Sicairos, et al., 2024)
	Y	Cu@Y	20 kHz, 60 W, 20 min, RT	(Jodłowski et al., 2021)
	USY	Cu@USY		
	Y	Fe@Y	20 kHz, 60 W, 20 min, 60 °C	(Chlebda et al., 2018)
	USY	Fe@USY		
	USY	Cu@USY	20 kHz, 60 W, 20 min ice bath	(Kuterasiński et al., 2019)
		Co@USY		
		Fe@USY		
		Cr@USY		
	Y	$Bi_2Sn_2O_7$-C_3N_4/Y NPs	150 W, 1 h	(Heidari et al., 2018)
	NaX	$NiFe_2O_4$/X	1 h, 25 °C	(Sadeghi et al., 2020b)
		CdS NRs/ $NiFe_2O_4$/X	30 min; 25 °C	
	NaX	$MgFe_2O_4$/CdO/X	1 h, RT	(Sadeghi et al., 2021)
	NaY	MoS_2 NFs/CdS NRs/ $MgFe_2O_4$/NaY	1h, RT	(Sadeghi et al., 2020a)
		$MgFe_2O_4$/NaY		
		MoS_2 NFs/CdS NRs/NaY		

Metal-Zeolite= exchanged metal ion, Metal/Zeolite= supported metal, Metal@Zeolite= fixed or encapsulated metal, NPs=nanoparticles, NFs= nanofibers, NRs=nanorods, RT= room temperature, USY= ultrastable Y zeolite.

Subsequently, post-synthesis studies were carried out for dealumination and sonochemical desilication of zeolite Y. Evidencing that sonochemistry is a more energy and time-efficient method, generating hierarchical mesoporous zeolites from a sequential dealumination-desilication process. Emphasizing that ultrasonic irradiation of a Y zeolite in an alkaline medium is a promising alternative for desilication compared to conventional hydrothermal heating, reducing the treatment time from 30 min to 5 min (R. Zhang et al., 2020). This ultrasound-assisted alkaline treatment creates mesopores and less destruction of the zeolitic structure (Oruji et al., 2018). Likewise, this sonochemistry-assisted desilication produces a hierarchical Y zeolite irradiating for 30 min, obtaining greater mesoporosity and improved catalytic properties (Kuterasiński et al., 2020). These works suggest the possibility of further investigations between ultrasonic irradiation and zeolites, such as irradiation power, solvent type, and solution temperature (Oruji et al., 2018).

In addition to the synthesis of zeolites, FAU-type structures are also benefited when modified by an ultrasound-assisted method with different ions (Li^+, Na^+, K^+, Mg^{2+}, Ca^{2+} and Ce^{3+}), metal species (Cr, Fe, Co, Ni, Cu, Mo, Pd, Cd, Pt and La) and compounds such as $Bi_2Sn_2O_7$-C_3N_4, TBAOH and NH_4NO_3 (Table 2). Applying ultrasound in different conditions increases efficiency, reduces treatment time, promotes homogeneous species distribution and increases catalytic activity and selectivity compared to the conventional methods evaluated.

Likewise, the modification of the ion exchange sites occupied by alkaline cations, alkaline earth metals and transition metals has been studied to improve the properties of FAU zeolites. When applying an ultrasound-assisted ion exchange to a FAU zeolite, the concentration of the alkali and alkaline earth metal cations (mono, di and trivalent) introduced in the cation exchange sites increased compared to the values achieved by the traditional method (Çakıcıoğlu-Özkan & Erten-Kaya, 2017; Erten-Kaya & Cakicioglu-Ozkan, 2012), as well as the catalytic performance (Rakmae et al., 2016). Equally, when introducing transition metal ions, the properties they obtained (resistance to halogens, low cost, high catalytic activity, and selectivity) made them substitutes for precious and rare metal zeolite catalysts. Introducing transition metals into zeolites can provide catalysts with oxidation properties, creating a synergistic effect with the inherent properties of the zeolitic framework itself (J. Zhang et al., 2022).

Based on these data, sonochemistry has facilitated and accelerated the synthesis of highly active composite metal zeolites, e.g., $Bi_2Sn_2O_7$-C_3N_4/Y nano photocatalysts (more uniform dispersion) (Heidari et al., 2018), CdS NRs/$NiFe_2O_4$/X (Sadeghi et al., 2020b), $MgFe_2O_4$/CdO/X (Sadeghi et al., 2021) and MoS_2 NFs/CdS NRs/$MgFe_2O_4$/NaY (Sadeghi et al., 2020a). Materials prepared by sonication at room temperature demonstrate the versatility and convenience of applying ultrasonic methods.

When modifying zeolite Y with different metals by ultrasound (Table 3), demonstrated that ultrasound irradiation has the potential to be an environmentally friendly, time-effective and low-cost approach (Dehghani & Tadjarodi, 2020), producing catalysts with a uniform distribution of nanoparticles while maintaining high catalytic selectivity (Dehghani et al., 2019). These improved properties are attributed to the effects of ultrasonic irradiation during the synthesis of the nanoparticles by the accelerated dispersion of the active species on the surface of the zeolite, which translates into most of the benefits observed compared to conventional methods (Oruji et al., 2019).

Table 3. Ultrasound-assisted zeolite Y modification with metals.

Me Zeolite	**Contribution**	**Effect of Sonication on Catalyst**	**Sonication Conditions**	**Ref.**
Pt@Y	US-P	Material for reactions in fuel cells	1 g FAU and 0.04 g H_2PtCl_6 for 60 min	(Dehghani & Tadjarodi, 2020)
Pd@Y	US-P	Improved Suzuki-Miyaura coupling reaction	1 g FAU and 0.04 g $PdCl_2$ for 60 min	(Tadjarodi et al., 2018)
$NiFe_2O_4$/ Pd@Y	US-S of $NiFe_2O_4$ and US-P of Pd-NP	Selective for Suzuki Miyaura coupling	1 g FAU, 0.05g $PdCl_2$ and 0.05 g $NiFe_2O_4$ for 60 min	(Dehghani et al., 2019)
La@Y	Comparison of IM and US-P	Selective for hydrocarbon fractionation	FAU and $La(NO_3)_36H_2O$ at pH 8.5 for 30 min	(Oruji et al., 2019)
NiCo@Y	US-P	Enhanced production for biogas	FAU, $Ni(NO_3)_2$ 0-7% and $Co(NO_3)_2$ 0-7% for 45 min	(Sharifi et al., 2015)
$MgFe_2O_4$/ NaY	US-S	Increased methylene blue degradation	0.15g of FAU and 0.15 g of $MgFe_2O_4$ for 30 min	(Sadeghi et al., 2020a)
$CoFe_2O_4$/Y	US-S	Synergistic effect for lead extraction	1.5 g FAU and 0.5 g of $CoFe_2O_4$ for 1 h	(Paris et al., 2020)
$MgFe_2O_4$/Y	US-S	Adsorbent for Co^{2+} and Mn^{2+}	1.5 g of FAU and 0.5 g $MgFe_2O_4$ for 30 min	(Meirelles et al., 2023)
Fe_3O_4@Y	US-P	Magnetic adsorbent for Cs^+ and Sr^{2+}	FAU in 1 M $FeCl_3$ and 2 M $FeCl_2$	(Shirani et al., 2014)
Cu@Y	Comparison of IE and US	Increased activity and selectivity for NO_x SCR	1.5 g of FAU in 100 mL at 0.5 M of $Cu(NO_3)$	(Jodłowski et al., 2021)
Fe@Y	Comparison of IE and US	Change in acid sites and higher efficiency for NO_x SCR	FAU in 0.5 M $Fe(NO_3)_3$ for 20 min <60 °C	(Chlebda et al., 2018)
Fe@Y	Comparison of IE and US	Increased BAS concentration	FAU in 5 mM of Fe nitrate for 20 min	(Kuterasiński et al., 2019)
Fe@Y	Comparison of pH by US	Increased selectivity and Fe speciation	30 g Y in 500 mL of 0.023 M Fe^{2+} at pH 3 for 5 min	(Reyes Villegas, De León Ramirez, Perez-Sicairos, et al., 2024)

continued on following page

Table 3. Continued

Me Zeolite	Contribution	Effect of Sonication on Catalyst	Sonication Conditions	Ref.
Fe@Y	Comparison of IE and US	Increased catalytic activity	30 g Y in 500 mL of 0.023 M Fe^{2+} at pH 3 for 120 min	(Reyes Villegas, De León Ramirez, Pérez-Cabrera, et al., 2024)
Fe@Y	Comparison of pH by US	Changes in Si/ Al ratio, pore distribution, and Fe speciation	2 g of FeY in 40 mL buffered at pH 5 or 9 with 132μL 0.5 M Fe^{2+} and 62 μL at 1M of Fe^{3+}	(Reyes Villegas, De León Ramirez, Pérez-Cabrera, et al., 2025)

Me@Y= Metal species inside zeolitic cavities, Me/Y= Metal species supported on zeolitic surface, IE = ion exchange, US= ultrasound, IM= impregnation, US-P= ultrasound assisted precipitation, US-S= ultrasound assisted support.

For the modification of HY with La nanoparticles, three different methods were compared: impregnation, precipitation, and a combination of ultrasonically assisted precipitation, revealing that incorporating ultrasound better dispersed La nanoparticles producing a zeolite-based catalyst with moderate acidity resistance, longer shelf life, and less coke formation (Oruji et al., 2019). Likewise, a quick and low-temperature sono-assisted method for the in-situ synthesis and support of Pt nanoparticles onto zeolite Y produced 14 nm metal crystals homogeneously distributed on the surface (Dehghani & Tadjarodi, 2020). Without being the exception, Pd nanoparticles (Tadjarodi et al., 2018) and Pd and nickel ferrite (Dehghani et al., 2019) have also benefited from using an ultrasound-assisted methodology.

These same sonochemistry-assisted methods have also been used to synthesize and optimize Ni/Co bimetallic catalysts over zeolite Y, observing the same improvements detected for Y zeolite catalysts with La, Pd and Pt metals (Sharifi et al., 2015). However, the results are more complex when the same methods are applied to iron (Fe) (Table 3[REMOVED REF FIELD]). Fe can coexist in different oxidation states in the same catalyst sample. Therefore, the analysis for identifying the active form of Fe in zeolite materials is challenging due to the wide distribution of possible structures of Fe species in zeolites. Within this wide speciation, ultrasonic-prepared Fe zeolite catalysts consist primarily of small Fe oxide clusters on the surface. Zeolite Y and USY-based catalysts are classified as highly charged and contain even more iron than the literature shows according to the cation exchange sites. Significant amounts of Fex3+Oy oligomeric groups can be observed for ultrasonically prepared catalyst samples. These $Fe_x^{3+}O_y$ species are located within the pores of zeolite, and it is suspected that they could serve as active sites (Chlebda et al., 2018).

However, these Fe zeolites' reaction mechanisms are not so straightforward, which means synergy between active species could influence the evaluated chemical reactions. These same effects were observed not only with Fe-modified zeolite Y but also for Cu, Co and Cr species by ultrasound-assisted ion exchange, concluding that Brønsted acid site concentrations are higher than samples modified by the conventional method (Kuterasiński et al., 2019). Following this method, a commercial USY zeolite with Fe was modified by controlling the pH with a buffer solution with the sonoassisted method, finding that several active Fe sites can be tuned (through pH, sonication time and amount of Fe), such as Fe nanoparticles and α-Fe, for selective hydrolysis of lactose at 30 or 60 °C, respectively (Reyes Villegas, De León Ramirez, Perez-Sicairos, et al., 2024). The activity for lactose hydrolysis was higher at 60 °C for the 1) sono-assisted Fe ion-exchanged sample and 30 °C for the 2) post ion-exchanged Fe-modified zeolite. When comparing this method for the selective oxidation of benzene to phenol (SOBP), the most promising material was a two-step Fe-modified zeolite Y, a Fe-exchange process followed by a pH-controlled Fe modification. When analyzing the pH-controlled post-ion-exchange Fe modification method for zeolite Y, the process allowed for the shift of speciation from oxide to hydroxide species and tuning of the Si/Al ratio and pore architecture. Interestingly, when evaluating the ultrasonic irradiation time in the Fe ion-exchange process, the catalytic activity for phenol increased by only treating the zeolitic suspension for 5 min. From these studies, we can conclude that 1) a 5 min sono-assisted Fe ion exchange and 2) a post-ion exchange pH-controlled Fe modification method produce promising catalysts for the SOBP.

4 FE-Y ZEOLITE CATALYST FOR SOBP: A PRELIMINARY STUDY

Focusing on the benzene-to-phenol reaction, industrially, 90% of the production of phenol (a value-added product) from benzene (a carcinogenic contaminant) follows a three-step process using cumene as the precursor. This method produces a product that needs purification to use phenol as a precursor molecule for the chemical synthesis of nylon-66, various polymers, and pharmaceuticals (Pérez & Manuel Sánchez-Sánchez Editors, 2018; Snyder, Bols, Schoonheydt, et al., 2018). This "Sunuco/UOP Phenol process" converts cumene into phenol, acetone and α-methyl styrene, demanding phenol purification and refining of α-methyl styrene (Schmidt, 2005). To solve this issue, research has been carried out on designing a one-step process for selective oxidation of benzene to phenol (SOBP) employing various catalysts (Table 4), including Fe-containing zeolites employing oxidants such as H_2O_2, N_2O, O_2, and H_2 in acetonitrile or water solutions (Mancuso et al., 2020).

Table 4. Comparison of the literature review for the oxidation of benzene to phenol over a Fe-zeolite catalyst in a water-based heterogeneous reaction.

Zeolite	Introduction Method	Drying or Calcination	Reaction Conditions					Benzene Conversion	Phenol Selectivity	By-products	Ref.
			Phase	Solvent	Oxidant	Time (h)	Temp (°C)				
ZSM-5	SSIE	900 °c for 2 h	Gas	He	N_2O	8	450	40	98	NM	(Wącław et al., 2004)
ZSM-5	HTS	550 °C in air	Gas	He	N_2O	25	400	-	>98	CAT, HQ, BQ	(Ivanov et al., 2003)
ZSM-5	SSIE	3-SP	Gas	He	N_2O^*	1 µL pulse	RT	ID	ID	-	(Xia et al., 2008)
ZSM-5	IMP	4-SP	Gas	He	N_2O^*	0.5	RT	>40	>95	Fe-PhOH	(Snyder, Bols, Rhoda, et al., 2018)
Beta	IMP	4-SP	Gas	He	N_2O^*	0.5	RT	>40	>95	Fe-PhOH	
Beta	IE	2-SP	Liquid	CH_3CN	H_2O_2	6	60	31	17	CAT, HQ	(Xiao et al., 2020)
Beta	SSIE	5 h 550 °C in air	Liquid	CH_3CN	H_2O_2	6	60	43	18	CAT, HQ	
S-1	IMP	5 h 550 °C in air	Liquid	CH_3CN	H_2O_2	6	60	8	100	-	(Xiao et al., 2019)
ZSM-5	IMP	5 h 550 °C in air	Liquid	CH_3CN	H_2O_2	6	60	47	99	HQ	
ZSM-5	HTS	5 h 550 °C in air	Liquid	CH_3CN	H_2O_2	6	60	60	95	CAT, HQ	
ZSM-5	HTS	5 h 550 °C in air	Liquid	CH_3CN	H_2O_2	8	60	12	97	CAT, BQ	(Yue et al., 2020)
Y	IE	80 °C overnight	Liquid	C_6H_6	H_2O_2	2	80	>80	>80	NM	(Ahmed & Thabet, 2015)
Y	IE and $TMCL_1$	RT	Liquid	CH_3CN	H_2O_2	20	70	<4	>96	ND	(Mori et al., 2008)

continued on following page

Table 4. Continued

Zeolite	Introduction Method	Drying or Calcination	Reaction Conditions					Benzene Conversion	Phenol Selectivity	By-products	Ref.
			Phase	Solvent	Oxidant	Time (h)	Temp (°C)				
Y	IE and $TMCL_1$	RT	Liquid	CH_3CN	H_2O_2	24	50	1	100	-	(Yamaguchi et al., 2018)
Y	IE and $TMCL_2$	RT	Liquid	CH_3CN	H_2O_2	24	50	1	100	-	
Y	IE and $TMCL_3$	RT	Liquid	CH_3CN	H_2O_2	24	50	1	100	-	
Y	IE and $TMCL_1$	RT	Liquid	CH_3CN/ H_2O	H_2O_2	24	50	14	~80	CAT, HQ	
Y	IE and $TMCL_1$	RT	Liquid	H_2O	H_2O_2	24	50	2	~57	CAT, HQ	
Y	IE and $TMCL_2$	RT	Liquid	H_2O	H_2O_2	24	50	11	~64	CAT, HQ	
Y	IE and $TMCL_3$	RT	Liquid	H_2O	H_2O_2	24	50	8	~67	CAT, HQ	
Y	IE	1h 150°C	Liquid	CH_3CN	H_2O_2	6	70	UR	UR	-	(Abbo & Titinchi, 2009)
Y	IE	1h 150°C	Liquid	H_2O	H_2O_2	6	70	23	100	-	
Y	IE	1h 150°C	Liquid	CH_3CN	H_2O_2	6	80	27^{***}	100^{***}	CAT, HQ	
Y	IMP	500 °C for 4 h	Liquid	H_2O	H_2O_2	1	50	50^{***}	70^{***}	CAT, HQ, BQ	(Long et al., 2009)
Y	SAIE	90 °C overnight	Liquid	H_2O	H_2O	0.5	90	76	25	NM	(Reyes Villegas, De León Ramirez, Pérez-Cabrera, et al., 2024)
Y	NFI	90 °C overnight	Liquid	H_2O	H_2O	0.5	90	75^{N}	4^{N}	NM	
Y	NFI	90 °C overnight	Liquid	H_2O	H_2O	0.5	90	39^{N**}	91^{N**}	ND	(Reyes Villegas, De León Ramirez, Perez-Cabrera, et al., 2025)
Y	SAM	90 °C overnight	Liquid	H_2O	H_2O	0.5	90	77	24	ND	
Y	SAM	90 °C overnight	Liquid	H_2O	H_2O	0.5	90	73^{**}	92^{**}	ND	
Y	SAM	90 °C overnight	Liquid	H_2O	H_2O	0.5	RT	91^{**}	100^{**}	-	

$X_{C_6H_6}$= Benzene conversion, $S_{C_6H_5OH}$= Phenol selectivity,S-1= Silicalite-1, IE= ion exchange, SSIE= solid-state ion exchange, HTS= hydrothermal synthesis, IMP= impregnation, SAIE= Sonoassisted ion exchange, SAM= Sonoassisted modification, TMC= Transition metal complex, L_1=ligand $[Fe(bpy)_3]^{2+}$, L_2=ligand $[Fe(phen)_3]^{2+}$, L_3=ligand $[Fe(terpy)_2]^{2+}$, NFI=No Fe introduced, 2-SP= two-step procedure: 1) 100 °C overnight, 2) 5 h 550 °C in air, 3-SP=three-step procedure: 1) 1h 550 °C in O_2, 2)1 h 900 °C in He, 3) N_2O/He 250 °C, 4-SP= four-step procedure: 1) 550 °C in air 2) 2 h 900 °C He, 3) 1 h 700 °C H_2, 4) N_2O/He 160 °C RT= room temperature,*=NO_2 activated,**=aqueous dichloromethane extraction ***Conversion and selectivity for phenol hydroxylation, CAT=catechol, HQ=Hidroquinone, BQ=Benzoquinone, NM= not measured, ID= $X_{C_6H_6}$ identified not measured, ND= not detected, UR= Unreacted benzene, ^{N}= No Fe introduced.

Among the most studied Fe-zeolite catalysts for the selective hydroxylation of benzene to phenol (SOBP) is the N_2O-activated Fe-ZSM-5, whose active site is characterized as an extraframework (EF) α-Fe(II) anchored in the center of a six-membered ring (6MR), which is activated by nitrous oxide and heat (above 300 °C) to form the reactive intermediate of α-O (Joyner & Stockenhuber, 1999; Wącław et al., 2004; Xia et al., 2008; Xiao et al., 2019). Fe-ZSM-5 is not the only zeolite that can stabilize α-Fe; this active site can also be introduced in zeolites with 6MR, such as BEA, FER, CHA, and FAU (Bols, Devos, et al., 2021). Therefore, theoretical calculations and simulations have shed insight into the structural configuration of α-Fe/O, although only ZSM-5, FER, CHA and BEA were considered (Bols et al., 2018; Bols, Snyder, et al., 2021; Hallaert et al., 2017; Rhoda et al., 2022; Snyder, Böttger, et al., 2018; Snyder et al., 2016, 2021). BEA zeolite was the most promising material selected for experimental research for the SOBP (Snyder, Bols, Rhoda, et al., 2018).

The stabilization of α-O achieved the activation of α-Fe in BEA zeolite through the decomposition of N_2O near 300 °C. This same decomposition reaction occurs in Fe-exchanged FAU zeolites near the same temperatures (Li et al., 2008), suggesting the stabilization of α-Fe. The interest in the Fe-FAU system becomes more attractive when considering that Fe-containing zeolites are enzyme-mimetic materials due to possessing Fe-active sites with high selectivity at ambient temperature and pressure for POBP (Snyder, Bols, Schoonheydt, et al., 2018; Y. Wang et al., 2023). The only well-studied catalytically active site in Fe zeolite is the α-Fe/α-O site, which is involved in low-temperature SOBP (Bols et al., 2020). It is not only necessary to distinguish between the types of iron species present and the precise active sites for specific reactions (van Bokhoven & Lamberti, 2014) but also to select the conditions for their reproducible preparation.

Apart from the ease of catalyst preparation, an ideal catalyst for heterogenous reactions should be able to catalyze any reaction, including the SOBP, in water, with air as the oxidizer, at room temperature and room pressure. This SOBP reaction can be catalyzed with a Fe-based non-zeolitic material in water as the solvent and H_2O_2 as the oxidant, successfully hydroxylating benzene to phenol with hydroxylated phenolic by-products such as catechol, resorcinol, hydroquinone, and benzoquinone (Mancuso et al., 2023). As expected, a Fe-Y zeolite catalyst under the same conditions (water and H_2O_2 as an oxidant) also produces the same hydroxylated phenolic compounds (Long et al., 2009). Suggesting that using H_2O_2 with water as a solvent may increase the complexity of the reaction by promoting the hydroxylation of phenol. This overoxidation of phenol decreases the selectivity. For evaluating the source of oxygen in phenol, an experiment was performed by changing the solvent from water to acetonitrile and by analyzing the isotopic distribution of oxygen with gas chromatography coupled to a mass spectrum (GC/MS), using labeled $H_2{}^{16}O_2$,

$H_2{}^{18}O_2{}^{16}O_2$ and $^{18}O_2$ as the oxidizer. The experiment showed that the oxygen atom in phenol comes from H_2O_2 (Abay et al., 2022). However, this same study showed that when H_2O_2 is used as the oxidant, phenol selectivity decreases by promoting the formation of hydroxylated phenolic compounds. Hence, by changing the oxidant from H_2O_2 to air and evaluating the reaction with labeled water ($H_2{}^{16}O$and $H_2{}^{18}O$) with a non-zeolitic photocatalyst, high phenol selectivity was obtained, and the source of oxygen in phenol was measured with a GC/MS to be from water, the solvent (Xie et al., 2023). This supports the idea that using H_2O_2 over oxidizes benzene and that high selectivity for phenol can be obtained in a water-based environment using water as both the solvent and oxidizer, extracting the extra oxygen from H_2O.

In previous work, α-Fe has been introduced in a series of Fe-FAU zeolites prepared by a sonoassisted method at 60 ℃ (Reyes Villegas, De León Ramirez, Pérez-Cabrera, et al., 2025), and adding to the evidence that FAU zeolites exchanged with Fe can produce phenol (Abbo & Titinchi, 2009), these sono-assisted treated Fe modified samples could catalyze the SOBP. When these ultrasound-assisted modified Fe-FAU zeolites were evaluated for the SOBP in a water-based system with H_2O as the oxidant and at room temperature, phenol production and selectivity were enhanced. For the sono-assisted ion-exchanged (SAIE) sample, the selectivity increased from 4 to 25% (Reyes Villegas, De León Ramirez, Pérez-Cabrera, et al., 2024), and when precipitating Fe species (sono-assisted modification, SAM) to a Fe-FAU (Fe exchanged FAU by a conventional ion-exchange method), the selectivity rose to near 100% (Reyes Villegas, De León Ramirez, Perez-Cabrera, et al., 2025). Combining the DFT calculations (Snyder, Bols, Rhoda, et al., 2018) and the above-mentioned studies, the following mechanism was proposed for the room-temperature α-Fe/O catalytic center in zeolite Y for the SOBP (Figure 1). When Benzene adsorbs to α-O, an NIH shift is crucial for phenol desorption, and if not occurring, the active site ends up poisoned. However, if phenol desorption is achieved, the remaining α-Fe can follow a single turnover or regenerate the α-O through a water nucleophilic attack (WNA). The exact mechanism for tilting the reaction towards phenol desorption and the LAS assisting the WNA regeneration remains to be elucidated. Suggesting further research on the dynamic effect of water and EFAl species in zeolite catalysts (Q. Liu & van Bokhoven, 2024; Ravi et al., 2020; Resasco et al., 2021). The ultrasound-assisted prepared Fe-zeolite should be tested under different solvents with different polarities in different water/solvent ratios to optimize the reaction conditions. Nevertheless, assisting the introduction of Fe in zeolites with ultrasound showed potential for producing active Fe-based zeolitic catalysts at relatively low temperatures that could catalyze the SOBP.

Figure 1. Gas chromatography A) of the extraction with dichloromethane for the partial oxidation of benzene to phenol (POBP) at 30 °C for 30 min with an initial benzene concentration of 1000 ppm by a sono-assisted modified Y zeolite with a molar ratio of 1:1 of divalent and trivalent iron catalyst (FeYII/III) with mass spectrum of the detected molecules at 4.27 min for benzene B) and 5.08 min for phenol C) and the proposed reaction mechanism D).

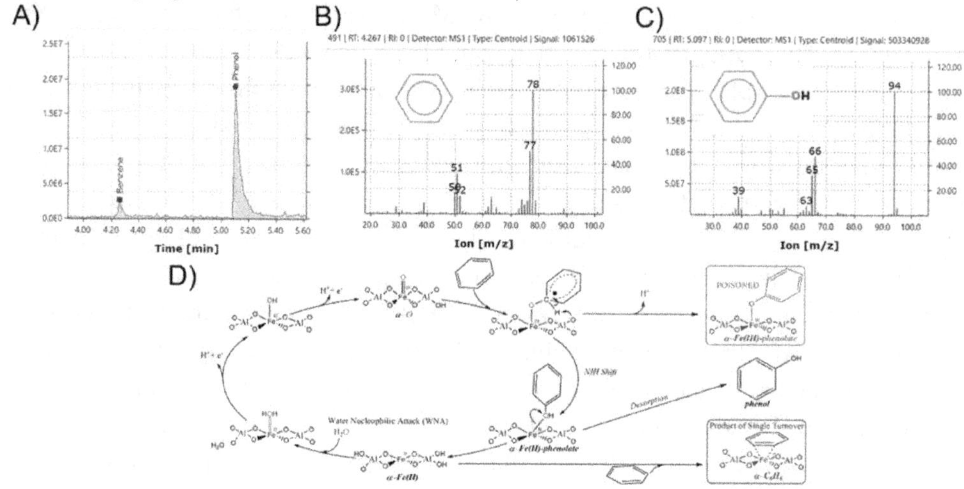

Reprinted from *(Reyes Villegas, De León Ramirez, Pérez-Cabrera, et al., 2025)* **with permission from Elsevier.**

Hence, modifying zeolite Y with Fe species using the sonoassisted method is proposed to develop more homogeneous, reproducible and selective conditions for preparing specific active sites of Fe to stabilize iron species in a zeolite (Chlebda et al., 2018). Also, considering the docile and variable properties of zeolite scaffolds, which allows for an easy modification of the materials, and the complex chemistry of iron oxides-hydroxides, these systems still need further research. In this preliminary work, zeolite catalyst sonochemically synthesized (SY), Fe ion-exchanged (Fe5SY), and post-IE Fe modification (Fe5SYFeII/III), with all three steps assisted by ultrasonic irradiation, were evaluated for the SOBP in an aqueous medium with water as an oxidizer at room temperature and room pressure. In addition, these samples were compared to the activity of their commercial zeolite Y counterparts (Y, FeY and FeYFeII/III).

4.1 Materials and Methods

4.1.1 Materials

Materials and reagents used in this work include commercial zeolite NH_4-Y (CBV 500, NH_4-Y, USY) with a nominal Si/Al ratio of 2.6 provided by Zeolyst International and $FeSO_4$ (Cat. No. 215422) and HNO_3 (Cat. No. 225711) from Sigma Aldrich. Materials and reagents used in this work include commercial faujasite in ammonium form with a nominal Si/Al ratio of 2.6, typical for Y-zeolite (CBV 500, NH_4-Y, USY), provided by Zeolyst International, and $FeSO_4$ (Cat. No. 215422), $Fe_2(SO_4)_3$ (Cat. No. 307718), NH_4OH (Cat. No. 05002), and HNO_3 (Cat. No. 225711) from Sigma Aldrich. The reactants used for this work such as NaOH (Catalog No. 221465), $Al(OH)_4Na$ (Catalog No. 13404), $Na_2O(SiO_2)_x{\cdot}xH_2O$ (Catalog No. 338443), C_6H_6O (Catalog No. 185450) and C_6H_6 (Catalog No. 319953) were supplied by Sigma Aldrich. A Branson Sonifier from the SLP Series (SLPt, 150W and 40 kHz) equipped with a tapered 1/8" Microtip delivered ultrasonic energy operating at a 15% amplitude in a continuous mode with a specified total processing time.

4.1.2 Sonoassisted Synthesis of Zeolite Y

To develop the sono-assisted synthesis of zeolite Y, a clear precursor solution with a molar ratio of 9 Na_2O: 0.7 Al_2O_3: 10 SiO_2: 160 H_2O was used. The synthesis consisted of three main stages: aging (called mixing due to the low time consumed), nucleation, and crystallization. The procedure conditions were as follows. Firstly, 25.33 mL of a sodium silicate solution was placed in a polypropylene (PP) bottle, referred to as solution A (Sol. A). In a separate PP bottle, 5.27 g of NaOH was dissolved in 21.53 g of distilled water and mixed under ultrasonic irradiation until transparent; then, 2.13 g of sodium aluminate was slowly added to this solution and continued treatment with ultrasound until a transparent suspension was reached (Sol. B). Both Sol. A and Sol. B was cooled for 10 minutes in a cryogenic freezer; then, the cooled Sol B was placed in an ice bath at a controlled temperature of 5 °C. Ultrasound was turned on, after which Sol A was added dropwise into Sol. B, over 10 minutes, and left to react for a further 20 minutes, stirring with ultrasound. The resulting clear suspension was changed to a 60 °C bath for the nucleation stage and sonicated again for 30 minutes. Next, the suspension was crystallized in an oven at 90 °C under static conditions for 1.5 h. Finally, the reaction mixture was filtered through a nitrocellulose membrane with a pore diameter of 0.45 μm under vacuum with excess rinse water, and the precipitate was dried in an oven at 100 °C for 24 h. The resulting material is named sono-synthesized zeolite Y (SY). All the prepared catalysts were filtered, washed, dried and grounded (FWDG) as explained above.

4.1.3 Sonoassisted Ion Exchange with Fe

Faujasite Y was modified with iron in two steps. Since zeolites are excellent ion exchangers, attempts to precipitate Fe species using ultrasonic irradiation on the initial zeolite would lead to competition between two processes - ion exchange and precipitation. Therefore, first, SY was modified by sono-assisted ion exchange, sonicating in a 0.23 M $FeSO_4$ solution, selecting an amount of Fe^{2+} ions in the liquid with three times exciding its theoretical cation exchange capacity, with a solid/liquid ratio of 1/20, for 5 minutes, at 60 °C (Fe5Y). This procedure was repeated for comparison using the commercial catalyst zeolite NH_4-Y zeolite Y, CBV-500; this Fe-exchanged zeolite was termed Fe5Y after the FWDG process (Reyes Villegas, De León Ramirez, Pérez-Cabrera, et al., 2024).

4.1.4 Sonoassisted Modification with Fe

This sono-assisted Fe-exchanged zeolite (Fe5SY) was then used as a scaffold to deposit different Fe species. In brief, 2 g of Fe5SY was suspended in an NH_4NO_3 buffer solution at pH 5 and sonicated for 30 min at 60 °C, with the addition of 132 µL of 0.5 M $FeSO_4$ (FeII) and 62 µL of 1 M $Fe_2(SO_4)_3$ (FeIII) immediately after starting sonication. The final concentration of Fe ions added to the zeolitic suspension was 1 wt. % of Fe with a molar ratio of 1:1 (II/III) of divalent (Fe^{2+}) and trivalent (Fe^{3+}) Fe ions, naming this sample as Fe5SYII/III after the FWDG process. For comparison, CBV-500 was ion exchanged as expressed above with slight changes such as stirring for 80 ºC for 2h, FWDG (FeY), then modified with different Fe species (FeYII/III) (Reyes Villegas, De León Ramirez, Pérez-Cabrera, et al., 2025).

4.1.5 Catalysis of SOBP

The catalytic activity was evaluated in a batch reaction; the OBP was assessed by placing 15 mg of the sample in 1 mL of solution with a benzene concentration of 1,000 ppm at 30 °C for 30 min (Figure 2). After the reaction time, the suspension was sonicated for less than 10 seconds and vigorously shaken to ensure a homogenous dispersion and disrupt any equilibrium if present. The suspension was centrifugated at 10,000 rpm for 10 min at 5°C to quench the reaction, and two aliquots were taken to measure the residual concentration of benzene and phenol in the supernatant. UV-Vis was used to measure the residual amount of benzene in the supernatant at 255 nm, calculating the moles of benzene consumed. Subsequently, moles of phenol formed were quantified using a colorimetric reaction with $FeCl_3$, measuring the violet phenoxide complex formed at 428 nm. Calibrations for benzene and phenol were done with 7-point linear regressions ranging from 5,000 to 0 ppm, linear model parameters for benzene with intercept = 0, slope = 3.9189 ± 6.13 E^{-4} and Adj. R^2 = 0.9985. As for the phenol model, the parameters were as follows: intercept = 0, slope = 3.4677 ± 4.522 E^{-5}, and Adj. R^2 = 0.9988.

Figure 2. Flow chart of the benzene to phenol reaction methodology measured by UV-Vis spectroscopy and gas chromatography coupled to mass spectra.

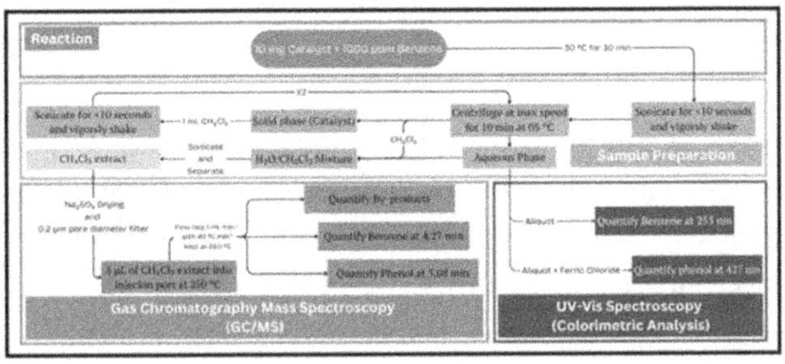

The conversion of benzene ($X_{C_6H_6}$) and selectivity for phenol ($S_{C_6H_5OH}$) for the POBP was calculated as follows:

$$X_{C_6H_6} = \frac{Moles\ of\ Benzene\ Consumed}{Inittial\ Moles\ of\ Benzene}$$

$$S_{C_6H_5OH} = \frac{Moles\ of\ Phenol\ Formed}{Moles\ of\ Benzene\ Consumed}$$

To confirm the presence of phenol and detect by-products of the POBP $X_{C_6H_6}$ and $S_{C_6H_5OH}$ was measured by employing gas chromatography coupled to a mass spectrometer (GC/MS). After the aliquot was taken for the colorimetric analysis, the suspension was recentrifuged (maximum speed for 1 min at 5 °C), recovering the supernatant (aqueous phase) and adding 1 mL of dichloromethane to the solid phase (catalyst). The dichloromethane suspension was homogenized by sonicating for less than 10 seconds and vigorous shaking and was centrifuged (10,000 rpm for 10 min at 5 °C), recovering the supernatant. The catalyst phase was extracted twice, and the obtained dichloromethane supernatant was mixed with the aqueous phase of the reaction. All the reaction mixtures were extracted twice. The dichloromethane extract was then dried with Na_2SO_4 (enough until no further agglomeration is seen) and filtered through a 0.2 μm pore diameter syringe filter. The mass spectra were obtained by an electron impact ionization technique in a Thermo Scientific Gas Chromatograph model TRACE 1310 coupled to a Thermo Scientific single quadrupole mass spectrometer model ISQ LT, where 3 μL of the sample were injected, with the injection port at 250 °C. The furnace temperature was 70 °C and the ion source 240 °C, with the transfer line at 250 °C. The heating ramp of the column in the furnace started at 70 ºC, maintaining this temperature for 3 min, rising at a speed of 40 ºC min^{-1} to 280 ºC, and kept at this temperature for 7 min. A flow rate of 1 mL min^{-1} was used, with a split of 20. As standards, 1000 ppm Benzene and phenol solutions were used to quantify the $X_{C_6H_6}$ and $S_{C_6H_5OH}$ of the reaction (Reyes Villegas, De León Ramirez, Perez-Cabrera, et al., 2025).

4.2 Results

Since the GC/MS technique is usually expensive, limited, and time-consuming, the detection of benzene and phenol was compared to a faster and less costly method, a colorimetric analysis with UV-Vis spectroscopy (Figure 3). The findings show no significant difference (p=0.9496) between these methodologies (GC/MS and Colorimetric analysis). Hence, UV-Vis spectroscopy is an attractive alternative to GC/MS for calculating phenol produced and residual benzene concentration. However, GC/MS is essential for elucidating a reaction mechanism for a novel catalyst.

Figure 3. Comparison of colorimetric and chromatographic measuring techniques for benzene conversion and phenol selectivity for the oxidation of benzene to phenol catalyzed by a commercial zeolite Y (CBV500) and sono-assisted synthesized zeolite Y (SY) modified with Fe by ultrasound-assisted ion exchange at 5 min (Fe5Y and Fe5SY) and pH 5 with Fe^{2+} *and* Fe^{3+} *(FeYII/III and Fe5SYII/III).*

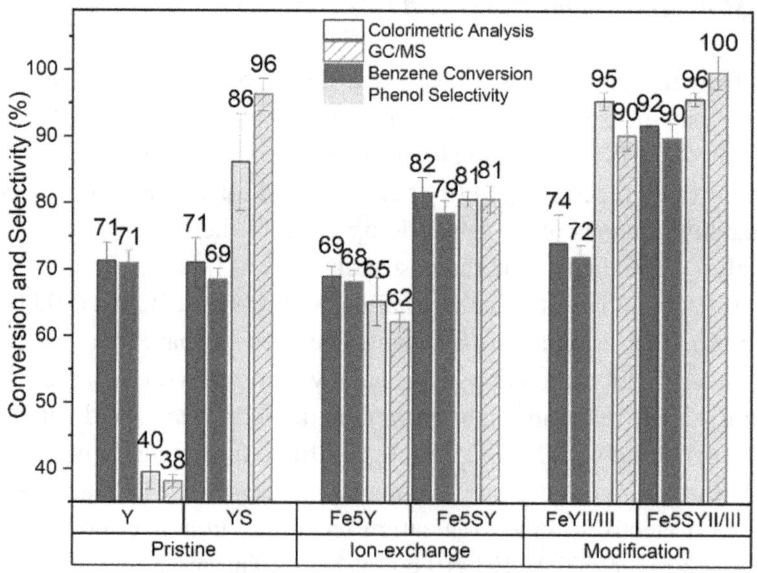

In these preliminary experiments, the commercial (CBV500) and sono-assisted synthesized zeolite Y (SY) were compared for the catalysis of benzene to phenol reaction. No by-products were detected for the pristine zeolites, and the selectivity for the modified FeYII/III catalyst was 55% higher, as expected from previous studies (Table 1). However, from the reaction from the ion exchange Fe5Y catalyst, more side-products were detected from the GC/MS, such as hydroxyphenol isomers (resorcinol, catechol and hydroquinone), maleic anhydride and maleic acid (a decomposition product of maleic anhydride). Suggesting different active sites involved in the reaction of benzene to phenol, activating a route to produce phenol and a minimum of two more routes, one that leads to hydroxyphenol and another to maleic anhydride. Nevertheless, the exact mechanism should be explored to elucidate the precise route of benzene to maleic anhydride and the active sites involved in the catalysis for the formation of phenol, hydroxyphenol and maleic anhydride since in the reaction with Y and FeYII/III, no by-products were detected.

To compare the efficiency between the commercial and sono-assisted synthesized zeolite Y (SY) as a catalyst for the SOBP, SY was sonochemically modified with Fe. The SY zeolite had high selectivity, indicating it to be a potential candidate for phenol production. When applying the sono-assisted Fe ion-exchange to SY (Fe5SY), selectivity increased compared to Fe5Y and decreased compared to SY. However, this selectivity improved when modifying with Fe species a buffer suspension at pH=5 in the presence of Fe^{2+} and Fe^{3+} cations (Fe5SYII/III). These samples confirm that the techniques used to synthesize a catalyst for the SOBF are an alternative to designing an even better catalyst and that the methodologies used to detect phenol are reliable.

4.3 Conclusions

This study successfully synthesized zeolite Y using a sonoassisted method from a clear solution in a significantly reduced total time of 4 hours. The sonoassisted synthesis of zeolite Y demonstrated a significant improvement in catalytic performance, with the 5 min sonication ion exchange process enhancing both conversion and selectivity, as shown by the comparison between commercial CBV500 (Fe5Y: Fe5Y: $X_{C_6H_6}$ 68%, $S_{C_6H_5OH}$=62%) and sonoassisted SY (Fe5YS: $X_{C_6H_6}$=79%, $S_{C_6H_5OH}$=81%). Furthermore, modifying the SY zeolite with Fe species enhanced the selectivity to values exceeding 99%, highlighting the potential of this approach for catalytic applications. While these findings underscore the efficacy of ultrasonic-assisted techniques and Fe modification in enhancing zeolite properties, this work remains in its preliminary stages. Verification and optimization of reaction conditions are essential to exploit this method's capabilities fully. Future research should focus on refining the synthesis parameters to achieve highly selective faujasite with shorter synthesis times and lower energy requirements, paving the way for more efficient and sustainable catalytic processes.

REFERENCES

Abay, T. A., Wanna, W. H., Natarajan, T., Tsai, Y.-F., Janmanchi, D., Jiang, J.-C., Abu-Reziq, R., & Yu, S. S.-F. (2022). Selective oxidation of benzene by an iron oxide carbonaceous nanocatalyst prepared from iron perchlorate salts and hydrogen peroxide in benzene and acetonitrile. *Molecular Catalysis*, *526*, 112397. DOI: 10.1016/j.mcat.2022.112397

Abbo, H. S., & Titinchi, S. J. J. (2009). Di-, tri- and tetra-valent ion-exchanged NaY zeolite: Active heterogeneous catalysts for hydroxylation of benzene and phenol. *Applied Catalysis A, General*, *356*(2), 167–171. DOI: 10.1016/j.apcata.2009.01.003

Ahmed, A. H., & Thabet, M. S. (2015). Physicochemical studies and heterogeneous hydroxylation of benzene on FeII, FeIII-Semicarbazone/zeolite-y clathrates. *Synthesis and Reactivity in Inorganic, Metal-Organic, and Nano-Metal Chemistry*, *45*(11), 1632–1641. DOI: 10.1080/15533174.2015.1031038

Ahmedzeki, N. S., Yilmaz, S., & Al-Tabbakh, B. A. (2016). Synthesis and Characterization of Nanocrystalline Zeolite Y. In *Journal Al-Khwarizmi. Engineering Journal (New York)*, *12*(1), •••.

Aldahri, T., Behin, J., Kazemian, H., & Rohani, S. (2016). Synthesis of zeolite Na-P from coal fly ash by thermo-sonochemical treatment. *Fuel*, *182*, 494–501. DOI: 10.1016/j.fuel.2016.06.019

Askari, S., & Halladj, R. (2012). Ultrasonic pretreatment for hydrothermal synthesis of SAPO-34 nanocrystals. *Ultrasonics Sonochemistry*, *19*(3), 554–559. DOI: 10.1016/j.ultsonch.2011.09.006

Askari, S., Miar Alipour, S., Halladj, R., & Davood Abadi Farahani, M. H. (2013). Effects of ultrasound on the synthesis of zeolites: A review. *Journal of Porous Materials*, *20*(1), 285–302. DOI: 10.1007/s10934-012-9598-6

Azizi, S. N., & Yousefpour, M. (2010). Static and ultrasonic-assisted aging effects on the synthesis of analcime zeolite. *Zeitschrift für Anorganische und Allgemeine Chemie*, *636*(5), 886–890. DOI: 10.1002/zaac.200900415

Bai, R., Song, Y., Lätsch, L., Zou, Y., Feng, Z., Copéret, C., Corma, A., & Yu, J. (2022). Switching between classical/nonclassical crystallization pathways of TS-1 zeolite: Implication on titanium distribution and catalysis. *Chemical Science (Cambridge)*, *13*(36), 10868–10877. DOI: 10.1039/d2sc02679a

Behin, J., Kazemian, H., & Rohani, S. (2016). Sonochemical synthesis of zeolite NaP from clinoptilolite. *Ultrasonics Sonochemistry*, *28*, 400–408. DOI: 10.1016/j.ultsonch.2015.08.021

Belviso, C., Cavalcante, F., & Fiore, S. (2013). Ultrasonic waves induce rapid zeolite synthesis in a seawater solution. *Ultrasonics Sonochemistry*, *20*(1), 32–36. DOI: 10.1016/j.ultsonch.2012.06.004

Belviso, C., Cavalcante, F., Lettino, A., & Fiore, S. (2011). Effects of ultrasonic treatment on zeolite synthesized from coal fly ash. *Ultrasonics Sonochemistry*, *18*(2), 661–668. DOI: 10.1016/j.ultsonch.2010.08.011

Bols, M. L., Devos, J., Rhoda, H. M., Plessers, D., Solomon, E. I., Schoonheydt, R. A., Sels, B. F., & Dusselier, M. (2021). Selective Formation of α-Fe(II) Sites on Fe-Zeolites through One-Pot Synthesis. *Journal of the American Chemical Society*, *143*(39), 16243–16255. DOI: 10.1021/jacs.1c07590

Bols, M. L., Hallaert, S. D., Snyder, B. E. R., Devos, J., Plessers, D., Rhoda, H. M., Dusselier, M., Schoonheydt, R. A., Pierloot, K., Solomon, E. I., & Sels, B. F. (2018). Spectroscopic Identification of the α-Fe/α-O Active Site in Fe-CHA Zeolite for the Low-Temperature Activation of the Methane C-H Bond. *Journal of the American Chemical Society*, *140*(38), 12021–12032. DOI: 10.1021/jacs.8b05877

Bols, M. L., Rhoda, H. M., Snyder, B. E. R., Solomon, E. I., Pierloot, K., Schoonheydt, R. A., & Sels, B. F. (2020). Advances in the synthesis, characterization, and mechanistic understanding of active sites in Fe-zeolites for redox catalysts. *Dalton Transactions (Cambridge, England)*, *49*(42), 14749–14757. DOI: 10.1039/D0DT01857K

Bols, M. L., Snyder, B. E. R., Rhoda, H. M., Cnudde, P., Fayad, G., Schoonheydt, R. A., Van Speybroeck, V., Solomon, E. I., & Sels, B. F. (2021). Coordination and activation of nitrous oxide by iron zeolites. *Nature Catalysis*, *4*(4), 332–340. DOI: 10.1038/s41929-021-00602-4

Boycheva, S., Zgureva, D., Miteva, S., Marinov, I., Behunová, D. M., Trendafilova, I., Popova, M., & Václaviková, M. (2020). Studies on the potential of nonmodified and metal oxide-modified coal fly ash zeolites for adsorption of heavy metals and catalytic degradation of organics for wastewater recovery. *Processes (Basel, Switzerland)*, *8*(7), 778. DOI: 10.3390/pr8070778

Bukhari, S. S., Behin, J., Kazemian, H., & Rohani, S. (2015). Conversion of coal fly ash to zeolite utilizing microwave and ultrasound energies: A review. *Fuel*, *140*, 250–266. DOI: 10.1016/j.fuel.2014.09.077

Çakıcıoğlu-Özkan, F., & Erten-Kaya, Y. (2017). Enhancement of Cerium exchange with Ultrasound. *Journal of the Turkish Chemical Society, Section A: Chemistry, 5*(sp.is.1), 53–64. https://doi.org/DOI: 10.18596/jotcsa.370769

Chlebda, D., Stachurska, P., Jędrzejczyk, R., Kuterasiński, Ł., Dziedzicka, A., Górecka, S., Chmielarz, L., Łojewska, J., Sitarz, M., & Jodłowski, P. (2018). DeNOx Abatement over Sonically Prepared Iron-Substituted Y, USY and MFI Zeolite Catalysts in Lean Exhaust Gas Conditions. *Nanomaterials (Basel, Switzerland), 8*(1), 21. DOI: 10.3390/nano8010021

Czuma, N., Zarebska, K., & Baran, P. (2017). Microwave and ultrasound energy in the process of fly ash zeolite synthesis and its application for CO<inf>2</inf> sorption. *2017 IEEE 7th International Conference Nanomaterials: Application & Properties (NAP)*, 03NNSA20-1-03NNSA20-5. https://doi.org/DOI: 10.1109/NAP.2017.8190267

Dehghani, M., & Tadjarodi, A. (2020). Formation and characterization of zeolite Y-platinum nanoparticles by rapid method of ultrasonic irradiation and investigation of its electrochemical property. *Journal of Nanostructures, 10*(3), 486–496. DOI: 10.22052/JNS.2020.03.005

Dehghani, M., Tadjarodi, A., & Chamani, S. (2019). Synthesis and Characterization of Magnetic Zeolite Y-Palladium-Nickel Ferrite by Ultrasonic Irradiation and Investigating Its Catalytic Activity in Suzuki-Miyaura Cross-Coupling Reactions. *ACS Omega, 4*(6), 10640–10648. DOI: 10.1021/acsomega.9b00666

Deneyer, A., Ke, Q., Devos, J., & Dusselier, M. (2020). Zeolite Synthesis under Nonconventional Conditions: Reagents, Reactors, and Modi Operandi. In *Chemistry of Materials* (Vol. 32, Issue 12, pp. 4884–4919). American Chemical Society. https://doi.org/DOI: 10.1021/acs.chemmater.9b04741

Erten-Kaya, Y., & Cakicioglu-Ozkan, F. (2012). Effect of ultrasound on the kinetics of cation exchange in NaX zeolite. *Ultrasonics Sonochemistry, 19*(3), 701–706. DOI: 10.1016/j.ultsonch.2011.10.010

Franko, I., Abrishamkar, M., Azizi, S. N., & Kazemian, H. (2010). Synthesis of borosilicate MFI type zeolite using different aging techniques. In *Chem. Met. Alloys* (Vol. 3, Issue 1/2, pp. 12–17). https://doi.org/DOI: 10.30970/cma3.0111

Grand, J., Awala, H., & Mintova, S. (2016). Mechanism of zeolites crystal growth: New findings and open questions. *CrystEngComm, 18*(5), 650–664. DOI: 10.1039/C5CE02286J

Gürbüz, H., Tokay, B., & Erdem-Şenatalar, A. (2012). Effects of ultrasound on the synthesis of silicalite-1 nanocrystals. *Ultrasonics Sonochemistry, 19*(5), 1108–1113. DOI: 10.1016/j.ultsonch.2012.01.008

Hallaert, S. D., Bols, M. L., Vanelderen, P., Schoonheydt, R. A., Sels, B. F., & Pierloot, K. (2017). Identification of α-Fe in High-Silica Zeolites on the Basis of ab Initio Electronic Structure Calculations. *Inorganic Chemistry, 56*(17), 10681–10690. DOI: 10.1021/acs.inorgchem.7b01653

Han, J., Ha, Y., Guo, M., Zhao, P., Liu, Q., Liu, C., Song, C., Ji, N., Lu, X., Ma, D., & Li, Z. (2019). Synthesis of zeolite SSZ-13 from coal gangue via ultrasonic pre-treatment combined with hydrothermal growth method. *Ultrasonics Sonochemistry, 59*. Advance online publication. DOI: 10.1016/j.ultsonch.2019.104703

Heidari, S., Haghighi, M., & Shabani, M. (2018). Ultrasound assisted dispersion of Bi2Sn2O7-C3N4 nanophotocatalyst over various amount of zeolite Y for enhanced solar-light photocatalytic degradation of tetracycline in aqueous solution. *Ultrasonics Sonochemistry, 43*, 61–72. DOI: 10.1016/j.ultsonch.2018.01.001

Hosseini, M., Zanjanchi, M. A., Ghalami-Choobar, B., & Golmojdeh, H. (2015). Ultrasound-assisted dealumination of zeolite Y. *Journal of Chemical Sciences, 127*(1), 25–31. DOI: 10.1007/s12039-014-0745-2

Huang, L., Qin, F., Huang, Z., Zhuang, Y., Ma, J., Xu, H., & Shen, W. (2016). Hierarchical ZSM-5 zeolite synthesized by an ultrasound-assisted method as a long-life catalyst for dehydration of glycerol to acrolein. *Industrial & Engineering Chemistry Research, 55*(27), 7318–7327. DOI: 10.1021/acs.iecr.6b01140

Ivanov, A. A., Chernyavsky, V. S., Gross, M. J., Kharitonov, A. S., Uriarte, A. K., & Panov, G. I. (2003). Kinetics of benzene to phenol oxidation over Fe-ZSM-5 catalyst. *Applied Catalysis A, General, 249*(2), 327–343. DOI: 10.1016/S0926-860X(03)00224-2

Jodłowski, P. J., Czekaj, I., Stachurska, P., Kuterasiński, Ł., Chmielarz, L., Jędrzejczyk, R. J., Jeleń, P., Sitarz, M., Górecka, S., Mazur, M., & Kurzydym, I. (2021). Experimental and Theoretical Studies of Sonically Prepared Cu–Y, Cu–USY and Cu–ZSM-5 Catalysts for SCR deNOx. *Catalysts, 11*(7), 824. DOI: 10.3390/catal11070824

Joyner, R., & Stockenhuber, M. (1999). Preparation, Characterization, and Performance of Fe–ZSM-5 Catalysts. *The Journal of Physical Chemistry B, 103*(29), 5963–5976. DOI: 10.1021/jp990978m

Jusoh, N., Yeong, Y. F., Mohamad, M., Lau, K. K., & Shariff, M., A. (2017). Rapid-synthesis of zeolite T via sonochemical-assisted hydrothermal growth method. *Ultrasonics Sonochemistry*, *34*, 273–280. DOI: 10.1016/j.ultsonch.2016.05.033

Khaleque, A., Alam, M. M., Hoque, M., Mondal, S., Haider, J. B., Xu, B., & Moni, M. A. (2020). Zeolite synthesis from low-cost materials and environmental applications: A review. *Environmental Advances*, *2*, 100019.

Khoshbin, R., & Karimzadeh, R. (2017). The beneficial use of ultrasound in free template synthesis of nanostructured ZSM-5 zeolite from rice husk ash used in catalytic cracking of light naphtha: Effect of irradiation power. *Advanced Powder Technology*, *28*(3), 973–982. DOI: 10.1016/j.apt.2017.01.001

Kuterasiński, Ł., Dymek, K., & Jodłowski, P. J. (2019). Sonically modified zeolites with MFI and USY type structure as catalysts for methane combustion. Preparation and physicochemical characterisation. *Science. Technology and Innovation*, *6*(3), 11–18. DOI: 10.5604/01.3001.0013.5388

Kuterasiński, Ł., Rojek, W., Gackowski, M., Zimowska, M., & Jodłowski, P. J. (2020). Sonically modified hierarchical FAU-type zeolites as active catalysts for the production of furan from furfural. *Ultrasonics Sonochemistry*, *60*, 104785. DOI: 10.1016/j.ultsonch.2019.104785

Li, L., Shen, Q., Li, J., Hao, Z., Xu, Z. P., & Lu, G. Q. M. (2008). Iron-exchanged FAU zeolites: Preparation, characterization and catalytic properties for N2O decomposition. *Applied Catalysis A, General*, *344*(1–2), 131–141. DOI: 10.1016/j.apcata.2008.04.011

Liang, D., Zhang, R., Liu, Y., Xie, Q., & Liu, J. (2023). A Review of High-Efficient Synthetic Methods for Zeolite Membranes and Challenges of Their Directional Growth Control. *Comments on Inorganic Chemistry*, *43*(4), 305–356. DOI: 10.1080/02603594.2022.2138864

Liu, Q., & van Bokhoven, J. A. (2024). Water structures on acidic zeolites and their roles in catalysis. *Chemical Society Reviews*, *53*(6), 3065–3095. DOI: 10.1039/D3CS00404J

Liu, Y., Zhang, H., Zhang, F., Zhu, M., Hu, N., & Chen, X. (2020). Facile synthesis of low-silica zeolite erionite by ultrasonic-assisted method. *Materials Letters*, *260*, 126934. DOI: 10.1016/j.matlet.2019.126934

Long, J., Wang, X., Ding, Z., Zhang, Z., Lin, H., Dai, W., & Fu, X. (2009). Binuclear μ-hydroxo-bridged iron clusters derived from surface organometallic chemistry of ferrocene in cavities of HY zeolite: Local structure, bound sites, and catalytic reactivity. *Journal of Catalysis*, *264*(2), 163–174. DOI: 10.1016/j.jcat.2009.04.002

Mancuso, A., Sacco, O., Sannino, D., Venditto, V., & Vaiano, V. (2020). One-Step Catalytic or Photocatalytic Oxidation of Benzene to Phenol: Possible Alternative Routes for Phenol Synthesis? *Catalysts*, *10*(12), 1424. DOI: 10.3390/catal10121424

Mancuso, A., Vaiano, V., Antico, P., Sacco, O., & Venditto, V. (2023). Photoreactive polymer composite for selective oxidation of benzene to phenol. *Catalysis Today*, *413–415*, 113914. DOI: 10.1016/j.cattod.2022.09.020

Meirelles, M. R., Malafatti, J. O. D., Escote, M. T., Pinto, A. H., & Paris, E. C. (2023). Magnetic Adsorbent Based on Faujasite Zeolite Decorated with Magnesium Ferrite Nanoparticles for Metal Ion Removal. *Magnetochemistry (Basel, Switzerland)*, *9*(5). Advance online publication. DOI: 10.3390/magnetochemistry9050136

Mori, K., Kagohara, K., & Yamashita, H. (2008). Synthesis of tris(2,2′-bipyridine)iron(II) complexes in zeolite y cages: Influence of exchanged alkali metal cations on physicochemical properties and catalytic activity. *The Journal of Physical Chemistry. C, Nanomaterials and Interfaces*, *112*(7), 2593–2600. DOI: 10.1021/jp709571v

Mu, Y., Zhang, Y., Fan, J., & Guo, C. (2017). Effect of ultrasound pretreatment on the hydrothermal synthesis of SSZ-13 zeolite. *Ultrasonics Sonochemistry*, *38*, 430–436. DOI: 10.1016/j.ultsonch.2017.03.043

Muraoka, K., Sada, Y., Miyazaki, D., Chaikittisilp, W., & Okubo, T. (2019). Linking synthesis and structure descriptors from a large collection of synthetic records of zeolite materials. *Nature Communications*, *10*(1), 4459. DOI: 10.1038/s41467-019-12394-0

Ng, E. P., Awala, H., Ghoy, J. P., Vicente, A., Ling, T. C., Ng, Y. H., Mintova, S., & Adam, F. (2015). Effects of ultrasonic irradiation on crystallization and structural properties of EMT-type zeolite nanocrystals. *Materials Chemistry and Physics*, *159*, 38–45. DOI: 10.1016/j.matchemphys.2015.03.047

Oruji, S., Khoshbin, R., & Karimzadeh, R. (2018). Preparation of hierarchical structure of Y zeolite with ultrasonic-assisted alkaline treatment method used in catalytic cracking of middle distillate cut: The effect of irradiation time. *Fuel Processing Technology*, *176*, 283–295. DOI: 10.1016/j.fuproc.2018.03.035

Oruji, S., Khoshbin, R., & Karimzadeh, R. (2019). Combination of precipitation and ultrasound irradiation methods for preparation of lanthanum-modified Y zeolite nano-catalysts used in catalytic cracking of bulky hydrocarbons. *Materials Chemistry and Physics*, *230*, 131–144. DOI: 10.1016/j.matchemphys.2019.03.038

Pal, P., Das, J. K., Das, N., & Bandyopadhyay, S. (2013). Synthesis of NaP zeolite at room temperature and short crystallization time by sonochemical method. *Ultrasonics Sonochemistry*, *20*(1), 314–321. DOI: 10.1016/j.ultsonch.2012.07.012

Paris, E. C., Malafatti, J. O. D., Musetti, H. C., Manzoli, A., Zenatti, A., & Escote, M. T. (2020). Faujasite zeolite decorated with cobalt ferrite nanoparticles for improving removal and reuse in Pb2+ ions adsorption. *Chinese Journal of Chemical Engineering*, *28*(7), 1884–1890. DOI: 10.1016/j.cjche.2020.04.019

Pérez, J., & Manuel Sánchez-Sánchez Editors, P. (2018). *Structure and Reactivity of Metals in Zeolite Materials* (J. Pérez Pariente & M. Sánchez-Sánchez, Eds.; Vol. 178). Springer International Publishing. https://doi.org/DOI: 10.1007/978-3-319-98905-1

Rakmae, S., Keawkumay, C., Osakoo, N., Montalbo, K. D., de Leon, R. L., Kidkhunthod, P., Chanlek, N., Roessner, F., Prayoonpokarach, S., & Wittayakun, J. (2016). Realization of active species in potassium catalysts on zeolite NaY prepared by ultrasound-assisted impregnation with acetate buffer and improved performance in transesterification of palm oil. *Fuel*, *184*, 512–517. DOI: 10.1016/j.fuel.2016.07.035

Ramirez Mendoza, H., Jordens, J., Valdez Lancinha Pereira, M., Lutz, C., & Van Gerven, T. (2020). Effects of ultrasonic irradiation on crystallization kinetics, morphological and structural properties of zeolite FAU. *Ultrasonics Sonochemistry*, *64*, 105010. DOI: 10.1016/j.ultsonch.2020.105010

Ravi, M., Sushkevich, V. L., & van Bokhoven, J. A. (2020). Towards a better understanding of Lewis acidic aluminium in zeolites. *Nature Materials*, *19*(10), 1047–1056. DOI: 10.1038/s41563-020-0751-3

Reinoso, D., Adrover, M., & Pedernera, M. (2018). Green synthesis of nanocrystalline faujasite zeolite. *Ultrasonics Sonochemistry*, *42*, 303–309. DOI: 10.1016/j.ultsonch.2017.11.034

Resasco, D. E., Crossley, S. P., Wang, B., & White, J. L. (2021). Interaction of water with zeolites: A review. *Catalysis Reviews*, *63*(2), 302–362. DOI: 10.1080/01614940.2021.1948301

Reyes Villegas, V. A., De León Ramirez, J. I., Perez-Cabrera, L., Estolano-Cobián, A., Chavez-Mendez, J. R., Huerta-Arcos, L., & Petranovskii, V. (2025). Sonochemical preparation of Fe-Y zeolite catalyst for the conversion of benzene to phenol. *Applied Catalysis A, General*, *689*, 120023. DOI: 10.1016/j.apcata.2024.120023

Reyes Villegas, V. A., De León Ramirez, J. I., Pérez-Cabrera, L., Pérez-Sicairos, S., Chávez-Méndez, J. R., & Petranovskii, V. (2024). Analysis of catalytic sites in FeY zeolite prepared by sono-assisted exchange of iron (II) ions. *Microporous and Mesoporous Materials*, *380*. Advance online publication. DOI: 10.1016/j.micromeso.2024.113306

Reyes Villegas, V. A., De León Ramirez, J. I., Pérez-Cabrera, L., Pérez-Sicairos, S., Yocupicio-Gaxiola, R. I., Chávez-Méndez, J. R., Huerta-Arcos, L., & Petranovskii, V. (2025). Sonochemical post-synthesis modification of Y zeolite with iron species. *Materials Chemistry and Physics*, *331*, 130199. DOI: 10.1016/j.matchemphys.2024.130199

Reyes Villegas, V. A., De León Ramirez, J. I., Perez-Sicairos, S., Yocupicio-Gaxiola, R. I., González-Torres, V., & Petranovskii, V. (2024). Catalyst for lactose hydrolysis based on zeolite Y modified with Fe species by ultrasound treatment. *Environmental Advances*, *15*, 100475. DOI: 10.1016/j.envadv.2023.100475

Rhoda, H. M., Heyer, A. J., Snyder, B. E. R., Plessers, D., Bols, M. L., Schoonheydt, R. A., Sels, B. F., & Solomon, E. I. (2022). Second-Sphere Lattice Effects in Copper and Iron Zeolite Catalysis. *Chemical Reviews*, *122*(14), 12207–12243. DOI: 10.1021/acs.chemrev.1c00915

Sadeghi, M., Farhadi, S., & Zabardasti, A. (2020a). Construction of magnetic $MgFe_2O_4$/CdS/MoS_2 ternary nanocomposite supported on NaY zeolite and highly efficient sonocatalytic degradation of organic pollutants. *RSC Advances*, *10*(72), 44034–44049. DOI: 10.1039/D0RA08831E

Sadeghi, M., Farhadi, S., & Zabardasti, A. (2020b). Fabrication of a novel magnetic CdS nanorod/$NiFe_2O_4$/NaX zeolite nanocomposite with enhanced sonocatalytic performance in the degradation of organic dyes. *New Journal of Chemistry*, *44*(20), 8386–8401. DOI: 10.1039/D0NJ01393E

Sadeghi, M., Farhadi, S., & Zabardasti, A. (2021). A NaX zeolite framework containing magnetic $MgFe_2O_4$/CdO nanoparticles: Synthesis, characterization and catalytic performance in the decontamination of 2-chloroethyl phenyl sulfide (2-CEPS) as a model of sulfur mustard agent. *New Journal of Chemistry*, *45*(45), 21315–21326. DOI: 10.1039/D1NJ04202E

Schmidt, R. J. (2005). Industrial catalytic processes—Phenol production. *Applied Catalysis A, General*, *280*(1), 89–103. DOI: 10.1016/j.apcata.2004.08.030

Sharifi, M., Haghighi, M., & Abdollahifar, M. (2015). Sono-dispersion of bimetallic Ni-Co over zeolite Y used in conversion of greenhouse gases CH4/CO2 to high valued syngas. *Journal of Natural Gas Science and Engineering*, *23*, 547–558. DOI: 10.1016/j.jngse.2015.03.006

Shirani, M., Semnani, A., Haddadi, H., & Habibollahi, S. (2014). Optimization of simultaneous removal of methylene blue, crystal violet, and fuchsine from aqueous solutions by magnetic NaY zeolite composite. *Water, Air, and Soil Pollution*, *225*(8). Advance online publication. DOI: 10.1007/s11270-014-2054-2

Snyder, B. E. R., Bols, M. L., Rhoda, H. M., Plessers, D., Schoonheydt, R. A., Sels, B. F., & Solomon, E. I. (2021). Cage effects control the mechanism of methane hydroxylation in zeolites. *Science*, *373*(6552), 327–331. DOI: 10.1126/science.abd5803

Snyder, B. E. R., Bols, M. L., Rhoda, H. M., Vanelderen, P., Böttger, L. H., Braun, A., Yan, J. J., Hadt, R. G., Babicz, J. T., Hu, M. Y., Zhao, J., Alp, E. E., Hedman, B., Hodgson, K. O., Schoonheydt, R. A., Sels, B. F., & Solomon, E. I. (2018). Mechanism of selective benzene hydroxylation catalyzed by iron-containing zeolites. *Proceedings of the National Academy of Sciences of the United States of America*, *115*(48), 12124–12129. DOI: 10.1073/pnas.1813849115

Snyder, B. E. R., Bols, M. L., Schoonheydt, R. A., Sels, B. F., & Solomon, E. I. (2018). Iron and Copper Active Sites in Zeolites and Their Correlation to Metalloenzymes. *Chemical Reviews*, *118*(5), 2718–2768. DOI: 10.1021/acs.chemrev.7b00344

Snyder, B. E. R., Böttger, L. H., Bols, M. L., Yan, J. J., Rhoda, H. M., Jacobs, A. B., Hu, M. Y., Zhao, J., Alp, E. E., Hedman, B., Hodgson, K. O., Schoonheydt, R. A., Sels, B. F., & Solomon, E. I. (2018). *Structural characterization of a non-heme iron active site in zeolites that hydroxylates methane.* *115*(18), 4565–4570. https://doi.org/DOI: 10.2307/26508720

Snyder, B. E. R., Vanelderen, P., Bols, M. L., Hallaert, S. D., Böttger, L. H., Ungur, L., Pierloot, K., Schoonheydt, R. A., Sels, B. F., & Solomon, E. I. (2016). The active site of low-temperature methane hydroxylation in iron-containing zeolites. *Nature*, *536*(7616), 317–321. DOI: 10.1038/nature19059

Tadjarodi, A., Dehghani, M., & Imani, M. (2018). Green synthesis and characterization of palladium nanoparticles supported on zeolite Y by sonochemical method, powerful and efficient catalyst for Suzuki-Miyaura coupling of aryl halides with phenylboronic acid. *Applied Organometallic Chemistry*, *32*(12). Advance online publication. DOI: 10.1002/aoc.4594

Vaičiukynienė, D., Jakevičius, L., Kantautas, A., Vaitkevičius, V., Vaičiukynas, V., & Dvořák, K. (2021). Conversion of silica by-product into zeolites by thermo-sonochemical treatment. *Ultrasonics Sonochemistry, 72*. Advance online publication. DOI: 10.1016/j.ultsonch.2020.105426

van Bokhoven, J. A., & Lamberti, C. (2014). Structure of aluminum, iron, and other heteroatoms in zeolites by X-ray absorption spectroscopy. *Coordination Chemistry Reviews, 277–278*, 275–290. DOI: 10.1016/j.ccr.2014.05.013

Wącław, A., Nowińska, K., & Schwieger, W. (2004). Benzene to phenol oxidation over iron exchanged zeolite ZSM-5. *Applied Catalysis A, General, 270*(1–2), 151–156. DOI: 10.1016/j.apcata.2004.04.032

Wang, B., Wu, J., Yuan, Z. Y., Li, N., & Xiang, S. (2008). Synthesis of MCM-22 zeolite by an ultrasonic-assisted aging procedure. *Ultrasonics Sonochemistry, 15*(4), 334–338. DOI: 10.1016/j.ultsonch.2007.07.007

Wang, R., Ma, N., Yan, Y., & Wang, Z. (2018). Ultrasonic-assisted fabrication of high flux T-type zeolite membranes on alumina hollow fibers. *Journal of Membrane Science, •••*, 676–684. DOI: 10.1016/j.memsci.2017.10.047

Wang, Y., Wang, J., Wei, J., Wang, C., Wang, H., & Yang, X. (2023). Catalytic Mechanisms and Active Species of Benzene Hydroxylation Reaction System Based on Fe-Based Enzyme-Mimetic Structure. *Catalysis Letters, 153*(11), 3311–3332. DOI: 10.1007/s10562-022-04238-2

Xia, H., Sun, K., Sun, K., Feng, Z., Li, W. X., & Li, C. (2008). Direct Spectroscopic Observation of Fe(III)–Phenolate Complex Formed From the Reaction of Benzene With Peroxide Species on Fe/ZSM-5 At Room Temperature. *The Journal of Physical Chemistry. C, Nanomaterials and Interfaces, 112*(24), 9001–9005. DOI: 10.1021/jp800455x

Xiao, P., Osuga, R., Wang, Y., Kondo, J. N., & Yokoi, T. (2020). Bimetallic Fe-Cu/beta zeolite catalysts for direct hydroxylation of benzene to phenol: Effect of the sequence of ion exchange for Fe and Cu cations. *Catalysis Science & Technology, 10*(20), 6977–6986. DOI: 10.1039/d0cy01216e

Xiao, P., Wang, Y., Kondo, J. N., & Yokoi, T. (2019). Consequences of Fe speciation in MFI zeolites for hydroxylation of benzene to phenol with H2O2. *Applied Catalysis A, General, 579*, 159–167. DOI: 10.1016/j.apcata.2019.04.025

Xie, J., Li, X., Guo, J., Luo, L., Delgado, J. J., Martsinovich, N., & Tang, J. (2023). Highly selective oxidation of benzene to phenol with air at room temperature promoted by water. *Nature Communications, 14*(1), 4431. DOI: 10.1038/s41467-023-40160-w

Yamaguchi, S., Miyake, Y., Takiguchi, K., Ihara, D., & Yahiro, H. (2018). Oxidation of cyclic hydrocarbons with hydrogen peroxide over iron complexes encapsulated in cation-exchanged zeolite. *Catalysis Today*, *303*, 249–255. DOI: 10.1016/j.cattod.2017.10.047

Yin, X., Long, Z., Wang, C., Li, Z., Zhao, M., & Yang, S. (2019). A time- and cost-effective synthesis of CHA zeolite with small size using ultrasonic-assisted method. *Ultrasonics Sonochemistry*, *58*. Advance online publication. DOI: 10.1016/j.ultsonch.2019.104679

Yue, M., Jiang, X., Zhang, H., Zhang, S., Xue, T., & Li, Y. (2020). Quasi-solid-phase synthesis of Fe-MFI zeolites by using Fe-containing zeolite seed sol for hydroxylation of benzene with H2O2. *Microporous and Mesoporous Materials*, *294*. Advance online publication. DOI: 10.1016/j.micromeso.2019.109891

Zang, J., Yu, H., Liu, G., Hong, M., Liu, J., & Chen, T. (2023). Research Progress on Modifications of Zeolite Y for Improved Catalytic Properties. *Inorganics*, *11*(1), 22. DOI: 10.3390/inorganics11010022

Zhang, J., Tang, X., Yi, H., Yu, Q., Zhang, Y., Wei, J., & Yuan, Y. (2022). Synthesis, characterization and application of Fe-zeolite: A review. *Applied Catalysis A, General*, *630*, 118467. DOI: 10.1016/j.apcata.2021.118467

Zhang, R., Zhong, P., Arandiyan, H., Guan, Y., Liu, J., Wang, N., Jiao, Y., & Fan, X. (2020). Using ultrasound to improve the sequential post-synthesis modification method for making mesoporous Y zeolites. *Frontiers of Chemical Science and Engineering*, *14*(2), 275–287. DOI: 10.1007/s11705-019-1905-1

Zhou, L., Wang, M., Yang, S., Guo, W., Pu, X., He, Y., Zhu, J., Wang, B., Zheng, M., Liu, S., & Zhang, Y. (2022). Facile synthesis of mesoporous ZSM-5 aided by sonication and its application for VOCs capture. *Ultrasonics Sonochemistry*, *88*. Advance online publication. DOI: 10.1016/j.ultsonch.2022.106098

Chapter 11
Ultrasonic Coating of LTA Zeolites by Antibacterial and Antibiofilm Nanoparticles

Jesus Isaias De Leon Ramirez
https://orcid.org/0000-0003-1393-610X
Centro de Nanociencias y Nanotecnologia UNAM, Mexico

Victor Alfredo Reyes-Villegas
Centro de Nanociencias y Nanotecnologia UNAM, Mexico

Sergio Perez Sicairos
Instituto Tecnologico de Tijuana Centro de Graduados, Mexico

José Román Chávez-Méndez
Universidad Autónoma de Baja California, Mexico

Luis Pérez-Cabrera
https://orcid.org/0000-0001-9896-3419
Centro de Nanociencias y Nanotecnologia UNAM, Mexico

Verónica González Torres
Universidad Autónoma de Baja California, Mexico

José Constantino González Crisóstomo
Universidad Autónoma de Baja California, Mexico

Ethiel Zavala-Flores
Universidad Autónoma de Baja California, Mexico

Fabian N. Murrieta-Rico
https://orcid.org/0000-0001-9829-3013
Universidad Politécnica de Baja California, Mexico

Vitalii Petranovskii
https://orcid.org/0000-0002-8794-0593
Centro de Nanociencias y Nanotecnologia UNAM, Mexico

DOI: 10.4018/979-8-3373-0320-8.ch011

ABSTRACT

Sonochemistry uses ultrasound to initiate chemical reactions under extreme conditions, difficult to achieve with conventional methods like hydrolysis or condensation. Ultrasound generates mechanical, transport, and chemical effects in liquids: microjets can erode surfaces, enhancing them for coatings, while localized high temperatures and radicals break molecular bonds. Ultrasound has been employed as an innovative approach to enhance the bactericidal properties of Zeolite A (LTA) by supporting nanoparticles. These improvements are attributed to cavitation effects, which promote adhesion and homogeneous distribution of nanoparticles onto the zeolite surface. Raising attention for environmental and medical applications by reducing the need for additional chemicals and shortening preparation time. The efficiency of ultrasound allows to explore how these modifications influence the bactericidal action of zeolitic materials. Ultrasound presents a more efficient, cleaner, and quicker method for enhancing zeolite-based materials.

INTRODUCTION TO ZEOLITE LTA

Zeolites are a family of microporous aluminosilicate materials widely found in nature. From a chemical composition perspective, these crystalline structures are composed primarily of Si and Al oxides, with fundamental building blocks consisting of tetrahedral atoms bonded to four oxygen atoms, forming a network where each oxygen links two tetrahedral atoms. In natural zeolites, the frameworks possess a three-dimensional structure and feature pores and cavities that can trap water molecules and alkali/alkaline earth metal cations. Despite their natural abundance, their use in industrial applications is limited due to impurities and high chemical composition variation. Various strategies have been developed to overcome these limitations, enhance the properties of natural zeolites, and expand their applications. This has led to the synthesis of new zeolitic structures with extended compositions surpassing nature's constraints.

Synthetic zeolites exhibit a higher diversity in the Si/Al ratio than their natural counterparts. They can incorporate additional structural elements such as zinc (Zn), gallium (Ga), germanium (Ge), phosphorus (P), boron (B), and beryllium (Be), among others. This ability to modify their composition has enabled the synthesis of materials with properties for specific industrial and scientific applications (Derakhshankhah et al., 2020). There are different types of zeolites, where the main differences lie in their structural arrangement and the silicon/aluminum ratio (Si/Al) within their crystalline framework.

Among the most important structures is the Linde Type A (LTA) zeolite, one of the first synthetic zeolites to be developed and commercialized. LTA zeolite was first synthesized in 1956 by Milton and his collaborators, marking a milestone in porous materials. Its structure consists of two cages: the β cage (or sodalite cage) and the α cage, interconnected through double four-membered rings, forming a central supercage accessible through eight-membered rings. This architecture provides a high cation exchange capacity, particularly in its sodium form (NaA), with a general formula of $Na_{12}[(AlO_2)_{12}(SiO_2)_{12}]\cdot 27H_2O$. Depending on the exchangeable cation, such as potassium, sodium or calcium cations, the variants were named 3A, 4A, and 5A, respectively. In addition to its versatility in industrial processes, LTA zeolite possesses key properties such as thermal stability, high selectivity, non-toxicity, and mechanical resistance. These characteristics have driven research into its structural and functional modification, aiming to enhance performance in specific applications. The study of LTA zeolite and its modifications remains an area of interest in materials science, offering new opportunities for developing advanced technologies (Collins et al., 2020; Petrov & Michalev, 2012).

Thanks to the properties of LTA zeolite, it can be modified with metals by various methods to enhance properties and even promote synergy between the introduced species; for example, a metal exchange can improve adsorption (da Silva et al., 2024). Studies have shown that incorporating Ag^+ into the zeolite structure provides bactericidal properties due to the controlled release and diffusion of the silver ions when in an aqueous media (Rieger et al., 2016). Many strategies have been developed to support these nanoparticles on LTA zeolite, pursuing an optimization of its properties and expanding its range of applications. One of the most promising materials is LTA zeolite with metallic nanoparticles and metal oxides due to their high efficiency as contaminant adsorption, photocatalysts, and antimicrobial activity (De León Ramirez et al., 2024; Ovchinnikov et al., 2023) Zinc, copper, and silver-based nanoparticles are among the most applied metal nanoparticles given the enhanced activity for degradation of organic compounds and antimicrobial properties(Azizi-Lalabadi et al., 2019; De León Ramirez et al., 2024; Ribeiro et al., 2024).

Ultrasound has emerged as an efficient and sustainable method within the various techniques available for zeolite modification. This approach enhances pore accessibility and promotes the interaction between the zeolite surface and metal precursors, optimizing nanoparticle deposition and reducing the agglomeration of supported materials (Zhang et al., 2020). Ultrasound, a wave with a frequency above the human audible range, can induce various physical and chemical effects applicable in chemistry, pharmaceuticals, and medicine (*Handbook of Ultrasonics and Sonochemistry*, 2016; Pollet & Ashokkumar, 2019). In recent years, ultrasound applications have increased considerably, driven by the development of sonochemistry, a field that studies chemical reactions induced by localized fields of high pressure and temperature (*Sonochemistry and the Acoustic Bubble*, 2015).

One of the fundamental phenomena of sonochemistry is acoustic cavitation, which occurs when ultrasound is applied to a reaction medium. This phenomenon is characterized by the formation, growth, and collapse of microbubbles in the liquid, creating extreme pressure and temperature conditions on very short time scales. The intensity and efficiency of cavitation depend on various factors, such as the reaction medium's nature and the reactor configuration (*Sonochemistry and the Acoustic Bubble*, 2015). The hot spot theory is the most widely accepted method to explain the sonochemical effects of ultrasound (Głowniak et al., 2023). It is based on the formation and collapse of cavitation bubbles, which generate extreme temperature and pressure conditions. This phenomenon occurs in three central regions: I) Bubble core: once a bubble collapses, it reaches temperatures of thousands of degrees and pressures of hundreds of atmospheres, promoting the formation of highly reactive radicals. II) Bubble/solution interface: an area where temperatures can reach up to 2000 K and non-volatile solutes can thermally decompose and participate in radical reactions.III) The radicals generated in the bubble can diffuse and react with solutes in the liquid phase (*Handbook of Ultrasonics and Sonochemistry*, 2016; Pollet & Ashokkumar, 2019).

The chemical effects are attributed to the production of •OH and •H radicals, which can generate or influence specific chemical reactions. It is well known that the intensity of these physical and chemical effects depends on various factors, such as the physical properties of the media, the emitted frequency, and the intensity (D. Chen et al., 2011a). Leading to the classification of sonochemistry into "true sonochemistry" and "false sonochemistry." The former occurs only when ultrasonic irradiation generates cavitation that produces reactive species such as radicals or radical ions. The latter include reactions that occur with mechanical agitation but whose rate increases significantly with ultrasound due to synergy between physical and mechanical effects like shock waves, microjets, and microcurrents (*Sonochemistry and the Acoustic Bubble*, 2015). Additionally, sonochemical reactions can be homogeneous (a single liquid phase) or heterogeneous (with suspended solids or immiscible liquids) depending on the phases present at the start of the reaction (Mason & Vinatoru, 2023). Induced reactions can occur directly in the reaction medium, which, when applied, can be carried out at different stages of material synthesis: pre-treatment, during and post-synthesis. The moment selected for the ultrasonic irradiation significantly influences the final properties of the obtained material, making sonochemical reactions a valuable tool for the structural modification of materials, including the modification of the zeolite surface with nanoparticles (D. Chen et al., 2011b).

This review focuses on the sono-assisted modification of LTA zeolite with nanoparticles, evaluating the impact of this technique on its physicochemical properties and its performance in antimicrobial applications. Implementing this method improves the efficiency of zeolitic materials and represents an eco-friendly and scalable alternative for designing new functional materials with applications in water decontamination and microbiological control.

SONOCHEMISTRY A SYNTHESIS METHOD FOR ZEOLITE LTA

Hydrothermal synthesis and sol-gel are the most applied zeolite synthesis methods. These processes have significant disadvantages, such as the difficulty in controlling nucleation. Synthesis time is one of the most common and vital challenges (Askari et al., 2013). However, nowadays, one of the most primitive techniques known, mechanochemistry, has been introduced, where it has been demonstrated that simply using milling reduces the crystallization times in zeolite syntheses (Natalya E. Gordina et al., 2017).

Ultrasound oscillates at frequencies higher than 20 kHz as a type of mechanical wave. When ultrasound propagates through a medium, it generates mechanical effects by causing vibrations. These vibrations create regions of high compression and rarefaction, increasing temperature and leading to radicals (Fig.1) (Zuo et al., 2024). The direct mechanical influence of ultrasound on zeolites results in modifications in particle size, pore opening, and structural defects (Ramishvili et al., 2017). It is an ideal environment for ultrasound to be used as a mechanochemical method for synthesizing or treating various zeolites. The application of ultrasound for material synthesis has attracted significant attention, given that ultrasonic irradiation achieves crystallization stages much faster than conventional methods. Additionally, many advantages come with this method, such as it is simple, does not require complex facilities, controls particle size, improves morphology, and increases control over nucleation (N. E. Gordina et al., 2017). Ultrasound can not only function in the synthesis stage of zeolites. However, it can also influence the properties of the zeolite once it is synthesized and even before it is synthesized, that is, in the pre-and post-synthesis stages (Cruz et al., 2020; Fu et al., 2016).

Figure 1. Ultrasonic vibrations generate regions of high compression and rarefaction, increasing temperature and leading to radical formation. Additionally, the direct mechanical influence of ultrasound on zeolites results in modifications in particle size, pore opening, and structural defects.

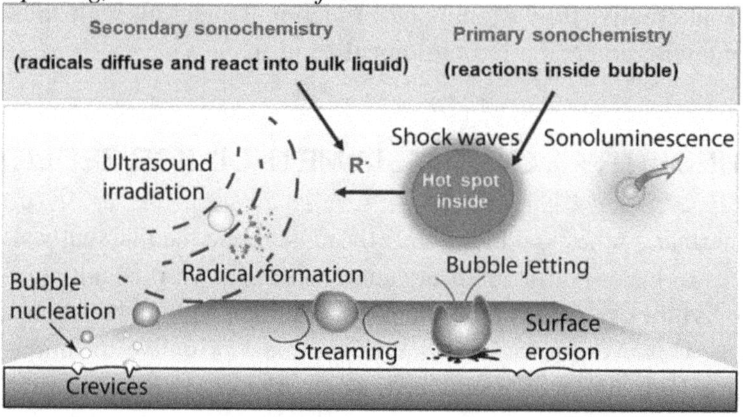

Image taken from Martínez, Cravotto, and Cintas (2021), "Organic Sonochemistry: A Chemist's Timely Perspective on Mechanisms and Reactivity," American Chemical Society, under CC-BY 4.0 license.

LTA zeolite is one of the most common due to its ease of synthesis and multiple uses in industry and medicine. Ultrasound has previously been used to synthesize LTA, and it has been shown to influence the physical properties of this zeolite (Vaičiukynienė et al., 2015). When applying ultrasound, crystalline increases, and greater microporosity is obtained, improving its adsorption capacity. Additionally, the particles obtained through ultrasound show a more uniform size distribution and morphology, which are critical characteristics for industrial applications. At elevated temperatures, ultrasound accelerates the transformation of LTA zeolite into undesirable phases such as hydroxy sodalite, highlighting the importance of temperature control. This transformation occurs above 60°C, so it is essential to maintain controlled conditions (Askari et al., 2013; Vaičiukynienė et al., 2015). The research conducted by Özlem Andaç in 2005 demonstrated the significant potential of ultrasound to influence hydrothermal crystallization and modify the properties of the final product (Andaç et al., 2005).

Ultrasound allows obtaining an LTA zeolite phase in significantly shorter times than conventional methods. Additionally, it reduces the formation of secondary phases, such as those generated in the agitation method, suggesting that ultrasound represents a promising technique for optimizing zeolite synthesis. The effect of ultrasound on the nucleation rate is related to the cavitation phenomenon, as each cavitation collapse generates a nucleation site like the effect of a trace particle or a surface imperfection. This results in an observed increase in the crystallization rate and the crystal growth speed due to the improvement in mass transfer near the crystal surface caused by acoustic transmission (Kim et al., 2010). Regarding particle aggregation, it occurs both in the absence and presence of ultrasound, being more evident at higher temperatures. In the initial synthesis stages, amorphous materials prepared with ultrasound exhibited stepped isotherms like mesoporous materials, with less sharp steps and higher P/P0 ratios. This indicates a more excellent pore size distribution and an increase in the average pore size. The results also demonstrated that the development of micropores begins earlier in the X-ray amorphous phase when ultrasound is used. This process is significantly accelerated, reflected in a higher adsorption capacity at low relative pressures and higher micropore volumes than samples synthesized without ultrasound (Andaç et al., 2005).

In the pre-synthesis stage, ultrasound has also been used to activate and generate nuclei for subsequent crystallization, with an optimal time of 10 minutes in the suspension of reagents. Beyond this time, certain impurities may appear (N. E. Gordina et al., 2017). When the ultrasonic activation method (UST) is compared to the traditional grinding method (MCA), both approaches lead to the formation of cubic sodium aluminate, but with significant differences in particle morphology and final product properties. While MCA produces spherical particles due to the mechanical action of the mill, UST generates angular and irregular particles through dispersion in an acoustic field.

This approach has demonstrated that ultrasound provides better crystallinity than other preparation methods, such as grinding. It highlights its ability to promote a homogeneous mixture and optimize the structural development of the material. The UST method significantly affects the deformation of chemical bonds (A–O(H) and Si–O) and the homogeneous dispersion of the mixture, which supports the subsequent synthesis of LTA zeolite with distinctive characteristics.

Although the thermal treatment following MCA results in larger and less defective crystals, UST stands out for producing LTA zeolite crystallites with more compact and uniform agglomerates, suggesting significant potential for controlling the morphology and size distribution of the final product. Sonication promotes a more homogeneous dispersion of reagents, facilitating the formation of nucleation sites and reducing structural irregularities. The crystallites obtained via UST exhibit specific properties, such as greater agglomerate compaction and improved adsorption characteristics. Ultrasound enables the production of materials with desirable characteristics through a process less reliant on intense mechanical forces, representing an efficient and reproducible option (N. E. Gordina et al., 2018).

Ultrasound has also been applied in the green synthesis of LTA zeolites using alum sludge, which often presents challenges due to the heterogeneity of the reagents. In these studies, preliminary ultrasonic gel treatment promoted nucleation and improved the crystallinity of zeolite. For example, 10 minutes of ultrasonication increased the crystallinity of the zeolite from 43% to 62% by weight, a significantly faster process than traditional methods. However, prolonged treatment times exceeding 40 minutes provided no additional benefits and, in some cases, even caused a slight reduction in the zeolite content due to the transformation into more stable phases, such as sodalite. Additionally, the particle size of the zeolite was observed to decrease with the application of ultrasound (Rozhkovskaya et al., 2021).

As for LTA zeolite membranes, these have also been synthesized by ultrasonic methods, enhancing the synthesis conditions, crystallinity, and morphology of the zeolite through acoustic cavitation (Sen et al., 2018). A study evaluating various combinations of methods for pretreatment and crystallization in preparing LTA membranes found that LTA membranes could be successfully prepared using all the examined combinations. However, the combination of ultrasonic processing and microwave crystallization produced membranes that were well-crystallized, defect-free, and exhibited high performance in ethanol/water pervaporation at 60°C (N. E. Gordina et al., 2021).

SONO-ASSISTED DEPOSITION OF NANOPARTICLES ON ZEOLITE LTA

Zeolites are widely recognized as highly effective support materials for various metals due to their well-defined microporous structures and the presence of intra-crystalline and intercrystalline mesopores that can arise during synthesis or post-synthetic modifications. These porosities enable metal species to be hosted within the microporous cavities and, in some cases, on the external surface of the zeolite crystals. Additionally, the coexistence of Brønsted and Lewis acid sites further enhances their functionality as support materials, facilitating metal encapsulation, impregnation, and ion exchange processes (Singh et al., 2020).

In metal@zeolite catalysts, metal nanoparticles are often located within the microporous framework of the zeolite, accessible only through its external surface. Molecules must diffuse through these micropores to reach the encapsulated nanoparticles, allowing the zeolite structure to act as a molecular sieve, thereby imparting selectivity to the catalytic process. However, it is also important to consider that, depending on the synthesis and preparation method, metal nanoparticles can be deposited on the external surface of the zeolite rather than being confined within its microporous structure. This dual distribution influences catalytic behavior by combining encapsulated active sites with those more accessible at the outer surface. A notable example of this effect is how the micropores of zeolites selectively influence molecular diffusion based on size differences, controlling the access of reactants to catalytic sites and impacting the overall reaction efficiency (Wang et al., 2021). To critically assess the effectiveness of zeolites as metal supports, it is necessary to consider both their structural advantages and potential limitations. While their microporous nature enables shape-selective catalysis, diffusion constraints can affect the accessibility of active sites. Moreover, metal deposition on the external surface may lead to unintended side reactions or reduced selectivity. These factors must be carefully controlled during catalyst design to optimize performance for specific applications.

Several post-synthesis treatments can be applied to zeolites, such as ion exchange, one of the most common methods to modify zeolite properties (Vaezi et al., 2025). In addition, other processes, such as thermal activation (Desales-Guzmán et al., 2025), metal impregnation (Asrori et al., 2024) or using zeolites as nanoparticle supports, can also be employed. These treatments enhance the performance of zeolites in various industrial processes, such as catalysis, gas or liquid separation, and water purification, tailoring them to the specific needs of each application (Ruíz-Baltazar, 2024). One method that has shown great promise in enhancing performance through optimizing structural parameters is supporting metallic nanoparticles on zeolites. This approach significantly improves the material's properties. Additionally, the solid support plays a key role in the performance of these materials across various applications, as it not only provides a large surface area for dispersing metallic species but also electronically and/or geometrically modifies the metallic nanoparticles through strong metal-support interactions (Wang et al., 2021).

The incorporation of nanoparticles into zeolites can occur during the synthesis stage, known as in situ synthesis or post-synthesis, after zeolite crystallization. In situ modification involves integrating metals during zeolite crystallization under hydrothermal conditions. Crystallization occurs at high temperatures, followed by calcination to remove organic residues and subsequent chemical reduction to form metallic nanoparticles. To prevent premature precipitation of metal cations, stabilizing ligands (e.g., NH_3, ethylenediamine) have been employed, allowing controlled encapsulation of noble metal nanoparticles (e.g., Pt, Pd, Rh, Ir, Re, Ag) within the micropores of LTA zeolites. These materials have demonstrated high selectivity in catalytic reactions, such as the oxidative dehydrogenation of alcohols and the hydrogenation of alkenes (Xu et al., 2018).

Incorporating nanoparticles into zeolite structures can also be achieved through post-synthesis modification techniques, such as impregnation, ion exchange, or precipitation. Specifically, post-synthesis impregnation involves depositing pre-synthesized nanoparticles onto the surface of zeolite crystals via wet impregnation or vapor-phase deposition. These techniques enhance the material's properties by improving efficiency in contaminant removal, preventing nanoparticle agglomeration and leaching, and ensuring long-term stability and recyclability. The close interaction between nanoparticles and the zeolite surface facilitates electron transfer, promoting key catalytic reactions in contaminant degradation. Furthermore, nanoparticle-supported zeolites have demonstrated antimicrobial properties, making them an effective environmental decontamination strategy. Notably, zeolites with supported nanoparticles exhibit a high capacity for eliminating pathogenic microorganisms due to the controlled release of metal ions that impair cellular viability (Ruíz-Baltazar, 2024)(Eom et al., 2022). Ion exchange is an intrinsic property of most zeolites, enabling the replacement of cations within their structure without significantly compromising their stability. This phenomenon is essential in various applications, such as catalysis, gas purification, and removing contaminants from aqueous media. Several factors influence the ion exchange process, including the cations' size and charge, the zeolite's structure, the medium's condition, and the competition between cations in the solution. The ion exchange mechanism is triggered when the zeolite encounters a solution containing other cations, allowing these to replace the original cations in the zeolite's structure. This process is widely utilized to modify zeolite properties and optimize their performance in diverse industrial and environmental applications (Townsend & Coker, 2001).

Various methods have been developed for synthesizing metal nanoparticles (NPs) supported on zeolites in recent years. Wet impregnation using metal salt solutions is one of the most suitable and widely used techniques to introduce different metals into zeolite structures. However, a significant drawback of this method is that it can lead to a non-uniform or disordered distribution of metals on the support material.

In contrast, ion exchange is a simpler and more convenient method for synthesizing transition metal cation-exchanged zeolite compounds. This process is conducted in an aqueous medium through continuous treatment with the target metal, followed by washing to remove residual ions (Singh et al., 2020). Zeolite-supported materials have gained prominence in various catalytic and adsorption applications due to the high thermal stability and specific surface area of these supports. The literature reports different strategies for impregnating active compounds into zeolites, allowing for modifying their physicochemical properties and enhancing their performance in industrial and environmental processes.

One of the most studied approaches involves impregnating ternary compounds into zeolites. For example, a study reports the synthesis of $CdS/TiO_2/CeO_2$ supported on zeolite through impregnation in deionized water, followed by stirring and solvent evaporation. The resulting material was sintered at 300 °C, ensuring the homogeneous integration of the nanocomposite into CBV400 zeolite (SiO_2/Al_2O_3 5.1). This procedure demonstrated the effectiveness of impregnation for obtaining composite materials with improved catalytic properties (Mahamud et al., 2023).

Additionally, the deposition of metallic nanoparticles on zeolites has been extensively explored. For instance, a study described the synthesis of Fe_2O_3-supported zeolite-Y by suspending the zeolite in a solution of $Fe(NO_3)_3{\cdot}9H_2O$, followed by pH adjustment with KOH to precipitate iron as hydroxide on the support's surface.

To remove impurities, the material was filtered, washed, and subjected to Soxhlet extraction. After drying at 327 °C and a final vacuum process, Fe_2O_3-Y with enhanced catalytic properties was obtained (Baruah et al., 2022). Another significant approach in synthesizing zeolite-supported materials is the incorporation of metal precursors, such as titanium isopropoxide, in various compositions on LTA and FAU zeolites. The chemical structure of LTA ($Na_{12}Al_{12}Si_{12}O_{48}\cdot37H_2O$) and FAU ($Na_{86}Al_{86}Si_{94\cdot6}O_{361}\cdot228.5H_2O$) allows for controlled metal impregnation, followed by calcination at 550 °C to stabilize the material's active phase (Al-Harbi et al., 2015). These studies highlight the importance of zeolite type and synthesis conditions in obtaining functional materials with applications in catalysis and selective adsorption.

Zeolites are widely used to support nanoparticles due to their porous structure and low synthesis cost. Various methods are currently used to synthesize nanomaterials supported on zeolites, including hydrothermal, sol-gel, microwave-assisted, template-directed, and sonochemical synthesis (Ruíz-Baltazar, 2024). The synthesis of nanoparticles for zeolite-supported materials has gained significant attention because it is a more eco-friendly, simple, and rapid method that does not require complex facilities compared to other techniques. This approach has effectively supported nanoparticles on natural (Liu et al., 2012) and synthetic zeolites (Talebi et al., 2010).

It has been demonstrated that using ultrasound for supporting nanoparticles is highly efficient, as it generates mechanical effects in the material, creating optimal conditions for nanoparticle synthesis and distribution on zeolites (Fu et al., 2016; Yosefi et al., 2015). When compared to conventional methods, ultrasound has proven to be more efficient. The ultrasound method produces a finer and more homogeneous distribution of nanoparticles on the zeolite surface (Bahiraei & Behin, 2020). Even with high nanoparticle loading, the efficiency does not decrease, unlike other methods where agglomeration reduces performance (Yosefi et al., 2015). This method has been employed to support various zeolite materials, including zinc oxides (Gicheva et al., 2021).

ANTIMICROBIAL ACTIVITY OF METAL MODIFIED ZEOLITES

Today, humanity faces various challenges, among which the fight against pathogens is one of the most significant. Several antimicrobial agents have been studied in this context, including natural compounds, synthetic substances, and advanced materials like zeolites. Zeolites are promising materials due to their adsorption capacity and ability to host metals within and outside their structure, granting them antimicrobial properties. For example, studies have shown that the ion exchange of sodium with silver in zeolite Y, followed by functionalization with APTES **(3-Aminopropil)trietoxisilano)**, significantly increases its antibacterial activity. In this case, *Escherichia coli* exhibited greater susceptibility to the functionalized zeolite (ZSA) compared to *Staphylococcus aureus* (Hanim et al., 2016).

Similarly, studies have indicated that modifying zeolite X with silver notably enhances its antimicrobial properties against *E. coli*, *Pseudomonas aeruginosa*, and *Staphylococcus aureus*. Likewise, natural zeolites loaded with silver have demonstrated significant efficacy against *E. coli* and *S. faecalis*, depending on the mineral-to-silver content ratio. For example, *E. coli* showed complete antibacterial effects after two hours of contact, while *S. faecalis* required similar exposure times but higher silver content (Kwakye-Awuah et al., 2008). However, silver is not the only metal investigated. A study demonstrated that zeolites exchanged with copper (Cu^{2+}), silver (Ag^{+}), and zinc (Zn^{2+}) exhibit different levels of antimicrobial activity. Silver stands out for its high efficacy against bacteria (MIC **(Minimal inhibitory concentration)** between 16 and 128 µg/mL) due to its higher free exchange energy and selectivity toward zeolites. In contrast, zinc and copper exhibit moderate efficacy against bacteria (MIC between 256 and 2,048 µg/mL) but are highly effective against fungi, broadening their range of applications (Demirci et al., 2014).

Natural Zeolites loaded with copper (CuNZ) and zinc (ZnNZ) have demonstrated high efficacy in eliminating *E. coli* and *S. aureus* in wastewater, reducing bacterial counts by six orders of magnitude with minimal metal leaching. In contrast, zeolites loaded with nickel (NiNZ) have shown high leaching and low efficacy, limiting their utility. These results highlight the potential of metal-exchanged zeolites as innovative materials for water disinfection and pathogen control (Hrenovic et al., 2012). As observed, antimicrobial activity is primarily related to the type of metal used rather than the zeolite itself. For example, most natural zeolites do not exhibit intrinsic bactericidal activity against *Escherichia coli* or *Enterococcus faecalis*. Highlighting ion exchange's role in promoting zeolitic antimicrobial properties for biological applications. However, the influence of zeolite architecture should not be underestimated, as its three-dimensional structure and adsorption capacity also contribute significantly to bacterial inhibition by facilitating contact between bacteria and active metals within its structure (Rivera-Garza et al., 2000).

Bacteria are much larger than the zeolite micropores, preventing direct contact with the active metals inside the structure. However, interactions can still occur at the external surface of the zeolite, where some of these metals are present. Additionally, the exchanged metals within the zeolite structure can be released into the surrounding medium, generating reactive oxygen species (ROS), which play a key role in antimicrobial activity. Clinoptilolite has been extensively studied as an antimicrobial agent due to its ability to exchange metal ions such as Ag^{+}, Zn^{2+}, and Cu^{2+}. Among these, Ag^{+} exhibits a high affinity for the zeolite structure, completely replacing Na^{+} and inducing strong activity against bacteria such as Pseudomonas aeruginosa and Escherichia coli. However, higher concentrations of Ag^{+} can lead to the formation of metallic silver and a decrease in porosity, reducing its effectiveness. The controlled release of Ag^+, essential for its antimicrobial action, can be modulated by the presence of other ions, such as Pb^{2+}, Cd^{2+}, and Zn^{2+}, which facilitate ion exchange and promote its mobility into the medium. Furthermore, clinoptilolites containing silver nanoparticles (AgNP) obtained by thermal reduction have demonstrated activity against E. coli and Salmonella typhi. However, higher concentrations are required for the elimination of the latter(Dutta & Wang, 2019).

Chabazite has been functionalized with AgNP via thermal annealing in air, achieving a silver load of 19 wt%. These materials eliminated six bacterial strains, including *S. epidermidis*, *Salmonella typhimurium*, and *E. coli*, within 48 hours at a zeolite concentration of 0.1 wt%, except for *S. aureus*, which showed higher resistance (Dutta & Wang, 2019). Zeolite EMT, with particle sizes of 10–20 nm, has been modified with Ag^+ and silver nanoparticles using microwave radiation. It shows high bactericidal efficacy against *E. coli* due to its mesoporosity, which facilitates silver mobility. In the case of ZSM-5, a zeolite with low aluminum content, samples exchanged with Ag^+ significantly reduced *S. aureus* populations, although biofilm formation can reduce effectiveness by blocking Ag^+ release. Combining Ag^+ with peracetic acid has shown a synergistic effect, breaking bacterial biofilms and enhancing antimicrobial action. However, silver nanoparticles (AgNP-ZSM5) without peracetic acid did not exhibit significant activity, possibly due to limited nanoparticle dissolution (Dutta & Wang, 2019).

Zeolite beta and Zeomic (commercial product based on zeolites modified with antimicrobial metals, primarily silver), with three-dimensional structures, have demonstrated significant differences in antimicrobial activity depending on their preparation method. Zeomic showed rapid Ag release with a silver content of 2 wt%, eliminating bacteria in under an hour. In contrast, zeolite beta showed variable action times depending on using a structuring agent. Surprisingly, samples without a structuring agent were more active due to higher Ag^+ release and possibly differences in particle size affecting bacterial interaction (Dutta & Wang, 2019). These findings show the importance of ionic selectivity, metal release kinetics, and structural properties for designing antimicrobial zeolites.

In addition to their metal-derived antimicrobial effects, zeolites can also serve as carriers for antibiotics. A study demonstrated that NaX, NaY, and HY zeolites could load clofazimine, a poorly water-soluble antibiotic, against gram-positive bacteria, leprosy, and certain multidrug-resistant cancers. Compared to mesoporous materials like SBA-15, MCM-41, and Al-MCM-41, which exhibited higher drug-loading capacities due to abundant silanol and OH groups, zeolites achieved stronger interactions with clofazimine molecules due to their smaller pore diameters, favoring retention. This finding highlights the potential of zeolites as carrier matrices in pharmaceutical applications (Souza et al., 2023).

As three-dimensional crystalline aluminosilicates, zeolites stand out as support materials due to their continuous channel structure, uniform pore size, and high specific surface area. Combined with their positive ion-exchange capacity, these properties make them ideal for supporting metallic nanoparticles and metal oxides with antibacterial properties. For example, zeolites loaded with silver nanoparticles (AgNP) show a superior antibacterial activity, attributed to the effect of zeolite on preventing agglomeration, showing a uniform and homogeneous particle dispersion, increasing the effective contact area for bacteria. Furthermore, particle morphology and size, such as nano zeolites, have an effect on ion exchange efficiency and metal ion release, being an alternative for eliminating resistant bacteria.

Metal oxides are highly valued and widely used in various applications. On the other hand, in their nanoparticulate form, metal oxides have shown great potential in fields such as catalysis, electronics, and biomedicine. Metal oxide nanoparticles (MO-NPs) have shown promising results in inhibiting bacterial growth and challenging antibiotic resistance. MO-NPs are increasingly sought as antimicrobial pharmaceutical products due to their durability, high stability, and low toxicity to mammalian cells, compared to organic nanoparticles. These properties make them an attractive option for combating bacterial infections, especially in the context of growing resistance to conventional. Zeolite LTA is an ideal support material for incorporating zinc oxide nanoparticles (ZnO) due to its high crystallinity, stable structure, and ion exchange capacity. A zeolite composite with supported ZnO particles and copper ions (Cu^{2+}). Remarkably, despite incorporating these components, the characteristic cubic structure of the zeolite remained intact. This composite material showed high antibacterial activity against Gram-positive and Gram-negative bacteria, including *E. coli*, demonstrating its ability to damage bacterial cell membranes, increase permeability, and cause cell deformation (Ma et al., 2019).

In another study, LTA zeolite was synthesized from waste materials, such as glass and aluminum scrap, through a low-temperature hydrothermal process. The highly crystalline zeolite supported ZnO nanoparticles via a combined ion exchange, precipitation, and calcination process. Analyses revealed that smaller ZnO nanoparticles formed within the zeolite's supercavities, while others grew on the surface without altering the material's structure. This zinc-rich nanocomposite exhibited significant antibacterial and antioxidant properties, making it a promising option for applications such as active packaging in the food industry. These results underscore the potential of zeolites as adequate supports for metallic nanoparticles and as a sustainable means of reusing solid waste and contributing to environmental protection (Mouna et al., 2024).

Metal oxides are highly valued and widely used in various applications. On the other hand, in their nanoparticulate form, metal oxides have shown great potential in fields such as catalysis, electronics, and biomedicine. Metal oxide nanoparticles

(MO-NPs) have shown promising results in inhibiting bacterial growth and challenging antibiotic resistance. MO-NPs are increasingly sought as antimicrobial pharmaceutical products due to their durability, high stability, and low toxicity to mammalian cells compared to organic nanoparticles. These properties make them an attractive option for combating bacterial infections, especially in the context of growing resistance to conventional antibiotics (Kadiyala et al., 2018).

Among the most well-known and studied metal oxide nanoparticles are silver oxide (Ag_2O), copper oxide (CuO), iron oxide (Fe_2O_3), magnesium oxide (MgO), titanium oxide (TiO2), and zinc oxide (ZnO). Some copper and silver nanoparticles have been shown to kill bacteria, likely by releasing Cu^{2+} and Ag^{+} ions. However, it is important to note that resistance to these particles has developed over time, although the mechanisms behind this are not fully understood. On the other hand, the antibacterial mechanism of TiO_2, MgO, and ZnO nanoparticles remains more controversial. Proposed mechanisms, which have been extensively reviewed, include the release of toxic ions, oxidative stress, lipid peroxidation, damage to the cell membrane, enzyme inhibition, and proteolysis (Kadiyala et al., 2018). Zinc oxide (ZnO) stands out as a particularly promising material due to its well-documented antibacterial properties, as well as its stability and relatively low toxicity. ZnO nanoparticles have gained significant attention for their ability to effectively inhibit bacterial growth, making them valuable in biomedical applications, including drug delivery and wound healing. Unlike some other metal oxide nanoparticles, ZnO has shown less risk of inducing resistance in bacteria, making it a more sustainable option for long-term antimicrobial use (Kadiyala et al., 2018).

The intrinsic properties of metallic nanoparticles, such as zinc oxide (ZnO), titanium dioxide (TiO2), and silver, are mainly characterized by their size, composition, crystallinity, and morphology. Reducing their size to the nanoscale can alter their chemical, mechanical, electrical, and optical properties. These modified characteristics allow nanoparticles to interact with biomolecules, facilitating the physical transfer of the nanoparticles to internal cellular structures. This enhanced interaction significantly contributes to their antimicrobial activity and their potential for use in various biomedical applications. (Kadiyala et al., 2018; Ovchinnikov et al., 2023) Zinc-based materials have emerged as up-and-coming agents, particularly in their oxide forms, such as zinc oxide (ZnO), zinc hydroxide, and zinc peroxide. These compounds have been extensively studied as antimicrobial agents when formulated at the nanoscale. Zinc oxide has demonstrated significant antibacterial activity across a broad spectrum of bacterial species, a property that numerous research groups have widely investigated (Sirelkhatim et al., 2015).

Nanoscale ZnO exhibits diverse morphologies and enhanced antimicrobial efficacy due to its reduced particle size. At the nanoscale, ZnO can interact directly with bacterial surfaces or penetrate bacterial cells, reaching the intracellular environment and triggering distinct bactericidal mechanisms. These interactions are predominantly toxic to bacteria, making ZnO nanoparticles highly suitable for antimicrobial applications, including their use in the food industry (Sirelkhatim et al., 2015). Interestingly, several studies have reported that ZnO nanoparticles exhibit low toxicity toward human cells. This unique combination of selective toxicity—harmful to microorganisms but biocompatible with human cells—has positioned ZnO as an ideal antibacterial agent. The antimicrobial mechanisms of ZnO nanoparticles are attributed to their high surface-area-to-volume ratio and distinctive physicochemical properties. However, these enhanced properties make elucidating the exact mechanisms underlying their antibacterial activity difficult, suggesting further research. So far, the proposed mechanisms include generating reactive oxygen species (ROS), membrane disruption, and intracellular damage. These mechanisms are actively explored to understand and fully optimize ZnO nanoparticles for various applications (Sirelkhatim et al., 2015). When zinc is in the form of zinc hydroxide ($Zn(OH)_2$), it exhibits remarkable physical properties and excellent biocompatibility. Several studies have demonstrated that $Zn(OH)_2$ shows effective antimicrobial activity, particularly against gram-positive bacteria such as *Staphylococcus aureus* and *Klebsiella pneumoniae*, as well as gram-negative bacteria like *Escherichia coli*. This highlights its potential as an antimicrobial agent for biomedical and environmental applications (Sanchez et al., 2024). Another effective species currently under study is zinc peroxide nanoparticles (ZnO_2-NPs), which have shown significant antimicrobial potential. The antimicrobial activity of metal oxides, including ZnO_2-NPs, is strongly linked to their morphology, particle size, and surface area. For instance, metal oxide particles smaller than 100 nm with larger surface areas exhibit stronger antimicrobial activity due to their enhanced ability to interact with and damage microbial membranes (Ali et al., 2017). In fact, *in vivo* histopathological results have demonstrated that ZnO_2-NPs can be promising antimicrobial agents against multidrug-resistant (*MDR*) species such as *Pseudomonas aeruginosa* and *Aspergillus niger*, commonly found in burn wound infections.

Further research involving human volunteers with skin burns is necessary to confirm the efficacy of ZnO_2-NPs as a novel antimicrobial drug for wound healing. This is especially relevant given their success in treating burn wounds in experimental animal models, where they outperformed traditional silver sulfadiazine ointments. For ZnO_2-NPs to be viable in biomedical applications, two critical requirements must be met: they must be small and uniform to ensure homogeneous distribution and controlled oxygen release, and they must be biofunctionalized to enable specific interactions with biological systems. Studies conducted on four bacterial species with varying oxygen susceptibilities have confirmed the antimicrobial efficacy of ZnO_2 nanoparticles. These findings suggest their strong potential for use as innovative antimicrobial agents in clinical and environmental settings (Bergs et al., 2017).

ANTIMICROBIAL ZEOLITE LTA WITH ZINC OXIDE NANOPARTICLES

The LTA zeolite has been evaluated after supporting different species of zinc oxides in studies against bacteria and fungi, demonstrating inhibition in several samples. In particular, samples containing ZnO_2 species show significant inhibition, mainly attributed to the generation and release of peroxide species into the medium. This process is also accompanied by the formation of microbubbles, further enhancing the antimicrobial effect (Balouiri et al., 2016; Parvekar et al., 2020).

Figure 2. Antimicrobial activity of Zn modified LTA with NaA A) and ZnA B) series against five different strains: Staphylococcus aureus (red line), Streptococcus mutans (blue line), Enterococcus faecalis (green line), Cryptococcus spp. (black line) and Candida albicans.(black dashed line).

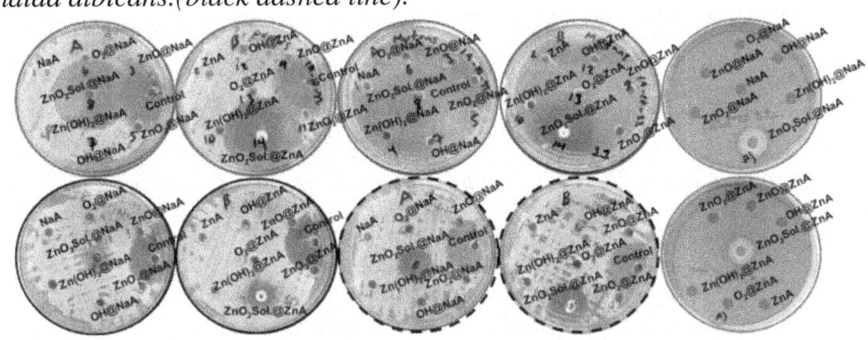

Reproduced with permission from Elsevier. Originally published in 'Antimicrobial activity of the LTA zeolite modified by zinc species', De León Ramírez J.I., Reyes Villegas V.A., Cadena-Nava R.D., Loredo-Garcia E., Chávez-Rivas F., González-Torres V., Petranovskii V., Microporous and Mesoporous Materials, Vol. 380, 2024, https://doi.org/10.1016/j.micromeso.2024.113295. © Elsevier 2024

Additionally, one of the factors responsible for bacterial inhibition is the presence of ions. However, as shown in Figure 1, these ions are not diffusing into the medium, as the evaluation was conducted in a solid medium (Pasquet et al., 2014). The activation of a Fenton-type reaction has been previously studied in various bacteria (Jannesari et al., 2023), and this effect is further enhanced by the solubility of the species in solution, which facilitates their interaction with water (Pasquet et al., 2014).

It is important to highlight that ultrasound significantly enhances the homogenization of ZnO_2 nanoparticles supported on LTA zeolites, resulting in a more uniform distribution across the zeolite surface. This improved uniformity optimizes the material's properties and enhances its antimicrobial performance. Specifically, ZnO_2 has demonstrated strong inhibition against resistant fungal strains, which are of critical clinical importance for immunosuppressed patients. As a result, innovative methods like ultrasound-assisted synthesis are being explored to combat these strains, aiming to save lives more efficiently (Pappas et al., 2015).

This study was conducted in a liquid medium to investigate the interaction mechanisms between bacteria and zeolite-supported nanoparticles in greater detail (fig.3). As is well known, zeolites are highly effective adsorbents, and this property plays a critical role in bacterial adhesion to their surfaces, subsequently enhancing antimicrobial activity. In the case of NaA zeolite, the adsorption of bacteria onto its surface is primarily attributed to the electrostatic interactions between the electrical charges of the bacterial membrane and the zeolite. Broth assays revealed that NaA exhibits a greater affinity for Gram-positive bacteria, which have a positive surface charge due to the amino acid residues in their peptidoglycans. This interaction is confirmed by the positive ζ potential measured for NaA. In contrast, Gram-negative bacteria, with a negative surface charge of the membrane phospholipids, are repelled by the NaA surface, limiting bacterial adhesion and avoiding antimicrobial activity (Kubota et al., 2008)(Oheix et al., 2022).

Figure 3. Schematic representation of the interaction mechanisms between bacteria and zeolite-supported nanoparticles in a liquid medium. Zeolites, known for their high adsorption capacity, facilitate bacterial adhesion to their surfaces, enhancing antimicrobial activity.

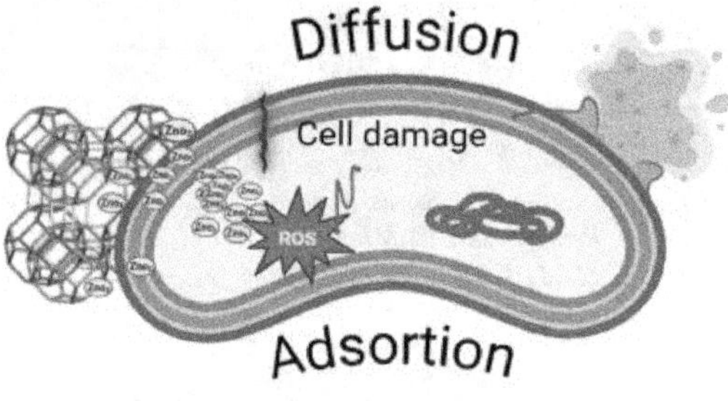

Created in BioRender. De Leon Ramirez, J. (2025). BioRender.

The results also suggest that the bacterial inhibition mechanism is closely linked to Zn species exchange within the zeolite structure. For instance, NaA ion exchange with Zn (ZnA) showed variations in bacterial growth inhibition based on the present type of Zn species. Ultrasound treatment during the deposition of these species, such as ZnO, $Zn(OH)_2$, and ZnO_2, significantly enhanced the homogeneity and distribution of the nanoparticles, optimizing surface interactions with bacteria. This improvement led to a treatment-dependent bactericidal activity, with ZnO_2 demonstrating particularly effective inhibition due to its ability to generate reactive oxygen species (ROS) and nanobubbles that induce oxidative stress in bacterial cells (Oheix et al., 2022).

*Figure 4. Antibacterial activity against E. coli and E. faecalis of zeolite NaA, modified with Zn species deposited by ultrasonic treatments done by triplicate in a broth seeding method. The asterisk denotes that the measurements are not significantly different ($p > 0.05$ for * and $p > 0.1$ for **) compared to the desired effect (positive control). Reproduced with permission from Elsevier.*

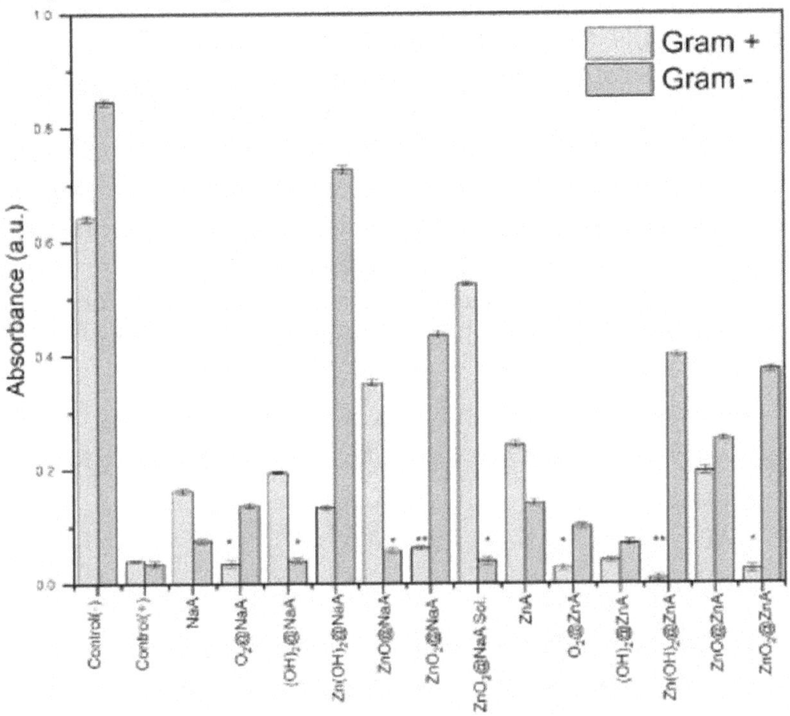

Originally published in 'Antimicrobial activity of the LTA zeolite modified by zinc species', De León Ramírez J.I., Reyes Villegas V.A., Cadena-Nava R.D., Loredo-Garcia E., Chávez-Rivas F., González-Torres V., Petranovskii V., Microporous and Mesoporous Materials, Vol. 380, 2024, https://doi.org/10.1016/j.micromeso.2024.113295. © Elsevier 2024

Interestingly, antimicrobial activity differs based on the bacterial type. Gram-negative bacteria, characterized by their thinner cell walls, were more susceptible to diffusion-based mechanisms, including ROS generation and Zn^{2+} release. In contrast, Gram-positive bacteria, with their thicker peptidoglycan layers, exhibited enhanced adsorption onto the zeolite surface. This adsorption mechanism restricted vital nutrients, creating a bacteriostatic effect. When ZnO_2 species were supported on NaA and treated with ultrasound, the material exhibited improved inhibitory activity, with the ultrasound facilitating a more uniform dispersion of Zn species (Sawyer et al., 2019)(Yusof et al., 2019).

Moreover, drying the ZnO_2@NaA particles activated membrane adsorption mechanisms, while the absence of drying favored diffusion-based mechanisms, further confirming the dual role of adsorption and diffusion in bacterial inhibition. The electrostatic interactions between the particles and the Gram-positive bacterial membrane enhanced the antimicrobial effect, making ultrasound-assisted Zn species deposition on zeolites a promising strategy for developing advanced antimicrobial agents (Yusof et al., 2019),(Bharti Sharma et al., 2022; Sengodan et al., 2022), (Mendes et al., 2022), (Jannesari et al., 2020),(Ameh et al., 2022).

One of the key benefits of loading nanoparticles onto zeolites is the ability to control and reduce their toxicity. A recent study compared different zinc peroxide species supported on LTA zeolite (type A), demonstrating that zeolites can moderate toxicity in fibroblasts. This effect is particularly relevant since controlled toxicity through zeolite support opens doors to various applications. Using ultrasound during the deposition process enhances nanoparticle dispersion and optimizes surface properties, such as adsorption capacity and ζ potential, critical factors for antimicrobial activity and toxicity regulation by ensuring better nanoparticle dispersion.

*Figure 5. Fibroblast cell viability conducted in triplicate, with a 48-h exposure of fibroblast to ZnO_2@NaA and ZnO_2@NaA Sol. at 1 mg/mL. The asterisk denotes that the measurements are not significantly different ($p > 0.05$ for *) compared to the undesired effect (negative control).*

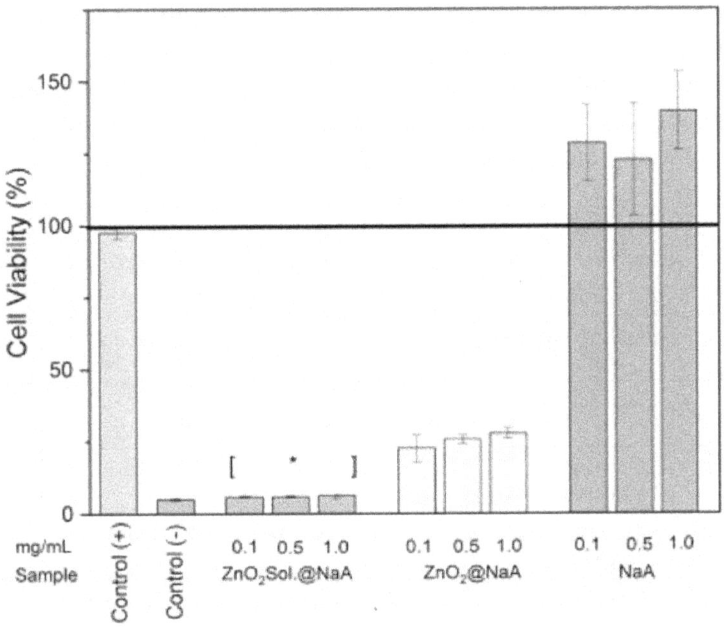

Reproduced with permission from Elsevier. Originally published in 'Antimicrobial activity of the LTA zeolite modified by zinc species', De León Ramírez J.I., Reyes Villegas V.A., Cadena-Nava R.D., Loredo-Garcia E., Chávez-Rivas F., González-Torres V., Petranovskii V., Microporous and Mesoporous Materials, Vol. 380, 2024, https://doi.org/10.1016/j.micromeso.2024.113295. © Elsevier 2024

Additionally, the results show an increasing trend in cell viability as the material concentration decreases. Conversely, cell viability decreases drastically when ZnO_2@NaA is not subjected to a drying process. This suggests that the observed activity is related to diffusion mechanisms involving increased release of reactive oxygen species (ROS) and Zn^{2+} into the solution.

These findings underscore the need for further research at lower concentrations (below 0.1 mg/mL) to enhance cell viability and better elucidate a concentration-dependent mechanism of action. Previous studies have shown that ZnO_2 exhibits highly toxic activities against DNA in cell lines at concentrations in the μg range. However, this toxicity can be significantly reduced when toxic ZnO_2 particles are supported on a non-toxic zeolitic material, such as NaA zeolite (De León Ramirez et al., 2024).

The synthesis and modification process of materials significantly influence their physicochemical and biological properties. In particular, the application of ultrasound during the synthesis of modified zeolitic materials has proven to be an effective strategy to enhance the deposition of metallic species, increase nanoparticle dispersion, and modify the material's surface properties. These modifications can directly impact on the antimicrobial activity of the material, making them relevant for applications in biomedicine and environmental remediation.

In this context, **Table 1** compares different materials based on zeolite A, synthesized or modified using various methods, focusing on the effect of ultrasound on their properties and antibacterial activity.

Table 1. Comparison of zeolite A-based materials with different synthesis/modification methods and their respective antibacterial activity.

Material	Synthesis/Modification Method	Effect of Ultrasound	Antibacterial Activity	Reference
Zeolite A	Conventional synthesis	Not applicable	Low	(Belaabed et al., 2016)
Zn-modified Zeolite A	Conventional precipitation	Not applicable	Moderate	(S. Chen et al., 2018)
Zn-modified Zeolite A	Conventional precipitation	Not applicable	Moderate	(Fox et al., 2010)
Ag/ZnO modified Zeolite A	Conventional precipitation	Not applicable	Moderate	(Wakweya & Jifar, 2023)
Zn-modified Zeolite A	Ultrasound-assisted precipitation	Improves the deposition of metal species and their interaction with zeolite and increases the surface area	High	(De León Ramirez et al., 2024)
Zeolite A modified with Ag	Ultrasound-assisted impregnation	Improves nanoparticle dispersion and increases surface area	Highter	(Asraf et al., 2024)
Zeolite A modified ZnO/TiO$_2$	Ultrasound-assisted doping	Improves metal species deposition and dispersion	Highter	(Azizi-Lalabadi et al., 2020)

The table summarizes various strategies for modifying zeolite A, including conventional and ultrasound-assisted techniques. The results indicate that ultrasound enhances the deposition of metallic species, increases nanoparticle dispersion, and improves the material's surface area, which translates into greater antimicrobial activity. Notably, materials synthesized using ultrasound-assisted techniques exhibit superior antibacterial activity compared to those obtained through conventional methods.

CONCLUSION

Sonochemistry through ultrasound emerges as an innovative, efficient, and environmentally friendly method for depositing nanoparticles of various zinc species onto the surface of LTA zeolites. This approach ensures proper dispersion of the nanoparticles and enhances their functional properties. Notably, zinc peroxide nanoparticles demonstrated remarkable effectiveness due to their unique mechanism of generating oxygen (O_2) microbubbles, significantly contributing to bacterial inhibition.

Furthermore, incorporating LTA zeolite optimizes antimicrobial activity and modulates toxicity, substantially reducing adverse effects on non-target cells. These findings highlight the potential of ultrasound for designing functional materials and exploring the application of zeolites with other metals, extending the research scope into catalytic and multifunctional activities. This study shows the importance of sustainable methods like sonochemistry in designing advanced solutions for technological and environmental challenges.

REFERENCE

Al-Harbi, L. M., Kosa, S. A., Abd El Maksod, I. H., & Hegazy, E. Z. (2015). The Photocatalytic Activity of TiO_2 -Zeolite Composite for Degradation of Dye Using Synthetic UV and Jeddah Sunlight. *Journal of Nanomaterials, 2015*(1). DOI: 10.1155/2015/565849

Ali, S., Morsy, R., El-Zawawy, N., Fareed, M., & Bedaiwy, M. (2017). Synthesized zinc peroxide nanoparticles (ZnO_2-NPs): A novel antimicrobial, anti-elastase, anti-keratinase, and anti-inflammatory approach toward polymicrobial burn wounds. *International Journal of Nanomedicine, 12*, 6059–6073. DOI: 10.2147/IJN.S141201 PMID: 28860766

Ameh, T., Gibb, M., Stevens, D., Pradhan, S. H., Braswell, E., & Sayes, C. M. (2022). Silver and Copper Nanoparticles Induce Oxidative Stress in Bacteria and Mammalian Cells. *Nanomaterials (Basel, Switzerland), 12*(14), 2402. DOI: 10.3390/nano12142402 PMID: 35889626

Andaç, Ö., Tatlıer, M., Sirkecioğlu, A., Ece, I., & Erdem-Şenatalar, A. (2005). Effects of ultrasound on zeolite A synthesis. *Microporous and Mesoporous Materials, 79*(1–3), 225–233. DOI: 10.1016/j.micromeso.2004.11.007

Askari, S., Miar Alipour, S., Halladj, R., & Davood Abadi Farahani, M. H. (2013). Effects of ultrasound on the synthesis of zeolites: A review. *Journal of Porous Materials, 20*(1), 285–302. DOI: 10.1007/s10934-012-9598-6

Asraf, M. H., Malek, N. A. N. N., Sani, N. S., & Matmin, J. (2024). Surface Morphology, Porosity and Antibiofilm Activity of Orthosiphon aristatus-Phytosynthesized–Silver Nanoparticles Supported Zeolite A. *Arabian Journal for Science and Engineering, 49*(7), 9121–9131. DOI: 10.1007/s13369-024-08832-x

Asrori, M. R., Santoso, A., & Sumari, S. (2024). Proofing the presence of metal oxide impregnated into zeolite A without calcination: XRD and FTIR studies. *Case Studies in Chemical and Environmental Engineering, 9*, 100676. DOI: 10.1016/j.cscee.2024.100676

Azizi-Lalabadi, M., Alizadeh-Sani, M., Khezerlou, A., Mirzanajafi-Zanjani, M., Zolfaghari, H., Bagheri, V., Divband, B., & Ehsani, A. (2019). Nanoparticles and Zeolites: Antibacterial Effects and their Mechanism against Pathogens. *Current Pharmaceutical Biotechnology, 20*(13), 1074–1086. DOI: 10.2174/1573397115666190708120040 PMID: 31309886

Azizi-Lalabadi, M., Ehsani, A., Ghanbarzadeh, B., & Divband, B. (2020). Polyvinyl alcohol/gelatin nanocomposite containing ZnO, TiO2 or ZnO/TiO2 nanoparticles doped on 4A zeolite: Microbial and sensory qualities of packaged white shrimp during refrigeration. *International Journal of Food Microbiology*, *312*, 108375. DOI: 10.1016/j.ijfoodmicro.2019.108375 PMID: 31669767

Bahiraei, A., & Behin, J. (2020). Sonochemical immobilization of MnO2 nanoparticles on NaP-zeolite for enhanced Hg (II) adsorption from water. *Journal of Environmental Chemical Engineering*, *8*(3), 103790. DOI: 10.1016/j.jece.2020.103790

Balouiri, M., Sadiki, M., & Ibnsouda, S. K. (2016). Methods for in vitro evaluating antimicrobial activity: A review. *Journal of Pharmaceutical Analysis*, *6*(2), 71–79. DOI: 10.1016/j.jpha.2015.11.005 PMID: 29403965

Baruah, M. J., Dutta, A., Biswas, S., Gogoi, G., Hoque, N., Bhattacharyya, P. K., & Bania, K. K. (2022). Fe2O3 Nanocatalysts Supported on Zeolite-Y for the Selective Synthesis of C2 Di-Indolyl Indolones and Isatins. *ACS Applied Nano Materials*, *5*(1), 1446–1459. DOI: 10.1021/acsanm.1c03987

Belaabed, R., Elabed, S., Addaou, A., Laajab, A., Rodríguez, M. A., & Lahsini, A. (2016). Synthesis of LTA zeolite for bacterial adhesion. *Boletín de la Sociedad Española de Cerámica y Vidrio*, *55*(4), 152–158. DOI: 10.1016/j.bsecv.2016.05.001

Bergs, C., Brück, L., Rosencrantz, R. R., Conrads, G., Elling, L., & Pich, A. (2017). Biofunctionalized zinc peroxide (ZnO_2) nanoparticles as active oxygen sources and antibacterial agents. *RSC Advances*, *7*(62), 38998–39010. DOI: 10.1039/C7RA06332F

Chen, D., Sharma, S. K., & Mudhoo, A. (Eds.). (2011a). *Handbook on Applications of Ultrasound*. CRC Press., DOI: 10.1201/b11012

Chen, D., Sharma, S. K., & Mudhoo, A. (Eds.). (2011b). *Handbook on Applications of Ultrasound*. CRC Press., DOI: 10.1201/b11012

Chen, S., Popovich, J., Zhang, W., Ganser, C., Haydel, S. E., & Seo, D.-K. (2018). Superior ion release properties and antibacterial efficacy of nanostructured zeolites ion-exchanged with zinc, copper, and iron. *RSC Advances*, *8*(66), 37949–37957. DOI: 10.1039/C8RA06556J PMID: 30555688

Collins, F., Rozhkovskaya, A., Outram, J. G., & Millar, G. J. (2020). A critical review of waste resources, synthesis, and applications for Zeolite LTA. *Microporous and Mesoporous Materials*, *291*, 109667. DOI: 10.1016/j.micromeso.2019.109667

Cruz, T. J. T., Melo, M. I. S., & Pergher, S. (2020). Optimization of Parameters and Methodology for the Synthesis of LTA-Type Zeolite Using Light Coal Ash. *Applied Sciences (Basel, Switzerland)*, *10*(20), 7332. DOI: 10.3390/app10207332

da Silva, A., Elias, E. B. C. D., Cruz, T. J. T., Pinto, F. G. H. S., de Mello, M. I. S., Bieseki, L., & Pergher, S. B. C. (2024). Synthesis and Cation Exchange of LTA Zeolites Synthesized from Different Silicon Sources Applied in CO2 Adsorption. *Coatings*, *14*(6), 680. DOI: 10.3390/coatings14060680

De León Ramirez, J. I., Reyes Villegas, V. A., Cadena-Nava, R. D., Loredo-Garcia, E., Chávez-Rivas, F., González-Torres, V., & Petranovskii, V. (2024). Antimicrobial activity of the LTA zeolite modified by zinc species. *Microporous and Mesoporous Materials*, *380*, 113295. DOI: 10.1016/j.micromeso.2024.113295

Demirci, S., Ustaoğlu, Z., Yılmazer, G. A., Sahin, F., & Baç, N. (2014). Antimicrobial Properties of Zeolite-X and Zeolite-A Ion-Exchanged with Silver, Copper, and Zinc Against a Broad Range of Microorganisms. *Applied Biochemistry and Biotechnology*, *172*(3), 1652–1662. DOI: 10.1007/s12010-013-0647-7 PMID: 24242073

Derakhshankhah, H., Jafari, S., Sarvari, S., Barzegari, E., Moakedi, F., Ghorbani, M., Shiri Varnamkhasti, B., Jaymand, M., Izadi, Z., & Tayebi, L. (2020). Biomedical Applications of Zeolitic Nanoparticles, with an Emphasis on Medical Interventions. *International Journal of Nanomedicine*, *15*, 363–386. DOI: 10.2147/IJN.S234573 PMID: 32021185

Desales-Guzmán, L. A., Pacheco-Sánchez, J. H., & Vazquez-Rivas, E. (2025). Enhanced hydrogen adsorption properties of Zeolite Templated Carbon via chemical activation: DFT study. *Fuel*, *379*, 133080. DOI: 10.1016/j.fuel.2024.133080

Dutta, P., & Wang, B. (2019). Zeolite-supported silver as antimicrobial agents. *Coordination Chemistry Reviews*, *383*, 1–29. DOI: 10.1016/j.ccr.2018.12.014

Eom, E., Song, M., Kim, J. C., Kwon, D., Rainer, D. N., Gołąbek, K., Nam, S. C., Ryoo, R., Mazur, M., & Jo, C. (2022). Confining Gold Nanoparticles in Preformed Zeolites by Post-Synthetic Modification Enhances Stability and Catalytic Reactivity and Selectivity. *JACS Au*, *2*(10), 2327–2338. DOI: 10.1021/jacsau.2c00380 PMID: 36311841

Fox, S., Wilkinson, T. S., Wheatley, P. S., Xiao, B., Morris, R. E., Sutherland, A., Simpson, A. J., Barlow, P. G., Butler, A. R., Megson, I. L., & Rossi, A. G. (2010). NO-loaded Zn2+-exchanged zeolite materials: A potential bifunctional anti-bacterial strategy. *Acta Biomaterialia*, *6*(4), 1515–1521. DOI: 10.1016/j.actbio.2009.10.038 PMID: 19861185

Fu, X., Sheng, X., Zhou, Y., Fu, Z., Zhao, S., Zhang, Z., & Zhang, Y. (2016). Ultrasonic/microwave synergistic synthesis of well-dispersed hierarchical zeolite Y with improved alkylation catalytic activity. *Korean Journal of Chemical Engineering, 33*(6), 1931–1937. DOI: 10.1007/s11814-016-0022-9

Gicheva, G., Mintcheva, N., & Panayotova, M. (2021). *Nanosized ZnO and ZnO@ zeolite composite prepared via sonication method.*

Głowniak, S., Szczęśniak, B., Choma, J., & Jaroniec, M. (2023). Recent Developments in Sonochemical Synthesis of Nanoporous Materials. *Molecules (Basel, Switzerland), 28*(6), 2639. DOI: 10.3390/molecules28062639 PMID: 36985612

Gordina, N. E., Prokof'ev, V. Yu., & Kochetkov, S. P. (2018). Use of Mechanochemical Activation and Ultrasonic Treatment for the Synthesis of LTA Zeolite. *Russian Journal of General Chemistry, 88*(9), 1981–1989. DOI: 10.1134/S1070363218090402

Gordina, N. E., Prokof'ev, V. Y., Kul'pina, Y. N., Hmylova, O. E., Gazahova, S. I., & Petuhova, N. V. (2017). Synthesis of granulated binder-free LTA zeolite from metakaolin using ultrasonic treatment. *Journal of Porous Materials, 24*(3), 667–678. DOI: 10.1007/s10934-016-0303-z

Gordina, N. E., Prokof'ev, V. Yu., Kul'pina, Yu. N., Petukhova, N. V., Gazakhova, S. I., & Khmylova, O. E. (2017). Use of Ultrasonic Processing at Early Stages of LTA Zeolite Synthesis from Metakaolin. *Glass and Ceramics, 73*(9–10), 334–337. DOI: 10.1007/s10717-017-9884-5

Gordina, N. E., Rumyantsev, R. N., Borisova, T. N., Kolobkova, A. E., Tsvetova, E. V., Afanas'eva, E. E., Severgina, E. S., & Prokof'ev, V. Y. (2021). Use of Combinations of Ultrasonic Treatment and Microwave Crystallization to Intensify the Synthesis of LTA Zeolite Membranes. *Petroleum Chemistry, 61*(3), 292–298. DOI: 10.1134/S096554412103021X

Handbook of Ultrasonics and Sonochemistry. (2016). Springer Singapore., DOI: 10.1007/978-981-287-278-4

Hanim, S. A. M., Malek, N. A. N. N., & Ibrahim, Z. (2016). Amine-functionalized, silver-exchanged zeolite NaY: Preparation, characterization and antibacterial activity. *Applied Surface Science, 360*, 121–130. DOI: 10.1016/j.apsusc.2015.11.010

Hrenovic, J., Milenkovic, J., Ivankovic, T., & Rajic, N. (2012). Antibacterial activity of heavy metal-loaded natural zeolite. *Journal of Hazardous Materials, 201–202*, 260–264. DOI: 10.1016/j.jhazmat.2011.11.079 PMID: 22178285

Jannesari, M., Akhavan, O., Madaah Hosseini, H. R., & Bakhshi, B. (2020). Graphene/CuO2Nanoshuttles with Controllable Release of Oxygen Nanobubbles Promoting Interruption of Bacterial Respiration. *ACS Applied Materials & Interfaces*, *12*(32), 35813–35825. DOI: 10.1021/acsami.0c05732 PMID: 32664715

Jannesari, M., Akhavan, O., Madaah Hosseini, H. R., & Bakhshi, B. (2023). Oxygen-Rich Graphene/ZnO2-Ag nanoframeworks with pH-Switchable Catalase/Peroxidase activity as O2 Nanobubble-Self generator for bacterial inactivation. *Journal of Colloid and Interface Science*, *637*, 237–250. DOI: 10.1016/j.jcis.2023.01.079 PMID: 36701869

Kadiyala, U., Kotov, N. A., & VanEpps, J. S. (2018). Antibacterial Metal Oxide Nanoparticles: Challenges in Interpreting the Literature. *Current Pharmaceutical Design*, *24*(8), 896–903. DOI: 10.2174/1381612824666180219130659 PMID: 29468956

Kim, W., Choi, D., & Kim, S. (2010). Sonochemical synthesis of zeolite a from metakaolinite in NaOH solution. *Materials Transactions*, *51*(9), 1694–1698. DOI: 10.2320/matertrans.M2010191

Kubota, M., Nakabayashi, T., Matsumoto, Y., Shiomi, T., Yamada, Y., Ino, K., Yamanokuchi, H., Matsui, M., Tsunoda, T., Mizukami, F., & Sakaguchi, K. (2008). Selective adsorption of bacterial cells onto zeolites. *Colloids and Surfaces. B, Biointerfaces*, *64*(1), 88–97. DOI: 10.1016/j.colsurfb.2008.01.012 PMID: 18313276

Kwakye-Awuah, B., Williams, C., Kenward, M. A., & Radecka, I. (2008). Antimicrobial action and efficiency of silver-loaded zeolite X. *Journal of Applied Microbiology*, *104*(5), 1516–1524. DOI: 10.1111/j.1365-2672.2007.03673.x PMID: 18179543

Liu, Z. F., Liu, Z. C., Wang, Y., Li, Y. B., Qu, L., e, L., Ya, J., & Huang, P. Y. (2012). Photocatalysis of TiO 2 nanoparticles supported on natural zeolite. *Materials Technology*, *27*(3), 267–271. DOI: 10.1179/1753555712Y.0000000011

Ma, X., Pei, Y., Ma, Y., Pu, T., & Lei, Y. (2019). Antibacterial Activity of Cu2+-ZnO-modified 13X Zeolite against E.coli and S.aureus. *Journal of Wuhan University of Technology. Materials Science Edition*, *34*(2), 481–486. DOI: 10.1007/s11595-018-2077-z

Mahamud, M., Taddesse, A. M., Bogale, Y., & Bezu, Z. (2023). Zeolite supported CdS/TiO2/CeO2 composite: Synthesis, characterization and photocatalytic activity for methylene blue dye degradation. *Materials Research Bulletin*, *161*, 112176. Advance online publication. DOI: 10.1016/j.materresbull.2023.112176

Mason, T. J., & Vinatoru, M. (2023). *Sonochemistry*. De Gruyter., DOI: 10.1515/9783110566178

Mendes, C. R., Dilarri, G., Forsan, C. F., Sapata, V., Lopes, P. R. M., de Moraes, P. B., Montagnolli, R. N., Ferreira, H., & Bidoia, E. D. (2022). Antibacterial action and target mechanisms of zinc oxide nanoparticles against bacterial pathogens. *Scientific Reports*, *12*(1), 2658. DOI: 10.1038/s41598-022-06657-y PMID: 35173244

Mouna, S., Hajji, S., & Tounsi, H. (2024). Waste to health: Green synthesis of Zn loaded LTA zeolite prepared from waste glass and aluminum scrap with high antioxidant and antimicrobial activities. *Journal of Cleaner Production*, *434*, 139946. DOI: 10.1016/j.jclepro.2023.139946

Oheix, E., Reicher, C., Nouali, H., Michelin, L., Josien, L., Daou, T. J., & Pieuchot, L. (2022). Rational Design and Characterisation of Novel Mono- and Bimetallic Antibacterial Linde Type A Zeolite Materials. *Journal of Functional Biomaterials*, *13*(2), 73. DOI: 10.3390/jfb13020073 PMID: 35735928

Ovchinnikov, N. L., Vinogradov, N. M., Gordina, N. E., & Butman, M. F. (2023). Preparation of Photoactive TiO2/LTA Zeolite Composites by Solution Technology in Hydrothermal Conditions. *Protection of Metals and Physical Chemistry of Surfaces*, *59*(4), 570–576. DOI: 10.1134/S2070205123700776

Pappas, P. G., Kauffman, C. A., Andes, D. R., Clancy, C. J., Marr, K. A., Ostrosky-Zeichner, L., Reboli, A. C., Schuster, M. G., Vazquez, J. A., Walsh, T. J., Zaoutis, T. E., & Sobel, J. D. (2015). Clinical Practice Guideline for the Management of Candidiasis: 2016 Update by the Infectious Diseases Society of America. In *Clinical Infectious Diseases* (Vol. 62, Number 4, pp. e1–e50). Oxford University Press. DOI: 10.1093/cid/civ933

Parvekar, P., Palaskar, J., Metgud, S., Maria, R., & Dutta, S. (2020). The minimum inhibitory concentration (MIC) and minimum bactericidal concentration (MBC) of silver nanoparticles against *Staphylococcus aureus.Biomaterial Investigations in Dentistry*, *7*(1), 105–109. DOI: 10.1080/26415275.2020.1796674 PMID: 32939454

Pasquet, J., Chevalier, Y., Pelletier, J., Couval, E., Bouvier, D., & Bolzinger, M.-A. (2014). The contribution of zinc ions to the antimicrobial activity of zinc oxide. *Colloids and Surfaces. A, Physicochemical and Engineering Aspects*, *457*, 263–274. DOI: 10.1016/j.colsurfa.2014.05.057

Petrov, I., & Michalev, T. (2012). Synthesis of Zeolite A. *RE:view*.

Pollet, B. G., & Ashokkumar, M. (2019). *Introduction to Ultrasound, Sonochemistry and Sonoelectrochemistry*. Springer International Publishing., DOI: 10.1007/978-3-030-25862-7

Ramishvili, T., Tsitsishvili, V., Chedia, R., Sanaia, E., Gabunia, V., & Kokiashvili, N. (2017). Preparation of Ultradispersed Crystallites of Modified Natural Clinoptilolite with the Use of Ultrasound and Its Application as a Catalyst in the Synthesis of Methyl Salicylate. *American Journal of Nano Research and Applications.Special Issue: Nanotechnologies*, *5*(1), 26–32. DOI: 10.11648/j.nano.s.2017050301.17

Ribeiro, A. C., de Oliveira, A. M., Beltran, L. B., Diório, A., Magalhães-Ghiotto, G. A. V., de Abreu Filho, B. A., de Almeida Duarte, E., & Bergamasco, R. (2024). Antibacterial activity of functionalized natural zeolites (NZ-AgNPs) and its application in bacteriological water treatment and commercial paints. *Environmental Nanotechnology, Monitoring & Management*, *22*, 101001. DOI: 10.1016/j.enmm.2024.101001

Rieger, K. A., Cho, H. J., Yeung, H. F., Fan, W., & Schiffman, J. D. (2016). Antimicrobial Activity of Silver Ions Released from Zeolites Immobilized on Cellulose Nanofiber Mats. *ACS Applied Materials & Interfaces*, *8*(5), 3032–3040. DOI: 10.1021/acsami.5b10130 PMID: 26788882

Rivera-Garza, M., Olguín, M. T., García-Sosa, I., Alcántara, D., & Rodríguez-Fuentes, G. (2000). Silver supported on natural Mexican zeolite as an antibacterial material. *Microporous and Mesoporous Materials*, *39*(3), 431–444. DOI: 10.1016/S1387-1811(00)00217-1

Rozhkovskaya, A., Rajapakse, J., & Millar, G. J. (2021). Optimisation of zeolite LTA synthesis from alum sludge and the influence of the sludge source. *Journal of Environmental Sciences (China)*, *99*, 130–142. DOI: 10.1016/j.jes.2020.06.019 PMID: 33183690

Ruíz-Baltazar, Á. de J. (2024). Advancements in nanoparticle-modified zeolites for sustainable water treatment: An interdisciplinary review. In *Science of the Total Environment* (Vol. 946). Elsevier B.V., DOI: 10.1016/j.scitotenv.2024.174373

Sanchez, J. A., Materon, L., Parsons, J. G., & Alcoutlabi, M. (2024). Synthesis, Characterization, and Antibacterial Activity of Graphene Oxide/Zinc Hydroxide Nanocomposites. *Applied Sciences (Basel, Switzerland)*, *14*(14), 6274. DOI: 10.3390/app14146274

Sawyer, W. S., Wang, L., Uehara, T., Tamrakar, P., Prathapam, R., Mostafavi, M., Metzger, L. E., Feng, B., & Baxter Rath, C. M. (2019). Targeted lipopolysaccharide biosynthetic intermediate analysis with normal-phase liquid chromatography mass spectrometry. *PLoS One*, *14*(2), e0211803. DOI: 10.1371/journal.pone.0211803 PMID: 30735516

Sen, M., Dana, K., & Das, N. (2018). Development of LTA zeolite membrane from clay by sonication assisted method at room temperature for H2-CO2 and CO2-CH4 separation. *Ultrasonics Sonochemistry*, *48*, 299–310. DOI: 10.1016/j.ultsonch.2018.06.007 PMID: 30080554

Sengodan, S., Matheswaran, B., Shanmugam, S., & Thangadurai Thangaian, D. (2022). Investigation on crystal facet-dependent antibacterial activity of Cu2O crystals – A structural, morphological, and spectroscopy studies. *Results in Chemistry*, *4*, 100358. DOI: 10.1016/j.rechem.2022.100358

Sharma, B., Tandon, R. P., & Mahapatro, A. K. (2022). Interaction of zinc peroxide nanoparticle with fibroblast cell. *Advanced Materials & Processes*, *2*(7), 440–444. DOI: 10.5185/amp.2017/708

Singh, B. K., Lee, S., & Na, K. (2020). An overview on metal-related catalysts: Metal oxides, nanoporous metals and supported metal nanoparticles on metal organic frameworks and zeolites. *Rare Metals*, *39*(7), 751–766. DOI: 10.1007/s12598-019-01205-6

Sirelkhatim, A., Mahmud, S., Seeni, A., Kaus, N. H. M., Ann, L. C., Bakhori, S. K. M., Hasan, H., & Mohamad, D. (2015). Review on Zinc Oxide Nanoparticles: Antibacterial Activity and Toxicity Mechanism. *Nano-Micro Letters*, *7*(3), 219–242. DOI: 10.1007/s40820-015-0040-x PMID: 30464967

Sonochemistry and the Acoustic Bubble. (2015). Elsevier. DOI: 10.1016/C2013-0-18886-1

Souza, I., García-Villén, F., Viseras, C., & Pergher, S. (2023). Zeolites as Ingredients of Medicinal Products. *Pharmaceutics*, *15*(5), 1352. DOI: 10.3390/pharmaceutics15051352 PMID: 37242594

Talebi, J., Halladj, R., & Askari, S. (2010). Sonochemical synthesis of silver nanoparticles in Y-zeolite substrate. *Journal of Materials Science*, *45*(12), 3318–3324. DOI: 10.1007/s10853-010-4349-z

Townsend, R. P., & Coker, E. N. (2001). *Ion exchange in zeolites.*, DOI: 10.1016/S0167-2991(01)80253-6

Vaezi, M., Noormohammadbeigi, M., Cruciani, G., & Zendehdel, M. (2025). Ion-exchange of copper into mordenite and clinoptilolite zeolites by molecular dynamics simulations and experimental investigations. *Microporous and Mesoporous Materials*, *382*, 113397. DOI: 10.1016/j.micromeso.2024.113397

Vaičiukynienė, D., Kantautas, A., Vaitkevičius, V., Jakevičius, L., Rudžionis, Ž., & Paškevičius, M. (2015). Effects of ultrasonic treatment on zeolite NaA synthesized from by-product silica. *Ultrasonics Sonochemistry*, *27*, 515–521. DOI: 10.1016/j.ultsonch.2015.06.001 PMID: 26186874

Wakweya, B., & Jifar, W. W. (2023). In vitro Evaluation of Antibacterial Activity of Synthetic Zeolite Supported AgZno Nanoparticle Against a Selected Group of Bacteria. *Journal of Experimental Pharmacology*, *15*, 139–147. DOI: 10.2147/JEP.S396118 PMID: 36941893

Wang, Y., Wang, C., Wang, L., Wang, L., & Xiao, F. S. (2021). Zeolite Fixed Metal Nanoparticles: New Perspective in Catalysis. *Accounts of Chemical Research*, *54*(11), 2579–2590. DOI: 10.1021/acs.accounts.1c00074 PMID: 33999615

Xu, D., Lv, H., & Liu, B. (2018). Encapsulation of Metal Nanoparticle Catalysts Within Mesoporous Zeolites and Their Enhanced Catalytic Performances: A Review. *Frontiers in Chemistry*, *6*, 550. Advance online publication. DOI: 10.3389/fchem.2018.00550 PMID: 30474024

Yosefi, L., Haghighi, M., Allahyari, S., & Ashkriz, S. (2015). Effect of ultrasound irradiation and Ni-loading on properties and performance of CeO2-doped Ni/clinoptilolite nanocatalyst used in polluted air treatment. *Process Safety and Environmental Protection*, *95*, 26–37. DOI: 10.1016/j.psep.2015.02.006

Yusof, N. A. A., Zain, N. M., & Pauzi, N. (2019). Synthesis of ZnO nanoparticles with chitosan as stabilizing agent and their antibacterial properties against Gram-positive and Gram-negative bacteria. *International Journal of Biological Macromolecules*, *124*, 1132–1136. DOI: 10.1016/j.ijbiomac.2018.11.228 PMID: 30496864

Zhang, R., Zhong, P., Arandiyan, H., Guan, Y., Liu, J., Wang, N., Jiao, Y., & Fan, X. (2020). Using ultrasound to improve the sequential post-synthesis modification method for making mesoporous Y zeolites. *Frontiers of Chemical Science and Engineering*, *14*(2), 275–287. DOI: 10.1007/s11705-019-1905-1

Zuo, M., Xiao, R., Du, F., Cheng, C., Rodriguez, R. D., Ma, L., Zhu, B., & Qiu, L. (2024). Ultrasound-activated mechanochemical reactions for controllable biomedical applications. *Smart Materials in Medicine*, *5*(4), 461–476. DOI: 10.1016/j.smaim.2024.09.001

Chapter 12
Theoretical Advances in Zeolites

Dipan Kumar Das
https://orcid.org/0000-0003-0224-3295
Centurion University of Technology and Management, India

Padmaja Patnaik
https://orcid.org/0000-0003-4468-7871
Centurion University of Technology and Management, India

Nibedita Nayak
Centurion University of Technology and Management, India

Sudip Kumar Das
https://orcid.org/0009-0003-0814-0884
Dr. C.V. Raman University, India

ABSTRACT

Zeolites, crystalline microporous materials with high thermal and chemical stability, have garnered significant research interest due to their diverse applications in catalysis, adsorption, and ion exchange. Recent theoretical advancements have expanded our understanding of their structural, electronic, and adsorption properties. Studies have demonstrated novel insights into CO_2 adsorption in non-Löwenstein zeolites, revealing enhanced uptake efficiency. Computational approaches, such as QM/MM methods, have provided accurate descriptions of diffusion behaviour within porous frameworks. Investigations into cation-loaded Mg^{2+}-zeolite layers have furthered developments in energy storage applications. Theoretical and experimental characterizations have elucidated the adsorption mechanisms of heavy metals and sulfur in zeolites. Advancements in zeolite-templated carbon structures have led to improved materials for batteries and gas separation. Theoretical studies

DOI: 10.4018/979-8-3373-0320-8.ch012

on Pb-exchanged zeolites and phosphorus-modified frameworks have refined our understanding of their electronic interactions.

1. INTRODUCTION

Zeolites are crystalline microporous materials with good adsorption, ion-exchange, and catalytic properties and are very important in various industrial applications. Advanced theoretical work done recently has gained deeper insight into their structural behavior, diffusion mechanism, and surface interaction, so that their use in gas separation, energy storage, and remediation of pollutants is improved significantly. Computational models and experimental approaches are further tailoring the structure of zeolite frameworks in order to provide better selectivity and stability towards new applications. It describes the contemporary developments made in zeolite theory while relating fundamental principles with practical innovation. Zeolite-templated carbons (ZTCs) are porous framework materials created by carbonization inside a zeolite template. They consist of a three-dimensional network of pores and polycyclic hydrocarbon walls, and are proposed as a potential synthetic pathway to carbon schwarzites (Taylor et al., 2020). Using ab initio data, a QM-MM technique is described to enhance the theoretical explanation of slow diffusion in microporous materials. The two-step ab initio adjustment to activation energy greatly improves force field model performance at a low computational cost. This method was used to study the diffusion of tiny hydrocarbons (C1–C3) in siliceous zeolites via 8MRs (8-membered rings). The technique that captures the trends in experimental data behaves more systematically as a result of the enhanced ab initio description of the hydrocarbon transition states in 8MRs. The suggested approach might be widely used for many different systems for which there aren't enough high-quality empirical potentials (Trachta et al., 2023). Using a highly flexible diquaternary ammonium cation (N,N′-bis-triethylpentanediyldiammonium) as a structure directing agent and tetramethylammonium as a minor co-structure directing agent, ferrierite zeolite was created by a co-structure-directing method. Strong changes in the bulky organic dication's 13C NMR bands during confinement within the ferrierite framework point to a conformational shift caused by host-guest interactions. The observed differences between the NMR bands of the dication in solution and when hosted in ferrierite were explained by a combination of molecular mechanics and DFT calculations of the theoretical 13C NMR chemical shifts. During confinement, the flexible dication must squeeze through the flexible pentyl chain that is located along the 10 MR ferrierite channels in order to host the two bulky triethyl ammonium groups in adjacent intersections with 8 MR channels. According to our research, highly flexible cations—which are typically regarded as less practical organic structure-directing agents due to their low specificity—have the benefit of being able to adapt to the various sizes of zeolite cells by stretching or squeezing their flexible chains, creating an accordion-like effect (Almeida et al., 2020).

The key kinetics-controlling parameter in crystallisation kinetics has not been identified, despite previous studies showing its influence on zeolite phase selectivity due to metastable species in synthesis solutions. The study reveals that the hydrolysis of tetraethyl orthosilicate (TEOS) before hydrothermal treatment affects the final zeolite phase selectivity, with heterogeneous nuclear development influenced by silica content, affecting MFI and TON zeolite production. Theoretical estimates on the creation of composite building units and silica polymerization modelling using the Reaction Ensemble Monte Carlo approach corroborate the experimental findings. The findings are useful for comprehending the zeolite crystallisation nucleation mechanism (Li et al.,2022). A lengthy amine called 1,8-diaminoctane (DAO) was used as an organic structure guiding agent to effectively create ferrierite zeolite crystals. In contrast, ethylenediamine (EN), a shorter amine, was also utilised as OSDA. According to the results, DAO shortens the crystallisation period and makes it possible to produce single FER-type crystals that are smaller and more acidic—two important factors to consider when using zeolites in catalysis. The materials were characterized using various techniques, with theoretical modeling suggesting a preferred placement in 10-MR ring channels and TGA indicating DAO may fit within the zeolite framework (Catizzone et al., 2020). Stepanyuk et al. (2020) investigates the interaction of zinc (II) with silica and alumina-silica surfaces for the creation of medicinal products. The study investigates adsorption patterns in d-elements with electronic structures using adsorption isotherm and thermal desorption of argon, and performs quantum chemical analysis using Hartree-Fock-Roothaan method. The study concludes that the high adsorption activity of zinc(II) is primarily due to the developed porous structure. The most probable interaction sites are determined through quantum chemical simulation (Stepanyuk et al., 2020). Because of its high theoretical specific capacity, lithium metal is perfect for lithium-ion batteries. Uncontrolled dendritic development presents difficulties, though. To allow for dendritic-free deposition on the Li metal anode, a porous Mg2+-Zeolite layer is suggested. With a cyclic lifetime of more than 2100 hours, this process enables high-efficiency deposition and homogeneous Li+ diffusion on the lithium substrate. A synergistic protection mechanism can direct the creation of Li metal batteries that are both safe and highly effective (Su et al.,2024). Particularly stiff membrane materials with customised designs have shown permeabilities and selectivities that exceed the polymer upper bound in gaseous and organic solvent separations for challenging combinations, including species of similar size and shape. Whether such membrane materials can sustain large guest diffusivities is one potential problem if the microporous sections are realistically filled with guest molecules. The diffusivity of water and small organics in a microporous membrane is estimated here by applying a simplified form of transition state theory. Variations in Maxwell-Stefan diffusivities and transport depending on guest species loading are revealed by the transition state theory model parameterized by experimental data from zeolites and carbon molecular sieve materials (White et al., 2022).

The research of metal-air batteries and fuel cells depends on the creation of non-noble metal catalysts with high catalytic activity and robust stability. A three-dimensional, hierarchically ordered porous carbon nanomaterial with zeolite-imidazole framework (ZIF-8) carbonization was produced by using the silica-template technique and carbon nanotube growth. Porous mFeNC-nitrogen-incorporated CNT is enhanced by an iron source, which creates dispersive Fe-Nx sites and Fe-based nanoparticles that increase catalytic activity for oxygen reduction on par with commercial Pt/C. The interlaced CNTs created by urea catalysis significantly narrow the ion and electron transport channel. According to both practical and theoretical assessments, the optimised mFeNC-CNT showed a robust ORR activity when utilised on zinc-air batteries, with a substantial open-circuit voltage (1.556 V) and a half-wave potential of 0.908 V. This work provides a feasible method for the rational design and straightforward synthesis of high-performing non-noble metal-based electrocatalysts for energy storage, conversion, and transport applications (Sun et al.,2022). Using density functional theory (DFT) calculations and in situ infrared (IR) spectroscopy, the oxidation of NO to NO2 and the subsequent reduction by NH3 via a NO+ intermediate over a proton-type chabazite zeolite were investigated (H-CHA). Compared to H-CHA and Cu-CHA zeolite, NO+ species generated by NO + O2 at 27–250 °C are more stable at lower temperatures. The activation barrier energy is negative, according to the Arrhenius plot. When NO is oxidised by O2, NO2 is converted to NO+ and NO3-. According to DFT simulations, N2O4 intermediates can dissociate into NO+ and NO3-molecules with a low activation barrier thanks to Brønsted acid sites in zeolite pores (15 kJ mol-1). Furthermore, NH3's predicted activation barrier to decrease NO+ species was extremely low (6 kJ mol-1). Cu-free H-CHA zeolites have a tremendous potential for enabling lean NOx capture to form NO+ species and the subsequent reduction by NH3 at room temperature, according to the study's theoretical and practical findings. (Yasumura et al., 2021).

2. ZEOLITES IN CATALYSIS AND REACTION MECHANISMS

Zeolites are of interest as catalysts due to their unique porous structure, high surface area, and tunable acidity. They allow for a wide range of reactions, such as hydrocracking, isomerization, and alkylation. They are characterized by well-defined channels and active sites, which result in selective adsorption and transformation of reactants. This in turn leads to efficient catalytic cycles with greater stability and reusability. Katada et al. (2024) uses first principles calculations to identify a pathway for methylating benzene with methane into toluene and dihydrogen, catalyzed by MFI type zeolite-supported Co species. The high activation energy for methane dissociation is suppressed by the significant benzene adsorption on Co^{2+} in the MFI framework, in conjunction with steric hindrance. The activation energy of methane dissociation is sensitive to the Al-Al distance; too long a distance prevents hopping H^+ to reach the base site, while too narrow a distance makes it difficult to break the Co-O bond. The high adsorption of benzene is what propels benzene methylation with methane, and the strong electron-withdrawing properties of Co species that are retained by the zeolite's ion exchange site are thought to be the source of catalytic activity (Katada et al., 2024).

Mahyuddin et al. (2020) uses a computationally efficient meta-GGA method and the D2 method of dispersion correction to calculate the homolytic C-H bond cleavage of methane on the periodic structures of Fe- and Cu-exchanged zeolites. The study emphasises precise Fe-OH bond treatment and O atom radical character, identifying FeIV=O, CuII2(μ-O)]2+, and CuII2CuIII(μ-O)2(μ-O·)]2+ as the optimum electronic structures for active species. Additionally, it highlights how crucial it is to accurately characterise weak contacts and Cu-O-Cu bond strengths in order to forecast an acceptable reaction energy trend (Mahyuddin et al., 2020). The dynamic nature of the copper cations acting as active sites for the selective catalytic reduction of nitrogen oxides with ammonia is investigated using a mix of theoretical and spectroscopic techniques. Molecular dynamics simulations of Cu-CHA catalysts in contact with reactants and intermediates under realistic operating conditions have shown that only ammonia can release Cu^+ and Cu^{2+} cations from their positions coordinated to the zeolite framework, forming mobile $Cu^+(NH_3)_2$ and $Cu^{2+}(NH_3)_4$ complexes that migrate to the centre of the cavity. Lattice asymmetry bands are disrupted and experimental monitoring of dynamic reconfiguration of active sites under operating circumstances is made possible by the mobilisation of copper cations, which changes the vibrational fingerprint in the infrared spectrum (Millan et al., 2020). Molybdenum carbides are promising alternatives to Pt-group metals for hydrogenation of unsaturated hydrocarbons. A systematic DFT study examined the interaction of MoCy nanoparticles with ethylene, revealing their potential to enhance catalytic activity in materials like zeolites and carbon or gold. The results showed that 1:1 stoichiometric Mo_6C_6, $Mo_{12}C_{12}$, and $Mo_{24}C_{24}$ and nonstoichiometric Mo_4C_6, Mo_8C_{12} (MetCar), and $Mo_{14}C_{13}$ (nanocube) were excellent systems for ethylene binding and activation. C_2H_4 binding on NPs with a Mo/C < 1.08 is advantageous due to low energy values for NPs deformation, C=C distortion, and C_2H_4 attachment (Jimenez-Orozco et al.,2021).

Because of its high hydrogen concentration, ammonia borane (AB), a liquid hydrogen storage substance, has drawn more interest for hydrogen consumption. The indefinite catalytic mechanism and the sluggish kinetics of AB hydrolysis, however, continue to be major obstacles to its widespread practical use. Therefore, it is crucial and essential to create effective AB hydrolysis catalysts and identify their catalytic processes. Graphite, metal-organic frameworks (MOFs), metal oxides, carbon nitride (CN), molybdenum carbide (MoC), carbon nanotubes (CNTs), boron nitride (h-BN), zeolites, carbon dots (CDs), and metal carbide and nitride are among the supported catalysts whose preparation and structural properties are summarised in this paper (MXene). With an emphasis on designing efficient AB hydrolysis catalysts and gathering possible paths, the study investigates the connection between electronic structure and catalytic performance. Three methods for AB stimulation from dehydrogenation by-products and the construction of possible hydrogen product-regeneration systems are also summarised. Finally, the remaining challenges and possible research directions for the effective development of AB catalysts are discussed (Guan et al., 2023).

The oxygen evolution reaction (OER) mechanism is crucial for improving water electrolysis efficiency. Because of their soft-lattice structure, methylammonium lead halide perovskites (MAPbX3) have demonstrated encouraging OER performance. Nevertheless, no research has been done on the function of A-site MA or X-site elements in electrochemical reconstruction and OER processes. Some study indicates that the OER process in perovskite@zeolite composites is dominated by lead-halide perovskites, and the kind of X-site halogen that might affect reconstruction kinetics is depicted in Figure 1 (Ren et al., 2025).

Zeolites are crucial in industrial catalysis because to their adjustable acidity, shape selectivity, and stability, enabling the design and manufacture of catalysts for specific processes as DME carbonylation. In order to increase activity in the DME carbonylation process, modified MOR containing stannum (Sn) was designed and acidity changes were predicted using density functional theory (DFT) calculations. By lowering activation barriers and facilitating DME adsorption, this increased acidity increases activity. This work presents a viable method for regulating zeolites' acidic strength (Li et al.,2024). The inability to regulate the size or chirality of catalyst nanoparticles, as well as their creation and development, is a barrier in the synthesis of single-wall carbon nanotubes (SWCNTs). Zeolite MFI nanosheets stabilize cobalt nanoparticles under extended annealing conditions. ETEM shows MFI-Ns control nanoparticle size and shape, resulting in SWCNTs of various chiral angles and a limited diameter distribution. The growth of these unfavourable tubes from unsupported catalysts causes the catalyst nanoparticles carrying them to encapsulate quickly, according to theoretical simulations; their presence in the growth products implies that the MFI-Ns inhibit nanoparticle encapsulation and prologue ZZ and near-ZZ SWCNT growth. The findings thus offer a way forward for regulating the production and development of nanoparticles, attaining size- and shape-selectivity at elevated temperatures, and regulating the synthesis of SWCNTs (Carpena-Núñez et al., 2020).

Figure 1. Preparation process of CsPbX3@AlPO-5

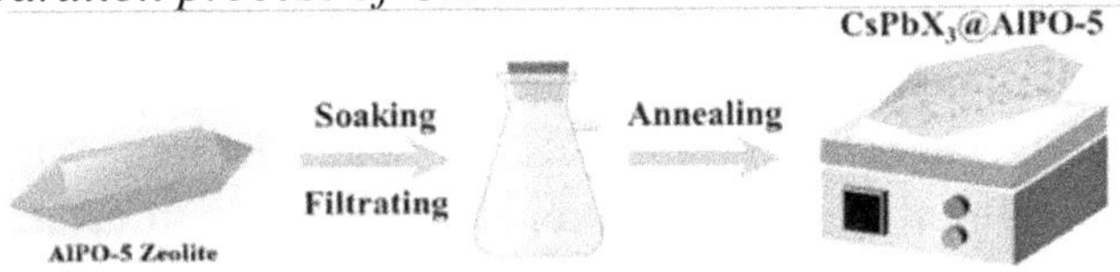

(CC. BY LICENSE 4.0, Ren et al., 2025)

3. ZEOLITES IN GAS ADSORPTION AND SEPARATION

The high importance of zeolites for applications in gas adsorption and separation is because of their porous structure, high molecular sieving properties, and tunable surface chemistry. Specificity in selective gas adsorption for CO_2, CH_4, and N_2, makes zeolites indispensable in industrial applications like carbon capture, natural gas purification, and air separation. The impact of aluminium distribution on carbon dioxide adsorption characteristics in MFI, MOR, and ITW zeolites was examined. The study generalizes a transferable force field for CO2 adsorption in non-Löwenstein zeolites, revealing that the number of Al atoms determines the efficiency of carbon dioxide adsorption in MFI, while the Al distribution impacts the heat of adsorption in MOR. Although there were notable variations between structures with a very high and very low number of non-Löwenstein links, the bonds themselves had no direct effect on the heat of adsorption. Rather, they discovered that a uniform distribution of Al atoms in the C-channel-forming sites is preferable. Lastly, the ITW zeolite's small-pore distribution produced broad error bars and high heat of adsorption values, making the analysis possible only at low aluminium concentrations (Romero-Marimon et al., 2023). Tsiotsias et al. (2023) investigated CO2 adsorption using a zeolitic industrial molecular sieve (IMS) and its effects on CO2 feed concentration and temperature. The study found a negative correlation between temperature and adsorption capacity, indicating physical adsorption, and suitable models like Langmuir isotherm, Bangham kinetic model, and Thomas and

Yoon-Nelson were used. The IMS adsorbent material maintained high capacity after multiple cycles (Tsiotsias et al., 2023).

The adsorption of CO2 on a number of distinct samples of high silica zeolites (silicalite, the aluminum-free version of ZSM-5, and DDR, the aluminum-free form of ZSM-58), made using various synthesis techniques, has been evaluated using equilibrium isotherms. Although there are notable variations in the isotherms for the different samples, a thorough examination shows that the normalised adsorption equilibrium constants are essentially the same for every sample. The observed isotherms for silicalite and DDR samples closely match theoretical ideal structures, confirming the Langmuir model's requirement that adsorption heats are almost independent of loading. The structural defects that result in an increase in the accessible micropore volume can account for some of the DDR samples' significantly higher capacities. Since a larger hydroxyl content suggests a faultier structure, the capacity sequence for silicalite appears to be indirect, following the hydroxyl content (Vidoni et al., 2020). Metal organic frameworks (MOFs) are being explored as gas adsorbents for environmental protection against toxic gases. A fiber-based gas filter was developed using zeolite imidazole framework-8 (ZIF-8) on polypropylene nonwoven. The sol-gel impregnation method demonstrated optimal ZIF-8 loading efficiency, affecting gas molecule accessibility to pores and active sites, highlighting the need for a multifaceted approach in gas sorption processes (Lee et al., 2021).

IR spectroscopy, calorimetry, and theoretical calculations based on DFT corrected to coupled-cluster accuracy (DFT/CC) have all been used to study the interaction of CO with alkali metal-exchanged mordenite. The low-frequency band at 2181 cm−1 and isosteric heat of 44 kJ/mol suggest partial Li+ transfer into MOR structure, while CO desorption changes Si/Al ratio, affecting band stabilities. It can be inferred from the kinetic data that diffusion constraints, which are strongly impacted by the Si/Al ratio and cation size, play a critical role in the MOR material. The "gate" effect on N2 adsorption is observed in the 2138 cm−1 band of Na/K-MOR materials with higher Al content, while the weak band at 2150 cm−1 reveals dual cationic sites (Bulánek et al.,2020). Passive NOxadsorbers (PNAs) are proposed for trapping NOx in automotive exhaust during cold start periods when the three-way convertor is not hot enough for NOx reduction. Pd-exchanged chabazite (Pd/H-CHA) is a good candidate due to its ability to store and retain NO at high temperatures. NO adsorbs on Pd2+ and Pd+cations, with desorption occurring at lower temperatures. Pd+cations form through NO adsorption and CO adsorption, thermodynamically feasible. (Kim et al., 2022). Within the subcategories of liquid theory, chromatography, Nano fluidics, energy storage, recovery, and dissipation, determining the molecular processes behind the wetting and drying of hydrophobic porous materials is a broad scientific and technological challenge. In this paper, we provide a novel approach to address this issue by investigating the impact of the architecture of zeolites, nanotubes, and pure silica nanoparticles. It was demonstrating through molecular dynamics simulations how secondary porosity influences a material's hydrophobicity and encourages water to enter micropores. Here, it is shown that altering the ratio of open to closed lateral holes of nanometer size may influence the hydrophobicity of nano-objects. When hydrophobicity has to be controlled without substantially altering the materials' chemistry or structure, this effect can be used to create novel materials for real-world uses. The study analyzed intrusion pressures in pure silica zeolites using simulations and theoretical considerations. It found a relationship between pore surface area ratio and incursion pressure, useful for energy storage, recovery, and dissipation (Bushuev et al., 2022). The Langmuir's nonlinear isotherm provides theoretical foundations for mathematical modeling of adsorption and desorption in nanoporous catalytic zeolite media, determining equilibrium mechanisms and utilizing efficient linearization and Hevisides operational approaches (Petryk et al., 2020).

Karamanis et al. (2023) used DFT calculations to screen faujasite zeolites for capturing NO and NO2 emissions from diesel engine exhaust gases. They investigated faujasite's adsorption selectivity, explored alkali and alkali-earth metals, and bond activation. Cu(i) and Fe(ii) based Y zeolites were identified as attractive candidates (Karamanis et al., 2023). Methane adsorption in bulk and thin layers of MFI zeolite, molecular models and experiments are employed (silicalite-1). Zeolite layers with varying crystallographic orientations and degrees of surface flexibility are taken into consideration after comparing the theoretical adsorption data for bulk MFI at different temperatures that were generated using Grand Canonical Monte Carlo simulations with experimental data. Archetypal systems reveal that methane adsorbs at low pressures in zeolite core, while exterior surface adsorption occurs at saturated internal porosity pressures. A comprehensive logical framework for fluid adsorption in finely split materials may be provided by extending this groundbreaking theory to explain both surface and volume adsorption, even if it has already been employed for gases in nanoporous solids. The suitability of this approach for gas adsorption data in supercritical environments is also covered (Kellouai et al., 2022). Findings from a combination of experimental and theoretical studies on the possible application of natural clinoptilolite zeolite (CLI) as an odor-adsorbing substance. The adsorption capacity of various granulations of CLI to determine if metals as active centers enhance odorous substance absorption through metallic forms. Its modifications were made by ion exchange using Sn and Fe metals. Hydrochloric acid-driven hierarchization of hydrogen zeolite for adsorption was utilized in an acidic environment, with synthetic zeolite MFI evaluated for CLI sorption impact, including metal additions (including acetaldehyde, dimethylamine, pentanoic acid and octanoic acid). The measurements show that clinoptilolite, a naturally occurring substance, is the most effective odorant trap. It has distinct abilities for different substances depending on the faction. Clinoptilolite, with grain sizes up to 2 mm, is an efficient material for acetaldehyde, and a 3-4 mm fraction with carboxylic acid hierarchization. According to data gathered from theoretical simulations, metallic cores are more stable. According to data from theoretical calculations, metallic centers—which are linked to the skeletal aluminium in clinoptilolite—are more stable above oxygen which is shown in Figure 2(Czekaj and Sobuś, 2024). The superior sulfur adsorption and selectivity of Cu and Ce exchanged Y zeolite (CuCeY) for hydrocarbon fuel removal was studied. X-ray diffraction confirmed the absence of large crystalline phases, and X-ray adsorption fine structure analysis revealed Cu and Ce cations within the Y zeolite framework. The results reveal CuCeY's structure and its superior sulfur adsorption performance, providing insight into its potential for desulfurization (Sokol et al.,2022).

Figure 2. In a clinoptilolite chamber (middle column), ammonia (left column) and dimethylamine (right column) are adsorbed: Adsorption on Brønsted acid centres (top) and metallic centres (bottom) (Na, Sn or Fe).

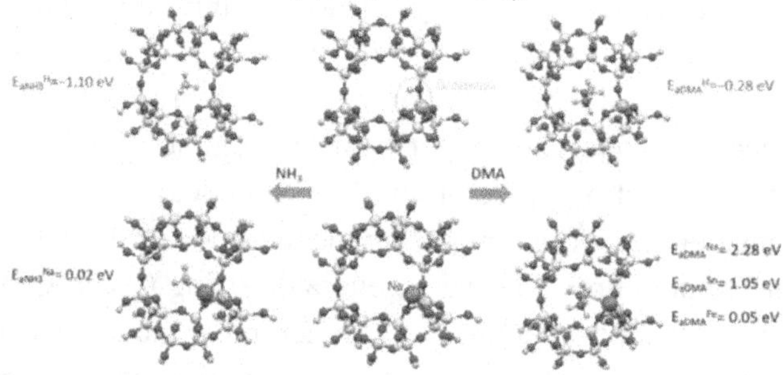

(CC BY LICENSE 4.0, Czekaj and Sobuś, 2024)

4. FUNCTIONAL PROPERTIES AND INDUSTRIAL APPLICATIONS OF ZEOLITES

Zeolites have unique functional properties, such as high surface areas, ion exchange capacity, and catalytic properties, and versatile applications for extensive use in varied industrial processes. They are being used in processes such as the refining of petrochemical products, treatment of wastewater, gas separation, materials for energy storage, and also as sensors. The thermal stability and high temperature structural evolution of highly crystallized ZSM-5 zeolite using organic species as structure directing agents (SDA), including tetra-n-propylammonium (TPA) and Na cations was studied. The study uses in situ synchrotron powder diffraction to study the structure of TPA-ZSM-5 with different Si/Al ratios (15, 25, 50, and 100) and their structural modifications induced by template calcination. The results show that tetrapropylammonium ions are located at the intersection of sinusoidal and straight channels, and the refined TPA+ content increases with the increasing Si/Al ratio. High temperature refinements confirm the high thermal stability of all four samples, and the study also provides insights into the effect of aluminum content on the interaction between SDA and zeolite structure (Mancinelli et al., 2023).

The poor mechanical performance and freeze-thaw resistance of recycled concrete are primarily due to the significant water absorption and porosity of recycled coarse particles. In this investigation, reclaimed concrete was mixed with varying amounts of zeolite powder. Zeolite powder-modified recycled concrete's workability and FT durability were assessed using a battery of macroscopic tests (ZPRC). X-ray diffraction and scanning electron microscopy (SEM) were used to identify the micro-mechanisms of FT resistance in ZPRC (XRD). According to the study, adding 10% zeolite powder to ZPRC enhances its mechanical qualities and FT condition tolerance while increasing its zeolite powder content reduces its droop and water absorption. The pozzolanic reaction of zeolite powder, which produces a denser microstructure, the particle-filling effect, and the water storage function are responsible for the ZPRC's increased strength and FT resistance. Zeolite powder's particle-filling action encourages recycled coarse aggregates to have fewer surface holes (RCAs). Zeolite powder's water storage reduces free water in recycled concrete, supplying water for cement particle hydration and compacting the microstructure, potentially promoting cold climate engineering use of recycled concrete (Yu et al., 2024).

Montesano et al. (2022) evaluates the petrographic, mineralogical, chemical, and technical characteristics of several zeolitic tuff samples from the Basin and Range Province in the western United States. These samples mostly include erionite, mordenite, clinoptilolite/heulandite, and phillipsite. In order to programme a potential usage as an additive for blended cements, the purpose of this characterization is to assess the pozzolanic activity of these samples in accordance with European regulatory UNI-EN 196/5 (Fratini test). Petrographic and mineralogical results reveal two phillipsite-bearing tuffs have higher theoretical Cation Exchange Capacity, with higher pozzolanic behavior. Blended cements could reduce CO2 emissions by 70-90% using these samples. (Montesano et al., 2022). Due to its exceptional light emission characteristics, including as their high quantum efficiency, small linewidths, and electron transport properties, perovskite quantum dots, or PQDs, have been the subject of much research. Poor stability, however, restricts their use in optical devices, particularly when operating in ambient temperatures with moisture present, which quickly reduces their performance. Zhang et al. (2022) created a composite EMT-CsPbBr3 by loading PQDs into nanosized EMT zeolite crystals made from template-free precursor systems. A tiny quantity of water molecules in the zeolites' pores can encourage the crystallization of perovskite nanocrystals. The energy and bond length of the perovskite $CsPbBr_3$ enclosed in the EMT zeolite cages were calculated with water molecules present, accounting for the effect of humidity. Crucially, the pore geometry of the EMT molecular sieve provides a helpful research paradigm. The remarkable stability and reliability of the EMT-$CsPbBr_3$ humidity sensor are shown. The luminous intensity efficiency of the EMT-$CsPbBr_3$ composite stayed about 90 percent after six months of constant usage. The luminescence of perovskite stabilised in the hydrophilic EMT zeolite is enhanced by a tiny quantity of water, according to both theoretical and experimental data (Zhang et al., 2022).

The packaging industry uses LDPE due to its affordability, durability, and flexibility, but its mechanical recycling is limited due to contamination and economic constraints. A variety of processing settings are used to assess two kinds of zeolite catalysts at various loading amounts. The extruded samples' structural, thermal, and rheological characteristics are examined and contrasted with those of virgin LDPE and LDPE extruded without a catalyst. NMR spectroscopy is employed to examine changes in the polymer's structure. The study reveals that microporous Y zeolite-extruded LDPE shows increased branching and reduced breakdown temperature, while mesoporous MCM-41 induces increased branching but maintains the same degradation temperature. A possible low-cost, straightforward method for reusing flexible packaging made of LDPE into coatings and adhesives is the hybrid reactive extrusion technique that has been shown (Martey et al., 2021).

In worldwide CCUS initiatives, CO_2 adsorption technology has grown in prominence as a key component of achieving net-zero emissions by 2050. To increase efficiency, new materials and post-modification methods are being created. A methodical understanding of the fundamental components of current research trends in terms of manufacturing specialised solid CO_2 adsorbent, with an emphasis on the next concerns was studied. When evaluating the candidates, which include amine-based adsorbents, metal-organic frameworks (MOFs), silica, alumina, metal oxides, polymers, zeolites, carbon-based materials, and other composite porous materials, the practical constraints imposed by industrial scale and economics are taken into account. Sustainable sorbents made from industrial waste and biomass are also being researched because of the circular economy's critical role and the increased need for affordable raw materials. Last but not least, a techno-economic analysis (TEA) is provided to highlight the most important considerations for industrial adsorbent use. Consequently, along with recommendations for more research in the creation of CO_2 capture materials. Additionally, it seeks to pinpoint and clearly state the most important directions for future study that may be beneficial to follow in the upcoming years (Dziejarski et al., 2023).

The heat capacity of partly dehydrated zeolite A (LTA), fully exchanged with Cu2+, Fe2+, and Mn2+ ions, was studied by Dickson et al. (2021) between 1.8 and 300 K. they ascribe the samples' broad excess heat capacity contribution, which is focused around 4 K, to local electric fields that divide the cations' magnetic moments. A combination of many Schottky anomalies is used to model the extra heat capacity. Based on these models, they deduce that the Cu2+ zeolite's cations inhabit a minimum of four different coordination habitats, with some of these environments being quite asymmetric in all three zeolites. They also include values of the conventional thermodynamic functions CP,m, Δ0KTSm°, Δ0KTHm°, and Φm° at smooth temperatures, as well as theoretical fits of the heat capacity data. At 298.15 K, the conventional molar entropies for Cu-zeolite A (Cu0.22Al0.49Si0.51O2·1.04 H2O), Fe-zeolite A (Na0.01Fe0.23Al0.50Si0.51O2·0.77 H2O), and Mn-zeolite A (Mn0.26Al0.49Si0.50O2·0.53 H2O) are 71.8 J·K−1·mol−1, 71.1 J·K−1·mol−1, and 66.0 J·K−1·mol−1 for each (Dickson et al., 2021).

Sun et al. (2023) investigates the moisture susceptibility of Zeolite-Foamed Asphalt (ZFA) from a mesoscopic perspective using the Surface Free Energy (SFE) concept. Sessile drop tests on aggregates and bitumen binders assessed zeolite incorporation, adhesion, debonding, spreading coefficient, and SFE-related energy ratio, confirming moisture susceptibility in ZFA systems. The findings show that, depending on the predominate failure mode and the aggregate's SFE, the zeolite foaming process's rise of the polar component and decrease of the dispersion component of the binder's SFE have a favourable or negative impact on the moisture susceptibility of the ZFA. When an interfacial adhesion failure occurs, it is more advised to prepare ZFA using an aggregate with a greater polar component of the SFE, such the granite used in this study. The study suggests that a ZFA composed of granite, 9% zeolite, and a rubber-modified binder can be optimized for improved moisture susceptibility (Sun et al., 2023). Researchers are exploring ballistic resistance architectures in aerospace and defense sectors, focusing on intricate, lightweight designs for increased mechanical energy absorption, despite limited manufacturing processes. However, automation, great design flexibility, and sustainability are made possible by additive manufacturing (AM) and 3D printing. The purpose of the zeolite-inspired 3D printed structures is to investigate deformation under low-velocity drop impact. The specific energy absorption of several zeolite-inspired structural types is compared in this paper. Additionally, it uses computer tomography and theoretical molecular dynamics to examine the impact depth at several scales in relation to velocity (experimental) (Ambekar et al.,2024). High-energy X-ray diffraction characterizes insulating and hydrophobic mordenite zeolite structures, while soft X-ray emission and absorption spectroscopies estimate partial densities of states in valence and conduction bands. While pre-shoulders are seen in mordenite, the SXAS spectra for the conduction band O 2p orbital shows distinctive features similar to those of crystalline SiO2.

SXES spectra were produced by selecting characteristic energies from the SXAS spectrum for the incident photon energies. These spectra show three minor peaks or shoulders and a big peak that can be attributed to bonding (σ) and lone pair orbital ones, respectively. The precise electronic states and atomic structures were found using a density functional theory, and they fit well with the related experiments. It is determined that the Al and Na atoms contribute just slightly to the O 2p partial DOS, which is mostly composed of OSi covalent connections. The study reveals that combining theoretical and experimental research can identify the active sites in functional zeolites, revealing the basic characteristics of complex zeolites (Hosokawa et al., 2024).

Fluid catalytic cracking (FCC) is a major source of fuels, specialized materials, polymers, and medications, with ZSM-5 as its active component stabilized by phosphorus compounds. There is still much disagreement on the mechanism of activation and the relationship between phosphate and zeolite aluminium species, even though this process is one of the most widely used catalytic processes in industry. In this study, they locate the phosphate groups inside the pore system of P-activated ZSM-5 and unambiguously confirm the activation mechanism of phosphorus-based promotion in FCC catalysis using synchrotron-based powder XRD, neutron diffraction, and subsequent pair distribution function analysis. Traditional XRD investigations were unable to detect the local disorder in the zeolite T–O coordination. To support these experimental results, they also offer comprehensive periodic quantum-mechanical modelling (QMM) of the crucial but usually overlooked combination of dealumination via phosphatation and hydrolysis (steaming) of the zeolite framework. We therefore show that partially dislodged aluminium species can react with phosphate as long as they remain stable and linked to their original framework site. The two models that result from this work rank among the three most relevant remaining models by taking into account the highly relevant inclusion of steaming prior to phosphatation. This finding is based on an energetic comparison of our acquired results with all accessible literature postulations evaluated using the same periodic QMM. The activation and promotion mechanism of propylene-selective FCC, one of the most widely used chemical processes in the world, is essentially explained by this combined experimental and theoretical effort (Louwen et al., 2020). The zeolite-templated carbon (ZTC) material was made via vapour phase impregnation. Sodium-bearing zeolites were used to create a new ZTC with a high specific surface area and pore volume. With a high initial discharge capacity of 1714 mAh·g−1 and a theoretical capacity of 552 mAh·g−1, the ZTC was employed as a negative electrode in a sodium battery (Tan et al., 2022).

5. ADVANCES AND LIMITATIONS IN THEORETICAL MODELING OF ZEOLITES

Improved theoretical modeling of zeolites has provided better understanding toward their properties of adsorption and diffusion as well as catalytic properties, hence providing impetus to the design of more environment-friendly as well as more efficient energy-using materials. Advances in the precision and predictability with which one could obtain using quantum mechanics and molecular dynamics together with advanced machine learning techniques have provided atomic-scale predictions of the behavior of these zeolite systems with considerable accuracy and precision. Of the 253 zeolites that are accessible, only a tiny fraction (21) are extra-large pore zeolites. The development of novel low-glycemic sugars is highly appealing as a healthy food ingredient. The 6-kestose is an example of this. In the current instance, it coexists with sucrose in an aqueous solution, necessitating the mixture's separation. Since 6-kestose has a larger molecular size than sucrose, it is required to encourage the sorption of the latter in order to purify the former. They have thus focused on using certain zeolites that have enough pore widths to allow for the separation of this mixture. After the micropores of the 253 IZA zeolites were computer-screened, eleven zeolites were selected. Among them, three extra-large pore zeolites (AET, DON, and ETR) have been proposed; a molecular dynamics study considering the external surface was used to extensively investigate them. The results show that DON provides the most promising theoretical results for a sucrose/6-kestose selective separation (Losada et al., 2021).

Single-atom catalysts (SACs) are a new class of catalysts with unique catalytic properties due to their easy tuneability. Zeolites are commonly used as support materials for SACs due to their steric restrictions, which suppress metal particle aggregation and provide exceptional stability. However, only a few transition metal atoms have been tested, and determining their precise locations remains a challenge. This study provides a comprehensive dispersion corrected density functional theory study of the energetic, structural, magnetic, and electronic properties of transition metal atoms encapsulated on MFI-type Silicalite-1 zeolite, TM@S-1. Most TM atoms adsorb in the center of S-1 channels, with the exception of Group 3 –Sc, Y, La–, Group 10 –Ni, Pd, Pt–, Ru, and Rh atoms (Prats et al., 2020). In order to investigate the structural alterations of phosphorus species in zeolites, the 31P MAS NMR spectra of phosphorus-modified chabazite (P-CHA) zeolites were recorded throughout the hydrothermal treatment. The hydrothermal structural changes in P-CHA zeolites are correlated with the characteristic variations of the spectra, which were found in the region of −27 to −42 ppm. To reveal the potential intra- and extra-framework phosphorus species as well as the structural alterations brought about by the hydrothermal treatment, theoretical calculations on the 31P and 27Al

NMR chemical shifts have been methodically carried out. For condensed phosphates and alumina-phosphates, the 31P resonances were shown to move toward higher fields. The initial hydrothermal treatment peak at -42 ppm is caused by SAPO species in the zeolite framework, while the later peak at -29 ppm is due to extra-framework condensed phosphate or alumina-phosphate species.The phosphates in the framework are identified by the prominent signal at -33 ppm in all 31P NMR spectra (Zhao et al., 2020).

The structural properties of Pb-exchanged zeolites, which are crucial for environmental remediation and industrial processes was studied. The zeolite, Pb-STI, has a chemical composition of Pb13.4(OH)10Al17.4Si54.6O144 •38H2O, with Pb2+, H2O, and OH− occupying the extra framework. The study reveals a strong positional-disorder in the extra framework, with Pb2+ ions coordinated by 2 OH− and 1H2O at distances <2.5 Å(Roos et al., 2021).

The whole visible spectrum may be illuminated by silver nanoclusters (Ag NCs), which are nanoscale aggregates with distinct energy levels that resemble molecules and are reliant on electronic arrangement. Zeolites are preferred for distributing and stabilizing Ag NCs due to their thermal and chemical properties, nanometer-sized cages, and efficient ion exchange capacity (Pan et al., 2023). The quantity and positioning of zeolites' acid sites—strong Brønsted and weak silanols—are essential to their uses. The mystery of the intricate H-bonded silanol networks in pure silica MFI-type zeolites was solved in this work by combining DFT calculations with 1H solid-state NMR and IR spectroscopy. In addition to being crucial in controlling the zeolite's acidity, stability, lifespan, and hydrophobicity, the silanol sites also determine phase selectivity during synthesis. In pure silica zeolites, the spectral signals of silanol flaws are seen. They discovered four distinct silanol types in pure silica zeolites based on theoretical and practical data, which differed according to the size of the crystallites: Isolated (free) silanols that do not participate in hydrogen bonds as either proton-acceptors or proton-donors; proton-acceptor silanols that only participate in hydrogen bonds as proton-acceptors; proton-donor silanols that participate in weak hydrogen bonds; and medium and strong hydrogen-bonding silanols that participate as proton-donors or concurrently as both proton-acceptors and proton-donors. The strength of hydrogen bonds formed by a silanol group is primarily determined by the proton's proximity to the oxygen center of the accepting silanol, affecting its spectrum characteristics (Dib et al., 2021). Dib et al. (2023) demonstrate how the 29Si NMR peaks of (≡Si-O-Si≡) (Q4(0Al)), Brønsted acid sites (≡Si-OH-Al≡) (Q4(1Al)), and silanols (≡Si-OH) (Q3) can be characterised in nanosized ZSM-5 zeolites using a combination of I symmetry-based recoupling in 29Si NMR, (ii) spin dynamics-based simulations, and (iii) theoretical calculations based on DFT. One can clearly assign the 29Si NMR peaks because of the significant variations in chemical shift anisotropy between silicon nuclei that are close to aluminium (Q4(1Al)) and those that are not, such as Q4(0Al) and Q3. These differences are computed theoretically and observed empirically. Such discrepancies cannot be resolved by the isotropic chemical shift alone (Dib et al., 2023).

Quantum chemistry methods (QMs) have been used to predict the structure of zeolite materials, even with low symmetry, by determining the location and coordination of NaI and CuI cations in Si-rich and Al-rich faujasites. The theoretical approach predicted the absence of symmetry at the atomic level and that sites I were not occupied by NaI nor CuI cations, even for Al-rich faujasites of X-type. The infrared CO probe was used to analyze the interaction of both NaI and CuI with the zeolite framework. The upshifting of the calculated vCO stretching frequency and topological analyses indicated that the interactions of NaI and CuI with the oxygen atoms of the hosted zeolite were mainly ionic with a weak covalent character in the case of CuI (Petitjean et al., 2020).

Some study presents a method for obtaining crystal structures at high pressures using geometric modelling, implemented in GASP, to overcome the difficulties of determining the structures of zeolite materials due to their intricate compositions and frameworks. Geometric Analysis of Structural Polyhedra software offers optimized structural models that serve as logical beginning points for comprehensive crystal structure investigation by simulating theoretical compression processes of zeolite frameworks at the atomic level. Price et al discuss the use of geometric modelling to predict high-pressure behavior in structures, highlighting the potential of GASP software for identifying crystal structures in high-pressure zeolite frameworks. (Price et al.,2021). Zeolite doped photopolymers can be used to create optical sensors, with holographic recording demonstrating dopant redistribution. An expanded mathematical model to describe nanozeolite redistribution dynamics and examines factors like starting load, dopant transport parameters, and recording conditions was studicd (Mackey et al., 2024).

Table 1. Methods used for zeolitic systems and key findings

Zeolitic System	Methods Used	Key Findings
Zeolite-Templated Carbons (ZTCs)	Ab initio modelling, QM-MM techniques	Improved force field models for hydrocarbon diffusion in microporous materials.
Ferrierite Zeolite (with flexible diquaternary ammonium cations)	Molecular Mechanics, DFT, 13C NMR	Flexible OSDAs adapt to zeolite pores, enabling better synthesis of zeolite frameworks.
Zeolite crystallization via TEOS hydrolysis	Reaction Ensemble Monte Carlo simulation, Experimental crystallization	Hydrolysis conditions control zeolite phase selectivity during synthesis.

continued on following page

Table 1. Continued

Zeolitic System	Methods Used	Key Findings
Ferrierite Zeolite (using DAO)	Theoretical modeling, TGA analysis, Crystallization experiments	DAO promotes faster crystallization and smaller, highly acidic crystals.
Zinc adsorption on silica/alumina-silica	Hartree-Fock-Roothaan quantum chemical analysis, Adsorption isotherms	High adsorption due to porous structure; key interaction sites identified.
Mg^{2+}-Zeolite layer for lithium anodes	Experimental (cycling tests), Theoretical Li^+ diffusion modeling	Enables dendrite-free lithium deposition, enhancing battery lifespan.
Microporous membranes for gas separation	Transition state theory, Experimental permeability data	High guest diffusivity predicted; guest loading affects transport.
mFeNC-CNT from ZIF-8 carbonization	Experimental synthesis, Theoretical ORR modeling	High ORR catalytic performance comparable to Pt/C in zinc-air batteries.
H-CHA Zeolite for NOx reduction	DFT calculations, in situ IR spectroscopy	NO^+ species formation and easy NH_3 reduction at room temperature.
Co-MFI Zeolite for methane activation	First principles DFT	Benzene adsorption facilitates methane activation on Co sites.
Cu-CHA Zeolite for NOx SCR	Molecular dynamics simulations, Infrared spectroscopy	Cu^+/Cu^{2+} cations mobilize under reaction conditions, enhancing SCR activity.
MoC nanoparticles for ethylene activation	Systematic DFT studies	Mo-rich nanoparticles bind and activate ethylene efficiently.
Zeolites for ammonia borane hydrolysis	Review of catalytic supports, DFT studies	Pathways for efficient hydrogen release and catalyst design outlined.
Zeolite-based composites for OER	DFT and electrochemical analysis	$MAPbX_3$ perovskites in zeolite matrices enhance OER performance.
Sn-modified MOR Zeolite	DFT calculations, Experimental catalysis	Enhanced acidity boosts dimethyl ether carbonylation activity.
Zeolite MFI nanosheets for SWCNTs	Theoretical simulations, ETEM	Nanosheets stabilize catalyst size and control SWCNT chirality.
Zeolite for CO_2 adsorption	Force field simulations, Experimental isotherms	Al distribution influences CO_2 adsorption strength and selectivity.
Zeolite fiber-based gas filters (ZIF-8)	Experimental adsorption, Theoretical modeling	Sol-gel method optimizes ZIF-8 gas filter efficiency.
Zeolite Clinoptilolite for odor adsorption	Experimental adsorption studies, Theoretical stability simulations	Hierarchized clinoptilolite effectively adsorbs different odorous compounds.
Pd/H-CHA Zeolite for NOx capture	Experimental adsorption-desorption studies, Theoretical thermodynamics	NOx trapping and Pd^+ formation processes explained.

continued on following page

Table 1. Continued

Zeolitic System	Methods Used	Key Findings
ZSM-5 phosphatation for FCC	Synchrotron XRD, Neutron diffraction, Quantum mechanical modeling	Localized phosphate activation mechanism clarified.
Natural clinoptilolite zeolite for CO_2 capture	Experimental adsorption and SEM	Effective CO_2 capture and stabilization by natural zeolites.
Ag nanoclusters in Zeolites	Luminescence studies, DFT modeling	Stable visible light emission enabled by zeolite encapsulation.
MFI Zeolite for methane adsorption	Grand Canonical Monte Carlo simulations, Experimental adsorption	External surface adsorption critical at high methane pressures.

6. LIMITATIONS OF THEORETICAL CALCULATIONS IN ZEOLITE RESEARCH

Although theoretical modeling has greatly boosted our knowledge on zeolitic systems, various limitations still hamper its entire potential. The limitations are:

i. Computational Time and Cost

Theoretical approaches, especially ab initio quantum mechanical calculations (e.g., Density Functional Theory, DFT), can be very computationally demanding. Simulating large and complex zeolite structures, particularly when modeling dynamic processes such as diffusion or catalytic reactions, typically demands thousands to millions of processor hours. The long computation time constrains the size of systems that can be realistically modeled and limits the number of potential reaction pathways or configurations that can be probed.

ii. Licensed Software Requirement

Most of the advanced computational chemistry packages (e.g., VASP, Gaussian, Materials Studio, CP2K) are proprietary software packages with high licensing costs. Such license costs can be exorbitant for small research centers or individual researchers. Additionally, the limitations in the availability of licenses (e.g., maximum numbers of users at a given time) can also decelerate progress when several simulations are required simultaneously.

iii. Availability of High-Performance Computing Facilities (Supercomputers)

To conduct simulations on big or realistic models of zeolites with an adequate degree of accuracy, scientists usually have access to supercomputer centers or HPC clusters. Such resources may not be accessible at all, especially in low-budget research environments. Even when available, quota restrictions or queuing mechanisms might extend project durations, and proper parallelization of intricate models still poses a technical challenge.

iv. Accuracy vs. Scalability Trade-off

Highly reliable quantum mechanical approaches (such as coupled-cluster theories) tend not to be tractable for large zeolite systems because they scale badly with system size. Researchers thus have to make do by applying quicker, less reliable approaches (e.g., semi-empirical methods, lower-level DFT functionals) in describing large frameworks. This makes predictions uncertain, particularly for systems that exhibit delicate electronic interactions such as transition metal centers.

v. Limited Transferability and Model Dependence

Theoretical models tend to be based on particular assumptions (e.g., stiff framework, idealized structures, periodic boundary conditions) that do not necessarily account for all of the real-world experimental complexities like defects, dynamic flexibility, or solvent effects. Thus, simulation results may not necessarily be transferable between different zeolite types, operating conditions, or scales (e.g., from atomic to macroscopic behavior). Rigorous validation against experimental data is still essential but not necessarily easy.

7. CONCLUSION

The theoretical development of zeolites has considerably expanded knowledge about their structural, catalytic, and adsorption properties. New synthesis methods, such as the use of flexible organic structure-directing agents, have been developed in order to control zeolite frameworks precisely, improving their functional versatility. It has furnished extensive information into the atomic and electronic mechanisms governing zeolite behavior through recent advances in computational modeling, including DFT studies, molecular dynamics simulations, and spectroscopic analysis. These approaches have given a good guide to experimentally designing ways for optimizing catalytic performance by predicting zeolite stability, cation interactions, and reaction pathways. The potential of employing zeolites in such novel applications such as impact-resistant 3D-printed structures, gas adsorption technologies, and catalytic systems for sustainable energy conversion encompasses their broad applicability. Theoretical studies on ion-exchanged zeolites, zeolite-encaged nanoclusters, and modified frameworks have revealed tunable properties that can be applied toward CO2 capture, selective adsorption, and enhanced catalytic efficiency. The ability to compute the computationally driven modifications that transform the acidity, hydrophobicity, and pore topology now allows the design of next-generation zeolite materials.

However, much remains to be done in optimizing zeolite synthesis for industrial applications as well as in the long-term stability under operational conditions, to perhaps unlock all the potential of hierarchical and nanoscale zeolites. Future research should focus on refining theoretical models to bridge the gap between simulations and real-world performance, developing more sustainable synthesis routes, and exploring multifunctional hybrid zeolite systems. Continuing to integrate theoretical insights with experimental validation, zeolites will remain at the forefront of materials science, driving innovations across catalysis, separation processes, and energy storage technologies.

REFERENCES

Almeida, R. K. S., Peréz-Pariente, J., & Gómez-Hortigüela, L. (2020). Accordion-like effect of flexible organic structure-directing agents in the synthesis of ferrierite zeolite. *Microporous and Mesoporous Materials, 300*, 110149. Advance online publication. DOI: 10.1016/j.micromeso.2020.110149

Ambekar, R. S., Oliveira, E. F., Pugazhenthi, P., Singh, S., Mahapatra, D. R., Galvao, D. S., & Tiwary, C. S. (2024). Drop-Weight Impact Resistance of 3D-Printed Complex Zeolite-Inspired Structures. *Advanced Engineering Materials, 26*(22), 2400035. Advance online publication. DOI: 10.1002/adem.202400035

Bulánek, R., Koudelková, E., Trachta, M., Bludský, O., & Rubeš, M. (2020). Re-examining the interpretation of CO adsorbed on Lewis acid sites of alkali metal-exchanged MOR zeolite. *Microporous and Mesoporous Materials, 296*, 110007. Advance online publication. DOI: 10.1016/j.micromeso.2020.110007

Bushuev, Y. G., Grosu, Y., Chorążewski, M. A., & Meloni, S. (2022). Effect of the Topology on Wetting and Drying of Hydrophobic Porous Materials. *ACS Applied Materials & Interfaces, 14*(26), 30067–30079. DOI: 10.1021/acsami.2c06039 PMID: 35730678

Carpena-Núñez, J., Rao, R., Kim, D., Bets, K. V., Zakharov, D. N., Boscoboinik, J. A., Stach, E. A., Yakobson, B. I., Tsapatsis, M., Stacchiola, D., Stacchiola, D., & Maruyama, B. (2020). Zeolite Nanosheets Stabilize Catalyst Particles to Promote the Growth of Thermodynamically Unfavorable, Small-Diameter Carbon Nanotubes. *Small, 16*(38), 2002120. Advance online publication. DOI: 10.1002/smll.202002120 PMID: 32812375

Catizzone, E., Migliori, M., Mineva, T., van Daele, S., Valtchev, V., & Giordano, G. (2020). New synthesis routes and catalytic applications of ferrierite crystals. Part 1: 1,8-Diaminooctane as a new OSDA. *Microporous and Mesoporous Materials, 296*, 109987. Advance online publication. DOI: 10.1016/j.micromeso.2019.109987

Czekaj, I., & Sobuś, N. (2024). Odors Adsorption in Zeolites Including Natural Clinoptilolite: Theoretical and Experimental Studies. *Materials (Basel), 17*(13), 3088. Advance online publication. DOI: 10.3390/ma17133088 PMID: 38998171

Dib, E., Costa, I. M., Vayssilov, G. N., Aleksandrov, H. A., & Mintova, S. (2021). Complex H-bonded silanol network in zeolites revealed by IR and NMR spectroscopy combined with DFT calculations. *Journal of Materials Chemistry. A, Materials for Energy and Sustainability, 9*(48), 27347–27352. DOI: 10.1039/D1TA06908J

Dib, E., Mintova, S., Vayssilov, G. N., Aleksandrov, H. A., & Carravetta, M. (2023). Chemical Shift Anisotropy: A Promising Parameter To Distinguish the 29Si NMR Peaks in Zeolites. *The Journal of Physical Chemistry. C, Nanomaterials and Interfaces*, *127*(22), 10792–10796. DOI: 10.1021/acs.jpcc.3c01533

Dickson, M. S., Rosen, P. F., Neilsen, G., Navrotsky, A., & Woodfield, B. F. (2021). Heat capacity and thermodynamic functions of transition metal ion (Cu2+, Fe2+, Mn2+) exchanged, partially dehydrated zeolite A (LTA). *The Journal of Chemical Thermodynamics*, *161*, 106556. Advance online publication. DOI: 10.1016/j.jct.2021.106556

Dziejarski, B., Serafin, J., Andersson, K., & Krzyżyńska, R. (2023). CO<inf>2</inf> capture materials: A review of current trends and future challenges. *Materials Today Sustainability*, *24*, 100483. Advance online publication. DOI: 10.1016/j.mtsust.2023.100483

Guan, S., Liu, Y., Zhang, H., Shen, R., Wen, H., Kang, N., Zhou, J., Liu, B., Fan, Y., Jiang, J., Jiang, J., & Li, B. (2023). Recent Advances and Perspectives on Supported Catalysts for Heterogeneous Hydrogen Production from Ammonia Borane. *Advancement of Science*, *10*(21), 2300726. Advance online publication. DOI: 10.1002/advs.202300726 PMID: 37118857

Hosokawa, S., Sato, H., Tezuka, Y., Adachi, J.-I., Kimura, K., Hayashi, K., Kohara, S., Tajiri, H., Kobayashi, K., Koura, A., Koura, A., & Shimojo, F. (2024). Atomic and Electronic Structures on a Mordenite Zeolite. *E-Journal of Surface Science and Nanotechnology*, *22*(1), 25–31. DOI: 10.1380/ejssnt.2023-063

Jimenez-Orozco, C., Figueras, M., Flórez, E., Vines, F., Rodriguez, J. A., & Illas, F. (2021). Size and Stoichiometry Effects on the Reactivity of MoC<inf>y</inf>Nanoparticles toward Ethylene. *The Journal of Physical Chemistry. C, Nanomaterials and Interfaces*, *125*(11), 6287–6297. DOI: 10.1021/acs.jpcc.0c10689

Karamanis, I., Daouli, A., Monnier, H., Dziurla, M.-A., Maurin, G., & Badawi, M. (2023). A systematic DFT screening of cationic faujasite-type zeolites for the adsorption of NO, NO<inf>2</inf> and H<inf>2</inf>O. *Molecular Systems Design & Engineering*, *8*(9), 1165–1181. DOI: 10.1039/D3ME00044C

Katada, N., Ozawa, N., Tsuji, E., Kanehara, K., Otsuka, A., Sakamoto, T., Umezawa, K., Matsubara, H., Suganuma, S., & Kubo, M. (2024). Methylation of benzene with methane induced by strong adsorption of benzene on Co ion at α-position in zeolite with moderate Al–Al distance. *Microporous and Mesoporous Materials*, *364*, 112855. Advance online publication. DOI: 10.1016/j.micromeso.2023.112855

Kellouai, W., Judeinstein, P., Plazanet, M., Baudoin, S., Drobek, M., Julbe, A., & Coasne, B. (2022). Gas Adsorption in Zeolite and Thin Zeolite Layers: Molecular Simulation, Experiment, and Adsorption Potential Theory. *Langmuir*, *38*(18), 5428–5438. DOI: 10.1021/acs.langmuir.1c03420 PMID: 35486814

Kim, P., van der Mynsbrugge, J., Head-Gordon, M., & Bell, A. T. (2022). Experimental and Theoretical Studies of Pd Cation Reduction and Oxidation during NO Adsorption on and Desorption from Pd/H-CHA. *The Journal of Physical Chemistry. C, Nanomaterials and Interfaces*, *126*(44), 18744–18753. DOI: 10.1021/acs.jpcc.2c06357

Lee, J., Lee, K., & Kim, J. (2021). Fiber-Based Gas Filter Assembled via in Situ Synthesis of ZIF-8 Metal Organic Frameworks for an Optimal Adsorption of SO<inf>2</inf>: Experimental and Theoretical Approaches. *ACS Applied Materials & Interfaces*, *13*(1), 1620–1631. DOI: 10.1021/acsami.0c19957 PMID: 33395254

Li, X., Li, T.-H., Zhou, W., Li, Y.-P., Sit, P. H.-L., Wu, Z., Curnow, O. J., Wu, K. C.-W., Choi, J., & Yip, A. C. K. (2022). Unveiling the elusive role of tetraethyl orthosilicate hydrolysis in ionic-liquid-templated zeolite synthesis. *Materials Today. Chemistry*, *23*, 100658. Advance online publication. DOI: 10.1016/j.mtchem.2021.100658

Li, Y., Yu, M., Qi, G., Liu, Y., Lv, J., Huang, S., & Ma, X. (2024). Designing a Mordenite Catalyst with Enhanced Acidity for Dimethyl Ether Carbonylation by Engineering Open Sn Sites. *Engineering*, *35*, 104–114. DOI: 10.1016/j.eng.2024.07.009

Losada, I. B., Grobas-Illobre, P., Misturini, A., Polaina, J., Seminovski, Y., & Sastre, G. (2021). Separation of an aqueous mixture of 6-kestose/sucrose with zeolites: A molecular dynamics simulation. *Microporous and Mesoporous Materials*, *319*, 111031. Advance online publication. DOI: 10.1016/j.micromeso.2021.111031

Louwen, J. N., van Eijck, L., Vogt, C., & Vogt, E. T. (2020). Understanding the Activation of ZSM-5 by Phosphorus: Localizing Phosphate Groups in the Pores of Phosphate-Stabilized ZSM-5. *Chemistry of Materials*, *32*(21), 9390–9403. DOI: 10.1021/acs.chemmater.0c03411

Mackey, D., Lyons, J., & Naydenova, I. (2024). Modeling redistribution of nanozeolites in holographic recording. *Journal of the Optical Society of America. A, Optics, Image Science, and Vision*, *41*(7), 1413–1419. DOI: 10.1364/JOSAA.523677 PMID: 39889130

Mahyuddin, M. H., Staykov, A., Saputro, A. G., Agusta, M. K., Dipojono, H. K., & Yoshizawa, K. (2020). Novel mechanistic insights into methane activation over Fe and Cu active sites in zeolites: A comparative DFT study using meta-GGA functionals. *The Journal of Physical Chemistry. C, Nanomaterials and Interfaces*, *124*(33), 18112–18125. DOI: 10.1021/acs.jpcc.0c04991

Mancinelli, M., Precisvalle, N., Ardit, M., Beltrami, G., Gigli, L., Catizzone, E., Migliori, M., Giordano, G., & Martucci, A. (2023). Thermal stability of templated ZSM-5 zeolites: An in-situ synchrotron X-ray powder diffraction study. *Microporous and Mesoporous Materials*, *362*, 112777. Advance online publication. DOI: 10.1016/j.micromeso.2023.112777

Martey, S., Addison, B., Wilson, N., Tan, B., Yu, J., Dorgan, J. R., & Sobkowicz, M. J. (2021). Hybrid Chemomechanical Plastics Recycling: Solvent-free, High-Speed Reactive Extrusion of Low-Density Polyethylene. *ChemSusChem*, *14*(19), 4280–4290. DOI: 10.1002/cssc.202100968 PMID: 34089238

Millan, R., Cnudde, P., Hoffman, A. E. J., Lopes, C. W., Concepción, P., van Speybroeck, V., & Boronat, M. (2020). Theoretical and Spectroscopic Evidence of the Dynamic Nature of Copper Active Sites in Cu-CHA Catalysts under Selective Catalytic Reduction (NH<inf>3</inf>-SCR-NO<inf>x</inf>) Conditions. *The Journal of Physical Chemistry Letters*, *11*(23), 10060–10066. DOI: 10.1021/acs.jpclett.0c03020 PMID: 33179925

Montesano, G., Cappelletti, P., Caputo, D., Liguori, B., Campanile, A., & Rispoli, C. (2022). Mineralogical and Technological Characterization of Zeolites from Basin and Range as Pozzolanic Addition of Cement. *Materials (Basel)*, *15*(7), 2684. Advance online publication. DOI: 10.3390/ma15072684 PMID: 35408016

Pan, L., Ye, S., Xv, X., Lin, P., Huang, R., & Wang, D. (2023). Zeolite-Encaged Luminescent Silver Nanoclusters. *Materials (Basel)*, *16*(10), 3736. Advance online publication. DOI: 10.3390/ma16103736 PMID: 37241363

Petitjean, H., Lepetit, C., Nour, Z., Poteau, R., del Rosal, I., & Berthomieu, D. (2020). How CuI and NaI interact with faujasite zeolite? A theoretical investigation. *The Journal of Physical Chemistry. C, Nanomaterials and Interfaces*, *124*(51), 28026–28037. DOI: 10.1021/acs.jpcc.0c06862

Petryk, M. R., Boyko, I. V., Petryk, M. M., Fraissard, J., & Mudryk, I. (2020). Modeling of adsorption and desorption of hydrocarbons in nanoporous catalytic zeolite media using nonlinear Langmuir isotherm. *Proceedings of SPIE - the International Society for Optical Engineering*, *11369*, 42. Advance online publication. DOI: 10.1117/12.2553926

Prats, H., Alonso, G., Sayós, R., & Gamallo, P. (2020). Transition metal atoms encapsulated within microporous Silicalite-1 zeolite: A systematic computational study. *Microporous and Mesoporous Materials*, *308*, 110462. Advance online publication. DOI: 10.1016/j.micromeso.2020.110462

Price, L. A., Ridley, C. J., Bull, C. L., Wells, S. A., & Sartbaeva, A. (2021). Determining the structure of zeolite frameworks at high pressures. *CrystEngComm*, *23*(33), 5615–5623. DOI: 10.1039/D1CE00142F

Ren, X., Zhai, Y., Yang, N., Wang, B., & Liu, S. (2025). Lattice Oxygen Redox Dynamics in Zeolite-Encapsulated CsPbBr<inf>3</inf> Perovskite OER Electrocatalysts. *Advancement of Science*, *12*(8), 2412679. Advance online publication. DOI: 10.1002/advs.202412679 PMID: 39783715

Romero-Marimon, P., Gutiérrez-Sevillano, J. J., & Calero, S. (2023). Adsorption of Carbon Dioxide in Non-Löwenstein Zeolites. *Chemistry of Materials*, *35*(13), 5222–5231. DOI: 10.1021/acs.chemmater.3c01258

Roos, D. P., Scheinost, A. C., Churakov, S. V., Nagashima, M., & Cametti, G. (2021). On the nature of Pb species in Pb-exchanged zeolite stellerite (STI): A combined experimental and theoretical study. *Microporous and Mesoporous Materials*, *327*, 111444. Advance online publication. DOI: 10.1016/j.micromeso.2021.111444

Sokol, H. J., Ebrahim, A. M., Caratzoulas, S., Frenkel, A. I., & Valla, J. A. (2022). In Situ XAFS, XRD, and DFT Characterization of the Sulfur Adsorption Sites on Cu and Ce Exchanged y Zeolites. *The Journal of Physical Chemistry. C, Nanomaterials and Interfaces*, *126*(3), 1496–1512. DOI: 10.1021/acs.jpcc.1c09128

Stepanyuk, K. O., Kravchenko, A. A., Gerashchenko, I. I., Grebenyuk, A. G., & Borysenko, M. V. (2020). Adsorption of zinc (II) on silica and alumina-silica surfaces: results of experiment and simulation | Адсорбция цинка (II) на кремнеземных и алюмосиликатных поверхностях: результаты эксперимента и моделирование | АДСОРБЦІЯ ЦИНКУ (II) НА КРЕМНЕЗЕМНИХ ТА АЛ. Himia. *Fizika Ta Tehnologia Poverhni*, *11*(3), 304–318. DOI: 10.15407/hftp11.03.304

Su, B., Wang, X., Chai, L., Huo, S., Qiu, J., Huang, Q., Li, S., Wang, Y., & Xue, W. (2024). Cation-Loaded Porous Mg2+-Zeolite Layer Direct Dendrite-Free Deposition toward Long-Life Lithium Metal Anodes. *Advancement of Science*, *11*(23), 2308939. Advance online publication. DOI: 10.1002/advs.202308939 PMID: 38600650

Sun, Q., Zhu, K., Ji, X., Chen, D., Han, C., Li, T.-T., Hu, Y., Huang, S., & Qian, J. (2022). MOF-derived three-dimensional ordered porous carbon nanomaterial for efficient alkaline zinc-air batteries | . *Science China Materials*, *65*(6), 1453–1462. DOI: 10.1007/s40843-021-1933-4

Sun, Z., Guo, Y., Liu, S., Li, S., & Sun, X. (2023). Moisture susceptibility evaluation of zeolite-foamed asphalts based on the Surface Free Energy (SFE) concept. *Construction & Building Materials*, *371*, 130762. Advance online publication. DOI: 10.1016/j.conbuildmat.2023.130762

Tan, S., Wang, C., Foucaud, Y., Badawi, M., Guo, H., Sun, K., Yang, G., & Mintova, S. (2022). Ordered sodium zeolite-templated carbon with high first discharge capacity for sodium battery application. *Microporous and Mesoporous Materials*, *336*, 111853. Advance online publication. DOI: 10.1016/j.micromeso.2022.111853

Taylor, E. E., Garman, K., & Stadie, N. P. (2020). Atomistic Structures of Zeolite-Templated Carbon. *Chemistry of Materials*, *32*(7), 2742–2752. DOI: 10.1021/acs.chemmater.0c00535

Trachta, M., Volný, T., Rubeš, M., Bulánek, R., & Bludský, O. (2023). QM-MM Approach to the Accurate Description of Slow Diffusion in Porous Materials. *The Journal of Physical Chemistry. C, Nanomaterials and Interfaces*, *127*(18), 8856–8863. DOI: 10.1021/acs.jpcc.3c01340

Tsiotsias, A. I., Georgiadis, A. G., Charisiou, N. D., & Goula, M. A. (2023). CO<inf>2</inf> Physisorption over an Industrial Molecular Sieve Zeolite: An Experimental and Theoretical Approach. *Materials (Basel)*, *16*(20), 6656. Advance online publication. DOI: 10.3390/ma16206656 PMID: 37895638

Vidoni, A., Ravikovitch, P. I., Afeworki, M., Calabro, D., Deckman, H., & Ruthven, D. (2020). Adsorption of CO<inf>2</inf> on high silica MFI and DDR zeolites: Structural defects and differences between adsorbent samples. *Microporous and Mesoporous Materials*, *294*, 109818. Advance online publication. DOI: 10.1016/j.micromeso.2019.109818

White, H. D., Huang, H. G., D'amaro, M. C., Mignone, E. A., Sholl, D. S., & Lively, R. P. (2022). Theoretical Pathway toward Improved Reverse Osmosis Membrane Selectivity for Neutral Solutes: Inspiration from Gas Separations. *The Journal of Physical Chemistry. C, Nanomaterials and Interfaces*, *126*(45), 19496–19506. DOI: 10.1021/acs.jpcc.2c06016

Yasumura, S., Liu, C., Toyao, T., Maeno, Z., & Shimizu, K.-I. (2021). Lean NO<inf>x</inf>Capture and Reduction by NH<inf>3</inf>via NO+Intermediates over H-CHA at Room Temperature. *The Journal of Physical Chemistry. C, Nanomaterials and Interfaces*, *125*(3), 1913–1922. DOI: 10.1021/acs.jpcc.0c10913

Yu, T., Zhang, Y., Cao, L., Cao, P., Zhou, C., & Gu, S. (2024). Freeze–Thaw Cycle Durability and Mechanism Analysis of Zeolite Powder-Modified Recycled Concrete. *Materials (Basel)*, *17*(11), 2671. Advance online publication. DOI: 10.3390/ma17112671 PMID: 38893935

Zhang, X., Lv, J., Liu, J., Xu, S., Sun, J., Wang, L., Xu, L., Mintova, S., Song, H., & Dong, B. (2022). Stable EMT type zeolite/CsPbBr<inf>3</inf> perovskite quantum dot nanocomposites for highly sensitive humidity sensors. *Journal of Colloid and Interface Science*, *616*, 921–928. DOI: 10.1016/j.jcis.2022.02.079 PMID: 35259722

Zhao, P., Boekfa, B., Nishitoba, T., Tsunoji, N., Sano, T., Yokoi, T., Ogura, M., & Ehara, M. (2020). Theoretical study on 31P NMR chemical shifts of phosphorus-modified CHA zeolites. *Microporous and Mesoporous Materials*, *294*, 109908. Advance online publication. DOI: 10.1016/j.micromeso.2019.109908

KEY TERMS AND DEFINITIONS

Zeolites: Crystalline microporous aluminosilicate materials known for their adsorption, ion-exchange, and catalytic properties.

Theoretical Modeling: Computational methods, including quantum mechanics and molecular dynamics, used to simulate the behavior and properties of materials at the atomic or molecular level.

Density Functional Theory (DFT): A quantum mechanical modeling method used to investigate the electronic structure of atoms, molecules, and solids.

Ab initio Methods: First-principles computational techniques that do not rely on empirical parameters, often used for high-accuracy simulations.

QM/MM (Quantum Mechanics/Molecular Mechanics): A hybrid computational approach that combines quantum mechanical and classical molecular mechanics calculations for large systems.

Structure-Directing Agent (SDA): Organic or inorganic molecules used during zeolite synthesis to guide the formation of a specific crystal framework.

Molecular Dynamics (MD): A computer simulation method that analyzes the physical movements of atoms and molecules over time.

Microporous Materials: Materials with pores less than 2 nanometers in diameter, critical for applications in catalysis, gas separation, and adsorption.

Adsorption Isotherm: A curve describing the relationship between the amount of a substance adsorbed on a surface and its pressure (or concentration) at constant temperature.

Transition State Theory: A theory that explains how chemical reactions occur and how reaction rates are calculated, focusing on the formation of an activated complex.

Extra-Framework Ions: Ions located within the pores or cavities of zeolites, but not part of the primary framework structure; crucial in ion-exchange and catalytic properties.

High-Performance Computing (HPC): The use of powerful supercomputers and parallel processing techniques for running advanced simulations and solving complex computational problems.

Catalytic Activity: A measure of a catalyst's ability to accelerate a chemical reaction without itself being consumed.

Single-Atom Catalysts (SACs): Catalysts consisting of isolated metal atoms dispersed on supports like zeolites, offering high activity and selectivity.

Passive NOx Adsorbers (PNAs): Materials, often based on zeolites, that trap nitrogen oxides (NOx) from vehicle emissions at low temperatures before they can be catalytically reduced.

Zeolite-Templated Carbon (ZTC): Carbon materials synthesized using zeolites as templates, resulting in a highly porous, structured carbon framework.

Grand Canonical Monte Carlo (GCMC) Simulation: A statistical mechanics method used to model adsorption processes by varying the number of particles, volume, and temperature.

Compilation of References

Abal, . (2021). Molecular fluid flow in MoS2 nanoporous membranes and hydrodynamics interactions. *The Journal of Chemical Physics*, *154*(18), 184704. Advance online publication. DOI: 10.1063/5.0039963 PMID: 33832278

Abatal, , Ruiz-Salvador, A. R., & Hernández, N. C. (2019). A DFT-based simulated annealing method for the optimization of global energy in zeolite framework systems: Application to natrolite, chabazite and clinoptilolite. *Microporous and Mesoporous Materials*, *294*, 109885. Advance online publication. DOI: 10.1016/j.micromeso.2019.109885

Abay, T. A., Wanna, W. H., Natarajan, T., Tsai, Y.-F., Janmanchi, D., Jiang, J.-C., Abu-Reziq, R., & Yu, S. S.-F. (2022). Selective oxidation of benzene by an iron oxide carbonaceous nanocatalyst prepared from iron perchlorate salts and hydrogen peroxide in benzene and acetonitrile. *Molecular Catalysis*, *526*, 112397. DOI: 10.1016/j.mcat.2022.112397

Abbo, H. S., & Titinchi, S. J. J. (2009). Di-, tri- and tetra-valent ion-exchanged NaY zeolite: Active heterogeneous catalysts for hydroxylation of benzene and phenol. *Applied Catalysis A, General*, *356*(2), 167–171. DOI: 10.1016/j.apcata.2009.01.003

Abdelwahab, , & Thabet, W. M. (2023). Natural zeolites and zeolite composites for heavy metal removal from contaminated water and their applications in aquaculture systems: A review. *Egyptian Journal of Aquatic Research*, *49*(4), 431–443. DOI: 10.1016/j.ejar.2023.11.004

Abdulridha, S., Jiao, Y., Xu, S., Zhang, R., Ren, Z., Garforth, A. A., & Fan, X. (2021). A comparative study on mesoporous Y zeolites prepared by hard-templating and post-synthetic treatment methods. *Applied Catalysis A, General*, *612*, 117986. DOI: 10.1016/j.apcata.2020.117986

Abukhadra, M. R., Ali, S. M., Nasr, E. A., Mahmoud, H. A. A., & Awwad, E. M. (2020). Effective Sequestration of Phosphate and Ammonium Ions by the Bentonite/Zeolite Na–P Composite as a Simple Technique to Control the Eutrophication Phenomenon: Realistic Studies. In ACS Omega (Vol. 5, Issue 24, pp. 14656–14668). American Chemical Society (ACS). https://doi.org/DOI: 10.1021/acsomega.0c01399

Abukhadra, M. R., Eid, M. H., El-Sherbeeny, A. M., Abd Elgawad, A. E. E., & Shim, J.-J. (2022). Effective desalination of brackish groundwater using zeolitized diatomite/kaolinite geopolymer as low-cost inorganic membrane; Siwa Oasis in Egypt as a realistic case study. In *Journal of Contaminant Hydrology* (Vol. 244, p. 103923). Elsevier BV., DOI: 10.1016/j.jconhyd.2021.103923

Adjal, , Guechtouli, N., Timón, V., Boussassi, R., Hammoutène, D., & Senent, M. L. (2024). DFT-based, Monte Carlo and Grand Canonical Monte Carlo simulations of nitro-organic pollutants 4-nitrophenol, 2-nitrophenol, 9-nitroanthracene and nitrogen trifluoride interacting with water in zeolite imidazole framework (ZIF-8). *Computational & Theoretical Chemistry*, *1243*, 114995. Advance online publication. DOI: 10.1016/j.comptc.2024.114995

Aguiar, E. F. S., Murta Valle, M. L., Silva, M. P., & Silva, D. F. (1995). Influence of external surface area of rare-earth containing Y zeolites on the cracking of 1,3,5-triisopropylbenzene. *Zeolites*, *15*(7), 620–623. https://doi.org/https://doi.org/10.1016/0144-2449(95)00069-I

Agustina, T. E., Rizkya, I., Utamaa, M. E. W., & Amalb, M. I. (2018). Characterization and Utilization of Zeolite for NPK Slow Release Fertilizer. *International Journal of Engineering TRANSACTIONS A: Basics*, *31*(4), 622–628.

Ahmed, A. H., & Thabet, M. S. (2015). Physicochemical studies and heterogeneous hydroxylation of benzene on FeII, FeIII-Semicarbazone/zeolite-y clathrates. *Synthesis and Reactivity in Inorganic, Metal-Organic, and Nano-Metal Chemistry*, *45*(11), 1632–1641. DOI: 10.1080/15533174.2015.1031038

Ahmed, , Mavukkandy, M. O., Giwa, A., Elektorowicz, M., Katsou, E., Khelifi, O., Naddeo, V., & Hasan, S. W. (2022). Recent developments in hazardous pollutants removal from wastewater and water reuse within a circular economy. *NPJ Clean Water*, *12*(1), 12. Advance online publication. DOI: 10.1038/s41545-022-00154-5

Ahmedzeki, N. S., Yilmaz, S., & Al-Tabbakh, B. A. (2016). Synthesis and Characterization of Nanocrystalline Zeolite Y. In *Journal Al-Khwarizmi. Engineering Journal (New York)*, *12*(1), •••.

Aili, , & Maruyama, I. (2020). Review of several experimental methods for characterization of micro- and nano-scale pores in cement-based material. *International Journal of Concrete Structures and Materials, 14*(1), 53. Advance online publication. DOI: 10.1186/s40069-020-00431-y

Akkoca, , Yılgın, M., Ural, M., Akçin, H., & Mergen, A. (2013). Hydrothermal and thermal treatment of the natural zeolite clinoptilolite from Bigadiç, Turkey: An experimental study. *Geochemistry International, 51*(4), 315–323. DOI: 10.1134/S0016702913040022

Akyalcin, , Akyalcin, L., & Ertugrul, E. (2024). Modification of natural clinoptilolite zeolite to enhance its hydrogen adsorption capacity. *Research on Chemical Intermediates, 1473*(3), 1455–1473. Advance online publication. DOI: 10.1007/s11164-023-05212-2

Alabdullah, M. A., Gomez, A. R., Vittenet, J., Bendjeriou-Sedjerari, A., Xu, W., Abba, I. A., & Gascon, J. (2020). A Viewpoint on the Refinery of the Future: Catalyst and Process Challenges. *ACS Catalysis, 10*(15), 8131–8140. DOI: 10.1021/acscatal.0c02209

Al-Ani, A., Freitas, C., & Zholobenko, V. (2020). Nanostructured large-pore zeolite: The enhanced accessibility of active sites and its effect on the catalytic performance. *Microporous and Mesoporous Materials, 293*, 109805. https://doi.org/https://doi.org/10.1016/j.micromeso.2019.109805. DOI: 10.1016/j.micromeso.2019.109805

Al-Ani, A., Haslam, J. J. C., Mordvinova, N. E., Lebedev, O. I., Vicente, A., Fernandez, C., & Zholobenko, V. (2019). Synthesis of nanostructured catalysts by surfactantlating of large-pore zeolites. *Nanoscale Advances, 1*(5), 2029–2039. DOI: 10.1039/C9NA00004F PMID: 36134217

Aldahri, T., Behin, J., Kazemian, H., & Rohani, S. (2016). Synthesis of zeolite Na-P from coal fly ash by thermo-sonochemical treatment. *Fuel, 182*, 494–501. DOI: 10.1016/j.fuel.2016.06.019

Alexandratos, S. D. (2008). Ion-Exchange Resins: A Retrospective from Industrial and Engineering Chemistry Research. In Industrial & Engineering Chemistry Research (Vol. 48, Issue 1, pp. 388–398). American Chemical Society (ACS). https://doi.org/DOI: 10.1021/ie801242v

Al-Harbi, L. M., Kosa, S. A., Abd El Maksod, I. H., & Hegazy, E. Z. (2015). The Photocatalytic Activity of TiO_2 -Zeolite Composite for Degradation of Dye Using Synthetic UV and Jeddah Sunlight. *Journal of Nanomaterials, 2015*(1). DOI: 10.1155/2015/565849

Ali, A., Chiang, Y. W., & Santos, R. M. (2022). X-ray diffraction techniques for mineral characterization: A review for engineers of the fundamentals, applications, and research directions. *Minerals (Basel)*, *12*(2), 205. DOI: 10.3390/min12020205

Ali, S., Morsy, R., El-Zawawy, N., Fareed, M., & Bedaiwy, M. (2017). Synthesized zinc peroxide nanoparticles (ZnO_2-NPs): A novel antimicrobial, anti-elastase, anti-keratinase, and anti-inflammatory approach toward polymicrobial burn wounds. *International Journal of Nanomedicine*, *12*, 6059–6073. DOI: 10.2147/IJN.S141201 PMID: 28860766

Almansouri, , Edokali, M., & Abu Seman, M. N. (2024). Flat sheet thin film composite forward osmosis membranes for water treatment and purification: A review on modification techniques and concepts. *Desalination and Water Treatment*, *320*, 100815. Advance online publication. DOI: 10.1016/j.dwt.2024.100815

Almeida, R. K. S., Peréz-Pariente, J., & Gómez-Hortigüela, L. (2020). Accordion-like effect of flexible organic structure-directing agents in the synthesis of ferrierite zeolite. *Microporous and Mesoporous Materials*, *300*, 110149. Advance online publication. DOI: 10.1016/j.micromeso.2020.110149

Alqarni, M., Ashour, A. A., Shafie, A., Alqarni, A., Felemban, M. F., Shukr, B. S., & Algahtani, F. S. (2024). Intelligence computational analysis of letrozole solubility in supercritical solvent via machine learning models. *Scientific Reports*, *14*(1), 21677.

Al-Zahrani, S. A., Patil, M. B., Mathad, S. N., Patil, A. Y., Al Otaibi, A., Masood, N., Mansour, D., Khan, A., Gupta, V., Topare, N. S., Somya, A., & Ayyar, M. (2023). Photocatalytic Azo Dye Degradation Using Graphite Carbon Nitride Photocatalyst and UV-A Irradiation. In Crystals (Vol. 13, Issue 4, p. 577). MDPI AG. https://doi.org/DOI: 10.3390/cryst13040577

Ambekar, R. S., Oliveira, E. F., Pugazhenthi, P., Singh, S., Mahapatra, D. R., Galvao, D. S., & Tiwary, C. S. (2024). Drop-Weight Impact Resistance of 3D-Printed Complex Zeolite-Inspired Structures. *Advanced Engineering Materials*, *26*(22), 2400035. Advance online publication. DOI: 10.1002/adem.202400035

Ambrosino, F., Gargiulo, N., Peruta, G. D., Gravino, C., Gagliardo, G., Pisciotta, F., Esposito, S., La Verde, G., Caputo, D., & Pugliese, M. (2024). Study and characterization of zeolites for the removal of artificial radionuclides in wastewater samples from nuclear power plants. In *Journal of Hazardous Materials Advances* (Vol. 16, p. 100458). Elsevier BV., DOI: 10.1016/j.hazadv.2024.100458

Ameh, E. S. (2019). A review of basic crystallography and X-ray diffraction applications. *International Journal of Advanced Manufacturing Technology*, *105*(7-8), 3289–3330. DOI: 10.1007/s00170-019-04508-1

Ameh, T., Gibb, M., Stevens, D., Pradhan, S. H., Braswell, E., & Sayes, C. M. (2022). Silver and Copper Nanoparticles Induce Oxidative Stress in Bacteria and Mammalian Cells. *Nanomaterials (Basel, Switzerland)*, *12*(14), 2402. DOI: 10.3390/nano12142402 PMID: 35889626

Amin, G., Konstantinovic, S., Jordanov, I., & Djordjevic, D. (2023). Adsorption study of textile dye Basic Red 46 on clinoptilolite. Iran. J. Chem. Chem. Eng. Research Article Vol, 42(11).

Amin, M., Ahmad, R., Ali, A., Aslam, M., & Lee, D. J. (2015). Silicon Fertilization Improves the Maize (Zea mays L.) Performance under Limited Moisture Supply. *Cereal Research Communications*. Advance online publication. DOI: 10.1556/0806.43.2015.035

Amoureux, J. P., & Pruski, M. (2008). MQMAS NMR: Experimental strategies and applications. In Harris, R. L., & Wasylishen, R. K. (Eds.), *eMagRes* (pp. 1–31). Wiley., DOI: 10.1002/9780470034590.emrstm0319.pub2

Andaç, Ö., Tatlıer, M., Sirkecioğlu, A., Ece, I., & Erdem-Şenatalar, A. (2005). Effects of ultrasound on zeolite A synthesis. *Microporous and Mesoporous Materials*, *79*(1–3), 225–233. DOI: 10.1016/j.micromeso.2004.11.007

Andana, T., Rappé, K. G., Gao, F., Szanyi, J., Pereira-Hernandez, X., & Wang, Y. (2021). Recent advances in hybrid metal oxide–zeolite catalysts for low-temperature selective catalytic reduction of NOx by ammonia. In Applied Catalysis B: Environmental (Vol. 291, p. 120054). Elsevier BV. https://doi.org/DOI: 10.1016/j.apcatb.2021.120054

Andrady, A. L. (2011). Microplastics in the marine environment. *Marine Pollution Bulletin*, *62*(8), 1596–1605.

Andronikashvili, T. G., Zautashvili, M., Eprikashvili, L., Pirtskhalava, N., & Dzagania, M. (2011). *New Substrate of Prolonging Action on the Basis of Natural Zeolite and Brown Coal for Growing Agricultural Crops* Bull. *Georg. Nat. Acad. Sci.*, *5*, 2.

Andronikashvili, T. G., Zautashvili, M., Eprikashvili, L., Pirtskhalava, N., & Dzagani, M. (2009). Zeolite containing substrates - A new way from plant-growing in plant production. *Annals of Agrarian Science*, *7*(4), 14–45.

Angell, C. L., & Schaffer, P. C. (1966). Infrared spectroscopic investigations of zeolites and adsorbed molecules. II. Adsorbed carbon monoxide. *Journal of Physical Chemistry*, *70*(5), 1413–1418. DOI: 10.1021/j100877a012

Angell, C. L., & Schaffer, P. C.Structural OH Groups. (1965). Infrared spectroscopic investigations of zeolites and adsorbed molecules. I. Structural OH groups. *Journal of Physical Chemistry*, *69*(10), 3463–3470. DOI: 10.1021/j100894a037

Anggoro, D. D., & Amin, N. A. S. (2006). Methane to Liquid Hydrocarbons over Tungsten-ZSM-5 and Tungsten Loaded Cu/ZSM-5 Catalysts. *Journal of Natural Gas Chemistry*, *15*(4), 340–347. DOI: 10.1016/S1003-9953(07)60016-4

Anis, , Hashaikeh, R., & Hilal, N. (2019). Flux and salt rejection enhancement of polyvinyl(alcohol) reverse osmosis membranes using nano-zeolite. *Desalination*, *463*, 18–27. DOI: 10.1016/j.desal.2019.114104

Antunes, F. P. N., Vaiss, V. S., Tavares, S. R., Chiaro, S. S., Souza, W. F., & Leitao, A. A. (2015). Vacancy formation in MoS2 supported on MgO: Electronic and energetic analysis. Chemical Physics Letters, 626, 55-62.

Antúnez-García, J., Galván, D., Petranovskii, V., Murrieta-Rico, F. N., Yocupicio-Gaxiola, R. I., Shelyapina, M. G., & Fuentes-Moyado, S. (2022). Aluminum distribution in mordenite-zeolite framework: A new outlook based on density functional theory calculations. *Journal of Solid State Chemistry*, *306*, 122725. DOI: 10.1016/j.jssc.2021.122725

Antúnez-García, J., Núñez-González, R., Petranovskii, V., Hmok, H., Reyes-Serrato, A., Murrieta-Rico, F. N., Xiao, M., & Zamora, J. (2025). Theoretical study of the electronic and optical properties of a composite formed by the zeolite NaA and a magnetite cluster. *Beilstein Journal of Nanotechnology*, *16*, 44–53. DOI: 10.3762/bjnano.16.5 PMID: 39834743

Antúnez-García, J., Petranovskii, V., Murrieta-Rico, F. N., Yocupicio-Gaxiola, R. I., Shelyapina, M. G., & Fuentes-Moyado, S. (2024a). Investigating patterns in aluminum distribution and stability in mordenite zeolite structure: Analysis and theoretical conclusions from existing experimental data. *Journal of Solid State Chemistry*, *331*, 124548. DOI: 10.1016/j.jssc.2024.124548

Antúnez-García, J., Ponce-Ruiz, J. L., Núñez-González, R., Petranovskii, V., Murrieta-Rico, F. N., Serrato, A. R., Shelyapina, M. G., Xiao, M., & Zamora, J. (2024b). Theoretical investigation of Fe and Cu cations hosted within the MOR zeolite framework. *Materials Today. Communications*, *110418*, 110418. Advance online publication. DOI: 10.1016/j.mtcomm.2024.110418

Araneta, M. J. P., Gaw, E. A. P., Seagan, C. G. V., Dolina, M. N. E., Bonto, A. P., Casalme, E. C. G., Salvador, A. A., & Nuñez, J. A. P. (2024). Enhanced Removal of Acetylsalicylic Acid Using Surfactant-Modified Natural Zeolites. In Trends in Sciences (Vol. 22, Issue 2, p. 8908). College of Graduate Studies, Walailak University. https://doi.org/DOI: 10.48048/tis.2025.8908

Aravind, S.. (2019). Nanomaterials: Synthesis, properties, and applications. *Journal of Materials Science and Technology*, *35*(6), 1092–1104.

Arias Cruz, C.L. and Martín Alonso, G.M. (2023) *personal report.*

Arribas, J., Corma, A., Fornés, V., & Melo, F. (1987). Influence of framework aluminum gradients on the catalytic activity of Y zeolites: Cracking of gas-oil on Y zeolites dealuminated by different procedures. *Journal of Catalysis*, *108*(1), 135–142. https://doi.org/https://doi.org/10.1016/0021-9517(87)90160-6. DOI: 10.1016/0021-9517(87)90160-6

Artioli, G. (2016). X-ray diffraction (XRD). *Encyclopedia of Earth Sciences Series*, 1019–1025. https://doi.org/DOI: 10.1007/978-1-4020-4409-0_29

Ashbrook, S. E., Dawson, D. M., & Seymour, V. R. (2014). Recent developments in solid-state NMR spectroscopy of crystalline microporous materials. *Physical Chemistry Chemical Physics*, *16*(18), 8223–8242. DOI: 10.1039/C4CP00578C PMID: 24675798

Askari, S., & Halladj, R. (2012). Ultrasonic pretreatment for hydrothermal synthesis of SAPO-34 nanocrystals. *Ultrasonics Sonochemistry*, *19*(3), 554–559. DOI: 10.1016/j.ultsonch.2011.09.006

Askari, S., Miar Alipour, S., Halladj, R., & Davood Abadi Farahani, M. H. (2013). Effects of ultrasound on the synthesis of zeolites: A review. *Journal of Porous Materials*, *20*(1), 285–302. DOI: 10.1007/s10934-012-9598-6

Asmi, D., & Low, I. M. (2014). Manufacture of graded ceramic matrix composites using infiltration techniques. In *Advances in Ceramic Matrix Composites* (pp. 109–140). Elsevier., DOI: 10.1533/9780857098825.1.109

Asraf, M. H., Malek, N. A. N. N., Sani, N. S., & Matmin, J. (2024). Surface Morphology, Porosity and Antibiofilm Activity of Orthosiphon aristatus-Phytosynthesized–Silver Nanoparticles Supported Zeolite A. *Arabian Journal for Science and Engineering*, *49*(7), 9121–9131. DOI: 10.1007/s13369-024-08832-x

Asrori, M. R., Santoso, A., & Sumari, S. (2024). Proofing the presence of metal oxide impregnated into zeolite A without calcination: XRD and FTIR studies. *Case Studies in Chemical and Environmental Engineering*, *9*, 100676. DOI: 10.1016/j.cscee.2024.100676

Azizi-Lalabadi, M., Alizadeh-Sani, M., Khezerlou, A., Mirzanajafi-Zanjani, M., Zolfaghari, H., Bagheri, V., Divband, B., & Ehsani, A. (2019). Nanoparticles and Zeolites: Antibacterial Effects and their Mechanism against Pathogens. *Current Pharmaceutical Biotechnology*, *20*(13), 1074–1086. DOI: 10.2174/1573397115666190708120040 PMID: 31309886

Azizi-Lalabadi, M., Ehsani, A., Ghanbarzadeh, B., & Divband, B. (2020). Polyvinyl alcohol/gelatin nanocomposite containing ZnO, TiO2 or ZnO/TiO2 nanoparticles doped on 4A zeolite: Microbial and sensory qualities of packaged white shrimp during refrigeration. *International Journal of Food Microbiology*, *312*, 108375. DOI: 10.1016/j.ijfoodmicro.2019.108375 PMID: 31669767

Azizi, S. N., & Yousefpour, M. (2010). Static and ultrasonic-assisted aging effects on the synthesis of analcime zeolite. *Zeitschrift für Anorganische und Allgemeine Chemie*, *636*(5), 886–890. DOI: 10.1002/zaac.200900415

Babaeva, M. A., & Tsyganenko, A. A. (1987). Infrared spectroscopic evidence for the formation of carbonite CO_2^{2-} ions in CO interaction with basic oxide surfaces. *Reaction Kinetics and Catalysis Letters*, *34*(1), 9–14. DOI: 10.1007/BF02069193

Babaeva, M. A., Tsyganenko, A. A., & Filimonov, V. N. (1984). Infrared spectra of the adsorbed SO_2. *Kinetics and Catalysis*, *25*(4), 787–792.

Babel, S., & Kurniawan, T. A. (2003). Low-cost adsorbents for heavy metals uptake from contaminated water: A review. *Journal of Hazardous Materials*, *97*(1-3), 219–243.

Badeenezhad, , & Azhdarpoor, A. (2019a). Efficiency of the activated carbon and clinoptilolite particles coated with iron oxide magnetic nanoparticles in removal of methylene blue. *Desalination and Water Treatment*, *154*, 347–355. DOI: 10.5004/dwt.2019.24039

Badeenezhad, , Azhdarpoor, A., Bahrami, S., & Yousefinejad, S. (2019b). Removal of methylene blue dye from aqueous solutions by natural clinoptilolite and clinoptilolite modified by iron oxide nanoparticles. *Molecular Simulation*, *45*(7), 564–571. DOI: 10.1080/08927022.2018.1564077

Baerlocher, C., & McCusker, L. B. (n.d.). *Database of zeolite structures,*http://www.iza-structure.org/databases/*; date accessed: May 30, 2024*. Retrieved May 30, 2024, from http://www.iza-structure.org/databases/

Bahiraei, A., & Behin, J. (2020). Sonochemical immobilization of MnO2 nanoparticles on NaP-zeolite for enhanced Hg (II) adsorption from water. *Journal of Environmental Chemical Engineering*, *8*(3), 103790. DOI: 10.1016/j.jece.2020.103790

Bahmanzadegan, , & Ghaemi, A. (2025). A comprehensive review on novel zeolite adsorbents for environmental pollutant. *Journal of Hazardous Materials Advances*, *100617*, 100617. Advance online publication. DOI: 10.1016/j.hazadv.2025.100617

Bai, R., Song, Y., Lätsch, L., Zou, Y., Feng, Z., Copéret, C., Corma, A., & Yu, J. (2022). Switching between classical/nonclassical crystallization pathways of TS-1 zeolite: Implication on titanium distribution and catalysis. *Chemical Science (Cambridge)*, *13*(36), 10868–10877. DOI: 10.1039/d2sc02679a

Balabinskaya, A. S., Ivanova, E. N., Ivanova, M. S., Kumzerov, Y. A., Pankova, S. V., Poborchii, V. V., Romanov, S. G., Solovyev, V. G., & Khanin, S. D. (2005). Investigation into the electrical and optical properties of sodium nitrite and sodium nitrate nanoparticles in regular porous matrices. *Glass Physics and Chemistry*, *31*(3), 330–336. DOI: 10.1007/s10720-005-0064-2

Balkan, N., & Erol, A. (2021). Intrinsic and extrinsic semiconductors. In *Semiconductors for Optoelectronics* (pp. 37–78). Springer., DOI: 10.1007/978-3-319-44936-4_2

Balouiri, M., Sadiki, M., & Ibnsouda, S. K. (2016). Methods for in vitro evaluating antimicrobial activity: A review. *Journal of Pharmaceutical Analysis*, *6*(2), 71–79. DOI: 10.1016/j.jpha.2015.11.005 PMID: 29403965

Bansiwal, A. K., Rayalu, S. S., Labhasetwar, N. K., Juwarkar, A. A., & Devotta, S. (2006). Surfactant-Modified Zeolite as a Slow Release Fertilizer for Phosphorus. *Journal of Agricultural and Food Chemistry*, *54*(13), 4773–4779. DOI: 10.1021/jf060034b PMID: 16787027

Baranov, D. A., Krauklis, I. V., & Tsyganenko, A. A. (2022). Quantum chemical study of isotopic isomerism of CO adsorbed on NiY zeolite. *Optics and Spectroscopy*, *130*(5), 530–537.

Barthomeuf, D. (1987). Zeolite acidity dependence on structure and chemical environment. Correlations with catalysis. *Materials Chemistry and Physics*, *17*(1–2), 49–71. DOI: 10.1016/0254-0584(87)90048-4

Barthomeuf, D. (1996). Basic zeolites: Characterization and uses in adsorption and catalysis. *Catalysis Reviews. Science and Engineering*, *38*(3), 521–612. DOI: 10.1080/01614949608006465

Baruah, M. J., Dutta, A., Biswas, S., Gogoi, G., Hoque, N., Bhattacharyya, P. K., & Bania, K. K. (2022). Fe2O3 Nanocatalysts Supported on Zeolite-Y for the Selective Synthesis of C2 Di-Indolyl Indolones and Isatins. *ACS Applied Nano Materials*, *5*(1), 1446–1459. DOI: 10.1021/acsanm.1c03987

Baskan, B.. (2021). Removal of arsenate using graphene oxide-iron modified clinoptilolite-based composites. *Journal of Analytical Science and Technology*, *12*(1), 14. Advance online publication. DOI: 10.1186/s40543-021-00274-6

Bauer, , Broome, S. T., Kibikas, W. M., & Wilson, J. E. (2024). Mechanical and poroelastic behavior of porous tuff under drained and undrained conditions. *Bulletin of Engineering Geology and the Environment*, *83*(10), 415. Advance online publication. DOI: 10.1007/s10064-024-03916-2

Baur, W. H., & Fischer, R. X. (2023). The flexibility of the T–X–T hinges between the coordination tetrahedra in various zeolitic frameworks: An empirical structural study. *Mineralogy and Petrology*, *117*(2), 165–179. DOI: 10.1007/s00710-023-00813-6

Beebe, T. P., Gelin, P., & Yates, J. T.Jr. (1984). Infrared spectroscopic observations of surface bonding in physical adsorption: The physical adsorption of CO on SiO_2 surfaces. *Surface Science*, *148*(2-3), 526–550. DOI: 10.1016/0039-6028(84)90597-1

Behin, J., Kazemian, H., & Rohani, S. (2016). Sonochemical synthesis of zeolite NaP from clinoptilolite. *Ultrasonics Sonochemistry*, *28*, 400–408. DOI: 10.1016/j.ultsonch.2015.08.021

Bein, T. (2007). Host–Guest interactions in zeolites and periodic mesoporous materials. In Studies in surface science and catalysis (pp. 611-657, XVII-XIX). https://doi.org/DOI: 10.1016/S0167-2991(07)80806-8

Belaabed, R., Elabed, S., Addaou, A., Laajab, A., Rodríguez, M. A., & Lahsini, A. (2016). Synthesis of LTA zeolite for bacterial adhesion. *Boletín de la Sociedad Española de Cerámica y Vidrio*, *55*(4), 152–158. DOI: 10.1016/j.bsecv.2016.05.001

Belviso, C., Cavalcante, F., & Fiore, S. (2013). Ultrasonic waves induce rapid zeolite synthesis in a seawater solution. *Ultrasonics Sonochemistry*, *20*(1), 32–36. DOI: 10.1016/j.ultsonch.2012.06.004

Belviso, C., Cavalcante, F., Lettino, A., & Fiore, S. (2011). Effects of ultrasonic treatment on zeolite synthesized from coal fly ash. *Ultrasonics Sonochemistry*, *18*(2), 661–668. DOI: 10.1016/j.ultsonch.2010.08.011

Belykh, R., Maevskaya, M., Krauklis, I., & Tsyganenko, A. (2015). Linkage isomerism of CO Adsorbed on alkali halides. *The Journal of Physical Chemistry A*, *119*(11), 2363–2370. Advance online publication. DOI: 10.1021/jp507394s PMID: 25222795

Belykh, R., Maugé, F., & Tsyganenko, A. (2019). Deep insight into CO adsorption on Ni-USY zeolite catalyst by FTIR spectroscopy. Evidence for isotopic isomerism. *Applied Catalysis A, General*, *583*, 117140. DOI: 10.1016/j.apcata.2019.117140

Bergs, C., Brück, L., Rosencrantz, R. R., Conrads, G., Elling, L., & Pich, A. (2017). Biofunctionalized zinc peroxide (ZnO_2) nanoparticles as active oxygen sources and antibacterial agents. *RSC Advances*, *7*(62), 38998–39010. DOI: 10.1039/C7RA06332F

Bernin, D., & Topgaard, D. (2013). NMR diffusion and relaxation correlation methods: New insights in heterogeneous materials. *Current Opinion in Colloid & Interface Science*, *18*(3), 166–172. DOI: 10.1016/j.cocis.2013.03.007

Bertaut, E. F. (1964). *La diffraction des rayons X*. Masson & Cie.

Bertsch, L., & Habgood, H. W. (1963). An Infrared Spectroscopic study of the Adsorption of water and Carbon Dioxide on Linde Molecular sieve X. *Journal of Physical Chemistry*, *67*(8), 1621–1628. DOI: 10.1021/j100802a013

Biaglow, A. I., Gorte, R. J., Kokotailo, G. T., & White, D. (1994). A probe of Brønsted site acidity in zeolites: ^{13}C chemical shift of acetone. *Journal of Catalysis*, *148*(2), 779–786. DOI: 10.1006/jcat.1994.1264

Biblioteca, I., Sambucci, M., & Valente, M. (2023). Zeolite-Clinoptilolite conditioning for improved heavy metals ions removal: A preliminary assessment. *Ceramics International*, *49*(23), 39649–39656.

Biesek, , Banaszak, M., & Adamski, M. (2021). Ducks' growth, meat quality, bone strength, and jejunum strength depend on zeolite in feed and long-term factors. *Animals (Basel)*, *11*(4), 1015. Advance online publication. DOI: 10.3390/ani11041015 PMID: 33916880

Bird, J. P. (2002). Semiconductors: An introduction. En *Encyclopedia of Materials: Science and Technology* (2nd ed., pp. 1-16). https://doi.org/DOI: 10.1016/B0-08-043152-6/01854-4

Bogomolov, V. N. (1978). Liquids in ultrathin channels (Filament and cluster crystals). *Soviet Physics - Uspekhi*, *21*(1), 77–83. DOI: 10.1070/PU1978v021n01ABEH005510

Bokov, D., Jalil, A. T., Chupradit, S., Suksatan, W., Ansari, M. J., Shewael, I. H., Valiev, G. H., & Kianfar, E. (2021). Nanomaterial by sol-gel method: Synthesis and application. *Advances in Materials Science and Engineering*, *5102014*(1), 5102014. Advance online publication. DOI: 10.1155/2021/5102014

Bols, M. L., Devos, J., Rhoda, H. M., Plessers, D., Solomon, E. I., Schoonheydt, R. A., Sels, B. F., & Dusselier, M. (2021). Selective Formation of α-Fe(II) Sites on Fe-Zeolites through One-Pot Synthesis. *Journal of the American Chemical Society*, *143*(39), 16243–16255. DOI: 10.1021/jacs.1c07590

Bols, M. L., Hallaert, S. D., Snyder, B. E. R., Devos, J., Plessers, D., Rhoda, H. M., Dusselier, M., Schoonheydt, R. A., Pierloot, K., Solomon, E. I., & Sels, B. F. (2018). Spectroscopic Identification of the α-Fe/α-O Active Site in Fe-CHA Zeolite for the Low-Temperature Activation of the Methane C-H Bond. *Journal of the American Chemical Society*, *140*(38), 12021–12032. DOI: 10.1021/jacs.8b05877

Bols, M. L., Rhoda, H. M., Snyder, B. E. R., Solomon, E. I., Pierloot, K., Schoonheydt, R. A., & Sels, B. F. (2020). Advances in the synthesis, characterization, and mechanistic understanding of active sites in Fe-zeolites for redox catalysts. *Dalton Transactions (Cambridge, England)*, *49*(42), 14749–14757. DOI: 10.1039/D0DT01857K

Bols, M. L., Snyder, B. E. R., Rhoda, H. M., Cnudde, P., Fayad, G., Schoonheydt, R. A., Van Speybroeck, V., Solomon, E. I., & Sels, B. F. (2021). Coordination and activation of nitrous oxide by iron zeolites. *Nature Catalysis*, *4*(4), 332–340. DOI: 10.1038/s41929-021-00602-4

Borai, E. H., Harjula, R., malinen, L., & Paajanen, A. (2009). Efficient removal of cesium from low-level radioactive liquid waste using natural and impregnated zeolite minerals. *Journal of Hazardous Materials*, *172*(1), 416–422. DOI: 10.1016/j.jhazmat.2009.07.033 PMID: 19656622

Borcuch, A., Rutkowska, M., Marzec, A., Kowalczyk, A., Michalik, M., Moreno, J. M., & Chmielarz, L. (2020). Selective ammonia oxidation over ZSM-5 zeolite: Impact of catalyst's support porosity and type of deposited iron species. *Catalysis Today*, *348*, 223–229. DOI: 10.1016/j.cattod.2019.08.054

Bordiga, S., Lamberti, C., Bonino, F., Travert, A., & Thibault-Starzyk, F. (2015). Probing zeolites by vibrational spectroscopies. *Chemical Society Reviews*, *44*(20), 7262–7341. DOI: 10.1039/C5CS00396B PMID: 26435467

Bornes, C., Fischer, M., Amelse, J. A., Geraldes, C. F. G. C., Rocha, J., & Mafra, L. (2021). What is being measured with P-bearing NMR probe molecules adsorbed on zeolites? *Journal of the American Chemical Society*, *143*(34), 13616–13623. DOI: 10.1021/jacs.1c05014 PMID: 34410690

Bornes, C., Sardo, M., Lin, Z., Amelse, J., Fernandes, A., Ribeiro, M. F., Geraldes, C., Rocha, J., & Mafra, L. (2019). ^{1}H-^{31}P HETCOR NMR elucidates the nature of acid sites in zeolite HZSM-5 probed with trimethylphosphine oxide. *Chemical Communications, 55*(84), 12635–12638. DOI: 10.1039/C9CC06763A PMID: 31580363

Bouzo, L., López, M., & Acosta, J. A. (1991) *Estudio de formas de aplicación de zeolita en caña de azúcar en suelos de textura ligera. Resultados de caña planta.* Memoirs 3rd Int. Conf. Occur. Prop. Util. Nat. Zeolites, Part II, 104-105

Boycheva, S., Zgureva, D., Miteva, S., Marinov, I., Behunová, D. M., Trendafilova, I., Popova, M., & Václaviková, M. (2020). Studies on the potential of nonmodified and metal oxide-modified coal fly ash zeolites for adsorption of heavy metals and catalytic degradation of organics for wastewater recovery. *Processes (Basel, Switzerland), 8*(7), 778. DOI: 10.3390/pr8070778

Bragg, W. H. (1913). The diffraction of X-rays by crystals. *Proceedings - Royal Society. Mathematical, Physical and Engineering Sciences, 88*(605), 428–438.

Braschinsky, A., & Steed, J. W. (2022). Molecular clusters in confined spaces. *Coordination Chemistry Reviews, 473*, 214840. DOI: 10.1016/j.ccr.2022.214840

Breck, D. W. (1974). Zeolite molecular sieves (9th ed.; Donald W. Breck, Ed.). Nashville, TN: John Wiley & Sons.

Breck, D. W. (1974). *Zeolite molecular sieves: structure, chemistry, and use.* John Wiley and Sons., DOI: 10.1093/chromsci/13.4.18A-c

Breiner, B., Clegg, J. K., & Nitschke, J. R. (2010). Reactivity modulation in container molecules. *Chemical Science (Cambridge), 2*(1), 51–56. DOI: 10.1039/C0SC00329H

Breysse, M., Geantet, C., Afanasiev, P., Blanchard, J., & Vrinat, M. (2008). Recent studies on the preparation, activation and design of active phases and supports of hydrotreating catalysts. *Catalysis Today, 130*(1), 3–13. DOI: 10.1016/j.cattod.2007.08.018

Brinker, C. J., Kirkpatrick, R. J., Tallant, D. R., Bunker, B. C., & Montez, B. (1988). NMR confirmation of strained "defects" in amorphous silica. *Journal of Non-Crystalline Solids, 99*(2-3), 418–428. DOI: 10.1016/0022-3093(88)90448-6

Brito, A., & Coutín, D. (2017). *Las zeolititas de Cuba: Results of 45 years dedicated to its study* Inst. Geol. Paleont., Geological Survey of Cuba Cairo Cairo, P., Machado de Armas, J., Torres Artiles, P., Diaz Martin, B., Jiménez Carrazana, R. and Rodríguez López, O. (2017) *Effects of zeolite and organic fertilizers on soil quality and yield of sugarcane.Australian Journal of Crop Science, 11*(06), 733–738. DOI: 10.21475/ajcs.17.11.06.p501

Brito, J. L., Barbosa, A. L., Albornoz, A., Severino, F., & Laine, J. (1994). Nickel molybdate as precursor of HDS catalysts: Effect of phase composition. *Catalysis Letters*, *26*(3), 329–337. DOI: 10.1007/BF00810606

Brouwer, D. H., Brouwer, C. C., Mesa, S., Semelhago, C. A., Steckley, E. E., Sun, M. P. Y., Mikolajewski, J. G., & Baerlocher, C. (2020). Solid-state ^{29}Si NMR spectra of pure silica zeolites for the International Zeolite Association Database of Zeolite Structures. *Microporous and Mesoporous Materials*, *297*(January), 110000-1–12. DOI: 10.1016/j.micromeso.2020.110000

Brown, D., & Green, R. (2015). Transmission electron microscopy: Principles and applications. *Journal of Microscopy*, *158*(1), 55–65.

Brus, J., Kobera, L., Schoefberger, W., Urbanová, M., Klein, P., Sazama, P., Tabor, E., Sklenak, S., Fishchuk, A. V., & Dědeček, J. (2015). Structure of framework aluminum lewis sites and perturbed aluminum atoms in zeolites as determined by ^{27}Al{^{1}H} REDOR (3Q) MAS NMR spectroscopy and DFT/molecular mechanics. *Angewandte Chemie International Edition*, *54*(2), 541–545. DOI: 10.1002/anie.201409635 PMID: 25393612

Bukhari, S. S., Behin, J., Kazemian, H., & Rohani, S. (2015). Conversion of coal fly ash to zeolite utilizing microwave and ultrasound energies: A review. *Fuel*, *140*, 250–266. DOI: 10.1016/j.fuel.2014.09.077

Bukka, , & Sarin, P. (2024). Effects of particle size reduction on the pore structure and accessibility in natural porous materials. *Energy & Fuels*, *38*(11), 9876. DOI: 10.1021/acs.energyfuels.4c00194

Bulánek, R., Koudelková, E., Trachta, M., Bludský, O., & Rubeš, M. (2020). Re-examining the interpretation of CO adsorbed on Lewis acid sites of alkali metal-exchanged MOR zeolite. *Microporous and Mesoporous Materials*, *296*, 110007. Advance online publication. DOI: 10.1016/j.micromeso.2020.110007

Bulanin, K. M., Mikheleva, A. Yu., Shchepkin, D. N., & Rudakova, A. V. (2022). Determination of the Extinction Coefficient of Carbon Monoxide Adsorbed on Titanium Dioxide. *Optics and Spectroscopy*, *130*(4), 248–256. DOI: 10.1134/S0030400X2204004X

Bunaciu, A. A., Udriştioiu, E. G., & Aboul-Enein, H. Y. (2015). X-ray diffraction: Instrumentation and applications. *Critical Reviews in Analytical Chemistry*, *45*(4), 289–299. DOI: 10.1080/10408347.2014.949616 PMID: 25831472

Buntkowsky, G., & Vogel, M. (2020). Small molecules, non-covalent interactions, and confinement. *Molecules (Basel, Switzerland)*, *25*(14), 3311. DOI: 10.3390/molecules25143311 PMID: 32708283

Busca, G. (2017). Acidity and basicity of zeolites: A fundamental approach. *Microporous and Mesoporous Materials*, *254*, 3–16. DOI: 10.1016/j.micromeso.2017.04.007

Busca, G., & Gervasini, A. (2020). Solid acids, surface acidity and heterogeneous acid catalysis. *Advances in Catalysis*, *67*, 1–90. DOI: 10.1016/bs.acat.2020.09.003

Bushuev, Y. G., Grosu, Y., Chorążewski, M. A., & Meloni, S. (2022). Effect of the Topology on Wetting and Drying of Hydrophobic Porous Materials. *ACS Applied Materials & Interfaces*, *14*(26), 30067–30079. DOI: 10.1021/acsami.2c06039 PMID: 35730678

Çakıcıoğlu-Özkan, F., & Erten-Kaya, Y. (2017). Enhancement of Cerium exchange with Ultrasound. *Journal of the Turkish Chemical Society, Section A: Chemistry*, *5*(sp.is.1), 53–64. https://doi.org/DOI: 10.18596/jotcsa.370769

Calderón-Magdaleno, M. Á., Mendoza-Nieto, J. A., & Klimova, T. E. (2014). Effect of the amount of citric acid used in the preparation of NiMo/SBA-15 catalysts on their performance in HDS of dibenzothiophene-type compounds. *Catalysis Today*, *220-222*(Supplement C), 78–88. DOI: 10.1016/j.cattod.2013.06.002

Callister, W. D., & Rethwisch, D. G. (2020). *Materials Science and Engineering: An Introduction* (10th ed.). Wiley.

Camacho, , Parra, R. R., & Deng, S. (2011). Arsenic removal from groundwater by MnO2-modified natural clinoptilolite zeolite: Effects of pH and initial feed concentration. *Journal of Hazardous Materials*, *189*(1–2), 286–293. DOI: 10.1016/j.jhazmat.2011.02.035 PMID: 21398033

Cardoza, L. A., Korir, A. K., Otto, W. H., Wurrey, C. J., & Larive, C. K. (2004). Applications of NMR spectroscopy in environmental science. *Progress in Nuclear Magnetic Resonance Spectroscopy*, *45*(3–4), 209–238. DOI: 10.1016/j.pnmrs.2004.06.002

Carolin, C. F., Kumar, P. S., Saravanan, A., Joshiba, G. J., & Naushad, Mu. (2017). Efficient techniques for the removal of toxic heavy metals from aquatic environment: A review. In Journal of Environmental Chemical Engineering (Vol. 5, Issue 3, pp. 2782–2799). Elsevier BV. https://doi.org/DOI: 10.1016/j.jece.2017.05.029

Carpena-Núñez, J., Rao, R., Kim, D., Bets, K. V., Zakharov, D. N., Boscoboinik, J. A., Stach, E. A., Yakobson, B. I., Tsapatsis, M., Stacchiola, D., Stacchiola, D., & Maruyama, B. (2020). Zeolite Nanosheets Stabilize Catalyst Particles to Promote the Growth of Thermodynamically Unfavorable, Small-Diameter Carbon Nanotubes. *Small, 16*(38), 2002120. Advance online publication. DOI: 10.1002/smll.202002120 PMID: 32812375

Carstens, , Splith, C., & Enke, D. (2019). Sol-gel synthesis of α-Al2O3 with enhanced porosity via dicarboxylic acid templating. *Scientific Reports, 9*(1), 19982. Advance online publication. DOI: 10.1038/s41598-019-56294-1 PMID: 31882695

Casentini, , Gallo, M., & Baldi, F. (2019). Arsenate and arsenite removal from contaminated water by iron oxides nanoparticles formed inside a bacterial exopolysaccharide. *Journal of Environmental Chemical Engineering, 7*(1), 102908. Advance online publication. DOI: 10.1016/j.jece.2019.102908

Cataldo, , Salvi, L., Paoli, F., Fucile, M., Masciandaro, G., Manzi, D., Masini, C. M., & Mattii, G. B. (2021). Application of zeolites in agriculture and other potential uses: A review. *Agronomy (Basel), 11*(8), 1547. Advance online publication. DOI: 10.3390/agronomy11081547

Catizzone, E., Migliori, M., Mineva, T., van Daele, S., Valtchev, V., & Giordano, G. (2020). New synthesis routes and catalytic applications of ferrierite crystals. Part 1: 1,8-Diaminooctane as a new OSDA. *Microporous and Mesoporous Materials, 296*, 109987. Advance online publication. DOI: 10.1016/j.micromeso.2019.109987

Cavani, F., Trifirò, F., & Vaccari, A. (1991). Hydrotalcite-type anionic clays: Preparation, properties and applications. *Catalysis Today, 11*(2), 173–301. DOI: 10.1016/0920-5861(91)80068-K

Cescon, , & Jiang, J.-Q. (2020). Filtration process and alternative filter media material in water treatment. *Water (Basel), 12*(12), 3377. Advance online publication. DOI: 10.3390/w12123377

Chaida-Chenni, F. Z., Grande Casas, M., Belhadj, F., Hamacha, R., & Pérez-Pariente, J. (2021). No Title. *Post-Synthesis Treatment of High-Silica Nanozeolite Beta with Cationic Surfactants for Enhancing Mesoporosity.*, 2.

Chakarova, K., & Hadjiivanov, K. (2011). H-bonding of zeolite hydroxyls with weak bases: FTIR study of CO and N_2 adsorption on H-D-ZSM-5. *The Journal of Physical Chemistry. C, Nanomaterials and Interfaces, 115*(11), 4806–4817. DOI: 10.1021/jp111961g

Chakarova, K., & Hadjiivanov, K. (2013). FTIR study of N_2 and CO adsorption on H-D-FER. *Microporous and Mesoporous Materials*, *177*, 59–65. DOI: 10.1016/j.micromeso.2013.04.022

Chal, R., Cacciaguerra, T., van Donk, S., & Gérardin, C. (2010). Pseudomorphic synthesis of mesoporous zeolite Y crystals. *Chemical Communications*, *46*(41), 7840–7842. DOI: 10.1039/c0cc02073g PMID: 20714567

Chang, W.-S., Hong, S.-W., & Park, J. (2002). Effect of zeolite media for the treatment of textile wastewater in a biological aerated filter. In Process Biochemistry (Vol. 37, Issue 7, pp. 693–698). Elsevier BV. https://doi.org/DOI: 10.1016/S0032-9592(01)00258-8

Chen, D., Sharma, S. K., & Mudhoo, A. (Eds.). (2011a). *Handbook on Applications of Ultrasound*. CRC Press., DOI: 10.1201/b11012

Cheng, W. P., Gao, W., Cui, X., Ma, J. H., & Li, R. F. (2016). Phenol adsorption equilibrium and kinetics on zeolite X/activated carbon composite. In Journal of the Taiwan Institute of Chemical Engineers (Vol. 62, pp. 192–198). Elsevier BV. https://doi.org/DOI: 10.1016/j.jtice.2016.02.004

Chen, L.-H., Sun, M.-H., Wang, Z., Yang, W., Xie, Z., & Su, B.-L. (2020). Hierarchically Structured Zeolites: From Design to Application. *Chemical Reviews*, *120*(20), 11194–11294. DOI: 10.1021/acs.chemrev.0c00016 PMID: 32915551

Chen, S., Popovich, J., Zhang, W., Ganser, C., Haydel, S. E., & Seo, D.-K. (2018). Superior ion release properties and antibacterial efficacy of nanostructured zeolites ion-exchanged with zinc, copper, and iron. *RSC Advances*, *8*(66), 37949–37957. DOI: 10.1039/C8RA06556J PMID: 30555688

Chen, T., Xia, G., Wu, Q., Zheng, J., Jin, Y., Sun, D., Wang, S., & Chi, D. (2017). The Influence of Zeolite Amendment on Yield Performance, Quality Characteristics, and Nitrogen Use Efficiency of Paddy Rice. *Crop Science*, *57*(5), 1–11. DOI: 10.2135/cropsci2016.04.0228

Chen, W., Maugé, F., van Gestel, J., Nie, H., Li, D., & Long, X. (2013). Effect of modification of the alumina acidity on the properties of supported Mo and CoMo sulfide catalysts. *Journal of Catalysis*, *304*, 47–62. DOI: 10.1016/j.jcat.2013.03.004

Chen, , Wang, X., Li, Q., Chen, J., Zhu, L., Wang, L., & Zhang, L. (2025). Soil fluoride enrichment process and the possible adaptation prevention principle in coal-burning fluorosis. *Scientific Reports*, *15*(1), 84381. DOI: 10.1038/s41598-024-84381-5 PMID: 39762348

Chen, Y.-H., Han, D.-M., Zhang, Q., & Cui, H.-X. (2020). In-situ synthesis of hierarchical lamellar ZSM-5 zeolite with enhanced MTP catalytic performance by a facile seed-assisted method. *Journal of Porous Materials, 27*(5), 1265–1275. DOI: 10.1007/s10934-020-00898-w

Chinobampo, R. (2008). *Report and Validation of Results for Zeolitic products to INIFAP*, state of Morelos, Mexico

Chizhik, V. I., Chernyshev, Y. S., Donets, A. V., Frolov, V. V., Komolkin, A. V., & Shelyapina, M. G. (2014). *Magnetic resonance and its applications*. Springer International Publishing., DOI: 10.1007/978-3-319-05299-1

Chlebda, D., Stachurska, P., Jędrzejczyk, R., Kuterasiński, Ł., Dziedzicka, A., Górecka, S., Chmielarz, L., Łojewska, J., Sitarz, M., & Jodłowski, P. (2018). DeNOx Abatement over Sonically Prepared Iron-Substituted Y, USY and MFI Zeolite Catalysts in Lean Exhaust Gas Conditions. *Nanomaterials (Basel, Switzerland), 8*(1), 21. DOI: 10.3390/nano8010021

Chmielewská, E., Tylus, W., Drábik, M., Majzlan, J., Kravčak, J., Williams, C., Čaplovičová, M., & Čaplovič, Ľ. (2017). Structure investigation of nano-FeO(OH) modified clinoptilolite tuff for antimony removal. *Microporous and Mesoporous Materials, 248*, 222–228. DOI: 10.1016/j.micromeso.2017.04.022

Christensen, C. H., Johannsen, K., Schmidt, I., & Christensen, C. H. (2003). Catalytic benzene alkylation over mesoporous zeolite single crystals: Improving activity and aelectivity with a new family of porous materials. *Journal of the American Chemical Society, 125*(44), 13370–13371. DOI: 10.1021/ja037063c PMID: 14583028

Chu, N., Wang, J., Zhang, Y., Yang, J., Lu, J., & Yin, D. (2010). Nestlike hollow hierarchical MCM-22 microspheres: Synthesis and exceptional catalytic properties. *Chemistry of Materials, 22*(9), 2757–2763. DOI: 10.1021/cm903645p

Chwatal, , Parizek, H., & Waldhauser, W. (2024). Simulation-based analysis of the atmospheric pressure plasma and its temperature distribution using Ansys Mechanical. *BHM. Berg- und Hüttenmännische Monatshefte, 169*(8), 351–356. DOI: 10.1007/s00501-024-01477-1

Cinar, S., & Beler-Baykal, B. (2005). Ion exchange with natural zeolites: an alternative for water softening? In Water Science and Technology (Vol. 51, Issue 11, pp. 71–77). IWA Publishing. https://doi.org/DOI: 10.2166/wst.2005.0392

Cinibulk, J., Kooyman, P. J., Vit, Z., & Zdražil, M. (2003). Magnesia-Supported Mo, CoMo and NiMo Sulfide Catalysts Prepared by Nonaqueous Impregnation: Parallel HDS/HDN of Thiophene and Pyridine and TEM Microstructure. *Catalysis Letters*, *89*(1), 147–152. DOI: 10.1023/A:1024748232539

Čižmek, A., Subotić, B., Šmit, I., Tonejc, A., Aiello, R., Crea, F., & Nastro, A. (1997). Dissolution of high-silica zeolites in alkaline solutions II. Dissolution of 'activated' silicalite-1 and ZSM-5 with different aluminum content. *Microporous Materials*, *8*(3–4), 159–169. DOI: 10.1016/S0927-6513(96)00082-X

Collins, F., Rozhkovskaya, A., Outram, J. G., & Millar, G. J. (2020). A critical review of waste resources, synthesis, and applications for Zeolite LTA. *Microporous and Mesoporous Materials*, *291*, 109667. DOI: 10.1016/j.micromeso.2019.109667

Conversa, G., Pacifico, S., La Rotonda, P., Lazzizera, C., Bonasia, A., & Elia, A. (2024). Foliar application of natural zeolites affects the growth and productivity of processing tomato. *European Journal of Agronomy*, *154*, 127100. DOI: 10.1016/j.eja.2024.127100

Conway, J. H., & Torquato, S. (2006). Packing, tiling, and covering with tetrahedra. *Proceedings of the National Academy of Sciences of the United States of America*, *103*(28), 10612–10617. DOI: 10.1073/pnas.0601389103 PMID: 16818891

Costa, D., Arrouvel, C., Breysse, M., Toulhoat, H., & Raybaud, P. (2007). Edge wetting effects of γ-Al2O3 and anatase-TiO2 supports by MoS2 and CoMoS active phases: A DFT study. *Journal of Catalysis*, *246*(2), 325–343. DOI: 10.1016/j.jcat.2006.12.007

Crespo, R. (1998), Producción *cañera más eficiente y ecológica.* Revista Cañaveral, Vol. 4, No. 1, Jan-Mar, MINAZ, p. 27.

Cruz, T. J. T., Melo, M. I. S., & Pergher, S. (2020). Optimization of Parameters and Methodology for the Synthesis of LTA-Type Zeolite Using Light Coal Ash. *Applied Sciences (Basel, Switzerland)*, *10*(20), 7332. DOI: 10.3390/app10207332

Cullity, B. D., & Stock, S. R. (2001). *Elements of X-ray Diffraction* (3rd ed.). Prentice Hall.

Cundy, C. S., & Cox, P. A. (2003). The hydrothermal synthesis of zeolites: History and development from the earliest days to the present time. *Chemical Reviews*, *103*(3), 663–701. DOI: 10.1021/cr020060i PMID: 12630849

Czekaj, I., & Sobuś, N. (2024). Odors Adsorption in Zeolites Including Natural Clinoptilolite: Theoretical and Experimental Studies. *Materials (Basel)*, *17*(13), 3088. Advance online publication. DOI: 10.3390/ma17133088 PMID: 38998171

Czuma, N., Zarebska, K., & Baran, P. (2017). Microwave and ultrasound energy in the process of fly ash zeolite synthesis and its application for CO<inf>2</inf> sorption. *2017 IEEE 7th International Conference Nanomaterials: Application & Properties (NAP)*, 03NNSA20-1-03NNSA20-5. https://doi.org/DOI: 10.1109/NAP.2017.8190267

D'Agostino, C., Bräuer, P., Zheng, J., Robinson, N., York, A. P. E., Song, L., & Fan, X. (2023). Adsorbate/adsorbent interactions in microporous zeolites: Mechanistic insights from NMR relaxation and DFT calculations. *Materials Today. Chemistry*, *29*, 101443. DOI: 10.1016/j.mtchem.2023.101443

D'Agostino, C., York, A. P. E., & Bräuer, P. (2022). Host-guest interactions and confinement effects in HZSM-5 and chabazite zeolites studied by low-field NMR spin relaxation. *Materials Today. Chemistry*, *24*, 100901. DOI: 10.1016/j.mtchem.2022.100901

da Silva, A., Elias, E. B. C. D., Cruz, T. J. T., Pinto, F. G. H. S., de Mello, M. I. S., Bieseki, L., & Pergher, S. B. C. (2024). Synthesis and Cation Exchange of LTA Zeolites Synthesized from Different Silicon Sources Applied in CO2 Adsorption. *Coatings*, *14*(6), 680. DOI: 10.3390/coatings14060680

Daage, M., & Chianelli, R. R. (1994). Structure-Function Relations in Molybdenum Sulfide Catalysts: The "Rim-Edge" Model. *Journal of Catalysis*, *149*(2), 414–427. DOI: 10.1006/jcat.1994.1308

Database of Zeolite Structures. (Accessed Feb. 4, 2025) (s. f.). http://www.iza-structure.org/databases

Dawson, D. M., Moran, R. F., & Ashbrook, S. E. (2017). An NMR crystallographic investigation of the relationships between the crystal structure and ^{29}Si isotropic chemical shift in silica zeolites. *The Journal of Physical Chemistry. C, Nanomaterials and Interfaces*, *121*(28), 15198–15210. DOI: 10.1021/acs.jpcc.7b03730 PMID: 28751927

De Antoni Migliorati, M., Leo, S., Millar, G. J., Bell, M. J., Strazzabosco, A., & Grace, P. R. (2024). Can zeolites be used in sugarcane cropping systems to help reduce nitrogen losses into the Australian Great Barrier Reef? *Plant and Soil*, *504*(1-2), 575–592. DOI: 10.1007/s11104-024-06643-5

de Jong, K. P. (2009). General Aspects *Synthesis of Solid Catalysts* (pp. 1-11).

De la Nuez, E. Y., Rodríguez Iznaga, I., Rodríguez Fuentes, G., Petranovskii, V., Martínez García, A., Calvino Gámez, J. J., Goma Jiménez, D., Cauqui, M. A., Rivero González, L. A., & Collazo García, O. (2024). Natural Clinoptilolite Materials Enriched in Nitrogen and Phosphorous for Agricultural Purposes: A Comprehensive Study on Their Development. *Journal of Inorganic and Organometallic Polymers and Materials*, *34*(6), 2522–2542. Advance online publication. DOI: 10.1007/s10904-023-02991-4

De la Rosa-Gómez, I., Olguín, M. T., & Alcántara, D. (2008). Antibacterial behavior of silver-modified clinoptilolite–heulandite rich tuff on coliform microorganisms from wastewater in a column system. *Journal of Environmental Management*, *88*(4), 853–863.

De León Ramirez, J. I., Reyes Villegas, V. A., Cadena-Nava, R. D., Loredo-Garcia, E., Chávez-Rivas, F., González-Torres, V., & Petranovskii, V. (2024). Antimicrobial activity of the LTA zeolite modified by zinc species. *Microporous and Mesoporous Materials*, *380*, 113295. DOI: 10.1016/j.micromeso.2024.113295

de Pablo, L., Doval, M., La Iglesia, A., & Soriano, J. (2014). CaK-clinoptilolite, KNa-chabazite, KNa-heulandite, KNa-erionite and Na-phillipsite from tuffaceous rocks, Province of the Mesa Central, Mexico. *Revista Mexicana de Ciencias Geológicas*, *31*(1), 116–126.

Dehghani, M., & Tadjarodi, A. (2020). Formation and characterization of zeolite Y-platinum nanoparticles by rapid method of ultrasonic irradiation and investigation of its electrochemical property. *Journal of Nanostructures*, *10*(3), 486–496. DOI: 10.22052/JNS.2020.03.005

Dehghani, M., Tadjarodi, A., & Chamani, S. (2019). Synthesis and Characterization of Magnetic Zeolite Y-Palladium-Nickel Ferrite by Ultrasonic Irradiation and Investigating Its Catalytic Activity in Suzuki-Miyaura Cross-Coupling Reactions. *ACS Omega*, *4*(6), 10640–10648. DOI: 10.1021/acsomega.9b00666

Demikhova, N. R., Rubtsova, M. I., Kireev, G. A., Cherednichenko, K. A., Vinokurov, V. A., & Glotov, A. P. (2023). Micro-mesoporous catalysts based on ZSM-5 zeolite synthesized from natural clay nanotubes: Preparation and application in the isomerization of C-8 aromatic fraction. *Chemical Engineering Journal*, *453*, 139581. DOI: 10.1016/j.cej.2022.139581

Demirci, S., Ustaoğlu, Z., Yılmazer, G. A., Sahin, F., & Baç, N. (2014). Antimicrobial Properties of Zeolite-X and Zeolite-A Ion-Exchanged with Silver, Copper, and Zinc Against a Broad Range of Microorganisms. *Applied Biochemistry and Biotechnology*, *172*(3), 1652–1662. DOI: 10.1007/s12010-013-0647-7 PMID: 24242073

Deneyer, , Ke, Q., Devos, J., & Dusselier, M. (2020). Zeolite synthesis under nonconventional conditions: Reagents, reactors, and modi operandi. *Chemistry of Materials*, *32*(12), 4884–4919. Advance online publication. DOI: 10.1021/acs.chemmater.9b04741

Denisenko, L. A., Tsyganenko, A. A., & Filimonov, V. N. (1984). Infrared study of the interaction between adsorbed molecules in the Co/ZnO system. *Reaction Kinetics and Catalysis Letters*, *25*(1–2), 23–26. DOI: 10.1007/BF02076533

Derakhshankhah, H., Jafari, S., Sarvari, S., Barzegari, E., Moakedi, F., Ghorbani, M., Shiri Varnamkhasti, B., Jaymand, M., Izadi, Z., & Tayebi, L. (2020). Biomedical Applications of Zeolitic Nanoparticles, with an Emphasis on Medical Interventions. *International Journal of Nanomedicine*, *15*, 363–386. DOI: 10.2147/IJN.S234573 PMID: 32021185

Derbe, T., Temesgen, S., & Bitew, M. (2021). A short review on synthesis, characterization, and applications of zeolites. *Advances in Materials Science and Engineering*, *6637898*(1), 6637898. Advance online publication. DOI: 10.1155/2021/6637898

Desales-Guzmán, L. A., Pacheco-Sánchez, J. H., & Vazquez-Rivas, E. (2025). Enhanced hydrogen adsorption properties of Zeolite Templated Carbon via chemical activation: DFT study. *Fuel*, *379*, 133080. DOI: 10.1016/j.fuel.2024.133080

Devautour, S., Henn, F., Giuntini, J., Zanchetta, J., & Vanderschueren, J. (1999). TSDC relaxation map analysis in a Na–mordenite zeolite. *Solid State Ionics*, *122*(1–4), 105–111. DOI: 10.1016/S0167-2738(99)00048-X

Díaz Coronel, G. T., Sánchez Mora, F. D., Llerena Ramos, L. T., & Vásconez, G. H. (2009). Use of natural zeolites in the fertilization and production of beans (Phaseolus vulgaris L.) in the Quevedo area. *Science and Technology*, *3*, 1–6.

Dib, E., Costa, I. M., Vayssilov, G. N., Aleksandrov, H. A., & Mintova, S. (2021). Complex H-bonded silanol network in zeolites revealed by IR and NMR spectroscopy combined with DFT calculations. *Journal of Materials Chemistry. A, Materials for Energy and Sustainability*, *9*(48), 27347–27352. DOI: 10.1039/D1TA06908J

Dib, E., Mintova, S., Vayssilov, G. N., Aleksandrov, H. A., & Carravetta, M. (2023). Chemical Shift Anisotropy: A Promising Parameter To Distinguish the 29Si NMR Peaks in Zeolites. *The Journal of Physical Chemistry. C, Nanomaterials and Interfaces*, *127*(22), 10792–10796. DOI: 10.1021/acs.jpcc.3c01533

Dickson, M. S., Rosen, P. F., Neilsen, G., Navrotsky, A., & Woodfield, B. F. (2021). Heat capacity and thermodynamic functions of transition metal ion (Cu2+, Fe2+, Mn2+) exchanged, partially dehydrated zeolite A (LTA). *The Journal of Chemical Thermodynamics*, *161*, 106556. Advance online publication. DOI: 10.1016/j.jct.2021.106556

Dight, L., García-Martínez, J., Valla, J., & Johnson, M. (2010). *No Title* (Patent No. US 2010/0190632.).

Dikić, , Hrenović, J., Durn, G., Kovačić, A., & Rajić, N. (2019). Antibacterial activity of metal-containing clinoptilolite in natural seawater. *Desalination and Water Treatment*, *169*, 135–142. DOI: 10.5004/dwt.2019.24712

Dobrotvorskaia, A. N., Kolomiitsova, T. D., Petrov, S. N., Shchepkin, D. N., Smirnov, K. S., & Tsyganenko, A. A. (2015). Effect of resonance dipole-dipole interaction on spectra of adsorbed SF_6 molecules. *Spectrochimica Acta. Part A: Molecular and Biomolecular Spectroscopy*, *148*, 271–279. DOI: 10.1016/j.saa.2015.04.002 PMID: 25897721

Dobrotvorskaia, A. N., Pestsov, O. S., & Tsyganenko, A. A. (2017). Lateral Interaction between Molecules Adsorbed on the Surfaces of Non-Metals. *Topics in Catalysis*, *60*(19–20), 1506–1521. Advance online publication. DOI: 10.1007/s11244-017-0835-8

Dong, H., Zhao, L., Zhang, L., Chen, H., Gao, C., & Winston Ho, W. S. (2015). High-flux reverse osmosis membranes incorporated with NaY zeolite nanoparticles for brackish water desalination. In *Journal of Membrane Science* (Vol. 476, pp. 373–383). Elsevier BV., DOI: 10.1016/j.memsci.2014.11.054

Dong, S., Liu, L., Ling, N. N. A., May, E. F., Johns, M. L., & Robinson, N. (2024). Toward quantifying the chemical sensitivity of nuclear spin surface relaxivity in mesoporous media. *Langmuir*, *40*(31), 16160–16171. DOI: 10.1021/acs.langmuir.4c01245 PMID: 39045723

Dryaz, , Shaban, M., AlMohamadi, H., Al-Ola, K. A. A., Hamd, A., Soliman, N. K., & Ahmed, S. A. (2021). Design, characterization, and adsorption properties of Padina gymnospora/zeolite nanocomposite for Congo red dye removal from wastewater. *Scientific Reports*, *11*(1), 21062. Advance online publication. DOI: 10.1038/s41598-021-00025-y PMID: 34702834

Duan, A., Wan, G., Zhao, Z., Xu, C., Zheng, Y., Zhang, Y., Dou, T., Bao, X., & Chung, K. (2007). Characterization and activity of Mo supported catalysts for diesel deep hydrodesulphurization. *Catalysis Today*, *119*(1), 13–18. DOI: 10.1016/j.cattod.2006.08.049

Duer, M. J. (Ed.). (2002). *Solid-state NMR spectroscopy. Principles and applications*. Blackwell Science.

Dutta, P., & Wang, B. (2019). Zeolite-supported silver as antimicrobial agents. *Coordination Chemistry Reviews, 383*, 1–29. DOI: 10.1016/j.ccr.2018.12.014

Dziejarski, B., Serafin, J., Andersson, K., & Krzyżyńska, R. (2023). CO<inf>2</inf> capture materials: A review of current trends and future challenges. *Materials Today Sustainability, 24*, 100483. Advance online publication. DOI: 10.1016/j.mtsust.2023.100483

Dzwigaj, S., de Mallmann, A., & Barthomeuf, D. (1990). Adsorption of benzene and ethylbenzene on the acidic and basic sites of beta zeolite. *Journal of the Chemical Society, Faraday Transactions, 86*(2), 431–435. DOI: 10.1039/ft9908600431

Egorova, M., & Prins, R. (2004). Hydrodesulfurization of dibenzothiophene and 4,6-dimethyldibenzothiophene over sulfided NiMo/γ-Al2O3, CoMo/γ-Al2O3, and Mo/γ-Al2O3 catalysts. *Journal of Catalysis, 225*(2), 417–427. DOI: 10.1016/j.jcat.2004.05.002

Eischens, R. P., Francis, S. A., & Pliskin, W. A. (1956). The effect of surface coverage on the spectra of chemisorbed CO. *Journal of Physical Chemistry, 60*(2), 194–201. DOI: 10.1021/j150536a015

El Hanache, L., Lebeau, B., Nouali, H., Toufaily, J., Hamieh, T., & Daou, T. J. (2019). Performance of surfactant-modified *BEA-type zeolite nanosponges for the removal of nitrate in contaminated water: Effect of the external surface. In *Journal of Hazardous Materials* (Vol. 364, pp. 206–217). Elsevier BV., DOI: 10.1016/j.jhazmat.2018.10.015

El-Azim, , & Mourad, F. (2018). Removal of heavy metals Cd (II), Fe (III) and Ni (II), from aqueous solutions by natural (clinoptilolite) zeolites and application to industrial wastewater. *Asian Journal of Environment & Ecology, 7*(1), 1–13. DOI: 10.9734/AJEE/2018/41004

Elgersma, S. V., Sederman, A. J., Mantle, M. D., & Gladden, L. F. (2022). Measuring the liquid-solid mass transfer coefficient in packed beds using T_2-T_2 relaxation exchange NMR. *Chemical Engineering Science, 248*, 117229. DOI: 10.1016/j.ces.2021.117229

Elgersma, S. V., Zheng, Q., Avrantinis, N., Sederman, A. J., Mantle, M. D., & Gladden, L. F. (2023). Quantifying liquid-solid mass transfer in a trickle bed using T_2-T_2 relaxation exchange NMR. *Applied Magnetic Resonance, 54*(11–12), 1423–1443. DOI: 10.1007/s00723-023-01607-x

Elsherbeni, , Youssef, I. M., Hamouda, R. E., Kamal, M., El-Gendi, G. M., El-Garhi, O. H., Alfassam, H. E., Rudayni, H. A., Allam, A. A., Moustafa, M., Alshaharn, M. O., & El Kholy, M. S. (2024). Performance and economic efficiency of laying hens in response to adding zeolite to feed and litter. *Poultry Science*, *103*(6), 103799. DOI: 10.1016/j.psj.2024.103799 PMID: 38759566

Eom, E., Song, M., Kim, J. C., Kwon, D., Rainer, D. N., Gołąbek, K., Nam, S. C., Ryoo, R., Mazur, M., & Jo, C. (2022). Confining Gold Nanoparticles in Preformed Zeolites by Post-Synthetic Modification Enhances Stability and Catalytic Reactivity and Selectivity. *JACS Au*, *2*(10), 2327–2338. DOI: 10.1021/jacsau.2c00380 PMID: 36311841

Erdoğan, N. N., & Başyiğit, A. B. (2023). An approach on determining micro-strain and crystallite size values of thermal spray barrier coated Inconel 601 super alloy. [TIJMET]. *The International Journal of Materials and Engineering Technology*, *6*(1), 21–25.

Erten-Kaya, Y., & Cakicioglu-Ozkan, F. (2012). Effect of ultrasound on the kinetics of cation exchange in NaX zeolite. *Ultrasonics Sonochemistry*, *19*(3), 701–706. DOI: 10.1016/j.ultsonch.2011.10.010

Espino, C., & Soca, M. (1991) Influencia de la Zeolita natural y modificada en tomate cultivado en invernadero. Memoirs 3rd Int. Conf. Occur. Prop. Util. Nat. Zeolites, Part II, 65-67.

Estrada Flores, S. L., Garcias-Morales, C., Mirón Velázquez, R. S., Pérez Berumen, C. M., & De la Cruz Duran, F. N. (2022). *An overview of organic electronics*. Revistas UNAM., DOI: 10.22201/fq.18708404e.2022.2.79879

Estrada-Cabrera, E., Torres-Ferrer, L. R., Luna-Barcenas, G., & Ramirez-Bon, R. (2021). Cellulose dialysis membrane containing raw clinoptilolite enhances the removal of Rhodamine 6G from aqueous solutions. *Microporous and Mesoporous Materials*, *321*, 111113. Advance online publication. DOI: 10.1016/j.micromeso.2021.111113

Evans, , & Coudert, F.-X. (2017). Predicting the mechanical properties of zeolite frameworks by machine learning. *Chemistry of Materials*, *29*(18), 7833–7839. Advance online publication. DOI: 10.1021/acs.chemmater.7b02532

Ezdakova, K. V., Rudakova, A. V., Krauklis, I. V., Shergin, Y. V., & Tsyganenko, A. A. (2023). The Influence of the Local Environment on the Electron Accepting Ability of the Cation: Induced Lewis Acidity. *Optics and Spectroscopy*, *131*(10), 964–974.

Ezugbe, . (2020). Membrane technologies in wastewater treatment: A review. *Membranes, 10*(5), 89. Advance online publication. DOI: 10.3390/membranes10050089 PMID: 32365810

Faisal, F., Rasul, M. G., Jahirul, M. I., & Chowdhury, A. A. (2023). Waste plastics pyrolytic oil is a source of diesel fuel: A recent review on diesel engine performance, emissions, and combustion characteristics. *The Science of the Total Environment, 886*, 163756. Advance online publication. DOI: 10.1016/j.scitotenv.2023.163756 PMID: 37146806

Farias da Silva, J. L., de Mello Prado, R., Alves, T. L., Lata-Tenesaca, L. F., & Bomfim Soares, M. (2023). *New strategy for silicon supply through fertigation in sugarcane integrating the pre-*sprouted *seedling phase and field cultivation.Scientific Reports, 13*(1), 1230. DOI: 10.1038/s41598-022-27323-3 PMID: 36681705

Farooq, , Muneer, M., Shahid, A., Rehman, M. A., Ullah, K., Murtaza, G., Iqbal, R., Iqbal, J., & Rahimi, M. (2024). Synthesis and characterization of fluorenone derivatives with electrical properties explored using density functional theory. *Scientific Reports, 14*(1), 80477. Advance online publication. DOI: 10.1038/s41598-024-80477-0 PMID: 39578658

Favvas, , Tsanaktsidis, C. G., Sapalidis, A. A., Tzilantonis, G. T., Papageorgiou, S. K., & Mitropoulos, A. C. (2016). Clinoptilolite, a natural zeolite material: Structural characterization and performance evaluation on its dehydration properties of hydrocarbon-based fuels. *Microporous and Mesoporous Materials, 225*, 385–391. Advance online publication. DOI: 10.1016/j.micromeso.2016.01.021

Febles (2006) Zeoponic, the antecedent of Organoponic. Report to Ministry of Agriculture of Cuba

Febles, J. A., Borsatto, F., Ferreira, J. P. & Ferreira, F. (2014) Celtonita: an environmentally friendly product from natural clinoptilolite for high technology agriculture. Zeolite 2014 Book of Abstract - 9th Int. Conf. Natural Zeolites, Belgrade, Serbia, 8 - 13 June 2014, 67-68

Fermoso, J., Hernando, H., Jana, P., Moreno, I., Přech, J., Ochoa-Hernández, C., Pizarro, P., Coronado, J. M., Čejka, J., & Serrano, D. P. (2016). Lamellar and pillared ZSM-5 zeolites modified with MgO and ZnO for catalytic fast-pyrolysis of eucalyptus woodchips. *Catalysis Today, 277*, 171–181. DOI: 10.1016/j.cattod.2015.12.009

Fichtner-Schmittler, H., Lohse, U., Miessner, H., & Maneck, H.-E. (1990). *Correlation between Unit-Cell Parameter, Skeletal Stretching Vibrations and Molar Fraction of Aluminium of Faujasite Type Zeolites for Si:Al = 1.1 - 1,000. 271O*(1), 69–80. https://doi.org/doi:10.1515/zpch-1990-27110

Field, , She, Q., Siddiqui, F. A., & Fane, A. G. (2021). Reverse osmosis and forward osmosis fouling: A comparison. *Discover Chemical Engineering*, *1*(1), 6. Advance online publication. DOI: 10.1007/s43938-021-00006-7

Filgueiras Dutra, A., Lima Leite, M. R., de Faria Melo, C. C., Silva Amaral, D., Farias da Silva, J. L., de Mello Prado, R., de Cássia Piccolo, M., de Souza Miranda, R., Barbosa da Silva Júnior, G., dos Santos Apollo Sousa, T. K., Mendes, L. W., Ferreira Araújo, A. S., Zuffo, A. M., & de Alcântara Neto, F. (2023). Soil and foliar Si fertilization alters elemental stoichiometry and increases yield of sugarcane cultivars. *Scientific Reports*, *13*(1), 16040. . DOI: 10.1038/s41598-023-43351-z PMID: 37749306

Fischer, . (2023). Density functional theory study of hydrophobic zeolites for the removal of triclosan from aqueous solution. *Environmental Science. Advances*, *2*(5), 678–689. DOI: 10.1039/D3VA00078H

Fleury, M., Pirngruber, G., & Jolimaitre, E. (2023). Probing diffusional exchange in mesoporous zeolite by NMR diffusion and relaxation methods. *Microporous and Mesoporous Materials*, *355*, 112575. DOI: 10.1016/j.micromeso.2023.112575

Fleury, M., & Soualem, J. (2009). Quantitative analysis of diffusional pore coupling from T_2-store-T_2 NMR experiments. *Journal of Colloid and Interface Science*, *336*(1), 250–259. DOI: 10.1016/j.jcis.2009.03.051 PMID: 19439315

Flores, et al. (2002). Ammonia removal in clarified wastewater by the ion exchange process with zeolites.

Foroutan, , Peighambardoust, S. J., Hemmati, S., Khatooni, H., & Ramavandi, B. (2021). Preparation of clinoptilolite magnetic nanocomposite powder and its elimination properties for cationic dyes from water and wastewater. *International Journal of Biological Macromolecules*, *189*, 432–442. Advance online publication. DOI: 10.1016/j.ijbiomac.2021.08.144 PMID: 34450143

Foroutan, R., Peighambardoust, S. J., Hemmati, S., Khatooni, H., & Ramavandi, B. (2021). Preparation of clinoptilolite/starch/CoFe2O4 magnetic nanocomposite powder and its elimination properties for cationic dyes from water and wastewater. *International Journal of Biological Macromolecules*, *189*, 432–442.

Forster, L., Kashbor, M. M. M., Railton, J., Chansai, S., Hardacre, C., Conte, M., & D'Agostino, C. (2023). Low-field 2D NMR relaxation and DRIFTS studies of glucose isomerization in zeolite Y: New insights into adsorption effects on catalytic performance. *Journal of Catalysis*, *425*, 269–285. DOI: 10.1016/j.jcat.2023.06.021

Fouad, M. M., Shihata, L. A., & Morgan, E. I. (2017). An integrated review of factors influencing the performance of photovoltaic panels. *Renewable and Sustainable Energy 23Reviews, 80*, 1499-1511. https://doi.org/DOI: 10.1016/j.rser.2017.05.141

Fox, S., Wilkinson, T. S., Wheatley, P. S., Xiao, B., Morris, R. E., Sutherland, A., Simpson, A. J., Barlow, P. G., Butler, A. R., Megson, I. L., & Rossi, A. G. (2010). NO-loaded Zn2+-exchanged zeolite materials: A potential bifunctional anti-bacterial strategy. *Acta Biomaterialia*, *6*(4), 1515–1521. DOI: 10.1016/j.actbio.2009.10.038 PMID: 19861185

Franko, I., Abrishamkar, M., Azizi, S. N., & Kazemian, H. (2010). Synthesis of borosilicate MFI type zeolite using different aging techniques. In *Chem. Met. Alloys* (Vol. 3, Issue 1/2, pp. 12–17). https://doi.org/DOI: 10.30970/cma3.0111

Fu, T., Chang, J., Shao, J., & Li, Z. (2017). Fabrication of a nano-sized ZSM-5 zeolite with intercrystalline mesopores for conversion of methanol to gasoline. *Journal of Energy Chemistry*, *26*(1), 139–146. DOI: 10.1016/j.jechem.2016.09.011

Fu, X., Sheng, X., Zhou, Y., Fu, Z., Zhao, S., Zhang, Z., & Zhang, Y. (2016). Ultrasonic/microwave synergistic synthesis of well-dispersed hierarchical zeolite Y with improved alkylation catalytic activity. *Korean Journal of Chemical Engineering*, *33*(6), 1931–1937. DOI: 10.1007/s11814-016-0022-9

Fyfe, C. A., Feng, Y., Grondey, H., Kokotailo, G. T., & Gies, H. (1991). One- and two-dimensional high-resolution solid-state NMR studies of zeolite lattice structures. *Chemical Reviews*, *91*(7), 1525–1543. DOI: 10.1021/cr00007a013

Gadore, V., Mishra, S. R., Yadav, N., Yadav, G., & Ahmaruzzaman, Md. (2024). Advances in zeolite-based materials for dye removal: Current trends and future prospects. In *Inorganic Chemistry Communications* (Vol. 166, p. 112606). Elsevier BV., DOI: 10.1016/j.inoche.2024.112606

Gagliano, E., Sgroi, M., Falciglia, P. P., Belviso, C., Cavalcante, F., Lettino, A., Vagliasindi, F. G. A., & Roccaro, P. (2022). Removal of ammonium from wastewater by zeolite synthetized from volcanic ash: Batch and column tests. In Journal of Environmental Chemical Engineering (Vol. 10, Issue 3, p. 107539). Elsevier BV. https://doi.org/DOI: 10.1016/j.jece.2022.107539

Gallet, J., Perez, M., Guillou, R., Ernould, C., Le Bourlot, C., Langlois, C., Beausir, B., Bouzy, E., Chaise, T., & Cazottes, S. (2023). Experimental measurement of dislocation density in metallic materials: A quantitative comparison between measurement techniques (XRD, R-ECCI, HR-EBSD, TEM). *Materials Characterization*, *199*, 112842. DOI: 10.1016/j.matchar.2023.112842

Galletti, . (2020). Zn2 and Cd2 removal from wastewater using clinoptilolite as adsorbent. *Environmental Sciences (Ruse)*, *27*(31), 38791–38802. DOI: 10.1007/s11356-020-08483-z PMID: 32212074

Gao, X., & Zhang, T. (2018). An overview: Facet-dependent metal oxide semiconductor gas sensors. *Sensors and Actuators. B, Chemical*, *277*, 604–633. DOI: 10.1016/j.snb.2018.08.129

Gao, Y., & Blümich, B. (2020). Analysis of three-site T_2-T_2 exchange NMR. *Journal of Magnetic Resonance (San Diego, Calif.)*, *315*, 106740. DOI: 10.1016/j.jmr.2020.106740 PMID: 32438312

García-Martínez , J., & Jonhson, M. (2010). (Patent No. US 2010/0021321).

García-Martínez, J., & Johnson, M. (2008). (Patent No. US 2008/0138274.).

García-Martínez, J., Johnson, M., Valla, J., Li, K., & Ying, J. Y. (2012). Mesostructured zeolite Y—High hydrothermal stability and superior FCC catalytic performance. *Catalysis Science & Technology*, *2*(5), 987–994. DOI: 10.1039/c2cy00309k

Garrone, E., Bonelli, B., Tsyganenko, A. A., Rodríguez Delgado, M., Turnes Palomino, G., Manoilova, O. V., & Otero Areán, C. (2003). Spectroscopic and thermodynamic characterization of strontium carbonyls formed upon carbon monoxide adsorption on the zeolite Sr-Y. *The Journal of Physical Chemistry B*, *107*(11), 2537–2542. DOI: 10.1021/jp0217841

Gattucci, . (2025). Tannic acid-assisted green functionalization of clinoptilolite A step-by- characterization. *Ceramics International*. Advance online publication. DOI: 10.1016/j.ceramint.2025.01.151

Gedeon, A., & Fernandez, C. (2007). Solid-state NMR spectroscopy in zeolite science. In Čejka, J., van Bekkum, H., Corma, A., & Schüth, F. (Eds.), *Studies in Surface Science and Catalysis* (Vol. 168, pp. 403–XII). Elsevier., DOI: 10.1016/S0167-2991(07)80800-7

Ghojavand, S., Coasne, B., Clatworthy, E. B., Guillet-Nicolas, R., Bazin, P., Desmurs, M., Jacobo Aguilera, L., Ruaux, V., & Mintova, S. (2022). Alkali Metal Cations Influence the CO2 Adsorption Capacity of Nanosized Chabazite: Modeling vs Experiment. In ACS Applied Nano Materials (Vol. 5, Issue 4, pp. 5578–5588). American Chemical Society (ACS). https://doi.org/DOI: 10.1021/acsanm.2c00537

Ghorbani, M. M., & Taherian, R. (2019). Methods of measuring electrical properties of material. En *Electrical Conductivity in Polymer-Based Composites: Experiments, Modelling, and Applications* (pp. 365-394). Plastics Design Library. https://doi.org/DOI: 10.1016/B978-0-12-812541-0.00012-4

Gicheva, G., Mintcheva, N., & Panayotova, M. (2021). *Nanosized ZnO and ZnO@ zeolite composite prepared via sonication method.*

Giraldo-Loaiza, C., Salazar-Loaiza, A. M., Sandoval-Barrera, M. A., Macías-Quiroga, I. F., Ocampo-Serna, D. M., & Sanabria-González, N. R. (2024). Integration of Ion Exchange—AOP—Biological System for the Treatment of Real Textile Wastewater. In ChemEngineering (Vol. 8, Issue 4, p. 76). MDPI AG. https://doi.org/DOI: 10.3390/chemengineering8040076

Gizatullin, B. I., Savinkov, A. V., Shipunov, T. V., Melnikova, D. L., Doroginitzky, M. M., & Skirda, V. D. (2018). Investigation of molecular mobility and exchange of n-hexane and water in silicalite-1 by 2D ^{1}H NMR relaxometry. *MRSej, 20*(1), 18102.

Glaser, R. H., Wilkes, G. L., & Bronnimann, C. E. (1989). Solid-state ^{29}Si NMR of TEOS-based multifunctional sol-gel materials. *Journal of Non-Crystalline Solids, 113*(1), 73–87. DOI: 10.1016/0022-3093(89)90320-7

Głowniak, S., Szczęśniak, B., Choma, J., & Jaroniec, M. (2023). Recent Developments in Sonochemical Synthesis of Nanoporous Materials. *Molecules (Basel, Switzerland), 28*(6), 2639. DOI: 10.3390/molecules28062639 PMID: 36985612

Godefroy, S., Korb, J. P., Fleury, M., & Bryant, R. G. (2001). Surface nuclear magnetic relaxation and dynamics of water and oil in macroporous media. *Physical Review E: Statistical Physics, Plasmas, Fluids, and Related Interdisciplinary Topics, 64*(2), 021605. DOI: 10.1103/PhysRevE.64.021605 PMID: 11497601

Gogoi, , Mazumder, P., Tyagi, V. K., Tushara Chaminda, G. G., An, A. K., & Kumar, M. (2018). Occurrence and fate of emerging contaminants in water environment: A review. *Groundwater for Sustainable Development, 6*, 169–180. DOI: 10.1016/j.gsd.2017.12.009

Gómez, J. M., Montes, I., Díez, E., & Rodríguez, A. (2022). Mesoporous low silica X (MLSX) zeolite: Mesoporosity in loewenstein limit? *Microporous and Mesoporous Materials, 330*, 111618. DOI: 10.1016/j.micromeso.2021.111618

Goñi Echave, I., & García Arnáez, I. (2023). *Incorporación de simvastatina a recubrimientos sol-gel para implantes dentales* [Tesis de licenciatura, Universidad del País Vasco]. https://addi.ehu.es/bitstream/handle/10810/62777/TFG_IreneToledo.pdf?sequence=2&isAllowed=y

González-Cortés, S. L., Xiao, T.-C., Lin, T.-W., & Green, M. L. H. (2006). Influence of double promotion on HDS catalysts prepared by urea-matrix combustion synthesis. *Applied Catalysis A, General, 302*(2), 264–273. DOI: 10.1016/j.apcata.2006.01.019

Góra-Marek, K., Glanowska, A., & Datka, J. (2012). Quantitative IR Studies of the Concentration of Different Nickel Sites in NiZSM-5 Zeolites. *Microporous and Mesoporous Materials*, *158*, 162–169. DOI: 10.1016/j.micromeso.2012.03.027

Gordina, N. E., Prokof'ev, V. Y., Kul'pina, Y. N., Hmylova, O. E., Gazahova, S. I., & Petuhova, N. V. (2017). Synthesis of granulated binder-free LTA zeolite from metakaolin using ultrasonic treatment. *Journal of Porous Materials*, *24*(3), 667–678. DOI: 10.1007/s10934-016-0303-z

Gordina, N. E., Prokof'ev, V. Yu., & Kochetkov, S. P. (2018). Use of Mechanochemical Activation and Ultrasonic Treatment for the Synthesis of LTA Zeolite. *Russian Journal of General Chemistry*, *88*(9), 1981–1989. DOI: 10.1134/S1070363218090402

Gordina, N. E., Prokof'ev, V. Yu., Kul'pina, Yu. N., Petukhova, N. V., Gazakhova, S. I., & Khmylova, O. E. (2017). Use of Ultrasonic Processing at Early Stages of LTA Zeolite Synthesis from Metakaolin. *Glass and Ceramics*, *73*(9–10), 334–337. DOI: 10.1007/s10717-017-9884-5

Gordina, N. E., Rumyantsev, R. N., Borisova, T. N., Kolobkova, A. E., Tsvetova, E. V., Afanas'eva, E. E., Severgina, E. S., & Prokof'ev, V. Y. (2021). Use of Combinations of Ultrasonic Treatment and Microwave Crystallization to Intensify the Synthesis of LTA Zeolite Membranes. *Petroleum Chemistry*, *61*(3), 292–298. DOI: 10.1134/S096554412103021X

Gowram, , Beulah, M., Ibrahim Mohamed, M. M., Ibrahim, S. A. M., Puneeth, V., Ben Moussa, S., & Nasr, S. (2024). Experimental and finite element studies on the mechanical properties of high-strength concrete using natural zeolite and additives. *Alexandria Engineering Journal*, *109*, 989–993. DOI: 10.1016/j.aej.2024.09.094

Grand, J., Awala, H., & Mintova, S. (2016). Mechanism of zeolites crystal growth: New findings and open questions. *CrystEngComm*, *18*(5), 650–664. DOI: 10.1039/C5CE02286J

Grass, , Pals, U., Inayat, A., Schwieger, W., Hartmann, M., & Dathe, W. (2021). Uptake and release characteristics of serotonin hydrochloride by natural Cuban zeolite containing clinoptilolite and mordenite. *Scientific Reports*, *11*(1), 14277. Advance online publication. DOI: 10.1038/s41598-021-93487-z PMID: 34253745

Grey, C. P. (2003). Nuclear magnetic resonance studies of zeolites. In Auerbach, S., Carrado, K., & Dutta, P. (Eds.), *Handbook of zeolite science and technology* (pp. 267–331). CRC Press., DOI: 10.1201/9780203911167.pt3

Grifasi, N., Ziantoni, B., Fino, D., & Piumetti, M. (2024). Fundamental properties and sustainable applications of the natural zeolite clinoptilolite. In *Environmental Science and Pollution Research*. Springer Science and Business Media LLC., DOI: 10.1007/s11356-024-33656-5

Groen, J. C., Jansen, J. C., Moulijn, J. A., & Pérez-Ramírez, J. (2004). Optimal aluminum-assisted mesoporosity development in MFI zeolites by desilication. *The Journal of Physical Chemistry B*, *108*(35), 13062–13065. DOI: 10.1021/jp047194f

Gualtieri, A. F., Lanci, L., & Tommasini, S. (2003). Crystallographic structure of natural magnetite. *Journal of Applied Crystallography*, *36*(3), 738–744.

Guan, S., Liu, Y., Zhang, H., Shen, R., Wen, H., Kang, N., Zhou, J., Liu, B., Fan, Y., Jiang, J., Jiang, J., & Li, B. (2023). Recent Advances and Perspectives on Supported Catalysts for Heterogeneous Hydrogen Production from Ammonia Borane. *Advancement of Science*, *10*(21), 2300726. Advance online publication. DOI: 10.1002/advs.202300726 PMID: 37118857

Guevara-Lara, A., Cruz-Pérez, A. E., Contreras-Valdez, Z., Mogica-Betancourt, J., Alvarez-Hernández, A., & Vrinat, M. (2010). Effect of Ni promoter in the oxide precursors of MoS2/MgO–Al2O3 catalysts tested in dibenzothiophene hydrodesulphurization. *Catalysis Today*, *149*(3), 288–294. DOI: 10.1016/j.cattod.2009.09.014

Gul, , Ullah, R., Sun, J., Munir, T., & Bai, S. (2021). The fabrication of TiO2-supported clinoptilolite via F– contained hydrothermal etching and a resultant highly energetic {001} facet for the enhancement of its photocatalytic activity. *RSC Advances*, *11*(29), 17849–17859. DOI: 10.1039/D1RA02269E PMID: 35480182

Güngör, , & Özen, S. (2021). Development and characterization of clinoptilolite-, mordenite-, and analcime-based geopolymers: A comparative study. *Case Studies in Construction Materials*, *15*, e00576. Advance online publication. DOI: 10.1016/j.cscm.2021.e00576

Gunther, W. R., Michaelis, V. K., Griffin, R. G., & Roman-Leshkov, Y. (2016). Interrogating the lewis acidity of metal sites in beta zeolites with ^{15}N pyridine adsorption coupled with MAS NMR spectroscopy. *The Journal of Physical Chemistry. C, Nanomaterials and Interfaces*, *120*(50), 28533–28544. DOI: 10.1021/acs.jpcc.6b07811 PMID: 28479940

Gürbüz, H., Tokay, B., & Erdem-Şenatalar, A. (2012). Effects of ultrasound on the synthesis of silicalite-1 nanocrystals. *Ultrasonics Sonochemistry*, *19*(5), 1108–1113. DOI: 10.1016/j.ultsonch.2012.01.008

Gurevich, M. (1930). Über eine rationelle klassifikation der lichtenstreuenden medien. *Physikalische Zeitschrift*, *31*, 753–763.

Gutiérrez, O. Y., Singh, S., Schachtl, E., Kim, J., Kondratieva, E., Hein, J., & Lercher, J. A. (2014). Effects of the Support on the Performance and Promotion of (Ni) MoS2 Catalysts for Simultaneous Hydrodenitrogenation and Hydrodesulfurization. *ACS Catalysis*, *4*(5), 1487–1499. DOI: 10.1021/cs500034d

Haag, W. O., Lago, R. M., & Weisz, P. B. (1984). The active site of acidic aluminosilicate catalysts. *Nature*, *309*(5969), 589–591. DOI: 10.1038/309589a0

Hadjiivanov, K. I. (1998). IR study of CO and NOx sorption on Ag-ZSM-5. *Microporous and Mesoporous Materials*, *24*(1–3), 41–49. DOI: 10.1016/S1387-1811(98)00147-4

Hadjiivanov, K. I., & Vayssilov, G. N. (2002). Characterization of oxide surfaces and zeolites by carbon monoxide as an IR probe molecule. *Advances in Catalysis*, *47*, 307–511. DOI: 10.1016/S0360-0564(02)47008-3

Hadjiivanov, K., Ivanova, E., & Knözinger, H. (2003). FTIR study of low-temperature CO adsorption on Y zeolite exchanged with Be2+, Mg2+, Ca2+, Sr2+ and Ba2+ cations. *Microporous and Mesoporous Materials*, *58*(3), 225–236. DOI: 10.1016/S1387-1811(02)00650-9

Hadjiivanov, K., & Knozinger, H. (1999a). FTIR study of the low-temperature adsorption and co-adsorption of CO and N_2 on NaY zeolite: Evidence of simultaneous coordination of two molecules to one Na^+ site. *Chemical Physics Letters*, *303*(5-6), 513–520. DOI: 10.1016/S0009-2614(99)00229-8

Hadjiivanov, K., & Knözinger, H. (1999b). FTIR spectroscopic evidence of formation of geminal dinitrogen species during the low-temperature N_2 adsorption on NaY zeolites. *Catalysis Letters*, *58*(1), 21–26. DOI: 10.1023/A:1019040825491

Hadjiivanov, K., & Knözinger, H. (2000). FTIR Study of Low-Temperature CO Adsorption on Cu-ZSM-5: Evidence of the Formation of $Cu^{2+}(CO)_2$ Species. *Journal of Catalysis*, *191*(2), 480–485. DOI: 10.1006/jcat.1999.2805

Hadjiivanov, K., Knözinger, H., Ivanova, E., & Dimitrov, L. (2001). FTIR study of low-temperature CO and $^{15}N_2$ adsorption on a CaNaY zeolite: Formation of site-specified and $Ca^{2+}(CO)_3$ complexes $Ca^{2+}(^{15}N_2)_3$. *Physical Chemistry Chemical Physics*, *3*, 2531–2536. DOI: 10.1039/b101782i

Hadjiivanov, K., Knözinger, H., & Mihaylov, M. (2002). FTIR study of CO adsorption on Ni-ZSM-5. *The Journal of Physical Chemistry B*, *106*(10), 2618–2624. DOI: 10.1021/jp0132782

Hagen, J. (2015). Economic Importance of Catalysts. In J. Hagen (Ed.) *Industrial Catalysis*. Wiley. doi: 9783527684625.ch17

Hallaert, S. D., Bols, M. L., Vanelderen, P., Schoonheydt, R. A., Sels, B. F., & Pierloot, K. (2017). Identification of α-Fe in High-Silica Zeolites on the Basis of ab Initio Electronic Structure Calculations. *Inorganic Chemistry*, *56*(17), 10681–10690. DOI: 10.1021/acs.inorgchem.7b01653

Hallajiqomi, , Mehdipourghazi, M., & Varaminian, F. (2019). Degradation of tetracycline using photocatalytic membrane reactor with nanocomposite Ag doped clinoptilolite zeolite photocatalyst. *Desalination and Water Treatment*, *139*, 111–122. DOI: 10.5004/dwt.2019.23306

Hama Aziz, K. H., Mustafa, F. S., Omer, K. M., Hama, S., Hamarawf, R. F., & Rahman, K. O. (2023). Heavy metal pollution in the aquatic environment: efficient and low-cost removal approaches to eliminate their toxicity: a review. In RSC Advances (Vol. 13, Issue 26, pp. 17595–17610). Royal Society of Chemistry (RSC). https://doi.org/DOI: 10.1039/D3RA00723E

Hamad, , Abdallah, H., Aly, S. T., Abobeah, R., & Amin, S. K. (2024). Fabrication and assessment of performance of clay based ceramic membranes impregnated with CNTs in dye removal. *Scientific Reports*, *77015*(1), 26728. Advance online publication. DOI: 10.1038/s41598-024-77015-3 PMID: 39496693

Handbook of Ultrasonics and Sonochemistry. (2016). Springer Singapore., DOI: 10.1007/978-981-287-278-4

Hanim, S. A. M., Malek, N. A. N. N., & Ibrahim, Z. (2016). Amine-functionalized, silver-exchanged zeolite NaY: Preparation, characterization and antibacterial activity. *Applied Surface Science*, *360*, 121–130. DOI: 10.1016/j.apsusc.2015.11.010

Han, J., Ha, Y., Guo, M., Zhao, P., Liu, Q., Liu, C., Song, C., Ji, N., Lu, X., Ma, D., & Li, Z. (2019). Synthesis of zeolite SSZ-13 from coal gangue via ultrasonic pretreatment combined with hydrothermal growth method. *Ultrasonics Sonochemistry*, *59*. Advance online publication. DOI: 10.1016/j.ultsonch.2019.104703

Hao, , Stavljenić Milašin, I., Batu Eken, Z., Mravak-Stipetic, M., Pavelić, K., & Ozer, F. (2021). Effects of zeolite as a drug delivery system on cancer therapy: A systematic review. *Molecules (Basel, Switzerland)*, *26*(20), 6196. Advance online publication. DOI: 10.3390/molecules26206196 PMID: 34684777

Harami, . (2020). Molecular simulation and computational modeling of gas separation through polycarbonate/p-nitroaniline/zeolite 4A mixed matrix membranes. *Industrial & Engineering Chemistry Research*, *59*(38), 16772–16785. Advance online publication. DOI: 10.1021/acs.iecr.0c02827

Hargreaves, J. S. J. (2016). Some considerations related to the use of the Scherrer equation in powder X-ray diffraction as applied to heterogeneous catalysts. *Catalysis. Structure & Reactivity*, *2*(1-4), 33–37. DOI: 10.1080/2055074X.2016.1252548

Hartmann, M., Thommes, M., & Schwieger, W. (2021). Hierarchically-Ordered Zeolites: A Critical Assessment. *Advanced Materials Interfaces*, *8*(4), 2001841. DOI: 10.1002/admi.202001841

Hassanzadeh-Tabrizi, S. A. (2023). Precise calculation of crystallite size of nanomaterials: A review. *Journal of Alloys and Compounds*, *968*, 171914. DOI: 10.1016/j.jallcom.2023.171914

He, . (2015). Removal of heavy metal ions from aqueous solution by zeolite synthesized from fly ash. *Environmental Science and Pollution Research International*, *4576*. Advance online publication. DOI: 10.1007/s11356-015-5422-6 PMID: 26446735

Heidari, S., Haghighi, M., & Shabani, M. (2018). Ultrasound assisted dispersion of Bi2Sn2O7-C3N4 nanophotocatalyst over various amount of zeolite Y for enhanced solar-light photocatalytic degradation of tetracycline in aqueous solution. *Ultrasonics Sonochemistry*, *43*, 61–72. DOI: 10.1016/j.ultsonch.2018.01.001

Helmi, , Hemmati, A., & Ghaemi, A. (2024). Clinoptilolite and MCM-41 impregnated with chitosan as a green novel adsorbent for CO2 capture. *Case Studies in Chemical and Environmental Engineering*, *9*, 100764. Advance online publication. DOI: 10.1016/j.cscee.2024.100764

Henriques, C., Marie, O., Thibault-Starzyk, F., & Lavalley, J.-C. (2001). NO^+ ions as IR probes for the location of OH groups and Na^+ ions in main channels and side pockets of mordenite. *Microporous and Mesoporous Materials*, *50*(2-3), 167–171. DOI: 10.1016/S1387-1811(01)00444-9

Hensen, E. J. M., de Beer, V. H. J., van Veen, J. A. R., & van Santen, R. A. (2002). A Refinement on the Notion of Type I and II (Co)MoS Phases in Hydrotreating Catalysts. *Catalysis Letters*, *84*(1), 59–67. DOI: 10.1023/A:1021024617582

Henson, N. J., Cheetham, A. K., & Gale, J. D. (1994). Theoretical Calculations on Silica Frameworks and Their Correlation with Experiment. *Chemistry of Materials*, *6*(10), 1647–1650. DOI: 10.1021/cm00046a015

Hermansson, K. (2002). Blue-Shifting Hydrogen Bonds. *The Journal of Physical Chemistry A*, *106*(18), 4695–4702. DOI: 10.1021/jp0143948

Hernández, M. A., Petranovskii, V., & Quiroz, K. (2018). Estimation of Nanoporosity of ZSM-5 Zeolites as Hierarchical Materials. In Rashed, M. N., & Palanisamy, P. N. (Eds.), *Zeolites and Their Applications*. IntechOpen. DOI: 10.5772/intechopen.73624

Hinnemann, B., Nørskov, J. K., & Topsøe, H. (2005). A Density Functional Study of the Chemical Differences between Type I and Type II MoS2-Based Structures in Hydrotreating Catalysts. *The Journal of Physical Chemistry B*, *109*(6), 2245–2253. DOI: 10.1021/jp048842y PMID: 16851217

Hitchcock, I., Holt, E. M., Lowe, J. P., & Rigby, S. P. (2011). Studies of freezing-melting hysteresis in cryoporometry scanning loop experiments using NMR diffusometry and relaxometry. *Chemical Engineering Science*, *66*(4), 582–592. DOI: 10.1016/j.ces.2010.10.027

Hobza, P., & Havlas, Z. (2000). Blue-shifting hydrogen bonds. *Chemical Reviews*, *100*(11), 4253–4264. DOI: 10.1021/cr990050q PMID: 11749346

Holder, C. F., & Schaak, R. E. (2019). Tutorial on powder X-ray diffraction for characterizing nanoscale materials. *ACS Nano*, *13*(7), 7359–7365. DOI: 10.1021/acsnano.9b05157 PMID: 31336433

Hosokawa, S., Sato, H., Tezuka, Y., Adachi, J.-I., Kimura, K., Hayashi, K., Kohara, S., Tajiri, H., Kobayashi, K., Koura, A., Koura, A., & Shimojo, F. (2024). Atomic and Electronic Structures on a Mordenite Zeolite. *E-Journal of Surface Science and Nanotechnology*, *22*(1), 25–31. DOI: 10.1380/ejssnt.2023-063

Hosono, H., Kajihara, K., Suzuki, T., Ikuta, Y., Skuja, L., & Hirano, M. (2002). Vacuum ultraviolet optical absorption band of non-bridging oxygen hole centers in SiO_2 glass. *Solid State Communications*, *122*(3–4), 117–120. DOI: 10.1016/S0038-1098(02)00118-7

Hosseinifard, , Aroon, M. A., & Dahrazma, B. (2020). Application of PVDF/HDTMA-modified clinoptilolite nanocomposite membranes in removal of reactive dye from aqueous solution. *Separation and Purification Technology*, *251*, 117294. Advance online publication. DOI: 10.1016/j.seppur.2020.117294

Hosseini, M., Zanjanchi, M. A., Ghalami-Choobar, B., & Golmojdeh, H. (2015). Ultrasound-assisted dealumination of zeolite Y. *Journal of Chemical Sciences*, *127*(1), 25–31. DOI: 10.1007/s12039-014-0745-2

Hossein, , Rwiza, M. J., Nyanza, E. C., Bakari, R., Ripanda, A., Nkrumah, S., Selemani, J. R., & Machunda, R. L. (2024). Fluoride contamination a silent global water crisis: A case of Africa. *Scientific African*, *25*, e02485. Advance online publication. DOI: 10.1016/j.sciaf.2024.e02485

Hovhannisyan, , Siposova, K., Musatov, A., & Chen, S.-J. (2021). Development of multifunctional nanocomposites for controlled drug delivery and hyperthermia. *Scientific Reports*, *11*(1), 5528. Advance online publication. DOI: 10.1038/s41598-021-84927-x PMID: 33750868

Hrenovic, J., Milenkovic, J., Ivankovic, T., & Rajic, N. (2012). Antibacterial activity of heavy metal-loaded natural zeolite. *Journal of Hazardous Materials*, *201–202*, 260–264. DOI: 10.1016/j.jhazmat.2011.11.079 PMID: 22178285

Huang, L., Qin, F., Huang, Z., Zhuang, Y., Ma, J., Xu, H., & Shen, W. (2016). Hierarchical ZSM-5 zeolite synthesized by an ultrasound-assisted method as a long-life catalyst for dehydration of glycerol to acrolein. *Industrial & Engineering Chemistry Research*, *55*(27), 7318–7327. DOI: 10.1021/acs.iecr.6b01140

Huang, S., Liu, X., Yu, L., Miao, S., Liu, Z., Zhang, S., Xie, S., & Xu, L. (2014). Preparation of hierarchical mordenite zeolites by sequential steaming-acid leaching-alkaline treatment. *Microporous and Mesoporous Materials*, *191*, 18–26. DOI: 10.1016/j.micromeso.2014.02.039

Huang, Z.. (2020). Quantum effects in nanomaterials: Insights and challenges. *Journal of Nanoscience*, *25*(4), 237–250.

Hu, , Han, J., Wei, Y., & Liu, Z. (2022). Dynamic evolution of zeolite framework and metal-zeolite interface. *ACS Catalysis*, *12*(9), 5060–5076. DOI: 10.1021/acscatal.2c01233

Huo, H., Peng, L., & Grey, C. P. (2009). Low temperature ^{1}H MAS NMR spectroscopy studies of proton motion in zeolite HZSM-5. *The Journal of Physical Chemistry. C, Nanomaterials and Interfaces*, *113*(19), 8211–8219. DOI: 10.1021/jp900313y

Hu, W., Li, J., Jiao, X., & Jiang, H. (2023). *Jiao*, X. and Jiang, H. (2023). *Zeolite as a Tool to Recycle Nitrogen and Phosphorus in Paddy Fields under Straw Returning Conditions.Agronomy (Basel)*, *13*(2), 327. DOI: 10.3390/agronomy13020327

Ibrahim, F. A., & El-Desoky, M. M. (2023). Erbium-doped (Pb0.9-Er0.01-Zr0.09) and (Pb0.9-Er0.01-Zr0.045-Ti0.045) nano-crystalline films and nano-rods ceramic synthesis by sol–gel technique for energy-storage application. *Scientific Reports*, *13*(1), 4192. Advance online publication. DOI: 10.1038/s41598-023-31279-3 PMID: 36918632

International Centre for Diffraction Data (ICDD). (2020). *PDF-4+ 2020*. ICDD.

International *Zeolite Corp (2022)* Greenhouse Gas Impact: Analysis & Estimation of NEREA on CO_2e reduction.

Islam, , & Mohr, B. J. (2023). Performance of clinoptilolite zeolite after milling as a pretreatment on hydration kinetics, shrinkage, and alkali-silica reaction of cementitious materials. *Cement*, *12*, 100069. Advance online publication. DOI: 10.1016/j.cement.2023.100069

Ivanova, I. I., Kuznetsov, A. S., Yuschenko, V. V., & Knyazeva, E. E. (2004). *Design of composite micro/mesoporous molecular sieve catalysts. 76*(9), 1647–1657. https://doi.org/doi:10.1351/pac200476091647

Ivanov, A. A., Chernyavsky, V. S., Gross, M. J., Kharitonov, A. S., Uriarte, A. K., & Panov, G. I. (2003). Kinetics of benzene to phenol oxidation over Fe-ZSM-5 catalyst. *Applied Catalysis A, General*, *249*(2), 327–343. DOI: 10.1016/S0926-860X(03)00224-2

Ivanova, I. I., & Knyazeva, E. E. (2013). Micro–mesoporous materials obtained by zeolite recrystallization: Synthesis, characterization and catalytic applications. *Chemical Society Reviews*, *42*(9), 3671–3688. DOI: 10.1039/C2CS35341E PMID: 23223641

Jackson, C. L., & McKenna, G. B. (1990). The melting behavior of organic materials confined in porous solids. *The Journal of Chemical Physics*, *93*(12), 9002–9011. DOI: 10.1063/1.459240

Jaime-Acuña, O. E., Villavicencio, H., Díaz-Hernández, J. A., Petranovskii, V., Herrera, M., & Raymond-Herrera, O. (2014). Atomic and Electronic structure of quaternary $Cd_xZn_yS_\delta O_\gamma$ nanoparticles grown on mordenite. *Chemistry of Materials*, *26*(21), 6152–6159. DOI: 10.1021/cm5024585

Jannesari, M., Akhavan, O., Madaah Hosseini, H. R., & Bakhshi, B. (2020). Graphene/CuO2Nanoshuttles with Controllable Release of Oxygen Nanobubbles Promoting Interruption of Bacterial Respiration. *ACS Applied Materials & Interfaces*, *12*(32), 35813–35825. DOI: 10.1021/acsami.0c05732 PMID: 32664715

Jannesari, M., Akhavan, O., Madaah Hosseini, H. R., & Bakhshi, B. (2023). Oxygen-Rich Graphene/ZnO2-Ag nanoframeworks with pH-Switchable Catalase/Peroxidase activity as O2 Nanobubble-Self generator for bacterial inactivation. *Journal of Colloid and Interface Science*, *637*, 237–250. DOI: 10.1016/j.jcis.2023.01.079 PMID: 36701869

Jarosz, et al. (2017). The effect of feed supplementation with Zakarpacki zeolite (clinoptilolite) on percentages of T and B lymphocytes. 2095–2100. DOI: 10.3382/ps/pex030

Jarosz, R., Szerement, J., Gondek, K., & Mierzwa-Hersztek, M. (2022). *The use of zeolites as an addition to fertilizers – A review.Catena*, *213*(June), 10612. DOI: 10.1016/j.catena.2022.106125

Javaid, , Munir, N., Abideen, Z., Siddiqui, Z. S., & Yong, J. W. H. (2024). The role of natural and synthetic zeolites as soil amendments for mitigating the harmful impacts of abiotic stresses to improve agricultural resilience. *Plant Stress*, *12*, 100627. Advance online publication. DOI: 10.1016/j.stress.2024.100627

Jiang, N., Shang, R., Heijman, S. G. J., & Rietveld, L. C. (2020). Adsorption of triclosan, trichlorophenol and phenol by high-silica zeolites: Adsorption efficiencies and mechanisms. In *Separation and Purification Technology* (Vol. 235, p. 116152). Elsevier BV., DOI: 10.1016/j.seppur.2019.116152

Jimenez-Orozco, C., Figueras, M., Flórez, E., Vines, F., Rodriguez, J. A., & Illas, F. (2021). Size and Stoichiometry Effects on the Reactivity of MoC<inf>y</inf>Nanoparticles toward Ethylene. *The Journal of Physical Chemistry. C, Nanomaterials and Interfaces*, *125*(11), 6287–6297. DOI: 10.1021/acs.jpcc.0c10689

Jiménez-Reyes, M., Almazán-Sánchez, P. T., & Solache-Ríos, M. (2021). Radioactive waste treatments by using zeolites. A short review. In *Journal of Environmental Radioactivity* (Vol. 233, p. 106610). Elsevier BV., DOI: 10.1016/j.jenvrad.2021.106610

Jin, , Fu, J., Yu, P., & Luo, D. (2023). Characterization and properties of manganese oxide film coated clinoptilolite as filter material in fixed-bed columns for removal of Mn(II) from aqueous solution. *Scientific Reports*, *13*(1), 17440. Advance online publication. DOI: 10.1038/s41598-023-44611-8 PMID: 37838822

Jodłowski, P. J., Czekaj, I., Stachurska, P., Kuterasiński, Ł., Chmielarz, L., Jędrzejczyk, R. J., Jeleń, P., Sitarz, M., Górecka, S., Mazur, M., & Kurzydym, I. (2021). Experimental and Theoretical Studies of Sonically Prepared Cu–Y, Cu–USY and Cu–ZSM-5 Catalysts for SCR deNOx. *Catalysts*, *11*(7), 824. DOI: 10.3390/catal11070824

Jones, A. (2018). Atomic force microscopy: Techniques and applications. *Surface Science Reports*, *71*(3), 123–136.

Joseph, J., Radhakrishnan, R. C., Johnson, J. K., Joy, S. P., & Thomas, J. (2020). Ion-exchange mediated removal of cationic dye-stuffs from water using ammonium phosphomolybdate. In *Materials Chemistry and Physics* (Vol. 242, p. 122488). Elsevier BV., DOI: 10.1016/j.matchemphys.2019.122488

Joyner, R., & Stockenhuber, M. (1999). Preparation, Characterization, and Performance of Fe–ZSM-5 Catalysts. *The Journal of Physical Chemistry B*, *103*(29), 5963–5976. DOI: 10.1021/jp990978m

Jusoh, N., Yeong, Y. F., Mohamad, M., Lau, K. K., & Shariff, M., A. (2017). Rapid-synthesis of zeolite T via sonochemical-assisted hydrothermal growth method. *Ultrasonics Sonochemistry*, *34*, 273–280. DOI: 10.1016/j.ultsonch.2016.05.033

Kabuba, , & Banza, M. (2020). Ion-exchange process for the removal of Ni (II) and Co (II) from wastewater using modified clinoptilolite: Modeling by response surface methodology and artificial neural network. *Results in Engineering*, *100189*, 100189. Advance online publication. DOI: 10.1016/j.rineng.2020.100189

Kadium, A. H., Sharif, J. A., & Akhgar, B. N. (2025). Evaluating microstructure and surface changes of clinoptilolite after mechanical activation. *Colloids and Surfaces. A, Physicochemical and Engineering Aspects*, *686*, 136523. Advance online publication. DOI: 10.1016/j.colsurfa.2025.136523

Kadiyala, U., Kotov, N. A., & VanEpps, J. S. (2018). Antibacterial Metal Oxide Nanoparticles: Challenges in Interpreting the Literature. *Current Pharmaceutical Design*, *24*(8), 896–903. DOI: 10.2174/1381612824666180219130659 PMID: 29468956

Kaluža, L., Gulková, D., Vít, Z., & Zdražil, M. (2015). High-activity MgO-supported CoMo hydrodesulfurization catalysts prepared by non-aqueous impregnation. *Applied Catalysis B: Environmental*, *162*(Supplement C), 430–436. DOI: 10.1016/j.apcatb.2014.07.016

Kanellopoulos, J., Gottert, C., Schneider, D., Knorr, B., Prager, D., Ernst, H., & Freude, D. (2008). NMR investigation of proton mobility in zeolites. *Journal of Catalysis*, *255*(1), 68–78. DOI: 10.1016/j.jcat.2008.01.028

Kannan, . (2021). Ammonia adsorption and recovery from swine wastewater permeate using naturally occurring clinoptilolite. *Journal of Water Process Engineering*, *43*, 102234. Advance online publication. DOI: 10.1016/j.jwpe.2021.102234

Karamanis, I., Daouli, A., Monnier, H., Dziurla, M.-A., Maurin, G., & Badawi, M. (2023). A systematic DFT screening of cationic faujasite-type zeolites for the adsorption of NO, NO<inf>2</inf> and H<inf>2</inf>O. *Molecular Systems Design & Engineering*, *8*(9), 1165–1181. DOI: 10.1039/D3ME00044C

Karimi, M., Mehrabadi, Z., Farsadrooh, M., Bafkary, R., Derikvandi, H., Hayati, P., & Mohammadi, K. (2021). Metal–organic framework. En *Interface Science and Technology, 33*, 279-387. https://doi.org/DOI: 10.1016/B978-0-12-818805-7.00010-2

Karimi, , & Azadmehr, A. (2017). Removal of sulfide from aqueous solution with transition metals ion impregnated clinoptilolite: Characteristics, equilibrium, kinetic and thermodynamic approach. *Desalination and Water Treatment*, *74*, 296–307. DOI: 10.5004/dwt.2017.20608

Kashi, . (2023). Electrocoagulation/flotation process for removing copper from an aqueous environment. *Scientific Reports*, *13*(1), 13334. Advance online publication. DOI: 10.1038/s41598-023-40512-y PMID: 37587185

Katada, N., Ozawa, N., Tsuji, E., Kanehara, K., Otsuka, A., Sakamoto, T., Umezawa, K., Matsubara, H., Suganuma, S., & Kubo, M. (2024). Methylation of benzene with methane induced by strong adsorption of benzene on Co ion at α-position in zeolite with moderate Al–Al distance. *Microporous and Mesoporous Materials*, *364*, 112855. Advance online publication. DOI: 10.1016/j.micromeso.2023.112855

Kato, M., Itabashi, K., Matsumoto, A., & Tsutsumi, K. (2003). Characteristics of MOR-framework zeolites synthesized in fluoride-containing media and related ordered distribution of Al atoms in the framework. *The Journal of Physical Chemistry B*, *107*(8), 1788–1797. DOI: 10.1021/jp022227h

Katsou, E., Malamis, S., Tzanoudaki, M., Haralambous, K. J., & Loizidou, M. (2011). Regeneration of natural zeolite polluted by lead and zinc in wastewater treatment systems. In Journal of Hazardous Materials (Vol. 189, Issue 3, pp. 773–786). Elsevier BV. https://doi.org/DOI: 10.1016/j.jhazmat.2010.12.061

Kavoosi, M. (2007). *Effect of zeolite application on rice yield, nitrogen recovery, and nitrogen use efficiency.* Communications in soil science and Plat Analysis Vo. 38, Issue 1-2, 69-76

Kellouai, W., Judeinstein, P., Plazanet, M., Baudoin, S., Drobek, M., Julbe, A., & Coasne, B. (2022). Gas Adsorption in Zeolite and Thin Zeolite Layers: Molecular Simulation, Experiment, and Adsorption Potential Theory. *Langmuir*, *38*(18), 5428–5438. DOI: 10.1021/acs.langmuir.1c03420 PMID: 35486814

Kerby, M. C., Degnan, T. F.Jr, Marler, D. O., & Beck, J. S. (2005). Advanced catalyst technology and applications for high quality fuels and lubricants. *Catalysis Today*, *104*(1), 55–63. DOI: 10.1016/j.cattod.2005.03.028

Kerstens, D., Smeyers, B., Van Waeyenberg, J., Zhang, Q., Yu, J., & Sels, B. F. (2020). State of the Art and Perspectives of Hierarchical Zeolites: Practical Overview of Synthesis Methods and Use in Catalysis. *Advanced Materials*, *32*(44), 2004690. https://doi.org/https://doi.org/10.1002/adma.202004690. DOI: 10.1002/adma.202004690 PMID: 32969083

Khaleque, A., Alam, M. M., Hoque, M., Mondal, S., Haider, J. B., Xu, B., Johir, M. A. H., Karmakar, A. K., Zhou, J. L., Ahmed, M. B., & Moni, M. A. (2020). Zeolite synthesis from low-cost materials and environmental applications: A review. In *Environmental Advances* (Vol. 2, p. 100019). Elsevier BV., DOI: 10.1016/j.envadv.2020.100019

Khaleque, A., Alam, M. M., Hoque, M., Mondal, S., Haider, J. B., Xu, B., & Moni, M. A. (2020). Zeolite synthesis from low-cost materials and environmental applications: A review. *Environmental Advances*, *2*, 100019.

Khan, S., Idrees, M., & Bilal, M. (2021). Revealing and elucidating chemical speciation mechanisms for lead and nickel adsorption on zeolite in aqueous solutions. In Colloids and Surfaces A: Physicochemical and Engineering Aspects (Vol. 623, p. 126711). Elsevier BV. https://doi.org/DOI: 10.1016/j.colsurfa.2021.126711

Khan, . (2021). *Experimental studies on arsenic removal efficiencies through adsorption using different natural adsorbents*. Water & Soil Pollution., DOI: 10.1007/s11270-020-04977-1

Khan, H., Yerramilli, A. S., D'Oliveira, A., Alford, T. L., Boffito, D. C., & Patience, G. S. (2020). Experimental methods in chemical engineering: X-ray diffraction spectroscopy—XRD. *Canadian Journal of Chemical Engineering*, *98*(6), 1255–1266. Advance online publication. DOI: 10.1002/cjce.23747

Khoshbin, R., & Karimzadeh, R. (2017). The beneficial use of ultrasound in free template synthesis of nanostructured ZSM-5 zeolite from rice husk ash used in catalytic cracking of light naphtha: Effect of irradiation power. *Advanced Powder Technology*, *28*(3), 973–982. DOI: 10.1016/j.apt.2017.01.001

Kianfar, . (2019). Nanozeolites: Synthesized, properties, applications. *Journal of Sol-Gel Science and Technology*, *91*(2), 415–429. DOI: 10.1007/s10971-019-05012-4

Kim, K. (1983). The integrated intensity of the carbon monoxide fundamental band. *Journal of Quantitative Spectroscopy & Radiative Transfer*, *30*(5), 413–416. DOI: 10.1016/0022-4073(83)90104-8

Kimmel, G., & Dayan, D. (2013). X-ray diffraction characterization of microstrain in some uranium alloys. *Powder Diffraction*, *13*(2), 89–95. DOI: 10.1017/S088571560000991X

Kim, P., van der Mynsbrugge, J., Head-Gordon, M., & Bell, A. T. (2022). Experimental and Theoretical Studies of Pd Cation Reduction and Oxidation during NO Adsorption on and Desorption from Pd/H-CHA. *The Journal of Physical Chemistry. C, Nanomaterials and Interfaces*, *126*(44), 18744–18753. DOI: 10.1021/acs.jpcc.2c06357

Kim, W., Choi, D., & Kim, S. (2010). Sonochemical synthesis of zeolite a from metakaolinite in NaOH solution. *Materials Transactions*, *51*(9), 1694–1698. DOI: 10.2320/matertrans.M2010191

Kittel, C. (2005). *Introduction to Solid State Physics*. Wiley.

Klaine, , Alvarez, P. J. J., Batley, G. E., Fernandes, T. F., Handy, R. D., Lyon, D. Y., Mahendra, S., McLaughlin, M. J., & Lead, J. R. (2008). Nanomaterials in the environment: Behavior, fate, bioavailability, and effects. *Environmental Toxicology and Chemistry*, *27*(9), 1825–1851. DOI: 10.1897/08-090.1 PMID: 19086204

Klicpera, T., & Zdražil, M. (2002). Preparation of High-Activity MgO-Supported Co–Mo and Ni–Mo Sulfide Hydrodesulfurization Catalysts. *Journal of Catalysis*, *206*(2), 314–320. DOI: 10.1006/jcat.2001.3488

Klimova, T., Solís Casados, D., & Ramírez, J. (1998). New selective Mo and NiMo HDS catalysts supported on Al2O3–MgO(x) mixed oxides. *Catalysis Today*, *43*(1), 135–146. DOI: 10.1016/S0920-5861(98)00142-4

Klinowski, J. (1984). Nuclear magnetic resonance studies of zeolites. *Progress in Nuclear Magnetic Resonance Spectroscopy*, *16*, 237–309. DOI: 10.1016/0079-6565(84)80007-2

Klinowski, J. (1988). Recent advances in solid-state nmr of zeolites. *Annual Review of Materials Science*, *18*(1), 189–218. DOI: 10.1146/annurev.ms.18.080188.001201

Klug, H. P., & Alexander, L. E. (1974). *X-ray Diffraction Procedures for Polycrystalline and Amorphous Materials* (2nd ed.). Wiley.

Kochubei, , Yaholnyk, S., Bets, M., & Malovanyy, M. (2020). Use of activated clinoptilolite for direct dye-contained wastewater treatment. *Chemistry & Chemical Technology*, *14*(3), 323–329. DOI: 10.23939/chcht14.03.386

Koekkoek, A. J. J., Xin, H., Yang, Q., Li, C., & Hensen, E. J. M. (2011). Hierarchically structured Fe/ZSM-5 as catalysts for the oxidation of benzene to phenol. *Microporous and Mesoporous Materials*, *145*(1), 172–181. DOI: 10.1016/j.micromeso.2011.05.013

Kondratieva, E. V., Manoilova, O. V., & Tsyganenko, A. A. (2008). Integrated absorption coefficient of adsorbed CO. *Kinetics and Catalysis*, *49*(3), 451–456. DOI: 10.1134/S002315840803021X

Konnov, S. V., Bruter, D. V., Pavlov, V. S., & Ivanova, I. I. (2024). State-of-the-art strategies for the synthesis of zeolite-encapsulated subnanometric metal clusters. *Inorganic Chemistry Frontiers*, *11*(13), 3669–3706. DOI: 10.1039/D4QI00703D

Korb, J. P. (2011). Nuclear magnetic relaxation of liquids in porous media. *New Journal of Physics*, *13*(3), 035016. DOI: 10.1088/1367-2630/13/3/035016

Król, M., Koleżyński, A., Mikuła, A., & Mozgawa, W. (2018). Vibrational spectroscopy of zeolites. In Challenges and advances in computational chemistry and physics (pp. 301–332). DOI: 10.1007/978-3-030-01355-4_10

Krylova, E. A., Shelyapina, M. G., Mazur, A., Baranov, D. A., Tsyganenko, A. A., & Petranovskii, V. P. (2022). Local structure of protonated mordenites with $SiO_2/Al_2O_3 \approx 15$ probed by multinuclear NMR. *Journal of Structural Chemistry*, *63*(6), 930–943. DOI: 10.1134/S0022476622060105

Krylova, E. A., Shelyapina, M. G., Nowak, P., Harańczyk, H., Chislov, M., Zvereva, I. A., Privalov, A. F., Becker, M., Vogel, M., & Petranovskii, V. (2018). Mobility of water molecules in sodium- and copper-exchanged mordenites: Thermal analysis and ^{1}H NMR. *Microporous and Mesoporous Materials*, *265*, 132–142. DOI: 10.1016/j.micromeso.2018.02.010

Kubelka, P., & Munk, F. (1931). Ein Beitrag zur Optik der Farbanstriche. Z. für techn. *Phys.*, *12*(Nr. 11a), 593–601.

Kubota, M., Nakabayashi, T., Matsumoto, Y., Shiomi, T., Yamada, Y., Ino, K., Yamanokuchi, H., Matsui, M., Tsunoda, T., Mizukami, F., & Sakaguchi, K. (2008). Selective adsorption of bacterial cells onto zeolites. *Colloids and Surfaces. B, Biointerfaces*, *64*(1), 88–97. DOI: 10.1016/j.colsurfb.2008.01.012 PMID: 18313276

Kukowska, . (2024). Management of the soil environment using biochar and zeolite in various combinations: Impact on soil condition and economical aspects. *Journal of Soils and Sediments*, *24*(10), 3456–3468. DOI: 10.1007/s11368-024-03927-2

Kuldeyev, , Seitzhanova, M., Tanirbergenova, S., Tazhu, K., Doszhanov, E., Mansurov, Z., Azat, S., Nurlybaev, R., & Berndtsson, R. (2023). Modifying natural zeolites to improve heavy metal adsorption. *Water (Basel)*, *15*(12), 2215. Advance online publication. DOI: 10.3390/w15122215

Kumar, M., Xiong, X., Sun, Y., Yu, I. K. M., Tsang, D. C. W., Hou, D., Gupta, J., Bhaskar, T., & Pandey, A. (2020). Critical Review on Biochar-Supported Catalysts for Pollutant Degradation and Sustainable Biorefinery. In Advanced Sustainable Systems (Vol. 4, Issue 10). Wiley. https://doi.org/DOI: 10.1002/adsu.201900149

Kumari, S., Chowdhry, J., Kumar, M., & Chandra Garg, M. (2024). Zeolites in wastewater treatment: A comprehensive review on scientometric analysis, adsorption mechanisms, and future prospects. In *Environmental Research* (Vol. 260, p. 119782). Elsevier BV., DOI: 10.1016/j.envres.2024.119782

Kumar, M., Ambika, S., Hassani, A., & Nidheesh, P. V. (2023). Waste to catalyst: Role of agricultural waste in water and wastewater treatment. In *Science of The Total Environment* (Vol. 858, p. 159762). Elsevier BV., DOI: 10.1016/j.scitotenv.2022.159762

Kumar, P., Sharma, S., & Patel, V. (2017). Calculation of interplanar distance in cubic crystals using X-ray diffraction. *Journal of Crystal Growth*, *461*, 38–42.

Kumar, S. (2023). Smart and innovative nanotechnology applications for water purification. *Hybrid Advances*, *3*, 100044.

Kumzerov, Y., & Vakhrushev, S. (2004). Nanostructures within Porous Materials. In Nalwa, H. S. (Ed.), *Encyclopedia of Nanoscience and Nanotechnology* (Vol. VII, pp. 811–849). American Scientific Publishers.

Kusrini, et al. (2018). The role of praseodymium oxide-impregnated clinoptilolite zeolite catalyst to increase octane number in gasoline. E3S Web of Conferences, 67, Article 03051. DOI: 10.1051/e3sconf/20186703051

Kustov, L. M. (1997). New trends in IR-spectroscopic characterization of acid and basic sites in zeolites and oxide catalysts. *Topics in Catalysis*, *1997*(4), 131–144. DOI: 10.1023/A:1019136121724

Kustov, L. M., Kazansky, V. B., Beran, S., Kubelková, L., & Jiru, P. (1987). Adsorption of carbon monoxide on ZSM-5 zeolites. Infrared spectroscopic study and quantum-chemical calculations. *Journal of Physical Chemistry, 91*(20), 5247–5251. Advance online publication. DOI: 10.1021/j100304a023

Kuterasiński, Ł., Dymek, K., & Jodłowski, P. J. (2019). Sonically modified zeolites with MFI and USY type structure as catalysts for methane combustion. Preparation and physicochemical characterisation. *Science. Technology and Innovation, 6*(3), 11–18. DOI: 10.5604/01.3001.0013.5388

Kuterasiński, Ł., Rojek, W., Gackowski, M., Zimowska, M., & Jodłowski, P. J. (2020). Sonically modified hierarchical FAU-type zeolites as active catalysts for the production of furan from furfural. *Ultrasonics Sonochemistry, 60*, 104785. DOI: 10.1016/j.ultsonch.2019.104785

Kwakye-Awuah, B., Williams, C., Kenward, M. A., & Radecka, I. (2008). Antimicrobial action and efficiency of silver-loaded zeolite X. *Journal of Applied Microbiology, 104*(5), 1516–1524. DOI: 10.1111/j.1365-2672.2007.03673.x PMID: 18179543

Kwon, , Yang, D. S., Koo, M. S., Ji, S. M., Jeong, J., Oh, S., Kuk, S. K., Heo, H., Ham, D. J., Kim, M., Choi, H., Lee, J.-M., Shur, J.-W., Lee, W.-J., Bin, C.-O., Timofeev, N., Wu, H., Wang, L., Lee, T., & Lee, H. C. (2023). Long-lifetime water-washable ceramic catalyst filter for air purification. *Nature Communications, 14*(1), 36050. Advance online publication. DOI: 10.1038/s41467-023-36050-w PMID: 36792592

Laffita Suárez, M. and Martínez Valdés, M. E. (2022). *Results of the Evaluation of NEREA® zeolitic products in the common beans (Phaseolus vulgaris L.) var BAT 304.Campaign* 2021. Report Grain Research Institute, MINAG

Laffita Suárez, M. and Martínez Valdés, M.E. (2024). *Results of the Evaluation of NEREA® products in the common beans (Phaseolus vulgaris L.) var BAT 304. Campaign* 2023. Report Institute of Grains Researchs, MINAG

Lappala, . (2024). The next revolution in computational simulations: Harnessing AI and quantum computing in molecular dynamics. *Current Opinion in Structural Biology, 89*, 102919. Advance online publication. DOI: 10.1016/j.sbi.2024.102919 PMID: 39306949

Latifian, B., Taghizade Firozjaee, T., & Abdi, J. (2023). Synthesis of biochar-iron oxide magnetic nanocomposite for efficient and rapid removal of Pb(II) heavy metal from water. In Water Practice & Technology (Vol. 19, Issue 1, pp. 181–199). IWA Publishing. https://doi.org/DOI: 10.2166/wpt.2023.223

Lavalley, J.-C. (1996). IR studies of the surface basicity of metal oxides and zeolites using adsorbed probe molecules. *Catalysis Today*, *27*(3-4), 377–401. DOI: 10.1016/0920-5861(95)00161-1

Leal, M., Martínez-Hernández, V., Meffe, R., Lillo, J., & de Bustamante, I. (2017). Clinoptilolite and palygorskite as sorbents of neutral emerging organic contaminants in treated wastewater: Sorption-desorption studies. *Chemosphere*, *175*, 534–542. DOI: 10.1016/j.chemosphere.2017.02.057 PMID: 28254718

Lee, D.-Y., Ko, W.-S., Kim, K.-N., Byun, J.-H., Kim, E.-G., Kwon, S.-Y., & Lee, G.-W. (2024). Analysis of metal and zinc oxide semiconductor interface resistance using transmission line method. *Solid-State Electronics*, *216*, 108916. DOI: 10.1016/j.sse.2024.108916

Lee, J., Lee, K., & Kim, J. (2021). Fiber-Based Gas Filter Assembled via in Situ Synthesis of ZIF-8 Metal Organic Frameworks for an Optimal Adsorption of SO<inf>2</inf>: Experimental and Theoretical Approaches. *ACS Applied Materials & Interfaces*, *13*(1), 1620–1631. DOI: 10.1021/acsami.0c19957 PMID: 33395254

Levitt, M. H. (2008). *Spin Dynamics: Basics of Nuclear Magnetic Resonance* (2nd Editio). Wiley.

Lewis, J. D., Ha, M., Luo, H., Faucher, A., Michaelis, V. K., & Román-Leshkov, Y. (2018). Distinguishing active site identity in Sn-Beta zeolites using ^{31}P MAS NMR of adsorbed trimethylphosphine oxide. *ACS Catalysis*, *8*(4), 3076–3086. DOI: 10.1021/acscatal.7b03533

Li, Z., & Bowman, R. S. (1998). Sorption of Perchloroethylene by Surfactant-Modified Zeolite as Controlled by Surfactant Loading. In Environmental Science & Technology (Vol. 32, Issue 15, pp. 2278–2282). American Chemical Society (ACS). https://doi.org/DOI: 10.1021/es971118r

Liang, D., Zhang, R., Liu, Y., Xie, Q., & Liu, J. (2023). A Review of High-Efficient Synthetic Methods for Zeolite Membranes and Challenges of Their Directional Growth Control. *Comments on Inorganic Chemistry*, *43*(4), 305–356. DOI: 10.1080/02603594.2022.2138864

Li, , Han, H., Evangelou, N., Wichrowski, N. J., Lu, P., Xu, W., Hwang, S.-J., Zhao, W., Song, C., Guo, X., Bhan, A., Kevrekidis, I. G., & Tsapatsis, M. (2023). Machine learning-assisted crystal engineering of a zeolite. *Nature Communications*, *14*(1), 3152. Advance online publication. DOI: 10.1038/s41467-023-38738-5 PMID: 37258522

Li, K., Valla, J., & García-Martínez , J. (2014). Realizing the Commercial Potential of Hierarchical Zeolites: New Opportunities in Catalytic Cracking. *ChemCatChem*, *6*(1), 46–66. https://doi.org/https://doi.org/10.1002/cctc.201300345. DOI: 10.1002/cctc.201300345

Li, L., Shen, Q., Li, J., Hao, Z., Xu, Z. P., & Lu, G. Q. M. (2008). Iron-exchanged FAU zeolites: Preparation, characterization and catalytic properties for N2O decomposition. *Applied Catalysis A, General*, *344*(1–2), 131–141. DOI: 10.1016/j.apcata.2008.04.011

Lima, , Silva, Y. F., & Lima, P. L. T. (2021). Iron removal efficiency in irrigation water by zeolite added to sand media filters. *Desalination and Water Treatment*, *220*, 134–141. DOI: 10.5004/dwt.2021.27024

Linares, C. F., Vásquez, M., Castillo, R., Bretto, P., Solano, R., & Rincón, A. (2015). Applications of CoMo/calcined quaternary hydrotalcites for hydrotreatment reactions. *Fuel Processing Technology*, *132*(Supplement C), 105–110. DOI: 10.1016/j.fuproc.2014.12.043

Linares, N., Sachse, A., Serrano, E., Grau-Atienza, A., De Oliveira Jardim, E., Silvestre-Albero, J., Cordeiro, M. A. L., Fauth, F., Beobide, G., Castillo, O., & García-Martínez, J. (2016). In Situ Time-Resolved Observation of the Development of Intracrystalline Mesoporosity in USY Zeolite. *Chemistry of Materials*, *28*(24), 8971–8979. DOI: 10.1021/acs.chemmater.6b03688

Lin, C. C. H., Dambrowitz, K. A., & Kuznicki, S. M. (2011). Evolving applications of zeolite molecular sieves. *Canadian Journal of Chemical Engineering*, *90*(2), 207–216. DOI: 10.1002/cjce.20667

Li, T., Cai, H., Li, C., Liu, X., & Huang, F. (2020). Rocksalt-zincblende–wurtzite mixed-phase ZnO crystals with high activity as photocatalysts for visible-light-driven water splitting. *Frontiers in Chemistry*, *8*, 351. DOI: 10.3389/fchem.2020.00351 PMID: 32411673

Liu, J., Guo, H., & Wang, F. (2015). Influence of crystallography on the properties of cementitious materials. *Cement and Concrete Research*, *70*, 36–44.

Liu, J., Huang, Z., Sun, J., Zou, Y., & Gong, B. (2020). Enhancing the removal performance of Cd(II) from aqueous solutions by NaA zeolite through doped thiourea reduced GO which is trapped within zeolite crystals. In *Journal of Alloys and Compounds* (Vol. 815, p. 152514). Elsevier BV., DOI: 10.1016/j.jallcom.2019.152514

Liu, , Kim, K.-H., Kumar, V., & Kim, S. (2019b). A review of functional sorbents for adsorptive removal of arsenic ions in aqueous systems. *Journal of Hazardous Materials, 393*, 121815. Advance online publication. DOI: 10.1016/j.jhazmat.2019.121815 PMID: 31831285

Li, , Ullah, R., Jiao, J., Sun, J., & Bai, S. (2020). Ion exchange of cations from different groups with ammonium-modified clinoptilolite and selectivity for methane and nitrogen. *Materials Chemistry and Physics, 256*, 123760. Advance online publication. DOI: 10.1016/j.matchemphys.2020.123760

Liu, Q., & van Bokhoven, J. A. (2024). Water structures on acidic zeolites and their roles in catalysis. *Chemical Society Reviews, 53*(6), 3065–3095. DOI: 10.1039/D3CS00404J

Liu, , Tian, R., Ding, W., He, Y., & Li, H. (2019a). Adsorption selectivity of heavy metals by Na-clinoptilolite in aqueous solutions. *Adsorption, 25*(4), 747–755. DOI: 10.1007/s10450-019-00081-x

Liu, Y., Zhang, H., Zhang, F., Zhu, M., Hu, N., & Chen, X. (2020). Facile synthesis of low-silica zeolite erionite by ultrasonic-assisted method. *Materials Letters, 260*, 126934. DOI: 10.1016/j.matlet.2019.126934

Liu, Z. F., Liu, Z. C., Wang, Y., Li, Y. B., Qu, L., e, L., Ya, J., & Huang, P. Y. (2012). Photocatalysis of TiO 2 nanoparticles supported on natural zeolite. *Materials Technology, 27*(3), 267–271. DOI: 10.1179/1753555712Y.0000000011

Li, W.. (2019). Nanoparticles with enhanced surface area for catalysis and energy storage. *Journal of Nanotechnology, 22*(3), 194–210.

Li, X., Li, T.-H., Zhou, W., Li, Y.-P., Sit, P. H.-L., Wu, Z., Curnow, O. J., Wu, K. C.-W., Choi, J., & Yip, A. C. K. (2022). Unveiling the elusive role of tetraethyl orthosilicate hydrolysis in ionic-liquid-templated zeolite synthesis. *Materials Today. Chemistry, 23*, 100658. Advance online publication. DOI: 10.1016/j.mtchem.2021.100658

Li, Y., Yu, M., Qi, G., Liu, Y., Lv, J., Huang, S., & Ma, X. (2024). Designing a Mordenite Catalyst with Enhanced Acidity for Dimethyl Ether Carbonylation by Engineering Open Sn Sites. *Engineering, 35*, 104–114. DOI: 10.1016/j.eng.2024.07.009

Li, Y., Zhu, G., Wang, Y., Chai, Y., & Liu, C. (2021). Preparation, mechanism and applications of oriented MFI zeolite membranes: A review. *Microporous and Mesoporous Materials, 312*, 110790. DOI: 10.1016/j.micromeso.2020.110790

Li, , & Yu, J. (2021a). Emerging applications of zeolites in catalysis, separation and host–guest assembly. *Nature Reviews. Materials, 6*(12), 1156–1174. DOI: 10.1038/s41578-021-00347-3

Li, Z. (2003). Use of surfactant-modified zeolite as fertilizer carriers to control nitrate release. *Microporous and Mesoporous Materials*, *61*(1), 181–188. DOI: 10.1016/S1387-1811(03)00366-4

Loewenstein, W. (1954). The distribution of aluminum in the tetrahedra of silicates and aluminates. *The American Mineralogist*, *39*, 92–96.

Loiola, , Bessa, R. A., Oliveira, C. P., Freitas, A. D. L., Soares, S. A., Bohn, F., & Pergher, S. B. C. (2022). Magnetic zeolite composites: Classification, synthesis routes, and technological applications. *Journal of Magnetism and Magnetic Materials*, *560*, 169651. Advance online publication. DOI: 10.1016/j.jmmm.2022.169651

Long, J., Wang, X., Ding, Z., Zhang, Z., Lin, H., Dai, W., & Fu, X. (2009). Binuclear µ-hydroxo-bridged iron clusters derived from surface organometallic chemistry of ferrocene in cavities of HY zeolite: Local structure, bound sites, and catalytic reactivity. *Journal of Catalysis*, *264*(2), 163–174. DOI: 10.1016/j.jcat.2009.04.002

López Paz, C. (2019). Fertilizantes completos con zeolita natural. Una alternativa sostenible en el cultivo del tomate. Tesis de Diploma. Facultad de Ciencias Agropecuarias Universidad de Matanzas, Cuba.

López, M., Bouza, L., & Acosta, A. (1991) Estudio de la interacción de la zeolita con diferentes dosis de nitrógeno en el cultivo de la caña de azúcar. I: resultados de caña planta var C266-70. Memoirs 3rd Int. Conf. Occur. Prop. Util. Nat. Zeolites, Part II, 102-103

López-Bastidas, C., Petranovskii, V., & Machorro, R. (2012). Optical response of Cu clusters in zeolite template. *Journal of Colloid and Interface Science*, *375*(1), 60–64. DOI: 10.1016/j.jcis.2012.02.036 PMID: 22440728

López-Bastidas, C., Smolentseva, E., Petranovskii, V., & Machorro, R. (2013). Plasmon features of coinage metal nanoparticles supported on zeolites. *Plasmonics*, *8*(4), 1551–1558. DOI: 10.1007/s11468-013-9571-5

López-Cruz, C., Guzman, J., Cao, G., Martínez, C., & Corma, A. (2021). Modifying the catalytic properties of hydrotreating NiMo–S phases by changing the electrodonor capacity of the support. *Catalysis Today*, *382*, 130–141. DOI: 10.1016/j.cattod.2021.08.002

López-Hernández, I., García, C., Truttmann, V., Pollitt, S., Barrabés, N., Rupprechter, G., Rey, F., & Palomares, A. E. (2020). Evaluation of the silver species nature in Ag-ITQ2 zeolites by the CO oxidation reaction. *Catalysis Today*, *345*, 22–26. DOI: 10.1016/j.cattod.2019.12.001

Losada, I. B., Grobas-Illobre, P., Misturini, A., Polaina, J., Seminovski, Y., & Sastre, G. (2021). Separation of an aqueous mixture of 6-kestose/sucrose with zeolites: A molecular dynamics simulation. *Microporous and Mesoporous Materials, 319*, 111031. Advance online publication. DOI: 10.1016/j.micromeso.2021.111031

Louwen, J. N., van Eijck, L., Vogt, C., & Vogt, E. T. (2020). Understanding the Activation of ZSM-5 by Phosphorus: Localizing Phosphate Groups in the Pores of Phosphate-Stabilized ZSM-5. *Chemistry of Materials, 32*(21), 9390–9403. DOI: 10.1021/acs.chemmater.0c03411

Luo, et al. (2025). Atomic-scale insights into topotactic transformations in an extra-large-pore zeolite using time-resolved 3D electron diffraction. 145–153. DOI: 10.1038/s44160-024-00715-1

Luque-Álvarez, L. A., Torres-Sempere, G., Romero-Sarria, F., Bobadilla, L. F., Ramírez-Reina, T., & Odriozola, J. A. (2024). Controlling copper location on exchanged MOR-type aluminosilicate zeolites for methanol carbonylation: In situ/operando IR spectroscopic studies. *Microporous and Mesoporous Materials, 378*, 113258. DOI: 10.1016/j.micromeso.2024.113258

Mackey, D., Lyons, J., & Naydenova, I. (2024). Modeling redistribution of nano-zeolites in holographic recording. *Journal of the Optical Society of America. A, Optics, Image Science, and Vision, 41*(7), 1413–1419. DOI: 10.1364/JOSAA.523677 PMID: 39889130

Maevsky, A. V., & Tsyganenko, A. A. (2012). Linkage isomerism in layered silicates [Paper presentation]. *International Symposium on Zeolites and Mycroporous Crystals ZMPC2012, Hiroshima, Japan.* Abstracts, p. 124

Mahamud, M., Taddesse, A. M., Bogale, Y., & Bezu, Z. (2023). Zeolite supported CdS/TiO2/CeO2 composite: Synthesis, characterization and photocatalytic activity for methylene blue dye degradation. *Materials Research Bulletin, 161*, 112176. Advance online publication. DOI: 10.1016/j.materresbull.2023.112176

Mahesh, . (2020). Machine learning algorithms - a review. [IJSR]. *International Journal of Science and Research (Raipur, India), 9*(1), 381–386. DOI: 10.21275/ART20203995

Mahmoudi, , Nasseri, S., Mahvi, A. H., Dargahi, A., Khubestani, M. S., & Salari, M. (2019). Fluoride removal from aqueous solution by acid-treated clinoptilolite: Isotherm and kinetic study. *Desalination and Water Treatment, 141*, 246–253. DOI: 10.5004/dwt.2019.23625

Mahyuddin, M. H., Staykov, A., Saputro, A. G., Agusta, M. K., Dipojono, H. K., & Yoshizawa, K. (2020). Novel mechanistic insights into methane activation over Fe and Cu active sites in zeolites: A comparative DFT study using meta-GGA functionals. *The Journal of Physical Chemistry. C, Nanomaterials and Interfaces, 124*(33), 18112–18125. DOI: 10.1021/acs.jpcc.0c04991

Maity, , Chen, C.-Y., Bhattacharya, P., Sharma, R. K., Ahmad, A., Patnaik, S., & Bundschuh, J. (2020). Advanced application of nano-technological and biological processes as well as mitigation options for arsenic removal. *Journal of Hazardous Materials, 405*, 123885. Advance online publication. DOI: 10.1016/j.jhazmat.2020.123885 PMID: 33183836

Makarova, M. A., Al-Ghefaili, K. M., & Dwyer, J. (1994a). Brønsted acid strength in US-Y: FTIR study of CO adsorption. *Journal of the Chemical Society, Faraday Transactions, 90*(2), 383–386. DOI: 10.1039/FT9949000383

Makarova, M. A., Zholobenko, V. L., Al-Ghefaili, K. M., Thompson, N. E., Dewing, J., & Dwyer, J. (1994b). Bronsted acid sites in zeolites. FTIR study of molecular hydrogen as a probe for acidity testing. *Journal of the Chemical Society, Faraday Transactions, 90*(7), 1047–1054. DOI: 10.1039/FT9949001047

Mallette, A. J., Seo, S., & Rimer, J. D. (2022). Synthesis strategies and design principles for nanosized and hierarchical zeolites. *Nature Synthesis, 1*(7), 521–534. DOI: 10.1038/s44160-022-00091-8

Mancinelli, M., Precisvalle, N., Ardit, M., Beltrami, G., Gigli, L., Catizzone, E., Migliori, M., Giordano, G., & Martucci, A. (2023). Thermal stability of templated ZSM-5 zeolites: An in-situ synchrotron X-ray powder diffraction study. *Microporous and Mesoporous Materials, 362*, 112777. Advance online publication. DOI: 10.1016/j.micromeso.2023.112777

Mancuso, A., Sacco, O., Sannino, D., Venditto, V., & Vaiano, V. (2020). One-Step Catalytic or Photocatalytic Oxidation of Benzene to Phenol: Possible Alternative Routes for Phenol Synthesis? *Catalysts, 10*(12), 1424. DOI: 10.3390/catal10121424

Mancuso, A., Vaiano, V., Antico, P., Sacco, O., & Venditto, V. (2023). Photoreactive polymer composite for selective oxidation of benzene to phenol. *Catalysis Today, 413–415*, 113914. DOI: 10.1016/j.cattod.2022.09.020

Manikandan, A., & Subramanian, K. S. (2014). Evaluation of Zeolite Based Nitrogen Nano-fertilizers on Maize Growth, Yield and Quality on Inceptisols and Alfisols. *African Journal of Agricultural Research, 9*(2), 276–284.

Manzoli, et al. (2021). New insights in the production of simulated moon agglutinates: The use of natural zeolite-bearing rocks. ACS, 5(10), 2878–2886. DOI: 10.1021/acsearthspacechem.1c00118

Mapes, J. E., & Eischens, R. P. (1954). The infrared spectra of ammonia chemisorbed on cracking catalysts. *Journal of Physical Chemistry*, *58*(12), 1059–1062. DOI: 10.1021/j150522a002

Martens, J. A., Perez-Pariente, J., Sastre, E., Corma, A., & Jacobs, P. A. (1988). Isomerization and disproportionation of m-xylene: Selectivities Induced by the Void Structure of the Zeolite Framework. *Applied Catalysis*, *45*(1), 85–101. https://doi.org/https://doi.org/10.1016/S0166-9834(00)82395-7. DOI: 10.1016/S0166-9834(00)82395-7

Martey, S., Addison, B., Wilson, N., Tan, B., Yu, J., Dorgan, J. R., & Sobkowicz, M. J. (2021). Hybrid Chemomechanical Plastics Recycling: Solvent-free, High-Speed Reactive Extrusion of Low-Density Polyethylene. *ChemSusChem*, *14*(19), 4280–4290. DOI: 10.1002/cssc.202100968 PMID: 34089238

Martínez Grillo, J. F., Martínez Valdés, M. and Rodríguez Reyes, M. (2022). *Results of the Evaluation of NEREA® zeolitic products in rice cultivation, 2021 winter campaign. Rice var. Selection-1.* Report Grain Research Institute, MINAG.

Martínez Grillo, J. F., Martínez Valdés, M. and Rodríguez Reyes, M. (2024). *Results of the Evaluation of NEREA® zeolitic products in rice cultivation, 2023 spring campaign. Rice var. IACuba 31.* Report Grain Research Institute, MINAG.

Martínez, C., & Corma, A. (2011). Inorganic molecular sieves: Preparation, modification and industrial application in catalytic processes. *Coordination Chemistry Reviews*, *255*(13–14), 1558–1580. DOI: 10.1016/j.ccr.2011.03.014

Martins, A., Amaro, B., Santos, M. S. C. S., Nunes, N., Elvas-Leitão, R., & Carvalho, A. P. (2024). Hierarchical Zeolites Prepared Using a Surfactant-Mediated Strategy: ZSM-5 vs. Y as Catalysts for Friedel–Crafts Acylation Reaction. In *Molecules* (Vol. 29, Issue 2). DOI: 10.3390/molecules29020517

Masai, H., Kohara, S., Wakihara, T., Shibazaki, Y., Onodera, Y., Masuno, A., Sukenaga, S., Ohara, K., Sakai, Y., Haines, J., Levelut, C., Hébert, P., Isambert, A., Keen, D. A., & Azuma, M. (2023). Siliceous zeolite-derived topology of amorphous silica. *Communications Chemistry*, *6*(1), 1–2. DOI: 10.1038/s42004-023-01075-1 PMID: 38071376

Mason, T. J., & Vinatoru, M. (2023). *Sonochemistry*. De Gruyter., DOI: 10.1515/9783110566178

Mastinu, , Kumar, A., Maccarinelli, G., Bonini, S. A., Premoli, M., Aria, F., Gianoncelli, A., & Memo, M. (2019). Zeolite clinoptilolite: Therapeutic virtues of an ancient mineral. *Molecules (Basel, Switzerland)*, *24*(8), 1517. Advance online publication. DOI: 10.3390/molecules24081517 PMID: 30999685

Matveeva, T. G., Ivanova, M. S., Solovyev, V. G., & Vanin, A. I. (2024). Electrophysical properties of the Rochelle salt dispersed in a porous dielectric matrix of type A zeolite. *Physics of Complex Systems.*, *5*(4), 195–201. DOI: 10.33910/2687-153X-2024-5-4-195-201

Maurin, G., Senet, P., Devautour, S., Henn, F., & Giuntini, J. (2002). Modelling of the cation motions in complex system: Case of Na-mordenites. *Journal of Non-Crystalline Solids*, *307–310*, 1050–1054. DOI: 10.1016/S0022-3093(02)01571-5

Ma, X., Pei, Y., Ma, Y., Pu, T., & Lei, Y. (2019). Antibacterial Activity of Cu2+-ZnO-modified 13X Zeolite against E.coli and S.aureus. *Journal of Wuhan University of Technology. Materials Science Edition*, *34*(2), 481–486. DOI: 10.1007/s11595-018-2077-z

McComb, D., & Howie, A. (1990). Characterisation of zeolite catalysts using electron energy loss spectroscopy. *Ultramicroscopy*, *34*(1–2), 84–92. DOI: 10.1016/0304-3991(90)90062-Q

McDonald, R. S. (1958). Surface functionality of amorphous silica by infrared spectroscopy. *Journal of Physical Chemistry*, *62*(10), 1168–1178. DOI: 10.1021/j150568a004

Mehrabadi, , & Faghihian, H. (2018). Elimination of highly consumed herbicide; 2,4-dichlorophenoxyacetic acid from aqueous solution by TiO2 impregnated clinoptilolite, study of degradation pathway. *Spectrochimica Acta. Part A: Molecular and Biomolecular Spectroscopy*, *204*, 676–684. DOI: 10.1016/j.saa.2018.06.047 PMID: 29936222

Meirelles, M. R., Malafatti, J. O. D., Escote, M. T., Pinto, A. H., & Paris, E. C. (2023). Magnetic Adsorbent Based on Faujasite Zeolite Decorated with Magnesium Ferrite Nanoparticles for Metal Ion Removal. *Magnetochemistry (Basel, Switzerland)*, *9*(5). Advance online publication. DOI: 10.3390/magnetochemistry9050136

Mendes, C. R., Dilarri, G., Forsan, C. F., Sapata, V., Lopes, P. R. M., de Moraes, P. B., Montagnolli, R. N., Ferreira, H., & Bidoia, E. D. (2022). Antibacterial action and target mechanisms of zinc oxide nanoparticles against bacterial pathogens. *Scientific Reports*, *12*(1), 2658. DOI: 10.1038/s41598-022-06657-y PMID: 35173244

Méndez Días, N. y Soca Núñez, M. (2016) *Evaluación agroindustrial del impacto de la zeolita en el cultivo de la caña de azúcar.* Revista electrónica de la Agencia de Medio Ambiente Año 16, No.31, jul-dic.

Mendoza-Castro, M. J., Serrano, E., Linares, N., & García-Martínez, J. (2021). Surfactant-Templated Zeolites: From Thermodynamics to Direct Observation. *Advanced Materials Interfaces*, *8*(4), 2001388. https://doi.org/https://doi.org/10.1002/admi.202001388. DOI: 10.1002/admi.202001388

Meneses, P., Méndez Parra, W., Hernández Gómez, A. and Bouza Magdaleno, Y. (2022). *Report on the results of the study of Nerea Plus foliar fertilizer in common bean cultivation*. Territorial Grain Research Station "Sur del Jíbaro", Grain Research Institute, MINAG.

Mengyao, L., Changging, F., Youliang, C., Xin, Z., Hanzhi, H., Jiarui, Z., & Yue, Z. (2024). A facile pathway to fabricate VO2 (M) nanoparticles via sol-gel method for flexible thermochromic films with efficient infrared stealth. *Vacuum*, *221*, 112885. DOI: 10.1016/j.vacuum.2023.112885

Mensah, J., Yan, P., Rawal, A., Lee, A. F., Wilson, K., Robinson, N., Johns, M. L., Kennedy, E., & Stockenhuber, M. (2024). Catalytic cracking of 1,3,5-triisopropylbenzene and low-density polyethylene over hierarchical Y zeolites and Al-SBA-15. *ChemCatChem*, *16*(3), e202300884. DOI: 10.1002/cctc.202300884

Milina, M., Mitchell, S., Crivelli, P., Cooke, D., & Pérez-Ramírez, J. (2014). Mesopore quality determines the lifetime of hierarchically structured zeolite catalysts. *Nature Communications*, *5*(1), 3922. DOI: 10.1038/ncomms4922

Millan, R., Cnudde, P., Hoffman, A. E. J., Lopes, C. W., Concepción, P., van Speybroeck, V., & Boronat, M. (2020). Theoretical and Spectroscopic Evidence of the Dynamic Nature of Copper Active Sites in Cu-CHA Catalysts under Selective Catalytic Reduction (NH<inf>3</inf>-SCR-NO<inf>x</inf>) Conditions. *The Journal of Physical Chemistry Letters*, *11*(23), 10060–10066. DOI: 10.1021/acs.jpclett.0c03020 PMID: 33179925

Minato, H. (1968). Characteristics and Uses of Natural Zeolites. *Koatsugasu*, *5*, 536–547.

Ming, D. W., & Allen, E. R. (2001) *Use of Natural Zeolites in Agronomy, Horticulture and Environmental Soil Remediation* Chapter 18. Nat. Zeol. Occur. Prop., Appl., Mineralogical Society of America. Geochemical Society ISSN 1529-6466

Ming, D. W., & Boettinger, J. L. (2001). Zeolites in Soil Environments, Chapter 11 of Natural Zeolites: Occurrence, Properties, Applications (Mineralogical Society of America, Geochemical Society ISSN 1529-6466

Ming, D.W. and Allen E.R. (1999). Zeoponic Substrates for Space Applications: Advances in the Use of Natural Zeolites for Plant Growth. Misailidis, P., Macásek, F., Pinnavaia, T.J., Colella, C. (eds) Natural Microporous Materials in Environmental Technology. NATO Science Series book series NSSE, vol 362, 157-176. Springer, Dordrecht. https://doi.org/doi:10.1007/978-94-011-4499-5_11

Min, G., Wei, W., Fan, Q., Wan, T., Ye, M., & Yun, S. (2025). High-throughput exploration of stable semiconductors using deep learning and density functional theory. *Materials Science in Semiconductor Processing*, *188*, 109150. DOI: 10.1016/j.mssp.2024.109150

Mitchell, J., Broche, L. M., Chandrasekera, T. C., Lurie, D. J., & Gladden, L. F. (2013). Exploring surface interactions in catalysts using low-field nuclear magnetic resonance. *The Journal of Physical Chemistry. C, Nanomaterials and Interfaces*, *117*(34), 17699–17706. DOI: 10.1021/jp405987m

Mitchell, J., Gladden, L. F., Chandrasekera, T. C., & Fordham, E. J. (2014). Low-field permanent magnets for industrial process and quality control. *Progress in Nuclear Magnetic Resonance Spectroscopy*, *76*, 1–60. DOI: 10.1016/j.pnmrs.2013.09.001 PMID: 24360243

Mitchell, J., Griffith, J. D., Collins, J. H. P., Sederman, A. J., Gladden, L. F., & Johns, M. L. (2007). Validation of NMR relaxation exchange time measurements in porous media. *The Journal of Chemical Physics*, *127*(23), 234701. DOI: 10.1063/1.2806178 PMID: 18154403

Mitchell, J., Webber, J. B. W., & Strange, J. H. (2008). Nuclear magnetic resonance cryoporometry. *Physics Reports*, *461*(1), 1–36. DOI: 10.1016/j.physrep.2008.02.001

Mittal, A., Nagpal, M., Sharma, D., Das, D., & Chahar, J. (2024). A Comprehensive Review of Natural and Surfactant- Modified Zeolites and Their Applications. *Letters in Applied NanoBioScience.*, *13*(4), 188. DOI: 10.33263/LIANBS134.188

Molina-Amat, J. C., Rodríguez-Iznaga, I., Rodríguez-Fuentes, G., de la Nuez-Pantoja, E. Y., & Caisés Ávalos, A. (2024, September-December). Industrial production of NEREA® zeolitic products for the agricultural industry: Alternatives for the curing-drying process. *Tecnología Química*, *44*(3), 529–543.

Molinari, R., Argurio, P., & Lavorato, C. (2015). Photocatalytic hydrogenation of organic compounds in membrane reactors. In *Membrane Reactors for Energy Applications and Basic Chemical Production* (pp. 605–639). Woodhead Publishing., DOI: 10.1016/B978-1-78242-223-5.00020-0

Möller, K., & Bein, T. (2013). Mesoporosity – a new dimension for zeolites. *Chemical Society Reviews*, *42*(9), 3689–3707. DOI: 10.1039/c3cs35488a PMID: 23460052

Monteilhet, L., Korb, J. P., Mitchell, J., & McDonald, P. J. (2006). Observation of exchange of micropore water in cement pastes by two-dimensional T2 - T2 nuclear magnetic resonance relaxometry. *Physical Review E: Statistical, Nonlinear, and Soft Matter Physics*, *74*(6), 061404. DOI: 10.1103/PhysRevE.74.061404 PMID: 17280070

Montesano, G., Cappelletti, P., Caputo, D., Liguori, B., Campanile, A., & Rispoli, C. (2022). Mineralogical and Technological Characterization of Zeolites from Basin and Range as Pozzolanic Addition of Cement. *Materials (Basel)*, *15*(7), 2684. Advance online publication. DOI: 10.3390/ma15072684 PMID: 35408016

Montes-Luna, . (2015). Characterization of natural and Ca2+-modified clinoptilolite by different physical-chemical methods for its possible application in gas separation processes. *Superficies y Vacío*, *28*(1). https://www.scielo.org.mx/scielo.php?script=sci_arttext&pid=S1665-35212015000100002

Moon, H. J., Sekiya, R. S., & Jones, C. W. (2023). Probing the morphology and mobility of amines in porous silica CO_2 sorbents by ^1H T_1-T_2 relaxation correlation NMR. *The Journal of Physical Chemistry. C, Nanomaterials and Interfaces*, *127*(24), 11652–11665. DOI: 10.1021/acs.jpcc.3c02441

Moradi, , Karimzadeh, R., & Moosavi, E. S. (2018). Modified and ion exchanged clinoptilolite for the adsorptive removal of sulfur compounds in a model fuel. *Fuel*, *217*, 467–477. DOI: 10.1016/j.fuel.2017.12.095

Morgan, , & Jacobs, R. (2020). Opportunities and challenges for machine learning in materials science. *Annual Review of Materials Research*, *50*(1), 71–103. DOI: 10.1146/annurev-matsci-070218-010015

Mori, K., Kagohara, K., & Yamashita, H. (2008). Synthesis of tris(2,2′-bipyridine) iron(II) complexes in zeolite y cages: Influence of exchanged alkali metal cations on physicochemical properties and catalytic activity. *The Journal of Physical Chemistry. C, Nanomaterials and Interfaces*, *112*(7), 2593–2600. DOI: 10.1021/jp709571v

Mortazavi, , Bahadori, M., Marandi, A., Tangestaninejad, S., Moghadam, M., Mirkhani, V., & Mohammadpoor-Baltork, I. (2021). Enhancement of CO2 adsorption on natural zeolite, modified clinoptilolite with cations, amines and ionic liquids. *Sustainable Chemistry and Pharmacy*, *22*, 100495. Advance online publication. DOI: 10.1016/j.scp.2021.100495

Mouna, S., Hajji, S., & Tounsi, H. (2024). Waste to health: Green synthesis of Zn loaded LTA zeolite prepared from waste glass and aluminum scrap with high anti-oxidant and antimicrobial activities. *Journal of Cleaner Production*, *434*, 139946. DOI: 10.1016/j.jclepro.2023.139946

Mukherjee, . (2021). Iron oxide-based nanomaterials for arsenic removal: A review. *Journal of Hazardous Materials*, *403*, 123558. Advance online publication. DOI: 10.1016/j.jhazmat.2020.123558

Mumpton, F. A. (1984). *Zeo-Agriculture: Use of Natural Zeolites in Agriculture & Aquaculture*. Westview.

Mumpton, F. A. (1999). *The magic rock: Uses of natural zeolites in agriculture and industry* colloquium "Geology, Mineralogy and Human Wellbeing" organized by the National Academy of Sciences of the United States of America on November 8-9, 1998, *Proc. Nat.* Science Academy of USA, Vol. 96, Issue 7, 3463-3470. DOI: 10.1073/pnas.96.7.3463

Muraoka, K., Sada, Y., Miyazaki, D., Chaikittisilp, W., & Okubo, T. (2019). Linking synthesis and structure descriptors from a large collection of synthetic records of zeolite materials. *Nature Communications*, *10*(1), 4459. DOI: 10.1038/s41467-019-12394-0

Murrieta-Rico, F. N., Antúnez-García, J., Yocupicio-Gaxiola, R. I., Galván, D. H., González, J. C., & Petranovskii, V. (2022). Zeolites as initial structures for the preparation of functional materials. *Journal of Applied Research and Technology*, *20*(1), 92–116. DOI: 10.22201/icat.24486736e.2022.20.1.1561

Murrieta-Rico, F. N., Lindner, L., Petranovskii, V., Galván, D. H., Antúnez-García, J., Yocupicio-Gaxiola, R. I., Sergiyenko, O., & Rivas-López, M. (2021) Zeolites with quantum dots as functional materials: current trends and perspectives for optical devices. *Proceedings of 2021 IEEE 30th International Symposium on Industrial Electronics (ISIE)*. Kyoto, Japan, P. 1–5. DOI: 10.1109/ISIE45552.2021.9576435

Murthi, M., & Destino, J. F. (2025). Sol-gel technologies for additive manufacturing glass materials. *Journal of Sol-Gel Science and Technology*. Advance online publication. DOI: 10.1007/s10971-025-06664-1

Mu, Y., Zhang, Y., Fan, J., & Guo, C. (2017). Effect of ultrasound pretreatment on the hydrothermal synthesis of SSZ-13 zeolite. *Ultrasonics Sonochemistry*, *38*, 430–436. DOI: 10.1016/j.ultsonch.2017.03.043

Nachtigall, P., Delgado, M. R., Nachtigallova, D., & Areán, C. O. (2012). The nature of cationic adsorption sites in alkaline zeolites - Single, dual and multiple cation sites. In *Physical Chemistry Chemical Physics* (Vol. 14, Issue 5). DOI: 10.1039/C2CP23237E

Na, K., Chol, M., Park, W., Sakamoto, Y., Terasakl, O., & Ryoo, R. (2010). Pillared MFI zeolite nanosheets of a single-unit-cell thickness. *Journal of the American Chemical Society*, *132*(12), 4169–4177. DOI: 10.1021/ja908382n PMID: 20136135

Na, K., Park, W., Seo, Y., & Ryoo, R. (2011). Disordered Assembly of MFI Zeolite Nanosheets with a Large Volume of Intersheet Mesopores. *Chemistry of Materials*, *23*(5), 1273–1279. DOI: 10.1021/cm103245m

Nami, . (2023a). Nitrate removal performance of different granular adsorbents using a novel Fe-exchanged nanoporous clinoptilolite. *Industrial & Engineering Chemistry Research*, *62*(8), 3659–3671. DOI: 10.1021/acs.iecr.2c03308

Narayanan, S., Tamizhdurai, P., Mangesh, V. L., Ragupathi, C., Santhana krishnan, P., & Ramesh, A. (2020). Recent advances in the synthesis and applications of mordenite zeolite - review. *RSC Advances*, *11*(1), 250–267. DOI: 10.1039/D0RA09434J PMID: 35423021

Ng, E. P., Awala, H., Ghoy, J. P., Vicente, A., Ling, T. C., Ng, Y. H., Mintova, S., & Adam, F. (2015). Effects of ultrasonic irradiation on crystallization and structural properties of EMT-type zeolite nanocrystals. *Materials Chemistry and Physics*, *159*, 38–45. DOI: 10.1016/j.matchemphys.2015.03.047

Nguyen, D.-K., Dinh, V.-P., Dang, N. T., Khan, D. T., Hung, N. T., & Thi Tran, N. H. (2023). Effects of aging and hydrothermal treatment on the crystallization of ZSM-5 zeolite synthesis from bentonite. In RSC Advances (Vol. 13, Issue 30, pp. 20565–20574). Royal Society of Chemistry (RSC). https://doi.org/DOI: 10.1039/D3RA02552G

Nguyen, , Loganathan, P., Nguyen, T. V., & Vigneswaran, S. (2019). Removing arsenic from water with an original and modified natural manganese oxide ore: Batch kinetic and equilibrium adsorption studies. *Environmental Science and Pollution Research International*, *27*(5), 5490–5502. DOI: 10.1007/s11356-019-07284-3 PMID: 31853842

Nidheesh, P. V., Kumar, M., Venkateshwaran, G., Ambika, S., & Bhaskar, S. Vinay, & Ghosh, P. (2024). Conversion of locally available materials to biochar and activated carbon for drinking water treatment. In Chemosphere (Vol. 353, p. 141566). Elsevier BV. https://doi.org/DOI: 10.1016/j.chemosphere.2024.141566

Novikov, R. G., Konopnitskii, R. & Tsyganenko, A. A. (2018). Distortions in IR Spectra Related to Registration Conditions: II. The Influence of Scattering. *Optics and Spectroscopy 2018 124:5, 124*(5), 655–659. DOI: 10.1134/S0030400X18050156

Núñez-González, R., Xiao, M., Antúnez-García, J., Ponce-Ruiz, J. L. A., Reyes-Serrato, A., Petranovskii, V., Murrieta-Rico, F. N., & Rakov, N. (2023). First-Principles Study of optical properties of Linde-Type A zeolite. *Physica Status Solidi. B, Basic Research, 260*(12), 2300378. Advance online publication. DOI: 10.1002/pssb.202300378

Ogorek, . (2024). Soil compaction sensing mechanisms and root responses. *Trends in Plant Science, 29*(12), 1345–1356. DOI: 10.1016/j.tplants.2024.10.014

Ogura, M., Shinomiya, S.-Y., Tateno, J., Nara, Y., Kikuchi, E., & Matsukata, M. (2000). Formation of uniform mesopores in ZSM-5 zeolite through treatment in alkaline solution. *Chemistry Letters, 8*(8), 882–883. DOI: 10.1246/cl.2000.882

Oheix, E., Reicher, C., Nouali, H., Michelin, L., Josien, L., Daou, T. J., & Pieuchot, L. (2022). Rational Design and Characterisation of Novel Mono- and Bimetallic Antibacterial Linde Type A Zeolite Materials. *Journal of Functional Biomaterials, 13*(2), 73. DOI: 10.3390/jfb13020073 PMID: 35735928

Oliveira, D. S., Lima, R. B., Pergher, S. B. C., & Caldeira, V. P. S. (2023). Hierarchical zeolite synthesis by alkaline treatment: Advantages and applications. *Catalysts, 13*(2), 316. DOI: 10.3390/catal13020316

Oliviero, L., Vimont, A., Lavalley, J.-C., Sarria, F. R., Gaillard, M., & Maugé, F. (2005). 2,6-Dimethylpyridine as a probe of the strength of Brønsted acid sites: Study on zeolites. Application to alumina. *Physical Chemistry Chemical Physics, 7*(8), 1861–1869. DOI: 10.1039/B500689A PMID: 19787950

Olla, C., & Carbonaro, C. M. (2024). The void side of silica: Surveying optical properties and applications of mesoporous silica. *Journal of Physics Condensed Matter, 36*(25), 253002. DOI: 10.1088/1361-648X/ad3474 PMID: 38545962

Onyutha, C., Okello, E., Atukwase, R., Nduhukiire, P., Ecodu, M., & Kwiringira, J. N. (2024). Improving household water treatment: using zeolite to remove lead, fluoride and arsenic following optimized turbidity reduction in slow sand filtration. In Sustainable Environment Research (Vol. 34, Issue 1). Springer Science and Business Media LLC. https://doi.org/DOI: 10.1186/s42834-024-00209-x

Oruji, S., Khoshbin, R., & Karimzadeh, R. (2018). Preparation of hierarchical structure of Y zeolite with ultrasonic-assisted alkaline treatment method used in catalytic cracking of middle distillate cut: The effect of irradiation time. *Fuel Processing Technology*, *176*, 283–295. DOI: 10.1016/j.fuproc.2018.03.035

Oruji, S., Khoshbin, R., & Karimzadeh, R. (2019). Combination of precipitation and ultrasound irradiation methods for preparation of lanthanum-modified Y zeolite nano-catalysts used in catalytic cracking of bulky hydrocarbons. *Materials Chemistry and Physics*, *230*, 131–144. DOI: 10.1016/j.matchemphys.2019.03.038

Ostrooumov, , Cappelletti, P., & de'Gennaro, R. (2011). Mineralogical study of zeolite from New Mexican deposits (Cuitzeo area, Michoacan, Mexico). *Applied Clay Science*, *55*, 156–160. DOI: 10.1016/j.clay.2011.09.011

Otero Areán, C., Manoilova, O. V., Tsyganenko, A. A., Turnes Palomino, G., Peñarroya Mentruit, M., Geobaldo, F. & Garrone, E. (2001). Thermodynamics of Hydrogen Bonding between CO and the Supercage Brønsted Acid Sites of the H-Y Zeolite – Studies from Variable Temperature IR Spectrometry. *European Journal of Inorganic Chemistry, 2001*(7). DOI: 10.1002/1099-0682(200107)2001:7<1739::AID-EJIC1739>3.3.CO;2-D

Otero Areán, C., Tsyganenko, A. A., Escalona Platero, E., Garrone, E., & Zecchina, A. (1998). Two Coordination Modes of CO in Zeolites: A Temperature-Dependent Equilibrium. *Angewandte Chemie International Edition*, *37*(22), 3161–3163. DOI: 10.1002/(SICI)1521-3773(19981204)37:22<3161::AID-ANIE3161>3.0.CO;2-B PMID: 29711312

Otero Areán, C., Turnes Palomino, G., Tsyganenko, A. A., & Garrone, E. (2002). Quantum Chemical and FTIR Spectroscopic Studies on the Linkage Isomerism of Carbon Monoxide in Alkali-Metal-Exchanged Zeolites: A Review of Current Research. *International Journal of Molecular Sciences*, *3*(7), 764–776. DOI: 10.3390/i3070764

Otitoju, T. A., Saari, R. A., & Ahmad, A. L. (2018). Progress in the modification of reverse osmosis membranes for enhanced performance. *Journal of Industrial and Engineering Chemistry*, *67*, 52–71. Advance online publication. DOI: 10.1016/j.jiec.2018.07.010

Ottoboni, , Coleman, S. J., Steven, C., Siddique, M., Fraissinet, M., Joannes, M., Laux, A., Barton, A., Firth, P., Price, C. J., & Mulheran, P. A. (2020). Understanding API static drying with hot gas flow: Design and test of a drying rig prototype and drying modeling development. *Organic Process Research & Development*, *24*(11), 2505–2520. DOI: 10.1021/acs.oprd.0c00035 PMID: 33250628

Ovchinnikov, N. L., Vinogradov, N. M., Gordina, N. E., & Butman, M. F. (2023). Preparation of Photoactive TiO2/LTA Zeolite Composites by Solution Technology in Hydrothermal Conditions. *Protection of Metals and Physical Chemistry of Surfaces*, *59*(4), 570–576. DOI: 10.1134/S2070205123700776

Ozaki, Y., Suzuki, Y., Hawai, T., Saito, K., Onishi, M., & Ono, K. (2020).. . *Computational Materials*, *6*, 1. DOI: 10.1038/s41524-019-0267-z

Ozbahcea, A., Fuat Tarib, A., Gönülalc, E., Simseklic, N., & Pademd, H. (2015). The effect of zeolitic zeolite applications on yield components and nutrient uptake of common bean under water stress. *Archives of Agronomy and Soil Science*, *61*(5), 615–626. DOI: 10.1080/03650340.2014.946021

Ozbahcea, A., Tarib, A. F., Gönülalc, E., & Simseklic, N. (2018). *Zeolite for Enhancing Yield and* Quality *of Potatoes Cultivated Under Water Deficit Conditions. Potato Research*, *61*(3), 247–259. DOI: 10.1007/s11540-018-9372-5

Pabiś-Mazgaj, , Gawenda, T., Pichniarczyk, P., & Stempkowska, A. (2021). Mineral composition and structural characterization of the clinoptilolite powders obtained from zeolite-rich tuffs. *Minerals (Basel)*, *1030*(10), 1030. Advance online publication. DOI: 10.3390/min11101030

Paixão, V., Carvalho, A. P., Rocha, J., Fernandes, A., & Martins, A. (2010). Modification of MOR by desilication treatments: Structural, textural and acidic characterization. *Microporous and Mesoporous Materials*, *131*(1–3), 350–357. DOI: 10.1016/j.micromeso.2010.01.013

Pal, P. (2017). Nanotechnology in water treatment. En *Industrial Water Treatment Process Technology* (pp. 513-536). https://doi.org/DOI: 10.1016/B978-0-12-810391-3.00007-2

Pal, P., Das, J. K., Das, N., & Bandyopadhyay, S. (2013). Synthesis of NaP zeolite at room temperature and short crystallization time by sonochemical method. *Ultrasonics Sonochemistry*, *20*(1), 314–321. DOI: 10.1016/j.ultsonch.2012.07.012

Pan, L., Ye, S., Xv, X., Lin, P., Huang, R., & Wang, D. (2023). Zeolite-Encaged Luminescent Silver Nanoclusters. *Materials (Basel)*, *16*(10), 3736. Advance online publication. DOI: 10.3390/ma16103736 PMID: 37241363

Pappas, P. G., Kauffman, C. A., Andes, D. R., Clancy, C. J., Marr, K. A., Ostrosky-Zeichner, L., Reboli, A. C., Schuster, M. G., Vazquez, J. A., Walsh, T. J., Zaoutis, T. E., & Sobel, J. D. (2015). Clinical Practice Guideline for the Management of Candidiasis: 2016 Update by the Infectious Diseases Society of America. In *Clinical Infectious Diseases* (Vol. 62, Number 4, pp. e1–e50). Oxford University Press. DOI: 10.1093/cid/civ933

Paris, E. C., Malafatti, J. O. D., Musetti, H. C., Manzoli, A., Zenatti, A., & Escote, M. T. (2020). Faujasite zeolite decorated with cobalt ferrite nanoparticles for improving removal and reuse in Pb2+ ions adsorption. *Chinese Journal of Chemical Engineering*, *28*(7), 1884–1890. DOI: 10.1016/j.cjche.2020.04.019

Parvekar, P., Palaskar, J., Metgud, S., Maria, R., & Dutta, S. (2020). The minimum inhibitory concentration (MIC) and minimum bactericidal concentration (MBC) of silver nanoparticles against *Staphylococcus aureus.Biomaterial Investigations in Dentistry*, *7*(1), 105–109. DOI: 10.1080/26415275.2020.1796674 PMID: 32939454

Pasquet, J., Chevalier, Y., Pelletier, J., Couval, E., Bouvier, D., & Bolzinger, M.-A. (2014). The contribution of zinc ions to the antimicrobial activity of zinc oxide. *Colloids and Surfaces. A, Physicochemical and Engineering Aspects*, *457*, 263–274. DOI: 10.1016/j.colsurfa.2014.05.057

Paukshtis, E. A., & Yurchenko, E. N. (1983). Study of the Acid–Base Properties of Heterogeneous Catalysts by Infrared Spectroscopy. *Russian Chemical Reviews*, *52*(3), 242–258. DOI: 10.1070/RC1983v052n03ABEH002812

Pavelić, . (2018). Critical review on zeolite clinoptilolite safety and medical applications in vivo. *Frontiers in Pharmacology*, *9*, 1350. Advance online publication. DOI: 10.3389/fphar.2018.01350 PMID: 30538633

Pavlak, , Kaufmann, C., Eyng, C., Carvalho, P. L. O., Pozza, P. C., Vieites, F. M., Rohloff Junior, N., Avila, A. S., Polese, C., & Nunes, R. V. (2023). Zeolite and corn with different compositions in broiler chickens feeding. *Poultry Science*, *102*(4), 102494. Advance online publication. DOI: 10.1016/j.psj.2023.102494 PMID: 36745958

Pavlov, A. Y., & Tsyganenko, A. A. (1994). Band profile of hydroxyl groups in the infrared spectrum of hydrogen-bonded surface complexes: Ammonia on silicon dioxide. [English Translation of Optika i Spektroskopiya]. *Optics and Spectroscopy*, *77*(1).

Pawelec, B., Halachev, T., Olivas, A., & Zepeda, T. A. (2008). Impact of preparation method and support modification on the activity of mesoporous hydrotreating CoMo catalysts. *Applied Catalysis A, General*, *348*(1), 30–41. DOI: 10.1016/j.apcata.2008.06.014

Pechar, F., & Rykl, D. (1983). Study of the complex vibrational spectra of natural zeolite mordenites. *Zeolites*, *3*(4), 329–332. DOI: 10.1016/0144-2449(83)90177-X

Pecharsky, V. K., & Zavalij, P. Y. (2009). Fundamentals of diffraction. Fundamentals of Powder Diffraction and Structural Characterization of Materials, 133-149.

Perego, , de Angelis, A., Pollesel, P., & Millini, R. (2021). Zeolite-based catalysis for phenol production. *Industrial & Engineering Chemistry Research*, *60*(18), 6379–6402. DOI: 10.1021/acs.iecr.0c05886

Pérez, J. E., Rivero, L., & Arozarena, N. (1991). *Estudio del contenido de nutrientes en sustratos NEREA*. Memoirs 3rd Int. Conf. Occur. Prop. Util. Nat. Zeolites, Part II, 3-7.

Pérez, J., & Manuel Sánchez-Sánchez Editors, P. (2018). *Structure and Reactivity of Metals in Zeolite Materials* (J. Pérez Pariente & M. Sánchez-Sánchez, Eds.; Vol. 178). Springer International Publishing. https://doi.org/DOI: 10.1007/978-3-319-98905-1

Pérez-Botella, , Valencia, S., & Rey, F. (2022). Zeolites in adsorption processes: State of the art and future prospects. *Chemical Reviews*, *17647*(24), 17647–17695. Advance online publication. DOI: 10.1021/acs.chemrev.2c00140 PMID: 36260918

Pérez-Cabrera, L., Díaz de León, J. N., Antúnez-García, J., Galván, D. H., Alonso-Núñez, G., & Fuentes-Moyado, S. (2019). Isomorphic substitution of Mg2+ by Al3+ on MgO: Effects on basicity, textural properties and microstructure. *Revista Mexicana de Ingeniería Química*, *18*(1), 9. DOI: 10.24275/uam/izt/dcbi/revmexingquim/2019v18n1/Perez

Pérez-Ramírez, J., Christensen, C. H., Egeblad, K., Christensen, C. H., & Groen, J. C. (2008). Hierarchical zeolites: Enhanced utilisation of microporous crystals in catalysis by advances in materials design. *Chemical Society Reviews*, *37*(11), 2530–2542. DOI: 10.1039/b809030k PMID: 18949124

Pérez-Ramírez, J., Verboekend, D., Bonilla, A., & Abelló, S. (2009). Zeolite Catalysts with Tunable Hierarchy Factor by Pore-Growth Moderators. *Advanced Functional Materials*, *19*(24), 3972–3979. https://doi.org/https://doi.org/10.1002/adfm.200901394. DOI: 10.1002/adfm.200901394

Peri, J. B. (1966). Infrared study of OH and NH_2 groups on the surface of a dry silica aerogel. *Journal of Physical Chemistry*, *70*(9), 2937–2945. DOI: 10.1021/j100881a037

Perkins, E. L., Lowe, J. P., Edler, K. J., Tanko, N., & Rigby, S. P. (2008). Determination of the percolation properties and pore connectivity for mesoporous solids using NMR cryodiffusometry. *Chemical Engineering Science*, *63*(7), 1929–1940. DOI: 10.1016/j.ces.2007.12.022

Petitjean, H., Lepetit, C., Nour, Z., Poteau, R., del Rosal, I., & Berthomieu, D. (2020). How CuI and NaI interact with faujasite zeolite? A theoretical investigation. *The Journal of Physical Chemistry. C, Nanomaterials and Interfaces*, *124*(51), 28026–28037. DOI: 10.1021/acs.jpcc.0c06862

Petkovich, N. D., & Stein, A. (2013). Controlling macro- and mesostructures with hierarchical porosity through combined hard and soft templating. *Chemical Society Reviews*, *42*(9), 3721–3739. DOI: 10.1039/C2CS35308C PMID: 23072972

Petranovskii, V., Gurin, V., & Machorro, R. (2005). Spectroscopic observation and ab initio simulation of copper clusters in zeolites. *Catalysis Today*, *107–108*, 892–900. DOI: 10.1016/j.cattod.2005.07.039

Petranovskii, V., Stoyanov, E., Gurin, V., Katada, N., Hernández, M.-A., Avalos, M., Pestryakov, A., Chávez Rivas, F., Zamorano Ulloa, R., & Portillo, R. (2013). Formation of copper nanoparticles in mordenites with variable SiO_2/Al_2O_3 molar ratios under redox treatments. *Revista Mexicana de Física*, *59*(2), 170–185.

Petrov, I., & Michalev, T. (2012). Synthesis of Zeolite A. *RE:view*.

Petrov, O. V., & Furó, I. (2009). NMR cryoporometry: Principles, applications and potential. *Progress in Nuclear Magnetic Resonance Spectroscopy*, *54*(2), 97–122. DOI: 10.1016/j.pnmrs.2008.06.001

Petrov, O., & Furó, I. (2006). Curvature-dependent metastability of the solid phase and the freezing-melting hysteresis in pores. *Physical Review E: Statistical, Nonlinear, and Soft Matter Physics*, *73*(1), 011608. DOI: 10.1103/PhysRevE.73.011608 PMID: 16486162

Petryk, M. R., Boyko, I. V., Petryk, M. M., Fraissard, J., & Mudryk, I. (2020). Modeling of adsorption and desorption of hydrocarbons in nanoporous catalytic zeolite media using nonlinear Langmuir isotherm. *Proceedings of SPIE - the International Society for Optical Engineering*, *11369*, 42. Advance online publication. DOI: 10.1117/12.2553926

Piccione, P. M., Laberty, C., Yang, S., Camblor, M. A., Navrotsky, A., & Davis, M. E. (2000). Thermochemistry of Pure-Silica Zeolites. *The Journal of Physical Chemistry B, 104*(43), 10001–10011. DOI: 10.1021/jp002148a

Pike, K. J., Malde, R. P., Ashbrook, S. E., McManus, J., & Wimperis, S. (2000). Multiple-quantum MAS NMR of quadrupolar nuclei. Do five-, seven- and nine-quantum experiments yield higher resolution than the three-quantum experiment? *Solid State Nuclear Magnetic Resonance, 16*(3), 203–215. DOI: 10.1016/S0926-2040(00)00081-3 PMID: 10868573

Poblete Palacios, G. E. (2017). *Estudio de la influencia de la densidad de dislocaciones en la resistencia mecánica del cobre* [Tesis de ingeniería civil mecánico, Universidad de Chile]. Facultad de Ciencias Físicas y Matemáticas, Departamento de Ingeniería Mecánica.

Pollet, B. G., & Ashokkumar, M. (2019). *Introduction to Ultrasound, Sonochemistry and Sonoelectrochemistry*. Springer International Publishing., DOI: 10.1007/978-3-030-25862-7

Ponce Cruz, L., Blanco Concepción, N., & García González, A. (2011). *Cultivos sin suelo empleando Zeolita en* Jagüey *Grande Matanzas.1st Workshop Use Industrial Minerals in Agricultural Branch*, Villa Clara, Cuba

Pour, . (2024). Rapid SO2 gas removal using MgO/AC/CaCO3/zeolite nanocomposite at room temperature. *Case Studies in Chemical and Environmental Engineering, 10*, 101042. Advance online publication. DOI: 10.1016/j.cscee.2024.101042

Prabhu, M. K., Louwen, J. N., Vogt, E. T. C., & Groot, I. M. N. (2025). Catalytic Hydrodesulfurization of Thiophene, Dibenzothiophene and 4,6-Dimethyldibenzothiophene on a CoMoS Catalyst. *Catalysis Letters, 155*(3), 113. DOI: 10.1007/s10562-025-04953-6

Prats, H., Alonso, G., Sayós, R., & Gamallo, P. (2020). Transition metal atoms encapsulated within microporous Silicalite-1 zeolite: A systematic computational study. *Microporous and Mesoporous Materials, 308*, 110462. Advance online publication. DOI: 10.1016/j.micromeso.2020.110462

Price, L. A., Ridley, C. J., Bull, C. L., Wells, S. A., & Sartbaeva, A. (2021). Determining the structure of zeolite frameworks at high pressures. *CrystEngComm, 23*(33), 5615–5623. DOI: 10.1039/D1CE00142F

Qasem, N. A. A., Mohammed, R. H., & Lawal, D. U. (2021). Removal of heavy metal ions from wastewater: a comprehensive and critical review. In npj Clean Water (Vol. 4, Issue 1). Springer Science and Business Media LLC. https://doi.org/DOI: 10.1038/s41545-021-00127-0

Qin, F., Li, J., Zhang, C., Zeng, G., Huang, D., Tan, X., Qin, D., Tan, H., 2022. Biochar in the 21st century: a data-driven visualization of collaboration, frontier identification, and future trend. Sci. Total Environ. 818, 151774 https://doi.org/. scitotenv.2021.151774.DOI: 10.1016/j

Qin, Z., Cychosz, K. A., Melinte, G., El Siblani, H., Gilson, J. P., Thommes, M., Fernandez, C., Mintova, S., Ersen, O., & Valtchev, V. (2017). Opening the cages of faujasite-type zeolite. *Journal of the American Chemical Society*, *139*(48), 17273–17276. DOI: 10.1021/jacs.7b10316 PMID: 29160696

Qiu, Zh.-Zh. Yu., Yu, Y.-X., & Mi, J.-G. (2012). Adsorption of carbon monoxide on Ag(I)-ZSM-5 zeolite: An ab initio density functional theory study. *Applied Surface Science*, *258*(24), 9629–9635. DOI: 10.1016/j.apsusc.2012.05.162

Rabia, , Ould Hamou, M., Kasperkiewicz, K., Brożek, J., & Augustyniak, M. (2019). Adhesion abilities and biosorption of Cd and Mg by microorganisms - first step for beneficiation of phosphate ore. *Scientific Reports*, *9*(1), 12929. Advance online publication. DOI: 10.1038/s41598-019-49406-4 PMID: 31506570

Rabo, J. A. (Ed.). (1976). *Zeolite Chemistry and Catalysis*. American Chemical Society.

Rahman, I., Vejayakumaran, P., Sipaut, C., Ismail, J., & Chee, C. (2008). Size-dependent physicochemical and optical properties of silica nanoparticles. *Materials Chemistry and Physics*, *114*(1), 328–332. DOI: 10.1016/j.matchemphys.2008.09.068

Rainho, J. P., Rocha, J., Carlos, L. D., & Almeida, R. M. (2001). ^{29}Si nuclear-magnetic-resonance and vibrational spectroscopy studies of SiO_2–TiO_2 powders prepared by the sol-gel process. *Journal of Materials Research*, *16*(8), 2369–2376. DOI: 10.1557/JMR.2001.0325

Rakanović, M., Vukojević, A., Savanović, M. M., Armaković, S., Pelemiš, S., Živić, F., Sladojević, S., & Armaković, S. J. (2022). Zeolites as Adsorbents and Photocatalysts for Removal of Dyes from the Aqueous Environment. In Molecules (Vol. 27, Issue 19, p. 6582). MDPI AG. https://doi.org/DOI: 10.3390/molecules27196582

Rakmae, S., Keawkumay, C., Osakoo, N., Montalbo, K. D., de Leon, R. L., Kidkhunthod, P., Chanlek, N., Roessner, F., Prayoonpokarach, S., & Wittayakun, J. (2016). Realization of active species in potassium catalysts on zeolite NaY prepared by ultrasound-assisted impregnation with acetate buffer and improved performance in transesterification of palm oil. *Fuel*, *184*, 512–517. DOI: 10.1016/j.fuel.2016.07.035

Ramasamy, , Senthil Kumar, R., Dharmaraj, R., & Kannan, M. (2025). Fluoride removal from contaminated water using natural bio-adsorbents: A numerical and experimental investigation. *Desalination and Water Treatment*, *324*, 100984. Advance online publication. DOI: 10.1016/j.dwt.2024.100984

Ramirez Mendoza, H., Jordens, J., Valdez Lancinha Pereira, M., Lutz, C., & Van Gerven, T. (2020). Effects of ultrasonic irradiation on crystallization kinetics, morphological and structural properties of zeolite FAU. *Ultrasonics Sonochemistry*, *64*, 105010. DOI: 10.1016/j.ultsonch.2020.105010

Ramishvili, T., Tsitsishvili, V., Chedia, R., Sanaia, E., Gabunia, V., & Kokiashvili, N. (2017). Preparation of Ultradispersed Crystallites of Modified Natural Clinoptilolite with the Use of Ultrasound and Its Application as a Catalyst in the Synthesis of Methyl Salicylate. *American Journal of Nano Research and Applications.Special Issue: Nanotechnologies*, *5*(1), 26–32. DOI: 10.11648/j.nano.s.2017050301.17

Ramkumar, T., Selvakumar, M., Vasanthsankar, R., Sathishkumar, A. S., Narayanasamy, P., & Girija, G. (2018). Rietveld refinement of powder X-ray diffraction, microstructural and mechanical studies of magnesium matrix composites processed by high energy ball milling. *Journal of Magnesium and Alloys*, *6*(4), 390–398. DOI: 10.1016/j.jma.2018.08.002

Rapado, , Garcés, D., Faba, L., & Ordóñez, S. (2024). Zeolite-assisted acid hydrolysis of cellulose: Optimization of reaction conditions and chemical pretreatments for enhancing HMF yields. *Industrial Crops and Products*, *212*, 118309. Advance online publication. DOI: 10.1016/j.indcrop.2024.118309

Rasmidi, R., Duinong, M., & Chee, F. P. (2021). Radiation damage effects on zinc oxide (ZnO) based semiconductor devices – a review. *Radiation Physics and Chemistry*, *184*, 109455. DOI: 10.1016/j.radphyschem.2021.109455

Rasyid, F., & Masri, M. (2019). Desalination of sea water using activated zeolite by cloride acid in tropical area based on column ion exchange method. *AMINA*, *1*(3), 114–119.

Rathi, , & Kumar, P. S. (2021). Application of adsorption process for effective removal of emerging contaminants from water and wastewater. *Environmental Pollution*, *116995*, 116995. Advance online publication. DOI: 10.1016/j.envpol.2021.116995 PMID: 33789220

Ravi, M., Sushkevich, V. L., & van Bokhoven, J. A. (2020). Towards a better understanding of Lewis acidic aluminium in zeolites. *Nature Materials*, *19*(10), 1047–1056. DOI: 10.1038/s41563-020-0751-3 PMID: 32958864

Ravi, M., Sushkevich, V. L., & van Bokhoven, J. A. (2021). On the location of Lewis acidic aluminum in zeolite mordenite and the role of framework-associated aluminum in mediating the switch between Brønsted and Lewis acidity. *Chemical Science (Cambridge)*, *12*(11), 4094–4103. DOI: 10.1039/D0SC06130A PMID: 34163680

Raymond, O., Villavicencio, H., Flores, E., Petranovskii, V., & Siqueiros, J. M. (2007). Spectral-Optic response of ZnS and mixed Cd_xZn_yS nanoclusters on synthetic mordenites. *The Journal of Physical Chemistry. C, Nanomaterials and Interfaces*, *111*(28), 10260–10266. DOI: 10.1021/jp072417y

Reid, , Klebe, S., van Zandwijk, N., & George, A. M. (2021). Asbestos and zeolites: From A to Z via a common ion. *Chemical Research in Toxicology*, *34*(4), 936–951. DOI: 10.1021/acs.chemrestox.0c00286 PMID: 33749247

Reinoso, D., Adrover, M., & Pedernera, M. (2018). Green synthesis of nanocrystalline faujasite zeolite. *Ultrasonics Sonochemistry*, *42*, 303–309. DOI: 10.1016/j.ultsonch.2017.11.034

Ren, X., Zhai, Y., Yang, N., Wang, B., & Liu, S. (2025). Lattice Oxygen Redox Dynamics in Zeolite-Encapsulated CsPbBr<inf>3</inf> Perovskite OER Electrocatalysts. *Advancement of Science*, *12*(8), 2412679. Advance online publication. DOI: 10.1002/advs.202412679 PMID: 39783715

Resasco, D. E., Crossley, S. P., Wang, B., & White, J. L. (2021). Interaction of water with zeolites: A review. *Catalysis Reviews*, *63*(2), 302–362. DOI: 10.1080/01614940.2021.1948301

Restiawaty, , Gozali, V. A., Wibisono, T. A. S. E., & Budhi, Y. W. (2024). Utilizing modified clinoptilolite for the adsorption of heavy metal ions in acid mine drainage. *Case Studies in Chemical and Environmental Engineering*, *9*, 100706. Advance online publication. DOI: 10.1016/j.cscee.2024.100706

Reyes Villegas, V. A., De León Ramirez, J. I., Perez-Cabrera, L., Estolano-Cobián, A., Chavez-Mendez, J. R., Huerta-Arcos, L., & Petranovskii, V. (2025). Sonochemical preparation of Fe-Y zeolite catalyst for the conversion of benzene to phenol. *Applied Catalysis A, General*, *689*, 120023. DOI: 10.1016/j.apcata.2024.120023

Reyes Villegas, V. A., De León Ramirez, J. I., Pérez-Cabrera, L., Pérez-Sicairos, S., Chávez-Méndez, J. R., & Petranovskii, V. (2024). Analysis of catalytic sites in FeY zeolite prepared by sono-assisted exchange of iron (II) ions. *Microporous and Mesoporous Materials*, *380*. Advance online publication. DOI: 10.1016/j.micromeso.2024.113306

Reyes Villegas, V. A., De León Ramirez, J. I., Pérez-Cabrera, L., Pérez-Sicairos, S., Yocupicio-Gaxiola, R. I., Chávez-Méndez, J. R., Huerta-Arcos, L., & Petranovskii, V. (2025). Sonochemical post-synthesis modification of Y zeolite with iron species. *Materials Chemistry and Physics*, *331*, 130199. DOI: 10.1016/j.matchemphys.2024.130199

Reyes Villegas, V. A., De León Ramirez, J. I., Perez-Sicairos, S., Yocupicio-Gaxiola, R. I., González-Torres, V., & Petranovskii, V. (2024). Catalyst for lactose hydrolysis based on zeolite Y modified with Fe species by ultrasound treatment. *Environmental Advances*, *15*, 100475. DOI: 10.1016/j.envadv.2023.100475

Rhoda, H. M., Heyer, A. J., Snyder, B. E. R., Plessers, D., Bols, M. L., Schoonheydt, R. A., Sels, B. F., & Solomon, E. I. (2022). Second-Sphere Lattice Effects in Copper and Iron Zeolite Catalysis. *Chemical Reviews*, *122*(14), 12207–12243. DOI: 10.1021/acs.chemrev.1c00915

Ribeiro, A. C., de Oliveira, A. M., Beltran, L. B., Diório, A., Magalhães-Ghiotto, G. A. V., de Abreu Filho, B. A., de Almeida Duarte, E., & Bergamasco, R. (2024). Antibacterial activity of functionalized natural zeolites (NZ-AgNPs) and its application in bacteriological water treatment and commercial paints. *Environmental Nanotechnology, Monitoring & Management*, *22*, 101001. DOI: 10.1016/j.enmm.2024.101001

Richardson, H. H., Baumann, C., & Ewing, G. E. (1987). Infrared spectroscopy and thermodynamic measurements of CO on NaCl films. *Surface Science*, *185*(1–2), 15–35. DOI: 10.1016/S0039-6028(87)80611-8

Rieger, K. A., Cho, H. J., Yeung, H. F., Fan, W., & Schiffman, J. D. (2016). Antimicrobial Activity of Silver Ions Released from Zeolites Immobilized on Cellulose Nanofiber Mats. *ACS Applied Materials & Interfaces*, *8*(5), 3032–3040. DOI: 10.1021/acsami.5b10130 PMID: 26788882

Rietveld, H. M. (2021). The Rietveld refinement method: Half of a century anniversary. *Crystal Growth & Design*, *21*(8), 4821–4822. DOI: 10.1021/acs.cgd.1c00854

Rivera-Garza, M., Olguín, M. T., García-Sosa, I., Alcántara, D., & Rodríguez-Fuentes, G. (2000). Silver supported on natural Mexican zeolite as an antibacterial material. *Microporous and Mesoporous Materials*, *39*(3), 431–444. DOI: 10.1016/S1387-1811(00)00217-1

Rivero González, L.A., Rodríguez Fuentes, G. and Rodríguez Iznaga, I. (2018). *Report on the field test of NEREA® zeolitic fertilizers in potato crop*, April 2018. NatZEng-IMRE

Rivero González, L. A., & Rodríguez-Fuentes, G. (1988). *Cuban experience with the use of natural zeolite substrates in soilless culture. Proceeding of Seventh International Congress on Soilless Culture*, The Netherlands, International Society for Soilless Culture. 405-416

Rivero-Gonzalez, L. A., & Rivero-Robaina, E. L. (2016). *Zeoecological Fertilization*, Proceedings of the 11th Meeting on Organic and Sustainable Agriculture. ACTAF, 24 – 26 mayo 2016, Matanzas, Cuba.

Rivero-González, L. A., Rivero-Robaina, E. L., Rodríguez-Iznaga, I., & Rodríguez-Fuentes, G. (2023). *Empleo de los productos NEREA® plus como fertilizante edáfico y foliar en el cultivo de papa var Rudol.* Report Institute of Material Science and Technology (IMRE), University of Havana.

Rivero-González, L. A., Rodríguez-Fuentes, G., Rivero-Robaina, E. L., & Rodríguez-Iznaga, I. (2018). *Report Validation of the effect of Cuban natural zeolite on sugarcane cultivation in Ingenio Tolulá S.A. Suchitepéquez.*

Rivero-Robaina, E.L., Coll, Y., Castillo C. R., Hansen, J., Köhler, S., Muskolus, A., Rodríguez-Iznaga. I. and Rodríguez-Fuentes, G. (2024). *NEREA® Plus-U and BIOBRAS® 16 as foliar fertilizer / biostimulant for potato and maize grown in a German field soil: study of concept.* Universidad de La Habana (IMRE, CEPN), Fundación Universitaria de Innovación y Desarrollo, Humboldt-Universität zu Berlin, IASP, WIWEX GmbH.

Robinson, N., Bräuer, P., York, A. P. E., & D'Agostino, C. (2021). Nuclear spin relaxation as a probe of zeolite acidity: A combined NMR and TPD investigation of pyridine in HZSM-5. *Physical Chemistry Chemical Physics*, *23*(33), 17752–17760. DOI: 10.1039/D1CP01515J PMID: 34190253

Robinson, N., Xiao, G., Connolly, P. R. J., Ling, N. N. A., Fridjonsson, E. O., May, E. F., & Johns, M. L. (2020). Low-field NMR relaxation-exchange measurements for the study of gas admission in microporous solids. *Physical Chemistry Chemical Physics*, *22*(24), 13689–13697. DOI: 10.1039/D0CP02002H PMID: 32525174

Rochard, , Olivet, L., Tannous, M., Poupin, C., Siffert, S., & Cousin, R. (2021). Recent advances in the catalytic treatment of volatile organic compounds: A review based on the mixture effect. *Catalysts, 11*(10), 1218. Advance online publication. DOI: 10.3390/catal11101218

Rodionova, T. A., Tsyganenko, A. A., & Filimonov, V. N. (1982). *Infrared study of low-temperature CO adsorption on metal oxides. Adsorbtsiya i adsorbenty, 10, 33-42*. Russ.

Rodríguez-Fuentes, G., & Rodríguez-Iznaga, I. (2009). Caracterización del mineral zeolítico para el desarrollo de materiales nanoestructurados. Rev. Cub. Física vol. 26, No. 1ª, 55-60

Rodríguez-Fuentes, G., Rivero González, L. A., Rodríguez-Iznaga, I., & Rivero Robaina, E. L. (2023). *Procedimiento para la obtención de sustratos y fertilizantes zeolíticos de liberación controlada y métodos de tratamiento de las plantas*. Patent CU20210017 A7. Patent application in Canada H8327186CA

Rodríguez-Fuentes, G., Rodríguez-Iznaga, I., Lariot-Sánchez, C. A., Concepción Rosabal, B., Molina-Amat, J. C., Rivero-González, L. A., Rivero-Robaina, E. L., De la Nuez-Pantoja, E. Y., & Caisés Ávalos, A. (2025). *Cuba Sedimentary Zeolite Deposits. Second report. Natural Zeolite Engineering Innovation: Industrial production of NEREA® materials for Agriculture*.https://www.iza-online.org/natural/Insights/Cuba_Insights

Rodríguez-Fuentes, G. (2015). Natural Zeolite Engineering: Development of New Materials and Utilization Technologies. [Associazione Italiana Zeoliti]. *Extended Abstract in Bollettino AIZ*, (45), 13–31.

Rodríguez-iznaga, I., Shelyapina, M. G., & Petranovskii, V. (2022). Ion exchange in natural clinoptilolite : Aspects related to its structure and applications. *Minerals, 12*, 1628-1–39. AcademicDOI: 10.3390/min12121628

Romanov, S. (1993). Electronic structure of the minimum-diameter Tl, Pb and Bi quantum wire superlattices. *Journal of Physics Condensed Matter*, *5*(8), 1081–1090. DOI: 10.1088/0953-8984/5/8/009

Romero-Marimon, P., Gutiérrez-Sevillano, J. J., & Calero, S. (2023). Adsorption of Carbon Dioxide in Non-Löwenstein Zeolites. *Chemistry of Materials, 35*(13), 5222–5231. DOI: 10.1021/acs.chemmater.3c01258

Roos, D. P., Scheinost, A. C., Churakov, S. V., Nagashima, M., & Cametti, G. (2021). On the nature of Pb species in Pb-exchanged zeolite stellerite (STI): A combined experimental and theoretical study. *Microporous and Mesoporous Materials*, *327*, 111444. Advance online publication. DOI: 10.1016/j.micromeso.2021.111444

Rosaline, D. R., Daphne, V., Srividya, G., Nivetha, P., Jarvin, M., Ashok Kumar, S., Manikandan, A., Inbanathan, S. S. R., Dinesh, A., Thanrasu, K., Raja, K. K., Almessiere, M. A., Slimani, Y., Baykal, A., Khan, A., Asiri, A. M., Topare, N. S., & Somya, A. (2024). Magnetic nanocomposites for biomedical and environmental applications. In *Magnetic Nanoparticles and Polymer Nanocomposites* (pp. 225–287). Elsevier., DOI: 10.1016/B978-0-323-85748-2.00011-6

Roth, W. J., Nachtigall, P., Morris, R. E., & Čejka, J. (2014). Two-dimensional zeolites : Current status and perspectives. *Chemical Reviews*, *114*(9), 4807–4837. DOI: 10.1021/cr400600f PMID: 24555638

Rozhkovskaya, A., Rajapakse, J., & Millar, G. J. (2021). Optimisation of zeolite LTA synthesis from alum sludge and the influence of the sludge source. *Journal of Environmental Sciences (China)*, *99*, 130–142. DOI: 10.1016/j.jes.2020.06.019 PMID: 33183690

Ruggiu, A., Carvalho, A. P., Rombi, E., Martins, A., Rocha, J., Parpot, P., Neves, I. C., & Cutrufello, M. G. (2024). Y and ZSM-5 Hierarchical Zeolites Prepared Using a Surfactant-Mediated Strategy: Effect of the Treatment Conditions. *Materials (Basel)*, *17*(17), 4401. Advance online publication. DOI: 10.3390/ma17174401 PMID: 39274788

Ruíz-Baltazar, Á. de J. (2024). Advancements in nanoparticle-modified zeolites for sustainable water treatment: An interdisciplinary review. In *Science of the Total Environment* (Vol. 946). Elsevier B.V., DOI: 10.1016/j.scitotenv.2024.174373

Saba, M., Esmaeil, M., & Saeed, H. (2024). The effect of apple pectin agent on the structural, magnetic, and optical properties of cobalt ferrite nanoparticles synthesized by the auto-combustion sol-gel method. *Materials Chemistry and Physics*, *315*, 129015. DOI: 10.1016/j.matchemphys.2024.129015

Saborit Reyes, R., Meneses Dartayet, P., & Cañizares Sierra, A. (2021). *Effect of the foliating Nerea* Plus - *U, on the rice's cultivation performances.Rev. Infociencia*, *25*(3), 36–48.

Sachse, A., & García-Martínez, J. (2017). Surfactant-Templating of Zeolites: From Design to Application. *Chemistry of Materials*, *29*(9), 3827–3853. DOI: 10.1021/acs.chemmater.7b00599

Sachse, A., Grau-Atienza, A., Jardim, E. O., Linares, N., Thommes, M., & García-Martínez, J. (2017). Development of Intracrystalline Mesoporosity in Zeolites through Surfactant-Templating. *Crystal Growth & Design*, *17*(8), 4289–4305. DOI: 10.1021/acs.cgd.7b00619

Sadeghi, M., Farhadi, S., & Zabardasti, A. (2020a). Construction of magnetic $MgFe_2O_4/CdS/MoS_2$ ternary nanocomposite supported on NaY zeolite and highly efficient sonocatalytic degradation of organic pollutants. *RSC Advances*, *10*(72), 44034–44049. DOI: 10.1039/D0RA08831E

Sadeghi, M., Farhadi, S., & Zabardasti, A. (2020b). Fabrication of a novel magnetic CdS nanorod/$NiFe_2O_4$/NaX zeolite nanocomposite with enhanced sonocatalytic performance in the degradation of organic dyes. *New Journal of Chemistry*, *44*(20), 8386–8401. DOI: 10.1039/D0NJ01393E

Sadeghi, M., Farhadi, S., & Zabardasti, A. (2021). A NaX zeolite framework containing magnetic $MgFe_2O_4$/CdO nanoparticles: Synthesis, characterization and catalytic performance in the decontamination of 2-chloroethyl phenyl sulfide (2-CEPS) as a model of sulfur mustard agent. *New Journal of Chemistry*, *45*(45), 21315–21326. DOI: 10.1039/D1NJ04202E

Saeed, , Abbas, G., Alagawany, M., Kamboh, A. A., Abd El-Hack, M. E., Khafaga, A. F., & Chao, S. (2019). Heat stress management in poultry farms: A comprehensive overview. *Journal of Thermal Biology*, *425*, 414–425. Advance online publication. DOI: 10.1016/j.jtherbio.2019.07.025 PMID: 31466781

Safarpour, , Vatanpour, V., Khataee, A., Zarrabi, H., Gholami, P., & Yekavalangi, M. E. (2017). High flux and fouling resistant reverse osmosis membrane modified with plasma treated natural zeolite. *Desalination*, *411*, 89–100. DOI: 10.1016/j.desal.2017.02.012

Sahu, , Pervez, S., Karbhal, I., Tamrakar, A., Mishra, A., Verma, S. R., Deb, M. K., Ghosh, K. K., Pervez, Y. F., Shrivas, K., & Satnami, M. L. (2024). Applications of different adsorbent materials for the removal of organic and inorganic contaminants from water and wastewater – A review. *Desalination and Water Treatment*, *317*, 100253. Advance online publication. DOI: 10.1016/j.dwt.2024.100253

Salahudeen, N. (2022). A Review on Zeolite: Application, Synthesis and Effect of Synthesis Parameters on Product Properties. In Chemistry Africa (Vol. 5, Issue 6, pp. 1889–1906). Springer Science and Business Media LLC. https://doi.org/DOI: 10.1007/s42250-022-00471-9

Samanta, C., & Das, R. K. (2021). C3-Based Petrochemicals: Recent Advances in Processes and Catalysts. In Pant, K. K., Gupta, S. K., & Ahmad, E. (Eds.), *Catalysis for Clean Energy and Environmental Sustainability: Petrochemicals and Refining Processes* - (Vol. 2, pp. 149–204). Springer International Publishing. DOI: 10.1007/978-3-030-65021-6_5

Sanchez, C.. (2018). Nanomaterials for drug delivery and diagnostics. *Advanced Drug Delivery Reviews, 56*(6), 705–747.

Sanchez, J. A., Materon, L., Parsons, J. G., & Alcoutlabi, M. (2024). Synthesis, Characterization, and Antibacterial Activity of Graphene Oxide/Zinc Hydroxide Nanocomposites. *Applied Sciences (Basel, Switzerland), 14*(14), 6274. DOI: 10.3390/app14146274

Sánchez-López, P., Kotolevich, Y., Yocupicio-Gaxiola, R. I., Antúnez-García, J., Chowdari, R. K., Petranovskii, V., & Fuentes-Moyado, S. (2021). Recent advances in catalysis based on transition metals supported on zeolites. *Frontiers in Chemistry, 9*, 1–25. DOI: 10.3389/fchem.2021.716745 PMID: 34434919

Sarioglu, M. (2005). Removal of ammonium from municipal wastewater using natural Turkish (Dogantepe) zeolite. In Separation and Purification Technology (Vol. 41, Issue 1, pp. 1–11). Elsevier BV. https://doi.org/DOI: 10.1016/j.seppur.2004.03.008

Sato, T., Tanaka, T., & Harada, H. (2019). Structural analysis of materials for performance optimization. *Materials Science and Engineering R Reports, 138*, 47–56.

Savant, N. K., Korndorfer, G. H., Datnoff, L. E., & Snyder, G. H. (1999). Silicon nutrition and sugarcane production: A review. *Journal of Plant Nutrition, 22*(12), 1853–1903. DOI: 10.1080/01904169909365761

Sawyer, W. S., Wang, L., Uehara, T., Tamrakar, P., Prathapam, R., Mostafavi, M., Metzger, L. E., Feng, B., & Baxter Rath, C. M. (2019). Targeted lipopolysaccharide biosynthetic intermediate analysis with normal-phase liquid chromatography mass spectrometry. *PLoS One, 14*(2), e0211803. DOI: 10.1371/journal.pone.0211803 PMID: 30735516

Schick, J., Caullet, P., Paillaud, J.-L., Patarin, J., & Mangold-Callarec, C. (2011). Nitrate sorption from water on a Surfactant-Modified Zeolite. Fixed-bed column experiments. In Microporous and Mesoporous Materials (Vol. 142, Issues 2–3, pp. 549–556). Elsevier BV. https://doi.org/DOI: 10.1016/j.micromeso.2010.12.039

Schmidt, R. J. (2005). Industrial catalytic processes—Phenol production. *Applied Catalysis A, General, 280*(1), 89–103. DOI: 10.1016/j.apcata.2004.08.030

Schulze, D., George, C., Cathline, K., & Germani Torrens, G. (2021). *Evaluation of the performance of zeolite-based NEREA® as a media amendment for greenhouse grown basil in Canada*. Agriculture & Environmental Technologies Innovation Centre, Niagara College.

Schwarzenbach, , Escher, B. I., Fenner, K., Hofstetter, T. B., Johnson, C. A., von Gunten, U., & Wehrli, B. (2006). The challenge of micropollutants in aquatic systems. *Science, 313*(5790), 1072–1077. DOI: 10.1126/science.1127291 PMID: 16931750

Schwieger, W., Machoke, A. G., Weissenberger, T., Inayat, A., Selvam, T., Klumpp, M., & Inayat, A. (2016). Hierarchy concepts: Classification and preparation strategies for zeolite containing materials with hierarchical porosity. *Chemical Society Reviews, 45*(12), 3353–3376. DOI: 10.1039/C5CS00599J PMID: 26477329

Seff, K. (2007). Doing chemistry in a one-nanometer test tube (in a zeolite). *Comments on Modern Chemistry. Part A, Comments on Inorganic Chemistry : a Journal of Critical Discussion of the Current Literature, 28*(5–6), 173–179. DOI: 10.1080/02603590701787415

Sellaoui, L., Hessou, E. P., Badawi, M., Netto, M. S., Dotto, G. L., Silva, L. F. O., Tielens, F., Ifthikar, J., Bonilla-Petriciolet, A., & Chen, Z. (2021). Trapping of Ag+, Cu2+, and Co2+ by faujasite zeolite Y: New interpretations of the adsorption mechanism via DFT and statistical modeling investigation. In *Chemical Engineering Journal* (Vol. 420, p. 127712). Elsevier BV., DOI: 10.1016/j.cej.2020.127712

Semeykina, V. S., Parkhomchuk, E. V., Polukhin, A. V., Parunin, P. D., Lysikov, A. I., Ayupov, A. B., & Glazneva, T. S. (2016). CoMoNi Catalyst Texture and Surface Properties in Heavy Oil Processing. Part II: Macroporous Sepiolite-Like Mineral. *Industrial & Engineering Chemistry Research, 55*(34), 9129–9139. DOI: 10.1021/acs.iecr.6b01208

Sengodan, S., Matheswaran, B., Shanmugam, S., & Thangadurai Thangaian, D. (2022). Investigation on crystal facet-dependent antibacterial activity of Cu2O crystals – A structural, morphological, and spectroscopy studies. *Results in Chemistry, 4*, 100358. DOI: 10.1016/j.rechem.2022.100358

Senila, , & Cadar, O. (2024). Modification of natural zeolites and their applications for heavy metal removal from polluted environments: Challenges, recent advances, and perspectives. *Heliyon, 10*(3), e25303. Advance online publication. DOI: 10.1016/j.heliyon.2024.e25303 PMID: 38352776

Sen, M., Dana, K., & Das, N. (2018). Development of LTA zeolite membrane from clay by sonication assisted method at room temperature for H2-CO2 and CO2-CH4 separation. *Ultrasonics Sonochemistry*, *48*, 299–310. DOI: 10.1016/j.ultsonch.2018.06.007 PMID: 30080554

Serrano, D. P., Escola, J. M., & Pizarro, P. (2013). Synthesis strategies in the search for hierarchical zeolites. *Chemical Society Reviews*, *42*(9), 4004–4035. DOI: 10.1039/C2CS35330J PMID: 23138888

Serykh, A. I., & Amirides, M. D. (2006). DRIFTS evidence for the formation of binuclear Cu(I)–dinitrogen complexes upon adsorption of N_2 on CuZSM-5. *Microporous and Mesoporous Materials*, *94*(1-3), 320–324. DOI: 10.1016/j.micromeso.2006.04.005

Shaba, E. Y., Jacob, J. O., Tijani, J. O., & Suleiman, M. A. T. (2021). A critical review of synthesis parameters affecting the properties of zinc oxide nanoparticle and its application in wastewater treatment. *Applied Water Science*, *11*(2), 48. DOI: 10.1007/s13201-021-01370-z

Shaban, , AbuKhadra, M. R., Nasief, F. M., & Abd El-Salam, H. M. (2017). Removal of ammonia from aqueous solutions, ground water, and wastewater using mechanically activated clinoptilolite and synthetic zeolite-A: Kinetic and equilibrium studies. *Water, Air, and Soil Pollution*, *228*(1), 20. Advance online publication. DOI: 10.1007/s11270-017-3643-7

Shankar, . (2023). Adsorption kinetics, equilibrium and thermodynamics of a textile dye V5BN by a natural nanocomplex material. *Energy Nexus*, *100197*, 100197. Advance online publication. DOI: 10.1016/j.nexus.2023.100197

Sharifi, M., Haghighi, M., & Abdollahifar, M. (2015). Sono-dispersion of bimetallic Ni-Co over zeolite Y used in conversion of greenhouse gases CH4/CO2 to high valued syngas. *Journal of Natural Gas Science and Engineering*, *23*, 547–558. DOI: 10.1016/j.jngse.2015.03.006

Sharma, B., Tandon, R. P., & Mahapatro, A. K. (2022). Interaction of zinc peroxide nanoparticle with fibroblast cell. *Advanced Materials & Processes*, 2(7), 440–444. DOI: 10.5185/amp.2017/708

Sharma, V., Javed, B., Byrne, H., Curtin, J., & Tian, F. (2022). *Zeolites as Carriers of Nano-Fertilizers: From Structures and Principles to Prospects and Challenges* Appl. *Nano*, *2022*(3), 163–187. DOI: 10.3390/applnano3030013

Shekarchi, , Ahmadi, B., Azarhomayun, F., Shafei, B., & Kioumarsi, M. (2023). Natural zeolite as a supplementary cementitious material – A holistic review of main properties and applications. *Construction & Building Materials*, *409*, 133766. Advance online publication. DOI: 10.1016/j.conbuildmat.2023.133766

Shelyapina, M. G. (2024). NMR relaxation to probe zeolites: Mobility of adsorbed molecules, surface acidity, pore size distribution and connectivity. *Molecules (Basel, Switzerland)*, *29*(22), 5432. DOI: 10.3390/molecules29225432 PMID: 39598821

Shelyapina, M. G., Krylova, E. A., Mazur, A. S., Tsyganenko, A. A., Shergin, Y. V., Satikova, E., & Petranovskii, V. (2022). Active sites in H-mordenite catalysts probed by NMR and FTIR. *Catalysts*, *13*(2), 344. DOI: 10.3390/catal13020344

Shelyapina, M. G., Nefedov, D. Y., Antonenko, A. O., Valkovskiy, G. A., Yocupicio-gaxiola, R. I., & Petranovskii, V. (2023). Nanoconfined water in pillared zeolites probed by ^{1}H nuclear magnetic resonance. *International Journal of Molecular Sciences*, *24*(21), 15898. DOI: 10.3390/ijms242115898 PMID: 37958879

Shelyapina, M. G., Yocupicio-Gaxiola, R. I., Zhelezniak, I. V., Chislov, M. V., Antúnez-García, J., Murrieta-Rico, F. N., Galván, D. H., Petranovskii, V., & Fuentes-Moyado, S. (2020). Local structures of two-dimensional zeolites – mordenite and ZSM-5 – probed by multinuclear NMR. *Molecules (Basel, Switzerland)*, *25*(20), 4678. DOI: 10.3390/molecules25204678 PMID: 33066351

Shi, et al. (2023). Adsorption properties of natural zeolite supported by chitosan on Cd(II) in micropolluted irrigation water. ACS, 12. DOI: 10.1021/acsomega.3c06041

Shimizu, , & Matubayasi, N. (2020). Fluctuation adsorption theory: Quantifying adsorbate–adsorbate interaction and interfacial phase transition from an isotherm. *Physical Chemistry Chemical Physics*, *22*(48), 28304–28316. DOI: 10.1039/D0CP05122E PMID: 33295900

Shintani, T., & Murata, Y. (2011). Evaluation of the dislocation density and dislocation character in cold rolled Type 304 steel determined by profile analysis of X-ray diffraction. *Acta Materialia*, *59*(11), 4314–4322. DOI: 10.1016/j.actamat.2011.03.055

Shirani, M., Semnani, A., Haddadi, H., & Habibollahi, S. (2014). Optimization of simultaneous removal of methylene blue, crystal violet, and fuchsine from aqueous solutions by magnetic NaY zeolite composite. *Water, Air, and Soil Pollution*, *225*(8). Advance online publication. DOI: 10.1007/s11270-014-2054-2

Shoinkhorova, , Dikhtiarenko, A., Ramirez, A., Dutta Chowdhury, A., Caglayan, M., Vittenet, J., Bendjeriou-Sedjerari, A., Ali, O. S., Morales-Osorio, I., Xu, W., & Gascon, J. (2019). Shaping of ZSM-5-based catalysts via spray drying: Effect on methanol-to-olefins performance. *ACS Applied Materials & Interfaces*, *11*(47), 44133–44143. Advance online publication. DOI: 10.1021/acsami.9b14082 PMID: 31612697

Shokrolahzadeh, . (2017). Modification of nano clinoptilolite zeolite using sulfuric acid and its application toward removal of arsenic from water sample. *Journal of Nanoanalysis*, *4*(1), 48–58. DOI: 10.22034/jna.2017.01.006

Siakpebru, O. K., Gurrala, L., Uchagawkar, A., Norris, O., Bartlett, J., & Morais, A. R. C. (2025). Application of Hierarchical H–Y Zeolites in Hydroxyalkylation–Alkylation of 2-Methylfuran with Furfural to Produce Jet Fuel Precursors. *Industrial & Engineering Chemistry Research*, *64*(11), 5830–5840. Advance online publication. DOI: 10.1021/acs.iecr.4c03317

Silletta, E. V., Franzoni, M. B., Monti, G. A., & Acosta, R. H. (2018). Influence of exchange in NMR pore size/relaxation correlation experiments. *Microporous and Mesoporous Materials*, *269*(April), 17–20. DOI: 10.1016/j.micromeso.2017.10.025

Silvy, R. P. (2004). Future trends in the refining catalyst market. *Applied Catalysis A, General*, *261*(2), 247–252. DOI: 10.1016/j.apcata.2003.11.019

Simmons, E. L. (1975). Diffuse reflectance spectroscopy: A comparison of the theories. *Applied Optics*, *14*(6), 1380. DOI: 10.1364/AO.14.001380 PMID: 20154834

Singh, B. K., Lee, S., & Na, K. (2020). An overview on metal-related catalysts: Metal oxides, nanoporous metals and supported metal nanoparticles on metal organic frameworks and zeolites. *Rare Metals*, *39*(7), 751–766. DOI: 10.1007/s12598-019-01205-6

Singh, S., Martínez-Ortigosa, J., Ortuño, N., Polshettiwar, V., & García-Martínez, J. (2024). Enhanced efficiency in plastic waste upcycling: The role of mesoporosity and acidity in zeolites. *Chemical Science (Cambridge)*, *15*(48), 20240–20250. DOI: 10.1039/D4SC05121A PMID: 39568899

Sing, K. S. W. (2014). *7 - Assessment of Surface Area by Gas Adsorption, Adsorption by Powders and Porous Solids* (2nd ed.). Academic Press.

Sirelkhatim, A., Mahmud, S., Seeni, A., Kaus, N. H. M., Ann, L. C., Bakhori, S. K. M., Hasan, H., & Mohamad, D. (2015). Review on Zinc Oxide Nanoparticles: Antibacterial Activity and Toxicity Mechanism. *Nano-Micro Letters*, *7*(3), 219–242. DOI: 10.1007/s40820-015-0040-x PMID: 30464967

Sivagami, M., & Asharani, I. V. (2022). Phyto-mediated Ni/NiO NPs and their catalytic applications: A short review. *Inorganic Chemistry Communications*, *145*, 110054. DOI: 10.1016/j.inoche.2022.110054

Skripov, A. V., & Shelyapina, M. G. (2016). Nuclear magnetic resonance. In Fritzsche, H., Huot, J., & Fruchart, D. (Eds.), *Neutron scattering and other nuclear techniques for hydrogen in materials* (pp. 337–376). Springer International Publishing., DOI: 10.1007/978-3-319-22792-4_12

Skuja, L., Kajihara, K., Hirano, M., & Hosono, H. (2011). Visible to vacuum-UV range optical absorption of oxygen dangling bonds in amorphous SiO_2. *Physical Review B: Condensed Matter and Materials Physics*, *84*(20), 205206. Advance online publication. DOI: 10.1103/PhysRevB.84.205206

Smiljanić, , de Gennaro, B., Izzo, F., Langella, A., Daković, A., Germinario, C., Rottinghaus, G. E., Spasojević, M., & Mercurio, M. (2020). Removal of emerging contaminants from water by zeolite-rich composites: A first approach aiming at diclofenac and ketoprofen. *Microporous and Mesoporous Materials*, *298*, 110057. Advance online publication. DOI: 10.1016/j.micromeso.2020.110057

Smirnov, E. P., & Tsyganenko, A. A. (1977). CNDO study of the properties of the OH group. *Reaction Kinetics and Catalysis Letters*, *7*(4), 425–429. DOI: 10.1007/BF02073831

Smith, D. (2010). X-ray diffraction methods for materials characterization. *Materials Science and Engineering*, *210*(2), 104–109.

Snyder, B. E. R., Böttger, L. H., Bols, M. L., Yan, J. J., Rhoda, H. M., Jacobs, A. B., Hu, M. Y., Zhao, J., Alp, E. E., Hedman, B., Hodgson, K. O., Schoonheydt, R. A., Sels, B. F., & Solomon, E. I. (2018). *Structural characterization of a non-heme iron active site in zeolites that hydroxylates methane. 115*(18), 4565–4570. https://doi.org/DOI: 10.2307/26508720

Snyder, B. E. R., Bols, M. L., Rhoda, H. M., Plessers, D., Schoonheydt, R. A., Sels, B. F., & Solomon, E. I. (2021). Cage effects control the mechanism of methane hydroxylation in zeolites. *Science*, *373*(6552), 327–331. DOI: 10.1126/science.abd5803

Snyder, B. E. R., Bols, M. L., Rhoda, H. M., Vanelderen, P., Böttger, L. H., Braun, A., Yan, J. J., Hadt, R. G., Babicz, J. T., Hu, M. Y., Zhao, J., Alp, E. E., Hedman, B., Hodgson, K. O., Schoonheydt, R. A., Sels, B. F., & Solomon, E. I. (2018). Mechanism of selective benzene hydroxylation catalyzed by iron-containing zeolites. *Proceedings of the National Academy of Sciences of the United States of America*, *115*(48), 12124–12129. DOI: 10.1073/pnas.1813849115

Snyder, B. E. R., Bols, M. L., Schoonheydt, R. A., Sels, B. F., & Solomon, E. I. (2018). Iron and Copper Active Sites in Zeolites and Their Correlation to Metalloenzymes. *Chemical Reviews*, *118*(5), 2718–2768. DOI: 10.1021/acs.chemrev.7b00344

Snyder, B. E. R., Vanelderen, P., Bols, M. L., Hallaert, S. D., Böttger, L. H., Ungur, L., Pierloot, K., Schoonheydt, R. A., Sels, B. F., & Solomon, E. I. (2016). The active site of low-temperature methane hydroxylation in iron-containing zeolites. *Nature*, *536*(7616), 317–321. DOI: 10.1038/nature19059

Soca Nuñez, M., & Constanza Daza, M. (2018). Zeolite and its effect on nitrogen efficiency in rice and corn Use of the zeolite in cultivations of economic importance. *INFOMIN*, *10*(1), 46–56.

Soca Nuñez, M., & Lorente, E. (2015). Efecto de las zeolitas naturales y activadas en la producción de tomate. *Agrotecnia de Cuba Volumen*, *39*(1), 67–78.

Sokol, H. J., Ebrahim, A. M., Caratzoulas, S., Frenkel, A. I., & Valla, J. A. (2022). In Situ XAFS, XRD, and DFT Characterization of the Sulfur Adsorption Sites on Cu and Ce Exchanged y Zeolites. *The Journal of Physical Chemistry. C, Nanomaterials and Interfaces*, *126*(3), 1496–1512. DOI: 10.1021/acs.jpcc.1c09128

Solovyev, V. G., Ivanova, M. S., Pankova, S. V., Trifonov, S. V., & Veisman, V. L. (2009). Preparation and physical properties of zeolite, zeolite-like single crystals and zeolite-based nanocomposite materials. In Wong, T. W. (Ed.), *Handbook of Zeolites: Structure, Properties and Applications* (pp. 77–99). Nova Science Publishers.

Somya, A. (2020). Hybrid Ion Exchangers. In Hybrid Nanomaterials - Flexible Electronics Materials. IntechOpen. https://doi.org/DOI: 10.5772/intechopen.92116

Somya, A. (2023). Metal Phosphates: Their Role as Ion Exchangers in Water Purification. In Metal Phosphates and Phosphonates (pp. 341–356). Springer International Publishing. https://doi.org/DOI: 10.1007/978-3-031-27062-8_19

Somya, A. (2024). Hybrid cation exchangers: A boon for analytical chemistry. Journal of Applied Optics, 263–270. Retrieved from https://appliedopticsjournal.net/index.php/JAO/article/view/111

Somya, A., & Maheria, K. (2020). Studies on polyoxyethylene octyl phenyl ether supported thorium(IV) phosphates: a new cation exchange material. In Bulletin of Materials Science (Vol. 43, Issue 1). Springer Science and Business Media LLC. https://doi.org/DOI: 10.1007/s12034-020-02153-z

Somya, A., & Singh, M. (2022). Studies on Sodium Lauryl Sulphate-supported Thorium (IV) Phosphate: A New Surfactant-supported Cation Exchange Resin, Useful in Water Purification. In Current Analytical Chemistry (Vol. 18, Issue 3, pp. 383–390). Bentham Science Publishers Ltd. https://doi.org/DOI: 10.2174/1573411017666210804102748

Somya, A., Kumar, S., & K., S. (2024). An exploration of AFLC-14-doped ceric phosphate, a novel liquid crystal hybrid ion exchanger: synthesis, ion exchange studies and analytical insights. In Materials Research Innovations (pp. 1–9). Informa UK Limited. https://doi.org/DOI: 10.1080/14328917.2024.2396655

Somya, A., Rafiquee, M. Z. A., Varshney, K. G., & Al-Lohedan, H. A. (2015). Synthesis, characterization and analytical applications of N-dodecyl pyridinium chloride-cerium (IV) phosphate fibrous ion exchanger: selective for Hg(II) and its binary separations. In Desalination and Water Treatment (Vol. 56, Issue 3, pp. 638–645). Elsevier BV. https://doi.org/DOI: 10.1080/19443994.2014.940388

Somya, A. (2019). Surfactant based hybrid ion exchangers. *Research Journal of Chemistry and Environment*, *23*(3), 96.

Somya, A. (2023). N-Cetyl-N N. N-tri methyl ammonium bromide based tin (IV) phosphate, a new surfactant based hybrid ion exchanger: Synthesis, ion exchange and physico-chemical characterization. *Research Journal of Chemistry and Environment*, *27*(6), 18–23. DOI: 10.25303/2706rjce018023

Somya, A. (2024). An investigation study on water quality of few lakes of Bengaluru. In *Industry 4.0 with Modern Technology* (pp. 200–204). CRC Press., DOI: 10.1201/9781003450924-37

Somya, A., Khan, A., Hussein, M. A., Al-Romaizan, A. N., Malash, A. M., Topare, N. S., & Asiri, A. M. (2023). Biological and environmental studies on pectin-zirconium (IV) phosphate, a biopolymer doped hybrid cation exchanger. In *International Journal of Biological Macromolecules* (Vol. 250, p. 125803). Elsevier BV., DOI: 10.1016/j.ijbiomac.2023.125803

Somya, A., Khan, A., & Varshney, A. P. (2025). Nanocomposites and Their Role in Wastewater Treatment. In Singh, L. (Ed.), *Design, Fabrication, and Significance of Advanced Nanostructured Materials* (pp. 75–100). IGI Global Scientific Publishing., DOI: 10.4018/979-8-3693-5320-2.ch004

Somya, A., Rafiquee, M. Z. A., & Upadhyaya, V. (2018). Role of surfactants in the adsorption studies of n-butyl acetate-cerium (IV) phosphate for some alkaline earths and heavy metal ions. *Research Journal of Chemistry and Environment*, *22*(6), 45.

Somya, A., & Upadhyaya, V. (2017). Synthesis, Characterization and Adsorption behaviour of N-dodecyl pyridinium chloride based Tin (IV) phosphate: A new surfactant based hybrid ion exchanger. *Research Journal of Chemistry and Environment, 21*(1), 18.

Somya, A., Varshney, A. P., & Thakur, A. (2025). Hybrid Ion Exchange Materials Useful in Water Purification. In Amrousse, R., Bachar, A., Mabrouk, A., & Azat, S. (Eds.), *Innovative Materials for Environmental and Aerospace Applications* (pp. 363–388). IGI Global Scientific Publishing., DOI: 10.4018/979-8-3373-0669-8.ch010

Song, R., Song, Y. Q., Vembusubramanian, M., & Paulsen, J. L. (2016). The robust identification of exchange from T_2-T_2 time-domain features. *Journal of Magnetic Resonance (San Diego, Calif.), 265*, 164–171. DOI: 10.1016/j.jmr.2016.02.001 PMID: 26905815

Song, Y. Q., Venkataramanan, L., Hürlimann, M. D., Flaum, M., Frulla, P., & Straley, C. (2002). T_1-T_2 correlation spectra obtained using a fast two-dimensional Laplace inversion. *Journal of Magnetic Resonance (San Diego, Calif.), 154*(2), 261–268. DOI: 10.1006/jmre.2001.2474 PMID: 11846583

Soni, et al. (2018). Synthesis of fly ash based zeolite-reduced graphene oxide composite and its evaluation as adsorbent for arsenic removal. DOI: 10.1016/j.chemosphere.2018.11.203

Sonochemistry and the Acoustic Bubble. (2015). Elsevier. DOI: 10.1016/C2013-0-18886-1

Sossou, , Prasad, S. B., Agbotsou, E. K., & Saidou Souley, H. (2024). Evaluation of the performance of magnetic zeolite nanocomposites in removing various water contaminants as heavy metals, organic pollutants, and emerging contaminants. *Article, 100075*, 100075. Advance online publication. DOI: 10.1016/j.nxnano.2024.100075

Sousa, T. G., Diniz, S. B., Pinto, A. L., & Brandão, L. P. (2015). Dislocation density by X-ray diffraction in α brass deformed by rolling and ECAE. *Materials Research, 18*(Suppl 2), 246–249. DOI: 10.1590/1516-1439.369214

Souza, , Borrego-Sánchez, A., Rigoti, E., Sainz-Díaz, C. I., Viseras, C., & Pergher, S. B. C. (2021). Experimental and molecular modelling study of beta zeolite as drug delivery system. *Microporous and Mesoporous Materials, 111152*, 111152. Advance online publication. DOI: 10.1016/j.micromeso.2021.111152

Souza, I., García-Villén, F., Viseras, C., & Pergher, S. (2023). Zeolites as Ingredients of Medicinal Products. *Pharmaceutics, 15*(5), 1352. DOI: 10.3390/pharmaceutics15051352 PMID: 37242594

Srivastava, D. J., Venetos, M. C., McCarthy-Carney, L., Baltisberger, J. H., Grandinetti, P. J., & Brouwer, D. (2025). Refining siliceous zeolite framework structures with ^{29}Si 2D J-resolved NMR spectroscopy. *Physical Chemistry Chemical Physics*, *27*(1), 419–435. DOI: 10.1039/D4CP03530E PMID: 39648968

Stanislaus, A., Marafi, A., & Rana, M. S. (2010). Recent advances in the science and technology of ultra low sulfur diesel (ULSD) production. *Catalysis Today*, *153*(1), 1–68. DOI: 10.1016/j.cattod.2010.05.011

Stepanov, A. G. (2016). Basics of solid-state NMR for application in zeolite science: Material and reaction characterization. In Sels, B. F., & Kustov, L. M. (Eds.), *Zeolites and Zeolite-like Materials* (pp. 137–188). Elsevier B.V., DOI: 10.1016/B978-0-444-63506-8.00004-5

Stepanyuk, K. O., Kravchenko, A. A., Gerashchenko, I. I., Grebenyuk, A. G., & Borysenko, M. V. (2020). Adsorption of zinc (II) on silica and alumina-silica surfaces: results of experiment and simulation | Адсорбция цинка (II) на кремнеземных и алюмосиликатных поверхностях: результаты эксперимента и моделирование | АДСОРБЦІЯ ЦИНКУ (II) НА КРЕМНЕЗЕМНИХ ТА АЛ. Himia. *Fizika Ta Tehnologia Poverhni*, *11*(3), 304–318. DOI: 10.15407/hftp11.03.304

Storozheva, E. N., Sekushin, V. N., & Tsyganenko, A. A. (2006). FTIR spectroscopy evidence for the basicity induced by adsorption. *Catalysis Letters*, *107*(3–4), 185–188. Advance online publication. DOI: 10.1007/s10562-005-0008-4

Storozhev, P. Y., Otero Otero Areán, C., Garrone, E., Ugliengo, P., Ermoshin, V. A., & Tsyganenko, A. A. (2003). FTIR spectroscopic and ab initio evidence for an amphipathic character of CO bonding with silanol groups. *Chemical Physics Letters*, *374*(5–6), 439–445. DOI: 10.1016/S0009-2614(03)00587-6

Storozhev, P. Y., Yanko, V. S., Tsyganenko, A. A., Palomino, G. T., Delgado, M. R., & Otero Areán, C. (2004). Isomeric states of polar molecules on ionic surfaces: Electrostatic model and FTIR studies. *Applied Surface Science*, *238*(1-4), 390–394. DOI: 10.1016/j.apsusc.2004.05.242

Strange, J. H., Rahman, M., & Smith, E. G. (1993). Characterization of porous solids by NMR. *Physical Review Letters*, *71*(21), 3589–3591. DOI: 10.1103/PhysRevLett.71.3589 PMID: 10055015

Suárez, N., Pérez-Pariente, J., Mondragón, F., & Moreno, A. (2019). Generation of hierarchical porosity in beta zeolite by post-synthesis treatment with the cetyltrimethylammonium cationic surfactant under alkaline conditions. *Microporous and Mesoporous Materials*, *280*, 144–150. https://doi.org/https://doi.org/10.1016/j.micromeso.2019.02.001. DOI: 10.1016/j.micromeso.2019.02.001

Su, B., Wang, X., Chai, L., Huo, S., Qiu, J., Huang, Q., Li, S., Wang, Y., & Xue, W. (2024). Cation-Loaded Porous Mg2+-Zeolite Layer Direct Dendrite-Free Deposition toward Long-Life Lithium Metal Anodes. *Advancement of Science*, *11*(23), 2308939. Advance online publication. DOI: 10.1002/advs.202308939 PMID: 38600650

Sun, , Abbass, R., Ghoroqi, M., Patra, I., Dwijendra, N. K. A., Uktamov, K. F., & Jasem, H. (2022). [Retracted] Optimization of dyes and toxic metals removal from environmental water samples by clinoptilolite zeolite using response surface methodology approach. *Scientific Reports*, *12*(1), 13218. Advance online publication. DOI: 10.1038/s41598-022-17636-8 PMID: 35918466

Sun, , Ding, Z.-H., Bi, Y.-Z., & Pu, S. (2023). Experimental investigation on water/gas resistance and strength properties of clinoptilolite/Na-PAA amended compacted clay cover. Geophysics for Geo-Energy and Geo-Resources. *Geomechanics and Geophysics for Geo-energy and Geo-resources*, *66*(1), 66. Advance online publication. DOI: 10.1007/s40948-023-00591-7

Sun, Q., Wang, N., & Yu, J. (2021). Advances in catalytic applications of zeolite-supported metal catalysts. *Advanced Materials*, *33*(51), 2104442. DOI: 10.1002/adma.202104442 PMID: 34611941

Sun, Q., Zhu, K., Ji, X., Chen, D., Han, C., Li, T.-T., Hu, Y., Huang, S., & Qian, J. (2022). MOF-derived three-dimensional ordered porous carbon nanomaterial for efficient alkaline zinc-air batteries | . *Science China Materials*, *65*(6), 1453–1462. DOI: 10.1007/s40843-021-1933-4

Sun, Y., Xie, J., Hou, H., Li, M., Wang, Y., & Wang, X. (2023). Effects of Zeolite on physiological characteristics and grain quality in rice under alternate wetting and drying irrigation. *Water (Basel)*, *15*(13), 2406. DOI: 10.3390/w15132406

Sun, Z., Guo, Y., Liu, S., Li, S., & Sun, X. (2023). Moisture susceptibility evaluation of zeolite-foamed asphalts based on the Surface Free Energy (SFE) concept. *Construction & Building Materials*, *371*, 130762. Advance online publication. DOI: 10.1016/j.conbuildmat.2023.130762

Suprapron, J., Pitayakon, L., Kamalapa, W., & Touchamon, P. (2002). *Effect of zeolite and chemical fertilizer on the change of physical and chemical properties of Lat Ya soil series for sugar cane*. 17th WCSS, 14-21 august, Thailand. 1897.

Suresh, C., Pérez-Cabrera, L., Díaz de León, J. N., Zepeda, T. A., Alonso-Núñez, G., & Fuentes-Moyado, S. (2017). Highly active CoMo/Al (10) KIT-6 catalysts for HDS of DBT: Role of structure and aluminum heteroatom in the support matrix. *Catalysis Today*, *296*(Supplement C), 214–218. DOI: 10.1016/j.cattod.2017.04.048

Suriya, P., Xu, S., Ding, S., Chansai, S., Jiao, Y., Hurd, J., Lee, D., Zhang, Y., Hardacre, C., Reubroycharoen, P., & Fan, X. (2024). Ethanol steam reforming over Ni/ZSM-5 nanosheet for hydrogen production. *Chinese Journal of Chemical Engineering*, *67*, 247–256. DOI: 10.1016/j.cjche.2023.12.006

Susarrey-Arce, A., Petranovskii, V., Hernández-Espinosa, M. A., Portillo, R., & de la Cruz, W. (2011). Optical properties of ZnO nanoparticles on the porous structure of mordenites and ZSM-5. *Journal of Nanoscience and Nanotechnology*, *11*(6), 5574–5579. DOI: 10.1166/jnn.2011.3426 PMID: 21770221

Sutapa, I. W., Wahid, A. W., Taba, P., & Nafie, N. L. (2018). Dislocation, crystallite size distribution, and lattice strain of magnesium oxide nanoparticles. *Journal of Physics: Conference Series*, *979*, 012021. DOI: 10.1088/1742-6596/979/1/012021

Tabacchi, G. (2018). Supramolecular organization in confined nanospaces. *ChemPhysChem*, *19*(11), 1249–1297. DOI: 10.1002/cphc.201701090 PMID: 29573368

Tadjarodi, A., Dehghani, M., & Imani, M. (2018). Green synthesis and characterization of palladium nanoparticles supported on zeolite Y by sonochemical method, powerful and efficient catalyst for Suzuki-Miyaura coupling of aryl halides with phenylboronic acid. *Applied Organometallic Chemistry*, *32*(12). Advance online publication. DOI: 10.1002/aoc.4594

Talebian-Kiakalaieh, A., & Tarighi, S. (2020). Synthesis of hierarchical Y and ZSM-5 zeolites using post-treatment approach to maximize catalytic cracking performance. *Journal of Industrial and Engineering Chemistry*, *88*, 167–177. https://doi.org/https://doi.org/10.1016/j.jiec.2020.04.009. DOI: 10.1016/j.jiec.2020.04.009

Talebi, J., Halladj, R., & Askari, S. (2010). Sonochemical synthesis of silver nanoparticles in Y-zeolite substrate. *Journal of Materials Science*, *45*(12), 3318–3324. DOI: 10.1007/s10853-010-4349-z

Tanimu, A., & Alhooshani, K. (2019). Advanced Hydrodesulfurization Catalysts: A Review of Design and Synthesis. *Energy & Fuels*, *33*(4), 2810–2838. DOI: 10.1021/acs.energyfuels.9b00354

Tan, S., Wang, C., Foucaud, Y., Badawi, M., Guo, H., Sun, K., Yang, G., & Mintova, S. (2022). Ordered sodium zeolite-templated carbon with high first discharge capacity for sodium battery application. *Microporous and Mesoporous Materials*, *336*, 111853. Advance online publication. DOI: 10.1016/j.micromeso.2022.111853

Tarach, K., Góra-Marek, K., Chrzan, M., & Walas, S. (2014). Quantification of silver sites in zeolites: Carbon monoxide sorption monitored by IR spectroscopy. *The Journal of Physical Chemistry. C, Nanomaterials and Interfaces*, *118*(41), 23751–23760. Advance online publication. DOI: 10.1021/jp506820v

Taylor, E. E., Garman, K., & Stadie, N. P. (2020). Atomistic Structures of Zeolite-Templated Carbon. *Chemistry of Materials*, *32*(7), 2742–2752. DOI: 10.1021/acs.chemmater.0c00535

Thakur, , Ojha, A., Kansal, S. K., Gupta, N. K., Swart, H. C., Cho, J., Kuznetsov, A., Sun, S., & Prakash, J. (2024). Advances in powder nano-photocatalysts as pollutant removal and as emerging contaminants in water: Analysis of pros and cons on health and environment. *Advanced Powder Materials*, *100233*(6), 100233. Advance online publication. DOI: 10.1016/j.apmate.2024.100233

Tian, Y., He, X., Chen, X., Qiao, C., Wang, H., Diao, Z., & Liu, G. (2024). Transition metal modified hierarchical ZSM-5 nanosheet for catalytic cracking of n-pentane to light olefins. *Fuel*, *363*, 130902. DOI: 10.1016/j.fuel.2024.130902

Timoshev, V., Haufe, L. A., Busse, O., Hamedi, H., Seifert, M., & Weigand, J. J. (2024). Recycling of Spent FCC Catalysts: Conversion of Leached Residues to Zeolite ZSM-5. *ChemSusChem*, *17*(14), e202301642. DOI: 10.1002/cssc.202301642 PMID: 38462539

Titova, T. I., Kosheleva, L. S., Zhdanov, S. P., & Shubaeva, M. A. (1993). IR spectroscopic study of structure-chemical aspects of the Na-Y zeolite dealumination with ethylene diamine tetraacetic acid. *Pure and Applied Chemistry*, *65*(10), 2231–2236. DOI: 10.1351/pac199365102231

Topsøe, H., Clausen, B. S., & Massoth, F. E. (1996). Hydrotreating Catalysis. In Anderson, J. R., & Boudart, M. (Eds.), *Catalysis: Science and Technology* (pp. 1–269). Springer Berlin Heidelberg. DOI: 10.1007/978-3-642-61040-0_1

Topsøe, H., Clausen, B. S., Topsøe, N. Y., Nørskov, J. K., Ovesen, C. V., & Jacobsen, C. J. H. (1995). The Bond Energy Model for Hydrotreating Reactions: Theoretical and Experimental Aspects. *Bulletin des Sociétés Chimiques Belges*, *104*(4-5), 283–291. DOI: 10.1002/bscb.19951040415

Torma, S., Vilcek, J., Adamisin, P., & Hronec, O. (2014). Influence of natural zeolite on the dynamics of nitrogen in the soil. *Turkish Journal of Agriculture and Forestry*. Advance online publication. DOI: 10.3906/tar-1311-13

Townsend, R. P., & Coker, E. N. (2001). *Ion exchange in zeolites.*, DOI: 10.1016/S0167-2991(01)80253-6

Trachta, M., Volný, T., Rubeš, M., Bulánek, R., & Bludský, O. (2023). QM-MM Approach to the Accurate Description of Slow Diffusion in Porous Materials. *The Journal of Physical Chemistry. C, Nanomaterials and Interfaces*, *127*(18), 8856–8863. DOI: 10.1021/acs.jpcc.3c01340

Tret'yakov, N. E., & Filimonov, V. N. (1972). An infrared spectroscopic study of the relative proton-donating power of the OH-groups of oxide surfaces. *Kinetics and Catalysis*, *13*(3), 735–737.

Tromp, M., van Bokhoven, J. A., Garriga Oostenbrink, M. T., Bitter, J. H., de Jong, K. P., & Koningsberger, D. C. (2000). Influence of the Generation of Mesopores on the Hydroisomerization Activity and Selectivity of n-Hexane over Pt/Mordenite. *Journal of Catalysis*, *190*(2), 209–214. https://doi.org/https://doi.org/10.1006/jcat.1999.2778. DOI: 10.1006/jcat.1999.2778

Trukhin, A. N. (1992). Excitons in SiO_2: A review. *Journal of Non-Crystalline Solids*, *149*(1–2), 32–45. DOI: 10.1016/0022-3093(92)90052-L

Tsiotsias, A. I., Georgiadis, A. G., Charisiou, N. D., & Goula, M. A. (2023). CO<inf>2</inf> Physisorption over an Industrial Molecular Sieve Zeolite: An Experimental and Theoretical Approach. *Materials (Basel)*, *16*(20), 6656. Advance online publication. DOI: 10.3390/ma16206656 PMID: 37895638

Tsyganenko, A. (2011). Low-temperature infrared spectroscopy of surface species. In Khriachtchev, L. (Ed.), *Physics and Chemistry at Low Temperatures* (pp. 267–296). Pan Stanford Publishing Pte., DOI: 10.1201/b11403-10

Tsyganenko, A. A. (1975). IR spectra and structure of the hydroxyl covering of oxides comparison with the spectra of hydroxides and silicates. *Journal of Structural Chemistry*, *16*(4), 536–540. DOI: 10.1007/BF00754034

Tsyganenko, A. A. (1980). A cell for the investigation of IR spectra of adsorbed molecules at liquid helium temperatures. *Приборы и техника эксперимента*, *1*, 265–267.

Tsyganenko, A. A. (2013). Variable temperature IR spectroscopy in the studies of oxide catalysts. *Topics in Catalysis*, *56*(11), 905–913. DOI: 10.1007/s11244-013-0054-x

Tsyganenko, A. A., Chizhik, A. M., & Chizhik, A. I. (2010). A FTIR search for linkage isomerism of CN^- ions on oxides and zeolites. *Physical Chemistry Chemical Physics*, *12*(24), 6387–6395. DOI: 10.1039/c003942j PMID: 20454717

Tsyganenko, A. A., Escalona Platero, E., Otero Areán, C., Garrone, E., & Zecchina, A. (1999). Variable temperature IR spectroscopic studies of CO adsorbed on Na-ZSM-5 and Na-Y zeolites. *Catalysis Letters*, *61*(3-4), 187–192. DOI: 10.1023/A:1019089309446

Tsyganenko, A. A., & Filimonov, V. N. (1972). Infrared Spectra of Surface Hydroxyl Groups and Crystalline Structure of Oxides. *Spectroscopy Letters*, *5*(12), 477–487. DOI: 10.1080/00387017208065418

Tsyganenko, A. A., & Filimonov, V. N. (1973). Infrared spectra of surface hydroxyl groups and crystalline structure of oxides. *Journal of Molecular Structure*, *19*(2), 579–589. DOI: 10.1016/0022-2860(73)85136-1

Tsyganenko, A. A., Khomenya, A. V., & Filimonov, V. N. (1976). *Infrared spectroscopic study of the proton-donating ability of the hydroxyl groups of oxide surfaces. Adsorbtsiya i adsorbenty, 4, 86-91.* Russ.

Tsyganenko, A. A., Kondratieva, E. V., Yanko, V. S., & Storozhev, P. Y. (2006). FTIR study of CO adsorption on basic zeolites. *Journal of Materials Chemistry*, *16*(24), 2358. Advance online publication. DOI: 10.1039/b514752b

Tsyganenko, A. A., & Mardilovich, P. P. (1996). Structure of alumina surfaces. *Journal of the Chemical Society, Faraday Transactions*, *92*(23), 4843–4852. DOI: 10.1039/ft9969204843

Tsyganenko, A. A., Shergin, Y. V., Shelyapina, M. G., Satikova, E. A., & Petranovskii, V. (2025). (in press). FTIR Study of CO Adsorbed on Ag-mordenite. *Kinetics and Catalysis*. Advance online publication. DOI: 10.1134/S0023158425600142

Tsyganenko, A. A., Storozheva, E. N., & Manoilova, O. V. (2001). Manifestations of the acidity of adsorbed molecules in H-bonded complexes with silanol groups: Lewis acidity of ozone. *Catalysis Today*, *70*(1–3), 59–71. DOI: 10.1016/S0920-5861(01)00407-2

Tsyganenko, A. A., Storozhev, P. Y., & Otero Areán, C. (2004). IR-spectroscopic study of the binding isomerism of adsorbed molecules. *Kinetics and Catalysis*, *45*(4), 530–540. Advance online publication. DOI: 10.1023/B:KICA.0000038081.43384.56

Tsyganenko, A. A., & Trusov, E. A. (1978). Infrared study of the interaction between SO_3 and functional groups of SiO_2 surface. *Colloid Journal*, *40*(5), 875–880.

Tsyganenko, A. A., Zakharov, N. V., & Murzin, P. D. (2014). CHF_3 as a probe for surface basicity. *Catalysis Today*, *226*, 73–80. DOI: 10.1016/j.cattod.2013.11.049

Tsyganenko, A. A., & Zverev, S. M. (1988). Mechanism of lateral interactions between molecules adsorbed on oxide surfaces. *Reaction Kinetics and Catalysis Letters, 36*(2), 269–274. DOI: 10.1007/BF02063817

Turan, N. G., & Ergun, O. N. (2009). Removal of Cu(II) from leachate using natural zeolite as a landfill liner material. In Journal of Hazardous Materials (Vol. 167, Issues 1–3, pp. 696–700). Elsevier BV. https://doi.org/DOI: 10.1016/j.jhazmat.2009.01.047

Ugliengo, P., Garrone, E., Ferrari, A. M., Zecchina, A., & Otero Areán, C. (1999). Quantum Chemical Calculations and Experimental Evidence for O-Bonding of Carbon Monoxide to Alkali Metal Cations in Zeolites. *The Journal of Physical Chemistry B, 103*(23), 4839–4846. DOI: 10.1021/jp982078w

Ugrina, M., Gabrilo, I., & Jurić, A. (2023). Modeling, simulation, optimization, and experimental verification of mercury removal onto natural and sulfur-impregnated zeolite clinoptilolite—Assessment of feasibility for remediation of mercury-contaminated soil. *Processes (Basel, Switzerland), 11*(2), 606. Advance online publication. DOI: 10.3390/pr11020606

Ullah, , Sun, J., Gul, A., & Bai, S. (2020b). One-step hydrothermal synthesis of TiO2-supported clinoptilolite: An integrated photocatalytic adsorbent for removal of crystal violet dye from aqueous media. Journal of Environmental Chemical Engineering. *Journal of Environmental Chemical Engineering, 103852*(4), 103852. Advance online publication. DOI: 10.1016/j.jece.2020.103852

Ullah, , Sun, J., Gul, A., Munir, T., & Wu, X. (2020a). Evaluations of physicochemical properties of TiO2/clinoptilolite synthesized via three methods on photocatalytic degradation of crystal violet. *Chinese Journal of Chemical Engineering, 33*, 181–189. DOI: 10.1016/j.cjche.2020.09.045

Uma, S., & Shobana, M. K. (2023). Metal oxide semiconductor gas sensors in clinical diagnosis and environmental monitoring. *Sensors and Actuators. A, Physical, 349*, 114044. DOI: 10.1016/j.sna.2022.114044

Usman Azam, M., Fernandes, A., João Ferreira, M., McCue, A. J., Graça, I., & Afzal, W. (2025). Unlocking the structure-activity relationship of hierarchical MFI zeolites towards the hydrocracking of HDPE. *Fuel, 379*, 132990. DOI: 10.1016/j.fuel.2024.132990

Vaezi, M., Noormohammadbeigi, M., Cruciani, G., & Zendehdel, M. (2025). Ion-exchange of copper into mordenite and clinoptilolite zeolites by molecular dynamics simulations and experimental investigations. *Microporous and Mesoporous Materials, 382*, 113397. DOI: 10.1016/j.micromeso.2024.113397

Vaičiukynienė, D., Jakevičius, L., Kantautas, A., Vaitkevičius, V., Vaičiukynas, V., & Dvořák, K. (2021). Conversion of silica by-product into zeolites by thermo-sonochemical treatment. *Ultrasonics Sonochemistry*, *72*. Advance online publication. DOI: 10.1016/j.ultsonch.2020.105426

Vaičiukynienė, D., Kantautas, A., Vaitkevičius, V., Jakevičius, L., Rudžionis, Ž., & Paškevičius, M. (2015). Effects of ultrasonic treatment on zeolite NaA synthesized from by-product silica. *Ultrasonics Sonochemistry*, *27*, 515–521. DOI: 10.1016/j.ultsonch.2015.06.001 PMID: 26186874

van Bokhoven, J. A., & Lamberti, C. (2014). Structure of aluminum, iron, and other heteroatoms in zeolites by X-ray absorption spectroscopy. *Coordination Chemistry Reviews*, *277–278*, 275–290. DOI: 10.1016/j.ccr.2014.05.013

van Vreeswijk, S. H., & Weckhuysen, B. M. (2022). Emerging analytical methods to characterize zeolite-based materials. *National Science Review*, *9*(9), nwac047. Advance online publication. DOI: 10.1093/nsr/nwac047 PMID: 36128456

Vasconcelos, A. A., Len, T., de Nazaré de Oliveira, A., da Costa, A. A. F., da Silva Souza, A. R., da Costa, C. E. F., Luque, R., da Rocha Filho, G. N., Noronha, R. C. R., & do Nascimento, L. A. S. (2023). Zeolites: A theoretical and practical approach with uses in (bio)chemical processes. *Applied Sciences (Basel, Switzerland)*, *13*(3), 1897. DOI: 10.3390/app13031897

Vázquez, M.. (2020). *Introduction to nanomaterials and nanotechnology: Structure, properties, and applications*. Elsevier.

Velarde, , Nabavi, M. S., Escalera, E., Antti, M.-L., & Akhtar, F. (2023). Adsorption of heavy metals on natural zeolites: A review. *Chemosphere*, *328*, 138508. Advance online publication. DOI: 10.1016/j.chemosphere.2023.138508 PMID: 36972873

Velasco, M. I., Franzoni, M. B., Chávez, F. V., & Acosta, R. H. (2023). Characterization of structure and functionality of porous materials. *Journal of Magnetic Resonance Open*, *14–15*, 100099. DOI: 10.1016/j.jmro.2023.100099

Verboekend, D., & Perez-Ramirez, J. (2011). Design of hierarchical zeolite catalysts by desilication. *Catalysis Science & Technology*, *1*(6), 879–890. DOI: 10.1039/c1cy00150g

Verboekend, D., & Pérez-Ramírez, J. (2011). Desilication mechanism revisited: Highly mesoporous all-silica zeolites enabled through pore-directing agents. *Chemistry (Weinheim an der Bergstrasse, Germany)*, *17*(4), 1137–1147. DOI: 10.1002/chem.201002589 PMID: 21243680

Vidal de la Fuente, A., Mekki, A., Boukoussa, B., Grande-Casas, M., Sastre, E., Márquez-Álvarez, C., Sánchez-Sánchez, M., & Pérez-Pariente, J. (2024). Generation of mesoporosity in crystals of zeolite omega by post-synthesis treatments with surfactants. *Catalysis Today*, *425*, 114364. https://doi.org/https://doi.org/10.1016/j.cattod.2023.114364. DOI: 10.1016/j.cattod.2023.114364

Vidoni, A., Ravikovitch, P. I., Afeworki, M., Calabro, D., Deckman, H., & Ruthven, D. (2020). Adsorption of CO<inf>2</inf> on high silica MFI and DDR zeolites: Structural defects and differences between adsorbent samples. *Microporous and Mesoporous Materials*, *294*, 109818. Advance online publication. DOI: 10.1016/j.micromeso.2019.109818

Vilcek, J., Torma, S., Adamisin, P., & Hronec, O. (2013). Nitrogen sorption and its release in the soil after zeolite application. *Bulgarian Journal of Agricultural Science*, *19*, 228–234.

Villarreal-Nuñez, L. A., Barahona-Amores, O. A., & Castillo-Ortiz, O. A. (2015). Effect of zeolite on the nitrogen fertilizer efficiency in rice crop. *Agronomía Mesoamericana*, *26*(2), 315–321. DOI: 10.15517/am.v26i2.19324

Vitale, , Barbato, C., Crispo, A., Habetswallner, F., De Martino, B. M., Riccardi, F., Maione, A., Eisenwagen, S., Vitale, G., & Cartenì, G. (2020). ZeOxaNMulti trial: A randomized, double-blinded, placebo-controlled trial of oral PMA-zeolite to prevent chemotherapy-induced side effects, in particular, peripheral neuropathy. *Molecules (Basel, Switzerland)*, *2297*(10), 2297. Advance online publication. DOI: 10.3390/molecules25102297 PMID: 32414185

Vogt, C., Wondergem, C. S., & Weckhuysen, B. M. (2023). Ultraviolet-Visible (UV-Vis) spectroscopy. In Springer Handbooks (pp. 237–264). https://doi.org/DOI: 10.1007/978-3-031-07125-6_11

Wącław, A., Nowińska, K., & Schwieger, W. (2004). Benzene to phenol oxidation over iron exchanged zeolite ZSM-5. *Applied Catalysis A, General*, *270*(1–2), 151–156. DOI: 10.1016/j.apcata.2004.04.032

Wahono, , Suwanto, A., Prasetyo, D. J., Hernawan, , Jatmiko, T. H., & Vasilev, K. (2019). Plasma activation on natural mordenite-clinoptilolite zeolite for water vapor adsorption enhancement. *Applied Surface Science*, *483*, 940–946. Advance online publication. DOI: 10.1016/j.apsusc.2019.04.033

Wahyuni, E. T., Diantariani, N. P., Kartini, I., & Kuncaka, A. (2022). Enhancement of the photostability and visible photoactivity of ZnO photocatalyst used for reduction of Cr (VI) ions. *Results in Engineering*, *13*, 100351. DOI: 10.1016/j.rineng.2022.100351

Wakchaure, S. G., Dabhade, G. B., Borhade, A. V., Malpure, A. A., Palde, S., Shewale, Y., & Yannam, R. (2025). Mordenite's mineralogical study and application as an efficient adsorbent for heavy metal detection from waste water. In Journal of the Indian Chemical Society (Vol. 102, Issue 1, p. 101502). Elsevier BV. https://doi.org/DOI: 10.1016/j.jics.2024.101502

Wakweya, B., & Jifar, W. W. (2023). In vitro Evaluation of Antibacterial Activity of Synthetic Zeolite Supported AgZno Nanoparticle Against a Selected Group of Bacteria. *Journal of Experimental Pharmacology*, *15*, 139–147. DOI: 10.2147/JEP.S396118 PMID: 36941893

Wang, S., & Peng, Y. (2010). Natural zeolites as effective adsorbents in water and wastewater treatment. In Chemical Engineering Journal (Vol. 156, Issue 1, pp. 11–24). Elsevier BV. https://doi.org/DOI: 10.1016/j.cej.2009.10.029

Wang, S., & Zhu, Z. (2006). Characterisation and environmental application of an Australian natural zeolite for basic dye removal from aqueous solution. In Journal of Hazardous Materials (Vol. 136, Issue 3, pp. 946–952). Elsevier BV. https://doi.org/DOI: 10.1016/j.jhazmat.2006.01.038

Wang, X., Yan, N., Xie, M., Liu, P., Bai, P., Su, H., Wang, B., Wang, Y., Li, L., Cheng, T., Guo, P., Yan, W., & Yu, J. (2021). The inorganic cation-tailored "trapdoor" effect of silicoaluminophosphate zeolite for highly selective CO2 separation. In Chemical Science (Vol. 12, Issue 25, pp. 8803–8810). Royal Society of Chemistry (RSC). https://doi.org/DOI: 10.1039/D1SC00619C

Wang, B., Wu, J., Yuan, Z. Y., Li, N., & Xiang, S. (2008). Synthesis of MCM-22 zeolite by an ultrasonic-assisted aging procedure. *Ultrasonics Sonochemistry*, *15*(4), 334–338. DOI: 10.1016/j.ultsonch.2007.07.007

Wang, C., Dai, W., Wu, G., Guan, N., & Li, L. (2022). Application of ammonia probe-assisted solid-state NMR technique in zeolites and catalysis. *Magnetic Resonance Letters*, *2*(1), 28–37. DOI: 10.1016/j.mrl.2021.10.001

Wang, , & Nguyen, A. V. (2015). Characterisation of electrokinetic properties of clinoptilolite before and after activation by sulphuric acid for treating CSG water. *Microporous and Mesoporous Materials*, *220*, 175–182. DOI: 10.1016/j.micromeso.2015.09.003

Wang, R., Ma, N., Yan, Y., & Wang, Z. (2018). Ultrasonic-assisted fabrication of high flux T-type zeolite membranes on alumina hollow fibers. *Journal of Membrane Science*, •••, 676–684. DOI: 10.1016/j.memsci.2017.10.047

Wang, Y., Wang, C., Wang, L., Wang, L., & Xiao, F. S. (2021). Zeolite Fixed Metal Nanoparticles: New Perspective in Catalysis. *Accounts of Chemical Research, 54*(11), 2579–2590. DOI: 10.1021/acs.accounts.1c00074 PMID: 33999615

Wang, Y., Wang, J., Wei, J., Wang, C., Wang, H., & Yang, X. (2023). Catalytic Mechanisms and Active Species of Benzene Hydroxylation Reaction System Based on Fe-Based Enzyme-Mimetic Structure. *Catalysis Letters, 153*(11), 3311–3332. DOI: 10.1007/s10562-022-04238-2

Wang, , Zhang, T., & Wang, X. (2025). Removal mechanism of lead by the composite of clinoptilolite, biochar, and Ca(OH)2 and its application in acid wastewater. *Environmental Technology & Innovation, 37*, 104024. Advance online publication. DOI: 10.1016/j.eti.2025.104024

Ward, J. W., & Habgood, H. W. (1966). The infrared spectra of carbon dioxide adsorbed on zeolite X. *Journal of Physical Chemistry, 70*(4), 1178–1182. DOI: 10.1021/j100876a034

Warren, B. E. (1969). *X-ray Diffraction*. Dover Publications.

Washburn, K. E., & Callaghan, P. T. (2006). Tracking pore to pore exchange using relaxation exchange spectroscopy. *Physical Review Letters, 97*(17), 25–28. DOI: 10.1103/PhysRevLett.97.175502 PMID: 17155481

Webber, J. B. W., Livadaris, V., & Andreev, A. S. (2020a). USY zeolite mesoporosity probed by NMR cryoporometry. *Microporous and Mesoporous Materials, 306*, 110404. DOI: 10.1016/j.micromeso.2020.110404

West, A. R. (2014). *Solid State Chemistry and its Applications*. Wiley.

White, H. D., Huang, H. G., D'amaro, M. C., Mignone, E. A., Sholl, D. S., & Lively, R. P. (2022). Theoretical Pathway toward Improved Reverse Osmosis Membrane Selectivity for Neutral Solutes: Inspiration from Gas Separations. *The Journal of Physical Chemistry. C, Nanomaterials and Interfaces, 126*(45), 19496–19506. DOI: 10.1021/acs.jpcc.2c06016

Williams, D. B., & Carter, C. B. (2009). *Transmission Electron Microscopy: A Textbook for Materials Science*. Springer Science & Business Media. DOI: 10.1007/978-0-387-76501-3

Wołowiec, , Muir, B., Zięba, K., Bajda, T., Kowalik, M., & Franus, W. (2017). Experimental study on the removal of VOCs and PAHs by zeolites and surfactant-modified zeolites. *Energy & Fuels, 31*(8), 8812. DOI: 10.1021/acs.energyfuels.7b01124

Wu, L. B., Jiao, D. M., Chen, L. F., Wang, J. A., & Cao, F. H. (2010). Comparative studies of the CoMo/MgO, CoMo/Al2O3 and CoMo/MgO-MgAl2O4 catalysts prepared by a urea-matrix combustion method. Advanced Materials Research, 132, 45-54.

Wu, L., Jiao, D., Wang, J., Chen, L., & Cao, F. (2009). The role of MgO in the formation of surface active phases of CoMo/Al2O3–MgO catalysts for hydrodesulfurization of dibenzothiophene. *Catalysis Communications*, *11*(4), 302–305. DOI: 10.1016/j.catcom.2009.10.021

Wu, Q. (2016). *Xia*, G., Chen, T, Wang, X., Chi, D. and Sun, D. (2016). *Nitrogen use and rice yield formation response to zeolite and nitrogen coupling effects: Enhancement in nitrogen use efficiency.Journal of Soil Science and Plant Nutrition*, *16*(4), 999–1009.

Xia, H., Sun, K., Sun, K., Feng, Z., Li, W. X., & Li, C. (2008). Direct Spectroscopic Observation of Fe(III)–Phenolate Complex Formed From the Reaction of Benzene With Peroxide Species on Fe/ZSM-5 At Room Temperature. *The Journal of Physical Chemistry. C, Nanomaterials and Interfaces*, *112*(24), 9001–9005. DOI: 10.1021/jp800455x

Xiang, , Han, Y., Jiang, C., Li, M., Wei, L., Fu, J., & Zhu, L. (2021). Composite biologically active filter (BAF) with zeolite, granular activated carbon, and suspended biological carrier for treating algae-laden raw water. Journal of Water Process Engineering. *Journal of Water Process Engineering*, *102188*, 102188. Advance online publication. DOI: 10.1016/j.jwpe.2021.102188

Xiao, P., Osuga, R., Wang, Y., Kondo, J. N., & Yokoi, T. (2020). Bimetallic Fe-Cu/beta zeolite catalysts for direct hydroxylation of benzene to phenol: Effect of the sequence of ion exchange for Fe and Cu cations. *Catalysis Science & Technology*, *10*(20), 6977–6986. DOI: 10.1039/d0cy01216e

Xiao, P., Wang, Y., Kondo, J. N., & Yokoi, T. (2019). Consequences of Fe speciation in MFI zeolites for hydroxylation of benzene to phenol with H2O2. *Applied Catalysis A, General*, *579*, 159–167. DOI: 10.1016/j.apcata.2019.04.025

Xie, . (2013). *Adsorption of organic pollutants by surfactant modified zeolite as controlled by surfactant chain length*. Microporous., DOI: 10.1016/j.micromeso.2013.05.027

Xie, J., Li, X., Guo, J., Luo, L., Delgado, J. J., Martsinovich, N., & Tang, J. (2023). Highly selective oxidation of benzene to phenol with air at room temperature promoted by water. *Nature Communications*, *14*(1), 4431. DOI: 10.1038/s41467-023-40160-w

Xin, S., Wang, Q., Xu, J., Chu, Y., Wang, P., Feng, N., Qi, G., Trébosc, J., Lafon, O., Fan, W., & Deng, F. (2019). The acidic nature of "NMR-invisible" tri-coordinated framework aluminum species in zeolites. *Chemical Science (Cambridge)*, *10*(43), 10159–10169. DOI: 10.1039/C9SC02634G PMID: 32055370

Xu, D., Lv, H., & Liu, B. (2018). Encapsulation of Metal Nanoparticle Catalysts Within Mesoporous Zeolites and Their Enhanced Catalytic Performances: A Review. *Frontiers in Chemistry*, *6*, 550. Advance online publication. DOI: 10.3389/fchem.2018.00550 PMID: 30474024

Xu, , Huang, Z., Ye, J., & Zheng, J. (2023). Hourly emissions of air pollutants and greenhouse gases from open biomass burning in China during. *Scientific Data*, *10*(1), 629. Advance online publication. DOI: 10.1038/s41597-023-02541-0 PMID: 37717027

Xu, , Liu, H., Zhang, J., Xu, M., Zhao, X., & Wang, J. (2025). Functionalized zeolite regulates bone metabolic microenvironment. *Materials Today. Bio*, *101558*, 101558. Advance online publication. DOI: 10.1016/j.mtbio.2025.101558 PMID: 40034985

Xu, Z., Wang, W., Li, Y., Meng, S., Liu, Y., Cao, Y., You, J., Zhang, B., & Zang, J. (2024). The influence of solvent-temperature parameters of sodium titanium phosphate anode materials prepared by sol-gel method. *Materials Chemistry and Physics*, *315*, 128946. DOI: 10.1016/j.matchemphys.2024.128946

Yakimov, A. V., Sushkevich, V. L., van Bokhoven, J. A., & Copéret, C. (2022). Probing acid sites in MOR zeolite using low-temperature ^{13}C solid-state NMR spectroscopy of adsorbed carbon monoxide. *The Journal of Physical Chemistry. C, Nanomaterials and Interfaces*, *126*(7), 3681–3687. DOI: 10.1021/acs.jpcc.1c10026

Yakout, S. M. (2020). Spintronics: Future technology for new data storage and communication devices. *Journal of Superconductivity and Novel Magnetism*, *33*(9), 2557–2580. DOI: 10.1007/s10948-020-05545-8

Yamaguchi, S., Miyake, Y., Takiguchi, K., Ihara, D., & Yahiro, H. (2018). Oxidation of cyclic hydrocarbons with hydrogen peroxide over iron complexes encapsulated in cation-exchanged zeolite. *Catalysis Today*, *303*, 249–255. DOI: 10.1016/j.cattod.2017.10.047

Yang, . (2024). Water-saving irrigation combined with N-loaded clinoptilolite enhances nutrient yield, and water productivity by improving rice root characteristics: A combined PCA-SEM analysis. *Agricultural Water Management*, *301*, 109203. Advance online publication. DOI: 10.1016/j.agwat.2024.109203

Yasumura, S., Liu, C., Toyao, T., Maeno, Z., & Shimizu, K.-I. (2021). Lean NO<inf>x</inf>Capture and Reduction by NH<inf>3</inf>via NO+Intermediates over H-CHA at Room Temperature. *The Journal of Physical Chemistry. C, Nanomaterials and Interfaces*, *125*(3), 1913–1922. DOI: 10.1021/acs.jpcc.0c10913

Ying, J. Y., & García-Martínez , J. (2005). (Patent No. US 2005/0239634).

Yin, Q., & McAughey, S. (2022). *IZC_NEREA Validation. Greenhouse basil, lettuce, gerbera production and chrysanthemum rooting*. Vineland Research & Innovation Centre.

Yin, X., Long, Z., Wang, C., Li, Z., Zhao, M., & Yang, S. (2019). A time- and cost-effective synthesis of CHA zeolite with small size using ultrasonic-assisted method. *Ultrasonics Sonochemistry*, *58*. Advance online publication. DOI: 10.1016/j.ultsonch.2019.104679

Yocupicio-Gaxiola, R. I., Petranovskii, V., Antúnez-García, J., & Fuentes Moyado, S. (2019). One-pot synthesis of lamellar mordenite and ZSM-5 zeolites and subsequent pillaring by amorphous SiO2. *Applied Nanoscience*, *9*(4), 557–565. DOI: 10.1007/s13204-018-0935-1

Yocupicio, R. I., Díaz de León, J. N., Zepeda, T. A., & Fuentes, S. (2017). Study of CoMo catalysts supported on hierarchical mesoporous zeolites for hydrodesulfurization of dibenzothiophene. *Revista Mexicana de Ingeniería Química*, *16*(2), 503–520.

Yon, V., Rochat, N., Charles, M., Nolot, E., & Gergaud, P. (2020). X-ray diffraction micro-strain analysis for extraction of threading dislocation density of GaN films grown on silicon, sapphire and SiC substrates. [b]. *Physica Status Solidi. B, Basic Research*, *257*(4), 1900579. Advance online publication. DOI: 10.1002/pssb.201900579

Yosefi, L., Haghighi, M., Allahyari, S., & Ashkriz, S. (2015). Effect of ultrasound irradiation and Ni-loading on properties and performance of CeO2-doped Ni/clinoptilolite nanocatalyst used in polluted air treatment. *Process Safety and Environmental Protection*, *95*, 26–37. DOI: 10.1016/j.psep.2015.02.006

Yue, M., Jiang, X., Zhang, H., Zhang, S., Xue, T., & Li, Y. (2020). Quasi-solid-phase synthesis of Fe-MFI zeolites by using Fe-containing zeolite seed sol for hydroxylation of benzene with H2O2. *Microporous and Mesoporous Materials*, *294*. Advance online publication. DOI: 10.1016/j.micromeso.2019.109891

Yu, , Rengasamy, K., Lim, K.-Y., Tan, L.-S., Tarawneh, M. A., Zulkoffli, Z. B., & Se Yong, E. N. (2019). Comparison of activated carbon and zeolites filtering efficiency in freshwater. Journal of Environmental Chemical Engineering. *Journal of Environmental Chemical Engineering, 103223*(4), 103223. Advance online publication. DOI: 10.1016/j.jece.2019.103223

Yusof, N. A. A., Zain, N. M., & Pauzi, N. (2019). Synthesis of ZnO nanoparticles with chitosan as stabilizing agent and their antibacterial properties against Gram-positive and Gram-negative bacteria. *International Journal of Biological Macromolecules, 124*, 1132–1136. DOI: 10.1016/j.ijbiomac.2018.11.228 PMID: 30496864

Yusuf, , Sodiq, A., Giwa, A., Eke, J., Pikuda, O., De Luca, G., Di Salvo, J. L., & Chakraborty, S. (2020). A review of emerging trends in membrane science and technology. *Journal of Cleaner Production, 266*, 121867. Advance online publication. DOI: 10.1016/j.jclepro.2020.121867

Yu, T., Zhang, Y., Cao, L., Cao, P., Zhou, C., & Gu, S. (2024). Freeze–Thaw Cycle Durability and Mechanism Analysis of Zeolite Powder-Modified Recycled Concrete. *Materials (Basel), 17*(11), 2671. Advance online publication. DOI: 10.3390/ma17112671 PMID: 38893935

Yu, Z., Li, S., Wang, Q., Zheng, A., Jun, X., Chen, L., & Deng, F. (2011). Brønsted/lewis acid synergy in H-ZSM-5 and H-MOR zeolites studied by ^{1}H and ^{27}Al DQ-MAS solid-state NMR spectroscopy. *The Journal of Physical Chemistry. C, Nanomaterials and Interfaces, 115*(45), 22320–22327. DOI: 10.1021/jp203923z

Yu, Z., Wang, Q., Chen, L., & Deng, F. (2012). Brønsted/lewis acid sites synergy in H-MCM-22 zeolite studied by ^{1}H and ^{27}Al DQ-MAS NMR spectroscopy. *Cuihua Xuebao. Chinese Journal of Catalysis, 33*(1), 129–139. DOI: 10.1016/S1872-2067(10)60287-2

Yu, Z., Zheng, A., Wang, Q., Chen, L., Xu, J., Amoureux, J. P., & Deng, F. (2010). Insights into the dealumination of zeolitehy revealed by sensitivity-enhanced ^{27}Al DQ-MAS NMR spectroscopy at high field. *Angewandte Chemie International Edition, 49*(46), 8657–8661. DOI: 10.1002/anie.201004007 PMID: 20931586

Zaki, M. I., & Knözinger, H. (1987). Characterization of oxide surfaces by adsorption of carbon monoxide — A low temperature infrared spectroscopy study. *Spectrochimica Acta. Part A: Molecular Spectroscopy, 43*(12), 1455–1459. DOI: 10.1016/S0584-8539(87)80030-2

Zang, J., Yu, H., Liu, G., Hong, M., Liu, J., & Chen, T. (2023). Research Progress on Modifications of Zeolite Y for Improved Catalytic Properties. *Inorganics, 11*(1), 22. DOI: 10.3390/inorganics11010022

Zasukhin, D. S., Kasyanov, I. A., Kolyagin, Y. G., Bulygina, A. I., Kharas, K. C., & Ivanova, I. I. (2022). Evaluation of zeolite acidity by ^{31}P MAS NMR spectroscopy of Aadsorbed phosphine oxides: Quantitative or not? *ACS Omega*, *7*(14), 12318–12328. DOI: 10.1021/acsomega.2c00804 PMID: 35449977

Zepeda, T. A. (2025). Ultrasonic-assistance influence on the impregnation method for high metal loading CoMo/Ti-HMS catalysts. *Applied Catalysis A, General*, *693*, 120135. DOI: 10.1016/j.apcata.2025.120135

Zepeda, T. A., Pawelec, B., Fierro, J. L. G., Olivas, A., Fuentes, S., & Halachev, T. (2008). Effect of Al and Ti content in HMS material on the catalytic activity of NiMo and CoMo hydrotreating catalysts in the HDS of DBT. *Microporous and Mesoporous Materials*, *111*(1), 157–170. DOI: 10.1016/j.micromeso.2007.07.025

Zhang, . (2020). *Electron microscopy studies of local structural modulations in zeolite crystals* (International Edition). Angewandte Chemie., DOI: 10.1002/anie.202007490

Zhang, J., Tang, X., Yi, H., Yu, Q., Zhang, Y., Wei, J., & Yuan, Y. (2022). Synthesis, characterization and application of Fe-zeolite: A review. *Applied Catalysis A, General*, *630*, 118467. DOI: 10.1016/j.apcata.2021.118467

Zhang, , Liu, N., Lei, Z., & Chen, B. (2016). Selective transformation of various nitrogen exhaust gases toward N2 over zeolite catalysts. *Chemical Reviews*, *117*(10), 6838–6880. DOI: 10.1021/acs.chemrev.5b00474 PMID: 26889565

Zhang, R., Zhong, P., Arandiyan, H., Guan, Y., Liu, J., Wang, N., Jiao, Y., & Fan, X. (2020). Using ultrasound to improve the sequential post-synthesis modification method for making mesoporous Y zeolites. *Frontiers of Chemical Science and Engineering*, *14*(2), 275–287. DOI: 10.1007/s11705-019-1905-1

Zhang, S., Liu, J., Zu, L., Wang, B., & Li, K. (2025). Analysis of real-world diesel sulphur in diesel engines in China's key regions. *Transportation Research Part D, Transport and Environment*, *139*, 104562. DOI: 10.1016/j.trd.2024.104562

Zhang, X., Lv, J., Liu, J., Xu, S., Sun, J., Wang, L., Xu, L., Mintova, S., Song, H., & Dong, B. (2022). Stable EMT type zeolite/CsPbBr<inf>3</inf> perovskite quantum dot nanocomposites for highly sensitive humidity sensors. *Journal of Colloid and Interface Science*, *616*, 921–928. DOI: 10.1016/j.jcis.2022.02.079 PMID: 35259722

Zhang, Y., Cao, B., Yin, H., Meng, L., Jin, W., Wang, F., Xu, J., & Al-Tabbaa, A. (2022). Application of zeolites in permeable reactive barriers (PRBs) for in-situ groundwater remediation: A critical review. In *Chemosphere* (Vol. 308, p. 136290). Elsevier BV., DOI: 10.1016/j.chemosphere.2022.136290

Zhang, Y., Han, W., Long, X., & Nie, H. (2016). Redispersion effects of citric acid on CoMo/γ-Al2O3 hydrodesulfurization catalysts. *Catalysis Communications*, *82*(Supplement C), 20–23. DOI: 10.1016/j.catcom.2016.04.012

Zhao, P., Boekfa, B., Nishitoba, T., Tsunoji, N., Sano, T., Yokoi, T., Ogura, M., & Ehara, M. (2020). Theoretical study on 31P NMR chemical shifts of phosphorus-modified CHA zeolites. *Microporous and Mesoporous Materials*, *294*, 109908. Advance online publication. DOI: 10.1016/j.micromeso.2019.109908

Zhao, R., Yin, C., Zhao, H., & Liu, C. (2003). Synthesis, characterization, and application of hydotalcites in hydrodesulfurization of FCC gasoline. *Fuel Processing Technology*, *81*(3), 201–209. DOI: 10.1016/S0378-3820(03)00012-2

Zhao, S., He, S., Du Kim, K., Wang, L., Ryoo, R., Wang, Z., & Huang, J. (2021). Influence of hierarchical ZSM-5 catalysts with various acidity on the dehydration of glycerol to acrolein. *Magnetic Resonance Letters*, *1*(1), 71–80. DOI: 10.1016/j.mrl.2021.100002

Zhao, Y., Zhu, L., Kang, Y., Shen, C., Liu, X., Jiang, D., Fu, L., Guselnikova, O., Huang, L., Song, X., Asahi, T., & Yamauchi, Y. (2024). Nanoengineering multilength-scale porous hierarchy in mesoporous metal–organic framework single crystals. *ACS Nano*, *18*(33), 22404–22414. DOI: 10.1021/acsnano.4c07119 PMID: 39108023

Zheng, M., Chu, Y., Wang, Q., Wang, Y., Xu, J., & Deng, F. (2024). Advanced solid-state NMR spectroscopy and its applications in zeolite chemistry. *Progress in Nuclear Magnetic Resonance Spectroscopy, 140–141*(March 2023), 1–41. DOI: 10.1016/j.pnmrs.2023.11.001

Zholobenko, V. L., Makarova, M. A., & Dwyer, J. (1993). Inhomogeneity of Brönsted acid sites in H-mordenite. *Journal of Physical Chemistry*, *97*(22), 5962–5964. DOI: 10.1021/j100124a030

Zholobenko, V., Hollamby, M., Al-Ani, A., Lebedev, O. I., Smith, A. J., & Snow, T. (2025). A nanoscale investigation of the formation of mesostructured zeolites FAU and LTL. *Microporous and Mesoporous Materials*, *381*, 113363. https://doi.org/https://doi.org/10.1016/j.micromeso.2024.113363. DOI: 10.1016/j.micromeso.2024.113363

Zhou, L., Wang, M., Yang, S., Guo, W., Pu, X., He, Y., Zhu, J., Wang, B., Zheng, M., Liu, S., & Zhang, Y. (2022). Facile synthesis of mesoporous ZSM-5 aided by sonication and its application for VOCs capture. *Ultrasonics Sonochemistry*, *88*. Advance online publication. DOI: 10.1016/j.ultsonch.2022.106098

Zhou, P., Liu, M. N., Luo, Q. X., Zhang, J., Chen, H., Ma, X., & Hao, Q. Q. (2022). Synthesis of hierarchical nanocrystalline β zeolite as efficient catalyst for alkylation of benzene with benzyl alcohol. *RSC Advances*, *12*(8), 4865–4873. DOI: 10.1039/D2RA00209D PMID: 35425505

Zhou, , Shen, Y., Zhao, S., Chen, Y., Gao, S., Liu, W., & Wei, D. (2020). Hydrothermal synthesis of novel ternary hierarchical MoS2/TiO2/clinoptilolite nanocomposites with remarkably enhanced visible light response towards xanthates. *Applied Surface Science*, *542*, 148578. Advance online publication. DOI: 10.1016/j.apsusc.2020.148578

Zhou, Z., Jiang, R., Chen, X., Wang, X., & Hou, H. (2021). One-step synthesis of hierarchical lamellar H-ZSM-5 zeolite and catalytic performance of methanol to olefin. *Journal of Solid State Chemistry*, *298*, 122132. DOI: 10.1016/j.jssc.2021.122132

Zhukov, Y. M., Kovalyov, A. N., Kultaeva, A. Y., Shelyapina, M. G., & Petranovskii, V. (2016). A comparative analysis of the protonated and copper exchanged mordenites with SiO_2/Al_2O_3 molar ratio equal to 10. *International Journal of Nanotechnology*, *13*(1/2/3), 136–146. DOI: 10.1504/IJNT.2016.074529

Zineb, C. B., Amina, K., Sanaa, E. K., Mohammed, R. R., Amel, B., Khaldoun, B., & Abderrahim, C. B. (2023). Enhancing structural properties of simple perovskite materials based on zirconate and titanate prepared by sol-gel method. *Materials Chemistry and Physics*, *310*, 128460. DOI: 10.1016/j.matchemphys.2023.128460

Zining, Y., Anlong, F., Kangyu, Z., Dmitry, V. K., Qi, G., Jiaojiao, Y., & Lisha, L. (2024). Great enhancement of ferroelectric properties of Al2O3-modified BiFeO3 thin films obtained by sol-gel method. *Journal of the European Ceramic Society*, *44*(1), 224–232. DOI: 10.1016/j.jeurceramsoc.2023.08.058

Znak, Z., Zin, O., Mashtaler, A., Korniy, S., Sukhatskiy, Y., Gogate, P. R., Mnykh, R., & Thanekar, P. (2021). Improved modification of clinoptilolite with silver using ultrasonic radiation. *Ultrasonics Sonochemistry*, *73*, 105496. DOI: 10.1016/j.ultsonch.2021.105496 PMID: 33636551

Zou, R., Chansai, S., Xu, S., An, B., Zainal, S., Zhou, Y., Xin, R., Gao, P., Hou, G., D'Agostino, C., Holmes, S. M., Hardacre, C., Jiao, Y., & Fan, X. (2023). Pt nanoparticles on Beta zeolites for catalytic toluene oxidation: Effect of the hydroxyl groups of Beta zeolite. *ChemCatChem*, *15*(21), e202300811. DOI: 10.1002/cctc.202300811

Zuo, M., Xiao, R., Du, F., Cheng, C., Rodriguez, R. D., Ma, L., Zhu, B., & Qiu, L. (2024). Ultrasound-activated mechanochemical reactions for controllable biomedical applications. *Smart Materials in Medicine*, *5*(4), 461–476. DOI: 10.1016/j.smaim.2024.09.001

About the Contributors

Fabian N. Murrieta-Rico received B.Eng. and M.Eng. degrees from Instituto Tecnológico de Mexicali (ITM) in 2004 and 2013 respectively. In 2017, he received his PhD in Materials Physics at Centro de Investigación Científica y Educación Superior de Ensenada (CICESE). He has worked as an automation engineer, systems designer, as a university professor, and as postdoctoral researcher at Facultad de Ingeniería, Arquitectura y Diseño from Universidad Autónoma de Baja California (UABC) and at the Centro de Nanociencias y Nanotecnología from Universidad Nacional Autónoma de México (CNyN-UNAM), currently he works as professor at the Universidad Politécnica de Baja California. His research has been published in different journals and presented at international conferences since 2009. He has served as reviewer for different journals, some of them include IEEE Transactions on Industrial Electronics, IEEE Transactions on Instrumentation, Measurement and Sensor Review. His research interests are focused on the field of time and frequency metrology, the design of wireless sensor networks, automated systems and highly sensitive chemical detectors.

Joel Antúnez García received his B.Sc. in Physics from the Universidad Autónoma de Baja California (UABC), his M.Sc. from the Centro de Investigación Científica y Educación Superior de Ensenada (CICESE), and his Ph.D. in Industrial Engineering Physics from the Universidad Autónoma de Nuevo León (UANL). From 2011 to 2013, he completed a postdoctoral fellowship at the Center for Nanosciences and Nanotechnology (CNyN) at UNAM. He was a full-time faculty member at the Centro de Enseñanza Técnica y Superior (CETYS), Tijuana campus, from 2013 to 2015. From 2016 to mid-2023, he worked as a postdoctoral researcher at CNyN. He currently teaches at the Universidad Autónoma de Baja California (UABC). As a researcher, his work has focused on molecular dynamics studies of metallic clusters and first-principles calculations of materials at both bulk and nanometer scales. Dr. Antúnez is currently researching catalysis, as well as the magnetic and optoelectronic properties of zeolitic materials and dichalcogenides.

Vitalii Petranovskii received the Ph.D. degree in physical chemistry from the Moscow Institute of Crystallography in 1988. From 1993 to 1994, he worked as a Visiting Fellow at the National Institute of Materials Science and Chemical Research, Japan. Since 1995, he has been working with the Center for Nanotechnology and Nanotechnology, National University of Mexico, as the Head of the Department of Nanocatalysis, from 2006 to 2014. He is a member of the Mexican Academy of Sciences, the International Association of Zeolites, and the Russian Chemical Society. He has published over 160 articles in peer-reviewed journals and five invited book chapters. He is also a coauthor of the monograph Clusters and Matrix Isolated Clustered Superstructures (St. Petersburg, 1995). His research interests include the synthesis and properties of nanoparticles deposited on zeolite matrices, and the modification of the zeolite matrices themselves for their high-tech use.

***

Julia Aguilar Pliego was born in Mexico City. She is a chemist from the Universidad Nacional Autónoma de México UNAM. She has a master's degree in chemistry from the Universidad Autonoma Metropolitana-Iztapalapa and a PhD from the Universidad Autónoma de Madrid. She is a full-time research professor at the Universidad Autónoma Metropolitana-Azcapotzalco. Her research interests are the synthesis, characterisation and application of micro/mesoporous materials for heterogeneous catalysis. She has published about 50 articles in indexed journals and book chapters. She has been the advisor of more than 20 postgraduate theses. She was president of the Mexican Academy of Catalysis (2015-2017). She holds a patent jointly with the ALFA group on catalytic depolymerisation of plastic waste (in use MX/a/2016/000917). He is a member of the National System of Researchers (SNI) at level II. Member of the American Chemical Society, ACAT.

Diana Barraza-Jiménez, a 2008 CIMAV PhD graduate in Materials Sciences, has taught materials science and nanotechnology at the master's level since 2015. Formerly with Interceramic, she now co-leads the nanotechnology and biotechnology research group and the scientific computing lab. Her research focuses on nanomaterial synthesis and simulation for biotechnology, clean tech (e.g., artificial photosynthesis, solar cells, purification), and machine learning. Dr. Barraza-Jimenez interests are nanotechnology simulation, metallic nanoparticles (Pt, Ag, Au, Fe), CFO, BFO, zeolites, 2D nanomaterials and plant pigments. She is a National Researcher Level I.

Manuel Flores-Hidalgo, a 2011 CIMAV PhD graduate in Materials Sciences, has been a research professor at FCQ-UJED since 2015. With over 15 years' experience in industry at firms like General Electric Aviation and Delphi (Aptiv), he now leads the nanotechnology and biotechnology research group and the scientific computing lab. His work focuses on nanomaterials simulation, synthesis, 3D printing, and machine learning applied in nanotechnology, alongside AI, robotics, and solar cells. Dr. Flores-Hidalgo is a National Researcher Level I.

Luis Perez Cabrera obtained his superior grade in chemistry at the Benemerita Universidad Autonoma de Puebla. Luis received his M.C. (with honors) and PhD from UNAM in the postgraduate program of Science and Engineering of Materials centralized to the Instituto de Investigación en Materiales of UNAM. He participates as a researcher-professor at the Center of Nanoscience and Nanotechnology, incorporated into UNAM through the postdoctoral program of SECIHTI. His research department is Nanocatalysis, particularly in synthesizing pure or mixed oxides such as magnesia, alumina, zeolites, and hydrotalcites, modified with different transition metals for application as catalysts for fuel refining and ambiental remediation reactions.

Joaquín Pérez Pariente (b. 1957) is Research Professor at the Institute of Catalysis and Petroleum Chemistry (ICP) of the Spanish Research Council (CSIC). He was the Director of ICP from 12/2005 until 01/2014. His research focuses on the synthesis of zeolites and porous materials and the study of their catalytic activity. He is also involved in research and teaching on the history of chemistry. He is the author of 290 publications, two books, 30 patents and has supervised 14 PhD. He is member of the Board of the History of Science Group of the Royal Spanish Society of Chemistry and of the Scientific Council of the journal Llull edited by the Spanish Society for the History of Sciences and Technology. He is corresponding academician of the Royal National Academy of Pharmacy of Spain, and was awarded in 2014 Doctor Honoris Causa of the Universidad Autónoma Metropolitana of Mexico.

Manuel Sánchez- Sánchez became PhD in 2000 at the Instituto de Tecnología Química (ITQ, UPV-CSIC, Valencia, Spain). He completed two postdoctoral stays in the United Kingdom: Chemistry Department at the University of Cambridge and The Royal Institution of Great Britain, where he held an Individual Marie Curie Fellowship. He returned to Spain to Universidad Rey Juan Carlos (Madrid) with a Ramón y Cajal contract in 2004, and then joined Instituto de Catálisis y Petroleoquímica (ICP, CSIC, Madrid), where he is Senior Scientist since 2015. His research career has focused on the use of nanoporous-based (mainly MOFs and zeolites) catalysts to address different energetic, chemical production and environmental challenges. He has participated in numerous research competitive projects, some of them as principal investigator, and he is author of 86 peer-review publications, three patents, two books and seven book chapters. He has also participated in various contracts with private companies.

Marina G. Shelyapina obtained a PhD in Physics (1999) and DrSc in Condenced Matter Physics (2018) at St Petersburg State University, Russia. She worked as invited reseacher at Neel Institute, National Center for Scientific Research, Grenoble, France. Since 2020 she is a full professor at St Petersburg State University. Her main research fields are related to local structure of solids and dynamics of guest atoms and molecules on their surface or in bulk probed by nuclear magnetic resonance and density functional theory

Index

Symbols

2D Relaxation 48

A

Acidity 23, 24, 27, 39, 40, 49, 51, 55, 58, 59, 62, 65, 66, 67, 69, 70, 71, 91, 92, 94, 97, 99, 182, 184, 186, 200, 201, 203, 204, 205, 206, 209, 315, 316, 319, 325, 326, 327, 329, 330, 333, 335, 337, 338, 339, 341, 342, 345, 346, 347, 351, 356, 358, 372, 430, 432, 442, 444, 447, 450
Acid Strength 94, 315, 325, 326, 329, 330, 333, 334, 342, 345, 351, 352
Adsorption 2, 25, 40, 42, 44, 45, 46, 54, 55, 65, 66, 67, 68, 69, 70, 71, 73, 74, 75, 76, 77, 78, 79, 80, 81, 83, 84, 86, 87, 88, 90, 91, 92, 93, 94, 95, 96, 97, 98, 99, 107, 124, 143, 153, 154, 156, 158, 159, 162, 165, 166, 167, 168, 169, 172, 173, 174, 175, 177, 178, 183, 188, 191, 205, 212, 215, 216, 217, 218, 219, 221, 222, 223, 224, 225, 226, 227, 228, 229, 230, 231, 232, 235, 238, 239, 240, 241, 242, 244, 247, 251, 252, 253, 255, 256, 257, 258, 260, 261, 263, 266, 269, 270, 271, 273, 275, 287, 320, 325, 327, 328, 329, 333, 336, 340, 344, 345, 347, 358, 363, 385, 390, 397, 400, 401, 404, 405, 406, 412, 413, 414, 419, 420, 422, 427, 428, 429, 431, 432, 433, 434, 435, 436, 438, 441, 444, 445, 446, 447, 448, 449, 450, 451, 452, 453, 454, 455
Adsorption Mechanisms 159, 175, 242, 414, 427
Aluminophosphate AFI 12, 15
Antimicrobial 223, 397, 399, 404, 405, 406, 407, 408, 409, 410, 411, 412, 413, 414, 416, 417, 418, 419, 420, 422, 423, 424

B

Basicity 65, 67, 68, 71, 83, 84, 85, 91, 93, 96, 98, 204, 212, 336, 339
Bean 280, 281, 282, 283, 285, 309, 310
Beta 25, 42, 55, 56, 62, 92, 166, 259, 318, 338, 341, 342, 343, 344, 353, 356, 374, 393, 407

C

Carbon Monoxide 61, 65, 72, 74, 90, 91, 92, 93, 95, 96, 98
Catalysis 2, 17, 19, 27, 51, 54, 55, 56, 59, 61, 86, 90, 91, 92, 93, 94, 96, 97, 98, 99, 102, 142, 146, 147, 157, 172, 183, 208, 209, 210, 211, 212, 213, 214, 227, 250, 253, 256, 316, 319, 326, 353, 354, 355, 357, 380, 383, 384, 385, 387, 388, 389, 390, 391, 392, 393, 394, 403, 404, 405, 408, 426, 427, 429, 430, 432, 440, 444, 447, 454
Catalyst 4, 23, 27, 30, 42, 62, 68, 90, 98, 174, 182, 183, 184, 185, 186, 187, 188, 189, 190, 191, 192, 193, 194, 195, 196, 197, 198, 199, 200, 201, 202, 203, 204, 205, 206, 207, 208, 211, 212, 218, 252, 317, 359, 360, 371, 372, 373, 374, 376, 378, 380, 381, 382, 383, 387, 391, 392, 403, 424, 432, 438, 444, 448, 450, 455
Clinoptilolite 59, 102, 108, 111, 124, 126, 128, 129, 138, 141, 142, 157, 162, 163, 164, 173, 215, 216, 217, 218, 219, 220, 221, 222, 223, 224, 225, 226, 227, 228, 229, 230, 231, 232, 234, 235, 236, 237, 238, 239, 240, 241, 242, 243, 244, 245, 246, 247, 248, 249, 250, 251, 252, 253, 254, 255, 256, 257, 258, 259, 260, 261, 262, 267, 270, 273, 274, 275, 276, 286, 292, 307, 308, 365, 385, 406, 424, 426, 435, 436, 437, 444, 445, 448
Coating 220, 395
CTA 37, 318, 320, 322, 323, 325, 326,

327, 330, 331, 332, 333, 334, 335, 336, 337, 338, 339, 340, 341, 342, 343, 344, 345, 346, 347, 348, 349, 358

D

Diffuse Reflectance Spectroscopy 20
Dislocation Density 102, 125, 126, 127, 145, 149, 150

E

Emerging pollutants 151, 232, 233, 263

F

Faujasite 25, 58, 78, 157, 177, 317, 318, 319, 320, 323, 324, 325, 326, 327, 331, 334, 337, 339, 354, 359, 368, 379, 380, 383, 389, 390, 435, 449, 451
Fertilizer 265, 266, 268, 269, 270, 272, 273, 276, 277, 278, 279, 280, 281, 282, 283, 284, 285, 286, 288, 289, 290, 291, 292, 293, 294, 295, 296, 297, 298, 299, 300, 302, 303, 304, 305, 306, 309, 311, 312, 361, 364
Fe species 196, 372, 377, 378, 380, 383, 391
Fluoroform 65, 67, 84, 85, 87
Framework Density 319, 332, 348, 350

H

HDS 182, 183, 184, 185, 201, 202, 204, 205, 206, 207, 208, 209, 210, 211, 213, 214
Hierarchical 23, 24, 25, 28, 29, 32, 33, 36, 38, 42, 43, 47, 48, 49, 52, 55, 56, 57, 58, 59, 60, 62, 158, 183, 186, 209, 210, 213, 214, 261, 271, 319, 327, 354, 355, 356, 357, 363, 368, 369, 387, 388, 389, 421, 447
Hydroxyl Groups 27, 33, 62, 65, 66, 78, 80, 95, 97

I

Infrared Spectra 75, 90, 94, 97, 98, 99, 304
Ion exchange 2, 24, 59, 107, 151, 152, 153, 154, 155, 156, 157, 158, 159, 161, 162, 163, 164, 165, 166, 167, 168, 169, 170, 172, 173, 176, 177, 178, 215, 217, 220, 248, 253, 257, 266, 271, 273, 274, 302, 317, 339, 370, 372, 373, 375, 380, 383, 393, 402, 403, 404, 405, 406, 407, 408, 412, 425, 427, 431, 435, 436, 442
Iron 115, 167, 175, 182, 183, 185, 186, 187, 191, 192, 193, 195, 196, 198, 203, 206, 208, 220, 221, 229, 245, 246, 253, 255, 269, 274, 372, 376, 378, 380, 384, 385, 386, 388, 389, 391, 392, 393, 394, 404, 409, 419, 430

L

Lattice Parameters 102, 128, 131, 132, 224
Linkage Isomerism 65, 75, 78, 79, 80, 81, 82, 90, 94, 95, 97, 99
Local Structure 30, 32, 36, 50, 56, 389
LTA 6, 395, 396, 397, 399, 400, 401, 402, 403, 405, 408, 411, 414, 417, 419, 420, 421, 423, 424, 425, 439, 449

M

Mesoporous Zeolites 23, 25, 30, 36, 49, 183, 214, 357, 369, 426
Mesostructured zeolites 330, 350, 351, 357
Metal 4, 18, 21, 25, 55, 60, 65, 66, 71, 79, 80, 81, 87, 93, 95, 98, 106, 145, 146, 147, 149, 156, 161, 162, 164, 165, 166, 172, 173, 174, 175, 176, 177, 178, 179, 182, 187, 188, 194, 202, 204, 206, 213, 214, 223, 229, 230, 231, 232, 236, 242, 243, 247, 250, 252, 257, 258, 262, 369, 370, 372, 375, 384, 385, 389, 396, 397, 402, 403, 404, 405, 406, 407, 408, 409, 410, 416, 418, 421, 422, 425, 426, 429, 430, 432, 434, 435, 438, 441, 446, 448, 449, 450, 452, 455
Microdeformations 102, 128, 129
Microporous Zeolites 27, 28, 30, 34, 36, 45, 53

Molecular Dynamics 226, 252, 426, 431, 435, 439, 441, 444, 446, 450, 454
Mordenite 1, 2, 3, 4, 5, 8, 10, 11, 12, 15, 16, 17, 25, 36, 37, 55, 58, 59, 69, 73, 87, 93, 97, 98, 102, 108, 111, 124, 128, 129, 130, 133, 136, 137, 138, 141, 142, 157, 162, 166, 179, 212, 214, 249, 260, 267, 270, 274, 301, 318, 335, 336, 337, 338, 355, 357, 426, 434, 437, 439, 449, 450

N

Natural Zeolite 19, 157, 173, 174, 175, 178, 179, 215, 217, 227, 228, 239, 240, 241, 244, 248, 249, 254, 255, 257, 258, 262, 265, 266, 269, 270, 275, 276, 277, 278, 279, 280, 281, 283, 284, 288, 289, 291, 292, 293, 294, 295, 296, 297, 298, 300, 301, 302, 304, 305, 306, 311, 312, 330, 421, 422
NEREA Products 265, 275, 285, 291, 302, 303
NMR Cryoporometry 23, 25, 44, 45, 46, 58, 61
Nuclear Magnetic Resonance 23, 24, 52, 54, 55, 56, 57, 58, 59, 60, 62, 63

O

Omega 62, 171, 318, 330, 333, 337, 357, 386

P

Particle Size 102, 109, 115, 123, 124, 125, 126, 204, 220, 221, 223, 245, 267, 302, 360, 362, 363, 364, 365, 367, 399, 402, 407, 409, 410
Pore Connectivity 25, 43, 46, 49, 50, 57
Pore Size distribution 30, 44, 45, 46, 49, 50, 59, 188, 263, 328, 340, 401
post-synthesis 315, 317, 319, 336, 348, 353, 356, 357, 368, 369, 391, 394, 398, 399, 403, 426
Potato 277, 292, 293, 294, 302, 303, 305, 310, 311
Pyridine 24, 41, 55, 58, 67, 68, 209, 325, 327, 329, 333, 341, 342, 345, 346, 351, 352

Q

Quantum Mechanics 441, 454

R

Remediation 159, 165, 169, 170, 179, 215, 216, 218, 239, 240, 242, 259, 262, 263, 309, 416, 428, 442
Rice 261, 277, 286, 287, 288, 289, 290, 291, 305, 307, 308, 309, 312, 313, 360, 362, 388
Rietveld Refinement 102, 136, 137, 138, 148

S

Semiconductor Oxides 106
SiO2 3, 11, 12, 17, 19, 20, 25, 29, 36, 37, 41, 56, 58, 62, 90, 98, 158, 183, 186, 187, 191, 195, 198, 203, 214, 274, 347, 363, 379, 397, 404, 439
Sol-gel Method 101, 108, 109, 144, 147, 148, 150, 444
Sonochemistry 262, 359, 361, 362, 369, 370, 372, 384, 385, 386, 387, 388, 389, 390, 393, 394, 396, 397, 398, 399, 417, 421, 423, 424, 425, 426
Substrate 104, 266, 267, 268, 269, 270, 272, 273, 276, 277, 278, 303, 304, 306, 425, 429
Sugarcane 277, 295, 296, 297, 298, 299, 300, 307, 308, 311, 312
Surface Acidity 23, 39, 40, 49, 59, 91
Surface Sites 33, 75, 86, 89
Sustainable water treatment 152, 153, 154, 156, 169, 170, 424
Synthesis 3, 4, 18, 23, 24, 29, 36, 49, 52, 57, 62, 106, 108, 109, 121, 129, 142, 144, 145, 149, 158, 169, 172, 174, 175, 176, 177, 178, 182, 183, 184, 186, 209, 210, 213, 214, 222, 223, 246, 247, 248, 251, 253, 259, 261,

263, 271, 315, 317, 319, 320, 336, 339, 348, 353, 354, 355, 356, 357, 358, 359, 360, 361, 362, 363, 364, 366, 367, 368, 369, 370, 372, 373, 375, 379, 383, 384, 385, 386, 387, 388, 389, 390, 391, 392, 393, 394, 396, 398, 399, 400, 401, 402, 403, 404, 405, 412, 416, 418, 419, 420, 421, 422, 423, 424, 425, 426, 429, 430, 432, 433, 434, 442, 443, 444, 446, 447, 448, 450, 454

T

Tomato 274, 277, 278, 279, 301, 305, 307

U

Ultrasound 189, 359, 360, 361, 362, 363, 364, 365, 366, 367, 368, 369, 370, 371, 372, 373, 377, 379, 384, 385, 386, 387, 388, 389, 390, 391, 394, 396, 397, 398, 399, 400, 401, 402, 405, 411, 412, 413, 414, 416, 417, 418, 419, 424, 426

V

VUV-UV-Vis 1, 2

W

Water pollution 151, 152, 153, 154, 161, 165, 168, 170, 216, 242

Water Treatment 148, 151, 152, 153, 154, 155, 156, 157, 158, 159, 161, 162, 163, 167, 169, 170, 175, 176, 178, 219, 222, 228, 230, 232, 236, 244, 245, 246, 247, 249, 251, 253, 254, 256, 257, 262, 424

X

X-ray Diffraction 101, 102, 112, 114, 117, 125, 128, 132, 133, 136, 142, 143, 144, 145, 146, 147, 148, 149, 150, 188, 195, 216, 224, 263, 327, 436, 437, 439

Z

Zeolite 1, 2, 3, 4, 5, 6, 8, 11, 12, 15, 16, 17, 18, 19, 20, 23, 24, 25, 26, 27, 28, 30, 32, 33, 34, 36, 37, 39, 40, 41, 42, 43, 45, 46, 48, 49, 50, 51, 52, 53, 54, 55, 56, 57, 58, 60, 61, 62, 63, 65, 69, 71, 72, 73, 75, 76, 78, 80, 83, 84, 85, 86, 87, 90, 91, 92, 95, 98, 99, 101, 107, 108, 110, 116, 121, 123, 125, 126, 127, 128, 130, 133, 136, 137, 138, 141, 142, 143, 147, 151, 152, 153, 154, 157, 158, 159, 162, 163, 164, 165, 166, 167, 168, 169, 170, 171, 172, 173, 174, 175, 176, 177, 178, 179, 183, 186, 187, 191, 196, 208, 209, 210, 211, 213, 214, 215, 216, 217, 219, 220, 222, 226, 227, 228, 229, 231, 232, 238, 239, 240, 241, 242, 243, 244, 245, 246, 247, 248, 249, 250, 251, 252, 253, 254, 255, 256, 257, 258, 259, 260, 261, 262, 265, 266, 267, 268, 269, 270, 271, 272, 273, 274, 275, 276, 277, 278, 279, 280, 281, 283, 284, 286, 287, 288, 289, 291, 292, 293, 294, 295, 296, 297, 298, 299, 300, 301, 302, 303, 304, 305, 306, 307, 308, 309, 310, 311, 312, 313, 315, 316, 317, 318, 319, 320, 322, 323, 324, 325, 326, 327, 329, 330, 331, 332, 333, 334, 335, 336, 337, 338, 339, 340, 342, 343, 345, 346, 347, 348, 349, 350, 351, 352, 353, 354, 355, 356, 357, 358, 359, 360, 361, 362, 363, 364, 365, 366, 367, 368, 369, 370, 371, 372, 373, 374, 376, 377, 378, 379, 380, 383, 384, 385, 386, 387, 388, 389, 390, 391, 392, 393, 394, 396, 397, 398, 399, 400, 401, 402, 403, 404, 405, 406, 407, 408, 411, 412, 413, 414, 415, 416, 417, 418, 419, 420, 421, 422, 423, 424, 425, 426, 427, 428, 429, 430, 431, 432, 433, 434, 435, 436, 437, 438, 439, 440, 441, 442, 443, 444, 445, 446, 447, 448, 449, 450, 451, 452,

453, 454, 455
Zeolite L 318, 327
Zeolites 1, 2, 3, 4, 5, 12, 17, 18, 19, 20, 23, 24, 25, 26, 27, 28, 29, 30, 32, 33, 34, 36, 37, 38, 39, 40, 41, 42, 45, 46, 48, 49, 50, 51, 52, 53, 54, 55, 56, 57, 58, 59, 60, 61, 62, 65, 66, 68, 70, 71, 72, 73, 75, 76, 77, 78, 79, 80, 81, 83, 86, 87, 90, 91, 92, 93, 94, 95, 96, 97, 98, 101, 102, 107, 108, 110, 111, 142, 143, 145, 148, 149, 151, 154, 156, 157, 158, 159, 160, 161, 162, 163, 164, 165, 166, 167, 168, 169, 170, 171, 172, 174, 175, 176, 179, 183, 184, 185, 191, 195, 206, 210, 212, 214, 216, 222, 231, 243, 246, 247, 248, 251, 252, 253, 256, 257, 258, 260, 261, 262, 265, 266, 267, 269, 270, 271, 272, 273, 275, 276, 286, 297, 301, 304, 305, 306, 307, 308, 309, 310, 312, 315, 316, 317, 318, 319, 320, 324, 330, 332, 335, 336, 338, 342, 345, 348, 349, 350, 351, 353, 354, 355, 356, 357, 358, 360, 361, 364, 365, 366, 367, 368, 369, 370, 372, 373, 376, 377, 380, 383, 384, 385, 386, 387, 388, 390, 392, 393, 394, 395, 396, 397, 399, 401, 402, 403, 404, 405, 406, 407, 408, 411, 412, 414, 417, 418, 419, 420, 422, 424, 425, 426, 427, 428, 429, 430, 431, 432, 433, 434, 435, 436, 438, 439, 440, 441, 442, 444, 445, 446, 447, 448, 449, 450, 451, 452, 453, 454, 455
ZSM-5 4, 5, 20, 25, 28, 41, 52, 59, 61, 62, 69, 70, 72, 73, 75, 78, 91, 92, 93, 95, 97, 175, 181, 182, 183, 184, 185, 186, 187, 188, 189, 190, 191, 192, 193, 195, 196, 198, 201, 202, 203, 204, 206, 208, 209, 210, 211, 212, 213, 214, 258, 317, 318, 345, 346, 355, 356, 362, 374, 376, 387, 388, 393, 394, 407, 434, 436, 437, 440, 442, 445, 450, 451

www.ingramcontent.com/pod-product-compliance
Lightning Source LLC
LaVergne TN
LVHW080615300926
849162LV00005B/1163

* 9 7 9 8 3 3 7 3 0 3 2 0 8 *